International Conference on Environmental, Energy and Earth Science (ICEEES)

IOP Conference Series: Earth and Environmental Science
Volume 1041

Online
22 – 23 September 2021

ISBN: 978-1-7138-5819-5
ISSN: 1755-1307

Printed from e-media with permission by:

Curran Associates, Inc.
57 Morehouse Lane
Red Hook, NY 12571

Institute of Physics
Dirac House, Temple Back
Bristol BS1 6BE UK

Phone: 44 1 17 929 7481
Fax: 44 1 17 920 0979

techtracking@iop.org

Additional copies of this publication are available from:

Curran Associates, Inc.
57 Morehouse Lane
Red Hook, NY 12571 USA
Phone: 845-758-0400
Fax: 845-758-2633
Email: curran@proceedings.com
Web: www.proceedings.com

TABLE OF CONTENTS

Preface International Conference on Environmental, Energy and Earth Science (ICEEES)

Peer Review Statement

The Effect of Chitosan Addition on Physicochemical Properties, Proximate and Antioxidant of Functional Drinks of Red and Purple Roselle (Hibiscus Sabdariffa L.) Extracts 1
Kusnadi, Aldi Budi Riyanta, Rizki Febriyanti

The Characteristics of Bio-Oil Hydrothermal Pyrolysis Empty Fruit Bunches Based on Powder Size and Hydrogen Gas Pressure 8
Rina Novia Yanti, Erliza Hambali, Gustan Pari, Ani Suryani

The Effect of Proportion of Ganyong Starch and Waste of Straw Rice on Biodegradable Foam Production as Sustainable Packaging 14
Gresita Fitaloka Sari

Physicochemical and Antioxidant Characteristics of Black Garlic from Indigenous Indonesian Garlic 22
K S Sasmitaloka, SM Widayanti, I Mulyawanti, ES Iriani

Efficiency Analysis of Vannamei Shrimp Supply Chain Performance in PT Sari Indonesia Group, Suak Pandan Village, West Aceh Regency, Aceh Province 29
Yoga Nugroho, Teuku Athaillah, Muhibbul Qadri

Nutrient Testing of Organic Liquid Fertilizer in Tempeh Factory in Pekanbaru City 36
Neng Susi, Vonny Indah Sari, Indra Purnama, Hazra Yuvendius

Extraction of Catfish Bone Waste (pangasius Hypophtalmus) by Utilizing an Organic Liquid of Pineapple Peel Waste (ananas Comosus) into Gelatin 39
Ira Oktaviani Rz, Rahimatul Uthia, Fathul Jannah, Alexsander Yandra

The Effect of Orange Peel Waste Addition on ZnO Characteristics and Its Ability to Degrade Methyl Orange 47
Aulia Dewi Rosanti, Fahmi Hidayat, Yuly Kusumawati, Arif Fadlan, Anggita R.K Wardani, Eva Umi Latifah

Carbon Sequestration of City Agriculture: Between Farming and Non-Farming Land 55
D.R.D. Hastuti, R. Darma, D. Salman, S. Santoso, A. Rahim

Comparison Predicted Using Nonlinear Model and in Vivo Studies Growth Production of Broiler After Fed Different Levels of Fermented Tamarind Flour as Replacing Soya Bean Meal 63
O Sjofjan, D N Adli, M H Natsir, Y F Nuningtyas, A D Budianto

Dry Matter (DM), Crude Protein (CP) and Water Consumption in Sheep Which Supplemented from Moringa Leaves Meal and Palm Oil in Elephant Silage-Based Feed 68
S Rizqiana, W Suryapratama, F M Suhartati

Effect of Various Packaging Materials for Storing Ground Yellow Corn of Hybrid C-1 Variety on Water and Amylum Content 74
Munasik, N Nurbaety, N Hidayat, E Susanti

Exploring the Herbaceous Plant Height and Dry Matter Relationship on a Sub-Humid Rangeland in Zimbabwe 80
T A Zezai, P H Mugabe, V E Imbayarwo-Chikosi

Potential Development Based on the Agricultural Waste in the Eastern Region of West Java Province 88
F. D. Perwitasari, A. R. S. Putra, B. Suwignyo, R. Widiati

Farming in Kediri Indonesia: Analysis of Cluster K-Means 96
N Solikin, B Hartono, Sugiono, Linawati

The Identification and Role Pf Mold in Lowering the HCN Levels in Black Cassava Based on the Immersion Duration 102
Mery Pattipeilohy, Alamanda Pelamonia

The Effects of Some Additives on Characteristics RVA Profile of Cassava Flour 112
H Herawati, E. Kamsiati

A Comprehensive Review on Power Tillers Usage – Bottlenecks and Prospects in India 118
Krishnadas R, Renganathan R

Android-Based Feasibility Study Transformation Analysis Case Study of the RASI Development in Cirendeu-Kampung Cimahi 124
H Herawati, A S Soemantri, E Kamsiati, M. Bachtiar

Water Pump Control System in Hydroponic Plants Using the Ebb and Flow Method 131
David Setiawan, Hamzah, Latifa Siswati, Anto Ariyanto, Guntoro

Resetting the Ground Fault Relay on a 28 MVA Power Transformer Due to Grounding Method Retrofit 139
Hamzah Eteruddin, Arlenny, Hadi Siswanto, Yanuar Z. Arief, Fathimah Hasanti

Intelligent Ubiquitous Technology as a Precision Agri-Food Framework: A Proposed Framework 145
T A Cahyanto, S Wahjuni, H Sukoco, H Rahmawan, S N Neyman

Development of Internet of Things-Based Monitoring System for Application Solar Energy Technology in Bunaken Island 152
M Rumbayan, I Pundoko, D Ruindungan, N V Panjaitan

Technical Analysis on Household Energy Consumption: LMDI Decomposition Index and Innovative 158
Yetty Murni, Eka sudarmaji, Sri Ambarwati, Ismiriati Nasip

Optimization of Electrical Energy in the Fan Using Fuzzy Logic Controllers 173
Arlenny, David Setiawan

The Effect of Giving Chia Seeds on the Bodyweight of Pregnant Mice (Mus Musculus L.) 184
R Uthia, I O Rz

Supplementing Mineral Selenium and Vitamin E in Diets on in Vivo Digestibility, Blood Glucose, and Urea Levels of Cows 191
D. Pamungkas, A.S. Putri, F. Firdaus, R. Widiyawati, D.M. Dikman

Effect of Saccharomyces Cerevisiae on Probiotic Properties of Goat Milk Kefir 197
L E Radiati, D L Hati, D Fardiaz, L R H Sari

Cattle Production System in Semi-Arid Area of Timor Island 203
Y Ngongo, S Ratnawaty, PR Matitaputty

Sustainable Indigenous Cattle Production in Brebes Regency, Indonesia: Opportunities and Threats During Pandemic 211
M Sugiarto, Y. N Wakhidati, O. E. Djatmiko, S. Nur, A Einstein

The Role of Philanthropy in Reducing Impact of Disaster on Economies of ASEAN-9 215
M S A Majid, R Nurdin, T Azhar, S Sartiyah

Community Participation in Disaster Management in Gondoriyo Village Administration, Ngaliyan Sub-District, Semarang City 222
Supratiwi, Yuwanto, Kushandajani

Assessing and Preparedness for Earthquake Disaster in Salaf-Khalaf Islamic Boarding School 228
Fatiya Rosyida, Neni Wahyuningtyas, Ardyanto Tanjung, Abdul Kodir

Disaster Education in Disaster-Prone Schools: A Systematic Review 235
R Yusuf, Razali, Sanusi, Maimun, I Fajri, S A Gani

LCA Methodology for Detecting Environmental Impacts on Natural Gas Drilling Process 242
Masayu Rosyidah, Andiyan Andiyan, Haniek Listyorini, Pranoto Hadi Prayitno, Yuswardi Yuswardi, Yuhanah Yuhanah

Community Based Ecotourism Management to Strengthen Environmental Ethics and Supports Sustainable Development in Pelalawan District, Indonesia 252
Nawari, Thamrin, Nofrizal, Almasdi Syahza, Juandi Muhammad, Nur Islami

Community Forestry for Environmental Sustainability and Ecotourism: The Context and Problems in Indonesia 262
R Sulistiowati, S Wahyuni, M K Yunanto, M Elsera, W E Yudiatmaja, T Samnuzulsari

Study on the Commitment of Oil Palm Companies to Achieve Sustainable Agriculture in Riau Province from the Perspective of Pesticide Use 268
S Syafrani, I Purnama, A Mutamima, W N Dewi

Assessing the Feasibility of Migration Policy from LPG Stoves to Induction Stoves in Indonesia 273
T Anggono, I Ruslan, C Anditya, D G Cendrawati, M I A Irsyad

The Innovation of Ecological Fiscal Transfers Policy at Siak Regency 280
Muhammad Ichsan Kabullah

Farmers' Perceptions of the Performance of Agricultural Instructors in Mangosteen Farmers' Groups in Segati Village, Langgam District, Pelalawan Regency 285
Asgami Putri, Latifa Siswati, Hamdan Yasid

Environment Conservation Responsibility in Community Development 289
N Sutisna, H Herdiansyah, P Sanjatmiko

Environmental Conservation in Ciletuh Geopark, West Java 295
R A Zuvara, H Herdiansyah, D Asteria

Effectiveness of Silica Gel from Palm Kernel Shell Ash as a Moisture Absorber of Bottle Packaging Medicine 301
Lidya Novita, Iswadi Idris

Comparative Study of Village Community Participation in Peat Restoration in the Peat Swamp Forest Management Unit (PSFMU) Tebing Tinggi Island, Indonesia 308
H. Malik, Ig. L. S. Purnama, Sudarmadji, L. W. Santosa, U. Widyanarko, M. Naufal

Swot Analysis of Utilization of Palm Oil Mill Effluent to Improve Soil Quality and Crop Productivity (case Study at PT X, Lampung, Indonesia) 313
Dela Febriana, Sylvia Madusari, Vira Irma Sari

Groundwater Level Fluctuations in Peatlands as Responses to Canal Block Construction in Sungai Tohor, Tebing Tinggi Island, Indonesia 319
H. Malik, Ig. L. S. Purnama, Sudarmadji, L. W. Santosa, M. Naufal

Identification of Land Use and Land Cover Using the Image Landsat 8 in Upstream Lematang Sub-Watershed by Support Vector Machine and Random Trees Methods 324
A Dinata, F Dhiniati, L E Diansari

Conservation and Utilization on Sawahlunto Ex Mining Area for Cultural Landscape 329
Rika Cheris, Amanda Rosetia

Measuring Cross-Sector Collaboration an Effective Solution for Solving Unlicensed Mining Problems in Kuantan Singingi, Indonesia 334
MY Tiyas Tinov, Aras Mulyadi, Sujianto, Zulkarnain, Ardi Putra, Tika Mutia, Tito Handoko

Helminthiasis of Rabbits on the Upland and Lowland Areas and the Risk Factors 343
D Indrasanti, M Indradji, Sufiriyanto, M Samsi, E Yuwono

Water Quality Analysis Based on Land Use in Sekampung River, Lampung, Indonesia 350
K Fitriani, T K Nufutomo, R Putra

Application of Oil Palm Empty Bunches as Organic Mulch in Oil Palm Plantation (Elaeis Guineensis Jacq.): An Evaluation and SWOT Analysis 357
Lala Novita Sari, Sylvia Madusari, Vira Irma Sari

Oil Palm Trunk Replanting as Brown Sugar Raw Materials 363
Latifa Siswati, Enny Insusanty, Neng Susi, Nopryanti

Prediction of Palm Oil Production in Riau Province Using the Single Exponential Smoothing Method 370
Guntoro, Lisnawita, Zamzami, David Setiawan

Study of Biogas Utilization as Waste-To-Energy Plant and Transport Modelling of Iron (Fe), Lead (Pb) and Copper (CU) in Leachate at Muara Fajar Landfill Pekanbaru 374
H Yuvendius, E Zondra, Zainuri, V I Sari

Utilization of Oil Palm Waste as Animal Feed for Bali Cattle in Margomulyo Village, Central Bengkulu 380
E Ramon, Z Efendi, H Kusnadi, S Yuliasari, A Ishak, WA Wulandari

Etnobotany of Medicine Plants in the Community of Rawang Kao Village, Siak, Indonesia 386
Sri Wahyuni, Mar'atul Afidah

Can Mangrove Intensive Policies in Indonesia Benefit More Coastal Communities? Environmental Policy Analysis 392
Anto Ariyanto, Afni Zulkifli, Jeni Wardi

Validation of UV-VIS Spectrophotometric Method for the Determination of Calcium in Maman (cleome Gynandra L) ... 399
Fathul Jannah, S Meri Agritubella, Ira Oktaviania, Lily Restusari

Utilization of Bamboo Charcoal to Decrease Tannin Content of Lindur Fruit Flour (Bruguiera Gymnorrhiza) ... 405
Niken Pujirahayu, Zakiah Uslinawaty, Nuhayati Hadjar, Muhtadir

Supplementing Safflower Oil (Carthamus Tinctori L) and Inocytol in Feed and Its Effect ... 411
Ning Iriyanti, Ismoyowati, Efka Aris Rimbawanto, Modawy Abdelgader Albasheer Altayb

The Effects of Nano Encapsulated Syzygium Polyanthum Leaves Extract Inclusion in Drinking Water ... 419
A N 'Afifah, Zuprizal, N D Dono

Effects of Plant Extracts Against Gastrointestinal Nematodes of Livestock: A Meta-Analysis ... 426
A Jayanegara, D M Sinaga, E B Laconi

Coconut Oil Effects on in Vitro Ruminal Fermentation and Methanogenesis ... 430
A Cusiayuni, RK Nurfatahillah, D Evvyernie, A Jayanegara, KG Wiryawan

Effect of Palm Oil Supplementation Level on in Vitro Ruminal Fermentability ... 436
RK Nurfatahillah, A Cusiayuni, A Jayanegara, KG Wiryawan, D Evvyernie

The Composition of Goat Milk Kefir Added with Moringa Oleifera Leaf Powder ... 443
S R N Endah, P D Wulasari, A Nofriyaldi, Nurliyani, E Harmayani

The Effect of Safflower Oil (Carthamus Tinctorius L.) and Inositol Supplementation on Egg Production ... 448
Ismoyowati, N Iriyanti, M Abdelgader

The Anti-Inflammatory Activity of Cherry Leaf Extract (Muntingia Calabura L.) Balm Stick ... 452
Fitria Nugrahaeni, Kriana Efendi, Abdul Kholik Aziz

The Effect of HPMC Concentration as a Gelling Agent on Color Stability of Copigmented Blush Gel Extract of Purple Sweet (Ipomoea Batatas (L.) Lam.) ... 458
Fitria Nugrahaeni, Nining, Redina Okvianida

Sunscreen Factor Formulation and Test of Gel Preparations of 70% Ethanol Extract on Arabica Coffee Leaf (Coffea Arabica L.) ... 465
Sofia Fatmawati, Fitria Nugrahaeni, Fith Khaira Nursal, Ana Fitriana

Antioxidant Activity and Sun Protection Factor (SPF) Graded Extract of Katuk Leaves (Sauropus Androgynus (L.) Merr.) ... 478
Sofia Fatmawati, Ni Putu Ermi Hikmawanti, Akbar Fadillah, Awlia Mustia Putri

Anti-Inflamatory Activity of Water Extract of Talinum Paniculatum. (Jact). Gaertn Leaves on Wistar Rat ... 487
A Emelda, Sukmawati, A L Dongke, I Marzuki

Assessing Food Safety Compliance for Food SMEs in Indonesia ... 492
Tri Fajarwaty, David Jukes

In-Silico Screening of Inhibitor on Protein Epidermal Growth Factor Receptor (EGFR) ... 498
Jeremi Ongko, Jesica Viona Setiawan, Alfina Gracia Feronytha, Agnes Juliana, Andy Effraim, Mariana Wahjudi, Yulanda Antonius

Effect of Various Feed Additives on Carcass and Meat Quality of Two Different Strains of Chickens 505
E Tugiyanti, E Susanti

Microbiological Profile of Concentrated Yoghurt Manufactured from Low and Full Fat Milk During Storage 510
T Setyawardani, J Sumarmono, D R Kusuma

Effect of Dietary Supplementation with Flying Fish (exocoetidae) Silage on the Egg Quality and Duck Performance 515
L Ambarwati, S P Syah

Production Profile and Microbiological Properties of Locally Produced Pork Sausage in the Philippines 521
MM Dealino, CM Bueno

Farming Sustainability: A Sensitivity Analysis 531
Y N Wakhidati, M Sugiarto, H Aunurrohman, S Mastuti, J A Suryani

Improving Genetic Quality of Local Livestock for Achieving Food Security 536
E Kurnianto

Current Goat Milk Production, Characteristics, and Utilization in Indonesia 542
J Sumarmono

The Impact of Agricultural Waste on River Water Quality of Kreo Watershed in Semarang City 550
W Setyaningsih, R S Sanjaya

Utilization of Plastic Waste as an Eco-Friendly Construction Material 563
Zainuri, G Yanti, S W Megasari

Biogas as Methane Mitigation Path on Dairy Cattle Waste Management 569
A Herliatika, Y Widiawati

Does Rebound Effect Influence the Factors of Carbon Emission in Indonesia? Kaya Index and LMDI Decomposition 575
Eka Sudarmaji, Noer Azam Achsani, Yandra Arkeman, Idqan Fahmi

Study of Phytoplankton Biology Index and Water Quality Parameters of Kali Surabaya River 585
O H Cahyonugroho, S Hariyanto, G Supriyanto

Factor and Perception Analysis of the Impact of Boat Parking on River Pollution 591
B Rahman, M S Aryanto

Author Index

IOP Conf. Series: Earth and Environmental Science 1041 (2022) 011001 doi:10.1088/1755-1315/1041/1/011001

Preface International Conference on Environmental, Energy and Earth Science (ICEEES)

Universitas Lancang Kuning Pekanbaru organized with **the International Conference on Environmental, Energy and Earth Science (ICEEES)** on September 22, 2021 in Pekanbaru, Indonesia. The conference is aims to exchange knowledge and research finding among academicians, researchers, professionals, policy makers, and postgraduate students.

The Awareness to increase the number of publications on the results of research that has been done, deserves appreciation by all parties. Because of that, Universitas Lancang Kuning Pekanbaru provide motivation and space for researcher to disseminate their research and accommodate the result of research that has been done. **The International Conference on Environmental, Energy and Earth Science (ICEEES)**, is another International Conference held by Universitas Lancang Kuning Pekanbaru.

The International Conference on Environmental, Energy and Earth Science (ICEEES)was chosen to be implemented virtually, this is because the cov-19 pandemic is still spreading. The conference was perform using zoom. **The International Conference on Environmental, Energy and Earth Science (ICEEES)** event is virtually implemented with a model that all invited speakers are given time to present their material for 30 minutes every invited speaker and after that a question and answer is carried out with the participants with a direct questioning system, through chat forums and Q&A forums provided by the zoom application. Overall, the conference took 5 hours the number of participants who joined the zoom room was recorded at participants. Participants came from few countries, namely Indonesia, Malaysia, India, Egypt and Australia.

The International Conference on Environmental, Energy and Earth Science (ICEEES) is implemented with the support of a stable internet network system and a zoom application. In the implementation there were several technical obstacles encountered by the participants, namely the difficulty of joining the zoom application due to the unstable internet signal. The holding of a virtual conference felt less meaningful, due to the lack of interaction between speakers and participants

The International Conference on Environmental, Energy and Earth Science (ICEEES) committee received 124 manuscripts and a total of 117 papers were presented and discussed. The papers were authored by researchers from Indonesia, Malaysia, India, Egypt, japan and Australia.

All papers have been scrutinized by a panel of reviewers who provide critical comments and corrections, and thereafter contributed to the improvement of the quality of the papers. Based on the reviewer's reports, 92 papers were selected and eligible to be published in the proceeding.

We sincerely express our gratitude to the international/national advisory committee, presenters, organizing committee members, session chairs, all members of organization, participants, contributors and all the members of **The International Conference on Environmental, Energy and Earth Science (ICEEES)**. Last but not the least, we are thankful to IOP EES Conference Series for producing the proceeding.

IOP Conf. Series: Earth and Environmental Science 1041 (2022) 011001 doi:10.1088/1755-1315/1041/1/011001

Organizing Committee

General Chair :
Dr. Anto Ariyanto, M.Si (Universitas Lancang Kuning, Indonesia)

General Co-Chair:
Dr. H. Jeni Wardi, S.E., M.Ak., Ak., CA (Universitas Lancang Kuning, Indonesia)

Secretary :
Dr. David Setiawan, M.T (Universitas Lancang Kuning, Indonesia)

Treasurer and Finance Chair :
Ambar Tri Ratnaningsih, S.Hut., M.Si (Universitas Lancang Kuning, Indonesia)

Publication Chair :
Robbi Rahim, S.Kom., M.Kom (STIM Sukma Medan, Indonesia)

Website and Social Media Chair :
Guntoro, S.T., M.Kom (Universitas Lancang Kuning, Indonesia)
Febrizal Alfarasy Syam, S.Kom., M.Kom (Universitas Lancang Kuning, Indonesia)
Ismar Husnudin, S.T (Universitas Lancang Kuning, Indonesia)

Editors

1. Prof. Dr. Joran (Linkoping University, Swedia)
2. Prof. Dr. Amran bin Hamzah (Universitas Teknologi Malaysia, Malaysia)
3. Prof. Dr. Khaerurijal (Institut Teknologi Bandung, Indonesia)
4. Prof. Dr. Asep Saefuddin (Institut Pertanian Bogor, Indonesia)
5. Prof. Dr. Mauridi (Institut Teknologi Sepuluh November, Indonesia)
6. Prof. Dr. Adi Soeprijanto (Institut Teknologi Sepuluh November, Indonesia)
7. Prof. Dr. Syafrani (Universitas Lancang Kuning, Indonesia)
8. Prof Dr. Sudi Fahmi (Universitas Lancang, Kuning, Indonesia)
9. Prof. Dr. Hariadi Kartodiharjo (Institut Pertanian Bogor, Indonesia)
10. Prof. Dr. Sumbangan Baja (Universitas Hasanuddin, Indonesia)
11. Prof. Wiendu Nuryanti Ph.D (Universitas Gajah Mada, Indonesia)
12. Prof. Dr. Chairul Saleh (Universitas Islam Indonesia, Indonesia)
13. Prof. Dr. Richardus Eko Indrajit (STMIK Perbanas, Indonesia)

Organizing Institutions

- Universitas Lancang Kuning

IOP Conf. Series: Earth and Environmental Science 1041 (2022) 011002 doi:10.1088/1755-1315/1041/1/011002

Peer Review Statement

All papers published in this volume have been reviewed through processes administered by the Editors. Reviews were conducted by expert referees to the professional and scientific standards expected of a proceedings journal published by IOP Publishing.

- **Type of peer review:** Single Anonymous
- **Conference submission management system:** https://iceees.unilak.ac.id/paper-submission/
- **Number of submissions received:** 129
- **Number of submissions sent for review:** 129
- **Number of submissions accepted:** 87
- **Acceptance Rate (Submissions Accepted / Submissions Received × 100):** 67.4
- **Average number of reviews per paper:** 2
- **Total number of reviewers involved:** 15
- **Contact person for queries:**
  Name: Robbi Rahim
  Email: robbirahim@ieee.org
  Affiliation: Sekolah Tinggi Ilmu Manajemen Sukma

The effect of chitosan addition on physicochemical properties, proximate and antioxidant of functional drinks of red and purple roselle (Hibiscus sabdariffa L.) extracts

Kusnadi*, Aldi Budi Riyanta, Rizki Febriyanti

Politeknik Harapan Bersama, Tegal, Indonesia

*kusnadi.adi87@gmail.com

Abstract This investigation aimed to determine the physicochemical, proximate, and antioxidant results of a functional drink made from red and purple roselle with the addition of chitosan. Six drink formulations were prepared based on different ratios for red, purple roselle extracts and chitosan, respectively; C0 (30: 0: 0)%; C1 (0: 30: 0)%; F1 (20: 10: 0.5)%, F2 (10: 20: 0.5)%, F3 (20: 10: 1)%, and F4 (10: 20: 1)%. Functional drink samples were made by the pasteurization method at 90 °C for 15 minutes. The scavenging activity of 2,2-diphenyl-1-picrylhydrazyl (DPPH) was used to measure antioxidant activity. The resulting pH values ranging from 2.62 to 3.20. Moisture content ranging from 90.45 to 93.10 g/100 ml, ash content ranging from 2.20 to 3.44 g/100 ml, protein content ranging from 1.42 to 2.62 g/100 ml, fat content ranging from 0.22 to 0.48 g/100 ml. The scavenging ability on DPPH radical is the highest on F4 with the addition of 1% chitosan ranging from 18.08% to 34.10% and the smallest at C0 is 10.05% to 27.15%. The results show that the addition of chitosan can improve the physicochemical, proximate, and antioxidant properties in functional drinks from red and purple roselle extracts.

1. Introduction

Utilization of foodstuffs into food and beverages can increase food diversification and is indispensable for the human body to meet nutritional needs. Trends in functional drinks as functional food are always being developed and can be consumed because they have good properties in improving health for the community [1]. One of the potentials possessed by functional drinks is that they have high antioxidant activity which can prevent degenerative diseases as well as for body fitness [2]. Functional drinks that are more practical and contain food active substances, have more food that can be processed. The basic ingredients in the manufacture of functional drinks, plants, both vegetables and fruits, contain biocative substances for health [3].

Rosella is a type of plant that has bioactive compounds and produces fiber which is useful for health. In Indonesia, there are two types of rosella varieties, namely red and purple. Rosella flower petals are very beneficial for health because they contain phenolic compounds which function as antioxidants that can cure degenerative diseases including lowering blood cholesterol levels, anticancer, lowering blood pressure, and antidiabetic [4–6]. This bioactive content causes rosella to be widely used as a food coloring, antioxidant, flavoring, preservative [7, 8].

At present, there has been research on drinks developed from a mixture of rosella with vegetables or fruits in beverage formulations such as roselle-pineapple [9], rosella-stevia-citric acid [1], and roselle-mango-orange drinks [3]. However, at this time, there has been no research on the functional drink of red and bronze roselle extracts combined with chitosan.

Chitosan is an oligosaccharide compound produced by the deacetylation process of alkaline chitin by converting N-acetyl-D-glucosamine units into D-glucosamine units at high temperatures. Chitosan has high economic value, and its processed products can be used for various purposes [10]. Chitosan can be used in food processing, medicine, biotechnology and become an attractive material in pharmaceutical biomedical applications, environmental, and agricultural wastewater treatment [11], [12]. Chitosan is non-toxic, has a biological activity; biocompatibility, biodegradability, and can be modified chemically and physically [13, 14].

There have been many studies regarding chitosan as a preservative supplement, juice purifier, gelling agent, edible formation, antioxidant, antimicrobial, anti-inflammatory, anti-cancer, and anti-tumor properties [15]–[17]. However, the use of chitosan in additional supplements in the functional drink mixture of red and purple rosella has not been studied before.

A study by Jin et al showed that giving chitosan to a drink formulation mixed with water can increase the value of antioxidants, as a coagulant and can eliminate bacteria [18]. The addition of 0.5% to 2% chitosan can increase the phenol content and antioxidant value in reducing free radicals using the DPPH method and can increase the thickness of the edible coating provided [19].

Another study shows that chitosan can also reduce water content and water vapor transfer rate in edible film due to the hydrophobic nature of chitosan [20]. Based on the stated explanatory criteria, The aim of this study was to determine the best formulation of functional drinks made from a combination of red and purple roselle extracts with the addition of chitosan by evaluating the physico-chemical properties, proximate levels, and free radical scavenging activity of the extracts.

2. Methodology

2.1 Material

Red and purple roselle flowers (Hibiscus sabdariffa) purchased at a local market in Yogyakarta, Indonesia. The chitosan used in this study was commercial chitosan which was purchased at the CHIM-Pharmacy Multipurpose Store Bandung-Indonesia with a deacetylation degree of 83.5% and a pH of 6.8.

2.2 Preparation of Rosella Extract

Rosella was extracted using the infusion method mentioned previously by Zannou et al with a slight modification, dried roselle petals and then immersed in water with a ratio of 1:10 and heated at 90 °C for 15 minutes [21]. Rosella extract filtering was done using a muslin cloth.

2.3 Preparation of mixed drink roselle extracts and chitosan

Six mixed drink formulations were experimented on based on the proportion of red and purple roselle extract and chitosan, respectively: C0/control (30: 0: 0)%; C1/control (0: 30: 0)%; F1 (20: 10: 0.5)%, F2 (10: 20: 0.5)%, F3 (20: 10: 1)%, and F4 (10: 20: 1)% and additional ingredients such as 5% sucrose, honey 5 %, citric acid 0.2% and water 59.8%. The functional drink formulation is made by mixing the amount of extract with other additives, then pasteurized at 90 °C for 15 minutes [22]. The resulting mixed drink in hot conditions is then poured into a polypropylene bottle and cooled rapidly at room temperature.

2.4 Physicochemical and proximate analysis

2.4.1 Degree of acidity (pH)

The pH measurement with a wireless pH meter (Hanna Instrument, HI 9126, California USA) by pipetting 30 ml of each formulation sample then pouring it into a beaker and dipping the pH measuring electrode into the sample then recording [23].

2.4.2 Moisture content

Moisture content was determined by filling a sterile chemical gala measuring 50 ml with water and weighing it (W1). A separate sample of 10 mL was weighed and placed in a pre-weighed beaker (W2), which was then dried in an oven at 110 for 2 hours of the AOAC [24]. After 2 hours of completion, the sample was placed into a desiccator and its weight was recorded (W3).

$$\text{Moisture content (g/100 ml)} = \frac{W2-W3}{W2-W1} \times 100$$

2.4.3 Total titrated acid
The titration method can be used to calculate the titrated acidity level by homogenizing 5 ml of the sample with 20 ml of distilled water, then filtering it with Whatman No.1 filter paper. 3 to 5 drops of 1 g/100 ml phenolphthalein (PP) indicator were applied to a total of 15 ml of sample and titrated with 0.1 N NaOH solution until it displayed a clear pink color [23].

$$\text{Titritable acidity (g/100ml)} = \frac{\text{The titrated sum of NaOH x N of NaOH x 9}}{\text{Volume of sample (ml)}}$$

2.4.4 Ash content
The ash content is determined by taking a 5 ml sample and pouring it into a jar for combustion in a furnace at 450 °C for 4 hours, after which the desiccator container is allowed to cool of the AOAC [24].

$$\text{Ash content (g/100ml)} = \frac{\text{(Container weight plus residue-empty container weight)x100}}{\text{The sample's weight}}$$

2.4.5 Fat content
To assess the fat content, 4 ml of the sample was dissolved in 2 ml of alcohol and then hydrolyzed for 30 minutes at 70 °C with 10 ml of concentrated HCl. Using the Soxhlet extractor, the hydrolyzed fat is extracted for at least 4 hours with petroleum ether 100 ml AOAC [24].

$$\text{Fat content (g/100 ml)} = \frac{W2-W1}{W} \times 100$$

Where W1 is the weight of the extraction flask; W2 is the weight of the extraction flask plus the weight of the dried crude fat (g); W is the sample's weight.

2.4.6 Protein content
Protein content analysis was carried out using the Kjeldahl method AOAC [24] with three stages (destruction, distillation, and titration). The results of protein content obtained directly from the experiment in percentage values (%).

2.4.7 Total carbohydrate
Measurement of total carbohydrate levels used the Carbohydrate by Difference of the method AOAC [24]. Total carbohydrate content = (100 – [(Moisture content + protein content + fat content + ash content) g/100 ml].

2.5 Antioxidant activity
The antioxidant activity of functional drinks of red and purple roselle extracts was determined using a modified version of Cittan et al [25] method of DPPH radical scavenging activity using spectrophotometry. In summary, the preparation of DPPH solution (1 mg/ml) as a stock solution. Each 1 ml of functional drinks of red and purple roselle extracts was diluted into a DPPH solution, then made several different concentrations (20, 40, 60, 80, and 100 µg/ml), then shaken until homogeneous and incubated in a room dark for 30 minutes.

Spectrophotometry was used to calculate the solution's light absorption at a wavelength of 510 nm. As a control, a DPPH solution with no sample and no standard mark was used. Ascorbic acid was also tested and used as a reference point. % Inhibition of free DPPH radical scavenging activity was calculated.

$$(\%)\text{Inhibition}=\frac{\text{Acontrol - Asample}}{\text{Acontrol}} \text{ x } 100$$

2.6 Statistical analysis

Statistical data analysis used analysis of variance to determine the level of significance. The smallest significant difference between treatments was used the least significant difference (LSD) test.

3. Result and Discussion

3.1 Physicochemical and proximate analysis

The physicochemical and proximate results on the formulation of functional drinks of red and purple roselle (Hibiscus sabdariffa L.) with the addition of chitosan and without chitosan are described in Table 1. The results show that giving chitosan can increase the pH and titratable acid values. This is because rosella extract also has a low glucose content which contributes to the low total titratable acid and pH [22].

Table 1. Physicochemical and proximate of functional drink of red-purple rosella and chitosan

Sample	C0	C1	F1	F2	F3	F4
pH	$2{,}68 \pm 0{,}02^a$	$2{,}62 \pm 0{,}02^a$	$2{,}92 \pm 0{,}01^b$	$2{,}86 \pm 0{,}01^b$	$3{,}20 \pm 0{,}11^c$	$3{,}05 \pm 0{,}21^c$
Moisture	$93{,}10 \pm 0{,}03^a$	$93{,}01 \pm 0{,}01^a$	$91{,}35 \pm 0{,}05^b$	$91{,}30 \pm 0{,}01^b$	$90{,}50 \pm 0{,}04^c$	$90{,}45 \pm 0{,}11^c$
Titratable	$0{,}47 \pm 0{,}05^a$	$0{,}53 \pm 0{,}02^a$	$1{,}17 \pm 0{,}11^b$	$1{,}22 \pm 0{,}21^b$	$1{,}32 \pm 0{,}01^b$	$1{,}36 \pm 0{,}03^b$
Ash content	$2{,}20 \pm 0{,}02^a$	$2{,}25 \pm 0{,}05^a$	$3{,}05 \pm 0{,}11^b$	$3{,}09 \pm 0{,}01^b$	$3{,}42 \pm 0{,}03^c$	$3{,}44 \pm 0{,}08^c$
Fat content	$0{,}22 \pm 0{,}01^a$	$0{,}26 \pm 0{,}11^a$	$0{,}30 \pm 0{,}01^b$	$0{,}33 \pm 0{,}04^b$	$0{,}46 \pm 0{,}05^c$	$0{,}48 \pm 0{,}01^c$
Protein	$1{,}42 \pm 0{,}08^a$	$1{,}47 \pm 0{,}01^a$	$2{,}12 \pm 0{,}03^b$	$2{,}21 \pm 0{,}01^b$	$2{,}60 \pm 0{,}02^c$	$2{,}62 \pm 0{,}05^c$
Carbohydrate	$3{,}06 \pm 0{,}03^a$	$3{,}01 \pm 0{,}08^a$	$3{,}08 \pm 0{,}01^a$	$3{,}07 \pm 0{,}08^a$	$3{,}02 \pm 0{,}04^a$	$3{,}01 \pm 0{,}03^a$

Data sharing similar letter in row are statistically nonsignificant ($p > 0.05$). All values represent the mean of three replication ± standard error. C0 = red/purple roselle/chitosan (30:0:0)%, C1 = red/purple roselle/chitosan (0:30:0)%, F1 = red/purple roselle/chitosan (20:10:0.5)%, F2 = red/purple roselle/chitosan (10:20:0.5)%, F3 = red/purple roselle/chitosan (20:10:1)%, F4 = red/purple roselle/chitosan (10:20:1)%.

The resulting functional drink formulation has a low pH with a range of 2.68 to 3.20 so that chitosan can dissolve in drinks made from red and purple rosella. This is because chitosan can dissolve in organic acid solutions with a pH <6 [26]. The difference in pH results in all formulations from C0 to F4 made from red roselle has a higher pH than formulations made from purple roselle, this is also influenced by differences in mineral content, organic acids and polyphenols found in rosella. A related study by Aryani et al showed that the anthocyanin content in purple roselle (75.60 mg/l) was higher than that of red roselle (63.45 mg/l) [27].

The addition of 0.5% and 1% chitosan to the beverage formula results in a decrease in moisture content at C0 and C1. This is due to chitosan's hydrophobic qualities, which can inhibit the rate of water vapor production, lowering the quantity of water content in the solution [28]. The value of titratable acid in beverage formulations also shows higher yields than those made from red roselle, this is also because the number of organic acids contained in purple roselle is higher than red roselle. Ningrum et al. (2019) also claimed that purple roselle flowers had 16 more volatile chemicals than red rosella flowers, which had 11 volatile compounds [29].

Likewise, the results of the proximate content of the ash, protein, and fat content resulted in the highest content obtained in the F3 and F4 formulations with the addition of 1% chitosan, followed by F2 and F1 with 0.5% chitosan, while the lowest proximate values were obtained in the control formulations C0 and C1 without chitosan. The administration of chitosan in rosella functional drink formulations can increase the amount of protein and fat levels compared to without giving chitosan. This is owing to chitosan's positively charged amine NH2 group, which allows it to bind bioactive molecules in the formulation as well as negatively charged compounds like protein and fat [30]. The ash content also indicates the presence of minerals contained in natural materials, the more minerals the more the ash content [31].

Based on the by difference formula, carbohydrate levels are strongly influenced by other nutrients. According to Matsuo et al. (2019) the higher the carbohydrate value, the lower the other nutritional components (water, protein, fat, and ash). Vice versa, the lower the carbohydrates, the greater the other nutrients [32]. The amount of nutrient content from water, protein, fat, and ash will affect the carbohydrate content in the roselle extract functional drink. Functional drinks with drink formulations made from purple roselle with 0.5% or 1% chitosan showed a higher ash content compared to C0 and C1 without chitosan, this was due to the mineral content in the simplicial material of purple roselle extract that was greater than that of red roselle.

3.2 Antioxidant activity

The effect of chitosan administration at different concentration levels on the scavenging activity of DPPH radicals from the formulation of red and bronze roselle extract drinks can be seen in Figure 1. Metode DDPH is an in vitro method used to determine free radical scavenging activity. The DPPH method can be used to assess the antioxidant activity of particular substances as well as to evaluate and quantify the dose utilized in vitro. The inhibitory activity value given as % inhibition can be used to demonstrate DPPH radical scavenging activity [33].

Antioxidant properties can be expressed in the scavenging activity of DPPH radicals by calculating % inhibition, the greater the % inhibition value, the greater the antioxidant activity [34]. The scavenging ability on radical DPPH is shown at the highest inhibitory power value on F4 with the addition of 1% chitosan ranging from 18.08% to 34.10%, and the activity below is at F3 of 15.98% to 30.05%, F2 ranging from 14.08% to 31%, 95%, F1 ranged from 12.90% to 30%, and the smallest at C0 without chitosan administration was 10.05% to 27.15%.

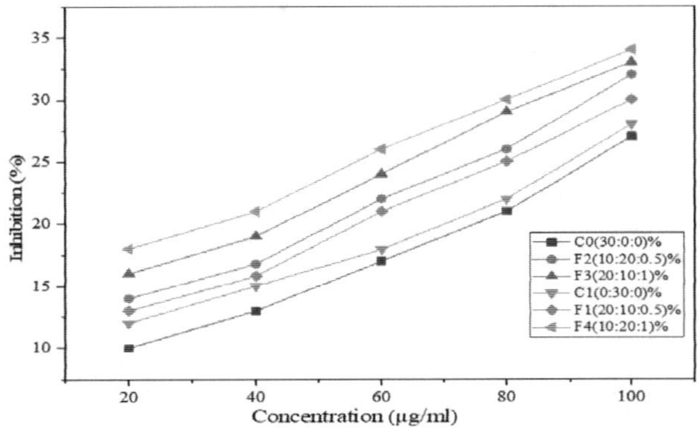

Figure 1. Antioxidant activity of functional drinks and control formulations from red and purple roselle with the addition of chitosan

According to studies, chitosan can boost the antioxidant value and the stability of electrostatic contacts in a ginger extract beverage compared to a beverage without chitosan [35]. Purple roselle-based drink formulations also have higher antioxidant activity than drinks containing red roselle ingredients, as evidenced by research that reveals purple roselle has higher antioxidant activity than red roselle [36].

4. Conclusion

The findings of this study show that adding chitosan to functional beverage formulations can boost physicochemical properties such as pH, moisture content, and titratable acidity. The results of the proximate content including protein, fat and ash content showed an increase with the addition of 0.5%

and 1% chitosan. The antioxidant test also showed an increase with the addition of chitosan. Drink formulations made from purple roselle show the results of the proximate and antioxidant content that is greater than the formula from red roselle, this is because the content of organic compounds and anthocyanins in purple roselle is greater than that of red roselle.

Acknowledgments
An institutional grant provided financial support for this study. The author would like to thank the Department of Pharmacy, Harapan Polytechnic, Tegal-Indonesia, for the supporting facilities that made this research possible.

References

[1] Pérez-Ramírez I F Castaño-Tostado E Ramírez-De León J A Rocha-Guzmán N E and Reynoso-Camacho R 2015 Food Chemistry 172 885
[2] Kamdaeng O and Singkaew K 2021 Food Research 5 366
[3] Ogundele O M A Awolu O O Badejo A A Nwachukwu I D and Fagbemi T N 2016 Food Sci. Nutr. 4 679
[4] Riaz G and Chopra R 2018 Biomed. Pharmacother. 102 575
[5] Da-Costa-Rocha I Bonnlaender B Sievers H Pischel I and Heinrich M 2014 Food Chemistry 165 424
[6] Mardiah Zakaria F R Prangdimurti E and Damanik R 2015 Procedia Food Sci. 3 182
[7] Wu H Y Yang K M and Chiang P Y 2018 Molecules 23 6
[8] Ifie I Marshall L J Ho P and Williamson G 2016 J. Agric. Food Chem. 64 4921
[9] Mgaya-Kilima B Remberg S F Chove B E and Wicklund T 2015 Food Sci. Nutr. 3 100
[10] Hafsa J Smach M A Charfeddine B Limem K Majdoub H and Rouatbi S 2016 Ann Pharm Fr. 74 27
[11] Xu T Gao C C Feng X Yang Y Shen X and Tang X 2019 Int. J. Biol. Macromol. 134 230
[12] Pu S Li J Sun L Zhong L and Ma Q 2019 Carbohydr. Polym. 211 161
[13] Genskowsky E Puente L A Pérez-Álvarez J A Fernandez-Lopez J Muñoz L A and Viuda-Martos M 2015 LWT - Food Sci. Technol. 64 1057
[14] Bertolo M R V Martins V C A Horn M M Brenelli L B and Plepis A M G 2020 Carbohydr. Polym. 228 115386
[15] Kim J H Hong W sung and Oh S W 2018 Int. J. Biol. Macromol.120 1468
[16] Zhang H He P Kang H and Li X 2018 Lwt 93 470
[17] Cheng J Zhu H Huang J Zhao J Yan B Ma S Zhang H and Fan D 2020 Food Sci. Nutr. 8 1987
[18] Jin Y Pei H Hu W Zhu Y Xu H Ma C Sun J and Li H 2017 Sci. Total Environ. 583 496
[19] Tesfay S Z and Magwaza L S 2017 Food Packag. Shelf Life. 11 40
[20] Yuan G Chen X and Li D 2016 food systems 89
[21] Zannou O Kelebek H and Selli S 2020 Food Res. Int. 133 109133
[22] Rasheed D M Porzel A Frolov A El Seedi H R Wessjohann L A and Farag M A, 2018 C Food Chemistry 250 236
[23] Nemo R and Bacha K 2020 Lwt 131 109713
[24] Latimer JrGW 2016 Official Methods of Analysis of AOAC International, twentieth ed. AOAC International, Gaithersburg, p. 3172.
[25] Cittan M Altuntaş E and Çelik A 2018 Ind Crops Prod. 122 553
[26] Rasweefali M K Sabu S Sunooj K V Sasidharan A and Xavier K A M 2021 Carbohydr. Polym. Technol. Appl. 2 100032
[27] Aryanti N Nafiunisa A and Wardhani D H 2019 Int Food Res J. 26 529
[28] Phisut N and Jiraporn B 2013 Int Food Res J. 20 1085
[29] Ningrum A Schreiner M Luna P Khoerunnisa F and Tienkink E 2019 Food Research 3 749
[30] Shen F Zhong H Ge W Ren J and Wang X 2020 Carbohydr. Polym. 234 115927
[31] Liu K 2019 Algal Research 40 101486

[32] Matsuo Y Miura L A Araki T and Yoshie-Stark Y 2019 Food Chemistry 279 356
[33] Liew S S Ho W Y Yeap S K and Bin Sharifudin S A 2018 PeerJ 8 1
[34] Goswami S Das R Ghosh P Chakraborty T Barman A and Ray S 2020 Ind Crops Prod. 154 112667
[35] Hu Y He C Jiang C Liao Y Xiong H and Zhao Q 2020 Food Hydrocoll. 104 1
[36] Kusnadi and Purgiyanti 2021 J. Phys. Conf. Ser. 1918 052007

The characteristics of bio-oil hydrothermal pyrolysis empty fruit bunches based on powder size and hydrogen gas pressure

Rina Novia Yanti[1], Erliza Hambali[2], Gustan Pari[3], Ani Suryani[4]

[1]Universitas Lancang Kuning, Pekanbaru, Indonesia
[2]IPB University, Bogor, Indonesia
[3]Indonesian Ministry of Natural Environment and Forestry Affairs, Indonesia
[4]IPB University, Bogor, Indonesia

*rinafahutan@unilak.ac.id

Abstract. Bio oil is the result of pyrolysis of biomass raw materials in the form of dark-colored liquids containing major chemical compounds such as benzene compounds, aromatic compounds, aliphatic, and other organic compounds. In the future, bio-oil is expected to be a substitute for fossil fuels as they are more environmentally friendly. Empty fruit bunches (EFB) from oil palm are one type of biomass that can be used as bio-oil products. The elements of pyrolysis that were analyzed were High Heating Value (MJ/kg), pH, viscosity, density and analysis of the components of hydrocarbon compounds formed using GCMS analysis. The results showed that HHV value 35.66 MJ/ - 39.09 MJ/kg, the pH value of bio-oil was somewhat similar to normal pH 4 - 6, and viscosity was 4 - 9 cp. The results of GCMS analysis showed that the content of the optimal percentage of hydrocarbon compounds was treated with 40 mesh powders with a hydrogen pressure of 15 bar. 19 types of hydrocarbon compounds were found with a total percentage of 13.79 %.

1. Introduction

Fossil fuel contributes negatively to the increase in air pollution such as increased CO2 gas pollution in the air, so that gradually fossil fuels will be replaced with environmentally friendly energy sources, such as energy with biomass raw materials. Research on energy sourced from the utilization of biomass such as biodiesel, and bioethanol is currently underway on the pyrolysis process, namely Bio oil. Indonesia is the country in the world which has the largest oil palm plantations which cover 12.3 million hectares with CPO production of 35.4 million tons (BPS, 2018).

One type of waste from CPO products is an empty fruit bunch (EFB) which has so far not been utilized to the maximum and has low economic value. Currently, EFB has only been used for compost that is disposed of in oil palm plantations. Empty fruit bunch (EFB) has a fairly high cellulose content of 52.93 %, hemicellulose 15.27 %, low lignin content of 17.43 % and has a calorific value of 18.14 KJ / gr 2 which is a suitable raw material for producing bio-oil through the hydrothermal pyrolysis process (Yanti et al, 2018). Solid waste from oil palm plantations has been utilized as a raw material in the pyrolysis process to produce bio-oil for this research. Pyrolysis results obtained cannot be directly called bio-oil as an alternative fuel, because the liquid resulting from this process still contains a lot of oxygen, water and acidic materials. The liquid product produced is called crude bio-oil (CBO) (Caprariis et al.2017) Kim et al.(2010).

Choi (2015) carried out rapid pyrolysis with palm kernel shells as raw materials and this produced CBO characteristics between 40-50%. Water content was still greater than 30% with a pH of 3-4. Tests conducted by Sulaiman and Abdulah (2011) from empty palm oil bunches resulted in 40% CBO and 30% water content. The utilization of waste stems, palm fronds and leaves has been studied by Abnisa et al.(2013) with a given pyrolysis temperature of 500 °C resulting in a CBO moisture content of 52% to 66% and a pH of 3.1–4.6.

Oramahi et al.(2015) also conducted research on palm oil trunks with a slow pyrolysis process resulting in a CBO yield of 42.05% with a water content of 9.26%. The results of research by Novianti et al.(2014) on bio-oil hydrothermal pyrolysis with empty fruit bunch (EFB) at 220 °C, with HHV value of 22.8 MJ/kg and with a high oxygen content of 32.94%. When comparing this biomass, the result of fast pyrolysis contains a lot of water (30-40%), high oxygen 30 % (Lehto et al.2014).

The CBO in this study used hydrothermal pyrolysis (HP) technology., where HP is carried out by adding liquid smoke (solvent) from solid waste of oil palm plantations as solvents. Bio oil belongs to a second generation of bioenergy, in the form of dark-colored liquid fuel, flavored like smoke and produced from biomass that can be used to substitute types of fuels such as gasoline, diesel (Bridgwater, 2012; Basu, 2013).

It is expected that the use of EFB as a raw material for bio–oil can be a substitution for fossil fuels and can reduce CO2 pollution in the air. The purpose of this study is to study the characteristics of EFB bio-oil produced through the hydrothermal pyrolysis (HP) process, comparing differences in powder size and amount of hydrogen gas provided. The provision of hydrogen gas and wet raw materials is expected to accelerate the disconnection reaction of carbon chain polymers.

2. Methodology

The materials used in this study are EFB powder size 40.60, 80, hydrogen gas pressure 0.10,15 bar and EFB liquid smoke as a solvent. Bio production in this study used the Hydrothermal Pyrolysis (HP) process, including slow pyrolysis. The HP process begins by inserting EFB powder mixed with liquid smoke (1 : 1) into the pyrolysis reactor, the reactor is closed-vacuumed for 30 minutes, after which hydrogen gas is pumped into the tube with a variation of pressure of 0.10 bar and 15 bar. Added solvents and hydrogen gases act as catalyst and can serve as a contribution to energy as depolymerization takes place faster when compared to the pyrolision of dry raw material without hydrogen (Akhtar and Ami, 2011).

The reactor is ignited with an initial temperature of 40 °C - 315 °C after that the temperature is maintained for 2 hours with the maximum pressure of the reactor is 100 bar and the reactor is cooled to a reactor temperature of 40 °C. The reactor is restarted to a temperature of 450 °C. During the heating process, at 100 °C, liquid comes out of the condenser pipe that is attached to the reactor.

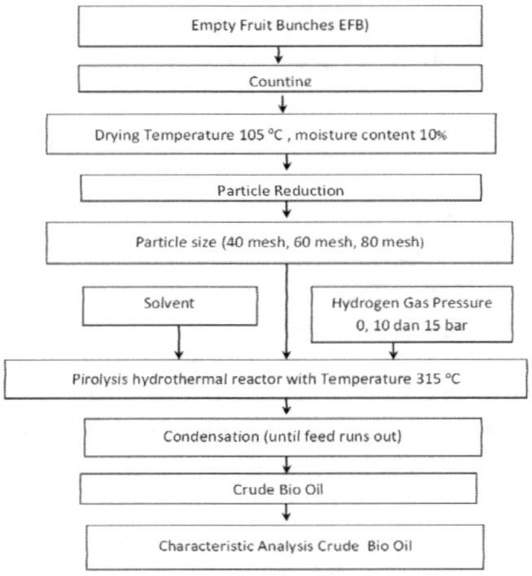

Figure 1. Bio Oil Production Process

IOP Conf. Series: Earth and Environmental Science 1041 (2022) 012002 doi:10.1088/1755-1315/1041/1/012002

Tools to measure the characteristics of bio oil are an automatic total acid number tester to measure pH, capillary-type viscometer and Visco Clock to measure viscosity, and Vario EL cube Elemental, Analyser GC – MS (Agilent Technology) for content of other chemical elements such as C, H, N, S, O. HHV (MJ/kg) is obtained using a formula.

HHV (MJ/kg) = 0.336C 1.418H - (0.153 – 0.00072)O 0.0941S (Maffut in Abnisa et al, 2013)

Table 1. Experimental conditions

Material	Treatment
EFB Powder	40, 60, 80 mesh
Hydrogen	0 bar, 10 bar, 15 bar
Temperature HP	40 °C – 315 °C
Temperature of condensation process	40 °C – 450 °C

3. Result and Discussion

Hydrothermal Pyrolysis is the conversion of biomass material into liquid fuel using wet raw material, using pressure to break down the lignocellulose polymers structure with an operating temperature of 270 °C – 340 °C and pressures of 40 – 220 bar 6-7. The HP process in this study worked at a maximum temperature of 315 °C and a maximum pressure of 100 bar (adjusted to reactor conditions). HP results from the combustion conversion process without oxygen produces 4 types of products, namely bio oil, liquid smoke, solid residues of charcoal and gas (Basu, 2013).

The characteristics of HP bio-oil including pH, density (kg/m3), viscosity (cp) and HHV (Mj/kg) can be seen in table 2. The pH value of bio-oil from this study is close to the normal pH of 4.1 – 6.56 with a powder treatment of 60 mesh hydrogen gas pressure 10 bar. The pH value of HP products is higher than the result of fast pyrolysis, namely pH 2-3 (Bridgwater, 2012; Chang, 2014) while the results of the study (Yang et al, 2016) of the HP process has a pH value of 4.8. The acidity level of bio-oil greatly affects the durability of machine tools because high acid pH will cause corrosion.

Table 2. Characteristics Of Crude Bio Oil Empty Fruit Bunches of Palm Oil

No	Analysis	Treatment								
		1.1	1.2	1.3	2.1	2.2	2.3	3.1	3.2	3.3
1	pH	4.85	4.18	4.90	6.49	6.56	6.38	5.81	5.79	5.
2	Viskositas (cp)	4.22	6.65	5.63	8.64	9.30	9.33	6.89	6.84	6.
3	Carbon (%)	71.40	76.20	74.10	70.85	71.60	75.67	72.49	70.70	69.
4	Hydrogen (%)	8.38	9.63	9.92	10.23	8.75	9.81	8.13	8.98	9.
5	Nytrogen (%)	2.38	2.38	3.27	0.47	2.57	1.77	1 .88	1.68	2.
6	Total Sulfur (%)	0.81	0.60	0.71	0.90	0.70	0.66	0.30	0.21	0.
7	Oxygen (%)	11.03	11.19	12.00	18.17	16.38	13.09	17.20	18.43	18.
8	HHV (MJ/kg)	35.66	39.06	38.75	38.09	36.26	39.13	35.72	36.33	36.
9	LHV (MJ/kg)	35.83	36.96	36.59	35.86	34.35	36.99	33.94	34.37	34.

Information:

1.1 : Particle zise 40 mesh, Hydrogen Gas Pressure 0 bar
1.2 : Particle zise 40 mesh, Hydrogen Gas Pressure 10 bar
1.3 : Particle zise 40 mesh, Hydrogen Gas Pressure 15 bar
2.1 : Particle zise 60 mesh, Hydrogen Gas Pressure 10 bar
2.2 : Particle zise 60 mesh, Hydrogen Gas Pressure 15 bar
2.3 : Particle zise 60 mesh, Hydrogen Gas Pressure 15 bar
3.1 : Particle zise 80 mesh, Hydrogen Gas Pressure 0 bar
3.2 : Particle zise 80 mesh, Hydrogen Gas Pressure 10 bar
3.3 : Particle zise 80 mesh, Hydrogen Gas Pressure 15 bar

The density (Kg/m3) of bio-oil is an influential factor in the quality of fuel ignition. The lowest bio-oil density value (Kg/m3) was for the EFB 40 mesh powder particle size treatment and 10 bar hydrogen gas pressure while the highest was the 80 bar powder particle size treatment. Viscosity is related to fuel storage properties. The lower the viscosity value, the better the quality of the fuel produced. The viscosity value of the results of this study was in the range of 4.22 cp – 9.33 cp.

The maximum calorific value (HHV) is influenced by high carbon content and low oxygen content. The results of this study showed the highest calorific value of 39.76 MJ/kg was produced in the particle size treatment of 40 mesh powder and 15 bar hydrogen gas and the lowest of 33.66 MJ/kg in the particle size treatment of 40 mesh powder without pressure. The effect of adding hydrogen gas and solvent during the HP processes is to speed up the depolymerization process of long-chain lignocellulosic polymer structures consisting of hydrogen, oxygen and long chain hydrocarbons into short-chain hydrocarbons.

The HP process is also affected by the final temperature and pressure of the Pyrolysis (Yang et al, 2016). This study was conducted with the same temperature for each treatment which was a maximum of 315 °C with the resulting final pressure differing, depending on the length of pyrolysis and the amount of hydrogen gas added.

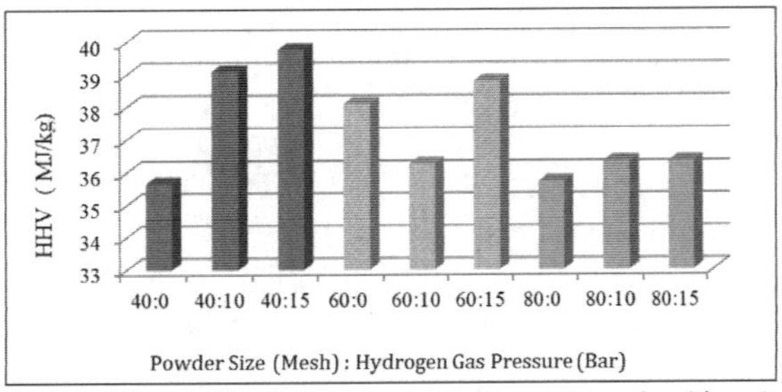

Figure 2. The Calorific Values Produced with Different Powder Size And Gas Pressure.

Particle size

The particle size during pyrolysis is also a determining factor in the resulting yield. The smaller the particle size, the more yield is produced. However, in this study, the main research is the physical and chemical characteristics of bio-oil, so yield is not the main factor being examined. The highest treatment in this study was a powder size of 60 mesh.

According to Gollakota et al.(2018) the size of the powder does not affect the property of bio oil from the HP process produced, but it is possible that the number of peak hydrocarbon compounds found is influenced by the amount of hydrogen gas pressure applied. This is because a powder of 80 mesh does not maximize the occurrence of depolymerization of cellulose and hemicellulose compounds due to the clumping of powder at the time of pyrolysis. Powder clumping during pyrolysis will cause incomplete combustion during the process (Lehto 2014).

Chromatography Mass Spectrometry (GC-MS) Analysis

GC-MS analysis shows the results of the content of chemical compounds contained in bio oil. Analysis of bio-oil samples undergoing 6 treatments, namely differences in powder size (40, 60, and 80 mesh) and differences in the amount of hydrogen pressure (0, 10, 15 bar). The results of GC-MS analysis found several groups of compounds consisting of aromatic aliphatic hydrocarbon compounds, acid compounds and other types.

This research focuses on the production of bio-oil as an alternative fuel substitute for gasoline or diesel, so the key emphasis in the results of GC-MS analysis is the presence of aliphatic hydrocarbon compounds as flammable compounds. The study found 43 types of aromatic compounds, 36 types of aliphatic compounds and 28 types of acid compounds.

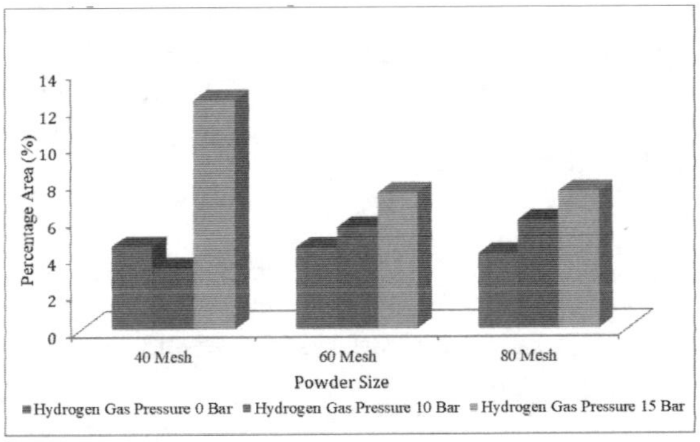

Figure 3. Graph Of the Percentage Area Of Hydrocarbon Compound Found From GCMS Bio Oil Analysis

Hydrocarbon compounds in table 3 show 19 types of hydrocarbon compounds that have the similarities above 70%, meaning that they have a series of straight-chain hydrocarbon functions. There are several peaks that are not reported because they are not identified, and the similarity is less than 70%. The number of peak areas of hydrocarbon compounds found, also has an effect on the value of HHV. The higher the HHV value the more hydrocarbon compounds. The presence of aromatic compounds and aliphatic compounds is also influenced by the decomposition of the cellulose and hemicellulose content found in raw materials (Caprariis et al.2017).

Table 3. The Percentage of Aliphatic Hydrocarbon Compounds From GCMS Bio Oil Analysis

No	Name of Compound	Treatment								
		1.1	1.2	1.3	2.1	2.2	2.3	3.1	3.2	3.3
		Area Percentage (%)								
1	Decane	-	-	0.88	-	-	0.68	-	-	0.56
2	Dodecane	0.45	-	1.59	-	0.52	0.25	-	-	-
3	Heptene	-	1.85	-	-	-	-	-	-	-
4	Heptadecane	1.87	-	0.25	-	1.54	1.87	1.05	0.5	0.47
5	Hexane	-	-	-	-	-	1.37	-	-	1.05
6	Hexadecane	-	-	0.44	0.43	0.4	0.76	0.77	1.12	0.72
7	Naphthalene	1.01	0.64	0.27	-	0.68	0.99	-	-	-
8	Nonadecane	-	0.21	-	0.34	0.73	-	-	-	-
9	Nonane	-	-	0.68	-	-	1.95	-	-	-
10	Octadecane	0.1	1.29	-	0.19	-	0.43	0.53	0.15	-
11	Pentadecane	-	0.11	2.32	-	0.91	0.12	0.77	1.02	1.31
12	Pentane	0.06	-	1.95	-	-	-	-	-	-
13	Pentadecanone	-	1.08	-	-	1.08	1.26	-	-	-
14	Tricosane	-	-	-	-	1.17	0.35	-	-	-
15	Tridecanone	-	-	-	-	-	1.35	-	-	-
16	Tridecane	-	2.28	2.27	0.51	-	-	-	0.62	0.91
17	Tetradecane	-	1.12	1.26	2.43	0.71	2..27	0.49	0.84	0.95
18	Tetradecanone	0.2	-	-	-	-	1.59	-	-	-
19	Undecane	-	-	1.86	-	-	0.88	0.31	-	-
	Total % Area	3.69	8,.58	11.91	3.9	7.74	15.24	3.61	4.25	5.97

Note:

1.1 : Powder Zise 40 mesh, Hydrogen Pressure 0 bar
1.2 : Powder Zise 40 mesh, Hydrogen Pressure 10 bar
1.3 : Powder Zise 40 mesh, Hydrogen Pressure 15 bar
2.1 : Powder Zise 60 mesh, Hydrogen Pressure 10 bar
2.2 : Powder Zise 60 mesh, Hydrogen Pressure 15 bar
2.3 : Powder Zise 60 mesh, Hydrogen Pressure 15 bar
3.1 : Powder Zise 80 mesh, Hydrogen Pressure 0 bar
3.2 : Powder Zise 80 mesh, Hydrogen Pressure 10 bar
3.3 : Powder Zise 80 mesh, Hydrogen Pressure 15 bar

4. Conclusion

Hydrothermal pyrolysis with EFB as raw materials will produce hydrocarbon compounds such as butanone, dodecane, heptadecane, tetradecane, hexadecane, tridecane and undecane. The larger the area of hydrocarbon compounds produced by HP, the higher the HHV values. The results of the study showed that the ph value of bio-oil had a normal tendency of 6.56. The particle size of the EFB material and the amount of hydrogen gas given will affect the characteristics of bio-oil. The optimal particle size is 40 mesh with a hydrogen gas pressure of 15 bar. There needs to be a continuation of research into the effects of upgrading bio-oil by using catalysts.

Acknowledgment

The authors would like to gratefully acknowledge support : (1) Ministry of research and Technology Directorate General of Higher Degree Education, Republic of Indonesia. (2) Surfactant Bioenergy Research Center IPB Bogor Indonesia. (3) Forestry research Center of Development and Innovation, Ministry of Environment and Forestry Bogor Indonesia.

References

[1] Abnisa F, Arami-Niya A, Wan Daud WMA, Sahu JN, Noor IM. 2013. Utilization of oil palm tree residues to produce bio-oil and bio-char via pyrolysis. Energy Convers. Manag. 76 1073–1082. doi:10.1016/j.enconman.2013.08.038

[2] [BPS] Badan Pusat Statistik. 2019. Statistik Kelapa Sawit Indonesia. Jakarta (ID).

[3] Basu P. 2013. Biomass Gasification, Pyrolysis and Torrefaction. Practical Design and Theory. (second Edition, ed.). India: Elsevier.

[4] Bridgwater, A, V. 2012. Review of fast pyrolysis of biomass and product upgrading. Biomass and Bioenergy,38, 68-94.

[5] Caprariis B, De Filippis P, Petrullo A, Scarsella M. 2017. Hydrothermal liquefaction of biomass: Influence of temperature and biomass composition on the bio-oil production. Fuel. 208: 618–625. doi:10.1016/j.fuel.2017.07.054.

[6] Choi GG, Oh SJ, Lee SJ, Kim JS. 2015. Production of bio-based phenolic resin and activated carbon from bio-oil and biochar derived from fast pyrolysis of palm kernel shells. Bioresour. Technol. 178 99–107. doi:10.1016/j.biortech.2014.08.053

[7] Chang SH. 2014. An overview of empty fruit bunch from oil palm as feedstock bio oil production. Biomass and Energy. 62 : 174 - 181. doi : 10.1016/j.biombiooe.2014.01.002.

[8] General D of EC.2017. Tree Crop Estate Statistic of Indonesia. Jakarta Indonesia.

[9] Kim SJ, Jung SH, Kim JS. 2010. Fast pyrolysis of palm kernel shells: Influence of operation parameters on the bio-oil yield and the yield of phenol and phenolic compounds. Bioresour. Technol. 101(23) 9294–9300. doi:10.1016/j.biortech.2010.06.110

[10] Lehto J, Oasmaaa A, Solantausta Y, Kyto M, Chiaramonti. 2014. Review of fuel oil quality and combustion of fast pyrolysis bio-oils from lignocellulosic biomass. ap2014.plied Energy 116 :b178–190.

[11] Novianti S, Biddinika MK, Prawisudha P, Yoshikawa K. 2014. Upgrading of Palm Oil Empty Fruit Bunch Employing Hydrothermal Treatment in Lab-scale and Pilot Scale. Procedia Environ. Sci. 20 46–54. doi:10.1016/j.proenv.2014.03.008

[12] Oramahi HA, Wahdina, Diba F, Nurhaida, Yoshimura T. 2015. Optimization of Production of Lignocellulosic Biomass Bio-oil from Oil Palm Trunk. Procedia Environ. Sci. 28(SustaiN 2014) 769–777. doi:10.1016/j.proenv.2015.07.090.

[13] Sulaiman F, Abdullah N. 2011. Optimum conditions for maximising pyrolysis liquids of oil palm empty fruit bunches. Energy. 36(5) 2352–2359. doi:10.1016/j.energy.2010.12.067.

[14] Yang, L. Nazari, L, Yuan, Z, Corscadden, K, Xu, C,C, He, Q.S. 2016. Hydrothermal liquefaction of spent coffee grounds in water medium for bio-oil production. Biomass and Bioenergy, 86, 191-198.

[15] Yanti RN, Hambali E, Pari G, Suryani A. 2018. The characteristics of palm oil plantation solid biomass wastes as raw material for bio oil. IOP Conf Ser Earth Environ Sci 141(Earth and Environmental Science),1-7.

The effect of proportion of ganyong starch and waste of straw rice on biodegradable foam production as sustainable packaging

Gresita Fitaloka Sari*

Universitas Internasional Semen Indonesia, Gresik, Indonesia

*gresita.sari17@student.uisi.ac.id

Abstract. Biodegradable foam (biofoam) packaging is a prospective packaging material that could developed from starch. Canna as a starch sources tuber of 93.30% consist of amylose content of 42.40% and amylopectin 50.90% is not utilized optimally. Nevertheless, starch-based foams are brittle, hydrophilic and rigid. Natural fiber-based material such a rice straw could be added as a cavity filler; thus, obtained a strong and sturdy product. The method used for producing biofoam was the baking process at a temperature of 170°C for 15 minutes. The aim of this study was to determine the effect of canna starch and rice straw fiber proportion on biofoam quality. The formulations used between canna starch and rice straw fiber were 4:0, 3:1, 2:2, and 1:3. Treatment with a ratio of 2:2 showed the best biodegradability for an average solubility and degradation of 18.900% and 82.693%, respectively. Meanwhile, the density parameter obtained from treatment with ratio of 1:3 of 0.225 gr/cm3 as the lightest density due contains more straw fiber. The results of the SEM test showed that the dough was not homogeneous, and no cavity formed in the biofoam structure. In addition, the results of the FTIR showed there was not alteration in the compound content from biofoam produced.

1. Introduction

People's lifestyles that are increasingly practical and consumptive encourage high consumption of plastic in various aspects of life, especially as packaging. The high dependence of humans on plastic packaging in everyday life cannot be separated as a packaging material for food and ready-to-eat food. Based on a report by Greenpeace Indonesia, the consumption of plastic packaging reaches 65% of the total national plastic consumption.

Indonesia is a consumptive country with the second largest plastic waste contributor in the world after China. Based on data obtained from the Indonesian Plastic Industry Association (INAPLAS) and the Central Statistics Agency (BPS), plastic waste in Indonesia reaches 64 million tons/year, of which 3.2 million tons are plastic waste dumped into the sea. According to the same source, 10 billion pieces of plastic bags are disposed of into the environment per year or as many as 85,000 tons of plastic bags. Meanwhile, plastic consumption per capita in Indonesia reaches 20 kilograms per year.

Styrofoam is one type of plastic that is often used. Due to its light weight, practicality and economical price, this material is widely used by the public as food packaging (Irawan, et al, 2018). Styrofoam is widely used as a disposable vegetable and fruit packaging container. However, styrofoam is not suitable for use as food packaging because it contains the chemical styrene which can migrate into the food being packaged. In addition to having a negative impact on health, Styrofoam also has a negative impact on the environment because of its non-biodegradable nature. Styrofoam waste treatment by burning below

1000°C will produce dangerous contaminants in the form of carcinogenic dioxins (Fikri & Veronica, 2018).

With the problems that arise due to the use of styrofoam, it is necessary to develop alternative packaging materials that are not harmful to human health and are more environmentally friendly. One of the bioplastic packaging that can be an alternative to styrofoam is biodegradable foam (biofoam). Biopolymer is a safe and environmentally friendly material to be used as a material for making biodegradable plastics (Iriani, 2013). Biopolymer raw materials that can potentially be used are starch and cellulose-based agricultural products or wastes. This is because the two materials have renewable properties, abundant availability and low price (Nofar, et al., 2015).

The use of starch must sometimes conflict with the interests of being used as food consumption so that government programs for food security can continue to be realized. One of the tuber materials with high starch potential and its utilization has not been explored optimally is canna tuber (Canna edulis Kerr). Canna is one of the typical Indonesian tubers that grows well in the tropics. Canna contains high starch of 93.30% with amylose content of 42.40% and amylopectin 50.90% (Harmayani, et al, 2011). Canna starch has high hydrophilic properties, biofoam products with a high hydrophilic level can cause biofoam to easily absorb water content in the material being packaged. Therefore, it is necessary to add material as a filler to fill empty cavities so that the resulting biofoam product has low hydrophilic properties.

The addition of cellulose fiber that can be used as a filler is rice straw. Rice straw was chosen because it has a cellulose content of 39.1%. The addition of materials that have a high cellulose content will produce biofoam products that have strong and sturdy properties. To be able to replace the use of styrofoam as ready-to-eat food packaging, the characteristics of the biofoam produced are as close as possible to the characteristics of styrofoam such as having light weight, flexibility, and water resistance. Therefore, it is necessary to add synthetic polymers such as Polyvinyl Alcohol (PVA) to reduce the water-absorbing properties and it is necessary to add a plasticizer in the form of glycerol to give the biofoam elastic properties so that it does not break easily.

2. Methodology

This research on making biofoam uses the formula of canna tubers and rice straw of 4:0, 3:1, 2:2 and 1:3. Making biofoam is done by mixing the ingredients using a mixer until all the dough is well mixed. After the ingredients are mixed, the dough is weighed 26 grams and put into a mold in the form of an acrylic plate/mold measuring 14x9 cm with a thickness of 2 mm. The baking process is heated in the oven for 15 minutes at a temperature of 170°C. The function of adding magnesium stearate to facilitate the removal of biofoam products from the mold. Meanwhile, the addition of water is carried out proportionally to the material. Table 1. shows the formulation of the materials used.

Table 1. Material Formulation

Sampel Code	UG : JP Ratio	UG (gr)	JP (gr)	PVA (%)	Glycerol (%)	Mg stearate (%)
V1	4 : 0	50	0	30 %	5 %	1,5%
V2	3 : 1	37,5	12,5	30 %	5 %	1,5%
V3	2 : 2	25	25	30 %	5 %	1,5%
V4	1 : 3	12,5	.37,5	30 %	5 %	1,5%

2.1 Test Parameters

a. Density Test (ASTM D 792-08)

Density testing is done by calculating the mass and volume of the sample. Measurement of mass by weighing the sample on an analytical balance, while the volume is calculated by multiplying the length, width and thickness of the sample. The purpose of conducting a density test is to determine the mass of the sample so that lightweight packaging can be produced (Ritongga, 2019).

$$\rho \text{ (density)} = \text{massa (gr)} / \text{volume (cm}^3\text{)} \quad (1)$$

b. Water Solubility Test

Solubility test was conducted to see the solubility of biofoam in water. The solubility test was carried out by weighing the initial weight of the biofoam sample (W0), then soaking it in distilled water for 24 hours. After 24 hours, the biofoam was then dried and re-weighed as the final weight (W1) (Harumarani, et al, 2016).

$$\text{Percent Solubility} = \frac{(W0-W1)}{W0}\ x\ 100\ \% \qquad (2)$$

c. Biodegrability Test (EN13432)

This test aims to determine the size of the sample that is decomposed by microorganisms in the soil. According to the International standard (ASTM 5336) the length of time for biofoam to degrade is at least 60 days. In the biodegradation test, the physical appearance and weight of the biofoam were observed by calculating the density of the biofoam before (W0) and after (W1) was placed in the soil (Ritonga, 2019).

$$\text{Weight loss (\%)} = \frac{(W0-W1)}{W0}\ x\ 100\ \% \qquad (3)$$

3. Result and Discussion

3.1 Biofoam Canna Starch and Straw Density

The morphological structure will also affect the physical characteristics of the biofoam such as the density of the biofoam. Structures with large and many cavity sizes will produce biofoam with low density. A low-density value indicates that the foam has a low weight (Warsiki et al., 2012). When compared with the density of commercial styrofoam with a density value of 0.035 g/cm3, the density of biofoam produced in this study is still quite high. Where the density of biofoam in this study ranged from 0.217 to 0.495 g/cm3. The average value of density in the biofoam sample with the ratio of canna starch and straw fiber 4:0, 3:1, 2:2, and 1:3 respectively was obtained at 0.472 gr/cm3, 0.441 gr/cm3, 0.265 gr/ cm3 and 0.225 gr/cm3.

In this study, samples with higher starch addition ratio resulted in higher density. Meanwhile, an increase in the ratio of straw fibers resulted in a lower density. The high density in the sample with a higher starch ratio is thought to be because canna starch absorbs more water than rice straw fiber. Absorption of more water causes the dough to become thick, and the expansion process becomes hampered because water as a blowing agent is reduced so that the resulting biofoam has a denser structure.

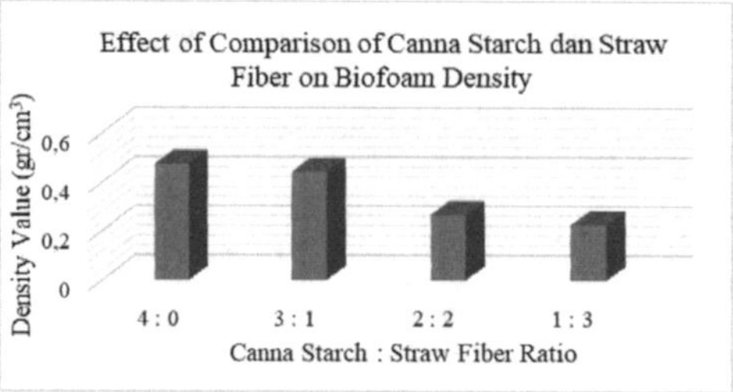

Figure 1. Average Density Value of Biofoam Canna Starch and Straw Fiber

While the low density value can be caused because the fiber from the straw has not been able to fill the empty spaces in the matrix so that it has not been able to increase the density of biofoam (Iriani, 2013). In addition, when adding straw fiber there will be times when the optimum point will be reached which makes the density value of the biofoam decrease. This is because the more fiber as a filler used, the more uneven the interaction between the filler and the matrix so that there is a filler that does not bind to the matrix (Clareyna & Mawarani, 2013). As a packaged product, biofoam is expected to have a low density because it affects the overall product weight. In this study, the ratio of canna starch and straw fiber 3:1 was the best result from the density test because it had the lowest density as can be seen in Figure 1.

3.2 Biofoam Canna Starch and Straw Water Solubility

The solubility test of biofoam in water was carried out to determine the strength of the structure of biofoam against water and also to determine the ability of biofoam to dissolve in water because in its application biofoam is a disposable product, where biofoam packaging waste is often found on the ground and in waters. With the solubility test in water, it is hoped that the biofoam waste found in the waters can dissolve in the water so that it does not cause buildup and pollution in the waters.

The results of the solubility test in this study can be seen in Figure 2, where the average % solubility of the biofoam sample with the ratio of canna starch and straw fiber 4:0, 3:1, 2:2, and 1:3 respectively was obtained at 65.656. %, 48.042%, 18.900% and 20.382%.

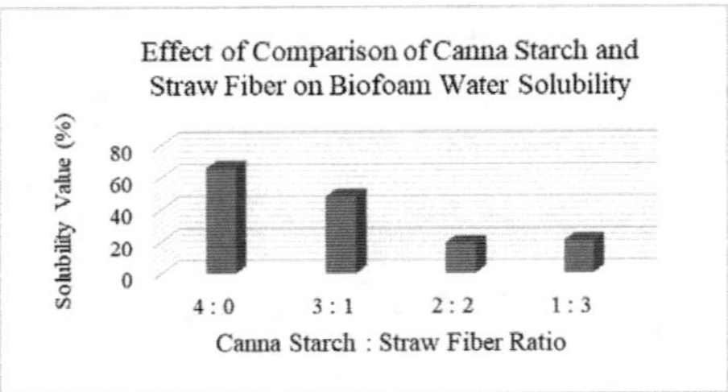

Figure 2. Average Water Solubility Value of Biofoam Canna Starch and Straw Fiber

Biofoam with the addition of more starch showed a higher solubility value. This could be due to canna starch having a high degree of swelling which causes biofoam to be easily degraded by water into small fragments. In the process of making biofoam, the expansion process that occurs will produce a hollow structure. If the biofoam is immersed in water, the water will fill the cavities. As a result, the solubility in water will increase.

However, if there is an addition of cellulose fiber, the resulting structure is more solid, and the cavity formed will be smaller. This causes less water absorbed to fill the cavity so that the solubility value of biofoam in water will also decrease. According to Zulfreriyeni et al (2014), films with low solubility values indicate that the film is good for use as packaging materials. In this study, the ratio of canna starch and straw fiber 2:2 was the best result because it had the lowest solubility value with an average value of 18.900%. Further testing showed that all samples of canna starch biofoam and straw fiber were completely dissolved after being soaked for approximately 2 weeks.

3.3 Biofoam Canna Starch and Straw Biodegrability

As environmentally friendly packaging, biofoam products must of course be easily degraded naturally, so it is necessary to observe the level of biodegradability of the biofoam. To determine the rate of decomposition of biofoam in the soil, a biodegradation test can be carried out. The faster the biofoam is decomposed by the soil, the better the biofoam can be. Cellulose has good water binding ability which allows microorganisms to live. Microorganisms will decompose the material well because cellulose is easily broken down compared to starch. This causes faster degradation of high cellulose materials.

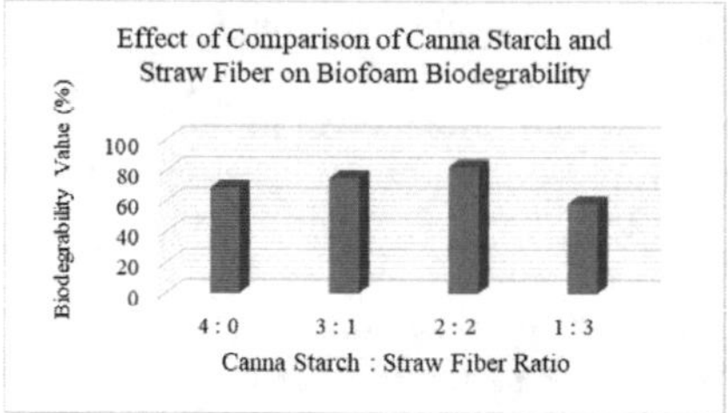

Figure 3. Average Biodegrability Value of Biofoam Canna Starch and Straw Fiber

The results of the biodegradation test showed that the biofoam samples with the ratio of canna starch and straw fiber 3:1 and 2:2 experienced an increase in % degradation along with the increase in the composition of the fiber contained. The average % degradation in biofoam samples with ratios of canna starch and straw fiber 4:0, 3:1, 2:2, and 1:3 respectively was obtained at 68.708%, 75.123%, 82.693% and 58.700%.

This shows that the average biodegradability of the canna starch and straw fiber biofoam products has been degraded at least more than 50% after 36 days or approximately 5 weeks. The best results in the biodegradation test were obtained in samples with the ratio of canna starch and straw fiber 2:2 as the highest degradation value.

3.4 Effect of Starch and Fiber Variations on Biofoam Characteristics by using Fourier Transform Infra-Red (FTIR)

Functional group analysis was carried out to find out what functional groups were contained in the biofoam produced based on the wave number where a peak appeared. From the results of the analysis shown in Figure 4, it can be seen that the functional groups contained in the biofoam sample are a combination of specific functional groups contained in the components such as starch, cellulose, PVA and glycerol.

Where the OH group comes from the constituent components of starch, namely amylose and amylopectin (Nurfitasari, 2018), cellulose which has OH, CO and CH groups, glycerol which has OH, CO and CH functional groups (Pratiwi, et al, 2016), and CH from alkanes derived from polyvinyl alcohol (PVA) compounds (Ritongga, 2019). The FTIR test results identify the presence of a typical functional group from the content of biofoam products in general, namely in the form of carbon (C) and hydrogen (H) elements which are common compounds of biofoam because most of the constituent materials consist of carbon and hydrogen elements. This indicates that the biofoam was produced from the physical blending process because no new functional groups were found.

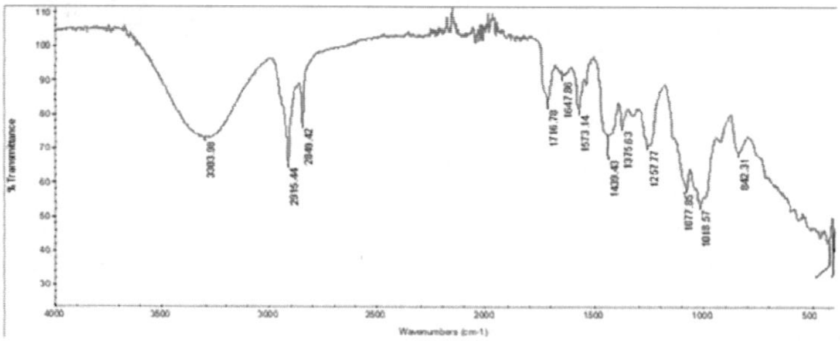

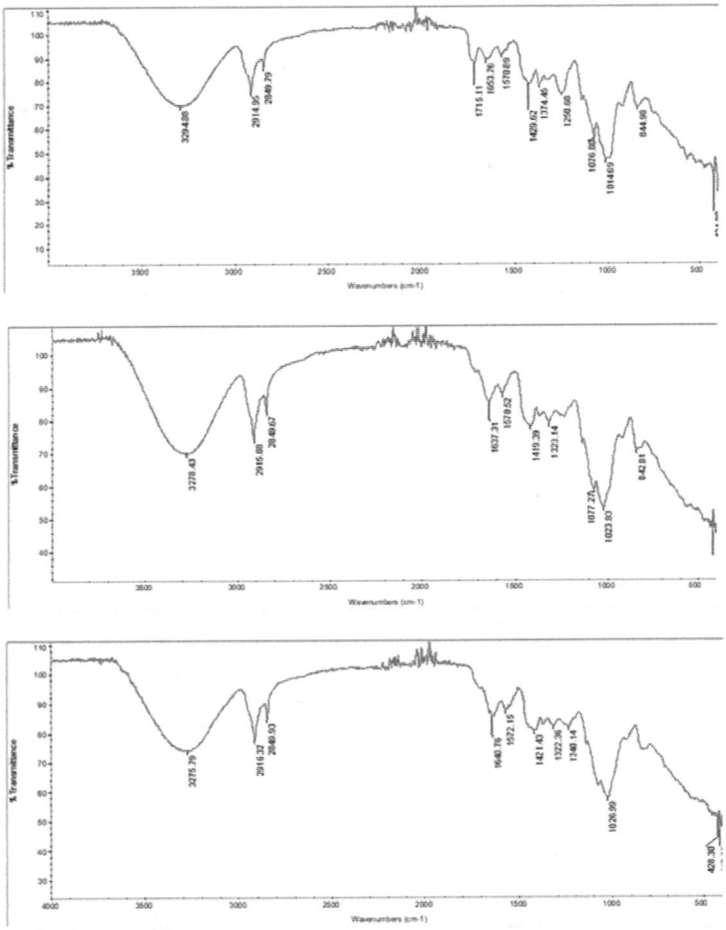

Figure 4. FTIR Spectrum of Biofoam with Ratio Canna Starch and Straw Fiber
(a) 4:0, (b) 3:1, (c) 2:2, (d) 1:3

3.5 The Effect of Starch and Fiber Variations on Biofoam Structure using Scanning Electron Microscope Analysis (SEM)

The results of SEM analysis are used to show the distribution of filler particles in the matrix so that it can be seen that the distribution of particles in the matrix is evenly distributed or not (Marbun, 2012). Based on SEM analysis conducted on biofoam with a magnification of 2,000x and 180x, it was found that the mixing of materials was not homogeneous in all biofoam samples. This can be due to the lack of mixing of the biofoam mixture and other ingredients that do not dissolve completely.

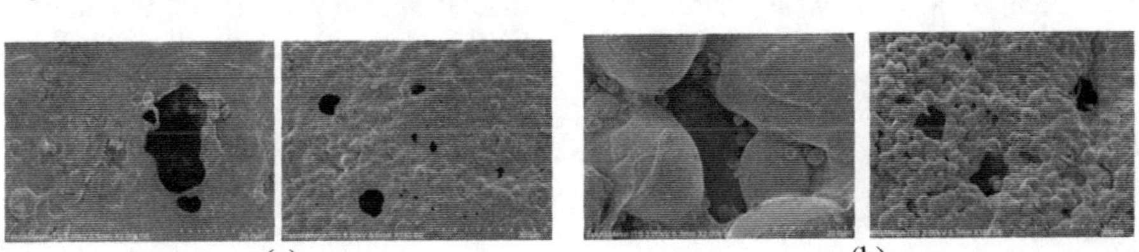

(a) (b)

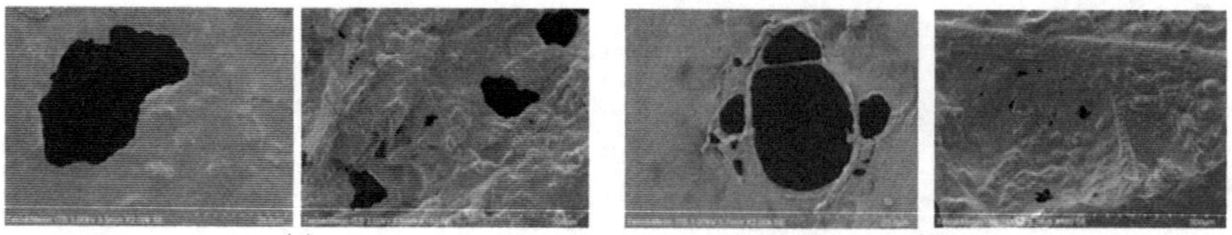

(c) (d)

Figure 5. Morphological Characteristics of Biofoam at Ratio of Canna Starch dan Straw Fiber (a) 4:0, (b) 3:1, (c) 2:2, (d) 1:3 with 2.000x andm180x magnification

When compared with the four samples of biofoam, there is a difference in biofoam with more starch or straw fiber composition. Biofoam with a higher starch composition still contains small lumps in the form of starch granules. While biofoam with more composition and straw, the lumps formed were not much but the fibers from the straw looked quite clear as can be seen in **Figure 5 (d)**. In the resulting biofoam structure, all biofoam samples still have no cavities in the biofoam structure. This can be caused because the heating process carried out in the oven is hampered.

4. Conclusion

The results showed that the ratio of addition of starch and fiber was able to affect the characteristics of the resulting biofoam product. Where the results of the density test show that the more fiber added, the smaller the density value of the biofoam. The best biodegradability was obtained from biofoam with a ratio of canna starch and straw fiber 2:2. In the water solubility test, the addition of cellulose causes the formation of a solid structure so that it is not easily broken down. While in the biodegradation test, the addition of cellulose serves to bind water, where microorganisms can live that function as decomposers. Thus, the degradation ability will be higher.

Meanwhile, several suggestions are needed for the development of further research, namely the need for updating the technology used to obtain biofoam with better quality, the need for an improvement in the composition of the biofoam manufacture of canna starch and straw fiber with fewer variations in the composition of straw fiber for future development and in the future. Further research is suggested to conduct several tests such as water absorption test, tensile strength test, modulus of elasticity test, elongation test and thermal conductivity in order to produce biofoam that meets the desired specifications and has better quality.

References

[1] Clareyna, E. D., & Mawarani, L. J. (2013). Pembuatan dan karakterisasi Komposit Polimer Berpenguat Bagasse. Jurnal Teknik Pomits Vol.2 No.2, 208-213.

[2] Fikri, E., & Veronica, A. (2018). Effectiveness of Carbon Monoxide Concentration Reduction on Active Carbon Contact System in Burning Polysteren Foam. Journal of Ecological Engineering, 19(4), 1-6.

[3] Harmayani, E., Agnes, M., & Griyaningsih. (2011). Karakteristik Pati Ganyong (Canna edulis) dan Pemanfaatan sebagai Bahan Pembuatan Mie. Program Studi Teknologi Pangan Fakultas Pertanian Universitas Gadjah Mada, 31(4), 3-4.

[4] Harumarani, S., Ma'aruf, W. F., & Romadhon. (2016). Pengaruh Perbedaan Konsentrasi Gliserol pada Karakteristik Edible Film Komposit Semirefined Karagenan Eucheuma Cottoni dan Beeswax. J. Peng. & Biotek. Hasil Pi. Vol.5 No.1, 101-105.

[5] Irawan, C., Aliah, & Ardiansyah. (2018). Biodegradable Foam dari Bonggol Pisang dan Ubi Nagara sebagai Kemasan Makanan yang Ramah Lingkungan. Jurnal Riset Industri Hasil Hutan Vol.10, No.1,, 33-42.

[6] Iriani, E. S. (2013). Pengembangan Produk Biodegradable Foam Berbahan Baku Campuran Tapioka dan Ampok. Bogor: Institut Pertanian Bogor.

[7] Marbun. (2012). Sintesis Bioplastik dari Pati Ubi Jalar Menggunakan Penguat Logam ZnO dan

Penguat Alami Selulosa, Skripsi Sarjana. Depok: Fakultas Teknik, Universitas Indonesia.
[8] Nofar, M. R., Ameli, A., & Park, C. B. (2015). Development of Polylactide Bead Foam with Double Crystal Melting Peak Structure. Polymer, 69(1), 83-94.
[9] Nurfitasari, I. (2018). Pengaruh Penambahan Kitosan dan Gelatin terhadap Kualitas Biodegradable Foam Berbahan Baku Pati Biji Nangka (Artocarpus heterophyllus). Makassar: Universitas Islam Negeri Alauddin Fakultas Sains dan Teknologi.
[10] Pratiwi, R., Rahayu, D., & Barliana, M. I. (2016). Pemanfaatan Selulosa dari Limbah Jerami Padi (Oryza sativa) sebagai Bahan Bioplastik. IJPST Vol.3, No.3, 83-91.
[11] Ritonga, A. U. (2019). Pembuatan dan Karakterisasi Biofoam Berbasis Komposit Serbuk Daun Keladi yang diperbkuat oleh Polivinil Asetat (PVAc). Medan: Program Studi Magister (S2) Fisika, Fakultas Matematika dan Ilmu Pengetahuan Alam, Universitas Sumatera Utara.
[12] Warsiki, E., Iriani, E. S., & Swandaru, R. (2012). Physical characteristics of microwave assisted moulded foam cassava starch-corn hominy. J Tek ki Indo. 10(2), 108-115.
[13] Zulferiyenni, Marniza, & Sari. (2014). Pengaruh Konsentrasi Gliserol dan Tepung Tapioka terhadap Karakteristik Biodegradable Film Berbasis Ampas rumpu Laut. Jurnal Teknologi dan Industri Hasil Pertanian 19(3), 257 - 273.

IOP Conf. Series: Earth and Environmental Science 1041 (2022) 012004 doi:10.1088/1755-1315/1041/1/012004

Physicochemical and antioxidant characteristics of black garlic from indigenous Indonesian garlic

K S Sasmitaloka, SM Widayanti, I Mulyawanti, ES Iriani

Indonesian Center for Agricultural Postharvest Research and Development, Bogor, Indonesia

*ks.sasmitaloka@gmail.com

Abstract. Black garlic is a processed garlic product which is obtained by heat treatment of the raw garlic at high temperature under high humidity for several days. It was reported that black garlic has stronger antioxidant activity than raw garlic. This research aimed to produce black garlic from indigenous Indonesian garlic and its characteristics. The raw material used was Sangga Sembalun garlic variety. The study was designed using a completely randomized design, consisted of one factor, namely the type of garlic clove (x1: single clove and x2: multi clove), with four replications. The results showed that when garlic undergoes heat treatment, various properties changes occur, including changes in physicochemical and antioxidant content. Single clove black garlic has antioxidant higher than multi clove black garlic. The single clove black garlic had moisture of 47.83%, fructose of 4.92%, glucose of 2.85%, sucrose of 1.36%, maltose of 0.63%, total flavonoid of 2054.9 ppm, total phenol of 335.43 ppm, and S-allyl cysteine of 635.09 mg/ 100 g. Meanwhile, the compound clove black garlic had total flavonoid of 1745.7 ppm, total phenol of 104.71 ppm, and S-allyl cysteine of 326.7 mg/ 100 g.

1. Introduction

Garlic (Allium sativum) is widely used worldwide as a spice, condiment, and traditional medicine for several diseases. Several studies observed that it had therapeutic effect as a anti-cancer, antibacterial, antiviral, anti-diabetic, anti-hypertensive, cardioprotective, hepatoprotective, hypolipidemic, and antioxidant effects because of its organosulfur and bioactive components [1] [2]. However, the consumption and application of fresh garlic as a traditional medicine is limited due to its pungent taste, odor, and tendency to cause gastrointestinal discomfort in some people [3]. Therefore, it had been developed for removing the unpleasant taste and odor of fresh garlic.

Black garlic is processed garlic that is produced by heating fresh garlic for a period of time under controlled high humidity. According to Liang et al., [4], heat processing is important for improving the taste of garlic and also critical to generate a functional food. Several studies report the changes of nutritional components, active composites and antioxidant capacity of black garlic. During heat processing to produce black garlic, many reactions occurred, including the Maillard reaction [5], which control the quality of black garlic. It may result in changes of the nutrient content, color, texture, taste, and flavor.

When compared with fresh garlic, black garlic does not release a strong off-flavor due to the reduced content of allicin, which was converted into antioxidant compounds such as bioactive alkaloids and flavonoid compounds during the aging process [6]. Several studies have demonstrated the health benefits of black garlic. It was reported that black garlic has stronger antioxidant activity than raw garlic.

Black garlic has stronger antioxidant activity compared with raw garlic and may have greater efficacy for preventing metabolic diseases and alcoholic hepatotoxicity [7] [8].

Indonesia has many varieties of local garlic, i.e Sangga Sembalun, Tawangmangu Baru, Lumbu Hijau, Lumbu Kuning, and others. However, limited information is available regarding changes in the quality of black garlic during thermal processing using Indonesian garlic as raw material. This research aimed to produce black garlic from indigenous Indonesian garlic and its characteristics. According to Tran et al., [9], single clove garlic is the product of atypical bulbing process of garlic under specific conditions, which exerted the higher antioxidant capacity and hepatoprotective effect than normal garlic, which was known as multi clove garlic. Therefore, this research also studied the characteristics of black garlic from different type.

2. Methodology

2.1. Materials

Raw material used in this research was Sangga Sembalun garlic variety obtained from Guci, Central Java. The main equipment used was magic com "Philips HD 3132", digital balance "OHAUS PA2202C" and others.

2.2. Black garlic production

Garlic was harvested approximately 115 days after planting [10] then cured until the moisture content 65-70%. After curing, 2 kg of garlic was heated in magic com at 50-60oC for 15 days [11] without addition of chemicals or microorganisms. The treatment consisted of one factor, namely the type of garlic clove (x1: single clove and x2: multi clove). After being heated, the black garlic is removed from the magic com, cooled, packaged in aluminum foil packaging to maintain its water content, and analysis the characteristics.

2.3. Analysis

Black garlic was analyzed their performance such as moisture content [12], protein content [12], lightness [13], glucose [14], fructose [14], maltose [14], sucrose [14], total phenol [15], total flavonoid [16], and S-allyl cysteine [7]. This research was conducted by Completely Random Design (CRD). All data reported as mean ± standard deviation (SD) of four replicates. The means were compared using t-student test with the significant level at 5% using SPSS software 21 version.

3. Result and Discussion

3.1. Lightness

Lightness is an important physical characteristic that affects the perception of consumer for black garlic product. Lightness represents the degree of brightness with the numbers 0-50 indicating dark colours and the number 51-100 indicating light colour [13]. Fresh garlic has lightness of 71.18-73.17, while black garlic has lightness of 10.75-13.91.

The lightness of fresh garlic is changed from white to dark brown and eventually became black with the increasing of thermal processing duration [17]. Therefore, the lightness of black garlic tends to decrease during heating. Choi et al., [18] mentioned that the colour changes of black garlic during heating are typically due to the Maillard reaction, known as non-enzymatic browning reaction. It can produce brown material due to the formation of melanoidin which affect the lightness of a material. The Maillard reaction was likely to occur in garlic that contains carbohydrates, amino acids and proteins [17].

The decrease of lightness is also affected by the moisture content. It appeared to affect the activation energy of non-enzymatic browning, and higher moisture content decreased the temperature dependence of browning [7]. Heat treatment in the black garlic production process increased the rate of moisture removal. Garlic heated at high temperatures is thought to require relatively higher activation energy for browning based on lower moisture content. There is no difference in lightness between single and multi-clove black garlic.

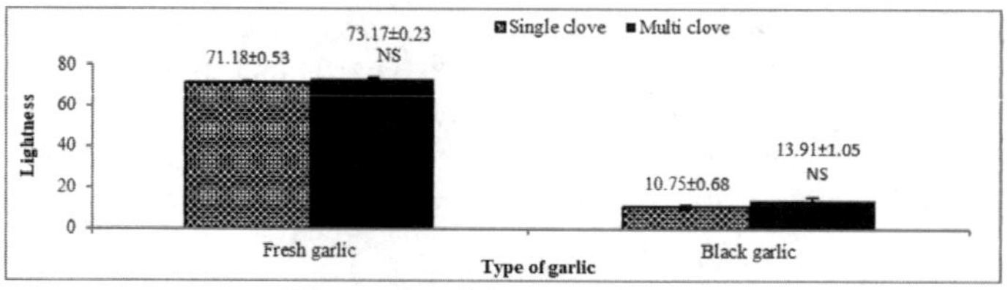

Figure 1. Lightness of garlic
NS = not significant; * = significant at the 5% level by t-student test

3.2. Moisture and protein content

Fresh garlic has moisture content in the range of 69.57-69.89%. The same condition of curing process cause fresh garlic has moisture content not much different. The heating process results in a decrease in moisture content of black garlic (Figure 2a). It is significantly decreased from 69.57-69.89% to 47.84-49.57%. Our result is similar with Zhang et al., [19] who stated that the moisture content decreased continuously during thermal processing. The decreased rate was faster at higher temperature.

The moisture content decreases due to evaporation. The longer of the heating causes the mass of water in the material to evaporate too. When it was decreasing, black garlic became soft and elastic. There is no difference in moisture content between single and multi-clove black garlic. This happens because the raw materials used, both single and multi-cloves, have moisture content that is not much different.

The heating process can increase the protein content of black garlic (Figure 2b). Fresh garlic had protein content of 6.98-9.47%. It was increase become 11.22-12.50% during processing. Heating is thought to increase protein synthesis. Some result reported by Sun and Wang [20], who mentioned that during heating process, the protein content dramatically increased within 6 days and then decreased. An increase in protein content will affect the color of the material because it can cause a Maillard reaction, namely the reaction between carbohydrates, especially reducing sugars with primary amine groups [21]. It will produce a brown to dark product.

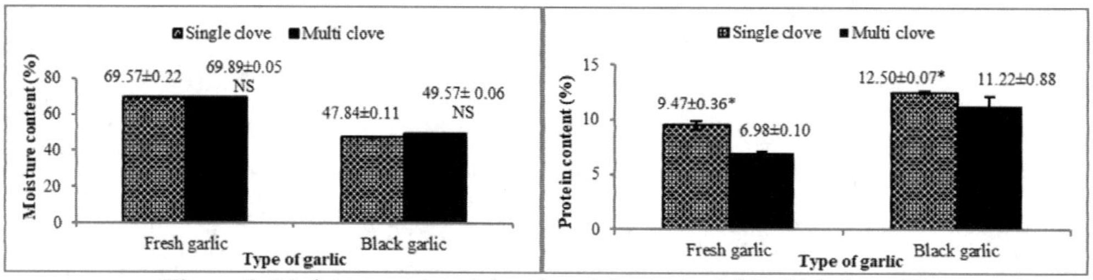

Figure 2. Moisture content (a) and protein content (b) of black garlic
NS = not significant; * = significant at the 5% level by t-student test

3.3. Glucose, fructose, maltose, and sucrose

Carbohydrates are the major constituents in fresh garlic bulb. According to Kimura et al., [6], fresh garlic contains approximately 63% of water and 28% of carbohydrate (fructans). In addition, the monosaccharide constituents of fresh garlic polysaccharide fraction were mainly fructose [22]. This statement is in line with this study, which the highest monosaccharide of fresh garlic is fructose (Figure 3).

Polysaccharides fractions were degraded to be mosaccharide or oligosaccharides during thermal processing. Zhang et al., [17] reported that the polysaccharide content in the black garlic decreased with the increasing of heating duration, whereas the monosaccharide and oligosaccharide increased gradually. In addition, during thermal processing of garlic showed that fructose was the major sugar in the largest quantity, followed by sucrose, glucose, arabinose, and maltose [14]. This statement is in line with this study, which the major sugar in black garlic is fructose (4.92%).

However, the second major sugar of black garlic in this research is glucose (2.85%), followed by sucrose (1.36%) and maltose (0.63%). The difference is thought to be caused by differences in the variety and location of the garlic used. The sugar content in black garlic was higher than fresh garlic. Black garlic consists of glucose increased 2.5 to 3.7-in fold, fructose increased 2.4 to 26.7-in fold, maltose 1.2 to 2 in-fold, and sucrose increased <1 in-fold the black garlic compared with fresh garlic (Figure 3).

This result in agreement with Kang [14], Zhang et al., [17], Martinez-casas et al., [23], and Yuan et al.,[5], who stated that sugar content (glucose, fructose, sucrose, and maltose) in black garlic was higher than fresh garlic. It was increased during thermal processing. Degradation of cell wall polysaccharides causes tissue softening. Therefore, the texture of black garlic is more gum-like than fresh garlic. It also could promote further hydrolysis of sucrose to be glucose and fructose. It tended to be supported by the increase in glucose and fructose content.

Furthermore, Generally, fructose is regarded as being 1.73 times as sweet as sucrose [17]. Moreover, the sweetness taste of black garlic might be attributed to the degradation of polysaccharides into fructose, glucose, and sucrose. Single clove black garlic have glucose, fructose, and sucrose higher than multi clove black garlic. Therefore, single clove black garlic have a sweeter taste than multi clove black garlic.

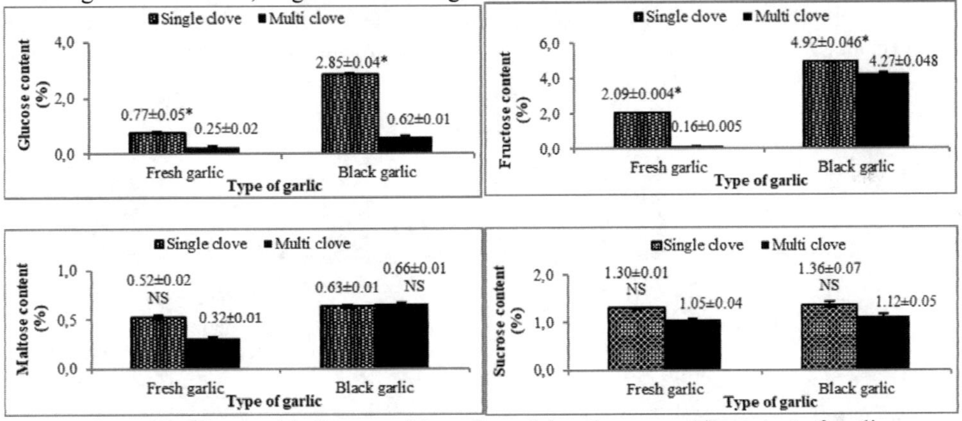

Figure 3. Glucose (a), fructose (b), maltose (c) and sucrose (d) content of garlic
NS = not significant; * = significant at the 5% level by t-student test

3.4. Total phenol and flavonoid content

Antioxidant content is the strongest property of black garlic. It representative by total phenols and flavonoid [3]. Phenolic and flavonoid content belong to the bioactive components of plant products and have good health-promoting activities [8]. It has interesting pharmacological properties and present in relatively high amounts.

Generally, vegetables processing caused in the breakdown of the cellulose structure of plant cell and improved the availability of bioactive component. Angeles [24] stated that during heat-treatment, unstable components in raw garlic are transformed into stable soluble components with a high antioxidant content. The total phenols and flavonoid content in garlic increased during the heating processes in this study (Figure 4). Heat treatments would lead to an increase in the free phenol forms from phenolic compounds, ester, glycoside or ester bound fractions or from other structures containing phenolic groups [22] [19]. Fresh garlic had total phenol of 25.19-46.88 ppm, while black garlic was 104.71-335.43 ppm.

The total phenol was increased by about 4 to 7-fold in the black garlic compared with the fresh garlic. These results are consistent with those obtained by Kim et al., [25], who reported that the concentration of the total phenol content in black garlic was significantly higher than fresh garlic. It was increased about 4-10-fold in the black garlic compared with the fresh garlic. The increase of total phenols content can improve the total antioxidant capacity of black garlic.

The higher concentration of total phenol was obtained in black garlic from single cloves (335.43 ppm). It increased 7-fold in single clove black garlic, compared to fresh garlic. Single clove fresh garlic had total phenol higher than multi clove fresh garlic. Therefore, single clove black garlic also had total phenol higher than multi black garlic (Figure 4a).

Single clove fresh garlic had total flavonoid higher than the multi one. After processing, single clove black garlic had total flavonoid 2054.9 ppm, while compound clove black garlic had 1745.7 ppm. Heat treatment also causes the disruption of the cell wall and produces changes in the extractability of flavonoids because high temperatures cause the disruption of the cell wall. The consequence is a release and therefore an increase of flavonoid compounds in the product that underwent the heat process [23].

In addition, heat treatment has a large influence on flavonoid availability, dependent on the magnitude and duration of treatment, their sensitivity to heat, and the physicochemical food environments [18]. Total flavonoids content of black garlic in this study increases around 1 to 2-fold compared with fresh garlic. The higher total flavonoid content was obtained in single clove black garlic (2054.9 ppm). The same as the total phenol, the total flavonoid content in single clove garlic is also higher than the multi one.

This is thought to cause the total flavonoid content in single clove black garlic is the highest in this study. Kim et al., [25] reported that the total flavonoid content was increased 1.1 to 3.5-fold in the black garlic compared with fresh garlic.

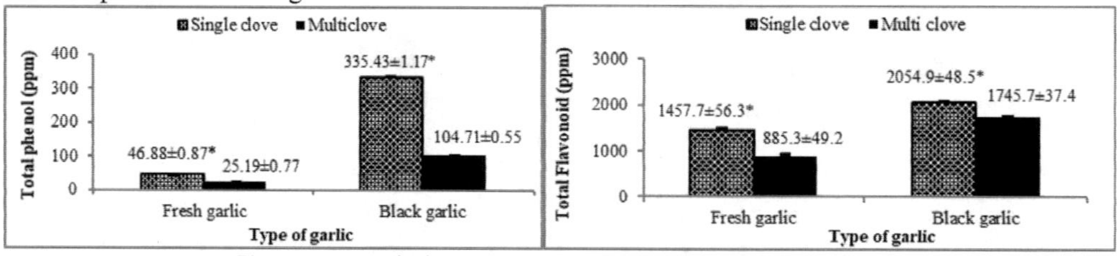

Figure 4. Total phenol (a) and total flavonoid (b) of garlic
NS = not significant; * = significant at the 5% level by t-student test

3.5. S-allylcysteine

S-allylcysteine (SAC) is a water-soluble sulphur compound with high antioxidant activity [7], which formed during the enzymatic hydrolysis of γ-glutamyl-S-allylcysteine, catalyzed by γ-glutamyl transpeptidase [27]. It is one of the It is one of the major Sulphur-containing amino acid components that is thought to be responsible for the beneficial of garlic [7]. S-allyl- cysteine content is an excellent indicator to evaluate the quality and bioactive value of black garlic, which can be used in nutraceutical and medical applications.

Single clove fresh garlic had SAC higher than multi clove garlic (Figure 5). This result is in line with Tran et al., [9]. It's increased during heat processing. This result is in line with Bae et al., [7], who stated that the amount of SAC increased significantly as the heat treatments process progressed. SAC content of black garlic in this study increases around 2 to 3-fold compared with fresh garlic, which single clove black garlic had SAC higher than multi clove black garlic.

There is not in line with our result. The amount of SAC in black garlic is generally five to six times higher than that in raw garlic [7] [28]. These differences are thought to be caused by different varieties and condition process of black garlic production. Some studies reported that SAC content influenced by temperature and time in black garlic manufacturing processing.

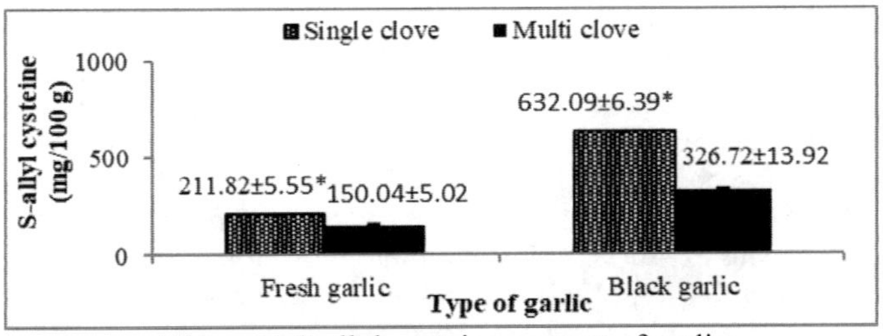

Figure 5. S-allyl cysteine content of garlic
NS = not significant; * = significant at the 5% level by t-student test

4. Conclusion

The results showed that when garlic undergoes heat treatment, various properties changes occur, including changes in physicochemical and antioxidant content. Lightness values of black garlic decreased during the heat treatment. Single clove black garlic has antioxidant higher than multi clove black garlic. The single clove black garlic had moisture of 47.83%, fructose of 4.92%, glucose of 2.85%, sucrose of 1.36%, maltose of 0.63%, total flavonoid of 2054.9 ppm, total phenol of 335.43 ppm, and S-allyl cysteine of 635.09 mg/ 100 g. Meanwhile, the multi clove black garlic had total flavonoid of 1745.7 ppm, total phenol of 104.71 ppm, and S-allyl cysteine of 326.72 mg/ 100 g.

References

[1] Jung Y, Lee S, Lee D, You M, Chung I K, Cheon W H et al. 2011 Fermented garlic protects diabetic, obese mice when fed a high-fat diet by antioxidant effects Nutr. Res. 31 5 387–396
[2] Santhosha S G, Jamuna P, Prabhavathi S N 2013 Bioactive components of garlic and their physiological role in health maintenance: A review Food Biosci. 3 59–74
[3] Ryu J H, Kang D 2017 Physicochemical properties, biological activity, health benefits, and general limitations of aged black garlic: A review Molecules 22 6 919–933
[4] Liang T, Wei F, Lu Y, Kodani Y, Nakada M, Miyakawa T et al. 2015 Comprehensive NMR analysis of compositional changes of black garlic during thermal processing J. Agric. Food Chem. 63 2 683–691
[5] Yuan H, Sun L, Chen M, Wang J 2016 The Comparison of the contents of sugar, amadori, and heyns compounds in fresh and black garlic J. Food Sci. 81 7 1662–1668
[6] Kimura S, Tung Y, Pan M, Su N, Lai Y, Cheng K 2017 Black garlic: A critical review of its production , bioactivity, and application J. Food Drug Anal. 25 1 62–70
[7] Bae S E, Cho S Y, Won Y D, Lee S H, Park H J 2014 Changes in S-allyl cysteine contents and physicochemical properties of black garlic during heat treatment LWT - Food Sci. Technol. 55 1 397–402
[8] Lu X, Li N, Qiao X, Qiu Z, Liu P 2016 Composition analysis and antioxidant properties of black garlic extract J. Food Drug Anal. 25 2 340–349
[9] Tran G, Dam S, Le N T 2018 Amelioration of single clove black garlic aqueous extract on dyslipidemia and hepatitis in chronic carbon tetrachloride intoxicated Swiss albino mice Int. J. Hepatol. 2018 1–10
[10] Sasmitaloka K S, Hidayat T, Arif A B, Jamal I B 2021 The dormancy breaking of garlic seeds through thermal shock storage methods and soaking in gibberellin acid in IOP Conf. Series: Earth and Environmental Science 653 12108
[11] Nelwida N, Berliana B, Nurhayati N 2019 Kandungan nutrisi black garlic hasil pemanasan dengan waktu berbeda J. Ilm. Ilmu-Ilmu Peternak. 22 1 53–64
[12] AOAC [Association of Official Analytical Chemist] 2006 Official Methods of Analytical of The

Association of Official Analytical Chemist. Washington, DC: AOAC
[13] Hunter Laboratories 2012 Measuring Color using Hunter L, a, b versus CIE L*, a*, b*. Reston, USA: Hunter Associates Laboratory Inc.
[14] Kang 2016 Physicochemical characteristics of black garlic after different thermal processing steps Prev. Nutr. Food Sci. 21 4 348–354.
[15] Lu X, Wang J, Al-Qadiri H M, Ross C F, Powers J R, Tang J et al. 2011Determination of total phenolic content and antioxidant capacity of onion (Allium cepa) and shallot (Allium oschaninii) using infrared spectroscopy Food Chem. 129 637–644
[16] Eghdami A, Sadeghi F 2010 Determination of total phenolic and flavonoids contents in methanolic and aqueous extract of Achillea millefolium Org. Chem. J. 2 81–84
[17] Zhang Z, Lei M, Liu R, Gao Y, Xu M, Zhang M 2015 Evaluation of alliin, saccharide contents and antioxidant activities of black garlic during thermal processing J. Food Biochem. 39 1 39–47
[18] Choi I S, Cha H S, Lee Y S 2014 Physicochemical and antioxidant properties of black garlic Molecules 19 10 16811–16823
[19] Zhang X, Li N, Lu X, Liu P, Qiao X 2015 Effects of temperature on the quality of black garlic J. Sci. Food Agric. 96 7 2366–2372
[20] Sun Y, Wang W 2018Changes in nutritional and bio-functional compounds and antioxidant capacity during black garlic processing J. Food Sci. Technol. 55 2 479–488
[21] Liu P, Lu X, Li N, Zheng Z, Zhao R, Tang X, Qiao X 2019 Effects and mechanism of free amino acids on browning in the processing of black garlic J. Sci. Food Agric. 99 10 4670–4676
[22] Ríos-ríos K L, Montilla A, Olano A, Villamiel M 2019 Physicochemical changes and sensorial properties during black garlic elaboration: A review Trends Food Sci. Technol. 88 459–467
[23] Martinez-casas L, Lage-yusty M, Lopez-hernandez J, 2017 Changes in aromatic profile , sugars and bioactive compounds when purple garlic is transformed into black garlic J. Agric. Food Chem. 65 49 10804–10811
[24] Angeles T M M 2016 Evolution of some physicochemical and antioxidant properties of black garlic whole bulbs and peeled cloves Food Chem. 199 135–139
[25] Kim J, Kang O, Gweon O 2013 Comparison of phenolic acids and flavonoids in black garlic at different thermal processing steps J. Funct. Foods 5 80–86
[27] Garcia-Villalon A L, Amor S, Monge L, Fernandez N, Prodanov M, Munoz M et al. 2016 In vitro studies of an aged black garlic extract enriched in S-allylcysteine and polyphenols with cardioprotective effects J. Funct. Foods 27 189–200
[28] Wang D, Feng Y, Liu J, Yan J, Wang M, Sasaki J, Lu C 2010 Black garlic (Allium sativum) extracts enhance the immune system Med. Aromat. Plant Sci. Biotechnol. 4 1 37–40

Efficiency analysis of vannamei shrimp supply chain performance in PT Sari Indonesia Group, Suak Pandan Village, West Aceh Regency, Aceh Province

Yoga Nugroho, Teuku Athaillah*, Muhibbul Qadri

Universitas Teuku Umar, Aceh, Indonesia

*athaillah.teuku@utu.ac.id

Abstract. Vannamei shrimp (Litopenaeus vannamei) is one of the perishable foodstuffs. Therefore, we need a good supply chain system so that vannamei shrimp can reach consumers while maintaining quality. This study aims to analyze the efficiency of supply chain performance at PT Sari Indonesia Group. The method used to measure the efficiency of supply chain performance is the Supply Chain Operation Reference (SCOR) method, starting with creating an initial hierarchy in the SCOR process (Plan, Source, Make, Delivery, and Return). The dimensions used are Reliability, Responsiveness, Flexibility, Cost, and Asset. The Analytical Hierarchy Process (AHP) and Objective Matrix (OMAX) methods were used to measure the supply chain performance efficiency. The results of this study indicated that the efficiency of supply chain performance at UD. Sari Inti Tani was in the GOOD category with a total index of 7,03. There are three categories included in the Red Zone, which the company must review immediately: the time to identify (check) employee performance, production quantities, and production costs.

1. Introduction

The fisheries sector has a significant influence on the development of the Indonesian economy and is one of the sectors relied upon to support national food security. In addition to contributing to Gross Domestic Product (GDP), the fisheries sector also plays a role in employment, increasing people's income and contributing to foreign exchange. Therefore, the fisheries sector becomes a superior sector and receives special attention from the government to continue developing. PT Sari Internasional Group is one of the companies that cultivate vannamei shrimp. The company is located in Suak Pandan Village, West Aceh Regency, and it has a large area on the coast. The choice of this type of shrimp is due to the characteristics of vannamei shrimp that are resistant to weather changes.

In addition, vannamei shrimp is one of Indonesia's leading fishery export commodities. Based on the export data of Indonesian fishery commodities in 2012, the exported shrimp commodities amounted to 96,406,844 kilograms consisting of tiger shrimp, vannamei shrimp, and giant prawns. The number of frozen vannamei shrimp exports was higher than that of giant prawns and tiger prawns. The enormous export potential makes the vannamei cultivation business quite promising.

PT. Sari Internasional Group (SIG) cultivates vannamei shrimp on a large scale. This company cooperates with other companies to support their production. Therefore, supply chain performance efficiency is needed to maximize every opportunity and minimize any potential losses. Research on supply chains should be the company's focus because the supply chain is the lifeblood of the company's business fluency. After all, it connects suppliers, companies, and customers (Harisnanda et al., 2012).

Not appropriately managed suppliers will have an impact on delays in procuring raw materials for the company. Ruteri and Qi yu (2009) say that strong relationships between members in the food industry supply chain are mandatory. Furthermore, the key to the performance level of multinational companies lies in the company's ability to cooperate with its business partners (Budiman 2013)

2. Literature Review

The supply chain is a network that works together to create and deliver a product to consumers. Prasetya (2009) states that a product will reach the end consumer after going through several processes, from the search for raw materials, the production process, and the distribution or transportation process. These processes involve various parties that are related to one another. Wuwung (2013) said the supply chain describes the entire process and activities to deliver products to consumers.

Assauri (2011) states that the supply chain concerns ongoing goods, money, and information relationships. Goods generally flow upstream to downstream, money flows from downstream to upstream, while information flows both ways from upstream to downstream and downstream to upstream. Hanugrani et al. (2011) said that determining supply chain performance measurement through the Supply Chain Operation Reference (SCOR) approach is needed. SCOR divides supply chain processes into five core processes, including Plan (planning process), Source (procurement process), Make (production process), Deliver (delivery process), and Return (return process).

Supply chain performance results from various efforts made by each supply chain member to meet the final supply chain goal, namely, customer satisfaction (Sari et al., 2014). Interdependence between partners in the supply chain network will strengthen the delivery of products and services upstream to downstream. Interdependence will also create long-term relationships that need each other. So that product availability and on-time delivery will be achieved (Rahmasari, 2011). A product will reach the end consumer after several processes from raw material search, production process, and distribution or transportation processes.

These processes involve various parties who are related to one another. Raw material providers (suppliers) supply the production needs of manufacturing companies that will process these raw materials into finished products. End-users received their finished product through distribution centers, retailers, small traders, and et cetera. This series of parties handling the product flow is called the Supply Chain (Prasetya, 2009).

In the supply chain, various types of information can be shared, including logistics, business, strategic, tactical, and others. These types of information are categorized as 1) Inventory Information; 2) Sales data; 3) Sales forecast; 4) Ordering information; 5) Information on product capabilities in the market; 6) New product exploitation information; and 7) Other Information. Inventory information such as information on expenditure, purchase of goods, receipt of goods, and others. Share sales data is also essential to represent customer demand and reduce losses caused by deficiency or excess of the products produced.

Information sharing will make a company quickly resolve a bottleneck in the supply chain and improve customer service quality. Product information can also help avoid the potential causes of the bullwhip effect. Information about the product allows consumers to receive timely supplies from suppliers when manufacturers get requests from retailers. There are also other types of information, such as product quality information, goods transportation information, plans, et cetera. (Lotfi, 2013).

3. Methodology

This research was conducted from February to April 2020 in Suak Pandan Village, Samatiga District, West Aceh Regency and was devoted to the vannamei shrimp supply chain at PT Sari Group Indonesia (SIG). The object of this research is vannamei shrimp cultivators and parties involved in the vannamei shrimp supply chain. The scope of this research is limited to the problem of efficiency in the supply chain performance of vannamei shrimp at PT SIG.

This study uses the Snowball sampling method. Snowball sampling technique is a method for identifying, selecting, and taking samples in a network or continuous chain of relationships (Nurdiani, 2014). Snowball takes several cases through the relationship of one person to another, or one case to another case, then looks for the next relationship through the same process, and so on (Athaillah, 2017). Snowball sampling is an approach to find critical informants who have much information. With this approach, several potential respondents were contacted and asked if they knew other people with the characteristics referred to for research purposes. Initial contact will help get other respondents through recommendations.

The types of data used in this study are primary data and secondary data. Researchers obtained primary data by conducting direct observations, interviews with related parties, and filling out questionnaires. The primary data in this study is AHP data obtained from the questionnaire recap. In comparison, secondary data was obtained from the Central Statistics Agency (BPS) in the form of the number of Production and Value of Aquaculture Production by District/City and Type of Cultivation in Aceh Province. Data on Pond Cultivation Production by District in Aceh Barat Regency 2018 were obtained from the Fisheries Service and times series data from PT. Sari International Group.

The data were analyzed using the SCOR (Supply Chain Operation Reference) method. This method was introduced by the Supply Chain Council (SCC) as a model for measuring supply chain performance. The SCOR model is a process reference model for supply chain operations developed by SCC, Pittsburgh, PA (Mardhiyah, 2008). SCOR divides supply chain processes into five processes, namely Plan (planning process), Source (procurement process), Make (production process), Deliver (delivery process), and Return (return process).

In addition to having five processes, SCOR also has a performance attribute. Performance attributes help assess supply chain processes from different perspectives. There are five attributes to assess the supply chain with the SCOR method. Several metrics can be used as performance measurement metrics (Supply Chain Council 2006 in Athaillah, 2017).

Performance attributes can be seen in Table 3.1 below:

Table 1. Performance Attribute

Performance Attribute	Definition
Supply Chain reliability	Supply chain reliability, supply chain performance in delivering the right product, at the right place, at the right time, and well documented
Supply chain responsiveness	Supply chain speed in providing products to consumers
Supply chain flexibility	The ability of the supply chain to adapt to market changes
Supply chain cost	The supply chain operations cost.
Supply chain asset	Value the effectiveness of an organization to manage its assets, to support demand satisfaction.

Source: Luthfiana and Perdana, (2012)

The results of the analysis are then scored using a performance indicator monitoring system. The following is a performance indicator monitoring system.

Table 2. Monitoring System

Monitoring System	Performance
<40	*Poor*
40-50	*Marginal*
50-79	*Average*
70-90	*Good*
>90	*Exellent*

Source: Wigaringtyas, 2013

The next stage is to score using the OMAX (Objective Matrix) method. OMAX combines productivity criteria into a unified and interrelated form. This model involves all levels in the company, from subordinates to superiors (Hanugrani, N et al., 2011). There are three assessment schemes based on the OMAX model, namely very low (red), medium (yellow), and very good (green).

4. Result and Discussion

PT SIG has pond area of 44,120 meters square, but only 30,000 meters square has been used as a pond area. The area is divided into 27 shrimp ponds and grouped into four blocks and three water reservoir ponds. Twenty-eight workers consist of a manager, officerss, security, technicians, and the pond crew. Cultivation carried out at PT SIG is rearing shrimp from Post Larva until the shrimp are ready to harvest, while shrimps fry are purchased from other companies. Sixty days after sowing the seeds, partial harvesting is carried out every two weeks (2-4 times) before the total harvest after 120 days. Vannamei shrimp cannot be harvested simultaneously because of the further growth of shrimp. The growth is strongly influenced by the quality of feed, drugs, feeding time, shrimp resistance to weather changes, and the cleanliness of the pond.

Based on field observations, preparatory activities such as checking the pond's condition, preparing nets to catch shrimp, shrimp storage places, and transport vehicles are prepared before shrimp are harvested. After thorough preparation, then the distributor of AAC (Aceh Aquaculture Cooperative) was contacted. AAC is an agent in charge of taking harvested shrimp, located in Lhokseumawe (approximately 350 kilometers), and is a branch of BMI in Medan.

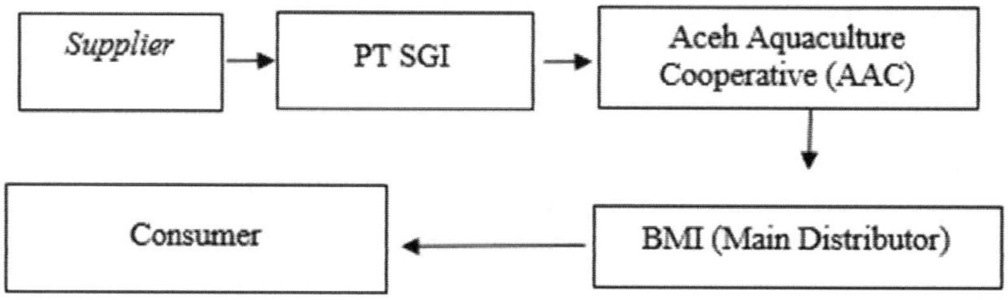

Figure 1. Supply Chain Management Vannamei Shrimp

After the AAC arrives, PT SIG will immediately harvest the vannamei shrimp and put it in the container. The vannamei shrimp sorting is performed to separate the different sizes of shrimp. Two sizes are separated here, namely size A (grade A), the bigger size of vannamei shrimp, with a higher price than size B (Grade B), which is smaller in size.

After sorting, the vannamei shrimp will be weighed first and then put into a container. After completion, the AAC will immediately return to Lhoksemawe to hand over the harvest. The next day, the shrimp will be brought to Medan (approximately 300 kilometers away). This action is done considering the nature of the shrimp that is easily damaged. Furthermore, the shrimp will become soft when in the container for a long time. This condition must be avoided because it can be detrimental. The price of shrimp will decrease along with changes in texture that occur in shrimp.

Weighting with Analytical Hierarchy Process (AHP)

Determination of AHP weight is done by selecting Key Performance Indicators (KPI). Key Performance Indicator (KPI) is an indicator that presents a series of measures that focus on the most critical aspects of organizational performance for the organization's success at this time and in the future (Adianto, 2014). There are 5 (five) dimensions that are seen as a reference for determining Key Performance Indicators on PT SIG's performance, namely:

a. Reliability: Supply chain reliability, supply chain performance in delivering the right product, at the right place, at the right time, and well documented.

b. Responsiveness: the speed of the supply chain in providing products to consumers.
c. Flexibility: The supply chain's ability to respond to market changes to win the market competition.
d. Cost: The costs associated with operating the supply chain.
e. Asset: The value of the effectiveness of an organization to manage its assets to support the satisfaction of demand.

PT SIG's supply chain performance measurement is carried out by selecting 15 Key Performance Indicators (KPI), which were selected based on the results of interviews. The weighting with the AHP method is based on the results of the questionnaire.

Table 3. Weighting with Analytical Hierarchy Process (AHP)

STEP	W	Dimension	W	KPI Code	Key Performance Indicator	W
Plan	0.05	Reliability	1	P.R 1	Time to identify employee performance	1
Source	0.45	Reliability	0.29	S.R 1	Quality of raw materials	0.8
				S.R 2	Fulfillment of raw materials	0.2
		Flexibility	0.14	S.F 1	Supply chain flexibility and adaptability	1
		Cost	0.57	S.C 1	Cost to source	1
Make	0.24	Reliability	0.32	M.R 1	Error in packing process	0.75
		Reliability		M.R 2	Number of Defective Products	0.25
		Responsiveness	0.12	M.Re 1	Production time	1
		Cost	0.56	M.C 1	Cost to make	1
Deliver	0.16		0.32	D.R 3	Perfect condition	0.56
		Reliability		D.R 4	Documentation accuracy	0.13
				D.R 5	Deliver cycle time	0.31
		Responsiveness	0.12	D.Re 1	Average document processing time	1.00
		Cost	0.56	D.C 1	Cost to deliver	1
Return	0.1	Reliability	1	R.R 1	Waranty and return	1

From Table 3, it can be seen that the highest weight on the core process is Source. This means that the Source process is considered the most important of other core processes. The Source process gets a weighted score of 0.45 or 45%. Meanwhile, the lowest weight score is Plan with a weighted score of 0.05 or 5% in the sense that Plan is considered the least important in the efficiency of the vannamei shrimp supply chain.

PT SIG scoring system

The scoring system is calculated using the Objective Matrix (OMAX). The calculation results of each SCOR perspective can be seen in the matrix below:

No	Perspektif	Dimensi	KPI	
1	Plan (1,83)	Reliability (2,44)	P.R.1	2,44
2	Source (6,53)	Reliability (9,78)	S.R1	10
			S.R2	8,89
		Flexibility (2,00)	S.F.1	2
		Cost (6)	S.C.1	6
3	Make (6.02)	Reliability (8,06)	M.R.1	10
			M.R.2	2,25
		Responsivenees (10)	M.Re.1	10
		Cost (4)	M.C.1	4
4	Deliver (9.76)	Reliability (9,23)	D.R.1	10
			D.R.2	10
			D.R.3	7,5
		Responsivenees (10)	D.Re.1	10
		Cost (10)	D.C.1	10
5	Return (10)	Reability (10)	R.R.1	10
Index Total				
7,03				

Figure 2. Supply Chain Performance Analysis

After measuring supply chain performance at PT SIG, overall, the total index value was 7.03. It can be concluded that the overall performance of PT SIG's Supply Chain has not yet reached the best performance because there are KPIs that are in the yellow and red zones. Based on the performance monitoring system table, the total index value of 7.03 is in a GOOD category.

Table 4. Performance Indicator Monitoring System

Monitoring System	Performance Indicator
<40	*Poor*
40-50	*Marginal*
50-70	*Average*
70-90	*Good*
>90	*Excellent*

5. Conclusion

Based on the results of measuring the efficiency of Supply Chain performance, PT Sari International Group is in a GOOD category, indicating that every supply chain activity carried out by UD Sari Inti Tani is overall good. The company must be able to improve efficiency in order to reach the EXCELLENT level. The total index value obtained by PT Sari International Group reached 7.03. There are three categories included in the red zone that the company must immediately fix: time to identify (check) employee performance, total production amount, and production costs.

References

[1] Adianto. 2014. *Analisis Pengukuran Kinerja Perusahaan Dengan Metode Analytical Hierarchy Process (AHP) Dan Objective Matrix (OMAX) Pada PT.Bpas.* Jurnal Teknik Industri Universitas Tarumanagara. Jakarta. SINERGI Vol. 18, No. 2, Juni 2014.

[2] Amrina I Fajar. 2014. IPB. *Analisi Rantai Pasok Jagung* Bogor.Program Studi Teknik Industri, Universitas Brawijaya.

[3] Athaillah, T., Hamid, A. H., dan Indra. 2018. Analisis Efisiensi Kinerja Rantai Pasok Ikan Tuna pada CV. Tuah Bahari dan PT. Nagata Prima Tuna di Banda Aceh. Marine Fisheries : Journal of Marine Fisheries Technology and Management, 9(2): 169-181.
[4] Atika. 2017. *Analisis Efisiensi dan Produktivitas Program Studi S-1.* Skipsi. UNY. Yogyakarta.
[5] Avianda *et* al. 2014. *Strategi Penigkatan Produktivitas Di Lantai Produksi Menggunakan Metode Objective Matrix (OMAX*). Jurnal Online Intitut Teknologi Nasional. No 4. Vol 01.
[6] Ayu Purbasari. 2017. *Analisis Rantai Pasok Beras Organik.* Skripsi. IPB. Bogor.
[7] Badan Pusat Statistik Kabupaten Aceh Barat (BPS), Tahun 2018.
[8] Dwita Adi Wicaksono. *Analisis Strategi Rantai Pasokan Udang Vaname.* Skripsi. IPB. Bogor.
[9] Febriarso, P. 2008. *Perancangan Sistem Pengukuran Kinerja Dengan Metode Analytical Hierarchy Process (AHP) Dan Objective Matrix (OMAX.* Program Studi Teknik Industri, Universitas Brawijaya
[10] Hanugrani, N. 2011. *Pengukuran Performasi Supply Chain Dengan Menggunakan Supply Chain Operation Reference (SCOR) Berbasis Performance Prism.* Universitas Muhammadiyah Surakarta.
[11] Imam Rabbani. 2018. *Analisis Efisiensi Rantai Pasok Komoditas Talas.* Skripsi. IPB. Bogor.
[12] Jacky S. B Sumarauw, Stevani flauren Kambey, Lotje Kawet. 2016. *Analisis Rantai Pasokan Kubis.* Manado: *Jurnal EMBA.* Vol.4 no.5 september 2016, hal. 303-408
[13] Kirana Pratiwi Aziz. 2016. *Analisis Supply Chain Komoditas Kedelai.* Skripsi. USU. Medan.
[14] Latifa Dinar Wigaringtyas. 2013. *Pengukuran Kinerja Supply Chain Management dengan pendekatan SCOR.* Universitas Muhammadiyah Surakarta. Surakarta.
[15] Luthfiana, Perdana. 2012. *Pengukuran Performasi Supply Chain Dengan Pendekatan Supply Chain Operation Reference (SCOR) Dan Analytical Hierarchy Process (AHP) Studi Kasus : PT. Indorma Global Medika.* Teknik UIN Sunan Kalijaga.
[16] Mardhiyah, n. 2008. *Kinerja Penyampaian Suku Cadang PT Toyota-Astra Motor Dengan Model Supply Chain Operations Reference.* Fakultas Ekonomi Dan Manajemen. Institute Pertanian Bogor.
[17] Muhammad Yogi Prayoga.2017. Rantai Pasok Ikan Tuna dan Cakalang. Skripsi. IPB.Bogor.
[18] Nina Hairiyah. 2014. *Model Peningkatan Kinerja Sistem Resi Gudang.* Skripsi. IPB.Bogor.
[19] Nur Khomsah. 2017. *Analisis Pendapatan Usaha Tambak Udang Vaname.* Skripsi. IPB. Bogor.
[20] Panjaitan. 2012. *Pemeliharaan Larva Udang Vaname.* Jurnal Manajemen Perikanan dan Kelautan vol.1 no.1. Jakarta : Universitas Terbuka.
[21] Pramudya Pratama Putra. 2012. *Peningkatan Produktivitas pada Reparasi di Dok Pembinaan UPT BTPI.* Skripsi. IPB. Bogor.
[22] Purbasari Indah. 2017. *Kajian Supply Chain Management.* Skripsi.IPB. Bogor.
[23] Rendiana.2015. *Analisis Pengaruh Efisiensi (BOPO) dan Capital Adequacy Ratio (CAR*). Skripsi. Unisba. Bandung
[24] Rijal Purwa Ilmiawan. 2014. *Kelayakan Bisnis Tambak Udang Vaname.* Skripsi. IPB. Bogor.
[25] Setiadi. 2018. *Kinerja Rantai Pasok Ikan Nila Pada Sriandoyo.* Skripsi. IPB. Bogor.
[26] T. Budiardi, T. Batara, D. Wahjuninggrum. 2005. *Tingkat Konsumsi Oksigen Udang Vaname.* Skripsi. IPB.Bogor
[27] Timbul rasoki. 2016. *Analisis Rantai Pasok Bawang Merah.* Skripsi. IPB. Bogor.

Nutrient testing of organic liquid fertilizer in tempeh factory in Pekanbaru City

Neng Susi*, Vonny Indah Sari, Indra Purnama, Hazra Yuvendius

Universitas Lancang Kuning, Pekanbaru, Indonesia

*neng_susi@ymail.com

Abstract. The increase of waste production has been affecting global warming, triggering the climate change. The right cultivation based on the environment and its existence could expectantly turn into a clean, sanitary and comfortable environment. Moreover, the appropriate waste could generate additional values that would increase income. Inedible factory waste of tempeh could be utilized for organic liquid fertilizer. Based on its premise, author has done research titled "Nutrient Testing of Organic Liquid Fertilizer in Tempeh Factory". The analysis done in a lab of Central Plantation Services resulted that N total 537 mg/Kg, P2O5 442 mg/Kg, K2O 2700 mg/Kg, Mg 175 mg/Kg, Cu 0,32 mg/Kg, Zn 0,95 mg/Kg, C-Organic 3800 mg/Kg pH H2O 4,88 and C/N Ratio 7,08Introduction.

1. Introduction

Land resources play an important part in succeeding in agriculture. Thus, many methods were tried and done to increase its productivities. Overuse of chemical fertilizer where the initial intention was to increase soil fertility could cause a decline in soil quality. The nutrient balance could be affected as well due to the absence of the restoration of soil organic. Good fertilization has to refer to the optimal effectivity and efficiency concepts. Economic, technical considerations influence the precision in deciding which fertilizer to use. The technical knowledge about plant physiology, fertilizer, and soil characteristics, where fertilizer would be applied, affects its efficiency.

Nowadays, people care even more about the quality of a product. Utilizing fertilizations from organic materials is believed to bring more benefits for agricultural products where they're healthier, environmentally friendly, and lessen the negative effects from chemical products. Waste is valueless materials, and its problem continues to be a pickle to solve, including tempeh waste as a consequence of population increase, in this case in Pekanbaru.

The uncontrollable heap of tempeh waste in factories affect many aspects of live, both directly and indirectly. Diseases, air pollutions, soil pollutions, even methane gas explosion can potentially be triggers of the global warming. The more waste production we have, the bigger the chance for climate change. The right cultivation based on the environment and its existence could expectantly turn into a clean, sanitary and comfortable environment. Moreover, the appropriate waste could generate additional values that would increase income.

Due to the fertilizer price that keeps getting higher and the subsidy abolishment from the government while less fertilizer materials to be imported, the waste utilization or organic waste could be an alternative, not to mention how much environmentally friendly they are. Inedible tempeh's factory waste can be utilized for organic liquid fertilizer. Based on this, author is interested to do research titled "Nutrient Testing of Organic Liquid Fertilizer in Tempeh Factory".

2. Methodology

This research is descriptive research, where the first process is the making process of organic fertilizer from tempeh's factory waste fermented for one month using bio activator EM4. POC that is ready will be sampled and tested in a lab to see both macro and micronutrient content and its C-organic. To measure its nutrient content, test method used was Kjeldahl (for nitrogen content) while Atomic Absorption Spectrophotometric (AAC) was used to check P,K, CaO, MgO and C-organic.

3. Result and Discussion

3.1 Result

Analysis result of parameters of organic liquid fertilizer of tempeh's factory waste tested in Kelapa Sawit Company "Mina Mas Research Centre" laboratory could be seen on Table 1 below.

Table 1. Nutrient content of tempeh's factory waste

No	Parameter Uji	Content	Methods
1	Phospat (mg/Kg)	442	*IKP-15 (Spectrophotometry)*
2	Calium (mg/Kg)	2700	*IKP-15 (Flamepphotometry)*
3	Nitrogen (mg/Kg)	537	*IKP-15 (Kjedhal)*
4	Magnesium / Mg (mg/Kg)	175	*IKP-15 (AAS)*
5	Copper / Cu (ppm)	0,32	*IKP-15 (AAS)*
6	Zinc / Zn (ppm)	0,95	*IKP-15 (AAS)*
7	C-Organic	3800	*IKP-15 (Spectrophotometry)*
8	C/N Ratio	7,08	*Calculation*
9	pH H_2O	4,88	*IKP-15 (pH Meter)*

Based on the result, it's shown that organic liquid fertilizer of tempeh's factory waste contained nutrient needed by plants, which are N total 537 mg/Kg, P2O5 442 mg/Kg, K2O 2700 mg/Kg, Mg 175 mg/Kg, Cu 0,32 mg/Kg, Zn 0,95 mg/Kg, C-Organik 3800 mg/Kg pH H2O 4,88 dan C/N Ratio 7,08. Plants need nutrient content, both macro and micro, such as Carbon (C), Hydrogen (H), Oxygen (O), Nitrogen, Phosphor (P), Potassium (K), Calcium (Ca), Magnesium (Mg), Sulfur (S), Iron (Fe), Manganese (Mn), Boron (B), Mo, Copper (Cu), Zinc (Zn) dan Chlore (Cl).

Some of the nutrients from organic liquid fertilizer of tempeh's factory waste are Phosphate, Potassium, Nitrogen, Calcium, and Magnesium. Phosphate is to carry the metabolism energy from plants, to stimulate flowering and insemination, root growth, seeding, plant cell division, and to enlarge network cells. Potassium is needed for photosynthesis, carrying assimilation result (including enzyme, water, and minerals), and increasing immune system. Nitrogen is for simulating plants' overall growth, synthesizing amino acid and protein and vegetative growth.

Nutrient content in organic liquid fertilizer of tempeh's factory waste is Iron (Fe), Manganese (Mn), Copper (Cu), and Zync (Zn). Iron is used for chlorophyll forming, protein, enzyme, and plays part in chloroplast development, and electron movement in metabolism process. Mn is the ribosome composer and also polymerase activator, protein synthesizer, and also carbohydrate. It plays the part as activator for a number of main enzymes in krebs cycle, where photosynthetic function is needed in chloroplast. Zinc functions are to activate enolase enzyme, aldolase, decarboxylated oxalic acid, lecithinase, d-cysteine desulfhydrase, superoxide dismutase (SOD), dehydrogenase, anhydrase carbon, proteinase and peptidase. Also, it is also in auxin biosynthesis, cell elongation and stem segment. Copper functions are to activate Cytochrome c oxidase enzyme, ascorbic acid, butyric-phenolase acid, and lactase. It also plays part in protein and carbohydrate metabolism, and in generative plant growth, N fixation by symbiotic and lignin forming.

4. Conclusion

Based on this research, it can be concluded that organic liquid fertilizer of tempeh's factory waste contains nutrient N total 537 mg/Kg, P2O5 442 mg/Kg, K2O 2700 mg/Kg, Mg 175 mg/Kg, Cu 0,32 mg/Kg, Zn 0,95 mg/Kg, C-Organik 3800 mg/Kg pH H2O 4,88 and C/N Ratio 7,08.

Acknowledgments

The author would like to thank all mothers who are under the Koran Yatama Foundation, Balai Raja Village, who have welcomed this activity, to the Faculty of Agriculture, Lancang Kuning University which has financed this service activity and UPM, Faculty of Agriculture, Lancang Kuning University who has supported and facilitate the running of this service activity.

References

[1] Arief, 1990. Horticulture. Penebar Swadaya. Jakarta.
[2] Asti, 2010. The Effect of Golden Conch Liquid on Rice Growth and Production (Oryza sativa).[Skripsi]. Faculty of Agricultural: Universitas Lancang Kuning Pekanbaru
[3] Ati P. 2013. rowth Response and Production of Caisim (Brassica juncea, L) Plants Due to Herbafarm Fertilizer Application. [Skripsi] Faculty of Agricultural Universitas Lancang Kuning Pekanbaru
[4] Habibi. 2013. Testing of Local Microorganisms (MOL) on Lettuce Growth and Production (Lactuca sativa). [Skripsi] Faculty of Agricultural: Universitas Lancang Kuning Pekanbaru
[5] Krisno. 2012. Biological Agents Bacillus sp. http://www.sehatcommunity.com/2011/06/agens - biology sp. html ≠ axzz2arcNLJD. Accessed date 29 October 2012.
[6] Musnawar E.I. 2003. Organic, Liquid and Solid Fertilizers, manufacture and application. Penebar Swadaya, Jakarta.
[7] Pepelakan, 2010. The Effect of Bacteria on Plant Growth. Penebar Swadaya. Jakarta
[8] Razak, Abdul. 2012. Introduction to Statistics, Malang: Intimedia.
[9] Sukarjo, 2010. The Effect of Provision of Rice Local Microorganisms on the Growth of Rubber Seeds (Heved brasilliensis) Sleep Eye Stump. [Skripsi]. Faculty of Agricultural. Universitas Lancang Kuning. Pekanbaru.
[10] Surtinah. 2013. Agricultural Cultivation Research Data Analysis. UNILAK Press. Pekanbaru.

Extraction of catfish bone waste (pangasius hypophtalmus) by utilizing an organic liquid of pineapple peel waste (ananas comosus) into gelatin

Ira Oktaviani Rz[1], Rahimatul Uthia[1], Fathul Jannah[1], Alexsander Yandra[2]

[1]Poltekkes Kemenkes Riau, Riau, Indonesia
[2]Universitas Lancang Kuning, Pekanbaru Indonesia

Email: ira@pkr.ac.id, rahimatul@pkr.ac.id, fathul@pkr.ac.id, alexsy@unilak.ac.id

Abstract. This study aims to determine how to extract the gelatin from catfish bones by using pineapple peel waste as a soaking material in the pre-treatment stage and examining the physicochemical properties of the gelatin produced. Kampung Patin in Kampar Regency, Riau Province, is one of the catfish cultivations in Indonesia. It is a center for producing catfish that produces fish bone waste. In this study, gelatin extraction will be undertaken in several stages: the preparation of pineapple waste liquid extract and gelatin extraction (pre-treatment and main extraction stages). Catfish bones will be soaked in 1:5 (m/v) pineapple waste liquid at the pre-treatment stage for 24 hours. While in the central extraction, ossein will be immersed in water at a temperature of 750C for 5 hours. The analysis results for physicochemical characteristics of gelatin obtained include yield of 2.55, pH of 4.58, water content; 11.66%, ash content; 17.71%, crude fat content; 0.17%, protein content; 64.71%. It can be inferred that it meets the established gelatin quality standards from various physicochemical characteristics of gelatin in this study.

1. Introduction

The demand for gelatin is intensifying every year; gelatin from pork is the primary source on the market. In 2007, sources of gelatin worldwide were derived from pork skins 46%, cow skins 29.4%, beef and pork bones 23.1%, and other sources 1.5% (Gomez-Guillen et al., 2009; Duconseille et al. al., 2015). Only about 1% comes from other sources (Ahmad & Benjakul, 2011). Kampar Regency, Riau Province, is one of the centers of catfish cultivation in Indonesia. Catfish processed into fillets have a higher economic value (Kemendag RI, 2013). This processing produces waste of fish skin and bones. It causes problems for the environment since its utilization has not been maximized (Agustin, 2013). Fish bones can be a source of raw material for the manufacture of gelatin. It has economic value and can overcome the problem of waste caused to the environment at the same time.

Research (Mahmoodani, 2004) reveals that catfish bones produce gelatin with a higher concentration than gelatin from other fish bones. It is in line with research from (Oktaviani, 2017) which mentions gelatin from catfish (Pangasius hyphothalmus) products produce gelatin compared with commercial gelatin. (Zhang et al., 2011) say that gelatin made from warm-water fish species has better physical characteristics than cold-water fishes. The necessary physical characteristics of gelatin include gel strength, texture, and viscosity (Badii & Howell, 2006; Ratnasari et al., 2013).

The gelatin manufacture using organic materials has not been widely used. One of the natural ingredients can be utilized for the manufacture is found in pineapple waste. Pineapple contains citric acid and bromelain enzymes have a high ability to break peptide bonds in proteins to convert collagen protein into gelatin (Nur & Rehalat, 2017). Utilizing pineapple waste for gelatin manufacture from fish bones, in addition to increasing value-added and minimizing agro-industrial wastes, can reduce the danger or risk of using chemicals. Pineapple waste liquid contains citric acid of 0.18-0.32% (Hajar et al., 2012).

On this occasion, the researchers would like to research the physicochemical properties of gelatin derived from the catfish bones (Pangasius hypopthalmus) taken from the Kampar area, precisely in Kampung Patin, which has the production of farming catfish in Riau. In this study, the pre-treated gelatin process will use pineapple waste as a solvent contains citric acid as a substitute for chemical solutions.

2. Methodology

2.1 Preparation of Pineapple Waste Liquid Extract (Atma et al., 2018)
The leftover peel and core of the pineapple obtained from the pineapple chips production in Kualu Nenas Villages, Kampar Regency, were taken to the laboratory. It is crushed using a blender to produce pulp. The pulp is then filtered and squeezed using a filter cloth to obtain pineapple waste liquid. The wastewater was then sterilized by autoclaving at 1210C, 2 atm, for 15 minutes and stored at room temperature before being used as an extracting solution.

2.2 Gelatin Extraction
Gelatin extraction was done based on the modified Atma (2018) method. Before conducting the extraction, the fish bones are first cleaned from the remnants of attached meat, boiled at a temperature of 1000C for approximately 15 minutes (Samsudin, 2018), dried, and the bone surface area is reduced. In the demineralization stage, the mashed catfish bones were soaked in 0.5 M EDTA solution (pH 7.5) for 24 hours (Theng, C.H., 2018), then washed with water. Subsequently, the catfish bones were soaked in pineapple waste liquid with a ratio of 1:5 (m/v) for 24 hours (pre-treatment stage). The bones were washed with water until the pH was neutral. The fish bone obtained is called ossein.

In the gelatin extraction stage, ossein was immersed in water with a ratio of 1:5 (m/v) at 750C for 5 hours (extraction stage). The primary extraction was done in each treatment with a mixture of fishbone ossein and water and then separated by filter paper. The liquid filtrate gained was collected in an Erlenmeyer flask. The liquid filtrate obtained is called the gelatin liquid extract. The liquid gelatin was then concentrated in a water bath at 550C and was dried at 550C in the oven. Dried gelatin, powdered.

2.3 Rendemen (AOAC, 1995)
The yield was attained from the comparison between the weight of dry gelatin.

2.4 Analysis of Physicochemical Characteristics
a. Organoleptic Examination
The organoleptic examination includes observing the shape, color, smell, and taste of the gelatin produced.
b. Gelatin pH (GMIA, 2019)
Gelatin powder is required as much as 7.5 ± 0.01 grams into a beaker and added 105 ± 0.2 grams of distilled water, then stirred until all the gelatin particles were suspended. Leave it for 1-3 hours at room temperature. Place the sample on a water bath at 650C for 10-15 minutes while stirring. Transfer the sample to a water bath with a temperature of 600C and let it remain. Calibrate the pH meter using pH 4 and pH 7 buffers at a temperature of 600C. Determine the gelatin pH liquid in accordance with the pH meter instructions. Rotate the liquid well using the pH probe to ensure the electrode is sufficiently saturated.

c. Proximate Analysis
The proximate analysis was conducted along with the procedures established by GMIA, 2019, and SNI 01-2891-1992. The proximate analysis includes water and ash content using GMIA, 2019, while protein and crude fat content using SNI-01-2891-1992.

3. Result and Discussion
The results of the study from gelatin of catfish bones with pineapple peel waste extract are:

3.1 Result
3.1.1 Rendemen
The yield resulting from this research is 2.55%.
Organoleptic
The organoleptic test results can be seen in table 1.

Table 1. Catfish Bone Gelatin Organoleptic Test Results

	Results	Standard (Farmakope V, Ditjen POM, 2012)
Form	Film sheet, coarse powder	Sheets, chips, moderately coarse powder
Color	Yellowish-brown	Weak yellow- light brown
Smells	Slight fishy smell	Weak broth

3.1.2 Physicochemical Properties
Table 2. Results of Physicochemical Properties of Catfish Bone Gelatin

	Results	GMIA (2012)	BSN (1995)
pH	4,58	3,8-5,5	4,5 -6,5
Water content (%)	11,66	< 10,5	<16
Ash content (%)	17,71	0,3-2	3,23
Crude Fat Content (%)	0,17	<5	Max 5
Protein Level (%)	64,71	>90	>87,65

3.2 Discussion
The gelatin formation from catfish bones with pineapple peel waste undergoes several stages, including demineralization, pre-treatment, and extraction—the first thing to do before the gelatin production process is making liquid pineapple waste. Pineapples are gained from Kualu Nenas, and the selected pineapples have the same maturity level. The peel and heads are taken, then cleaned from dirt and dust. After cleaning from impurities, the pineapple skin and hump are roughly chopped, then blended to minimize the surface, so the water is easily squeezed out. From 30 pineapples taken from the hump and skin, 12.5 liters of liquid pineapple waste were gained. This liquid pineapple waste was then sterilized in an autoclave at 1050C, with a pressure of 1 atm for 15 minutes. Next, it was allowed to stand at room temperature before the liquid was used as a soaking material in the pre-treatment stage.

Before the gelatin is extracted from the catfish bones, it is necessary to clean the bones from the remaining fat and meat attached, known as the degreasing process. From 30 kilograms of catfish bones from Kualu Nenas village, it was cleaned by boiling the bones at a temperature of 80-1000C for approximately 15 minutes. As claimed by Nurilmala (2017), the optimum temperature for removing fat from bone is 800C due to the melting point temperature of fat and the coagulation temperature of bone albumin, which is between 32-800C, resulting in optimum fat solubility. If the temperature exceeds this limit, it is feared that the fat in the extracted bone will run out, then boiling will decompose the protein contained, and the protein will be damaged first due to boiling. It can cause the percentage yield of gelatin to be small.

IOP Conf. Series: Earth and Environmental Science 1041 (2022) 012007 doi:10.1088/1755-1315/1041/1/012007

Furthermore, demineralization is done where there is a process of removing minerals in the bone so that ossein is acquired. In this process, the catfish bones were soaked in 0.5 M EDTA solution (pH 7.5) for 24 hours to diminish the bones' ash content. This process intensely determines the ash content in catfish bone gelatin (Theng, C.H., 2018). After soaking with EDTA solution, the bones were cleaned with distilled water, then soaked with liquid pineapple waste with a 24-hour soaking treatment. All of these soaking processes are done to remove calcium and mineral salts from the bones to become soft or called ossein contains collagen.

During the process, the catfish bones are stirred occasionally. In gelatin manufacture, the treatment of animal collagen raw materials with dilute acid or base affects the cross-linking of proteins to be cut, the structure to be broken, and the pieces to be dissolved in water. These water-soluble pieces of protein chains are called gelatin. The quality of gelatin produced depends on the concentration of acid or base used, temperature, and length of immersion time (GMIA, 2012).

In this study, pineapple peel waste liquid was used as an ingredient for soaking acidic fish bones; it is a natural source comprises high citric acid (Hajar et al., 2012). Citric acid is an organic acid has the most role in the acidity level of pineapple waste. The content of citric acid in pineapple waste is about 2.18 g per liter (Abdullah & Mat, 2008). Citric acid as a solvent in the pre-treatment stage of gelatin extraction is better than other organic acids (Mariod & Adam, 2013). After soaking with pineapple solution, the solution is removed and washed until the pH is neutral; the bone resulting from this washing is called ossein.

In the foremost extraction step, ossein was extracted with distilled water in a ratio of 1:5 (m/v) at 750C for 5 hours. Based on research by Atma (2017), which identified gelatin extracted from catfish bones using pineapple peel were using two extraction temperature conditions, namely 650C and 750C, it was found that at 750C extraction temperature showed high gelatin content, in line with (Pertiwi's research, 2018) which extracted fish bone using citric acid with various extraction temperatures of 45, 55, 65, and 750C for 5 hours, gained the best fishbone gelatin with a temperature of 750C for 48 hours of immersion. While the central extraction time of 5 hours is the optimum time because if it is more than 5 hours, the ossein will be destroyed and dissolved with distilled water (Rahayu & Fithriyah, 2015). The gelatin liquid was filtered using filter paper. It was concentrated in a water bath at 550C. The concentrated liquid was put into an oven at 550C to dry on gelatin sheets, then powdered.

3.3 The Yield

Yield is a imperative parameter in the gelatin isolation process. The greater the yield produced, the more influential the resulting treatment. The yield produced in this research is 2.55. These results indicate that the longer the immersion, the higher the yield. As said by Park (2013), the longer the immersion in acid, the weaker the bonds in the collagen, which causes more significant swelling, and more gelatin is produced. The yield value produced is still low. The acid used in the demineralization process of catfish bone may be weak. The activity of the acid solution is not optimal in loosening the collagen chain bonds and cross-linking between the collagen (Ramdani, 2014).

In line with the statement of Courts and Johns (1977) that the process of bone demineralization into ossein affects the success rate (yield) of collagen extraction in ossein. This collagen will then be converted into gelatin. Furthermore, the concentration factor of the acid solution and the immersion time are essential factors in the demineralization process. In a study conducted by Jaya (2020), who extracted snakehead fish bone gelatin with a variety of acids, namely synthetic citric acid, starfruit acid, and lime acid, the average yield values were 6%, 2.32%, and 3.33%, where Synthetic citric acid has a higher yield value than the yield value of organic acids. It is in line with the yield value of the pineapple peel waste used.

When compared with the study results of extracting gelatin from tilapia bones with several immersion times in 5% HCl, the yield percentage was between 7.34 to 10.20%. It presents the yield percentage is not too far from the yield of the research results.

3.4 Acidity (pH)

The pH value is the acidity level of the gelatin, which is vital for analysis. Gelatin pH affects other gelatin properties and also affects the application in the product. The pH value of gelatin is related to the process used. Acid processes tend to produce low pH values, while alkaline tends to produce high pH values. Gelatin with a neutral pH tends to be preferred, so the neutralization process has a vital role in neutralizing the remaining acid after soaking.

The pH value in this study was 4.58. It still meets the gelatin standard required by GMIA (2012); acid-made gelatin has a value range of 3.8-5.5, and the standard required by BSN (1995) is 4.5-6,5. In line with previous research, in catfish bone gelatin (Pangasius sutchi) with pineapple waste immersion, the pH value was 4.52 (Atma, 2018), snakehead fish bone gelatin with citric acid, lime acid, and starfruit acid immersion, respectively. i.e., 6.13; 6.80; 6.83 (Jaya, 2020), and the pH of catfish bone gelatin with citric acid immersion is 4.462 (Pertiwi, 2018).

3.5 Water content

Water content is one of the crucial tests in determining the quality of gelatin. It will affect the shelf life of gelatin because it is closely related to metabolic activities that occur while the gelatin is stored, such as enzyme activity, microbial activity, chemical activity, namely the occurrence of rancidity and non-enzymatic reactions, resulting in changes in gelatin properties quality and the value of gelatin quality (Darwin, 2018).

In this study, the value of the water content was 11.66%. The results of this study are included in the requirements set by SNI 06-3735-1995 (BSN, 1995) and the European Pharmacopoeia (European Pharmacopoeias Commission, 2014), which are a maximum of 16% and 15% but have not yet entered the requirements determined by the GMIA (2012) which should not exceed 10.5%. The value of this concentration is influenced by the amount of collagen formed in gelatin, which causes hydrogen bonds from non-collagen to bind to water molecules so the drying process will evaporate along with water. The longer the immersion, the lower the water content produced. The more open collagen structure causes the binding capacity of gelatin to adsorb water to be more significant, and the binding capacity to free water is low. The weaker binding capacity to free water causes the water in gelatin to evaporate quickly during drying and produces low water content (Ulfah, 2011).

Previous research on the extraction of catfish bone gelatin (Pangasius sutchi) by soaking pineapple waste is 8.59% (Atma, 2018). Research on extracting snakehead fish bone gelatin by soaking citric acid, lime acid, and wuluh starfruit acid, respectively. 9.21%, 11.29%, and 18.96%, while the catfish bone gelatin with citric acid immersion was 7.72%.

3.6 Ash Level

Measurement of ash content aimed to determine the mineral content in gelatin. Ash content affects gel strength and protein content. The purer the gelatin, the higher the protein content. The high ash content is due to calcium, potassium, sodium, iron, and magnesium in bones (Yenti, 2016). The ash content produced in this study was 17.71%. This value exceeds the requirements set by GMIA (2012), which is 0.3-2%, and BSN (1995), which is 3.25%.

Ash content exceeds the standard indicates the gelatin is not pure. Factors affecting the high ash content are the filtering process, raw materials content, and the extraction process (Du, 2013). Meanwhile, concerning Panjaitan (2016), the high ash content of fish bone gelatin may be caused by the minerals are still bound to collagen due to an imperfect demineralization process.

3.7 Crude Fat Content

Determination of fat content affects the material quality during storage. High-quality gelatin has a low-fat content (Shyni, 2014). In this study, the fat content value was 0.17%. It meets the standard by GMIA (2012) less than 5% and BSN (1995), a maximum of 5%. As attested by Winarno (2004), in general, the use of acids can accelerate hydrolyzing fats into glycerol and fatty acids. The higher the acid, the more the fat will decompose, thereby decreasing the fat content of the gelatin produced. The fat content value of this study was lower than that of catfish bone gelatin with citric acid immersion, which was 2.79% (Pertiwi, 2018). Likewise, the fat content of catfish bone gelatin (Pangasius sutchi) by soaking pineapple skin waste is 7.71%.

3.8 Protein Level

Gelatin is a type of conversion protein produced through the hydrolysis of collagen. The test shows how much protein content is in a food ingredient (Ulfah, 2011). The results of this study revealed that the protein content is 64.71%. This value is still far from the requirements determined by GMIA (2012), protein content of more than 90% and conforming to SNI (1995) of 87.65%. The low protein content results in the gelatin sample can be caused by several factors, including the high extraction temperature, which can cause further hydrolysis, so that the protein in the gelatin sample can be broken down into more superficial elements such as its constituent elements, namely C, N, O., and H (Yenti, 2015).

From a previous study conducted by Pertiwi (2018), catfish bone gelatin with citric acid immersion, the protein content was 57.80%, and research on mackerel bone gelatin by soaking in some acid solutions obtained a protein content of 58.37 %. Atma's (2018) research, namely the extraction of catfish bone gelatin (Pangasius sutchi) by soaking pineapple skin waste, obtained a protein content of 47.60%.

4. Conclusion

Catfish bone gelatin (Pangasius hypophtalmus) extracted by soaking pineapple peel waste has physicochemical properties, among others; yield 2.55, pH 4.58, water content 11.66%, ash content 17.71%, crude fat content 0.17%, protein content 64.71%, some physicochemical characteristics of gelatin in this study meets the standard of gelatin quality.The Promethee-based decision support framework assists decision-makers in determining the performance of lecturers based on the guidelines presented.

Acknowledgment

Thank you to the Health Polytechnic of the Riau Ministry of Health for facilitating researchers in terms of funds and research facilities so this research can run well.

References

[1] Mat, H.B., 2017. The Characteristic of pineapple waste from canning industry. Advanced Science Letters, 23(6), pp.5691-5693.
[2] Agustin, A.T., 2013. Gelatin ikan: sumber, komposisi kimia dan potensi pemanfaatannya. Media Teknologi Hasil Perikanan, 1(2).
[3] Ahmad, M. and Benjakul, S., 2011. Characteristics of gelatin from the skin of unicorn leatherjacket (Aluterus Monoceros) as influenced by acid pre-treatment and extraction time. Food Hydrocolloids, 25(3), pp.381-388.
[4] Atma, Y., Ramdhani, H., Mustopa, A.Z., Pertiwi, M. and Maisarah, R., 2018. Physical-chemicals Characteristic of Fish Bone Gelatin from Pangasius catfish Extracted Using Pineapple Waste. AGRITECH-JURNAL TEKNOLOGI PERTANIAN, 38(1), pp.56-63.
[5] Nur Azira, T., Amin, I. and YB, C.M., 2012. Differentiation of bovine and porcine gelatins in processed products via Sodium Dodecyl Sulphate-Polyacrylamide Gel Electrophoresis (SDS-PAGE) and principal component analysis (PCA) techniques.
[6] Badii, F. and Howell, N.K., 2006. Fish gelatin: structure, gelling properties and interaction with egg albumen proteins. Food hydrocolloids, 20(5), pp.630-640.
[7] BSN. (1995). Standar Nasional Indonesia SNI 06-3735:1995, Mutu dan Cara Uji Gelatin. Badan Standarisasi Nasional
[8] Darwin, D., Ridhay, A. and Hardi, J., 2018. Kajian ekstraksi gelatin dari tulang ikan mujair (Oreochromis mossambicus). KOVALEN: Jurnal Riset Kimia, 4(1), pp.1-15.
[9] Du, L., Khiari, Z., Pietrasik, Z. and Betti, M., 2013. Physicochemical and functional properties

of gelatin extracted from turkey and chicken heads. Poultry Science, 92(9), pp.2463-2474.

[10] Gómez-Guillén, M.C., Pérez-Mateos, M., Gómez-Estaca, J., López-Caballero, E., Giménez, B. and Montero, P., 2009. Fish gelatin: a renewable material for developing active biodegradable films. Trends in Food Science & Technology, 20(1), pp.3-16.

[11] GMIA., 2012. Gelatin Handbook. USA: Gelatin Manufacturers Institute of America.

[12] Grobben, A.H., Steele, P.J., Somerville, R.A. and Taylor, D.M., 2004. Inactivation of the bovine-spongiform-encephalopathy (BSE) agent by the acid and alkaline processes used in the manufacture of bone gelatine. Biotechnology and Applied Biochemistry, 39(3), pp.329-338.

[13] Hafidz, R.M.R.N., Yaakob, C.M., Amin, I. and Noorfaizan, A., 2011. Chemical and functional properties of bovine and porcine skin gelatin. International Food Research Journal, 18(2), pp.813-817.

[14] Hajar, N., Zainal, S., Nadzirah, K.Z., Roha, A.S., Atikah, O. and Elida, T.T., 2012. Physicochemical properties analysis of three indexes pineapple (Ananas Comosus) peel extract variety N36. APCBEE Procedia, 4, pp.115-121.

[15] Jaya, F.M., Utpalasari, R.L. and Kurniawan, R., 2020, November. Karakteristik Organoleptik Permen Jelly Gelatin Tulang Ikan Gabus (Ophioceplaus striatus) dengan Penambahan Lendir Okra. In Seminar Nasional Lahan Suboptimal (No. 1, pp. 364-372).

[16] Shakila, R.J., Jeevithan, E., Varatharajakumar, A., Jeyasekaran, G. and Sukumar, D., 2012. Functional characterization of gelatin extracted from bones of red snapper and grouper in comparison with mammalian gelatin. LWT-Food Science and Technology, 48(1), pp.30-36.

[17] Kemendag RI. (2013). Ikan patin hasil alam bernilai ekonomi dan berpotensi ekspor tinggi Warta ekspor, Ditjen PEN/MJL/004/10/2013: 3-11

[18] Ketnawa, S., Chaiwut, P. and Rawdkuen, S., 2012. Pineapple wastes: A potential source for bromelain extraction. Food and bioproducts processing, 90(3), pp.385-391.

[19] Khiari, Z., Rico, D., Martin-Diana, A.B. and Barry-Ryan, C., 2011. The extraction of gelatine from mackerel (Scomber scombrus) heads with the use of different organic acids. Journal of FisheriesSciences. com, 5(1), p.52.

[20] Koli, J.M., Basu, S., Nayak, B.B., Patange, S.B., Pagarkar, A.U. and Gudipati, V., 2012. Functional characteristics of gelatin extracted from skin and bone of Tiger-toothed croaker (Otolithes ruber) and Pink perch (Nemipterus japonicus). Food and bioproducts processing, 90(3), pp.555-562.

[21] Mahmoodani, F., Ardekani, V.S., Fern, S.S., Yusop, S.M. and Babji, A.S., 2014. Optimization of extraction and physicochemical properties of gelatin from pangasius catfish (Pangasius sutchi) skin. Sains Malaysiana, 43(7), pp.995-1002.

[22] Mariod, A.A. and Fadul, H., 2013. Gelatin, source, extraction and industrial applications. Acta Scientiarum Polonorum Technologia Alimentaria, 12(2), pp.135-147.

[23] Nur, S., Surati, S. and Rehalat, R., 2017. Aktifitas enzim bromelin terhadap peningkatan protein tepung ampas kelapa. Biosel: Biology Science and Education, 6(1), pp.84-93.

[24] Nurilmala, M., Jacoeb, A.M. and Dzaky, R.A., 2017. Quality of cultured wader pari during storage at different temperature. Jurnal Pengolahan Hasil Perikanan Indonesia, 20(2), pp.339-350.

[25] Oktaviani, I., Perdana, F. and Nasution, A.Y., 2017. Perbandingan sifat gelatin yang berasal dari kulit ikan patin (Pangasius hypophthalmus) dan gelatin yang berasal dari kulit ikan komersil. JOPS (Journal Of Pharmacy and Science), 1(1), pp.1-8.

[26] Panjaitan, T.F.C., 2017. Optimasi ekstraksi gelatin dari tulang ikan tuna (Thunnus albacares). Jurnal Wiyata: Penelitian Sains dan Kesehatan, 3(1), pp.11-16.

[27] Kim, H.W., Hwang, K.E., Song, D.H., Lee, S.Y., Choi, M.S., Lim, Y.B., Choi, J.H., Choi, Y.S., Kim, H.Y. and Kim, C.J., 2013. Effects of dietary fiber extracts from brewer's spent grain on quality characteristics of chicken patties cooked in convective oven. Food Science of Animal Resources, 33(1), pp.45-52.

[28] Pertiwi, M., Y. Atma, A. Z. Mustopa, R. Maisarah. (2018). Karakteristik Fisik dan Kimia Gelatin

dari Tulang Ikan Patin dengan Pre-Treatment Asam Sitrat. Jurnal Aplikasi Teknologi Pangan, 7(2): 83-91 Pertiwi, M., Atma, Y., Mustopa, A.Z. and Maisarah, R., 2018. Karakteristik fisik dan kimia gelatin dari tulang ikan patin dengan pre-treatment asam sitrat. Jurnal Aplikasi Teknologi Pangan, 7(2).

[29] Arima, I.N. and Fithriyah, N.H., 2015. Pengaruh waktu perendaman dalam asam terhadap rendemen gelatin dari tulang ikan nila merah. Prosiding Semnastek.

[30] Bali, T.S. and Ramadani, D., Pengaruh Perbedaan Jenis Asam dan Waktu Demineralisasi pada Nilai Rendemen dan Sifat Fisiko Kimia Gelatin.

[31] Ratnasari, I., Yuwono, S.S., Nusyam, H. and Widjanarko, S.B., 2013. Extraction and characterization of gelatin from different freshwater fishes as alternative sources of gelatin. International Food Research Journal, 20(6), p.3085.

[32] Saputra, R.H., Widiastuti, I. and Supriadi, A., 2015. Karakteristik fisik dan kimia gelatin kulit ikan patin (Pangasius pangasius) dengan kombinasi berbagai asam dan suhu. Jurnal FishtecH, 4(1), pp.29-36.

[33] Shyni, K., Hema, G.S., Ninan, G., Mathew, S., Joshy, C.G. and Lakshmanan, P.T., 2014. Isolation and characterization of gelatin from the skins of skipjack tuna (Katsuwonus pelamis), dog shark (Scoliodon sorrakowah), and rohu (Labeo rohita). Food hydrocolloids, 39, pp.68-76.

[34] Schrieber, R. and Gareis, H., 2007. Gelatine handbook: theory and industrial practice. John Wiley & Sons.

[35] Theng, C.H., Huda, N., Nik Muhammad, N.A., Wariyah, C. and Hashim, H., 2018. Physicochemical properties of duck feet collagen with different soaking time and its application in surimi. Int J Adv Sci Eng Inf Technol, 8, pp.832-41.

[36] Ulfah, M., 2011. Pengaruh konsentrasi larutan asam asetat dan lama waktu perendaman terhadap sifat-sifat gelatin ceker ayam. Agritech, 31(3).

[37] Wang, Y. and Regenstein, J.M., 2009. Effect of EDTA, HCl, and citric acid on Ca salt removal from Asian (silver) carp scales prior to gelatin extraction. Journal of food science, 74(6), pp.C426-C431.

[38] Winarno, F.G., 2004. Kimia Pangan dan Gizi, Jakarta: PT. Gramedia Pustaka Utama.

[39] Yenti, R., Nofiandi, D. and Rosmaini, R., 2015. Pengaruh Beberapa Jenis Larutan Asam pada Pembuatan Gelatin Dari Kulit Ikan Sepat Rawa (Trichogaster trichopterus) Kering sebagai Gelatin Alternatif. SCIENTIA: Jurnal Farmasi dan Kesehatan, 5(2), pp.114-121.

[40] Zhang, F., Xu, S. and Wang, Z., 2011. Pre-treatment optimization and properties of gelatin from freshwater fish scales. Food and Bioproducts Processing, 89(3), pp.185-193.

The effect of orange peel waste addition on ZnO characteristics and its ability to degrade methyl orange

Aulia Dewi Rosanti[1*], Fahmi Hidayat[1], Yuly Kusumawati[2], Arif Fadlan[2], Anggita R.K Wardani[1], Eva Umi Latifah[1]

[1]Universitas Islam Kadiri, Kediri, Indonesia
[2]Institut Teknologi Sepuluh Nopember, Surabaya, Indonesia

*aulia.dewi.r@uniska-kediri.ac.id

Abstract. The development of the textile industry in Indonesia is rapid. It cause serious environmental problems because the dyes used in the textile industry are synthetic that is azo dyes. Azo dyes are non-biodegradable substances that pollute the environment if not processed. One of the azo dyes that is often used in the textile industry is Methyl Orange. Photodegradation is an inexpensive and environmentally friendly method of removing synthetic dyes. This study will discuss the effect of adding orange peel on the characterization of ZnO and the ability to degrade Methyl Orange. The results of the FTIR spectra showed an absorption in the area of 1430-1455 cm-1 that means the orange peel had been impregnated into ZnO and the more ZnO as added on orange peel, it may reduce the number of -OH and CH groups of the ZnO/OP composite. Based on the results of characterization using XRD, it is known that the addition of orange peel into ZnO did not damage the crystal structure and crystal form. The crystal structure of ZnO/OP formed was wurtzite with a hexagonal shape. The results of the characterization using DR-UV showed that the band gap energy of ZnO/OP was smaller than ZnO and the more ZnO was added, the wider the band gap of energy. ZnO/OP with a variation of 1:25 can degrade Methyl Orange more than ZnO that was 73% in 60 minutes.

1. Introduction

Every year the textile industry in Indonesia develops rapidly. Although textile industry is beneficial for human life, it has a negative impact on the environment. It is because the dye used during the dyeing process in the textile industry is a textile dye [1]. One of the textile dyes that is often used is Methyl Orange. Methyl orange is an anionic molecule that is an azo dye with the charaterized by very stable and not easily biodegradable, so it is difficult to remove from wastewater [2], with the development of the textile industry in Indonesia, the wastewater containing methyl orange is increasing to disturbs the surrounding environment, and public health.

Photodegradation is an alternative technology for processing organic substances and dyes in liquid waste that is cheap and environmentally friendly. This photodegradation technology is able to treat wastewater pollutants with the assistance of UV light and semiconductor photocatalyst materials [3]. Photocatalyst is a material with catalytic ability after absorption of electromagnetic radiation. In the case of dyestuffs in the textile industry, photocatalysts with mesoporous pore types show high photocatalytic yields [4]. Some of the photocatalyst materials that can be used for the photocatalyst process from the oxide group are TiO2, SnO2, SrTiO3, Fe2O3, ZnO, WO3 [5] [6]. It is because the oxide group material has semiconductor properties where the material has a band gap energy in the UV region, so it is able to absorb UV rays [6].

ZnO semiconductor has an advantage compared to TiO2 because ZnO material has a larger band gap (3.4 eV) than TiO2 (3.2 eV), so ZnO has a higher photocatalytic efficiency than TiO2 due to the strong absorption of UV rays from the solar spectrum[7], [8]. The photocatalytic activity of a semiconductor can be increased by modifying the recombination of charge carriers and can be carried out by adding matrix or dopants and making a nanocomposite material [6], [9] Dopants used can be in the form of metallic, non-metallic or natural materials, one example of natural materials that can be added as dopants is orange peel.

Various research studies proved that orange peel is a good adsorbent for the purification of dyes in textile industry waste. Research conducted by Arami, Yousefi and Mohammad, (2005)proved that orange peel with 8 and 4 g/L concentration showed adsorption efficiencies of 92 and 91% for dyes DR23 and DR80, respectively. Knowledge of orange peel as an adsorbent as well as ZnO as a semiconductor, encourages further research to combine the two as composites to be applied as photocatalysts. In this study, the manufacture of natural material-based photocatalyst composite materials using orange peel was carried out.

2. Methodology

The material used in this study were orange peel (OP) (Orange peel was obtained from a local fruit field in the north of Sumatra, Indonesia), Aqua DM (Bratachem), Methyl Orange(MO)(Merck) NaOH (Merck), Acetone (Merck), dan Zn Acetate (Merck)

2.1 Preparation Of Orange Peel

Preparation of adsorbent was adapted from the method of Arami et al (2005) [10] with modification. The orange peels were washed several times with water and were sun-dried for 48 hours. The dried materials were blended and sieved through 60-mesh sieve

2.2 Synthesis of ZnO using Batch Impregnation

0.2 M Zn Acetate was heated at 60^0C and added 0.01 M NaOH to pH 12 and then stirred for 1 hour. The mixed solution was allowed to stand at room temperature for 24 hours. After being allowed to stand, the sample mixture was separated using the decantation method and after that it was washed using distilled water while heated at a temperature of 60^0C. Then the mixed solution was allowed to stand for 24 hours and then centrifuged at 2000 rpm for 1 hour. The solids obtained were washed using acetone and baked in an oven at 75^0C for 4 hours. The dry solids obtained were ground until smooth and then calcined for 4 hours at 450^0C.

2.3 Synthesis of ZnO/OP Composites using Batch Impregnation

Orange peel powder was added with the solution of 0.2 M Zn Acetate with the ratio of 1:25; 1:50 and 1:100 and then stirred for 30 minutes. Then the mixed solution was heated at 60^0C and added 0.01 M NaOH to pH 12 and then stirred for 1 hour. The mixed solution was allowed to stand at room temperature for 24 hours. After being allowed to stand, the sample mixture was separated using the decantation method and after that it was washed using distilled water while heated at a temperature of 60^0C. Then the mixed solution was allowed to stand for 24 hours and then centrifuged at 2000 rpm for 1 hour. The solids obtained were washed using acetone and baked in an oven at 75^0C for 4 hours. The dry solids obtained were ground until smooth and then calcined for 4 hours at 450^0C.

2.4 Composite Characterization of ZnO/OP

The synthesized ZnO/OP results were characterized by using an IR spectrophotometer to determine the functional groups, XRD characterization to determine the level of crystallinity, size and type of crystals, DR-UV to determine the band gap energy of the composite material. To determine the value of the band gap energy (Eg) can be calculated using the Kubelka-Munk equation with the Tauc equation then it is t substituted with the Kubelka-Munk coefficient[11].

$$(Khv)^{1/n} = A(hv - Eg) \qquad (1)$$

where h is Planck's constant = 6.34x10-34 Js; v is the Vibration frequency; Eg is the band gap energy; A is a proportional constant and the value is n=1/2, for direct transition (direct semiconductor), n=2, for indirect transition (indirect semiconductor).

2.5 Degradation of orange methyl using ZnO/OP

ZnO/OP was weighed as much as 0.01 grams then added Methyl Orange 3 mg/L as much as 25 mL and stirred at 400 rpm at 25°C under UV light. The solution was centrifuged at 3000 rpm for 10 minutes, and the resulting solution was observed using a UV-Vis Spectrophotometer at a wavelength of 465 nm. The level of decolorization efficiency (% decolorization efficiency) on MO can be calculated using the following equation [12] :

$$\%\ \text{decolorization efficiency} = \frac{(A_0 - A_t)}{Ao} \times 100$$

Where A_0 is the initial absorbance of MO before irradiation and A_t is the final concentration of MO after irradiation at t-time.

3. Result and Discussion

3.1 Material Characterization of ZnO and ZnO/OP using IR Spectroscopy

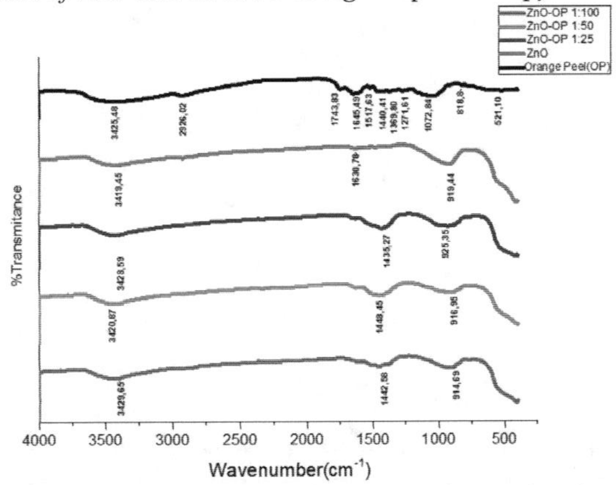

Figure 1. FTIR Spectra from OP, ZnO, ZnO-OP 1:25; ZnO-OP 1:50; ZnO-OP 1:100

The spectra of the FTIR characterization of orange peel and ZnO/OP composite can be seen in Figure 1 and the intensity that appears in the FTIR spectra of ZnO/OP can be seen in Table 1. In the spectra of orange peel and ZnO/OP composite there was a wide absorption at a wavenumber around 3400 cm $^{-1}$ that indicates the presence of -OH group strain vibration [3], [13]. The peaks of the orange peel FTIR spectra at wavenumbers 2926.02 cm^{-1} and 1743.83 cm^{-1} respectively showed stretching vibrations of C-H and carbonyl groups (C=O). The vibrations in the area around 1430-1455 cm^{-1} were C-H vibrations of methyl, methylene, methoxy aromatic and aliphatic groups.

In addition, the C-O stretching vibration of the carboxylic acid and alcohol groups was indicated by the presence of an absorption band in the wavenumber range of 1300-1000 cm^{-1}[14], [15]. The presence of ZnO in the ZnO/OP composite was indicated by the absorption at wavenumber 400-500 cm^{-1} as the ZnO strain area [2], [3], [13]. There were similar peaks in the ZnO/OP composite 1:25, 1:50 and 1:100. Based on Figure 1, it is known that there was absorption of methyl, methylene and methoxy groups in the area around 1430-1455 cm-1 at ZnO/OP 1:25; 1:50 and 1:100. It showed that the orange peel had entered into ZnO. Based on Table 1, the greater the addition of ZnO may increase the intensity of transmittance or decrease the intensity of absorption at wavelengths 3420-3430 cm^{-1} and 1435-1449 cm^{-1}. It indicates the more addition of ZnO to the orange peel may decrease the number of -OH and C-H groups of the ZnO/OP composite.

Table 1. Vibration intensity of functional groups in orange peel and ZnO/OP

Vibration	**Wavelength (cm^{-1})**	**Transmittance Intensity**			
		OP	**ZnO/OP 1:25**	**ZnO/OP 1:50**	**ZnO/OP 1:100**
-OH	3420-3430	18.70	41.26	47.40	54.71
C-H (mehyl, methylene and methoxy)	1435-1449	30.50	48.49	56.29	69.41

3.2 Material Characterization of ZnO and ZnO/OP Using XRD

Characterization using XRD is to determine the effect of adding orange peel on the ZnO/OP crystallinity with characterization using XRD. It can be seen the type of crystal, crystal size, and crystallinity level of ZnO/OP and ZnO. The information about changes in the sample structure can be obtained.

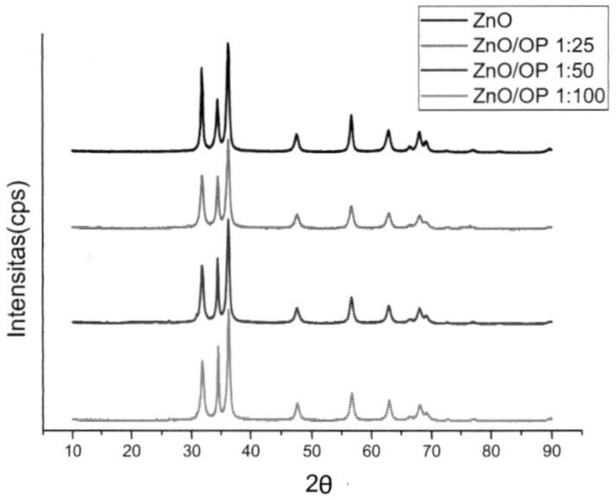

Figure 2. XRD diffractogram of ZnO and ZnO/OP with various concentrations : 1:25;1:50;1:100

Zinc Oxide is an N-type semiconductor with wide band gap of 3.37 eV [16]. The crystal structure of zinc oxide has three forms such as wurtzite, zinc blende and cubic rock salt. The most dominant form of crystal structure is wurtzite because the structure is the most stable at room temperature and pressure. The wurtzite structure is hexagonal in shape that has two sub-lattices, namely Zn^{2+} and O_{2-} that alternately form the base plane [17]. The coordination number of zinc changes to the coordination number of oxygen and it is surrounded by four cations and four anions at a corner of the tetrahedron and vice versa.

The arrangement of the tetrahedral atoms in the wurtzite structure is a noncentrosymmetric structure that causes a dipole moment[17]. This is a characteristic of zinc oxide wurtzite that rises to some unique properties such as piezoelectricity [17]. Based on the data obtained and matched using the Match! 3, it is known that the crystal structure formed in ZnO/OP was hexagonal and no amorphous phase found. Based on the comparison of the ZnO/OP diffractogram with the Crystalography Open Database and the Joint Committee on Powder Diffraction Standard (JCPDS no 36-1451), known that the crystal structure formed the ZnO/OP material was s a wurtzite structure with a hexagonal shape.

Based on Figure 2, it can be observed that there was a spectral peak with a relatively high intensity, that was the 36° region, this peak appears in both ZnO/OP and ZnO with a diffraction plane d101 (2.4Å).

The other peaks which had relatively greater intensity than the others were at 31 ° and 34 ° with their respective diffraction planes being d100(2.8 Å) and d002(2.6Å).

Table 2. Identification of ZnO and ZnO/OP Cristal Strcture

Material	2θ(°)	d(Å)	Cristal Structure	Cristal Form
	31.770	2.814		
JCPDS	34.422	2.602	wurzite	hexagonal
	36.253	2.475		
	31.797	2.811		
ZnO	34.451	2.601	wurzite	hexagonal
	36.271	2.475		
	31.808	2.811		
ZnO/OP 1:25	34.485	2.598	wurzite	hexagonal
	36.283	2.474		
	31.810	2.810		
ZnO/OP 1:50	34.480	2.599	wurzite	hexagonal
	36.287	2.474		
	31.843	2.808		
ZnO/OP 1:100	34.534	2.595	wurzite	hexagonal
	36.337	2.470		

Comparison JCPDS data (36-1451) with the synthesized XRD diffractogram data in Table 2, known that the crystal structure of ZnO and ZnO/OP was hexagonal, it can be assumed that the addition of OP into the ZnO crystal lattice did not change the crystal structure formed that was hexagonal. From the diffractogram, it can be seen that there was a difference in the absorption intensity, where the absorption intensity increased the addition of ZnO.

Meanwhile, if we see the diffractogram of ZnO and ZnO/OP 1:25, it is known that with the addition of orange peel the absorption intensity decreases. It is because the more OP was added to the ZnO/OP synthesis, the OP will enter the ZnO crystal lattice, causing the crystal structure to become more irregular and the crystallinity to decrease. Unlike the case for ZnO/OP 1:50 and 1:100 the absorption intensity increased again; it is due to the addition of more ZnO than OP so that less OP entered the crystal lattice causes the increase of crystallinity.

3.3 Characterization of ZnO and ZnO/OP Using DR-UV (Diffuse Refectance UV-Visible)

The responsiveness of ZnO/OP was observed using Diffuse Refectance UV-Visible (DR-UV) spectrometer carried out at a wavelength of 200-800 nm. From the results of the synthesis, it was found that the addition of orange peel into ZnO, the Wavelength shifted toward the larger Wavelength (visible light). It is clarified by the absorbance spectra (DR-UV) of ZnO and ZnO/OP 1:25 as seen in Figure 3. However, the ZnO increases, namely at ZnO/OP 1:50 and 1:100, the absorption shifts back to Wavelength lower (UV).

It is because the addition of more ZnO causes the band gap energy to widen causing the absorption to shift towards UV. Band gap energy (Eg) is the energy required for the process of excitation of electrons from the valence band to the conduction band. If the band gap energy is getting smaller, the energy required for the photocatalyst to excite electrons from the valence band to the conduction band will also be smaller.

It is very advantageous in photocatalyst applications because it is able to work at a wider range of Wavelengths so that it can also be applied to the visible light region [18][19]. Photocatalysts with the light with the right Wavelength, then the electrons in the valence band absorb photon energy then experience excitation and move to the conduction band. The result of this electron excitation is the formation of positively charged holes in the valence band (h^+) and electrons in the conduction band (e^-)[19].

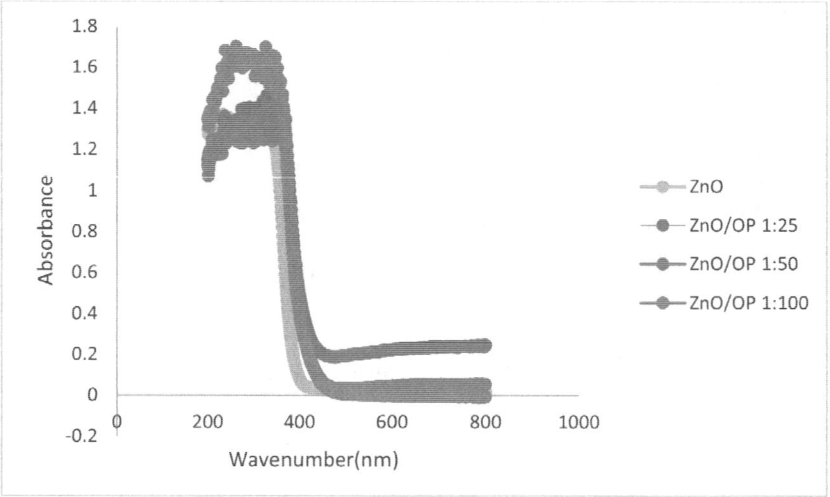

Figure 3. DR-UV Spectra from ZnO;ZnO/OP 1:25; ZnO/OP 1:50 and ZnO/OP 1:100

Based on Table 3, it is known that the greater the addition of ZnO, the higher the band gap energy. The addition of orange peel into ZnO causes the narrower band gap energy . Compared with ZnO, ZnO/OP had a narrower Eg value, it can be assumed that the orange peel had successfully entered the ZnO crystal lattice. However, with increasing of ZnO, the recovered Eg becomes even greater because with the increase in ZnO, the ZnO contained in ZnO/OP is more than orange peel, causing Eg to become larger. The more ZnO added, it may affect the agglomeration process [17], so that ZnO/OP with the addition of larger ZnO will have a greater Eg value.

There was a shift in absorption to the visible light region and a narrower Eg value at ZnO/OP 1:25 showed that the process of inserting orange peel on the ZnO crystal lattice can increase the responsiveness in visible light. In Table 3, it can be seen that the smallest Eg value was at ZnO/OP 1:25 at 3.171 eV. Based on these results, it can be concluded that the addition of orange peel can increase the responsiveness or ability of ZnO to respond to visible light, and with the greater addition of ZnO to ZnO/OP, the responsiveness will decrease caused by the agglomeration process.

Table 3. ZnO and ZnO/OP band gap energy

Material	Band Gap (eV)	λ edge
ZnO	3.331	372.636
ZnO/OP 1:25	3.171	391.438
ZnO/OP 1:50	3.188	389.229
ZnO/OP 1:100	3.192	388.628

Based on the results of IR, XRD and DR-UV characterization, it can be concluded that the material with the best character was ZnO/OP 1:25 material.

3.4 Photocatalytic Activity Test under UV light

Based on the results of the characterization, it is known that the material with the best character was ZnO/OP 1:25. The photodegradation process was carried out under UV light for 10 to 60 minutes. The results of the photodegradation of MO by ZnO and ZnO/OP 1:25 were presented in Figure 4. From Figure 4. it can be seen that the ability of ZnO/OP 1:25 can degrade MO better than ZnO.

The longer the contact time, the more degraded dyes , and it can be observedt in a contact time of 60 minutes ZnO/OP 1:25 can degrade 3ppm MO dyes up to 73% while ZnO is only able to degrade up to 34%. This data shows that the addition of orange peel increased the ability of ZnO in degrading Methyl Orange dye under UV light. It is because ZnO/OP 1:25 had a narrower band gap energy than ZnO so that the degradation process was maximized than ZnO . Methyl Orange is one of the azo dyes used in the textile industry in Indonesia. This dye waste can pollute the environment, by combining ZnO material and orange peel that may help reduce environmental pollution due to these dyes while utilizing orange peel waste.

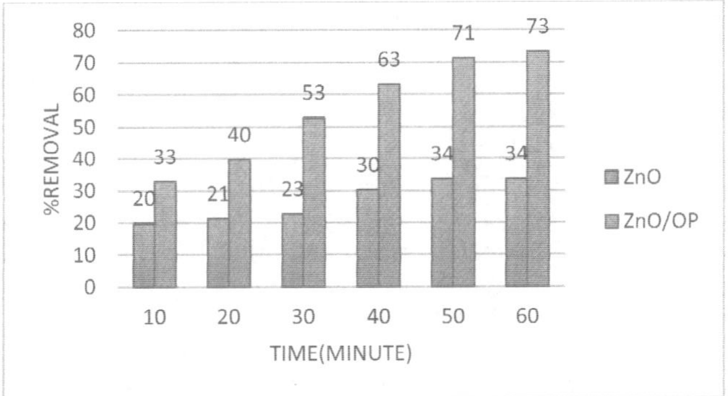

Figure 4. MO degradation efficiency using ZnO and ZnO/OP 1:25

4. Conclusion

Based on the results , it can be concluded that the greater the addition of ZnO the greater intensity of transmittance or less intensity of wavelengths absorption so that the more addition of ZnO on the orange peel will decrease the number of -OH and C-H groups of the ZnO/OP composite. The addition of orange peel to ZnO did not damage the crystal structure that is wurtzite with a hexagonal shape. The addition of orange peel can narrow the band gap energy so that it can increase the responsiveness of ZnO in the visible light region. Based on the results , it is known that the material with the best character was ZnO/OP material with the addition of 25 mL of ZnO (1:25). The results of the photocatalytic activity test under UV light showed that ZnO/OP material was better at degrading MO than ZnO. ZnO/OP can degrade 3 ppm MO by 73% in 60 minutes.

Acknowledgements
The authors gratefully acknowledge to Indonesian Ministry Education, Culture, Research and Technology for its financial support for this research and to Universitas Islam Kadiri for the facilities.

Conflict of interest statement
Authors declare no conflict of interest for this article.

References
[1] W. Muniroh, “Studi Fotodegradasi-Adsorpsi Methyl Orange Menggunakan Komposit TiO2-Kitosan,” Universitas Islam Negeri Sunan Kalijaga Yogyakarta, 2014.
[2] O. C. Sitepu, O. Ratnayani, and I. E. Suprihatin, “Sintesis Komposit ZnO-Bentonit dan Penggunaannya Dalam Proses Degradasi Methyl Orange,” Cakra Kim., vol. 4, no. 2, pp. 153–160, 2016.
[3] D. A. Wismayanti, N. P. Diantariani, and S. R. Santi, “Pembuatan Komposit ZnO-Arang Aktif Sebagai Fotokatalis Untuk Mendegradasi Zat Warna Metilen Blue,” J. Kim., vol. 9, no. 1, pp. 109–116, 2015.
[4] D. V. Wellia, Y. Kusumawati, L. J. Diguna, N. Pratiwi, R. A. Putri, and M. I. Amal, “Mesoporous Materialsfor Degradation of Textile Dyes,” in Green Methods for Wastewater Treatment. Environmental Chemistry for a Sustainable World, Springer, 2019, pp. 255–288.

[5] A. D. Rosanti, "Pengaruh Variasi Konsentrasi Garam FeCl3.6H2O,Suhu Kalsinasi, dan pH terhadap Karakter Fe-TiO2 yang disintesis Menggunakan Metode Sol-Gel," Universitas Gadjah Mada, 2014.
[6] D. V. Wellia, Y. Kusumawati, L. J. Diguna, and M. I. Amal, "Introduction of Nanomaterials for Photocatalysis," in Nanocomposites for Visible Light-induced Photocatalysis, Springer, 2017, pp. 1–17.
[7] N. Dony, H. Azis, and Syukri, "Studi Fotodegradasi biru metilen di Bawah Sinar Matahari Oleh ZnO-SnO2 Yang Dibuat Dengan Metoda Solid State Reaction," in Semirata FMIPA Universitas Lampung, 2013, pp. 297–304.
[8] H. Sutanto et al., "Deposisi Lapisan Tipis Foto Katalis Seng Oksida (ZnO) Berukuran Nano dengan Teknik Penyemprotan dan Aplikasinya untuk Pendegradasi Pewarna Methylene Blue," J. Fis., vol. 3, no. 1, pp. 69–75, 2013.
[9] E. W. P. Dini and S. Wardhani, "Degradasi metilen blue menggunakan fotokatalis ZnO-zeolit," vol. 7, no. 1, pp. 29–33, 2014.
[10] M. Arami, N. Yousefi, and N. Mohammad, "Removal of dyes from colored textile wastewater by orange peel adsorbent : Equilibrium and kinetic studies," J. Colloid.Interf.Sci, vol. 288, pp. 371–376, 2005, doi: 10.1016/j.jcis.2005.03.020.
[11] P. Pongwan, B. Inceesungvorn, K. Wetchakun, S. Phanichphant, and N. Wetchakun, "Highly efficient visible-light-induced photocatalytic activity of Fe-doped TiO 2 nanoparticles," Eng. J., vol. 16, no. 3, pp. 143–151, 2012, doi: 10.4186/ej.2012.16.3.143.
[12] T. Karnan and S. A. S. Selvakumar, "Biosynthesis of ZnO nanoparticles using rambutan (Nephelium lappaceumL.) peel extract and their photocatalytic activity on methyl orange dye," J. Mol. Struct., vol. 1125, pp. 358–365, 2016, doi: 10.1016/j.molstruc.2016.07.029.
[13] W. Raza, M. M. Haque, and M. Muneer, "Synthesis of visible light driven ZnO: Characterization and photocatalytic performance," Appl. Surf. Sci., vol. 322, pp. 215–224, 2014, doi: 10.1016/j.apsusc.2014.10.067.
[14] N. Feng, X. Guo, and S. Liang, "Adsorption study of copper (II) by chemically modified orange peel," J. Hazard. Mater., vol. 164, no. 2–3, pp. 1286–1292, 2009, doi: 10.1016/j.jhazmat.2008.09.096.
[15] N. A. Asyiqin Anas, Y. W. Fen, N. A. Yusof, N. A. S. Omar, N. S. Md Ramdzan, and W. M. E. M. Mohd Daniyal, "Investigating the properties of cetyltrimethylammonium bromide/hydroxylated graphene quantum dots thin film for potential optical detection of heavy metal ions," Materials (Basel)., vol. 13, no. 11, 2020, doi: 10.3390/ma13112591.
[16] M. Kumar and S. S. Sahu, "Zinc Oxide Nanostructures Synthesized by Oxidization of Zinc," 2010.
[17] A. T. Surono and H. Sutanto, "SIFAT OPTIK Zinc Oxide (ZnO) YANG DIDEPOSISI DI ATAS SUBSTRAT KACA MENGGUNAKAN METODE CHEMICAL SOLUTION DEPOSITION (CSD) DAN APLIKASINYA UNTUK DEGRADASI ZAT WARNA METHYLENE BLUE," Youngster Phys. J., vol. 3, no. 1, pp. 7–14, 2014.
[18] T. V. Surendra, S. M. Roopan, N. A. Al-Dhabi, M. V. Arasu, G. Sarkar, and K. Suthindhiran, "Vegetable Peel Waste for the Production of ZnO Nanoparticles and its Toxicological Efficiency, Antifungal, Hemolytic, and Antibacterial Activities," Nanoscale Res. Lett., vol. 11, no. 1, 2016, doi: 10.1186/s11671-016-1750-9.
[19] A. R. Hakim and A. Haris, "Sintesis Fotokatalis ZnO-Al dan Aplikasinya pada Degradasi Fenol dan Reduksi Cd(II) secara Simultan," J. Kim. Sains dan Apl., vol. 19, no. 1, p. 7, 2016, doi: 10.14710/jksa.19.1.7-10.

Carbon sequestration of city agriculture: between farming and non-farming land

D.R.D. Hastuti[1*], R. Darma[1], D. Salman[2], S. Santoso[2], A. Rahim[1]
[1]Universitas Negeri Makassar, Makassar, Indonesia
[2]Universitas Hasanuddin, Makassar, Indonesia

*diah.retno@unm.ac.id

Abstract. Urban agriculture is multifunctional. The environmental function is one of the agricultural functions. This research focuses on the level of carbon sequestration in food crops and vegetables. Furthermore, we estimate carbon sequestration differences in farming and non-farming land. This study calculates carbon sequestration using destructive methods. Furthermore, food crops carbon sequestration better than vegetable crops. In the same crop, there is more carbon sequestration in farming land when compared to non-farming land. Economically, this carbon sequestration can provide substantial economic value if it is better managed by considering urban spatial planning.

1. Introduction

The primary function of urban agriculture is to fulfill food needs [1], [2] as a substitute for urban and rural areas [3]. Therefore, it is necessary to increase food production sustainably with food security [4], [5], especially in developing countries [6]. Moreover, food is a fundamental problem for human civilization's continuity [7], whose population is increasing [8], [9]. Apart from these primary functions, urban agriculture has other multi-functions [3], [10], [11]. One of them is green open space [12], [13] that is productive and has economic value [14]. Also, make healthy [15], [16], reducing the carbon footprint [17]–[20]. This environmental function [21]–[23] can help overcome climate change [24]. Besides agricultural sustainability [25], especially for farmers small [26], even creating more sustainable cities [9], [16], [27]–[29].

Several studies have shown that trees [30] and tall cover crops [31] on agricultural land or green spaces [28], [32] helps the carbon sequestration of other sectors [20]. Also, organic agriculture [33], [34]. This contribution is limited quantities [35], [36]. Some studies show carbon loss due to high urbanization [37]–[39]. Besides, intensive land management [40], and changes in pastureland [41], [42] and forest [43] become agricultural land, but not the other way around [41], [44]. However, so far, there has not been a calculation of urban agriculture environmental services on farming land and non-farming land.

2. Methodology

a. Study Area and Plot Area

This study analyzes environmental services from carbon absorption limited to the soil surface in four potential sub-districts in Makassar City, namely, Tamalate, Manggala, Tamalanrea, and Biringkanaya sub-district with the purposive sampling method. The count was carried out destructively for understory

IOP Conf. Series: Earth and Environmental Science 1041 (2022) 012009 doi:10.1088/1755-1315/1041/1/012009

[45] using a sample plot measuring 0.5m x 0.5m, both on farming land and non-farming land, carried out randomly.

b. Carbon Stocks Estimation

The first step in calculating the biomass estimate is using the equation:

$$DW\ total = \frac{subsample\ DW}{subsample\ WW} \times total\ weight \tag{1}$$

where DW is dry weight (g) and WW is wet weight (g) [45]. Furthermore, by multiplying the conversion factor by 46% [45], [46], the biomass carbon content can be obtained so that the estimated carbon content is:

$$C = DW \times 0{,}46 \tag{2}$$

where C is carbon stocks (t/ha) and DW is dry weight of biomass (t/ha).

c. Estimated CO_2 Sequestered and O_2 Released

The amount of absorbed carbon dioxide is calculated using the conversion of the C atom to the CO_2 molecule [47], [48] as follows:

$$CO_2 = C \times \frac{Mw\ CO_2}{Aw\ C} \tag{3}$$

where CO_2 is adsorbed carbon dioxide (t/ha), $Mw\ CO_2$ is relative molecular weight of CO_2 compound, and $Aw\ C$ is relative weight of atom C. Furthermore, the release of O_2 can be estimated from CO_2 sequestered using the development formula of the CO_2 sequestered formula through the conversion of the element CO_2 to O_2 [49] as follows:

$$O_2 = CO_2 \times \frac{Mw\ O_2}{Mw\ CO_2} \tag{4}$$

where O_2 is oxygen produced (t/ha) and $Mw\ O_2$ is relative molecular weight of the compound O_2.

d. The Value of Environmental Services from CO_2 Sequestration

The value of environmental services from or the calculated economic valuation is a direct benefit of CO_2 sequestration and is calculated by the equation:

$$V_{ES} = P_C \times CO_2 \tag{5}$$

Where V_{ES} is the value of environmental services (USD/ha) and P_C is the selling price of carbon (USD/t), valued at the 10 USD [50].

3. Result and Discussion

Biomass calculation is needed in estimating carbon absorbed because biomass is organic material stored from photosynthesis [51]. Previously, it was necessary to calculate the plant's wet and dry weight to be calculated its biomass. Figure 1 shows the average wet weight.

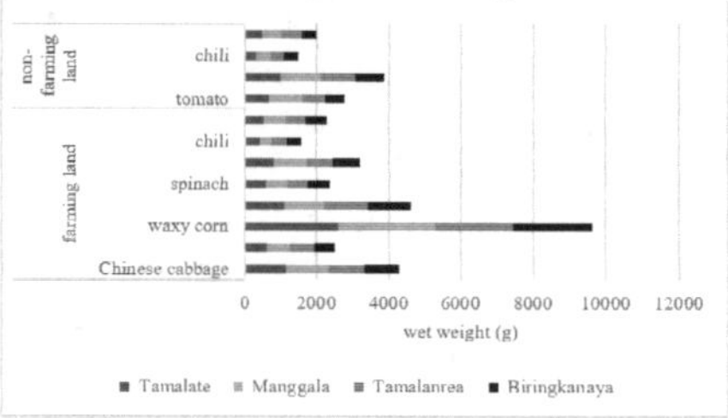

Figure 1. Crop wet weight

The whole plant's average wet weight was 3,379.73 g, where the highest average wet weight of waxy corn located in Manggala sub-district was 2,686.4 g, where 44.31% was in the stem section. Okra plants also had the highest average wet weight for the plant group on non-farming land of 1.097.7 g in Manggala sub-district with the most significant weight on the fruit part, namely 399.8 g. In contrast, the lowest wet weight was in chili plants, both on farming land and non-farming land, namely 1.570.40 g and 1.494.50 g, respectively.

The lowest weight on the Manggala sub-district farming land was 314.10 g and 323.7 g on the non-farming land of Tamalate sub-district with the most negligible weight in the leaf area from 25.95% - 26.57%. In comparison, Figure 2 shows the dry weight.

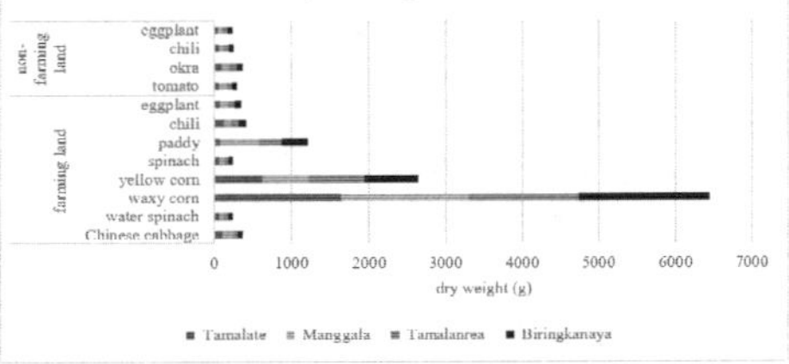

Figure 2. Crop' dry weight

Not different from the wet weight, the whole plant's average dry weight was 1,083.26 g, with the highest weight also in waxy corn. The highest weight is not in the Manggala sub-district but the Biringkanaya sub-district of 1,691.08 g with a composition of 47.7%, which is also in the stem section. Okra also has the highest dry weight for the plant group on a non-farming land of 100.13 g, located in the Manggala sub-district. However, in contrast to the wet weight, the stem's highest dry weight reaches 39%—the fruit's highest water content than to the stems and leaves.

Meanwhile, the lowest average dry weight was for water spinach on farming land and eggplant on non-farming land, 230.84 g and 229.14 g, respectively. Thus, the wet weight is not directly proportional to the dry weight, except for food crops.

Table 1. Crop' carbon stocks

Crop	Carbon stocks (t/ha)			
	Tamalate	Manggala	Tamalanrea	Biringkanaya
1. farming land				
Chinese cabbage	1.96	2.31	1.31	1.17
water spinach	1.06	1.23	1.10	0.85
waxy corn	30.27	30.53	26.51	31.12
yellow corn	11.51	10.97	13.26	12.83
spinach	1.14	1.19	1.08	0.86
paddy	7.37	9.05	5.50	6.42
chili	2.25	1.60	1.85	1.83
eggplant	1.40	1.62	1.61	1.67
2. non-farming land				
tomato	1.20	1.70	1.17	1.29
okra	1.69	1.84	1.70	1.42
chili	1.05	1.12	1.06	1.25
eggplant	0.92	0.98	1.27	1.04
sum	61.83	64.15	57.42	61.76
average	5.15	5.35	4.79	5.15

Overall, based on Table 2, it can be seen that the food crop groups that make a significant contribution to carbon stock, namely corn, and paddy. This contribution is due to the longer harvest life than vegetable crops. The highest carbon stock was obtained from waxy corn at 118.42 t/ha, followed by yellow corn and paddy. The three sequences are the same for the four sub-districts studied. Biringkanaya sub-district supplies the most significant carbon stocks for waxy corn, namely 31.12 t/ha or equivalent to 50.39% of the sub-district's total carbon stock, followed by Manggala and Tamalate sub-districts (Table 1).

Meanwhile, the most significant contribution to yellow corn comes from the Tamalanrea sub-district. Even though the Tamalate sub-district is not a minor contributor, the cost required is too large because it has to pump more water than other commodities. It gives a reluctance to plant yellow corn if not as a

IOP Conf. Series: Earth and Environmental Science 1041 (2022) 012009 doi:10.1088/1755-1315/1041/1/012009

member of a farmer group. The paddy commodity, as the third-largest carbon store, on average, stores carbon of 7.09 t/ha, which is not much different from the findings of [25] in Thailand with the ability to store carbon by 7.08 t/ha and has the most significant carbon footprint in China [52].

The most significant contribution came from the Manggala sub-district, amounting to 9.05 t/ha. Paddy field area of 801 ha [53] with the most significant rice production support this contribution. Thus, we can estimate that paddy carbon stock's contribution in this sub-district will reach 7,249.05 tons for one planting season. Likewise, the sub-districts of Biringkanaya, Tamalanrea, and Tamalate have 639 ha, 632 ha, and 509 ha of paddy fields, whose carbon reserves can reach 3,476 t - 4,102.38 t per planting season.

Table 2. Estimates of biomass, carbon sequestration, and economic value

type of land	crops	wet weight (g)	dry weight (g)	Carbon stocks (t/ha)	CO_2 sequestration (t/ha)	O_2 released (t/ha)	Economic value (USD/ha)
farming land	Chinese cabbage	4,277.50	366.88	6.75	24.75	4.91	247.52
	water spinach	2,504.60	230.84	4.25	15.57	3.09	155.74
	waxy corn	9,635.40	6,435.86	118.42	434.21	86.12	4,342.06
	yellow corn	4,609.20	2,640.01	48.58	178.11	35.33	1,781.12
	spinach	2,366.90	232.27	4.27	15.67	3.11	156.70
	paddy	3,190.90	1,216.02	28.35	103.94	20.62	1,039.37
	chili	1,570.40	409.02	7.53	27.60	5.47	275.95
	eggplant	2,280.20	342.84	6.31	23.13	4.59	231.30
non-farming land	tomato	2,761.80	291.71	5.37	19.68	3.90	196.81
	okra	3,874.20	361.30	6.65	24.38	4.83	243.75
	chili	1,494.50	243.30	4.48	16.41	3.26	164.15
	eggplant	1,991.10	229.14	4.22	15.46	3.07	154.59
sum		40,556.70	12,999.17	245.16	898.91	178.30	8,989.07
average		3,379.73	1,083.26	20.43	74.91	14.86	749.09

Meanwhile, the vegetable crop group can store carbon ranging from 4.22 t/ha - 7.53 t/ha. On the farming land of commodity, chilies, Chinese cabbage, and eggplants can store more carbon than spinach and water spinach. The highest carbon storage in chili commodity comes from Tamalate sub-district, while Manggala sub-district for Chinese cabbage and eggplant commodities in Biringkanaya sub-district. Furthermore, in the crop group on non-farming land, the highest carbon stock came from okra, namely 6.65 t/ha and tomatoes, which ranged from 1.17 t/ha - 1.7 t/ha. Compared to the same two commodities, namely chili and eggplant, the carbon stock is more significant in farming land due to the absorption of nutrients from the soil and above the soil surface and the more excellent absorption of sunlight on the farming land.

The most significant average value of carbon stock was in the Manggala sub-district, namely 5.35 t/ha, and the smallest was in the Tamalanrea sub-district. However, the Tamalanrea sub-district contributed carbon reserves of 24,304.92 g [51] or the equivalent of 121.52 t/ha as one of the urban forests. It is not like [32] findings that the carbon in urban park soils in winter is greater than that of forest soils. This contribution is smaller than the contribution of urban agriculture of 245.16 t/ha. The findings of [54] stated that urban agriculture absorbs more carbon than urban green open space, such as parks and forests.

Likewise, [28] calculates urban agricultural carbon stock, identical to seasonal crops of 123.22 t/ha. The four sub-districts have the largest carbon reserves originating from waxy corn, but not the minor contribution, except for Tamalate and Manggala sub-districts with the same commodity eggplant on non-farming land. Meanwhile, in the Tamalanrea sub-district is chili and water spinach on non-farming land in the Biringkanaya sub-district.

The uptake of CO_2, O_2 released, and environmental services can be obtained estimates by the obtained carbon stocks. The ability to absorb CO_2 in the atmosphere is estimated to reach 898.91 t/ha with the order of contributions such as carbon stocks, namely Manggala, Tamalate, Biringkanaya, and Tamalanrea sub-district. This sequence also applies to O_2 and the environmental services it produces. Food crops contribute 79.68% to CO_2 absorption, while plants on non-agricultural land only support 8.45%. This carbon absorption is still smaller than Seoul's agricultural city, which can reduce CO_2 emissions by 11,670 tons each year [18].

Table 3. Estimated population oxygen sufficiency

sub-district	Population* (people)	O_2 needs (kg/day)	availability of O_2** (kg/day)
Tamalate	180,824	151,892.16	706.33
Manggala	146,724	123,248.16	741.27
Tamalanrea	103,177	86,668.68	640.34
Biringkanaya	209,048	175,600.32	675.30
sum	639,773	537,409.32	2,763.25
average	159,943	134,352.33	690.81

**based on the results of the 2020 population census,*

*** based on planted area per hectare of the twelve commodities studied*

If the oxygen demand is 0.84 kg/day/person [55], then the amount of oxygen sufficiency in the four sub-districts can be estimated (Table 3). Manggala sub-district, as the agricultural center of Makassar City, has the most significant O_2 availability, but only 0.60% of the O_2 needs of its population due to its large population. Another case that happened in the Tamalanrea sub-district was the lowest available O_2 but was able to meet 0.74% of the O_2 needs of the population. Overall, the available O_2 can meet the O_2 needs of the four sub-districts residents by 0.51% or equivalent to 3,290 residents.

The availability of O_2 is inseparable from the carbon stocks of food crops, Chinese cabbage, chilies, and eggplant in farming land and okra and tomatoes on non-stretched soils (Table 1 and Table 2). The economic value obtained reaches 8,989.07 USD with a range of 156 - 4,342 USD/ha for commodities on farming land. Non-farming land also contributes 155 - 244 USD/ha. This indirect economic value is often neglected because of the tendency to leave agriculture which is not financially economical. This will affect carbon absorption [56].

However, better agricultural management will have various environmental and economic benefits in addition to climate mitigation potential [57]. Therefore, by estimating the economic value, it is hoped that urban agriculture will become policy makers' attention in the further development of urban spatial planning.

4. Conclusion

Makassar City agriculture has higher carbon sequestration in the food crop group than in the vegetables crop group. In the same types of plants, namely chilies and eggplants, the value of carbon uptake was higher in farming land than in non-farming land. This carbon sequestration provides the highest oxygen adequacy in the Tamalanrea sub-district. This carbon sequestration also has potential economic value. Therefore, in making policies, especially in urban spatial planning, it is necessary to pay attention to these economic values.

References

[1] C. Aubry, J. Ramamonjisoa, M.-H. Dabat, J. Rakotoarisoa, J. Rakotondraibe, and L. Rabeharisoa, “Urban agriculture and land use in cities: An approach with the multi-functionality and sustainability concepts in the case of Antananarivo (Madagascar),” *Land use policy*, vol. 29, no. 2, pp. 429–439, Apr. 2012, doi: 10.1016/j.landusepol.2011.08.009.

[2] S. Yusuf, O. Balogun, and O. Falegbe, “Effect of urban household farming on food security status in Ibadan metropolis, Oyo State, Nigeria,” *J. Agric. Sci. Belgrade*, vol. 60, no. 1, pp. 61–

75, 2015, doi: 10.2298/JAS1501061Y.

[3] W. Valley and H. Wittman, "Beyond feeding the city: The multifunctionality of urban farming in Vancouver, BC," *City, Cult. Soc.*, vol. 16, pp. 36–44, Mar. 2019, doi: 10.1016/j.ccs.2018.03.004.

[4] D. Baulcombe *et al.*, *Reaping the Benefits: Science and the Sustainable Intensification of Global Agriculture*. London: Royal Society, 2009.

[5] D. Cordell, J.-O. Drangert, and S. White, "The story of phosphorus: Global food security and food for thought," *Glob. Environ. Chang.*, vol. 19, no. 2, pp. 292–305, May 2009, doi: 10.1016/j.gloenvcha.2008.10.009.

[6] M. Thornbush, "Urban agriculture in the transition to low carbon cities through urban greening," *AIMS Environ. Sci.*, vol. 2, no. 3, pp. 852–867, 2015, doi: 10.3934/environsci.2015.3.852.

[7] E. Dal Bó, P. Hernández, and S. Mazzuca, "The Paradox of Civilization: Pre-Institutional Sources of Security and Prosperity," Cambridge, MA, Dec. 2015. doi: 10.3386/w21829.

[8] A. Burns, R. Gleadow, J. Cliff, A. Zacarias, and T. Cavagnaro, "Cassava: The Drought, War and Famine Crop in a Changing World," *Sustainability*, vol. 2, no. 11, pp. 3572–3607, Nov. 2010, doi: 10.3390/su2113572.

[9] L. Li, X. Li, C. Chong, C.-H. Wang, and X. Wang, "A decision support framework for the design and operation of sustainable urban farming systems," *J. Clean. Prod.*, vol. 268, p. 121928, 2020, doi: 10.1016/j.jclepro.2020.121928.

[10] N. Cohen and K. Reynolds, "Urban Agriculture Policy Making in New York's 'New Political Spaces,'" *J. Plan. Educ. Res.*, vol. 34, no. 2, pp. 221–234, Jun. 2014, doi: 10.1177/0739456X14526453.

[11] B. Pölling, W. Sroka, and M. Mergenthaler, "Success of urban farming's city-adjustments and business models—Findings from a survey among farmers in Ruhr Metropolis, Germany," *Land use policy*, vol. 69, pp. 372–385, Dec. 2017, doi: 10.1016/j.landusepol.2017.09.034.

[12] S. Hallett, L. Hoagland, and E. Toner, "Urban Agriculture: Environmental, Economic, and Social Perspectives," in *Horticultural Reviews*, Hoboken, NJ, USA: John Wiley & Sons, Inc., 2016, pp. 65–120.

[13] H. S. Mulyani, H. Agustin, and I. Zulfan, "Implementation of Community Empowerment Communication in the Need of Environmental Food Security through the Program 'Bandung Agri Market,'" *Int. J. Psychosoc. Rehabil.*, vol. 23, no. 2, pp. 289–297, 2019, doi: 10.37200/IJPR/V23I2/PR190290.

[14] J. Kaufman and M. Bailkey, "Farming Inside Cities: Entrepreneurial Urban Agriculture in the United States," Cambridge, 2000. [Online].

[15] K. Benis and P. Ferrão, "Commercial farming within the urban built environment – Taking stock of an evolving field in northern countries," *Glob. Food Sec.*, vol. 17, pp. 30–37, Jun. 2018, doi: 10.1016/j.gfs.2018.03.005.

[16] L. Ibrahim and S. A. Salim, "The Framework of Urban Farming Towards Enhancing Quality of Life In Malaysia," *Int. J. Supply Chain Manag.*, vol. 9, no. 1, pp. 520–526, 2020, [Online].

[17] C. Gómez *et al.*, "Controlled Environment Food Production for Urban Agriculture," *HortScience*, vol. 54, no. 9, pp. 1448–1458, Sep. 2019, doi: 10.21273/HORTSCI14073-19.

[18] G.-G. Lee, H.-W. Lee, and J.-H. Lee, "Greenhouse gas emission reduction effect in the transportation sector by urban agriculture in Seoul, Korea," *Lanscape urban Plan.*, vol. 140, pp. 1–7, 2015, doi: 10.1016/j.landurbplan.2015.03.012.

[19] H.-F. Mok, V. G. Williamson, J. R. Grove, K. Burry, S. F. Barker, and A. J. Hamilton, "Strawberry fields forever? Urban agriculture in developedcountries: a review," *Agron. Sustain. Dev*, vol. 34, no. 1, pp. 21–43, 2013, doi: 10.1007/s13593-013-0156-7.

[20] A. J. VandenBygaart and D. A. Angers, "Towards accurate measurements of soil organic carbon stock change in agroecosystems," *Can. J. Soil Sci.*, vol. 86, no. 3, pp. 465–471, May 2006, doi: 10.4141/S05-106.

[21] D. Despommier, "Farming up the city: the rise of urban vertical farms," *Trends Biotechnol.*, vol.

31, no. 7, pp. 388–389, Jul. 2013, doi: 10.1016/j.tibtech.2013.03.008.

[22] M. M. Khan, M. T. Akram, R. Janke, R. W. K. Qadri, A. M. Al-Sadi, and A. A. Farooque, "Urban Horticulture for Food Secure Cities through and beyond COVID-19," *Sustainability*, vol. 12, no. 22, p. 9592, Nov. 2020, doi: 10.3390/su12229592.

[23] E. Mohareb, M. Heller, P. Novak, B. Goldstein, X. Fonoll, and L. Raskin, "Considerations for reducing food system energy demand while scaling up urban agriculture," *Environ. Res. Lett.*, vol. 12, no. 12, p. 125004, Dec. 2017, doi: 10.1088/1748-9326/aa889b.

[24] J. Demenois *et al.*, "Barriers and Strategies to Boost Soil Carbon Sequestration in Agriculture," *Front. Sustain. Food Syst.*, vol. 4, Apr. 2020, doi: 10.3389/fsufs.2020.00037.

[25] N. Gnanavelrajah, R. P. Shrestha, D. Schmidt-Vogt, and L. Samarakoon, "Carbon stock assessment and soil carbon management in agricultural land-uses in Thailand," *L. Degrad. Dev.*, vol. 19, no. 3, pp. 242–256, May 2008, doi: 10.1002/ldr.838.

[26] T. Tennigkeit, K. Solymosi, M. Seebauer, and B. Lager, "Carbon Intensification and Poverty Reduction in Kenya: Lessons from the Kenya Agricultural Carbon Project," *F. ACTions Sci. Reports*, vol. 7, 2013, [Online].

[27] M. W. Adidja, J. Mwine, J. G. M. Majaliwa, and J. Ssekandi, "The Contribution of Agro-ecology as a Solution to Hunger in the World: A Review," *Asian J. Agric. Extension, Econ. Sociol.*, vol. 33, no. 2, pp. 1–22, 2019, doi: 10.9734/ajaees/2019/v33i230170.

[28] Helen, M. P. Jarzebski, and A. Gasparatos, "Land use change, carbon stocks and tree species diversity in green spaces of a secondary city in Myanmar, Pyin Oo Lwin," *PLoS One*, vol. 14, no. 11, p. e0225331, Nov. 2019, doi: 10.1371/journal.pone.0225331.

[29] N. H. Yusoff, M. R. Mohd Hussain, and I. Tukiman, "roles of community towards urban farming activities," *Plan. Malaysia J.*, vol. 15, no. 1, May 2017, doi: 10.21837/pmjournal.v15.i6.243.

[30] Y. Gebrewahid, T.-B. Gebre-Egziabhier, K. Teka, and E. Birhane, "Carbon stock potential of scattered trees on farmland along an altitudinal gradient in Tigray, Northern Ethiopia," *Ecol. Process.*, vol. 7, no. 1, p. 40, Dec. 2018, doi: 10.1186/s13717-018-0152-6.

[31] C. Poeplau and A. Don, "Carbon sequestration in agricultural soils via cultivation of cover crops – A meta-analysis," *Agric. Ecosyst. Environ.*, vol. 200, pp. 33–41, 2015, doi: 10.1016/j.agee.2014.10.024.

[32] L. Lindén, A. Riikonen, H. Setälä, and V. Yli-Pelkonen, "Quantifying carbon stocks in urban parks under cold climate conditions," *Urban For. Urban Green.*, vol. 49, p. 126633, 2020, doi: 10.1016/j.ufug.2020.126633.

[33] P. García-Palacios *et al.*, "Crop traits drive soil carbon sequestration under organic farming," *J. Appl. Ecol.*, vol. 55, no. 5, pp. 2496–2505, Sep. 2018, doi: 10.1111/1365-2664.13113.

[34] A. Gattinger *et al.*, "Enhanced top soil carbon stocks under organic farming," *Proc. Natl. Acad. Sci.*, vol. 109, no. 44, pp. 18226–18231, Oct. 2012, doi: 10.1073/pnas.1209429109.

[35] B. Goldstein, M. Hauschild, J. Fernández, and M. Birkved, "Testing the environmental performance of urban agriculture as a food supply in northern climates," *J. Clean. Prod.*, vol. 135, pp. 984–994, Nov. 2016, doi: 10.1016/j.jclepro.2016.07.004.

[36] L. G. Smith, G. J. D. Kirk, P. J. Jones, and A. G. Williams, "The greenhouse gas impacts of converting food production in England and Wales to organic methods," *Nat. Commun.*, vol. 10, no. 1, p. 4641, Dec. 2019, doi: 10.1038/s41467-019-12622-7.

[37] D. Murtala, L. A. Manaf, M. F. Ramli, M. R. Yacob, and A. A. Makmom, "quantifying the aboveground biomass and carbon storage of urban tree species in Sokoto Metropolis, North-Western Nigeria," *Plan. Malaysia J.*, vol. 17, no. 2, pp. 179 –190, 2019, doi: 10.21837/pm.v17i10.639.

[38] Y. Tao, F. Li, X. Liu, D. Zhao, X. Sun, and L. Xu, "Variation in ecosystem services across an urbanization gradient: A study of terrestrial carbon stocks from Changzhou, China," *Ecol. Modell.*, vol. 318, pp. 210–216, 2015, doi: 10.1016/j.ecolmodel.2015.04.027.

[39] Y. Tao, F. Li, R. Wang, and D. Zhao, "Effects of land use and cover change on terrestrial carbon stocks in urbanized areas: a study from Changzhou, China," *J. Clean. Prod.*, vol. 103, pp. 651–

657, 2015, doi: 10.1016/j.jclepro.2014.07.055.

[40] Y. L. Zinn, R. Lal, and D. V. S. Resck, "Changes in soil organic carbon stocks under agriculture in Brazil," *Soil Tillage Res.*, vol. 84, no. 1, pp. 28–40, Nov. 2005, doi: 10.1016/j.still.2004.08.007.

[41] E. Lugato, F. Bampa, P. Panagos, L. Montanarella, and A. Jones, "Potential carbon sequestration of European arable soils estimated by modelling a comprehensive set of management practices," *Glob. Chang. Biol.*, vol. 20, no. 11, pp. 3557–3567, Nov. 2014, doi: 10.1111/gcb.12551.

[42] A. Stevens, B. van Wesemael, G. Vandenschrick, S. Touré, and B. Tychon, "Detection of Carbon Stock Change in Agricultural Soils Using Spectroscopic Techniques," *Soil Sci. Soc. Am. J.*, vol. 70, no. 3, pp. 844–850, May 2006, doi: 10.2136/sssaj2005.0025.

[43] R. Priyadarshini, A. Hamzah, and B. W. Widjajani, "Carbon Stock Estimates due to Land Cover Changes at Sumber Brantas Sub-Watershed, East Java," *Caraka Tani J. Sustain. Agric.*, vol. 34, no. 1, p. 1, Feb. 2019, doi: 10.20961/carakatani.v34i1.27124.

[44] J. A. R. Martín *et al.*, "Assessment of the soil organic carbon stock in Spain," *Geoderma*, vol. 264, pp. 117–125, Feb. 2016, doi: 10.1016/j.geoderma.2015.10.010.

[45] K. Hairiah, A. Ekadinata, R. R. Sari, and S. Rahayu, *Petunjuk Praktis Pengukuran Cadangan Karbon dari Tingkat Lahan ke Bentang Lahan*, Kedua. Bogor: World Agroforestry Centre, 2011.

[46] Badan Standardisasi Nasional, "Pengukuran dan Penghitungan Cadangan Karbon: Pengukuran Lapangan untuk Penaksiran Cadangan Karbon Hutan (Ground Based Forest Carbon Accounting)," Jakarta, 2011. [Online].

[47] C. T. Stevanus and Sahuri, "Potensi Peningkatan Penyerapan Karbon di Perkebunan Karet Sembawa Sumatera Selatan," *Widyariset*, vol. 17, no. 3, pp. 363–371, 2014, doi: 10.14203/widyariset.17.3.2014.363-371.

[48] D. Usmadi, S. Wahyuni, and M. K. Riswati, "Potensi Dipterocarpaceae sebagai Penyerap CO2 dan Penyimpan Karbon di Kebun Raya Bogor," in *Prosiding Seminar Nasional Konservasi Tumbuhan Tropika: Kondisi Terkini dan Tantangan ke Depan*, 2011, p. 46, [Online].

[49] F. B. Salisbury and C. W. Ross, *Plant Physiology*. Belmont: Wadsworth Publishing Company, 1978.

[50] S. Juita, A. Lumangkun, and I. Dewantara, "Penilaian Ekonomi Jasa Lingkungan Hutan Kota pada Kawasan Universitas Tanjungpura Pontianak," *J. Hutan Lestari*, vol. 4, no. 3, pp. 380–386, 2016, [Online].

[51] S. Santosa, M. R. Umar, D. Priosambodo, and R. A. P. Santosa, "Estimation of biomass, carbon stocks and leaf litter decompo_sition rate in teak Tectona grandis Linn plantations in city forest of Hasanuddin University, Makassar," *Int. J. Plant Biol.*, vol. 11, no. 1, pp. 32–35, 2020, doi: 10.4081/pb.2020.8541.

[52] Q. Yue, X. Xu, J. Hillier, K. Cheng, and G. Pan, "Mitigating greenhouse gas emissions in agriculture: From farm production to food consumption," *J. Clean. Prod.*, vol. 149, pp. 1011–1019, Apr. 2017, doi: 10.1016/j.jclepro.2017.02.172.

[53] Badan Pusat Statistik Kota Makassar, "Makassar dalam Angka," Makassar, 2019. [Online].

[54] M. Kulak, A. Graves, and J. Chatterton, "Reducing greenhouse gas emissions with urban agriculture: A Life Cycle Assessment perspective," *Landsc. Urban Plan.*, vol. 111, pp. 68–78, 2013, doi: 10.1016/j.landurbplan.2012.11.007.

[55] E. lrwan Afrizal, I. S. Fatimah, and B. Sulistyantara, "studi potensi produksi oksigen hutan kota di Kampus Universitas Indonesia, Depok," *J. Lanskap Indones.*, vol. 2, no. 1, pp. 23–29, 2010, doi: 10.29244/jli.2010.2.1.%25p.

[56] J. A. R. Martín *et al.*, "Soil organic carbon stock on the Majorca Island: Temporal change in agricultural soil over the last 10 years," *CATENA*, vol. 181, p. 104087, Oct. 2019, doi: 10.1016/j.catena.2019.104087.

[57] P. Smith, "Carbon sequestration in croplands: the potential in Europe and the global context," *Eur. J. Agron.*, vol. 20, no. 3, pp. 229–236, 2004, doi: 10.1016/j.eja.2003.08.002.

Comparison predicted using nonlinear model and in vivo studies growth production of broiler after fed different levels of fermented tamarind flour as replacing soya bean meal

O Sjofjan*, D N Adli, M H Natsir, Y F Nuningtyas, A D Budianto

Universitas Brawijaya, Malang, Indonesia

*osofjan@yahoo.com

Abstract. This study compared two nonlinear models for growth performance of broiler after fed different levels of fermented tamarind flour as replacing soya bean meal. Two experiments were designed for this purpose. In first experiment, 20 chickens (Lohmann MB202) that had been fed on the different level of tamarind flour and weighed to determine best model suitable. In second experiment the treatment of the study was basal feed (control): basal feed substitute soy bean meal with tamarind flour 25%: basal feed substitute soy bean meal with tamarind flour 50%: basal feed substitute soy bean meal with tamarind flour 100%. The results showed that the used of tamarind flour did not give a significant effect ($p > 0.05$) on the production performance. It concluded the comparison of growth models in terms of goodness revealed that gompertz is the most appropriate model for describing between age and body weight in this experiment.

1. Introduction

The increasing population has an impact on increasing feed consumption, one of which is chicken. Some experts state that the level of consumption of purebred chicken per capita / year in 2017 is 5.68 kg per capita / year, an increase of 573 g (11.2%) compared to the previous year, because that is why there is sufficient meat to meet animal protein needs for the population Indonesia, one of which is broilers. Broilers experience rapid development in body weight gain and are relatively short for the age of the harvest. Rapid development, one of which is influenced by genetic improvement of broilers and feeding with good quality feed and effective for digestion.

One of them is influenced by the composition and nutrition of the feed ingredients used in the feed mixture so that it can accelerate the weight gain of broilers. Feed material is any food that can be eaten, preferred, partially or completely digested, absorbable and beneficial for livestock [1]. Poultry feed is a mixture of several kinds of feed ingredients given to poultry to meet their needs, development and reproduction [2].

Feed functions as the body's development and maintenance, energy sources, production, and regulating processes in the body. The nutrient content that must be present in feed is protein, fat, carbohydrates, minerals, vitamins, and water [3]. The cost of feed in the business of poultry farming (broilers) is the largest component, which is about 70% of the total production costs [4]. One protein source material commonly used for poultry feed mix is soybean meal, but its availability is still limited. Tamarind (Tamarindus indica L.) is a type of tree legume. This plant is easily found in the environment around us, especially on the roadside.

Tamarind seeds are native to Indonesia, easily cultivated, can be harvested many times (more than 5 times in one crop) with dry seed yields more than 1 ton / ha once harvest can grow and produce on less

IOP Conf. Series: Earth and Environmental Science 1041 (2022) 012010 doi:10.1088/1755-1315/1041/1/012010

fertile land, more cultivation costs cheap compared to soybeans, as well as cheap prices [5]. Every part of this plant, from roots to leaves, can be used for various purposes. Tamarind seeds have the potential to be used as poultry feed ingredients because they have a relatively high protein content of 2.8 g / 100 g seeds [6]. Based on the description above, it is necessary to do comparison and predicted using non-linear model and in vivo studies.

2. Methodology

Table 1. Nonlinear model to predict growth of broiler

	Mathematical expression	Weight at Inflection (W*)	Time to Inflection (t*)	Growth Rate (dw/ dt)
Gompertz	*w=w0*exp((1-exp(-b*t))*(log(wf/w0)))*	$0.368W_f$	[illegible]	*bWln* [illegible]

2.1 First experimental with Mathematical expression

Gompertz were fitted to data using NLIN procedure of the SAS studio university version 4.0 Red hat 64bit for the evaluation of the growth parameters (Table 1). The models and equations are shown in. In all models, Wt refers to live body weight (g) at age t (day), W0 is the initial body weight (g), b is the coefficient of relative growth or maturing index (smaller b indicates later maturity, while larger b indicates earlier maturity); t is the age of bird (day) and Wf is the mature body weight (g). The derived parameters were, then, used to estimate the inflection point Ti (day); body weight at the inflection point (g; Wi) and growth rate (GR; g/day).

While in vivo studies using GLM procedure of *SAS University version 4.0 red hat* (64-bit) with code encryption algorithm http://localhost:10080/SASStudio/38/index license owned by Danung adli and the differences among treatment means ($p<0.05$) were determined using Duncan's multiple range test according to [4] method.

Code algorithm in SAS as follows:

```
Data Q1;
Set pre.Q1;
Run;
Proc anova data=Q1;
Title'one way anova with a moringa oleifera on one factor';
Class calib;
Mode shape_1shape_2 shape_3 shape 4= calib/nouni;
```

Table 2. Estimated parameters for each model in the study of the biological growth curve

Models	W0	B	Wf	K\n	Wi	Ti	Iterations	R2	AIC
Gompertz	31.73 (1.90)	0.053 (0.001)	3623 (68)	-	1334	29.28	5	0.9970	648

W0: the initial body weight (g);Wf: mature body weight (g);Wi: body weight at the infection point (g); b: coefficient of relative growth;Ti: inflection point (day)

Table 3. Approximate correlation matrix of the Gompertz parameters

Treatments	Day 7	Day 14	Day 21	Day 28	Day 35
T0	88.2	263.7	577.2	982.1^{a}	1516^{a}
T1	81.2	255.4	556.1	979.4^{b}	$1499,8^{b}$
T2	82.4	265.4	541.2	981.7^{b}	1466.7^{b}
T3	86.4	261.3	549.1	$902.^{b}$	1459.5^{b}

T4	85.6	255.7	557.2	988.6	1447.8[b]
SEM	2.23	6.37	11.23	16.75	55.10
P-Value	0.212	0.238	0.223	0.223	0.027

Table 4. Approximate correlation matrix of the Gompertz parameters

Parameter	**W0**	**B**	**Wf**
W0	1	-0.9410502	0.8341997
B	-0.9410502	1	-0.9652989
Wf	0.8341997	-0.9652989	1

Gompertz were widely used to represent changes in size of the growth of body weight with days or age. In order, can be evaluated and nutrition can be matched to possibility. Regarding the goodness of fit criteria, all models were suitable for describing the growth of the broiler chicken. Among all nonlinear models used, the smallest AIC and the lowest number of iterations were calculated for the Gompertz function.

The Gompertz model best described the growth curves for body weight of broiler chicken, with R2, AIC values and number of iterations being 0.8341, 997, respectively. The Gompertz model is one of the most common models to describe broiler growth curve. In this model, the growth curve is asymmetric around the point of maximum growth rate [7]

Table 5. Experimental diet

Feedstuff	T0 (%)	T1 (%)	T2 (%)	T3 (%)	T4 (%)
Yellow maize	58	58	58	58	58
Rice bran	10	10	10	10	10
Soya bean	25	18.75	12.50	6.25	-
Fermented Tamarind Flour	-	6.25	12.50	18.75	25
DL Methionine	0.3	0.3	0.3	0.3	0.3
Premix	0.7	0.7	0.7	0.7	0.7
MBM	6	6	6	6	6
Total	100	100	100	100	100

Nutrient content	T0 (%)	T1 (%)	T2 (%)	T3 (%)	T4 (%)
GE (kcal/kg)	2932	3018.25	3103.70	3189.95	3275.20
CP (%)	20.04	19.81	19.58	19.35	19.12
Fat (%)	3.04	3.27	3.50	3.73	3.97
CF (%)	4.73	4.76	4.94	5.12	5.3
Ca (%)	0.86	0.87	0.89	0.91	0.93
P (%)	0.54	0.52	0.49	0.44	0.42

***Mineral premix (Per kg of diet); Fe 70 mg, Zn, 90 mg; CU, 10 mg; Mn, 80 mg*

Table 6. Nutrient content of treatment

Feedstuff	EM (Kcal/kg)	CP (%)	Fat (%)	CF (%)	Ca (%)	P (%)
Yellow maize	3370	9	2.61	4.76	0.02	0.1
Rice bran	2860	12	7	3	0.04	0.16
Soya bean	2240	42	0.9	6	0.29	0.65

Fermented tamarind flour	3611	23.12	4.61	8.9	0.59	0.23
DL Methionine	0	0	0	0	0.3	0
Premix	0	0	0	0	25	0
MBM	2190	52	10	2.8	10	5.1

Notes: formulated using *UBfeed software*

Table 7. Growth Performance of Broiler

	Treatment				
Variabel	T0	T1	T2	T3	T4
FI (g/head)	2.994,50	2.917,50	2.904,25	2.902,75	2.889,75
BWG (g/head)	1477,7	1.461,5	1.428,5	1.421,5	1.409,5
Feed/gain (g/head)	1.99	2.00	2.03	2.04	2.05
Production index	211.92	209.23	200.79	198.88	196.45
IOFC (IDR/head)	11.172,00	11.533	11.488,26	11.877,37	12.225,02

3. Result and Discussion

The result showed on the table 6 that in vivo studies on the tamarind doesn't give significant effect ($p > 0.05$) on feed consumption between all treatments. The protein and energy content in each treatment feed is in a nearly balanced state, so the consumption of food for each treatment obtained is identical. As is well the balance between protein-influenced known that feed consumption and energy in feed, thus consumption of feed in the treatment produces results that are not significantly different.

[8] states that the lower the protein-energy content of feed, the higher feed consumption so that the need for protein-energy for growth can be achieved subsequently in feed that meets quality based on the need for growth, then feed consumption will be adjusted to differences in protein-concentration feed energy [1]0. That high protein consumption will affect the intake of protein in meat and sufficient amino acids in the body so that the metabolism of cells in the body takes place normally [9].

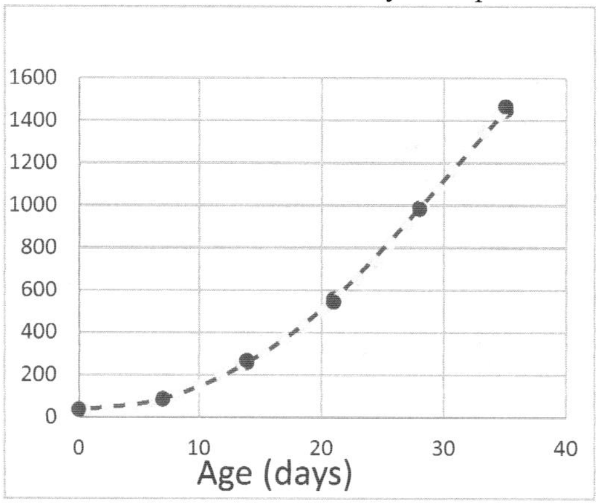

Figure 1. comparison between gompertz and in vivo studies

4. Conclusion

It can be concluded the comparison of growth models in terms of goodness revealed that gompertz is the most appropriate model for describing between age and body weight in this experiment.

References

[1] Sjofjan O and Adli D N. 2020. April Effect of Dietary of Supplementation Mannan-Riched Fraction (MRF) and Probiotic-Enhanced Liquid Acidifier on the Growth Performance, Serum

IOP Conf. Series: Earth and Environmental Science 1041 (2022) 012010 doi:10.1088/1755-1315/1041/1/012010

Blood Biochemistry, and Intestinal Properties of Broilers in IOP Conference Series: Earth and Environmental Science 478 012066 IOP Publishing

[2] Sjofjan O, Natsir M H, Adli D N, Adelina, D D, and Triana, L M Effect Of Symbiotic Flour (Lactobacillus Sp. And FOS) To the Egg Quality and Performance of Laying Hens in IOP Conference Series: Earth and Environmental Science 465 012033 IOP Publishing

[3] Adli D N, Sjofjan O, Natsir M H, Nuningtyas Y F, Sholikah N, and Marbun A C 2020 The effect of replacing maize with fermented palm kernel meal (FPKM) on broiler performance. Livestock Research for Rural Development. 32, Article #120. http://www.lrrd.org/lrrd32/7/danung32120.html

[4] Adli D N and Sjofjan O Growth performance, serum blood biochemistry, and intestinal properties of Arbor Acres Broiler fed diets containing mannan-riched fraction (MRF) and probiotic-enhanced liquid acidifier Buletin Peternakan 44

[5] Mirnawati, Ciptaan G and Ferawati 2020Broiler performance on a diet containing palm kernel meal fermented with Bacillus subtilis. Livestock Research for Rural Development. 32 Article #21. Retrieved May 21, 2020, from http://www.lrrd.org/lrrd32/2/mirna32021.html

[6] Widiyawati I and Sjofjan O 2020 Peningkatan kualitas dan persentase karkas ayam pedaging dengan subtitusi bungkil kedelai menggunakan tepung biji asam (Tamarindus indica L) fermentasi Jurnal Nutrisi Ternak Tropis 3 35-40

[7] Narinc D, Karaman E, Fira M Z, and Aksoy T 2010 Comparison of non-linear growth models to describe the growth in Japanese quail Journal of Animal and Veterinary Advances 9 1961-1966

[8] Masoudi A and Azarfar A 2017 Comparison of nonlinear models describing growth curves of broiler chickens fed on different levels of corn bran International Journal of Avian and Wildlife Biology 2 1-7

[9] Natsir M H, Djunaidi I, Sjofjan O, Suwanto A, Puspitasari E, and Virginia L J 2018 The Effect of Corn Substitution with Palm Kernel Meal Treated by Enzyme on Production Performance and Carcass Quality of Broiler Buletin Peternakan 42

[10] Leke, J R, Sjofjan, O, and Najoan, M 2013 Egg Internal Quality and n-3 Fatty Acids of Native Chicken Fed on Skipjack Fish (Katsuwonus pelamis. I) Industrial Waste Containing Feed. International Journal of Poultry Science, 12 484.

IOP Conf. Series: Earth and Environmental Science 1041 (2022) 012011 doi:10.1088/1755-1315/1041/1/012011

Dry matter (DM), crude protein (CP) and water consumption in sheep which supplemented from *Moringa* leaves meal and palm oil in elephant silage-based feed

S Rizqiana*, W Suryapratama, F M Suhartati

Universitas Jenderal Soedirman, Purwokerto, Indonesia

*sista.rizqiana@gmail.com

Abstract. This study aimed to examine the effect of supplementation on Moringa leaf meal and palm oil on the dry matter (DM), crude protein (CP) and water consumption in sheep. The material research used 18 thin tailed of 8-month-old sheep. The supplementation feed used was Moringa leaf flour and palm oil. The design used in this study was a Completely Randomized Design (CRD) with 3 treatments and 6 replications. The treatments include: P0: basal feed; P1: P0 + 2% Moringa leaf meal; P2: P1 + 0.5% palm oil. The variables measured were dry matter (DM), crude protein (CP) and water consumption. Data were analyzed by Analysis of Variance (ANOVA). The results of the analysis of variance showed that the treatment had no significant effect on dry matter (DM) and water consumption but had a significantly effect on crude protein consumption (CPI). The results showed that the dry matter consumption (DMI) of each treatment were P0 (control): 615.95 ± 98.50 g / head / day; P1: 728.37 ± 75.16 g / head / day; and P2: 616.62 ± 99.37 g / head / day. Crude protein consumption (CPI) from each treatment were P0 (control): 80.28 ± 12.84 g / head / day; P1: 122.55 ± 12.65 g / head / day; and P2: 93.64 ± 15.09 g / head / day. Water consumption from each treatment were P0 (control): 1535.67 ± 798.01 ml / head / day; P1: 1572.17 ± 855.73 ml / head / day; and P2: 1512.50 ± 532.67 ml / head / day. Based on the research results, concluded that the use of 2% Moringa leaf meal in silage-based feed supplementation increases the crude protein consumption (CPI).

1. Introduction

Efforts to increase sheep production must be followed by the provision of feed, especially quality forage, both in terms of quantity and quality. Indonesia is a tropical region that causes erratic forage production throughout the year, especially during the dry season. From the problems mentioned above, it is necessary to make a strategy to provide adequate forage with grass preservation technology, namely silage. Silage is animal feed made with the aim of preserving animal feed with the help of microorganisms and processed under anaerobic or airless conditions. Giving silage to sheep feed is better balanced with concentrate feed. Concentrate can be used as a source of energy and protein feed that can increase livestock growth. However, the concentrate used in this study was a low-quality concentrate, so it is necessary to provide feed supplementation.

Supplementation is needed as an effort to increase the value of nutrient consumption in livestock. The feed used for supplementation in this study was Moringa leaf meal and palm oil. Moringa trees are easily found in various regions in Indonesia. There have been many studies that use the Moringa plant mainly on its leaves. Moringa leaves have been used as animal feed for both ruminants and non-ruminants. Moringa oleifera leaves contain high nutrients, namely protein, carbohydrates, ash, fat and fiber, each of 28.44; 57.01; 7.95; 2.74; and 12.63%. Moringa leaves have a high protein source that can increase livestock productivity [1]. Apart from Moringa, palm oil can also be used as a feed supplement. The use of palm oil other than for home industries, can also be used as feed for livestock. Palm oil is used for animal feed usually as a source of energy feed. [2] stated that palm oil is a vegetable oil that can be used as a source of energy for livestock. However, the maximum use of palm oil in animal feed is 3%. This is due to the unsaturated fat content in palm oil which can affect the work of bacteria.

Supplementation using Moringa leaf meal and palm oil in grass silage and concentrate feed for sheep has not been widely used. However, there are previous studies that have utilized Moringa leaves and palm oil supplementation. A study from [3] stated that giving Moringa leaves can improve the detrimental effects of oxidized palm oil. Based on the description above, the use of Moringa for feed supplements has the potential to increase feed consumption and digestibility, so it is necessary to further study the use of Moringa and palm oil on dry matter, crude protein and water consumption in sheep.

2. Methodology

2.1 Experimental livestock

The livestock used were 18 local male sheep aged about 8 months and body weight of 18.9 ± 1.9 kg, obtained from breeders in Banyumas Regency and its surroundings. Prior to the research, all experimental livestock were treated with deworming and shaved their hair. Research sheep are placed in individual cages. The cage is made of wood, equipped with a feed and water bucket for each individual.

2.2 Experimental ration

The rations used during the study consisted of grass silage and concentrate with a ratio of 40%:60%. Silage is made from elephant grass (*Pennisetum purpureum*) while the concentrate consists of bran and coconut cake (2:1 balance). The supplementation feed used was *Moringa* leaf flour and palm oil. Water is given *albidum* but measured. The experimental design used in the study was a completely randomized design (CRD) with 3 treatments and 6 replications. The treatments used include:
P0: basal feed.
P1: P0 + 2% *Moringa* leaf flour.
P2: P1 + 0.5% palm oil.
The composition of feed ingredients and nutrient content of each ration can be seen in Table 1.

Table 1. Composition of feedstuff and nutrient content.

Feedstuff	Treatments		
	P0	P1	P2
	-------------------%-------------------		
Elephant grass silage	40	40	40
Concentrate	60	60	60
***Moringa* leaf flour**	-	2	2
Palm oil	-	-	0.5
Total	100	102	102.5
Nutrient content*			
Dry matter	52.04	55.85	58.53
Ash	11.27	13.08	12.39
Crude Protein	13.03	16.83	15.19
Crude Fiber	31.22	38.95	39.87
Crude Fat	7.38	9.30	10.99
BETN	37.10	21.84	21.56

*) Analysis results of the Laboratory of Animal Nutrition and Forage, UNSOED (2020)

2.3 *Parameters observed*
The variables measured were of dry matter (DM), crude protein (CP) and water consumption. The research was conducted at Experimental Farm and the Laboratory of Animal Nutrition and Forage, Faculty of Animal Husbandry, Jenderal Soedirman University, Purwokerto. The research method that will be used in this research is the experimental method and the experiment is carried out *in vivo*.

2.4 *Maintenance period*
Maintenance of sheep is carried out for 2 months in an individual pen. Feeding during adaptation was 3.5% of body weight. Sheep were adapted to feed and environment for two weeks prior to starting treatment. Then continued the preliminary period for two weeks. Furthermore, the collection period is carried out for 7 days. Data measurement of dry matter (DM), crude protein (CP) and water consumption were carried out for 7 days during the collection period.

2.5 *Feed and water consumption*
The sheep are given treatment feed twice a day (morning and evening). The feed is given in accordance with the calculation of the body weight of the livestock. The remaining feed is weighed every morning before feeding. Feed consumption is calculated from the difference between the amount of feed given and the remaining feed that is not consumed.
Dry Matter Consumption (DMI) = given feed (DM) - leftover feed (DM) (gr)
Crude Protein Consumption (CPI) = % CP feed x DM consumption (gr)
Water consumption = water provision - remaining water (ml)

2.6 *Data analysis*
The data obtained were analyzed using analysis of variance (ANOVA). If the results of the analysis of variance have a significant effect, it will be followed by a test using *Honest Real Difference* (HRD). Data analysis was performed using the SPSS ver. 16 program.

3. Result and Discussion
Consumption is the amount of feed eaten and consumed by livestock. The average consumption observations of local sheep that were observed during the study are presented in Table 2.

Table 2. Average consumption of sheep.

Treatments	Dry matter consumption (g/head/day)	Crude protein consumption (g/head/day)	Water consumption (ml/head/day)
P0 = Basal feed	637.56^{a}±94.79	86.08^{a}±18.20	1535.67^{a}±798.01
P1 = P0 + 2% *Moringa* leaf flour	757.15^{a}±108.62	129.54^{b}±18.28	1572.17^{a}±855.73
P2 = P1 + 0.5% palm oil	683.58^{a}±84.03	109.76^{a}±13.60	1512.50^{a}±532.67

3.1 *Dry matter consumption*
The dry matter consumption in research sheep ranged from 637.56 to 757.15 g / head / day. P1 treatment with the addition of 2% *Moringa* leaf powder supplement treatment showed the highest dry matter consumption was 757.15 ± 108.62 g / head / day. The results of the analysis of variance showed that the treatment had no significant effect (P> 0.05) on the consumption of dry matter in sheep. This means that the addition of supplementary feed containing 2% *Moringa* leaf meal and 0.5% palm oil does not interfere with the consumption of dry matter in thin tail sheep but addition of supplementary feed containing 2% *Moringa* leaf meal able to increase dry matter consumption. However, despite this there is an increasing trend in dry matter consumption.

Research results from [4] stated that supplemental feeding in the form of *Moringa* leaves can increase feed nutrients that are not available in forage feed. In addition, supplementary feed containing *Moringa* leaves can increase palatability and feed consumption of ruminants. It is reinforced by the statement from [5] which states that the level of feed consumption is influenced by palatability. Palatability can affect the amount of consumption and the ability of livestock to consume dry matter contained in rations. Besides that, the palatability of the feed is influenced by the taste, smell and color of the feed.

The presence of anti-nutritional substances contained in *Moringa* leaf meal, in the form of tannins and saponins, does not affect the dry matter consumption in sheep. According to [6] that the presence of saponin content in cattle feed at a level of 2.5% reduces the palatability of cows because the taste of the feed given becomes bitter. Whereas the results of research from [7] reported that feeding protein sources from *Moringa* leaves up to a level of 75% did not have an effect on dry matter consumption in Jawarandu goats because the goats were more able to tolerate the bitter taste in the feed.

3.2 Crude protein consumption

Consumption of crude protein in the research sheep ranged from 86.08 to 129.54 g / head / day. The P1 treatment feed with the addition of 2% *Moringa* leaf meal treatment showed the highest crude protein consumption was 129.54 ± 18.28 g / head / day. The results of the analysis of variance showed that the treatment had a very significant effect ($P < 0.01$) on the consumption of crude protein in sheep. This means that the addition of feed containing *Moringa* leaf meal as much as 2% affects the consumption of crude protein in local sheep.

This is in line with the opinion [8] which states that the increased consumption of crude protein can be caused by the increase in dry matter consumption. Dry matter consumption in research sheep tends to increase, so that consumption of crude protein also increases. It is quite possible that dry matter consumption has a positive correlation with crude protein consumption.

The results of feed consumption from this study were higher than the results of [9] stating that the protein consumption in sheep fed water *hyacinth* based silage was 87.97-107.14 g / day. This is related to the level of palatability of livestock used as experimental animals in the study. It is possible that the addition of *Moringa* leaf meal and palm oil to the feed gives a slightly different smell or aroma. The protein content of *Moringa* leaf flour used in the study was 27.33%. Apart from the increased consumption of dry matter, the high content of crude protein in the feed given to livestock also affects the consumption of animal crude protein.

It is reinforced by the statement from [10] which states that high levels of protein in the feed ration will increase the proliferation of microbes in the digestive tract so that livestock can digest feed better. [11] stated that a diet with sufficient protein content and a finer size structure can increase the amount of food consumption. The feed that was carried out during the study was the *Total Mixed Ration* (TMR) method, so that the feed could not be separated between silage and concentrate because it had been mixed.

3.3 Water consumption

The research data showed that drinking water consumption in sheep ranged from 1512.50–1572.17 ml / head / day. The highest drinking water consumption was found in treatment P1 by giving basal feed supplemented with 2% *Moringa* leaf powder, namely 1572.17 ± 855.73 ml / head / day. The results of the analysis of variance showed that the treatment had no significant effect ($P > 0.05$) on drinking water consumption in sheep (Table 2), although there was an increasing trend in drinking water consumption. [12] stated that drinking water consumption in livestock is influenced by several factors, including water quality and the place where livestock adapt (dry, semi-dry, wet areas).

In addition, sheep have a higher water consumption level than goats. It was also reported that the higher consumption of dry matter will be accompanied by higher consumption of drinking water. However, this is different from the results of the study which showed that the dry matter content in the feed used in the study did not affect drinking water consumption.

The results of the drinking water consumption of sheep in this study were lower than the study conducted by [13] reporting that the drinking water consumption of sheep fed with hay and concentrate during the day ranged from 6.75 to 7.80 liters / head / day. This is due to heat stress due to livestock being kept in semi-arid areas. In addition, water quality in semi-arid areas is not good for consumption. Research from [14] reported that drinking water consumption in sheep was influenced by the level of cleanliness and freshness of the water given. It is reinforced by the opinion of [15] that water quality greatly affects the level of water consumption, but each sheep has its own adaptation characteristics to the water consumed.

Research conducted [16] reported that drinking water consumption in sheep fed in the form of kumpai grass was 100.94-751.72 ml / head / day, which means it is lower than the research results. It was also explained that the water consumption of sheep given dry kumpai grass was higher than that of sheep given fresh kumpai grass. This is because the grass that is given fresh still contains high enough water. In contrast to the research conducted, the feed given to sheep during the study was in the form of silage and concentrate, causing the texture of the feed to tend to be dry so that the livestock tended to drink more.

[17] stated that apart from being influenced by dry matter consumption, the fat content in the ration could also affect drinking water consumption. A study conducted by [18] showed that the addition of saturated fatty acids in the ration led to decreased drinking water consumption. In this study, it was also explained that the fat content contained in the ration produced high metabolic water so that some of the water needs of the sheep's body were fulfilled. This is in line with research conducted that the drinking water consumption in treatment P2 was the lowest, namely 1512.50 ± 532.67 ml / head / day. It is possible that this is influenced by the addition of a feed supplement in the form of 0.5% palm oil.

4. Conclusion

Supplementary feeding of 2% Moringa leaf meal and 0.5% palm oil did not interfere with the consumption of dry matter in sheep. Likewise, the consumption of drinking water in each treatment was observed to be not much different. However, in the feed treatment with the addition of 2% Moringa leaf powder supplement feed, increased consumption of crude protein.

References

[1] Aminah, S., T. Ramdhan dan M. Yanis. 2015. Kandungan Nutrisi dan Sifat Fungsional Tanaman Kelor (Moringa oleifera). Buletin Pertanian Perkotaan, 5(2): 35–44.

[2] Abqoriyah., S.D. Widyawati dan Lutojo. 2013. Penggunaan Minyak Ikan Lemuru (Sardinella longiceps) dan Minyak Kelapa Sawit Diproteksi dalam Ransum Domba Lokal Jantan Terhadap Daya Guna Pakan Serat. Zoo Indonesia, 22(2): 39–46.

[3] Okwari, O.O., K. Dasofunjo, A.A. Asuk, E.A. Alagwu and C.M. Mokwe. 2013. Anti-Hypercholesterolemic and Hepatoprotective Effect of Aqueous Leaf Extract of Moringa oleifera in Rats Fed with Thermoxidized Palm Oil Diet. IOSR J. Pharm. Biol. Sci, 8(2): 57–62.

[4] Soetanto., E. Marhaeniyanto dan S. Chuzaemi. 2011. Penerapan Teknologi Suplementasi Berbasis Daun Kelor dan Molases pada Peternakan Kambing Rakyat. Buana Sains, 11(1): 25–34.

[5] Syam, J., M. Nur, A.L. Tolleng dan S.T. Aisyah. 2018. Konsumsi Pakan Sapi Bali yang Diberikan Pakan Daun Kelor (Moringa oleifera). Prosiding Seminar Nasional Megabiodiversitas Indonesia. Gowa, Sulawesi Selatan: 8-14.

[6] Suharti, S., D.A. Astuti dan E. Wina. 2009. Kecernaan Nutrien dan Performa Produksi Sapi Potong Peranakan Ongole (PO) yang Diberi Tepung Lerak (Sapindus rarak) dalam Ransum. JITV, 14(3): 200–207.

[7] Rohmah, A.N., F. Wahyono dan J. Achmadi. 2020. Pengaruh Substitusi Bungkil Kedelai dengan Daun Kelor (M. oleifera) Terhadap Profil Darah Merah Kambing Pra-Sapih. Jurnal Sain Peternakan Indonesia, 15(1): 29–36. doi: https://doi.org/10.31186/jspi.id.15.1.29-36.

[8] Robo, M.M., M.M. Kleden dan L.S. Enawati. 2019. Pengaruh Pemberian Konsentrat yang Mengandung Tepung Daun Kelor dengan Level yang Berbeda Terhadap Penggunaan Nitrogen Kambing Lokal. Jurnal Trop. Anim. Sci. Technol, 1(1): 7–13.

[9] Hanun, L., A. Muktiani dan L.K. Nuswantara. 2018. Kecernaan Protein dan Retensi Nitrogen

pada Domba yang Diberi Silase Pakan Komplit Berbahan Eceng Gondok dengan Starter Lactobacillus plantarum. Jurnal Pengembangan Penyuluhan Pertananian, 15(27): 41–47.

[10] Ayuningsih, B., I. Hernaman, D. Ramdani dan Siswoyo. 2018. engaruh Imbangan Protein dan Energi Terhadap Efisiensi Penggunaan Ransum pada Domba Garut Betina. Jurnal Ilmiah Peternakan Terpadu, 6(1): 97–100. doi: 10.23960/jipt.v6i1.p97-100.

[11] Wulandari, S., A. Agus, M. Soejono, M. N. Cahyanto dan R. Utomo. 2014. Performa Produksi Domba yang Diberi Complete Feed Fermentasi Berbasis Pod Kakao Serta Nilai Nutrien Tercernanya Secara In Vivo. Buletin Peternakan, 38(1): 42-50. doi: 10.21059/buletinpeternak.v38i1.4615.

[12] Attia-Ismail, S.A., A. R. Abdo and A. R. T. Asker. 2008. Effect of Salinity Level in Drinking Water on Feed Consumption, Nutrient Utilization, Water Consumption and Turnover and Rumen Function in Sheep and Goats. Egypt. Sheep Goat Sci. Journal, 2(31): 77–92.

[13] De, K., D. Kumar, S. Sharma, P. Kumawat, A. Mohapatra and A. Sahoo. 2020. Effect of Drinking Earthen Pot Water on Physiological Response and Behavior of Sheep Under Heat Stress. Thermal Biology Journal, 87: 1-7. doi: 10.1016/j.jtherbio.2019.102476.

[14] Ghanem, M., M. Zeineldin, A. Eissa, E.E. Ebissy, R. Mohammed and Y. Abdelraof. 2018. The effects of Saline Water Consumption on the Ultrasonographic and Histopathological Appearance of the Kidney and Liver in Barkisheep. Veterinary Medical Science Journal, 80(5): 741–748. doi: https://doi.org/10.1292/jvms.17-0596.

[15] Hussein, A., R. Puchala, I. Portugal, B.K. Wilson, T.A. Gipson and A.L. Goetsch. 2019. Effects of Restricted Availability of Drinking Water on Body Weight and Feed Consumption by Dorper, Katahdin, and St. Croix Sheep From Different Regions of the USA. Animal Science, 98(1): 1–47. doi: 10.1093/jas/skz367.

[16] Nasution, A. 2009. Pengaruh Pengganti Rumput Gajah (Pennisetum purpureum) dengan Rumput Kumpai (Hymenachne amplixicaulis) Terhadap Kecernaan Bahan Kering dan Bahan Organik dan Konsumsi Air Minum Domba Lokal Jantan. Junral Ilmiah Ilmu-ilmu Peternak, 12(2): 78–82.

[17] Sudarman, A., K.G. Wiryawan dan H. Markhamah. 2008. Penambahan Sabun-Kalsium dari Minyak Ikan Lemuru dalam Ransum: Pengaruhnya terhadap Tampilan Produksi Domba. Media Peternakan, 31(3): 166–171.

[18] Harvatine, K. J. and M.S. Allen. 2006. Effects of Fatty Acid Supplements on Feed Consumption, and Feeding and Chewing Behavior of Lactating Dairy Cows. Journal of Dairy Science, 89(3): 1104–1112. doi:10.3168/jds.s0022-0302(06)72178-6.

Effect of various packaging materials for storing ground yellow corn of hybrid C-1 variety on water and amylum content

Munasik*, N Nurbaety, N Hidayat, E Susanti

Universitas Jenderal Soedirman, Purwokerto, Indonesia

*munasik2011@gmail.com

Abstract. This research aims to investigate the effect of various packaging materials (cotton sack and perforated plastic sack) and the storing time on ground corn quality that were assessed from water and amylum contents. The research method was experimental using Completely Randomized Design that consisted of: A (control), B (cotton sack, stored for 20 days), C (cotton sack, stored for 40 days), D (cotton sack, stored for 60 days), E (perforated plastic sack, stored for 20 days), F (perforated plastic sack, stored for 40 days), and G (perforated plastic sack, stored for 60 days). The treatments were replicated 3 times. The results showed that the longer storing time on cotton sack and perforated plastic sack resulted in higher water content and lower amylum content.

1. Introduction

Corn is a popular ration material for all types of livestock, especially poultry. This corn foodstuffs material could come from domestic production and imports. The usage of corn in poultry can reach 51 percent of the total ration [1]. [1] further stated that in 2019 the national feed production was 20.8 million tons, in other words the need of corn for animal feed production in Indonesia was 700,000 tons per month. According to [2], corn production in Indonesia, on the one hand, has a good market potential, but in fact many corn products at the farmer level are not absorbed by the industry due to several things such as: high water content, damaged corn grain, non-uniform grain color, broken grains and other impurities which cause the low quality of corn produced.

This condition is affected by the weather conditions of our country which relatively in high humidity (70-80 percent), so it is support the growth of fungi, especially of the Aspergillus, spp. In the grain storage, the most important biological conditions that affect the growth and production of fungal metabolites (mycotoxins) are water activity, temperature, and gas composition [3]; [4]. With the good corn storage techniques by maintaining a low water content, safety sacks and floor mats are needed to prevent from animal attack damage and mold growth. [5] stated that storing corn in sack packaging with a base can reduce the aflatoxin content stored for 21 days.

This is in line with the statement of [6] that aflatoxin contamination occurs because of inappropriate post-harvest handling, for example when corn is stored. One of the efforts to improve post-harvest handling with good corn storage techniques is to store the product by placing it on a mat. For packaged products, it is recommended to use a package that has pores for air circulation and placed on a board base. The corn that used as animal feed is generally yellow corn because it has a high carotene content. Carotene affects poultry as a form of egg yolk color, shank color and skin fat color. Yellow corn used for poultry feed is generally ground corn. The smaller particle size of the corn allows it to be easily digested.

Ground corn that is used as concentrate feed for livestock will have disadvantages when it is stored compared to whole corn, such as reduced nutrient content and palatability, so it is necessary to pay attention on how to overcome this problem. One aspect that influences the storage of ground corn is the packaging material. This packaging itself aims to protect the material against mechanical forces from outside, reduce evaporation or water absorption, reduce the possibility of microbial contamination or pest attack, thus the packaging can prolong the shelf life of corn.

During storage the ground corn will undergo physical and chemical changes. These changes can reduce the quality of ground corn. Physical changes, namely by increasing water content, while chemical changes, namely by decreasing amylum. One of the ways to inhibit the change process requires storage using the appropriate type of packaging. Packages that can be used include cotton sacks and perforated plastic sacks. The characteristic of cotton sacks is a large porosity while perforated plastic sacks have a smaller porosity. Therefore, which type of packaging can maintain water and amylum content to be developed.

2. Methodology

The materials used in the study were ground yellow corn varieties C-1 hybrids, cotton sacks, perforated plastic sacks, spectrophotometer, stirrer, desiccator, analytical scales, oven, room thermometer, amylum standard, absolute alcohol, 1 N NaOH, 1 N iodine solution and 1 N acetic acid. The research method was experimental using Completely Randomized Design consisted of A (control), B (cotton sack, stored for 20 days), C (cotton sack, storage for 40 days), D (cotton sack, stored for 60 days), E (perforated plastic sack, stored for 20 days), F (perforated plastic sack, stored for 40 days), and G (perforated plastic sack, stored for 60 days). The treatments were replicated 3 times. Determination of water content of ground corn using the [7] method and measurement of amylum content using a spectrophotometer [8].

3. Result and Discussion

3.1 The physical condition of ground corn before and after storage

The physical condition of ground corn before storage is fresh, which has characteristics of corn, fresh yellow color, does not clot and no insects were found. Stored for 60 days with storage warehouse temperatures ranging from 24-28 °C and relative humidity of 75-80 percent. This condition is still suitable for grain storage, this is in line with the statement by [9] that a good storage temperature for grain is 25-27 °C with an air humidity of 70-75 percent. However, there will be limitations in maintaining the quality of the materials during the storage period. Therefore, although the ground corn that stored was previously has a low water content, the water content of the stored corn will increase as well.

In the process of storing feed raw materials, there will be a balance between the water content in the feed ingredients and the water content in the room, meaning that if the water content of the feed ingredients is low while the water content in the storage room is high, the water content of the feed ingredients will gradually increase so that at the end will be balanced. Physical characteristics of ground corn during storage including discoloration, odor, clumping and insect attack are shown in Table 1.

Table 1. Physical characteristics of ground yellow corn during storage

Kinds of packaging	Physical characteristics	Long of storage (days)			
		0	20	40	60
Cotton sack	Color	-	+	++	+++
	Odor	-	-	+	++
	Clot	-	+	++	+++
	Insect	-	-	++	+++
Perforated plastic sack	Color	-	+	+++	++++
	Odor	-	+	+	++
	Clot	-	+	+++	++++

Insect	-	+	+++	++++

Note: - nothing changes, + little change, ++ moderate change, +++ a lot of change, ++++ very changed

The physical characteristics of ground corn (Table 1), the longer of the storage period, using cotton or perforated plastic sacks, tended to experience a decrease in the quality of ground corn. The storage time for ground corn still shows good physical characteristics when stored for 20 days.

3.2 Water content

The water content of ground corn during 60 days of storage is increased, both in cotton and perforated plastic sacks. The increase in water content of stored ground corn with cotton sacks was not different with the increase in water content of ground corn that occurred in storage with perforated plastic sacks. The average of water content of ground corn during storage (Table 2).

Table 2. The water content of ground corn during storage

No	Treatment	Water content (%)
1	A (control)	12.23 ± 0.20
2	B (cotton sack, stored for 20 days)	12.67 ± 0.25
3	C (cotton sack, storage for 40 days)	13.77 ± 1.21
4	D (cotton sack, stored for 60 days)	14.46 ± 0.49
5	E (perforated plastic sack, storage for 20 days)	12.62 ± 0.06
6	F (perforated plastic sack, stored for 40 days)	13.71 ± 0.25
7	G (perforated plastic sack, stored for 60 days)	14.82 ± 0.16
	Average	13.47 ± 1.03

The average water content of ground corn during storage for each type of packaging was increased, at storage with cotton sacks the water content of ground corn was from 12.23-14.46 percent and in storage with perforated plastic sacks the water content of ground corn was from 12.23-14.82 percent. The water content of stored ground corn is still low because the value of equilibrium moisture content (EMC) reach the equilibrium number of grains when stored at a temperature of 32 °C with 80 percent humidity indicates a water content of 15.5 percent [10].

Therefore, the longer of round corn is stored, the water content will increase. The increase in water content of ground corn during storage was caused by a balance between water content of ground corn and water in the storage warehouse. Based on the research results, temperature and humidity of the storage warehouse showed between 24-28 °C temperatures and 76-82 percent humidity. The water content of grain ingredients such as corn must be dried to a 14% water content to be safe during storage [4]. Other opinions from [11] stated that the water content of feedstuffs depends on the amount of main ingredients used. The water content that meets the quality requirements of corn is a maximum of 14%, before storage until the 21st day of storage [12].

The high humidity will also lead absorption of water vapor from the air which will result in moist material which affects the increase of water content [13]. The humidity is inversely related to air temperature. The higher of air temperature caused the smaller of humidity [14]. According to [15] 23-31 °C is the optimum growth temperature for *Aspergillus flavus*, aflatoxins can be produced with 7 days of an incubation periods. According to [10] that the water content of ground corn is 13.47 percent and the storage room temperature is 24-28 °C, the storage of ground corn can last up to 6 months. This is because a close relationship between storage room temperature of 27 °C and corn water content of 14 percent, so the corn can be stored for 6 months.

In line with the statement of [10], [10] further stated that the water content of the materials stored in the warehouse is closely related to humidity or water migration, feed ingredients which originally had a low water content if stored in a room with higher humidity then their water content would increase up to the point of equilibrium. This increase in water content will continue to gradually occur on the following days.

The increase in water content is caused by corn were being stored in the form of ground corn state so that it is easier to absorb water from the air. If seen physically, ground corn has smaller particles than the original size and has a hollow structure so that the ground corn has the higher ability to absorb water from the air and makes ground corn easily invest by insects. [16] stated that corn kernels are hygroscopic, so they always adjust to humidity and environmental temperature.

When viewed from the humidity of the warehouse during storage, which is 76-82 percent, it causes the ground corn to respire, thus the water content of ground corn will increase so that there will be an increasing of insect investment and microbial growth during storage. Storage of ground corn that packed with cotton sacks, water content is relatively still the same as ground corn that packed in perforated plastic sacks.

Shrinkage of the material during storage cannot be avoided because of the heating process and the discharge of water from the material. When the water leaves the material, there is an imbalance pressure inside the material and the outside which causes contraction and thus triggers shrinkage, changes of shape and occasionally breaks or cracks the material [17].

3.3 Amylum content

The amylum content of ground corn that stored for 60 days in cotton sacks or in perforated plastic sacks has decreased, this is presumably due to the damage to the carbohydrates. The average amylum content of ground corn during storage is shown in Table 3. The decrease in amylum content is caused by the presence of enzymatic activity which hydrolyzes carbohydrates into starch and then reducing sugars such as glucose and fructose, resulting in sugar accumulation. Amylum of ground corn contains 28 percent amylose and 72 percent amylopectin [18]. Amylum of corn is a fine powder, has a large surface area.

This is because natural amylum contains a lot of amylose so it is dry, less sticky and tends to absorb a lot of water [19]. Storing ground corn in cotton sacks causes a decreasing of amylum content lower than those stored in perforated plastic sacks. This is presumed by the greater humidity in the perforated plastic sack which causes a greater change of carbohydrates, among others, due to oxidation from incoming light, the large number of microorganisms that live at high humidity. The optimum temperature for the growth of microorganisms such as fungi is 25-30 °C with an air humidity of 65-93 percent [10]. In the storage process, corn grain still undergoes a respiratory process and produce carbon dioxide, water vapor, and heat [20].

Table 3. The Amylum content of ground yellow corn during storage

No	Treatment	Amylum content (mg/ml)
1	A (control)	5.50 ± 0.21
2	B (cotton sack, stored for 20 days)	4.73 ± 0.53
3	C (cotton sack, storage for 40 days)	3.69 ± 0.46
4	D (cotton sack, stored for 60 days)	2.73 ± 0.40
5	E (perforated plastic sack, stored for 20 days)	3.71 ± 0.30
6	F (perforated plastic sack, stored for 40 days)	2.82 ± 0.35
7	G (perforated plastic sack, stored for 60 days)	1.66 ± 0.12
	Average	3.55 ± 1.26

Chemical damage is caused by chemical decomposition during storage such as a decrease in carbohydrate and protein levels due to metabolic processes by both insects and microbes [21]. Furthermore, according to [21], the damage that occurs during storage and the main factors causing quality degradation include : Physical damage due to changes in water content during storage caused by changes in weather, granules become broken and susceptible to pests, biological damage caused by biological activities during storage such as pests, fungi and microbes, chemical damage is caused by chemical decomposition during storage such as decreased levels of carbohydrates and protein due to metabolic processes by both insects and microbes. The amylose content in carbohydrates also affects the reduction of starch levels in ground corn. This is because amylose is able to absorb water so that it affects the starch development process [22].

4. Conclusion

The storage of ground yellow corn with cotton sack is relatively the same as perforated plastic sack to maintain water and amylum contents.

References

[1] Istiadi 2020 Sistem Logistik Pakan yang Efektif dan Efisien PT. Charoen Pokphand Indones.
[2] Kristanto A 2018 Teknologi Pascapanen untuk Peningkatan Mutu Jagung https://scholar.google.co.id/scholar?hl=id&as_sdt=0%2C5&q=Teknologi+Pascapanen+untuk+Peningkatan+Mutu+Jagung&btnG=
[3] Ramos M E G S M V S A J 2003 Modified Atmosphere Packaging for Prevention of Mold Spoilage of Bakery Products with Different pH and Water Activity Levels J Food Prot 66 1864–72
[4] Naresh Magan D A 2007 Post-harvest control strategies: Minimizing mycotoxins in the food chain Int. J. Food Microbiol. 119 131–9
[5] Hasnani, S P J D R F 2019 Effect Of Storage Technique On Aflatoxin Control Of Corn (Zea Mays L) During Storage J. Pendidik. Teknol. Pertan. 5 S37–47
[6] Maryam R 2006 Pengendalian Terpadu Kontaminasi Mikotoksin Wartazoa 16 21–30
[7] AOAC 2005 Official methods of analysis of the Association of Analytical Chemist Virginia USA Assoc. Off. Anal. Chem. Inc
[8] Ferdiaz, D.A., Apriyantono, S. Yasni S B dan N L P 1986 Penuntun Praktikum Analisa Pangan Teknol. Pangan dan Gizi. IPB Bogor
[9] HP Imdad, Nawangsih A A 1999 Menyimpan bahan pangan PT Penbar Swadaya. Jakarta
[10] Sukria H A 2020 Penyimpanandan Pergudangan (Storage and Warehousing) Feed Manuf. Ind. Div. Dep. Nutr. Feed Technol. IPB Univ. Bogor
[11] Nurlaila, Andi Sukainah A 2016 PENGEMBANGAN PRODUK SOSIS FUNGSIONAL BERBAHAN DASAR IKAN TENGGIRI (Scomberomorus sp.) DAN TEPUNG DAUN KELOR (Moringa oleifera L) J. Pendidik. Teknol. Pertan. 2 105–13
[12] Nasional B S 2012 SNI 01- 3920-1995. Standar mutu jagung Bahan Baku Pakan Badan Standar Nas. (BSN). Jakarta
[13] Fauziah K R 2013 Pengaruh Lama Penyimpanan Terhadap Kualitas Jagung Kuning Dan Jagung Putih Balai Penelit. Tanam. Serealia Maros, Maros
[14] WANITA, Yeyen Prestyaning; BUDIARTI S W 2015 Pengaruh Teknik Pengemasan dan Penyimpanan Terhadap Mutu Kacang Tanah Pada Dua Musim Panen Berbeda Pros. Semin. Has. Penelit. Tanam. Aneka Kacang dan Umbi 466–73
[15] Pratiwi, C., Rahayu W P, Lioe, H. N., Herawati D, Broto W, And and Ambarwati S 2015 The effect of temperature and relative humidity for Aspergillus flavus BIO 2237 growth and aflatoxin production on soybeans Int. Food Res. J. 22 82–7
[16] S. Xu, Q. Zhang M G B 1997 A microstructural mechanics model for predicting hygroscopic loads in grain storage bins Am. Soc. Agric. Enginer 40 1435–9
[17] L Mayor A S 2004 Modelling shrinkage during convective drying of food materials: a review J. Food Eng. 61 373–86
[18] Wicaksono A 2008 Suksinilasi Pati Singkong Pregelatinasi sebagai penghancur dan Pengikat pada tablet Amoxicillin FMIPA Univ. Indones. Jakarta
[19] Hasibuan M 2009 Pembuatan Film Layak Makan dari Pati Sagu Menggunakan Bahan Pengisi Serbuk Batang Sagu dan Gliserol Sebagai Plastisiser Univ. Sumatera Utara. Medan
[20] Champ, B.R. H 1986 Technological change in postharvest handling and transportation of grains in humid tropics Int. Semin. Bangkok, Thail.
[21] Kushartono B 2011 Pengendalian Jasad Pengganggu Bahan Selama Penyimpanan Balai Penelit.

Ternak. Ciawi
[22] Mahdi Jufri R D F 2006 STUDI KEMAMPUAN PATI BIJI DURIAN SEBAGAI BAHAN PENGIKAT DALAM TABLET KETOPROFEN SECARA GRANULASI BASAH Pharm. Sci. Res. 3 78–86

Exploring the herbaceous plant height and dry matter relationship on a sub-humid rangeland in Zimbabwe

T A Zezai[1*], P H Mugabe[2] and V E Imbayarwo-Chikosi[2]

[1]Women's University in Africa, Mount Pleasant, Harare, Zimbabwe
[2]University of Zimbabwe, Mount Pleasant, Harare, Zimbabwe

*vazezai@gmail.com

Abstract. A study was carried out at the University of Zimbabwe Farm, to assess and quantify the correlation between plant canopy height and biomass in the rangelands of Zimbabwe. Two range sites, bush grassland and grassland, were selected and three paddocks within each range site were sampled. Four 25 m long transects were drawn in the four cardinal directions from the paddock center. Five 0.25 m2 quadrants were located at 5 m intervals on each transect line. Common plant species in both bush grassland and grassland range sites were Sporobolus pyramidalis, Hyparrhenia filipendula, Cynodon dactylon, and Eragrostis curvula. There was a significant (P<0.05) linear relationship between plant height and biomass for both range sites. There was a significant (P>0.05) quadratic relationship between plant height and biomass for grassland but not for the bush grassland. Pearson Correlation Coefficient (r) was 0.929 for grassland and 0.717 for bush grassland. The regression equation was y = 15.82x – 0.02x2 for bush grassland and y = 15.88x – 0.27x2 for grassland. The coefficient of determination (r2) for grassland was 0.863 and 0.514 for bush grassland. Variation in plant height explained 86.3% of the variation in biomass for grassland and 51.4% for bush grassland accordingly.

1. Introduction

Rangeland is defined as land on which the indigenous vegetation, climax or natural potential, is predominantly grasses, grass-like plants for example sedges, forbs or shrubs (1). Rangelands do not therefore include cropland and irrigated pastures but consist of extensively grazed native or naturalized plant species representative of historic climax vegetation. Biomass is a collective term for all material that is derived from the product of photosynthesis (2) hence biomass can be of plant or animal origin. In range science, biomass usually refers to plant mass only (3).

When measuring biomass, the above ground standing biomass is normally measured as it is seasonally consistent in weight because it is not affected by agents of erosion (4). There are direct and indirect methods of measuring biomass. Direct methods involve clipping and weighing fresh forage or weighing after drying the forage to obtain dry matter yield (5)(6). Indirect measurement methods involve developing a relationship between plant weight and an easier-to-measure attribute such as plant height, and electronic capacitance (7). Some of the examples of indirect methods of pasture biomass assessment are capacitance measurement, pasture height/biomass measurement and the spectral and remote sensing vegetation measurement.

In Zimbabwe, the conventional method used to estimate biomass so as to establish feed available for livestock is the direct method of clipping and weighing (8) . This is time consuming and requires a lot of labor. above all, it is destructive to the rangeland. some producers with experience are able to quantify their rangeland pasture biomass visually (9). visual assessment of biomass can however be inaccurate due to the assessor's subjectivity and also unforeseen natural environmental changes on rangelands.

The direct method of clipping and weighing in biomass assessment is time consuming, laborious and harmful to rangelands as it destroys both increaser and decrease plant species, thereby increasing bareness, reducing forage production and increasing soil compaction and erosion (10) (7). Alternate double sampling approaches such as spectral and remote–sensing options, capacitance measurement, and spectral reflectance developed in countries such as New Zealand (7) cannot be applied in Zimbabwe because of the advanced technology and costly equipment.

This research will determine easier biomass assessment through ascertaining the relationship between biomass and plant canopy height for the grassland and bush grassland range sites of the university of Zimbabwe farm. The ecological characteristics of the farm are representative of the agro-ecological zone ii a of the country. a more efficient means of biomass assessment will result in better decision making for livestock production and rangeland management. the practice of assessing rangeland pasture biomass for easier forage and stocking rates determination will also be promoted among Zimbabwean farmers through this research.

The main objective is to assess and quantify the correlation between herbaceous plant height and biomass in Grassland and Bush Grassland range sites of the University of Zimbabwe farm which is representative of the agro-ecological zone II a of the country. The specific objectives are to determine the relationship between herbaceous plant height and dry matter yield for the grassland and bush grassland range sites using regression analysis in Zimbabwe's Agro ecological zone II a and to compare the relationship between standing biomass height and dry matter yield between grasslands and bush grasslands in Zimbabwe's Agro ecological zone II a.

2. Materials and Methods

2.1 Environment

University of Zimbabwe (UZ) farm is located about 13km northwest of the capital city of Zimbabwe, Harare. The farm lies at longitude and latitude lines 3105' E and 1705' S respectively, at an altitude of 1420m above sea level (11). The farm is in agro-ecological zone II a which has an average annual temperature of 190 C and it consists of deep to moderately deep well drained red granular clay soils. Soil colour changes from red in well drained areas to grey and black in vleis. The black soils are deficient in phosphate and potash (12). University of Zimbabwe farm occupies a total area of 1805 hectares (ha) of which 800ha are rangelands.

The most common tree species on the rangelands include Brachystegia Spiciformis (Msasa), Julbernadia Globiflora (Munondo) and Combretum Molle (Mupembere) (12). University of Zimbabwe farmland types are differentiated into arable and grazing land. Arable land is used for crop production, mainly maize, sugar beans, groundnuts and potatoes and the grazing land is used for livestock production, currently beef and goat . Rotational grazing system with 13 paddocks is the grazing system being utilized. The UZ farm is divided into two sections on either side of the Harare-Bindura main road. The main farming section is on the eastern side of the Harare-Bindura Highway and the teaching unit area is on the western side.

2.2 Sampling framework

There are three range sites at the UZ farm, namely Grassland, Bush Grassland and Woodland. In this study, the first two range sites namely Grassland and Bush Grassland were assessed .The Bush Grassland has 40% tree cover and 60% grass cover dominated by Sporobolus Pyramidalis, Hyparrhenia Filipendula, Cynodon Dactylon and Eragrostis Curvula. The Woodland is composed of thick trees which do not allow sufficient sunlight to pass through for good undergrowth, hence Woodland was not assessed as it did not have sufficient herbaceous biomass for assessment.

IOP Conf. Series: Earth and Environmental Science 1041 (2022) 012013 doi:10.1088/1755-1315/1041/1/012013

2.3 Sampling frame

The basis for stratification was the range site. Within each range site, three paddocks were selected based on the statistical concept that at least 20% of the selected sites should be sampled (13). Line transects of 25m in length were systematically positioned in the four cardinal directions from the center point of each paddock. Quadrants measuring 0.5m x 0.5m were used as the sampling plots and were placed at intervals of 5m apart along the transects. This resulted in 5 quadrants along the 'north' transect, 5 quadrants along the 'south' transect, 5 quadrants along the 'east' transect and 5 quadrants along the 'west' transect. The total number of samples upon which data was collected was therefore calculated as below, 2 range sites x 3 paddocks per range site x 4 transects per paddock x 5 quadrants per transect = 120 samples.

A falling plate meter (14) was constructed at the UZ Mechanical Engineering Department for the purpose of measuring plant canopy height during data collection. Plant canopy height and species type and their number per quadrant were collected and recorded whilst in the field. The fresh forage was then cut and packed in plastic bags for dry matter determination at the UZ Department of Animal Science laboratory.

2.4 Data collection

Procedures for identifying species, measuring height and collecting fresh biomass were in such an order that the non-destructive technique was applied first followed by the destructive technique as recommended by (15). Plant species identification and counting procedures were therefore carried out first. The 0.25 m2 frequency quadrat was placed on the transect location (16), followed by species identification, counting and recording on the data form. After species data collection procedures per quadrant were completed, height measuring and biomass cutting, and packing followed.

This involved replacing the frequency quadrat with the 0.25 m2 biomass quadrat on the same transect location. The meter rule of the falling plate meter was then placed in the middle of the quadrant and the acrylic plate lowered to rest on the grass for the height to be recorded (17);(14);(5). The arcrylic plate was left to fall on the pasture from a height of 30cm on all quadrants such that the same force was exerted on the pasture for uniformity of conditions. The preceding procedures described were for the non-destructive techniques.

The falling plate meter was removed, and biomass clipped using a sickle, as the destructive technique performed last. All standing biomass in the 0.25m2 quadrant was cut to ground level, packed in polythene bags, and then taken to the laboratory for drying. The polythene bags were pre coded with an identification code to identify each plot from which data was collected. The code identified the quadrant number, the transect in which the quadrant occurred, the paddock and the range site. Grass plant height, plant species type and quantity data were recorded on a data form whilst in the field.

One hundred and twenty samples were taken to the laboratory for drying. The polythene bags with fresh forage were placed in forced-air ovens set at 60oC for 48 hours (18);(19);(20). The temperature level and period of drying ensured that the polythene bags did not melt due to high temperatures. After drying, the dry grass samples were weighed on a digital scale and the weight of the dried biomass for each quadrant was recorded in grams, next to the plant canopy height record for that same quadrant on the data form.

2.5 Data analysis

Data collected were processed through the Microsoft Office Excel 2007 software programmed, giving columnar information on the day, paddock, site, transect, quadrant, pasture height and dry matter yield for each of the 120 plots. The Statistical Analysis System (SAS) Version 9.3 (SAS, 2002-2010) was used to analyze the data. PROC REG fitted a regression model comprising both the linear and quadratic effects of plant height on biomass yield. Analysis was carried out by site (site 1=bush grassland; site 2= grassland). Initial analysis showed non-significant effect ($P<0.05$) of the intercept (β_0).

The final regression equation therefore excluded the intercept from the following model (no-intercept model):

$$y=\beta_1 x+ \beta_2 x^{\wedge 2}$$

Where:
y is the dry matter yield (biomass);
x is the plant height;
β_1 and β_2 are the respective partial linear and quadratic regression coefficients of plant height on biomass;

The PROC CORR procedure of SAS was used to analyze the herbaceous plant height and dried biomass data for the estimation of the Pearson Correlation Coefficients between dry matter yield and plant height. Plant species data was analyzed separately to obtain relative frequencies for species ranking. Microsoft Office Excel 2007 was used for the mathematical calculation of species frequency first and then species relative frequencies. The frequency a particular species in the two range sites was determined by diving the number of quadrants in which that species occurred by the total quadrants assessed, and then multiplying by 100 to convert into a percentage (21). The relative frequency of each plant species was calculated by dividing its frequency by the sum of frequencies for all species.

3. Result and Discussion

3.1 Descriptive Statistics

Table 4.1 and Table 4.2 show the descriptive statistics for herbaceous plant height and the dried biomass data and the common plant species respectively. The bush grassland (Site 1) had a higher mean dry matter yield (biomass) and plant height for equal number of observations than the grassland (Site 2). This could have been due to the presence of the Hyparrhenia filipendula species among the common species in the bush grassland site. Hyparrhenia filipendula forms thick sods (22) at the base which resulted in more biomass accumulation than other grass species in the two range sites. Hyparrhenia filipendula was not amongst the common grass species in grasslands.

3.2 Species frequency assessment

Table 4.2 shows the common species ranked by relative frequencies. The first five most common species per range site, out of twenty-two identified from both range sites, were selected. Plant morphological characteristics have a relationship with plant growth traits like biomass and height (23), hence dominant species in a range help to partly explain the correlation between biomass and plant canopy height in this study. Three of the most frequent grasses were common to both study sites, that is, Sporobolus pyramidalis, Cynodon dactylon and Setaria pumila. The total relative frequency of the common species was 62.2% for the bush grassland and 74.4% for the grassland. At least two thirds of the species frequency must be used for rangeland biomass characteristics analysis (24).

Table 1. Descriptive statistics for plant height and biomass data collected during field

Range Site	Variable	N	Mean	Std Dev	Sum	Minimum	Maximum
Bushgrassland (Site 1)	Plant height(cm)	60	10.24	4.18	614.4	3.00	23.00
	Biomass(g/m^2)	60	158.53	94.12	9512	23.10	337.00
Grassland (Site 2)	Plant height(cm)	60	6.15	3.41	369	1.50	14.00
	Biomass(g/m^2)	60	84.40	44.42	5064	11.70	192.20

Table 2. Shows the five most common plant species by site ranked according to relative frequency (February, mid-season, 2016)

Site	Plant species	Species relative frequency (%)
Bush grassland	*Sporobolus pyramidalis*	20
	Hyparrhenia filipendula	11.8
	Urochloa panicoides	10.9
	Setaria pumila	10.0
	Cynodon dactylon	9.5

Grassland	*Sporobolus pyramidalis*	23.1
	Eragrostis curvula	20.6
	Cynodon dactylon	15.0
	Setaria pumila	9.4
	Cyperus rotundus	6.3

3.3 Regression analysis

Regression analysis with plant height as the independent variable and dried biomass as the dependent was carried out by site and Table 4.3 presents the results of the model parameter estimates and significance level test figures. Within the grassland site plant height showed a significant linear and quadratic relationship ($P<0.05$) with biomass yield. The negative quadratic relationship between plant height and biomass may be due to plants grown beyond the vegetative stage where biomass rate of growth starts to decline in relation to pasture height (25) (26). Within the bush grassland, the relationship between plant height and biomass yield was predominantly linear with non-significant relationship between the biomass and the quadratic function of plant height.

The linear function parameter estimates show that for a unit increase in pasture height, there is a 15.82g increase in the bush grassland biomass and a 15.88g increase in the biomass yield for grassland. The quadratic function parameter estimates show a negative relationship between biomass and the quadratic function of pasture height. The non-significant relationship in the bush grassland showed that for a unit increase in pasture height squared, biomass yield rate of growth decreases by 0.02g.

For grassland ,the relationship between biomass and the quadratic function was significant with a unit increase in pasture height squared has a corresponding rate of decrease of 0.27g in biomass. Below are the regression analysis equations for the calibration of pasture height to biomass for UZ farm bush grassland and grassland rangeland sites and surrounding areas of similar ecological characteristics.

For bush grassland:

$$y = 15.82x - 0.02x^2; \quad r^2 = 0.51$$

and for grassland:

$$y = 15.88x - 0.27x^2; \quad r^2 = 0.86$$

where:
y = dry matter yield (biomass),
x = plant canopy height
r^2 = coefficient of determination

Table 3. Model parameter estimates and significance level test values from the regression analysis of plant height vs dried biomass in grassland and bush grassland sites (February, mid-summer, 2016)

Variable	Bush grassland (site 1)		Grassland (site 2)	
	Estimate (β)	P-value	Estimate (β)	P-value
Linear function of pasture height (cm)	15.82	<0.0001	15.88	<0.0001
Quadratic function of pasture height (cm²)	-0.02	0.9067	-0.27	0.0099

Figures 4.1 and 4.2 shows the scatter graphs for herbaceous pasture height and dried biomass yield for the grassland and bush grassland sites. Figure 4.1 shows the scatter plot for pasture height and biomass yield for the bush grassland range site. The relationship was largely linear especially at lower pasture height range of 2cm to about 12cm. At low plant height, which is the vegetative state, the plant physiology of all species is generally uniform resulting in similar biomass weight in relationship height for that stage (27).

The quadratic relationship was non-significant ($P>0.05$) for plant height and biomass in the bush grassland range site, hence the figure 4.1 shows a straight line without the curving effect of the quadratic function. However, figure 4.2 which shows a significant relationship between plant canopy height and biomass shows a more curvi-linear regression curve. The negative quadratic coefficient parameter (β_2) is greater for the grassland (0.27) than that of the bush grassland (0.02) showing that biomass decreases at a rate 13.5 times higher in grassland than bush grassland .It is important to note that for both range sites, linearity of the variables is largely observed in the lower plant canopy height ranges of 2 to 8cm for grassland and 3 to 12 cm for bush grassland.

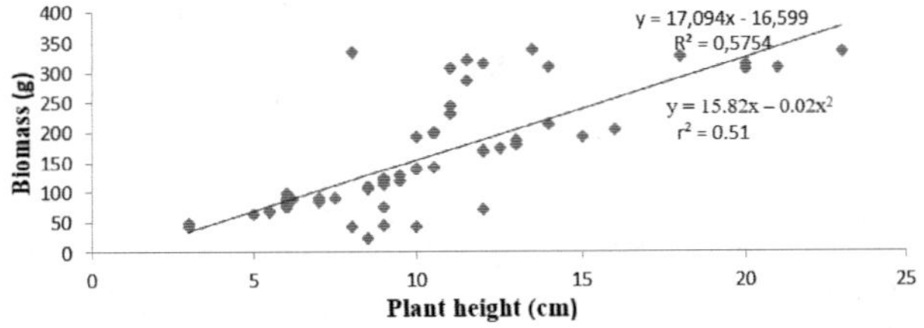

Figure 1. Plant height versus biomass yield for bush grassland

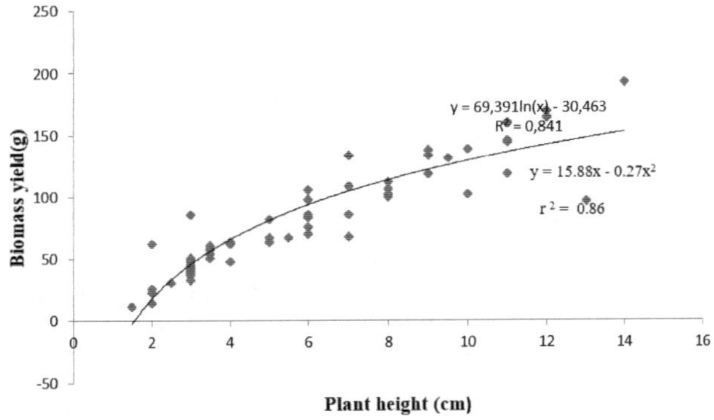

Figure 2. Plant height versus biomass yield for grassland

3.4 Correlation analysis results
The significant ($P<0.05$) positive correlation between dry matter and pasture height in the grassland range site was confirmed by a highly positive Pearson's correlation coefficient ($r = 0.929$) which shows a strong association between the variables. However, as height increases, biomass association with pasture height becomes less and less linear in concurrence with (5). This lack of linearity was more pronounced in the grassland than bush grassland probably due to the less sod forming grasses like Setaria pumila, Cyperus rotundus and Cynodon dactylon .Non sod forming grasses in the rangeland contribute to inconsistency in the pasture heights and biomass pattern (28). Hence the trendline was more curving in the grassland graph (Figure 4.2) than the bush grassland (Figure 4.1).

Grasslands range site's stronger correlation coefficient ($r = 0.929$) than that of Bush grassland ($r=0.717$) similar to results by (29) shows a stronger association between biomass and plant canopy height in the grassland biomass. The coefficient of determination, worked out by squaring the correlation coefficient of grassland (r2) is 0.863 and that of bush grassland is 0.514. According to the coefficient of determination, 86.3% of the variation in biomass was explained by variation in pasture height for grassland and 51.4% of variation in biomass was explained by variation in pasture height. Both the correlation coefficient(r) and the coefficient of determination (r2) indicate that the developed regression equations estimate biomass using pasture height more accurately in grassland range site rather than the bush grassland.

Table 4. Results of the correlation procedure for Bush grassland showing the coefficient of determination between dry matter biomass and pasture height and significance level test figures.

Pearson Correlation Coefficients, N=60	Prob > |r| under H0:Rho=0	
	DM	Past height
DM	1.000	0.717 <.0001
Pasture Height	0.717 <.0001	1.000

Table 5. Results of the correlation analysis procedure for Grassland showing the coefficient of determination between dry matter biomass and pasture height and significance level test figures.

Pearson Correlation Coefficients, N=60	Prob > |r| under H0:Rho=0	
	DM	Pasture height
DM	1.000	0.929 <.0001
Pasture Height	0.929 <.0001	1.000

4. Conclusion

There was an adequately strong association between herbaceous plant height and dry matter yield in the bush grassland and grassland range sites as indicated by the Pearson correlation coefficient values which were both above 0.70 as shown by tables 4 and 5. Hence the regression equations established for each range site can be accepted as a means to estimate biomass from plant height at UZ farm and surrounding grassland and bush grassland sites. Grassland had a stronger correlation coefficient and higher coefficient of determination as shown by the tables meaning that under the UZ farm ecological conditions, biomass is more accurately estimated through plant height in grassland than bush grassland. Common plant species in both range sites were Sporobolus pyramidalis, Hyparrhenia filipendula, Urochloa panicoides, Setaria pumila, Cynodon dactylon, Cyperus rotundus and Eragrostis curvula as shown by table 2. The dominancy of Sporobolous pyramidalis, an increaser, indicates degrading rangelands at the UZ farm.

References

[1] SocietyforRangeManagment. Society for Range Management Policy Statements, Position Statements and Resolutions. 2002;(February).

[2] Pittman JJ, Arnall DB, Interrante SM, Moffet CA, Butler TJ. Estimation of biomass and canopy height in bermudagrass, alfalfa, and wheat using ultrasonic, laser, and spectral sensors. Sensors (Switzerland). 2015;15(2):2920–43.

[3] Bunning S, McDonagh J, Rioux J. Vegetation assessment. Man Local Lev Assess L Degrad Sustain L Manag Part 2. 2011;39–66.

[4] Schucknecht a., Meroni M, Kayitakire F, Rembold F, Boureima a. Biomass estimation to support pasture management in Niger. ISPRS - Int Arch Photogramm Remote Sens Spat Inf Sci [Internet]. 2015;XL-7/W3(May):109–14. Available from: http://www.int-arch-photogramm-remote-sens-spatial-inf-sci.net/XL-7-W3/109/2015/

[5] Harmoney KR, Moore KJ, George JR, Brummer EC, Russell JR. Determination of pasture biomass using four indirect methods. Agron J. 1997;89(4):665–72.

[6] Smith R, Panciera M. Using a Grazing Stick for Pasture Management. World Wide Web Internet Web Inf Syst. 2007;1–6.

[7] Barnhart S. Estimating available pasture forage. 2009;(April). Available from: http://scholar.google.com/scholar?hl=en&btnG=Search&q=intitle:Estimating+Available+Pasture+Forage#2

[8] Gadzirayi CT, Mutandwa E, Mupangwa JF. Veld Condition Trend of Grazing Areas: Why poor livestock production in the tropics? Rangelands. 2007;29(1):17–21.

[9] Vibart, R. E., White-Bennett, S. L., Green, J. T., & Washburn SP. Visual assessment versus compressed sward heights as predictors of forage biomass in cool-season pastures. In: JOURNAL OF DAIRY SCIENCE. AMER DAIRY SCIENCE ASSOC 1111 N DUNLAP AVE, SAVOY, IL 61874 USA; 2004. p. 36.

[10] Rinehart L. Pasture, Rangeland and Grazing Management. Management. 2008;20.

[11] Moyo S, Rutherford B, Amanor-Wilks D. Land reform & changing social relations for farm workers in Zimbabwe. Rev Afr Polit Econ. 2000;27(84):181–202.

[12] Mapiye C, Foti R, Chikumba N, Poshiwa X, Mwale M, Chivuraise C, et al. Constraints to adoption of forage and browse legumes by smallholder dairy farmers in Zimbabwe. Livest Res Rural Dev. 2006;18(12).

[13] Ghorbania A, Bruce D, Tivera F. Stratification: A problem in rangeland monitoring. 1st Int Conf Object-based Image Anal. 2006;

[14] Rayburn E, Lozier J. A Falling Plate Meter for Estimating Pasture Forage Mass. System. 2003;3.

[15] Farajollahi A. Assessment and Comparison of Different Methods for Estimating Forage Production (Case Study: Rangeland of Kurdistan Province). J Rangel Sci. 2012;2(2):483–9.

[16] Johnson J, Newman M. A Comparison of Two Methods for Sampling Biomass of Aquatic Plants. J Aquat Plant Manag. 2011;49:1–8.

[17] Shewmaker G, Gillaspy B, Fransen S, Griggs T, Hooper L. Estimating Forage Production , Monitoring „ and Evaluating the Grazing System Estmathg forage procitori. 2010;

[18] Shinners KJ, Boettcher GC, Muck RE, Wiemer PJ, Casler MD. Drying , Harvesting and Storage Characteristics of Perennial Grasses as Biomass Feedstocks. 2006;0300(061012).

[19] Farnham DE, Moore KJ, George JR. Influence of Drying Method and Temperature on Ruminal Degradable Protein of Switchgrass. 1990;(988):21–2.

[20] Eversole DE, Fike JH, Lewis RM, Mccann M a. Estimating Forage Mass of Tall Fescue Pastures and Dry. 2005;

[21] Römermann C, Tackenberg O, Scheuerer M, May R, Poschlod P. Predicting habitat distribution and frequency from plant species co-occurrence data. J Biogeogr. 2007;34(6):1041–52.

[22] Storrie A. Hyparrhenia hirta – Coolatai grass. Weedwatch. 2003;12.

[23] Zhang Y, Shao X, Chen C, Zhang YJ, Wang K. Morphological characteristics and biomass allocation of Leymus chinensis (Poaceae) (Trin.) responses to long-term overgrazing in agro-pastoral ecotone of Northern China. J Anim Plant Sci. 2013;23(3):933–8.

[24] Madjelia CED, Kadidiatou HK. Morphological characteristic variation of eleven provenances of Moringa oleifera seedlings grown in the Northern Sudanese area of Burkina Faso. African J Plant Sci [Internet]. 2015;9(10):401–11. Available from: http://academicjournals.org/journal/AJPS/article-abstract/87F240955772

[25] Pettorelli N, Pelletier F, Hardenberg A von, Festa-Bianchet M, Côté SD. EARLY ONSET OF VEGETATION GROWTH VS. RAPID GREEN-UP: IMPACTS ON JUVENILE MOUNTAIN UNGULATES. Ecology. 2007;88(2):381–90.

[26] Angerer JP. Methodologies for Forage Evaluation in Grasslands and Rangelands. 2008;

[27] Karl MG, Nicholson RA. Society for Range Management Evaluation of the Forage-Disk Method in Mixed-Grass Rangelands of Kansas Published by : Society for Range Management Linked references are available on JSTOR for this article : Evaluation of the Forage-disk Method in Mixed-gra. 2016;40(5):467–71.

[28] Zou J, Rogers WE, Siemann E. Differences in morphological and physiological traits between native and invasive populations of Sapium sebiferum. Funct Ecol. 2007;21(4):721–30.

[29] Ahmed J, Bonham CD, Ahmed J, Bonham CD. Society for Range Management Optimum Allocation in Multivariate Double Sampling for Biomass Estimation Optimum Allocation in Multivariate Double Sampling for Biomass Estimation. 2016;35(6):777–9.

Potential development based on the agricultural waste in the eastern region of West Java Province

F. D. Perwitasari[1,2], A. R. S. Putra[2], B. Suwignyo[2] and R. Widiati[2*]

[1]Universitas Muhammadiyah Cirebon, Cirebon, Indonesia
[2]Universitas Gadjah Mada, Yogyakarta, Indonesia

*rini_w@ugm.ac.id

Abstract. Development of ruminant livestock requires to a stable availability of feed because feed is a determining factor for the continuity and success of businesses in livestock production The aimed of this study was to analyses the potential for beef cattle development based on the availability of forage from agricultural waste in the eastern region of West Java province. The sample locations were taken from 3 regencies based on low of population density per unit of agricultural land, namely Majalengka, Indramayu, and Ciamis Regency. Time series data on land use and agricultural waste production at the Regency's from 2015 to 2019 were used in this research. Data analysis used LQ (location question) and IDD (carrying capacity index). The results showed that the total sub-Regency's from 3 Regency's was 84 sub-districts, where there were 24 sub-districts that have LQ> 1, indicates that these sub- districts were the base area for beef cattle development. Furthermore, based on the IDD analysis, all of the 24 sub-districts have IDD values > 2, which means that based on the availability of forage from agricultural land, it was in the safe category to increase of the beef cattle population. The twenty-four sub-districts in West Java Province can serve as the basis for the development of beef cattle that need to be considered by farmers and the local government, namely increasing production (especially quality forage feed even though the availability of agricultural waste is sufficient), productivity and efficiency (marketing infrastructure) supported with policy government.

1. Introduction

The plan for the distribution and development of beef cattle in an area of Indonesia has received the attention of the government. The beef cattle development program has obstacles, namely the land where the land is getting narrower, due to the increasing competition in land use for human needs and farming. This affects decreasing the carrying capacity of natural resources, especially forage for livestock due to the conversion of agricultural land, as well as changes in cultivation patterns which are one of the causes of the decline in the beef cattle population.

This potential analysis is used to plan and develop livestock in an area that is closely related to the availability of land resources and carrying capacity, the commodities to be developed, and the supporting facilities and infrastructure. This analysis aims to establish a livestock area, balance development between regions, optimize resources to increase farmer income, population, and production, to empower farmer communities (Yuniar et.al, 2015)[1].

The eastern region of West Java Province, namely the Majalengka Regency, Indramayu Regency, and Ciamis Regency has several considerations, namely 1) seen from the low population density per agricultural land area, so that the agricultural land area will produce agricultural waste. 2) beef import is low compared to regencies and municipalities in West Java Province. West Java Province has several features, namely 1) as a center of consumption, 2) close to the center of production (Central

IOP Conf. Series: Earth and Environmental Science 1041 (2022) 012014 doi:10.1088/1755-1315/1041/1/012014

Java), 3) Characteristics of the community in running existing livestock businesses where beef cattle fattening is more developed. For this reason, it is necessary to carry out a potential analysis of the development of beef cattle fattening business based on agricultural waste resources in the eastern region of West Java Province.

2. Materials and Methods

Material. The materials used are secondary data, time-series data on land use, and agricultural waste production at the Regency level from 2015 - 2019 from the Regency BPS in numbers and the Agricultural Service, the Animal Husbandry Service per Regency in the eastern region of West Java Province. The research was conducted from July to August 2020.

Research methods. Selection of research sample locations through purposive sampling. The location of the research sample was taken from 3 Regencys based on low population density per area of agricultural land, namely the Majalengka Regency, Indramayu Regency, and Ciamis Regency. Methods of data analysis using descriptive.

The parameters observed

2.1 The potential of agricultural wastes (straw) [2]

Rumus : (Ps x 0,4) + (pi x 3 x 0,4) + (jg x 3 x 0,5) + (kd x 3 x 0,55) + (((kh +kt) x 2 x 0,55) + (uj x 0,25/6) + (uk x 0,25/40) x 0,65

Note: (Ps = rice fields; Pl = paddy fields; jg = corn; kd = soybean; kh = green beans; kt = peanuts; uj = yams; uk = cassava).

2.2 Requirement Minimum feed

The need for feed for one livestock unit (1 AU) beef cattle in one year requires a dry matter of 6.25 kg/day or equivalent to 2.80 tons/year (NRC 1984). There are various types of cattle in the field, for example, Po cattle, Simmental breeds, limousine breeds, assuming an average body weight of up to 400 kg. Therefore, for the minimum feed requirement for ruminants, the unit of livestock (1 AU) is calculated according to Thahar et al. (1991) in [2] as follows: (with modification)

K = 2,5% x 50% x 365 x 400 kg = 1,83 ton BKC/years/AU

Note:

K = Minimum feed requirement for 1 (AU) in tonnes of digestible dry matter or DDM (Digestible dry matter) for one year.

2,5 % = The minimum requirement ration the amount of forage (dry matter) of body weight of cattle

50% = the average value of the digestibility of various plants

365 = the number of days in a year

400= live weight 1 AU, assuming various breeds of cattle, Limousin, Simmental and PO

Carrying Feed Forage

This forage carrying capacity aims to determine the carrying capacity of livestock by using the following equation: [2]

$$\text{Carrying Feed Forage (ST)} = \frac{forage\ dry\ matter\ production\ (Kg)}{Minimum\ requirement\ BK\ Cattle\ Feed\ Adults\ (\frac{Kg}{AU})}$$

IDD Forage

This index of forage carrying capacity has the aim of knowing the safety level of animal feed in an area, namely the following equation: [2]

$$\text{IDD Forage} = \frac{forage\ dry\ matter\ production\ (Kg)}{Number\ of\ livestock\ population\ (ST)\ x\ Minimum\ requirement\ BK\ Cattle\ Feed\ Adults\ (\frac{Kg}{AU})}$$

Atau

$$\text{IDD Forage} = \frac{Capacity\ of\ Livestock\ Forage(AU)}{Number\ of\ livestock\ population\ (AU)}$$

Table 1. Criteria for forage supporting capacity based on IDD

No	(IDD)	Criteria	Explanation
1	≤ 1	Verry Critical	There is no choice of utilizing the available resources. There is resource depletion in the agroecosystem. No natural forage and waste is recycled.
2	> 1 – 1,5	Critical	Livestock have had the option to utilize resources but the conservation aspects have not yet fulfilled
3	>1,5 – 2	Prone	The availability of organic matters derived from waste does not have a significant role in fulfilling animal feed need
4	> 2	Safe	The availability of feed resources from waste is functionally sufficient for animal feed needs and is available in the environment efficiently

Sumber : [2]

Location Quotient (LQ) Analysis

The LQ method is an analysis to identify and identify the economic concentration in a relatively broader location or is called an economic base analysis [3], with the following equation:

$$LQ_{ij} = \frac{\frac{x_{ij}}{X_{i.}}}{\frac{X_{.j}}{X_{..}}}$$

Note:
Xij = The economic value of (j) the beef cattle commodity in the(i)-Regency
Xi. = total economic value of all livestock commodities in Regency i
X.j = the total economic value (j) the beef cattle in all Regencys
X = total economic value of all livestock commodities in the entire region

Criteria Value:
LQ> 1 means that the concentration of the activity concentration of commodity beef cattle in a region compared to other livestock commodities in the region.
LQ = 1 means that economic activity in the area of commodity cattle equivalent to the economic activity of other farm commodities.
LQ <1 means indicate that economic activity in the area of commodity cow is smaller proportion compared to other livestock commodities.

3. Result and Discussion

3.1 The potention of agriculture waste

The feed is a very important factor in the maintenance of cattle farming. The availability of continuous feed will determine the success of cattle farming. Therefore, to solve the problem in the availability of animal feed, the use of agricultural waste as an alternative feed [4]. The production of agricultural waste calculated in the study consists of the production of rice, corn, peanuts, green beans, soybeans, cassava, and sweet potatoes which are then converted into dry matter.

The results of this study indicate that most agricultural waste production is in Indramayu Regency with an average of 476,586 tons, the most agricultural waste product is produced from rice straw. Indramayu Regency is the second-largest rice producer after Cianjur with an average rice harvest area of 234,532.40 hectares and an average production of 649,883.15 tons. The results research [4] said that the high and low production of feedstuffs originating from food plant waste is caused by the harvest area supported by the production per m2. Each region has the potential for different agricultural waste, this is due to the harvested area which will affect the amount of agricultural waste produced [3].

Table 2. Total Production Agricultural waste (BK)

Regency	Total Production Agricultural Waste BK (Ton/TH)					Total	Average
	2015	2016	2017	2018	2019		
Majalengka	294.251,10	347.596,23	350.455,83	349.551,09	338.318,73	1.680.172,98	336.034,60
Indramayu	538,506.41	512,897.00	389,689.93	475,194.49	466,642.14	2,382,929.98	476,586.00
Ciamis	150,279.14	178,725.48	168,717.09	176,327.16	156,107.32	830,156.19	150,279.14

Processed Data Sources Secondary data from BPS Ciamis Regency, Indramayu Regency and Majalengka Regency (2015-2019)

The results of potensial agricultural waste in these three regencys, the dominant agricultural waste is rice straw. Rice straw is more widely used by farmers due to the more easily searchable, continuously available, nothing cost. Rice straw is predominantly used by farmers for cattle forage because of the high production of rice straw that led to serving as a source of animal feed. Rice straw has constraints in the nutrient content of high fiber, low protein but livestock can convert rice straw for growth, livestock living needs, and reproduction of ruminants.

Therefore, to overcome the lack of nutritional content in rice straw, it is necessary to process feed both physically, chemically, and biologically [4]. Utilization of agricultural waste for animal feed is one of the solutions to overcome the shortage of animal feed, diversified utilization of waste by-products for the creation zero waste [4]. Rice straw predominates over other agricultural wastes in the Indramayu Regency because rice can be harvested 3-4 times a year given the wide agricultural area and diversification allows, especially in the region's southern direction [5].

3.2 Carrying capacity agriculture waste and IDD

Carrying capacity has the aim of measuring the availability of forage in an area by calculating the area, grass production, or agricultural waste that has the potential for feed [1]. The carrying capacity of agricultural waste calculated in the study came from rice straw, corn, soybeans, peanuts, green beans, cassava, and sweet potatoes. The most dominant agricultural waste is sawed from the production of agricultural waste and the use of forage provided by farmers, namely rice straw and corn. Assuming an average weight of cattle ranchers in Ciamis, Indramayu, and Majalengka Regency is 400 kg so that the needs of mature cattle over the past year was 1.83 Ton/years.

The results of the study in three regency of West Java Province only 24 sub-districts which has the status of carrying capacity of agricultural waste by categories Secure consists of Ciamis Regency (11 Subdistrict), Indramayu Regency (12 subdistrict), and Majalengka Regency (1 Subdistrict), for more details can be seen in Table 3. The area of Ciamis Regency has a SAFE status with the widest carrying capacity for animal feed and the capacity to add livestock, there are only ten sub-district in Ciamis Regency that have the carrying capacity of livestock, namely 1) Banjarsari (5,701.94 AU), 2) Banjaranyar (792,49 AU), 3) Lakbok (6,315.14 AU), 4) Cieunjing (2,224.24 AU), 5) Cisaga (3,036.19 AU), 6) Tambaksari (7,034.65 AU), 7) Rajadesa (3,548.68 AU), 8) Cikoneng (2,095.08 AU), 9) Sadanaya (1,902.33 AU), 10) Barabeg (1,917.34 AU), 11) Sukamantri (2,664.08 AU).

The addition of livestock capacity in the Ciamis Regency is 792.49 AU to 5,613.28 AU. The beef cattle population in Ciamis Regency is influenced by beef cattle income, beef cattle output, grass gardens, grazing land, forest land, number of livestock ownership, population and beef prices [6].

Table 3. Average carrying capacity of agricultural waste Ciamis Regency, Majalengka Regency and Indramayu Regency (2015-2019) years

No	Subdistrict	Average production	Feed Needs for Mature Cows	Average livestock	average DD	Average Increase

		waste TON	BK/Ton/Year/AU	population AU		livestock
Kabupaten Ciamis						
1	Banjarsari	10.434,54	1,83	1.060,68	5.701,94	4.641,25
2	Banjaranyar	1.918,51	1,83	255,88	1.048,37	792,49
3	Lakbok	11.556,70	1,83	701,86	6.315,14	5.613,28
4	Cijeungjing	4.070,35	1,83	758,91	2.224,24	1.465,33
5	Cisaga	5.556,23	1,83	812,70	3.036,19	2.223,49
6	Tambaksari	12.873,42	1,83	2.831,70	7.034,65	4.202,96
7	Rajadesa	6.494,09	1,83	1.169,74	3.548,68	2.378,94
8	Cikoneng	3.833,99	1,83	857,03	2.095,08	1.238,05
9	Sadananya	3.481,26	1,83	335,14	1.902,33	1.567,19
10	Baregbeg	3.508,74	1,83	433,56	1.917,34	1.483,79
11	Sukamantri	4.875,28	1,83	1.124,27	2.664,08	1.539,81
Kabupaten Majalengka						
12	Majalengka	32.905,50	1,83	8.407,88	17.981,15	9.573,26
Kabupaten Indramayu						
13	Haurgeulis	16.190,20	1,83	260,15	8.847,11	8.586,95
14	Gantar	41.862,89	1,83	5.243,29	22.875,90	17.632,61
15	Gabuswetan	20.308,11	1,83	555,19	11.097,33	10.542,14
16	Cikedung	23.690,25	1,83	5.594,00	12.945,49	7.351,49
17	Terisi	37.642,01	1,83	2.404,42	20.569,41	18.164,99
18	Bangodua	12.635,34	1,83	401,47	6.904,56	6.503,09
19	Kertasemaya	11.478,97	1,83	423,91	6.272,66	5.848,75
20	kedokanbunder	7.685,99	1,83	433,59	4.200,00	3.766,41
21	Juntinyuat	14.639,27	1,83	366,96	7.999,60	7.632,64
22	Sliyeg	15.881,79	1,83	1.920,10	8.678,57	6.758,47
23	sindang	6.753,22	1,83	210,34	3.690,29	3.479,95
24	patrol	10.220,60	1,83	400,87	5.585,03	5.184,16

Processed Data Sources Secondary data from BPS Ciamis Regency, Indramayu Regency and Majalengka Regency (2015-2019)

Indramayu district (12 sub-districts) with an average carrying capacity of 3,690,29 – 22,875.90 AU consists of 1) Haurgelis, 2) Gantar, 3) Gabuswetan, 4) Cikedung, 5) Filled, 6) Bongdua, 7) Kertasemaya, 8) Kedokanbunder, 9) Juntinyuat, 10) Sliyeg, 11) Sindang, 12) Patrol. Gantar Subdistrict has the highest carrying capacity of 22,875.90 AU and can still accommodate livestock of 17,632.61 AU, while the lowest carrying capacity is Sindang District of 3,690,29 AU which can accommodate livestock of 3,479.95 AU. As for Majalengka Regency, Majalengka District has a carrying capacity of 17,981.15 which can still accommodate livestock of 9,573.26 AU.

The difference in the results of this research is caused by 1. Area and population density, 2. Agricultural area, 3. agricultural waste production 4. livestock population. This is following the opinion [7] that said several factors influence the carrying capacity of forage as follows 1) the level of livestock density, 2) production of agricultural and forage waste, 3) land area for forage, and agricultural land.

The IDD values generated in these three Regencys resulted in IDD values ranging from 2,14 to 34,01 AU, the SAFE category. The potential of agricultural waste available in Three Regencys can accommodate and provide for the needs of cattle feed at 2,14 to 34,01 AU. The results of the calculation of the livestock carrying capacity index of all Regencys in Ciamis Regency are in the safe category, meaning that the existing agricultural waste needs can meet the forage needs of livestock in accordance with the livestock population based on livestock units (AU). Ciamis Regency consists of 27 sub-district but there were only twelve sub-district in Ciamis Regency have value IDD and LQ> 2, which are described in the table below. The resulting IDD values were 2,37 – 9,00 ; means that the IDD value generated is in the Safe category.

The results of the calculation of the land carrying capacity index of all sub-district in the Indramayu Regency, Ciamis Regency and Majalengka Regency are in the safe category, meaning that the existing agricultural waste needs can meet the forage needs of livestock according to the livestock population-based on livestock units (AU). Indramayu Regency has the highest IDD value ranging from 4,36 - 34,01 compared to Majalengka and Ciamis Regency. Indramayu Regency has the highest development potential, followed by the Regency of Garut, Karawang, Subang, Majalengka postscript is rice-producing areas are high, so the availability of its rice straw can meet the needs of livestock feed. Other areas of considerable potential are the area of Cianjur, Sukabumi, and Cirebon to be developed because it is still possible to be developed in a large enough population [8].

3.3 Analisis location quotient (LQ)

The results of the Location quotient analysis for the three Regencys (Majalengka, Indramayu, and Ciamis) in the five years 2015 - 2019 period explaining below. The value of LQ in table 3 below is a choice of Regencys that have LQ> 1 and IDD> 2 values as follows, Ciamis Regency has 27 subdistricts but the results of the calculation LQ analysis only twelve sub-district is Banjasari, banjaranyar, Cijeungjing, Cikoneng, Sadananya, Baregbeg.

That is the twelve this sub-district has the potential basis for the development of cattle breeding business. Cijeungjing and Tambaksari have high LQ value, because in this region close to the scaffolding market, the market of large and small livestock for Ciamis Regency, and close enough to the animal market in Tasikmalaya. The base areas in Ciamis Regency for beef cattle commodities are the sub-districts of 1) Banjarsari, 2) Banjaranyar, 3) Lakbok, 4) Cieunjing, 5) Cisaga, 6) Tambaksari, 7) Rajadesa, 8) Cikoneng, 9) Sadanaya, 10) Barabeg, 11) Sukamantri the base areas in Ciamis Regency for beef cattle commodities are the districts of Banjarsari, Lakbok, Pamarican, Cidolog, Cijeungjing, Cisaga, Tambaksari, Rancah, Sukadana, Ciamis, Cikoneng, Sadananya, Panawangan, Panjalu, Barebeg , Sukamantri, can meet their own needs for beef cattle commodities, and can even export these commodities to other sub-districts or even outside the Ciamis Regency area[6].

Table 4. Carrying value index (IDD) of Ciamis Regency, Indramayu Regency, Majalengka Regency (2015-2019) years

No	Subdistrict	Average production waste TON	Feed Needs for Mature Cows BK/Ton/Year/AU	Average livestock population AU	average DD	average IDD	Category
Kabupaten Ciamis							
1	Banjarsari	10.434,54	1,83	1.060,68	5.701,94	5,38	SAFE
2	banjaranyar	1.918,51	1,83	255,88	1.048,37	4,10	SAFE
3	Lakbok	11.556,70	1,83	701,86	6.315,14	9,00	SAFE
4	Cijeungjing	4.070,35	1,83	758,91	2.224,24	2,93	SAFE
5	Cisaga	5.556,23	1,83	812,70	3.036,19	3,74	SAFE
6	Tambaksari	12.873,42	1,83	2.831,70	7.034,65	2,48	SAFE
7	Rajadesa	6.494,09	1,83	1.169,74	3.548,68	3,03	SAFE
8	Cikoneng	3.833,99	1,83	857,03	2.095,08	2,44	SAFE
9	Sadananya	3.481,26	1,83	335,14	1.902,33	5,68	SAFE
10	Baregbeg	3.508,74	1,83	433,56	1.917,34	4,42	SAFE
11	Sukamantri	4.875,28	1,83	1.124,27	2.664,08	2,37	SAFE
Kabupaten Majalengka							
12	Majalengka	32.905,50	1,83	8.407,88	17.981,15	2,14	SAFE
Kabupaten Indramayu							
13	Haurgeulis	16.190,20	1,83	260,15	8.847,11	34,01	SAFE
14	Gantar	41.862,89	1,83	5.243,29	22.875,90	4,36	SAFE
15	Gabuswetan	20.308,11	1,83	555,19	11.097,33	19,99	SAFE
16	Cikedung	23.690,25	1,83	5.594,00	12.945,49	2,31	SAFE
17	Terisi	37.642,01	1,83	2.404,42	20.569,41	8,55	SAFE
18	Bangodua	12.635,34	1,83	401,47	6.904,56	17,20	SAFE
19	Kertasemaya	11.478,97	1,83	423,91	6.272,66	14,80	SAFE
20	kedokanbunder	7.685,99	1,83	433,59	4.200,00	9,69	SAFE
21	Juntinyuat	14.639,27	1,83	366,96	7.999,60	21,80	SAFE
22	suyeg	15.881,79	1,83	1.920,10	8.678,57	4,52	SAFE
23	sindang	6.753,22	1,83	210,34	3.690,29	17,54	SAFE
24	patrol	10.220,60	1,83	400,87	5.585,03	13,93	SAFE

Indramayu Regency has 31 Sub-Regencys but the results of the calculation of LQ analysis are only twelve sub-Regencys is Haurgelis, Gantar, Gabuswetan, Cikedung Terisi, Bangodua, Kertasemaya, Kedokanbunder, Juntinyuat, Sliyeg, Sindang and Patrol. This means that these Indramayu Regency has begun to focus on these twelve sub-districts to order to be prepared the development of an increase in the number of ruminant livestock populations by paying attention to various aspects, such as cultivation techniques, farmer knowledge, and skills, and socioeconomics through a more focused approach by analyzing distinctive institutional functions both capital and non-capital institutions [5].

This sub-district is mostly located in the geographical position of the southern part of Indramayu where this area borders Subang, Majalengka, and Sumedang, then not far from the animal markets in Majalengka and Subang, making it easier for breeders to choose feeders or sell livestock. Majalengka Regency consists of 26 sub-Regencys however the results of the analysis calculation LQ only one is Majalengka (2.92) subRegency. That means Majalengka sub-Regencys have the potential for the development of cattle breeding business and increase agricultural waste to compared with population in Majalengka sub-district.

The results of this research are three Regencys in West Java Province with 84 sub-Regencys, only twenty-three sub-Regencys with LQ value> 1, which means these Regencys are the basis for beef cattle development. Factors that support the twenty-three sub-Regencys as the basis for the development is 1. geographical conditions border with other regions so as to facilitate access to marketing, 2. The area, 3. The number of populations, 4. Close to the infrastructure as a sales market mainly cattle and feeder cattle.

The results of this study are by the opinion of [9] said that an area has an LQ value > 1 due to the potential for land carrying capacity and the number of workers in an area, so that the ratio of arable land area to total population determines the distribution pattern of livestock in an area. Apart from that, several other factors have closely influenced the development of beef cattle farms in Taluditi Regency in addition to the availability of adequate facilities and infrastructure, also because cattle farming is a family business that has been carried out from generation to generation, so that almost every household in the Regency has cattle. The high and low LQ value can be influenced by land specificity, this geographical condition will produce various types of green forage resources and provision of the amount of food, land tenure per individual causes communal land use to be very limited [10] .

4. Conclusion

The conclusion from this research is that a total of 84 sub-Regencys from 3 sample Regencys, there are 24 sub-district with LQ value > 1, which means that these Regencys are the basis for beef cattle development. IDD analysis, all 24 sub-district which are the base areas have IDD value > 2, which means that based on the availability of forage from agricultural land, it is in the safe category to increase the beef cattle population.

The potential for beef cattle development in the eastern region of West Java needs to be prioritized in 24 sub-Regencys with LQ > 1 and IDD> 2 accompanied by investment support policies for smallholder farmer, the quality forage land, the number of livestock holdings supported with farmers' access to credit. Efforts were increased productivity and efficiency can be done through technical counseling and guidance.

Table 5. LQ Analysis of Ciamis Regency, Indramayu Regency, Majalengka Regency (2015-2019)

No	Kecamatan	Average Population Beef Cattle Subdistrict AU	Average Population Ruminansia Subdistrict AU	Average Total Population Beef Cattle Regency (AU)	Average Total Population Ruminansia Regency (AU)	LQ	Category
Kabupaten Ciamis							

1	Banjarsari	408,60	1.060,68	10.015,20	34.215,50	1,32	basis
2	banjaranyar	187,00	255,88	10.015,20	34.215,50	2,50	basis
3	Lakbok	249,60	701,86	10.015,20	34.215,50	1,21	basis
4	Cijeungjing	580,60	758,91	10.015,20	34.215,50	2,61	basis
5	Cisaga	357,00	812,70	10.015,20	34.215,50	1,50	basis
6	Tambaksari	1.942,60	2.831,70	10.015,20	34.215,50	2,34	basis
7	Rajadesa	417,80	1.169,74	10.015,20	34.215,50	1,22	basis
8	Cikoneng	353,40	857,03	10.015,20	34.215,50	1,41	basis
9	Sadananya	112,80	335,14	10.015,20	34.215,50	1,15	basis
10	Baregbeg	153,20	433,56	10.015,20	34.215,50	1,21	basis
11	Sukamantri	425,20	1.124,27	10.015,20	34.215,50	1,29	basis
	Kabupaten Majalengka						
12	Majalengka	1.786,00	5.572,50	13.918,80	108.814,33	2,51	basis
	Kabupaten Indramayu						
13	Haurgeulis	125,60	260,15	11.421,60	42.352,52	1,79	basis
14	Gantar	3.263,60	5.243,29	11.421,60	42.352,52	2,31	basis
15	Gabuswetan	246,40	555,19	11.421,60	42.352,52	1,65	basis
16	Cikedung	1.515,80	5.594,00	11.421,60	42.352,52	1,00	basis
17	Terisi	2.017,60	2.404,42	11.421,60	42.352,52	3,11	basis
18	Bangodua	111,60	401,47	11.421,60	42.352,52	1,03	basis
19	Kertasemaya	172,60	423,91	11.421,60	42.352,52	1,51	basis
20	kedokanbunder	212,20	433,59	11.421,60	42.352,52	1,81	basis
21	Juntinyuat	247,60	366,96	11.421,60	42.352,52	2,50	basis
22	suyeg	937,00	1.920,10	11.421,60	42.352,52	1,81	basis
23	sindang	88,60	210,34	11.421,60	42.352,52	1,56	basis
24	patrol	263,20	400,87	11.421,60	42.352,52	2,43	basis

Processed Data Sources Secondary data from BPS Ciamis Regency, Indramayu Regency and Majalengka Regency (2015-2019)

Acknowledgement
Thank you University Gadjah Mada for funding my Dissertation Research.

References

[1] Yuniar, P. S., Widiatmaka, & Fuah, A. M. 2015. Analysis of the Potential of Beef Cattle Development in South Tangerang City. Journal of Animal Production Science and Technology, 03(2), 106–112.

[2] Juarini E, Sumanto, Budiarsana IGM, Wibowo B. 2007. Estimation of Potential Local Feed Sources and Improvement Efforts to Increase Milk Production in Smallholder Dairy Cattle Business in Sukabumi. proceedings of the National Seminar on Animal Husbandry and Veterinary. Bogor, 21-22 August. Bogor (ID): Center for Animal Husbandry Research and Development. 301 - 308. Potong di Kabupaten Pesawaran," J. Imiah Peternak. Terpadu, vol. 4, no. 2, p. 233172, 2016.

[3] Panuju DR, Rustiadi E. 2012. Regional Planning Development Analysis Techniques. Bogor (ID): Department of Land Sciences and Land Resources. Institut Pertanian Bogor.

[4] Saputra, J., Liman, L., & Widodo, Y. 2016. Analysis of the Potential of Beef Cattle Development in Pesawaran Regency. Integrated Animal Husbandry Imiah Journal, 4(2), 233172.

[5] Perwitasari, F.D. A. Romadhoni, S. Putra, B. Suwignyo, and R. Widiati, "Analysis of the Potential of Beef Cattle Business Development in Indramayu District," vol. 23, no. 32, pp. 127–137, 2021.

[6] Isyanto, A.Y. and Sudrajat . 2019. "Keragaan Usaha Penggemukan Sapi Potong Di Kabupaten Ciamis." MIMBAR AGRIBISNIS: Jurnal Pemikiran Masyarakat Ilmiah Berwawasan Agribisnis 5(1): 33.

[7] Sari. A, Liman, & Muhtarudin. 2016. Supporting of Agricultural by Product as Ruminant Feed in Pringsewu Dictrict. Jurnal Imiah Peternakan Terpadu, 4(2), 100–107.

[8] Arief H and L. Khaerani. R.Z. Islami, "PETA POTENSI KAWASAN PETERNAKAN BERBASIS DAYA DUKUNG LOKAL DI JAWA BARAT," Pros. Semin. Nas. Agroinovasi Spesifik Lokasi Untuk Ketahanan Pangan Pada Era Masy. Ekon. ASEAN, pp. 1085–1097, 2012.

[9] Arsyad, A. H. 2012. Research Report Analysis of Potential Supporting Capacity of Beef Cattle Development in Pohuwato Regency, State University of Gorontalo. Gorontalo.

[10] Dewi, K.D. 2018. Potential Analysis of Ruminant Livestock Development Areas in Lamongan Regency. Animal Journal, 09(02), 5–11

Farming in Kediri Indonesia: analysis of cluster k-means

N Solikin[1*], B Hartono[2], Sugiono[1] and Linawati[1]

[1]Universitas Nusantara PGRI Kediri, Kediri, Indonesia
[2]Universitas Brawijaya Malang, Malang, Indonesia

*gatotkoco.80@gmail.com

Abstract. Beef cattle farming is a source of income preferred by the people of Kediri Regency to fulfill their daily needs. Various components related to beef cattle business in Kediri Regency need to be assessed to understand the real facts related to the business. This study aims firstly to analyze the density cluster of cattle and the second perception of beef cattle breeders in Kediri Regency regarding the components of strength, weakness, threat, and opportunity in the beef cattle business. The research was conducted in June - December 2019. Data collection was carried out through a survey of 150 breeders and FGD (Focus Group Discus) involving 15 experts was conducted. The data obtained were then analyzed by using the K-means cluster. The results of the first study regarding the density level of cluster 1 (one) solid category with successive central values (13239, 7658, 88190, 4759), there are 6 (six) districts. Cluster 2 (two) with moderate category with central value (9966, 1870, 59872, 3618) totaling 17 (seventeen) districts. Cluster 3 (three) categories rarely consist of three districts. The second result is based on the perception that the strength component of the government support aspect must be optimized, small amount of farmer capital became the weakness aspect of the business. The increasing demand for beef was the opportunity aspect. The rampant slaughter of productive female cattle was the threat aspect of this business. Based on the results of the K means cluster analysis, it is concluded that the beef cattle business in Kediri Regency is very potential to be developed because of the dominance of the area in clusters 1 and 2. The Kediri Regency Government and its stakeholders are actively involved in the success of the beef cattle breeder business. The results show that the strength component of the government support aspect must be optimized.

1. Introduction

In determining the development of beef cattle business in rural areas, it is necessary to have a mapping based on the similarity of characters possessed by the community and factual conditions. Various components will be linked to one another. The complexity of the problem is important to study using an approach that focuses on the perspective and the level of importance expected. Heterogeneity of opinion is deemed necessary to be classified in order to suppress the spread of the problem, grouping is indispensable with the aim of making it easier to take action.

Internal factors (strengths, weaknesses) and external factors (threats, opportunities) are an important part that must be considered in determining the progress of a business, especially beef cattle farming. A strong internal factor component can become the farmer's strength to solve problems caused by external factors. In the beef cattle business, it is important to identify internal and external factors as a SWOT unit, referring to [1] [2] [3] [4] strategic decisions can be obtained if strengths, weaknesses, opportunities, and threats are known. Indicators that are part of each component are then formulated to analyze the real conditions felt and expected by beef cattle breeders and stakeholders.

K means cluster analysis is a statistical analysis that aims to determine complex data structures by placing the similarities of observed objects into one group. K means cluster analysis is used as a tool to facilitate grouping and the tendency of many perceptions. Research [5] [6] [7] [8] [9] illustrates that strategic decisions are made based on grouping according to known cluster data. Not many have used it in the animal husbandry sector, the use of analysis can be an alternative to 1) knowing segmentation and determining potential areas, 2) knowing the positioning and determining the business of beef cattle development in Kediri Regency.

Kediri Regency is one of the regions in East Java that contributes (producers) to beef cattle. Referring to [10] as the basis for area and livestock development is the authority of each region. Mapping of livestock density in the area needs to be taken into consideration so that the decisions taken can be an effective solution. Areas with different cattle populations require different treatment measures. The varied socio-economic characteristics of breeders also influence the breeders' perspective on the various internal and external components that affect the progress and development of beef cattle breeders, especially in Kediri Regency.

2. Materials and Methods

The research focuses on examining 2 (two) important things, first about the density of cattle in the Kediri district by grouping them into cluster 1 (dense), cluster 2 (medium), and cluster 3 (rare), secondly about breeders' perceptions of strengths, weaknesses. , opportunities, and threats in beef cattle business. The analysis used is the K means cluster analysis as a decision-making tool. The k-mean cluster analysis step or using a partitional clustering approach, each cluster is associated with a centroide value (center point).

Each point is placed in a cluster with the closest centroid. calculate the distribution of each component, in this case the researcher uses the Excel microsof application by = SQRT (first data in column and row 1 minus the cetroid 1 value) to the power of 2 (two) plus (data in row 1 column 2 minus the value of centroid 2) to power 2 , so on until the nth data. After knowing the results, it is continued with repetition (iteration1,2, etc.) to find out the new cluster by first calculating the centroid (central value) until there is no change in the cluster.

3. Result and Discussion

Kediri Regency is one of the districts that has contributed to providing the needs of cattle in East Java and Indonesia. Kediri Regency ranks 6th in producing beef cattle in East Java [11] supported by data from the Central Statistics Agency for Kediri Regency in 2018, which recorded a cow population of 223,213 with a distribution in 26 (twenty-six) sub-districts included in the administrative area of Kediri Regency. The distribution of the cattle population in each district is different depending on the trends of the population in the area.

Beef cattle business in rural areas is dominated by traditional small-scale breeders, because the beef cattle business is a side business and serves as a filler of the remaining time from the main job (farming, public, private, traders, coolies, etc.). Beef cattle business in the area of Kediri Regency is still popular with the community because it can increase income even though it is not too much. According to Solikin's research [12], social capital affects the development of beef cattle business. It is confirmed by [13] stating that the increase in income of cattle farmers can be obtained greater if they process existing natural resources as feed and fertilizer. The increase in farmer's income can be a leverage factor as well as a motivation for breeders to increase the livestock population.

Each sub-district with the character of the population, land area, and cattle population as shown in table 1. Referring to the data in table 1 by grouping cluster 1 (dense), cluster 2 (medium), and cluster 3 (rare). The initial cluster was determined randomly by having 3 districts as representatives of each cluster. In this stage, cluster 1 of Wates sub-district, cluster 2 of Plemahan sub-district, and cluster 3 of Ngasem sub-district were selected. After determining the value of each centroid, then calculating and repeating (iteration) until the cluster does not change. The distribution of beef cattle population, population, and land area in Kediri Regency is as shown in table 1 below:

Table 1. Distribution of beef cattle, population, land area in Kediri Regency

No	sub-district	Number of beef cattle (tail)	Total population (soul)	Area (km^2)
1.	Mojo	12.726	75.764	102.73
2.	Semen	8.550	53.494	80.42
3.	Ngadiluwih	7.394	77.551	41.85
4.	Kras	8.793	59.351	44.81
5.	Ringinrejo	8.943	52.287	42.38
6.	Kandat	8.958	59.098	51.96
7.	Wates	13.915	87.113	76.58
8.	Ngancar	11.272	47.178	94.05
9.	Plosoklaten	12.468	69.937	88.59
10.	Gurah	13.142	81.234	50.83
11.	Puncu	11.773	60.980	68.25
12.	Kepung	7.108	81.841	105.65
13.	Kandangan	3.579	48.304	41.67
14.	Pare	6.630	101.861	47.21
15.	Badas	3.724	62.081	39.21
16.	Kunjang	6.689	34.297	29.98
17.	Plemahan	10.475	58.721	47.88
18.	Purwoasri	9.528	55.559	42.50
19.	Papar	10.688	50.006	36.22
20.	Pagu	9.482	38.635	24.86
21.	Kayenkidul	11.635	44.529	35.58
22.	Gampengrejo	2.474	34.054	16.76
23.	Ngasem	3.295	67.310	21.83
24.	Banyakan	4.966	54.848	72.55
25.	Grogol	5.080	46.385	34.50
26.	Tarokan	10.302	58.974	47.20
	Jumlah	223.213	1.561.392	1386.05

Source: BPS Kediri 2018

The results of the research on the existing conditions of beef cattle business density in Kediri Regency through two repetitions (iterations) obtained sub-district results for category 1 (one) with successive central values (13239, 7658, 88190, 4759) there are 6 (six) districts namely Mojo, Ngadiluwih , Wates, Gurah, Kepung, and Pare. Cluster 2 (two) medium categories with central values (9966, 1870, 59872, 3618) with a total of 17 (seventeen) districts namely Semen, Kras, Ringinrejo, Kandat, Ngancar, Plosoklaten, Puncu, Kunjang, Plemahan, Purwoasri, Papar, Pagu, Kayen kidul, Gampengrejo, Banyakan, Grogol, and Tarokan. Cluster 3 (three) rare categories consist of three sub-districts, namely Kandangan, Badas and Ngasem.

The difference in density for each region is an important factor that must be considered in implementing development strategies. One example in areas where the density category is urgently needed is increasing the speed of adding cattle weight by providing quality feed. The hope is that rapid weight gain will affect the cattle harvest period. In areas with rare densities, the strategy is to spur the increase in the population of cattle.

In addition to analyzing livestock density, this study also analyzes breeders' perceptions of the components of strengths, weaknesses, opportunities, and threats that are happening. The results were obtained as follows:

3.1 The results of the strength component cluster

On the strength component with indicators, farming experience, geographic location and land availability, farmer skills, and government support. The results showed that most of the breeders' perceptions knew that the position of the strength they had was very potential to develop a beef cattle business, especially if it was supported by the government to optimize mentoring and training for farmers in improving technical skills in processing agricultural waste as animal feed and waste. livestock as fertilizer that can be applied in farming. The results of the assessment are presented in table 2 below:

Table 2. The results of the cluster perceptions of the strength components

strength components question items	Respondents 1	2	3	4	5	6	7	8	9	10	11	12	13	14	15
Breeding experience	2	2	3	3	2	3	2	2	2	2	2	3	2	2	2
Geographical location and land availability	2	1	2	2	2	3	2	2	2	2	2	2	2	2	2
Breeding skills	3	4	2	3	3	4	3	3	3	3	2	4	4	4	3
Government support	4	3	4	4	4	4	4	4	4	3	3	3	3	4	4
Sub total	16	15	15	16	16	19	16	15	15	15	14	16	15	18	17

With the centroid value for cluster A: 15.25 and cluster B: 17.75, the results in table 2 show that 70 percent of respondents rated the strength component as not urgent to handle, meaning that Kediri Regency has the power to develop beef cattle business. There are 30 percent of respondents who fall into cluster B, with the perception that the strength in Kediri Regency is still urgent to handle, especially in relation to points of government support.

3.2 The results of the cluster component weaknesses

Weakness components include business capital, market access and prices, livestock control, availability of feed and feeder cattle, and access to capital. The centroid values for the weakness component were 19.5 and 22.5. The k-mean test for weakness component clusters was obtained by 30 percent of respondents in cluster A, which means that they think that the urgency value of weakness is not urgent or not problematic and 70 percent of respondents occupy cluster B, meaning that perceptions regarding the urgency of subscribing to the weakness component need to be subscribed. As in table 3 below:

Table 3. The results of the perception cluster component of weakness

component of weakness question items	Respondents 1	2	3	4	5	6	7	8	9	10	11	12	13	14	15
startup capital	4	3	2	4	4	3	4	4	4	4	4	4	4	4	3
market access and prices	3	4	4	4	4	4	4	4	4	4	4	4	3	4	3
control of livestock	4	4	3	4	4	4	3	4	4	4	3	4	4	4	3
animal feed	3	3	3	4	3	4	4	4	3	3	4	4	4	4	4
feeder cattle	3	4	3	3	3	4	3	3	4	4	3	4	3	4	4
access to capital	4	3	3	3	4	4	4	4	4	4	3	4	4	4	4
sub total	21	21	18	22	22	23	22	23	23	23	21	24	22	24	21

The results of the study based on the weakness component analysis show that the business capital of beef cattle breeders in Kediri Regency is relatively small. The breeder has a working capital ranging from IDR 15,000,000 - 20,000,000, this figure is equivalent to one male going 18 months. Limited business capital causes breeders to be unable to develop their businesses and the income they earn from raising cattle, especially if business is not a priority for their livelihoods. Strategies in business development through additional capital must find solutions by all stakeholders.

3.3 The results of the opportunity component cluster

Opportunity components consisting of beef prices, beef demand, government support, geographic location and availability of agricultural waste, the role of banks in providing loans, and the existence of a cattle market are analyzed using the k-mean cluster, the centroid values of 18 (A) and 20 (B) it is known that the responses of expert respondents were 30 percent including cluster A (bottom) and 70 percent cluster B (top). The dominant perception of cluster B shows that the business opportunity for beef cattle is wide open as a potential business option to be increased.

Table 4. The results of the opportunity component perception cluster

Opportunity component	Respondents														
Question item	1	2	3	4	5	6	7	8	9	10	11	12	13	14	15
Price of beef	3	4	4	4	4	4	4	4	3	3	3	3	4	4	3
Beef needs	3	3	3	3	3	3	3	3	3	3	3	4	4	3	4
Government support	4	3	3	4	4	4	4	4	4	4	4	4	4	3	4
Geographical location and availability of agricultural waste	2	3	3	3	3	3	4	3	2	2	2	2	2	2	3
The role of the Bank in providing credit	3	4	4	4	4	4	4	3	3	3	3	3	3	4	3
Market presence	2	3	3	2	3	3	2	2	2	2	3	3	3	3	3
sub total	17	20	20	20	21	21	21	19	17	17	18	19	20	19	20

The results show that the main opportunity component is the need for beef which continues to increase along with the population growth, which is a potential and promising situation for breeders in developing their business. This opportunity should be used as motivation for farmers to increase the scale of their business.

3.4 Results of the threat component cluster

Threat components include competition for substituted meat prices, market access, policy changes, land use change, livestock disease, and slaughter of productive females. The results of the k-means test for threat component clusters with centroid values of 13 (A) and 17 (B), it is known that the expert's view of the urgency of threat management faced in beef cattle development is 40 percent of respondents are included in cluster A (bottom) and 60 percent of cluster respondents. B (above). This means that most respondents think that the threat component should be considered as a determining factor for the development of the beef cattle farmer's business and conditions.

Table 5. Results of the threat component perception cluster

threat component	Respondents														
Question item	1	2	3	4	5	6	7	8	9	10	11	12	13	14	15
Competition for substitute meat prices	3	3	3	2	3	3	3	4	3	3	2	2	2	3	2
Market access	3	2	2	2	2	2	2	2	2	2	2	2	3	3	3
Policy changes	3	1	1	1	3	2	2	2	2	2	3	2	2	2	2
Change of land function	1	1	1	2	2	1	1	1	1	1	2	1	1	1	2
Farm animal disease	3	3	4	3	3	3	3	2	3	3	4	3	3	3	3
Productive slaughter of female cows	2	3	4	3	4	4	4	4	4	4	3	3	3	4	4
sub total	**15**	**13**	**15**	**13**	**17**	**15**	**15**	**15**	**15**	**15**	**16**	**13**	**14**	**16**	**16**

The result of the calculation of the threat component shows that productive female slaughter items have the highest position in the threat component of beef cattle development. Cows are an asset that should be safeguarded, in order to keep reproducing so as to meet the needs of animal food in

Indonesia. The government has banned the slaughter of productive females in order to prevent a reduction in productive female cattle. Supervision in policy implementation needs to be improved.

4. Conclusion

The results of the study using K means cluster beef cattle business in Kediri Regency are: first, the density level of the cluster 1 (one) category is dense, there are 6 (six) districts. Cluster 2 (two) with medium category 17 (seventeen) districts. Cluster 3 (three) categories rarely consist of three districts. The second result is based on the perception that the strength component of the government support aspect must be optimized, the weakness aspect is the small amount of farmer capital, the opportunity aspect is the increasing demand for beef, and the threat aspect is the rampant slaughter of productive female cattle. Based on the results of the K means cluster analysis, it is concluded that the beef cattle business in Kediri Regency is very potential to be developed because of the dominance of the area in clusters 1 and 2. The Kediri Regency Government and its stakeholders are actively involved in the success of the beef cattle farmer business.

References

[1] Valentin E K 2005 Away With SWOT Analysis : Use Defensive / Offensive Evaluation Instead J. Appl. Bus. Res.
[2] Manage T 2005 SWOT analysis Twelve Manag.
[3] Ghazinoory S, Abdi M and Azadegan-Mehr M 2011 Swot Methodology: A State-of-the-Art Review for the Past, A Framework for the Future J. Bus. Econ. Manag.
[4] Ommani A R 2010 Strengths, weaknesses, opportunities and threats (SWOT) analysis for farming system businesses management: Case of wheat farmers of Shadervan District , Shoushtar Township, Iran African J. Bus. Manag.
[5] Prayoga B S and Fatriani N N 2014 Penerapan Metode K-Means Cluster Analysis Untuk 73–8
[6] Darmi Y and Setiawan A 2016 Penerapan metode clustering k-means dalam pengelompokan penjualan produk J. Media Infotama Univ. Muhammadiyah Bengkulu 12 148–57
[7] Sani A 2018 Penerapan Metode K-Means Clustering Pada Perusahaan J. Ilm. Teknol. Inf. 1–7
[8] Siregar M H 2018 Data Mining Klasterisasi Penjualan Alat-Alat Bangunan Menggunakan Metode K-Means (Studi Kasus Di Toko Adi Bangunan) J. Teknol. Dan Open Source 1 83–91
[9] Soepomo P 2013 Perancangan Sistem Pendukung Keputusan Penentuan Rumah Sehat Peranc. Sist. Pendukung Keputusan Penentuan Rumah Sehat 1 584–96
[10] Menteri Pertanian 2001 Surat Keputusan No. 417/Kpts/OT.210/7/2001 tanggal 20 Juli 2001 tentang Pedoman Umum Penyebaran dan Pengembangan Ternak.
[11] BPS Jawa Timur 2019 Propinsi Jawa Timur Dalam Angka 2019
[12] Nur Solikin, Budi H and Fanani, Z, Ihsan M N 2019 The Contribution of Social Capital to the Income and Development of Beef Cattle Farmers The Contribution of Social Capital to the Income and Development of Beef Cattle Farmers pp 0–6
[13] Nur Solikin, Sugiono, Erna Y 2019 Kontribusi Perguruan Tinggi Tehadap Kemajuan dan Peningkatan Ekonomi Peternak (Studi Kasus di Desa Ngino Kecamatan Plemahan Kabupaten Kediri) Ekuivalensi 5 161–73

The Identification and Role pf Mold in Lowering the HCN Levels in Black Cassava Based on the Immersion Duration

Mery Pattipeilohy* and Alamanda Pelamonia

Universitas Pattimura, Ambon, Indonesia

*pattipeilohymery@gmail.com

ABSTRACT. This study aimed to provide information for people in Banda, Central Maluku, about types and role of mold in lowering the HCN levels in black cassava. It is expected that the availability of alternative choices of quality food ingredients and affordable prices can improve food security and nutrition at the family level. The Identification of mold was performed through 16S rRNA sequence analysis. The analysis of HCN levels in black cassava was conducted using argentometric method. Five mold isolates, namely J1, J2, J3, J4, and J5 were successfully identified from black cassava samples. The mold species were classified into Rhizopus, Aspergillus, Penicillium, Aspergillus flavus and Aspergillus clavatus. The results of the analyses showed that mold and immersion duration played a significant role in lowering the HCN levels in black cassava. In fact, the HCN levels in Manihot aipi decreased significantly from 0.3024% (302.4 mg HCN/kg) to 0.1296 (129.6 mg HCN/kg), 0.135 % (135 mg HCN/kg), 0.162% (162 mg HCN/kg), and 0.1782 (178.2 mg HCN/kg) after 12, 24, 36, and 48-hour immersion, respectively. Similarly, at 12h, 24h, 36h, and 48h fermentation duration, the HCN levels in black cassava were reduced from 0.135% (135 mg HCN/kg) to 0.108% (108 mg HCN/kg), 0.0918% (91.8 mg HCN/kg), and 0.0864% (86.4 mg HCN/kg). The analysis of mold species and HCN levels in black cassava from Manihot aipi Phol. is important because mold species and HCN levels can determine the number of foodborne pathogens that cause many diseases and, in extreme cases, death. The analysis of mold species and HCN levels is also vital to food security. The results of this study can provide insights to the people in Banda regarding the immersion and fermentation duration of Manihot aipiPhol., which is safe for black cassava processing.

1. Introduction

The uniqueness of Indonesia's natural environment influences the uniqueness of its natural inhabitants such as microorganisms [13]. Mold is the most diverse microorganisms that can live anywhere such as in the soil, in plant parts (leaves, roots and fruits), in water sources, even in the food we consume [1]. Mold can affect the level of food decomposition. Mold is saprophytes that have no chlorophyll. Mold can grow on agricultural products before they are harvested, when they are stored, after they are processed into food, or after they are sold in the market.

Black cassava is a fermented cassava from *Manihot aipi* Phol. *Manihot aipi* Phol. is a cassava species which contains high levels of cyanide acid that, if processed without special procedures, are poisonous for human body. On Banda Island, black cassava is processed in a very unique way through two-day fermentation process after soaking it in salt water (sea water). The cassava is dried by hanging it on the drain of the house that is protected from sunlight for approximately 3 weeks. Mold flourishing on the surface of the cassava are blackish brown in color. Therefore, the cassava that is processed in this way is called black cassava. Black cassava will later be processed into a traditional food called *sunu-sunu*. Our observations show that the quality of black cassava is determined by the number of molds attached on the cassava during the drying process where a darker color cassava tastes better. This process particularly leads to decreased HCN levels in *Manihot aipi* Phol. so that the cassava is no longer hazardous for human consumption.

Food that experiences decomposition by molds can rot and change color. Mold can cause diseases that are divided into two groups, namely (1) Mycosis (mold infection) and (2) Mycotoxicosis, which is a symptom of toxicity caused by the deletion of a toxic metabolic product of mold or fungus [23]. Toxic compounds produced by fungi are called mycotoxins. These toxins can trigger painful symptoms that are sometimes fatal. Some of them even have carcinogenic properties that can cause cancers. The main species of mold that can contaminate food ingredients include *Aspergilus flavus, A. oryzae, A. ochraceus, A. tamarii, Penicillium puberulum, P. Citrinum, P. italicum, P. chrysogenum, P. expansum, A. wentii, Alternaria alternata, A. melleus, A. terreus, dan A. Niger.* These mold species are able to produce toxic substances namely mycotoxins which cause damage to food [22].

Identification of mold on black cassava made from *Manihot esculenta* Crantz has been conducted [10]. Three genuses of mold, *Rhizopus, Aspergilus* and *Penicillium* were discovered. The two genuses, *Apergilus* and *Penicillium* can contaminate food materials. This study aimed to further identify mold species that can grow on black cassava made from *Manihot aipi* Phol., a cassava species that contains high levels of HCN (100 mg HCN/kg of fresh bulbs) [24, 25, 26]. The Ministry of Health of the Republic of Indonesia in 1990 stated that poisonous cassava contains 0.08-0.10 % HCN (80-100 mg HCN/kg), highly poisonous cassava contains more than 0.1% HCN (100 mg/kg). The cyanide acid level of each cassava variety is different and can affect the taste, so the reduction of cyanide acid levels in cassava should be a major concern in the utilization of the cassava (27, 28, 29).

Underlines the importance of mold research and exploration in revealing the potential and benefits of mold for humans [7]. In natural settings, mold has different roles and potentials. Each type of mold has distinctive characteristics. Some types of mold are known to have high economic value because they are needed in industrial activities. The economic potentials of mold include food and medicine, land fertility agents, biopesticides, enzyme and other bioactive basic materials [1]. Mold also plays a key role in nutrition and the process of improving plant health and development [30].

The results of this study are expected to provide information to the public, especially to the people of Banda in Central Maluku Regency, about the type of mold that grows on black cassava and the HCN levels contained in the black cassava. This study provides benefits for the community to improve food security and nutrition at the family level, where alternative choices of quality food at affordable prices are available. In the end, the dependence on rice can be reduced by food diversification [31, 33].

2. METHODS

2.1 Sampling Locations

A purposive sampling technique was employed to select the samples. Sampling was conducted on Banda Island, Central Maluku Regency. The selected *Manihot aipi* Phol. cassava samples were ready to harvest. The cassava tubers were collected. The isolation and purification of the mold were conducted at the basic laboratory of the Department of Biology Education, FKIP, Universitas Pattimura. Meanwhile, the 16S rRNA sequence analysis was performed at *PT. Genetika Science Indonesia.*

2.2 Equipment and Materials

The main materials of this study were 4 kg of black cassava samples made from *Manihot aipi* Phol., salt water (sea water), 1000 ml aquadest, alcohol 70%, 10 L sea water for immersion, ropes, label paper, aluminum foil, cotton and PDA media. The equipment used in the experiment included a scale, a knife, a basin, a sample box, a measuring cup, an erlenmeyer, a glass spatula, a hot plate, an autoclave, tweezers, petri dishes, an ose needle, a reaction tube, an oven, a microscope and a bunsen burner. The materials used for HCN level analysis were aquadest, alcohol, NaOH solution, AgNO3 solution.

2.3 Procedures

2.3.1. Black Cassava Sample Preparation

Manihot aipi Phol. cassava was peeled, washed, and immersed in a basin containing sea water at 12h, 24h, 36h and 48h. After immersion, the cassava was hanged by raffia rope on a house drain. During this process, the cassava must not be exposed to direct sunburn. The cassava was left to dry for 3 weeks until it was overgrown with mold, marked by a change in color from gray to black. When the cassava had changed color, the mold on the black cassava was ready for isolation.

2.3.2. PDA Media Preparation

All equipment and materials such as the Potato Dextrose Agar (PDA) media powder, aquadest, and antibiotics were prepared. The PDA media powder was weighed as much as 5.85 g based on the following formula: v 1 /w 1 = v 2 /w 2 and then transferred into a beaker glass. It was added with 150 ml and poured into an erlenmeyer. The solution was heated and stirred until homogeneous (stirring was stopped before it boiled and after it dissolved completely or until no more crystals were left on the bottom). The solution pH was checked (5.6 ± 0.2) at 25° C. The solution was added with 0.01N NaOH if the pH was acidic, or added with 0.01N HCl if the pH was basic. The medium was sterilized in an autoclave for 15 minutes at 121° C. After sterilization, the medium was left to cool until it reached 50° C and added with chloramphenicol dissolved in 10 ml aquadest (every 100 ml PDA=1 ml of chloramphenicol suspension). The medium was homogenized and poured into petri dishes using an aseptic technique. The medium was left to frezee perfectly and then stored in an incubator.

2.3.3. Mold Isolation

Mycelium from black cassava that had been overgrown with mold was taken and planted on agar plates made from PDA in petri dishes. Each agar medium was added with 0.2 g/L Chloramphenicol to prevent the growth of endophytic bacteria. The medium for mold growth in a petri dish was first sterilized in an autoclave. The sample inoculation was carried out on petri dishes where each dish contained 1 mycelium of mold. The petri dishes were sealed and stored at the room temperature (25°C) for 1-2 X 24 hours. After 2-3 days, mold would grow around the samples that were put on the agar media [9].

2.3.4. Mold Purification

Mycelium of the mold was collected by using sterile ose wire then transferred to other PDA media. This purification process aimed to separate the colonies with different morphologies to be used as isolates. The mycelium was then stored for 3-7 x 24 hours. Mold mycelium was isolated using the streak-plating method and then incubated at the room temperature for 3-5 days [5].

2.3.5. Mold Identification

The isolated and purified mold was identified. The identification of mold was conducted based on the morphological characteristics of each colony, including the color of the colony, the diameter of the colony, the microscopic properties of the colony (i.e. the color of the hypha, the diameter of the hypha, the color of the conidiophor, the diameter of the conidiophor, the color of the conidium and the diameter of the conidium). The mold identification was performed on the observed preparation using the slide culture method [4, 1].

A specimen preparation was provided for microscopic morphology observation using a binocular microscope with a magnification of 400x. Mold inoculums on the media were taken using an ose needle. The media that had been grown with mold were cut to the size of about half a centimeter and placed on a sterile glass object. The glass object was sealed with a glass cover and then pressed slowly. After that, observations were made. To identify the mold, the results of the observations were compared with the relevant literature, including Introduction to Food and Airborne-Fungi by [14] and Introduction to Common Tropical Mold by [22]

2.3.6. HCN Level Analysis

The analysis of the HCN levels in black cassava was conducted before and after the immersion and drying processes using the following quantitative testing procedures: (1) a small sample of flour (10-20 gr) was added into 100 ml aquadest in an erlenmeyer and left for two hours. didiamkan selama 2 jam; (2) the suspension was added with 100 ml aquadest and separated using steam distillation. The distillate was collected in an erlenmeyer that had been filled with 20 ml NaOH 2.5%. The distillation process was stopped after the distillate reached a volume of 150 ml. The distillate was then added with 5 ml KI 5% and 8 ml NH4OH. The distillate compounds were titrated with 0.02 N AgNO3 solution until it went turbid. The HCN level was measured using the following formula [15]:

$$\mathrm{HCN} = \mathrm{ml}\ \frac{(AgNO3\ x\ 0{,}54)}{Berat\ Bahan}\ \mathrm{x}\ 1.000\ \mathrm{mg/kg}$$

3. Results and Discussion

3.1. Mold Identification

Based on the macroscopic and microscopic morphological analyses, five species of mold with codes according to the time of immersion were identified. They were respectively Isolate 1 at 12h immersion, Isolate 2 at 24h immersion, Isolate 3 at 36h immersion and Isolate 4 at 48h immersion. All of which were repeated five times. Three species of mold were obtained at 12h immersion, 4 species of mold were obtained at 24h immersion, 4 species of mold were obtained at 36h and 5 species of mold were obtained at 48h immersion.

The results of the morphological observations (hyphal color, colony texture, topography, radial line, concentric line and spore shape) revealed that there were 5 (five) types of mold that grew predominantly on *Manihot aipi* (black cassava) such as presented in figure 1.

No.	Hyphal color	Colony texture	Topography	Radial line	Concentric Line
1.	White	*Cottony*	*Rugose*	-	-
2.	Black	*Cottony*	*Rugose*	-	-
3.	Gray	*Glabrous or Waxy*	*Umbunate*	-	Yes
4.	Dark green	*Velvety*	*Umbunate*	-	-
5.	Light green	*Velvety*	*Umbunate*	-	Yes

Figure 1. The Results of the Mold Morphological Observations

Based on figure 1, the identified mold possessed the following characteristics:

Identification of Isolate J1

Based on the results of the colony morphological and microscopic analyses, Isolate 1's colony was white in color and cottony in texture (the colony had long and thick hyphae similar to cotton). The colony had a rugose topography (the furrows had an irregular height which appeared to be a radial line from the reverse side). The colony had no radial line that resembled the colony fingers and no concentric line (circles created in a colony).

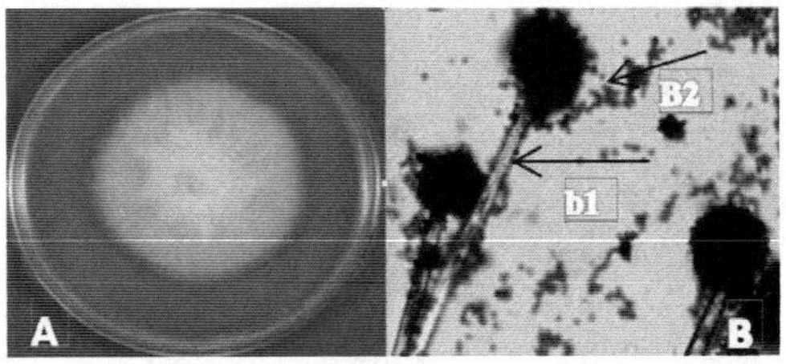

Figure 2. A. Isolate J1, B. Microscopic Shape: b1.Hypha, b2. Spore.

Identification of Isolate J2

Based on the results of the colony morphological and microscopic analyses, Isolate 2's colony was black in color and cottony in texture (the colony had long and thick hyphae similar to cotton). The J2 colony had a rugose topography (the furrows had an irregular height which appeared to be a radial line from the reverse side). The colony had no radial line that resembled the colony fingers and no concentric line (circles created in a colony).

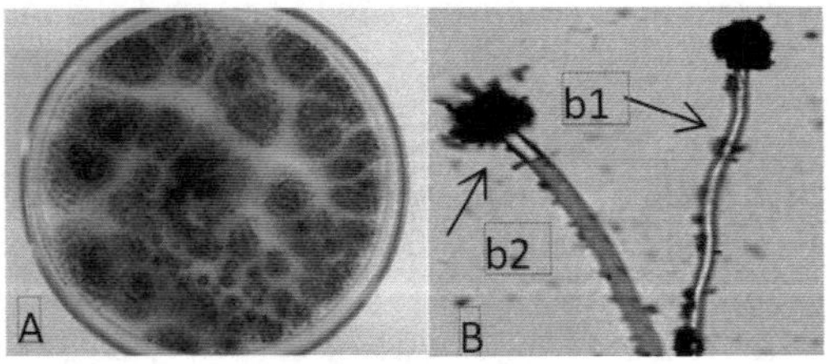

Figure 3. A. Isolate J2, B. Microscopic Shape: b1.Hypha, b2. Spore.

Identification of Isolate J3

Unlike Isolate J1 and J2, Isolate J3's colony was gray in color and glabrous and waxy in texture (the colony had a smooth surface; no aerial hypha was found). The J3 colony had an umbunate topography (a protrusion in the middle of the colony resembled a button). It had no radial line that resembled the colony fingers, but it had concentric lines that were created inside the colony.

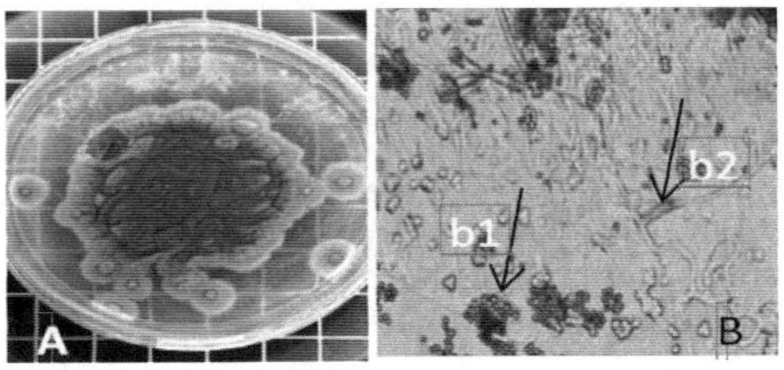

Figure 4. A. Isolate J3, B. Microscopic Shape: b1.Hypha, b2. Spore.

Identification of Isolate J4

Isolate J4's colony was dark green in color and velvety in texture (the colony had short aerial hyphae resembling velvet). The J4 colony had an umbunate topography (a protrusion in the middle of the colony resembled a button). Often This colony also had radial line furrows. The colony had no radial line that resembled the colony fingers and no concentric line (circles created in a colony).

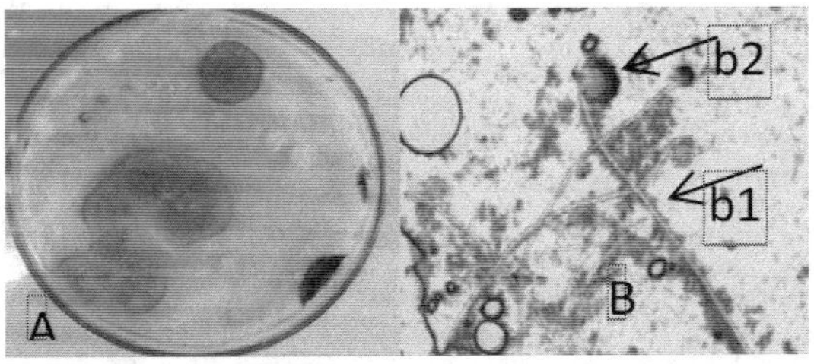

Figure 5. A, Isolate J4, B, Microscopic Shape: b1.Hypha, b2. Spore.

Identification of Isolate J5

Isolate J5's colony was light green in color and velvety in texture (the colony had short aerial hyphae resembling velvet). The J5 colony had an umbunate topography (a protrusion in the middle of the colony resembled a button). This colony also had radial line furrows. It had no radial line that resembled the colony fingers, but it had concentric lines that were created inside the colony.

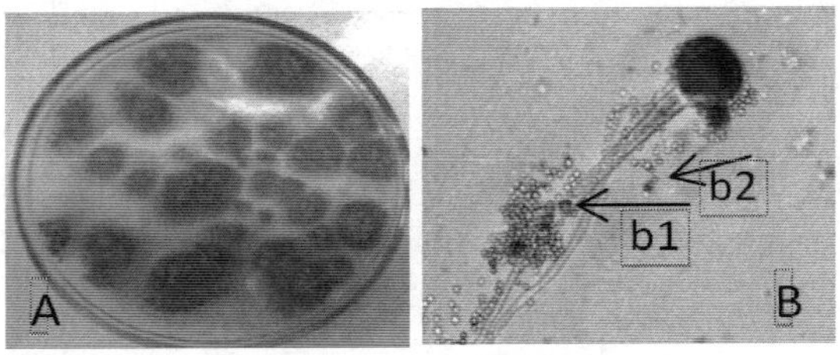

Figure 6. A. Isolate J5, B. Microscopic Shape: b1.Hypha, b2. Spore.

Based on the results of the mold morphological observations, a specimen preparation was done for 16S rRNA sequence analysis. The identification of the mold species based on the immersion and drying duration was presented in figure 7.

Isolate Code	Species Name	Immersion Duration			
		12 h	24 h	36 h	48 h
1.	*Rhizopus oryzae,*	✓	✓	✓	✓
2.	*Rhizopus stolonifer*	✓	✓	✓	✓
3.	*Penicillium sclerotigenum*	✓	✓	✓	✓
4.	*Aspergillus flavus*		✓	✓	✓
5.	*Aspergilus clavatus*				✓

Figure 7. Mold Species that Flourish on Black Cassava

Figure 7 showed that three mold isolates were found at 12h immersion, four isolates were obtained at 24h immersion, four isolates were found at 36h immersion and 5 isolates were obtained from 48h immersion.

A study has identified the genuses of mold found in the black cassava produced from *Manihot esculenta* Crantz [10]. They classified the mold into *Rhizopus, Aspergillus*, and *Penicillium*. This study, however, aimed to identify mold species in the black cassava fermented from *Manihot aipi* Phol using 16S rRNA sequence analysis. The results showed that the five isolates found in this study could be identified as *Rhizopus oryzae, Rhizopus stolonifer, Penicillium sclerotigenum, Aspergillus flavus* and *Aspergilus clavatus. Rhizopus oryzae*, *Rhizopus oryzae* and *Penicillium notatum* appear to be non-pathogenic to human [2], while *Aspergillus flavus* and *Aspergillus clavatus* are categorized as pathogenic mold to human body [16]. *Aspergillus flavus* and *Rhizopus oryzae* are mostly used in *tempeh* processing and are safe for consumption due to its zero toxicity and ability to produce lactic acids [17, 19].

Penicillium chrysogenum used to be known as *Penicillium notatum*. This species is rarely reported as a cause of disease in humans. This species is the source of several β-lactam antibiotics, especially penicillin [2]. On the other hand, *Aspergillus flavus* and *Aspergillus clavatus* produce aflatoxin which can cause cancer in humans. Both types of mold are part of the *Aspergillus* group that is well known for its role as a pathogen in plants and its ability to produce aflatoxin in infected plants. In addition, both species are also toxin producers from the *Aspergillus* group that can contaminate agricultural products [16]. These mold species cannot survive at 60°C if heated for 5-10 minutes [6, 32].

3.2. HCN Levels in the Black Cassava Samples

The HCN levels in the black cassava samples were measured before and after the immersion in salt water and fermentation by drying (Figure 8 and Figure 9).

No.	Sample	Duration	HCN (%)
1.	Control		0.3024
3.	Immersion in salt water	12 h	0.1296
		24 h	0.135
		36 h	0.162
		48 h	0.1782
2.	After fermentation helped by mold	12 h	0.135
		24 h	0.108
		36 h	0.0918
		48 h	0.0864

Figure 8. The HCN Levels Dissolved before and after Immersion and Fermentation

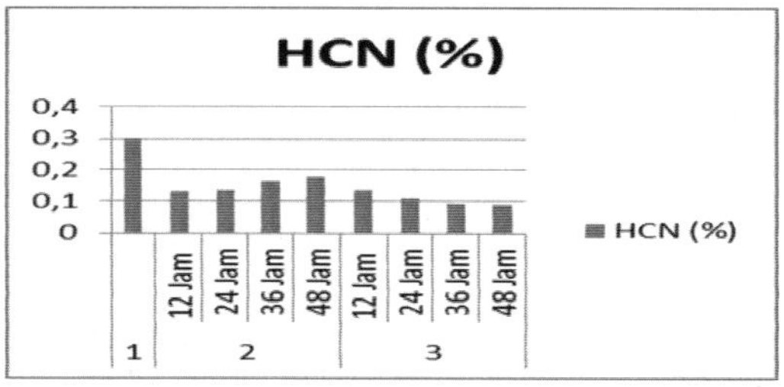

Notes: 1. Control, 2. Immersion, 3. Fermentation

Figure 9. HCN Levels

Based on the results of identification, it was obvious that mold played an important role in decreasing the HCN levels in cassava. This is consistent with the results of the research conducted [20], stating that bitter cassava which was fermented using *Rhizopus* had reduced levels of HCN. This is also in line with the results of the study in Table 3, where there was a decrease in the HCN levels in the black cassava samples in line with the immersion duration. In fact, the HCN levels in *Manihot aipi* decreased significantly from 0.3024% (302.4 mg HCN/kg) to 0.1296 (129.6 mg HCN/kg), 0.135 % (135 mg HCN/kg), 0.162% (162 mg HCN/kg), and 0.1782 (178.2 mg HCN/kg) after 12, 24, 36, and 48-hour immersion, respectively. Similarly, at 12h, 24h, 36h, and 48h fermentation duration, the HCN levels in black cassava were reduced from 0.135% (135 mg HCN/kg) to 0.108% (108 mg HCN/kg), 0.0918% (91.8 mg HCN/kg), and 0.0864% (86.4 mg HCN/kg). However, compared to the threshold for HCN recommended by The Ministry of Health of the Republic of Indonesia (1990) (0.08-0.10 % or 80-100 mg HCN/kg), these figures are considered relatively high.

The HCN levels in the black cassava after fermentation process (drying by hanging) experienced a decrease in line with the duration of the process (the longer the process, the lower the HCN level). For example, at 12h, 24h, 36h, and 48h, the HCN levels were reduced to 135 mg HCN/kg, 108 mg

HCN/kg, 91.8 mg HCN/kg, 86.4 mg HCN/kg. This finding explains that although they are still in the toxic category, the HCN levels in the black cassava had experienced a significant decrease. So, the immersion and fermentation processes indeed had an effect on reducing the levels of HCN dissolved in the black cassava. The HCN levels in the black cassava could be decreased because the HCN in the form of glycoside bonds in the cassava was hydrolyzed and decomposed into glucose, acetone and HCN.

The mold species that grew on the black cassava also had an effect on the HCN levels. Used *Rhizopus oryzae* to lower the HCN levels in four species of bitter cassava and the results demonstrated the ability of the mold to remove cyanide from bitter cassava [11]. Another research finding also showed that fermentation using *Rhizopus stolonifer* strain LAU 07 was able to reduce the HCN levels in cassava substrates. Furthermore argues that mold is one of the microorganisms that play a key role in degradation [21].

Black cassava is a traditional food from Banda that has been long consumed from generation to generation. All this time, processing cassava in this way has had no deadly effect. This is in line with the results of the studies described above where immersion, hanging, and fermentation duration as well as mold species have important roles in reducing the HCN levels in cassava. Therefore, the traditional processing of black cassava by the Banda community is quite effective in removing the poison from *Manihot aipi* Phol.

To avoid poisoning due to mycotoxins produced by mold, black cassava should be soaked for 48 hours and hung for a maximum of 2-3 weeks, then heated to temperatures above 60°C. In this way, the spores on the black cassava will die. Therefore, *sunu-sunu* (Banda's traditional food) prepared from black cassava is safe for consumption.

4. Conclusion

In this study, five species of mold in black cassava were successfully identified. They were *Rhizopus, Aspergillus*, *Penicillium*, *Aspergillus flavus* and *Aspergillus clavatus.* Mold species played a role in reducing the HCN levels in black cassava. Future research may wish to consider analyzing the HCN levels in boiled black cassava.

5. Acknowledgment

Authors wishing to acknowledge assistance or encouragement from colleagues, special work by technical staff or financial support from DIPA DIKTI Universitas Pattimura for the funding and kind assistance they have provided so that this research could be carried out successfully.

References

[1] Alsohaili, S. A. and Bayan M. B.H 2018 Morphological and Molecular Identification of Fungi Isolated from Different Environmental Sources in the Northern Eastern Desert of Jordan. *Jordan Journal of Biological Sciences*. **11(3),** *329 – 337.*

[2] Asan, A 2004 *Aspergillus,Penicillium* and Related Species Reported from Turkey. *Mycotaxon*. **89 (1),** 155 – 157.

[3] Puwastien P 2002 Issues in the development and use of food composition databases. Public Health Nutrition: 5(6A), 991–999

[4] Fitriarni, D. Kasiamdari, R S. 2018 Isolation and Identification of Endophytic Fungi from Leave and Stem of *Calopogonium mucunoides*. *J. Trop. Biodiv. Biotech*. **3,** 30-36.

[5] Sanders E. R 2012 Aseptic Laboratory Techniques: Plating Methods. Journa of Visualized Experiments. (**63**), e3064.

[6] Kumar D, Barad S, Yongchen, Xingyuluo, Tannous J, Dubey A, Matana N. G, ,Shipingtian, Boqiang Li, Keller N, And Prusky D 2017 LaeA regulation of secondary metabolism

modulates virulence in Penicillium expansum and is mediated by sucrose. Molecular Plant Pathology **18**(8), 1150–1163

[7] Ilyas. M 2007 Isolation and Identification of Fungal Mycophlora in Plant Leaf Litter Samples in the Gunung Lawu Area, Surakarta, Central Java. *Biodiversitas.* **8** (2).

[8] Leplepem 2005 Analysis of carbohydrate levels and identification of types of mushrooms that are thought to neutralize toxic compounds in the seeds of the hajj fern (Cycas rumphii Miq) traditionally by the people of the Aru Islands. Thesis is not published. FMIPA. Ambon

[9] Milon M.A. Abdul Muhid, md, Durajan, Goshwami, Mehedi Masud, Begum, B Md 2012 Antioxidant, Cytotoxic and Antimicrobial Activity 0f Sonneratia Alba Bark. International *Journal Pharmaceutical Sciences and Research.* **3** (7).

[10] Pattipeilohy, M dan Radjulan 2012 Isolation and Identification of Isolation Types of Fungi on Cassava (*Manihot esculenta Crantz*) in the Traditional Process of Making Black Cassava by the Banda People. Proceedings of the National Seminar on Biology and Biology of Mathematics and Natural Sciences. Yogyakarta State University.

[11] Akindahunsi A.A., Oboh G., Oshodi A.A 1999 Effect Of Fermenting Cassava with Rhizopus Oryzae On The Chemical Compositionof Its Flour And Gari Products. United Nations Educational Scientific and Cultural OrganizationandInternational Atomic Energy Agenc

[12] Basha S, Ogbaghebrie A, Yemane K, Zenebe M 2012 Isolation and screening of endophytic fungi from Eritrean traditional medicinal plant Terminalia brownii leaves for antimicrobial activityn.. Volume 6 Issue 5 ISSN 0973-8258

[13] Miguel Ángel G. G 2012 Presence Of Aspergillusand Other Fungal Symbiontsin Coffee Beans From Colombia. Acta biol. Colomb., Vol. **17** no 1, 2012 39 - 5

[14] Samson. R, Ellen Hoekstra, and Jens Frisvad 2004 Introduction to Food and Airborne Fungi. Ed 7th. Centraalbureau voor schimmelcultures. *Utrecht.*

[15] Nebiyu A and Getachew E 2011 Soaking and drying of cassava roots reduced cyanogenic potential of three cassava varieties at Jimma, Southwest Ethiopia. African Journal of Biotechnology Vol. **10**(62), pp. 13465-13469.

[16] Wang, J. Huang, Y, Fang, M, Zhang,Y, Zheng, Z, Zhao, Y, Su, W. 2006 Brefeldin A, a cytotoxin produced by *Paecilomyces* sp. and *Aspergillus clavatus* isolated from *Taxus mairei* and *Torreya grandis. Pathogens and Disease.volume **34** issue.*

[17] Battaglia E, Benoit I, Joost van den Brink, Wiebenga A, Coutinho P. M, Henrissat B, de Vries R. P 2011 Carbohydrate-active enzymes from thezygomycete fungusRhizopus oryzae: a highlyspecialized approach to carbohydratedegradation depicted at genome level. BMC Genomics,**12**:38

[19] Yuji Oda, Katsuichi Saito, Hiroaki Yamauchi & Motoyuki Mori 2002 Lactic Acid Fermentation of Potato Pulp by the Fungus *Rhizopus oryzae.* Current Microbiology. volume **45**, pages1–4.

[20] Prasetyo T., Ardhianto F., and Pawitra M, Sumardiono S 2018 Improvement the Chemical Properties of Cassava Through Microbial Fermentation Using Rhizopus Oligosporus. MATEC Web of Conferences 156, 01025 https://doi.org/10.1051/matecconf/201815601025

[21] Machida M, Yamada O, and Gomi K 2008 Genomics of Aspergillus oryzae: Learning from the History of Koji Moldand Exploration of Its Future. DNA Research **15**, 173–183

[22] Mahendra Rai 2009 *Applied Mycology* First Edition. Publisher : CABI. ISBN-**10** : 9781845935344

[23] Tournas, V et.al 2001 Yeast, Molds, and Mycotoxins. Washington, D.C: US Food Administration Center for Safety & Applied Nutrition.

[24] Apeh Akwu Omede, Emmanuel Uchenna Ahiwe, Ze Yuan Zhu, FidelisFru-Nji and Paul Ade Iji 2018 Improving Cassava Quality for Poultry Feeding Through Application of Biotechnology. Chapter · January 2018

[25] Okoli. I C., Okparaocha. C.O., Chinweze C. E and Udedibie 2012 Physicochemical and Hydrogen Cyanide Content of Three ProcessedCassava Products Used for Feeding Poultry in Nigeria. Asian Journal of Animal and Veterinary Advances 7(4) 334-340

[26] Amarachi D. Uchechukwu A & Oluwafemi J. Calebn & Opara U L 2015 Postharvest Handling and Storage of Fresh Cassava Rootand Products: a Review. Food Bioprocess Technol 8:729–748

[27] Adamafio A, Sakyiamah M and Josephyne 2010 Fermentation in cassava (Manihot esculentaCrantz) pulp juice improves nutritive value of cassava peelNaa. African Journal of Biochemistry Research Vol. **4**(3), pp. 51-56.

[28] Cliff, J., Muquingue, H., Nzwalo, H. and Bradbury, J.H 2011 Konzo and continuingcyanide intoxication from cassava in Mozambique. Workshop on toxico-nutritional neurodegenerations, IPBO–Ugent, Belgium. FoodChem. Toxicol. **49**, 631–635.

[29] Barrett, B 2007 Konzo, cassava foodborne paralysis-Cameroon: refugees. CCDNNews, No **10**, p. 2.

[30] Mulani K R.M., Shahim S 2013 Diversity of Rhizosphere Fungi and Soil Nutrient Properties of Amorphophallus sylvaticus (Roxb.) International Journal of Science and Research (IJSR) ISSN (Online): 2319-7064

[31] Imelda, Kusrini N, Hidayat R 2017Development Strategy 0f Local Food DiversificationI. Journal of Economics and Policy. Jejak Vol**10**(1):62-79

[32] Ralph E. Moo 2006 High Temperatueffects On Micro. Cleaning & Restoratoin. The Official Publication of ASCR International. Vol. **43** No. 5

[33] Mlingi N.V. and Ndunguru G.T. 2007 A review of post-harvest activities for cassava in Tanzania. Proceedings of the 13th ISTRC Symposium,pp. 506 – 513.

The Effects of Some Additives On Characteristics RVA Profile of Cassava Flour

H Herawati* and E. Kamsiati

Indonesia Center For Agricultural Postharvest Research and Development
Jl. Tentara Pelajar No 12, Cimanggu-Bogor, Indonesia
Email: herawati_heny@yahoo.com

Abstract. The use of wheat flour as a raw material in the manufacture of other food products will cause dependence on imports of wheat flour in Indonesia. One of the opportunities to suppress wheat imports is to develop local flour, including cassava flour. To be applied further, one of the characteristic parameters that can be observed from local flour is its viscosity profile. This study object to determine the effect of several additives on the RVA (Rapid Visco Analysis) profile of cassava flour. Several modifications were made by adding several types of additives, including: salt, sugar, developer, emulsifier and various hydrocolloids using various concentrations of 0.5 to 4%. Based on the results of the analysis, the presence of types and variations in concentration resulted in varying characteristics based on the results of the RVA analysis. With increasing levels of guar gum and xanthan gum, the peak viscosity increases. However, it is different with the addition of concentration in the developer and malto dextrin forms a peak at the concentration of additives added. These results can be used further as consideration for the formulation process and application of further processed products.

1. Introductions

The use of wheat flour as a raw material in the manufacture of other food products will cause dependence on imports of wheat flour in Indonesia. One of the opportunities to suppress wheat imports is to develop local flour, including cassava flour. One of the potential flours that can be developed and does not require specific planting habitat criteria with uncomplicated environmental conditions is cassava.

Cassava flour is widely used into several processed products. Some of the processed cassava products include making bread [1-3], noodles [4-6], and other processed products. In the implementation of some of these products, it is necessary to mix some additional ingredients to improve the quality of the resulting product. The addition of these ingredients include sugar, salt, eggs, emulsifiers, hydrocolloids and other additives.

The addition of some of these additional ingredients will affect the characteristics of the dough and the resulting final product. The type and concentration of these food additives varies greatly depending on the characteristics of the final quality produced. Some researchers add certain types and concentrations to the dough such as 1% monoglyceride [7]; addition of eggs to spaghetti, noodles and pasta [8-10] in order to improve elasticity and reduce cooking loss.

To be applied further, one of the characteristic parameters that can be observed from local flour is its viscosity profile. This study object to determine the effect of several additives on the RVA (Rapid Visco Analysis) profile of cassava flour.

2. Materials and Methods

The material used cassava flour with Prodes Brand. Other additional ingredients in the form of: salt, sugar, baking powder, and eggs obtained from commercial markets in Bogor. While the cellulose material from the BTP factory. Sodium alginate, maltodextrin and CMC from Setia Guna Store, Bogor.

The tools used are the Digestor Stove HYP-1008, the KDN-103F Automatic Nitrogen Determinator distillation device (Shanghai, Qianjian Instrument), the Fat Determinator and Soxtech SZC-D (Shanghai, Qianjian Instrument) tool, stirring rod, canister, cup. aluminum, porcelain cup, desiccator, fat-free cotton, analytical balance (Precisa, Swissmade), Memmert UF 110 Open Lab 108 oven, cup clamp, 10.00 mL measuring pipette, paddle, RVA Tecmaster (Perten Instrument), spoon, spatula, test tube large, furnace (Lenton Furnace).

Several modifications were made by adding several types of additives, including: salt, sugar, developer, emulsifier and various hydrocolloids using various concentrations of 0.5 to 4%. The local flour is uniform in size, namely 100 mesh and added with water and food additives and physical characterization is carried out by performing RVA analysis and proximate analysis. The analysis method for proximate was carried out according to the AOAC method ([11] and the RVA profile analysis method according to the method described in Herawati et al. [4].

3. Results and Discussions

To determine the characteristics of cassava flour as the main raw material, a proximate analysis of cassava flour was carried out. Based on the results of the proximate analysis which includes water, ash, fat, protein and carbohydrate content of cassava flour, the results are as shown in Table 1 below.

Table. Proximate Analysis Results of Cassava Flour

No	Analysis	Content (%)
1	Moisture	9,80
2	Ash	1,98
3	Fat	0,73
4	Protein	2,10
5	Carbohydrate	85,39

Cassava flour contains lower levels of fat and protein than wheat flour. The gelatinization profile and viscosity of cassava flour are quite different from other local flours. It is possible that the proximate content contained in cassava flour is different from other local flours.

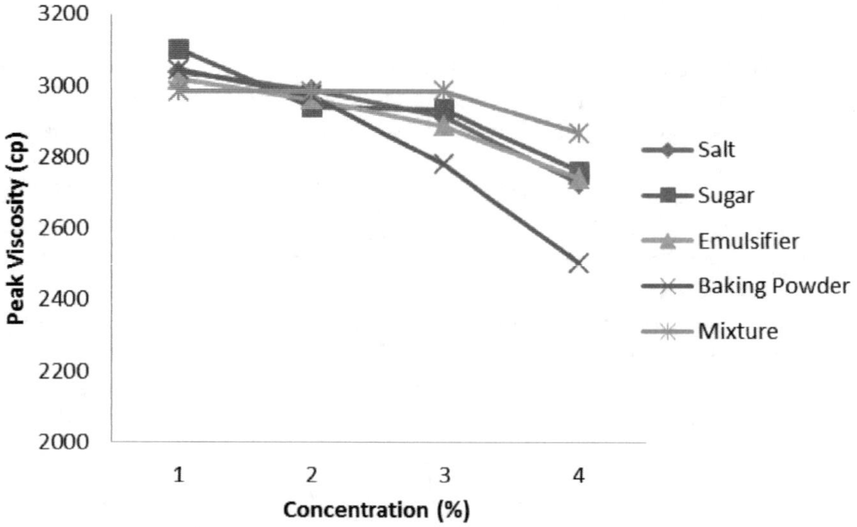

Figure 1. The influences of several additives to the peak viscosity of cassava flour

The addition of ingredients such as salt, sugar, emulsifier, baking powder and a mixture of all of them in cassava flour in the concentration range of 1-4% indicates that the peak viscosity of the dough is decreasing. Based on the literature, the peak viscosity of cassava flour was 1200–1440 cP, and cassava starch ranged from 2400–3000 cP [12]. This shows that with the addition of additional ingredients, there is an increase or decrease in the peak viscosity of cassava flour.

The presence of a high peak viscosity will indicate that starch or its additive material can expand more [13]. However, it must be noted that the possibility of too high a sedback can also cause the dough to stall. . In the final process, the viscosity indicates that the paste or gel affects the cooking and cooling process of the product. While the retrogradation of starch can be observed from the calculation results of the sedback viscosity [14].

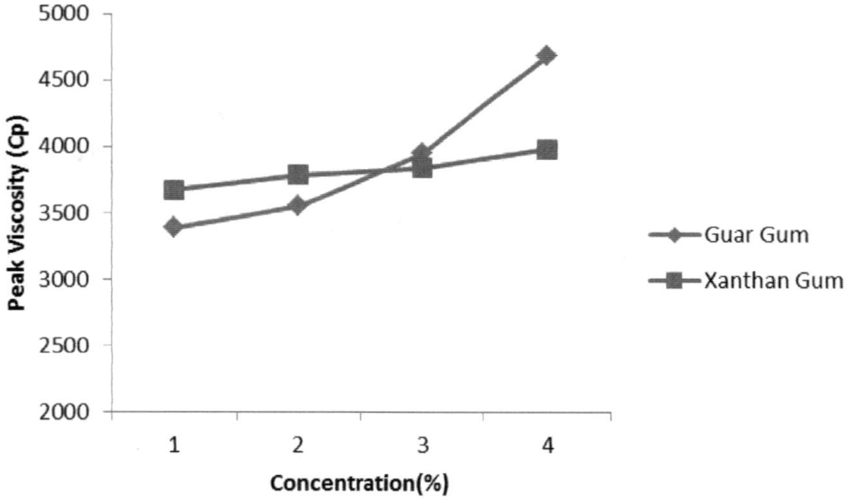

Figure 2. The influences of gums to the peak viscosity of cassava flour

The addition of gum in the form of guar gum and xanthan gum in a concentration range of 1-4% indicates an increasing peak viscosity. The increase in the peak viscosity profile from the addition of the gum group is possible because gum is a hydrocolloid group and can increase the peak viscosity of cassava flour.

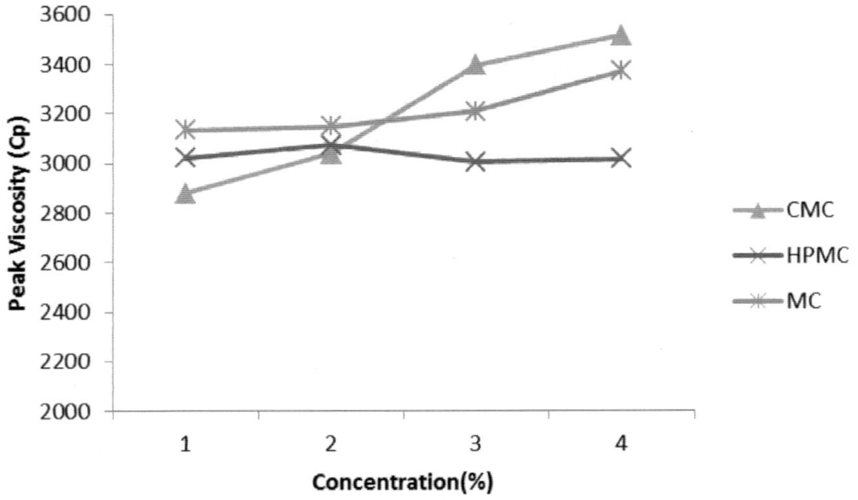

Figure 3. The influences of derivatives cellulose to the peak viscosity of cassava flour

The addition of food additives from the cellulose group relatively increased the peak viscosity profile of the cassava flour dough. This is possible, the hydrocolloid component of the cellulose group can synergize with the starch component contained in the cassava flour mixture and increase the peak viscosity produced.

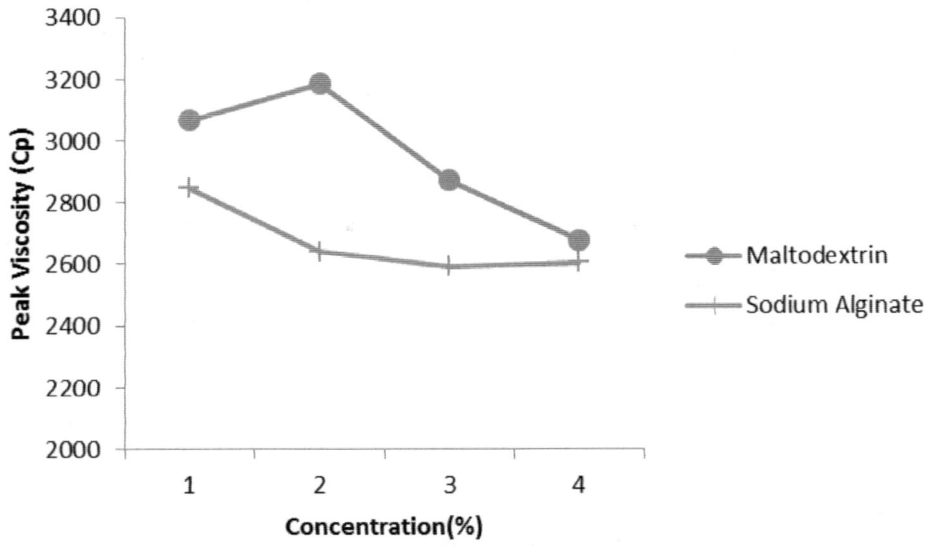

Figure 4. The influences maltodextrin and sodium alginate to the peak viscosity of cassava flour

Meanwhile, the addition of maltodextrin and sodium alginate actually decreased the peak viscosity of the cassava flour dough. Maltodextrin and sodium alginate may not be able to synergize as in the gum and cellulose groups. The interaction between these components is a consideration for further implementation.

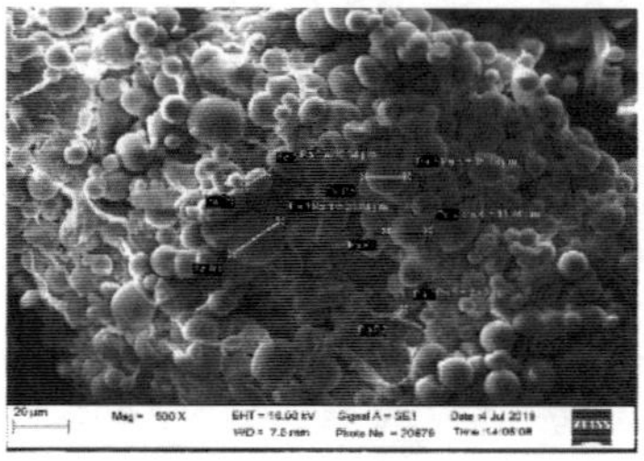

Figure 5. SEM analysis result of cassava flour

The microstructural profile of cassava is generally round. The granules that were still intact indicated that the starch had not undergone the gelatinization process. This is of course because cassava starch has not been subjected to further heating processes. Based on the literature, cassava starch granules are spherical or dome shaped [15]. Some literature mentions the possibility of a physical process so that the granules are close to each other.

4. Conclusion

Based on the results of the analysis, the presence of types and variations in concentration resulted in varying characteristics based on the results of the RVA analysis. With increasing levels of guar gum and xanthan gum, the peak viscosity increases. However, it is different with the addition of concentration in the developer and malto dextrin forms a peak at the concentration of additives added. These results can be used further as consideration for the formulation process and application of further processed products.

References

[1] A Pasqualone, F Caponio1, C Summo, VM Paradiso, G Bottega, MA Pagani. 2010. Gluten-Free Bread Making Trials from Cassava (Manihot Esculenta Crantz) Flour and Sensory Evaluation of the Final Product. International Journal of Food Properties, 13: 562–573.

[2] MA Ammar, MM Abd El Razik. 2013. Quality Charaxteristics of Gluten Free Produced From Cassava, Pumkin and Potato Flours. J. Food and Dairy Sci., Mansoura Univ., Vol. 4 (8): 401 – 412.

[3] Y Gao, ME Janes, B Chaiya, MA Brennan, CS Brennan, W Prinyawiwatkul. 2017. Gluten-free bakery and pasta products: prevalence and quality improvement. International Journal of Food Science and Technology 2017.

[4] H Herawati, E Kamsiati1, Sunarmani. 2021. Formulation of Food Ingredients (Peanut Flour, Egg Yolks, Egg Whites, and Guar Gum) to the Characteristics of Gluten-Free Noodles. International Journal of Technology 12(3) 602-612.

[5] Purwandari, U., Hidayati, D., Tamam, B., Arifin, S., 2014a. Gluten-free Noodle Made from Gathotan (an Indonesian Fungal Fermented Cassava) Flour: Cooking Quality, Textural, and Sensory Properties. *International Food Research Journal*, Volume 21(4), pp. 1615– 1621.

[6] Omeire, G.C., Kabuo, N.O., Nwosu, J.N., Peter-Ikechukwu, A., Nwosu, M.O., 2015. Enrichment of Wheat/Cassava Noodles with Partially Defatted Protein-Rich Flours. *IOSR Journal of Environmental Science, Toxicology and Food Technology*, Volume 9(5), pp. 121–125.

[7] Padalino, L., Mastromatteo, M., Sepielli, G., Del Nobile, M.A., 2011. Formulation Optimization of Gluten-Free Functional Spaghetti based on Maize Flour and Oat Bran Enriched in B-Glucans. *Materials*, Volume 4(12), pp. 2119–2135.

[8] Garcia, L.G.C., Silva, A.H.S., Cunha, P.D., Damiani, C., 2016. Preparation of Gluten-Free Noodles Incorporated of Jabuticaba Peel Flour. *Journal of Food and Nutrition Research*, Volume 4(2), pp. 82–87

[9] Schoenlechner, R., Wendner, M., Siebenhandl-Ehn, S., Berghofer, E., 2010. Pseudocereals as Alternative Sources for High Folate Content in Staple Foods. *Journal of Cereal Science*, Volume 52(3), pp. 475–479

[10] Matsuo, R.R., Bradley, J.W., Irvine, G.N., 1972. Effect of Protein Content on the Cooking Quality of Spaghetti. *Cereal Chemistry*, Volume 49, pp. 707–711.

[11] *AOAC*, 2005. Official Methods of Analysis of AOAC International. AOAC International, Gaithersburg, Maryland, USA.

[12] Charoenkul, N., Uttapap, D.W., Pathipanawat, W., Takeda, Y., 2011. Physicochemical Characteristics of Starches and Flours from Cassava Varieties Having Different Cooking Root Texture. *LWT-Food Science and Technology*, Volume 44(8), pp. 1774–1781.

[13] Fari, M.J.M., Rajapaksa, D., Ranaweera, K.K.D.S., 2011. Quality Characteristics of Noodles Made from Selected Varieties of Sri Lankan Rice with Different Physicochemical Characteristics. *Journal of the National Science Foundation of Sri Lanka*, Volume 39(1), pp. 53–60

[14] Afifah, N., Ratnawati, L., 2017. Quality Assessment of Dry Noodles Made from Blend of Mocaf Flour, Rice Flour and Corn *Flour. In:* IOP Conference Series: Earth and Environmental Science, Volume 101, The International Conference on Natural Products and Bioresource Science 2017 (ICONPROBIOS 2017) 23–24 October 2017, Balai Kartini Convention Center, Jakarta, Indonesia.

[15] Akuzawa, S., Okada, N., Tamaki, Y., Ikegami, A., Fujita, N., Vilpoux, O., Cereda, M.P., Jane, J., 2012. Physicochemical Properties of Starches Isolated from Five Cassava (*Manihot esculenta Crantz*) Landraces of Brazil. *Journal of Applied Glycoscience*, Volume 59(3), pp. 131–138.

IOP Conf. Series: Earth and Environmental Science 1041 (2022) 012018 doi:10.1088/1755-1315/1041/1/012018

A comprehensive review on power tillers usage – bottlenecks and prospects in India

Krishnadas R, Renganathan R

SASTRA Deemed University, Tamil Nadu, India

*r.krishnadas@sastra.ac.in

Abstract. Agriculture mechanization is essential to increasing agricultural productivity and lowering production costs via timely agricultural operations.. The vast majority of farmers in India (86 percent) are small and marginal. Due to their limited financial resources, they are unable to acquire heavy machinery like tractors and tractor-drawn implements. Power tillers, on the other hand, are less expensive and more adaptable tools that may be used for a wide range of farm tasks. Since power tillers are more suited to small-scale farms than tractors, they have been a key component of the plan. Plows and dry-tillage machines, as well as weeding and intercultural tasks, pesticide application (such as spraying and dusting), crop harvesting (such as thinning), and post-harvest handling (such as bagging) may all be accomplished with the help of power tillers. It is easy to maintain, and the operator does not require any special training to fix it on the spot. A variety of external attachments are available, or can be constructed, depending on the nature of the task at hand for the purposes of creating a realistic picture of India's power tiller market and examining previous studies on the topic, this research was carried out to determine the current state of the power tiller in India and to determine the potential for growth in the future as well as any obstacles that may stand in the way of widespread adoption. The study shows that a power tiller can accomplish agricultural chores in a timely way, with higher production and less drudgery, thanks to its adaptability.

1. Introduction

Indian agriculture is dominated by small- and medium-sized farmers. Agricultural power, which encompasses a variety of inputs including seeds, irrigation, and fertilizer, has a significant impact on farm productivity. There are also concerns about accessibility, viability, and sustainability on small and disadvantaged farms due to the high traction costs. Despite abundant water, good land, and a temperate climate, India's agricultural output is disappointingly poor. Two to three times the overall productivity may be achieved by using precision agriculture and automation (Tiwari et al., 2019a).

Farmers will be able to harvest three harvests a year as a result of this. It is essential to ensure timely operation and appropriate farm power supply in order to maximize production and yield (Singh and Sahni, 2019). Small and marginal landowners (those with a plot size of less than two hectares, or 2.0 ha) account for around forty-seven percent of India's total land area (DoACFW, 2018).

To increase agricultural output and reduce production costs, agricultural mechanization is a significant component. Because of the increased efficiency with which important inputs are managed, the yield of crops and the cash generated by farms are increased. As analysts point out, the progress of agricultural technology over conventional human- and animal-drawn activities has served as a barometer for the country's rural economy (Government of India, 2018a).

Farmers' incomes have a significant impact on the demand for agricultural machinery. (Morehouse, 1982; Lal & Singh, 2016). Land fragmentation in India has stifled the country's mechanization progress. Individual ownership of agricultural machinery is becoming increasingly challenging as the average holding size decreases from 2.28 ha in 1970-71 to 1.08 ha in 2015-16. Against the so-called "scale economies," tiny and non-contiguous groupings of small farms are automated for processes such as land production and harvesting, which are often done by hand (Mehta, Chandel, & Senthilkumar, 2014).

2. Power Tillers in India

As of 2020–21, India's agricultural industry would account for only 19.9% of the country's GDP, down from 52% in 1950–51. (Subramanian, 2021). The agriculture industry employs more than half of the country's workers; therefore, the GDP drop has not been matched by a decrease in employment possibilities. In 2016-17, the country's total agricultural power supply was 2.24 kW/ha, with animate power accounting for just 0.2 kW/ha (9.9%). (Mehta. et al., 2019). Furthermore, it is predicted that the percentage of animate power would fall to 4.1 percent in 2032–33 (Tiwari et al., 2019b), indicating that agricultural practices in the country are shifting towards greater reliance on machines due to advancements in the tractor, power tiller, and diesel engines.

The size of a farm's land, the availability of an irrigation water supply, and the availability of bank financing all have a role in whether or not a farmer can afford to buy machinery. Machinery in agriculture saves seeds and fertilizers (15–20 percent), labor and operational time (20–30 percent), and boosts crop production (10–15 percent) and cropping intensity (5–20 percent), consequently increasing profitability (Tiwari et al., 2017; Sahni et al., 2018).

In the coastal, hilly, and mountainous regions of the nation, mechanical farming methods are the least common. While water-scarce regions have the largest purchasing power for water lifting machines as well as irrigation-related hardware and tractors, the irrigated region has the highest purchasing power for power tillers and self-propelled equipment (Sarkar, 2020). Small and marginal farmers make up the vast majority of the farming population. A tractor or tractor-drawn implements are out of the question due to their limited financial resources. Power tillers, on the other hand, are less expensive and more adaptable tools that may be used for a wide range of agricultural tasks (Raghavan, 2008).

It was first thought that power tillage would be useful for seedbed preparation and puddling activities, but it performed well with all agricultural cultivation operations except for deep tillage. In all farming operations, such as puddle and dry tillage, sowing and transplanting, weeding and intercultural operations, irrigation, spraying and dusting pesticides, harvesting, and post-harvest operations, as well as transportation and appropriate technology for small farms, Varshney et al. (1990, 1995) found power tillers to be an extremely good resource of farm power. It has also been used to generate energy, pump water, and move people (Ademiluyi et al., 2008, Thomas et al., 2006).

The advantages of power tillers include simple maneuverability, a small turning radius, easy inter-row operation in crops such as maize, peach, and oranges, and suitable for small farms (Salokhe and Hendriadi, 1995; Panwar et al., 2001; Fasola et al., 2007; Ademiluyi et al. 2008; Mandal and Maity, 2011; Adamu et al. 2014). Since it may be used all year round, the power tiller is well-liked by farmers for its adaptability. This lowers the operating cost per unit of production (Mahapatra et al., 2010). The tiller may be used as a multi-purpose adaptable equipment due to the accessories that can be applied to it depending on the type of work (Cherian et al., 2016).

To boost agricultural productivity in nations such as Nepal and India, power tillers have been widely used (Dhakal et al. 2001; Singh, 2002). Power tiller tractors operate better and are more accurate on mountains than four-wheel tractors (Sakai, 2000). This page focuses on extensive assessments of power tillers' wide range of applications over the years, in accordance with the aforementioned topic. Power tiller-operated tools for farm-related operations, disaster management and agroforestry management have been the focus of research. Power tiller adoption and research will benefit from this study, as would field officers, subject-matter experts, administrators, and policymakers.

2.1 Entry of Power tillers in India

Power tillers were invented in the 1920s and widely utilized in Japan, where they were initially developed. Local farmers in Japan employed diesel engines to pump irrigation water in the early years of the twentieth century. It wasn't long before garden tractors were fitted with a diesel engine and renamed "power tillers." In compared to animal tillage, farmers immediately grasped the usefulness of power tillers. Power tiller sales soared as a result of the widespread acceptance of the machine by Japanese farmers.

It contributed to Japan's increased rice production. In the 1960s, India began investing in power tillers as a result of the country's growing interest in them. Several firms, including Iseki, Sato, Krishi, Kubota, Yanmar, and Mitsubishi, began producing power tillers in India in 1963. The Indian government invested in power tiller research and development, as well as a government subsidy scheme. In the late 1970s and early 1980s, the Indian Council of Agricultural Research (ICAR) began a series of coordinated initiatives around the country to test the appropriateness of power tillers.

For rice, wheat, and other crops a variety of cropping patterns and their performance were studied. The government of India initiated a subsidy program in 2007–2015 to encourage farmers to acquire and utilize power tillers, in light of a lack of agricultural labor and a rising demand from the land for increased output. The rise in sales of power tillers may be seen as a result of this. Sales of power tillers have climbed from 15529 in 2004–05 to 55000 in 2015–16 at a compound annual growth rate of almost 12 percent (Tiwari et al., 2019b).

2.2 Power Tiller Usage and Productivity

To prepare seedbeds and transport seedlings, power tillers have been a key component of the plan, as they are more appropriate for smaller farms than tractors (Mandal and Maity 2013). It is sometimes referred to as a garden tractor, a hand tractor, or a two-wheel tractor because of its wide range of applications (Mandal et al. 2016). The tiller may be used as a multi-purpose machine since it can be fitted with external attachments depending on the task at hand (Cherian et al. 2016). Farm labor shortages and rural youth's lack of interest in farming pose a serious threat to the long-term viability of agricultural output in India.

To compensate for a lack of human labor, agricultural mechanization is essential, as it helps to maximize the use of farm inputs, as well as transportation assistance. To ensure that farm mechanization was utilized to its best potential in 2007, India's government launched a farm mechanization program that ran until 2015, when it was replaced by a sub-mission on agricultural mechanization. Across states, power tiller costs, subsidies, and farmers' portion of the cost varied widely. A power tiller costs an average of Rs. 1.5 lakhs, including subsidies, across states, according to the people who use them.

During the dry season, soil moisture content and water availability are the most important determinants of crop intensity. With motorized tillers, ground preparation may be completed more quickly. In a plot, there may be additional items that can aid in the growth of crops. High-value inputs like potatoes, vegetables, and mustards are more likely to be grown by those who use power tillers. Because of the ability to discharge clods for breaking manual labor behind plowing, farmers employing tillers can plant their crops considerably more quickly and at a lower cost than those utilizing animals.

2.3 Prospects Of Power Tillers for Agri Operations In India

An estimated 320 billion rupees (5.74%) would be spent on farm mechanization in India year 2015–16. (CAGR). One of the most important agricultural machines is the power tiller (Oluka, 2000). Farmers in low-mechanized nations, such as India, face a difficult task in raising their productivity and boosting their incomes via agriculture. It demands constant technical advancement and innovation to be available at a fair cost to current inputs.

Many former subsistence farmers have been transformed into dynamic commercial farmers employing machined farm power from human and animal power as a catalytic tool for greater output and productivity. Farming policies, socio-economic conditions, climate policies, and government policies all have an impact on the need for agricultural gear (Vieweg, 2012).

Customized renting of power tillers is also possible. Power Tiller Operated Seeder (PTOS) machines are very profitable in Bangladesh, according to research done by the state's Department of Agriculture and Rural Development (DARD) (Mia and Haque 2015). Farmers in the mountainous region of north-eastern India who used motorized tillers felt that using the equipment would result in higher area coverage by individual farmers and save time and labor, according to new research. Additional revenue was also made owing to the bespoke hiring of the power tillers' owners (Thomas et al. 2006). Mechanization policies must be tailored to the specific demands of each location to maximize output (Mehta, Chandel, & Senthil Kumar, 2014).

2.4 Challenges Of Power Tiller Operations

Farmers' investment tendency to purchase new equipment has decreased in recent years as a result of structural changes in agricultural systems (Bonati and Gelb, 2005; Vieweg, 2012). The bulk of Indian farmers own a modest amount of marginal land. If you're going to spend money on farming equipment, you're going to be stuck with it for a lengthy period of time. When it comes to owning agricultural gear, the procurement of stand-alone devices is another common problem.

As a result, farmers with small or marginal holdings would be unable to afford to acquire every machine and technology. The farm mechanization ownership model is not particularly inclusive since agricultural machinery are not as totally divisible as seeds and fertilizers. In addition, there are huge technology gaps in meeting the needs of different crop patterns in each location. Investment in mechanization may not provide the intended outcomes if sufficient planning is not in place. Chinese power tillers have a market share of 32 percent at the moment, according to the research, because they are roughly 10% to 20% less expensive than indigenous tillers. According to the paper, the Indian power tiller sector has difficulties because of the free imports from China, subsidy uncertainty, and the difficulties related with the state-level execution of the programs.

Policymakers in India must keep an eye on the intricate price systems involved with farm gear. Lack of study into the production, standardization, and quality control of farm machinery are the main barriers to farm mechanization according to Verma et al. (1994). Farmers with low socioeconomic and educational status are constrained in their use of machinery by a lack of training and maintenance facilities (Singh 2006).

The adoption of a power tiller is considerably hampered by a lack of understanding of operating and maintenance procedures (Patra 2014). Training programs for agricultural machinery, such as tractors, power tillers, and so on, must be strengthened in order to meet the needs of a wide range of users (Mehta et al. 2014). It has been observed that involvement in agricultural extension activities and the availability of repair facilities have a favorable and significant impact on the usage of farm machinery (Kumar 2014).

3. Conclusion

The usage of farm gear in India, such as tractors and power tillers, has raised worries about the shrinking land holdings and rising fragmentation. In the vast majority of cases, the properties and areas of activity are little and insignificant. The land fragmentation is also likely to worsen, especially for small and marginal farmers. A healthy planting stand can't be established in insufficient soil moisture without timely farm activities, especially when it comes to seed preparation and seed preparation procedures. Wetland cultivation, hill agriculture, small farm cultivation, horticultural operation, and forestry are the primary uses of power tillers in Indian agriculture.

Furthermore, it is a tried-and-true method for a wide range of agricultural tasks, including plowing, wet puddling, sowing and transplanting, weeding and intercultural activities, irrigation, pesticide spraying and dusting, harvesting and post-harvest operations, and hauling. The use of a power tiller expedites agricultural activities, resulting in higher output, productivity, and cropping intensity. Nevertheless, a number of obstacles stand in the way of the widespread adoption of power tiller mechanization, including a lack of public awareness of the corresponding implement, the arduous nature of its use, the scarcity of adequate training on these topics, as well as inadequate support from financial institutions and the government. That is why the future of Indian agriculture lies with power tillers and other agricultural equipment.

References

[1] Adamu, F. A., Jahun, B. G., and Babangida, B., (2014). Performance evaluation of power tiller in Bauchi State Nigeria. Journal of Biology, Agriculture and Healthcare, 4 (9), 10–14.

[2] Ademiluyi, S. Y., Oladele, O. I., and Wakatsuki, T., 2008. Socio-economic factors influencing power tiller use among sawah farmers in Bida, Nigeria. Journal of Food, Agriculture & Environment, 6 (3&4), 387–390.

[3] Bonati, G., Gelb, E., 2005. Evaluating internet for extension in agriculture. In: Gelb, E., Offer, A. (Eds.), ICT in Agriculture: Perspectives of Technological Innovation. European Federation for Information Technologies in Agriculture, Food and the environment, Paris.

[4] Cherian A, Punnan J and Varghese A. 2016. A review on power tiller attachments. Paper presented at Kerala Technological Congress (KETCON-2016 Technology for Sustainability).

[5] DoACFW. (2018). Pocketbook of agricultural statistics 2017. Directorate of Economics and Statistics, Dept. of Agriculture Cooperation and Farmers Welfare, Ministry of Agriculture, Cooperation & Farmers Welfare, GOI.

[6] Dhakal, N.H., Pariyar, M.P. and Shrestha, K.B. (2001). Baseline Study in Agricultural Mechanization Needs in Nepal. New Delhi, India. 84.

[7] Government of India. (2018a). Doubling Farmers' Income (Volume VIII): Input Management for Resource Use Efficiency & Total Factor Productivity. New Delhi: Department of Agriculture, Cooperation and Farmers' Welfare, Ministry of Agriculture & Farmers' Welfare.

[8] Fashola, O. O., Ademiluyi, S. Y., Faleye, T., James, D., & Wakatsuki, T. (2007). Machinery systems management of walking tractor (power tiller) for rice production (sawah) in Nigeria. Journal of food agriculture and environment, 5(3/4), 284.

[9] Kumar S. 2014. A study on the agricultural mechanization in Karimnagar district of Andhra Pradesh. M Sc. thesis, Acharya N. G. Ranga Agricultural University. Rajendranagar, Hyderabad.

[10] Lal, B., & Singh, A. (2016). Understanding the Tractor Market: An Exploratory Study in Haryana (India). Global Journal of Management and Business Research: E-Marketing, 28-34.

[11] Mahapatra; M; Maitra; R; & Chattoupadhyay, P.S. (2010). J. Power tiller-an appropriate technology for rice-based small farms. A review, 4(1), 126–134.

[12] Mandal S K and Maity A. 2011. Development and performance evaluation of a lightweight power tiller. (In) Proceedings of National Conference on Machines and Mechanism, NACOMM 2011. IIT Madras, India, pp 319–25.

[13] Mandal, S., Kumar, A., and Prasad, A. K. (2016). Performance evaluation of a lightweight power tiller. Agriculture for Sustainable Development 3-4, 126–129.

[14] Mehta C, Chandel N.S., Jena, P. C. and Jha, A. (2019). Indian Agriculture Counting on Farm Mechanization. Agricultural Mechanization in Asia, Africa, and Latin America. 50(1), 84–89

[15] Mehta, C. R., Chandel, N. S., & Senthil Kumar, T. (2014). Status, Challenges, and Strategies for Farm Mechanization in India. Agricultural Mechanization in Asia, Africa & Latin America, Volume 45, No.4, 43

[16] Mia M A and Haque M E. 2015. Farm-level impact study of power tiller operated seeder on service providers' livelihood in some selected sites of Bangladesh. Bangladesh Journal of Agricultural Research 40(4): 669–82.

[17] Morehouse, W. (1982). Dilemma of Technological Choice Revisited: Case of Combine Harvester. Economic and Political Weekly, Volume 17, No.35, 83-84.

[18] Oluka, S. I. Cost of Tractor Ownership under Different Management Systems in Nigeria. Nigerian Journal of Technology, Vol. 19, No. 1, 2000, pp. 15.

[19] Panwar J S; Srivastava A P; and Indramani (2001). Power Tillers in Indian Agriculture. Technical Report. Division of Agricultural Engineering, Indian Agricultural Research Institute, NewDelhi, India.

[20] Patra S. 2014. Study the extent of farm mechanization with respect to tractor and power tiller in rice growing belt of Orissa. (In) Report of the Summer Internship Program at Odisha Farm Machinery Research and Development Centre (OFMRDC), Bhubaneswar.
[21] Payne, C.A., Chelson, J.B. and Reavill, R.P. Maintenance Management. Chapter in Management for Engineers, John Wiley and Sons, Chichester, 1996.
[22] Sahni, R. K., Patel, A., Kumar, M., Thorat, D. S., and Kumar, V. (2018). Status of Farm Mechanization in Indian Agriculture. Biotech Article.
[23] Raghavan, M. (2008). Changing pattern of input use and cost of cultivation. Economic and Political Weekly, 123–129.
[24] Sakai, J. (2000). Two-wheel Tractor Engineering: For Asian Wet Land Farming: Shin-norinsha.
[25] Salokhe, V. M., Hendriadi, A., 1995. Power tiller industry in Indonesia. Agricultural Mechanization in Asia, Africa, and Latin America, 26 (4), 29–32.
[26] Sarkar A., (2020). Agricultural Mechanization in India: A Study on the ownership and Investment in Farm Machinery by Cultivator Households across Agro-ecological Regions.
[27] Singh, G. (2002). Agricultural Machinery industry in India (Manufacturing, Marketing and Mechanization Promotion).
[28] Singh J. 2006. Scope, progress and constraints of farm mechanization in India, Status of Farm Mechanization in India, pp.48–56. ICAR. Indian Council of Agricultural Research, New Delhi.
[29] Singh, R. S., & Sahni, R. K. (2019). Transformation of Indian agriculture through mechanization. Economic Affairs, 64(2), 297–303.
[30] Subramanian K. V. (2021). State of the Economy 2020-21: A Macro View. Economic Survey 2020-21. 2, GOI
[31] Thomas, E.V. C.S. Sahay, K.K Satapathy, (2006). Prospects and problems of power tillers in selected districts of northeastern hilly region of India – A case study. Agricultural Mechanization in Asia, Africa and Latin America, pp 39–45.
[32] Tiwari, P. S., Gurung, T. R., Sahni, R. K., & Kumar, V. (2017). Agricultural mechanization trends in SAARC region. Mechanization for Sustainable Agricultural Intensification in SAARC Region. 1, 1-40
[33] Tiwari, P.S., Sahni, R.K., Kumar, S.P., Kumar, V. and Chandel, N.S., 2019a. Precision agriculture applications in horticulture. Pantnagar Journal of Research, 17(1),1–10.
[34] Tiwari, P. S., Singh, K. K., Sahni, R. K., & Kumar, V. (2019b). Farm mechanization–trends and policy for its promotion in India. Indian Journal of Agricultural Sciences, 89(10),1555–1562.
[35] Varshney, A.C.; Saxena, D.C. and Narang, Suresh (1990). Power tiller potential in forestry. Agricultural Mechanization in Asia, Africa and Latin America, 2l (2), 7l–72, Japan.
[36] Verma M and Tripathy A. 2015. Perspective of agricultural mechanization in Supaul district of North Bihar- a Research. Journal of Agriculture and Veterinary Science 8(8): 4–12.
[37] Vieweg, H.G. (Ed.), 2012. An introduction to mechanical engineering: study on the competitiveness of EU mechanical engineering industry. European Commission. <http://ec.europa.eu/enterprise/sectors/mechanical/files/competitiveness/comp-mech-eng-2012-frep_en.pdf> (accessed January 2013).

Android-based feasibility study transformation analysis case study of the RASI development in Cirendeu-Kampung Cimahi

H Herawati[1], A S Soemantri[1], E Kamsiati[2], M. Bachtiar[2]

[1]Indonesia Center for Agricultural Postharvest Research and Development, Indonesia
[2]IPB University, Bogor, Indonesia

*herawati_heny@yahoo.com

Abstract. Development of rasi or rice made from cassava flour was developed in Cirendeu Village-Cimahi City. The activity requires a feasibility study analysis that is mobile and fast so that it can be implemented more easily in the field. Nowadays, information technology is developing very fast to support decision support systems that are mobile. Analysis of economic-based feasibility studies has been widely developed in the field, but often is still based on virtual laptops. The purpose of this research activity is to conduct a trial-based Android-based feasibility study for the development of constellations in Cirendeu Village, Cimahi City. The research activity method includes several stages including optimizing and developing constellation technology in the field, analyzing the laptop-based feasibility study, transforming the results of an Android-based feasibility study and testing the implementation of performance in the field. Based on the results of the analysis in the field, the results were obtained that the software was able to identify the feasibility and optimum capacity of the production business based on the results of input data for several parameters that had been identified based on the system. The feasibility of the constellation development efforts developed in Cirendeu Village-Cimahi City can be identified more easily by using android for further implementation at the field level. This technology can be used and replicated as a mobile software-based decision-making system approach.

1. Introduction

Processing technology of Cassava Rice (RASI) began to develop in the village of Cirendeu, Cimahi City. Cirendeu village people consume traditional RASI from generation to generation, however they still have an unequal shape and are easily destroyed. One technology for uniforming the form so that it is not easily destroyed by processing it into an analog constellation or artificial rice with cassava flour raw material. In making rice granule or artificial rice can come from cereals or tubers which are a source of carbohydrates as the main raw material [1].

To increase the protein content, Herawati et al [2] added soy flour to the production of artificial rice from cassava flour raw material. Furthermore, Herawati et al. [3] made and analyzed the characteristics of artificial rice from corn starch added with tapioca and hydrocolloid raw materials to produce instant artificial rice used to extrusion technology method. Some other terms such as artificial rice [2-4]; analog rice [5]; pearl rice [6]; Smart Rice[7]; siger rice [8-9]; sorghum rice [10]; artificial rice [11-18].

Herawati et al. [1] conducted a rice processing technique from cassava raw materials and with the treatment of the addition of groundnut flour to increase protein content. Some rice processing technology as done by researchers using several granulation techniques [6; 2]. Furthermore, Herawati et al [4;17] carried out rice processing techniques using hot extruders. Technology of artificial rice production process can be produced from some raw based material of carbohydrate sources [19-21]. Hydrocolloids

IOP Conf. Series: Earth and Environmental Science 1041 (2022) 012019 doi:10.1088/1755-1315/1041/1/012019

as much as 0.2 to 2.5% can be added before the extrusion process in the manufacture of artificial rice [14]. To increase the water absorption capacity and artificial instant rice process with an optimal result, other food additives can be added [15].

Development of rasi or rice made from cassava flour was developed in Cirendeu Village-Cimahi City. The activity requires a feasibility study analysis that is mobile and fast so that it can be implemented more easily in the field. Nowadays, information technology is developing very fast to support decision support systems that are mobile. Analysis of economic-based feasibility studies has been widely developed in the field, but often is still based on virtual laptops. The purpose of this research activity is to conduct a trial-based Android-based feasibility study for the development of constellations in Cirendeu Village, Cimahi City.

2. Methodology

The research activity method includes several stages including optimizing and developing constellation technology in the field, analyzing the laptop-based feasibility study, transforming the results of an Android-based feasibility study and testing the implementation of performance in the field. Implementation of the activities began in the 2017-2019 FY.

For business feasibility analysis began to be carried out in 2018 and in 2019 transformed into the form of android. Techno-economic analysis of the feasibility of operational units in the field, data is collected and in the preliminary analysis described in the form of input and output flows in the form of mass balance produced. The components in the mass balance produced are presented on the basis of the unit of product produced.

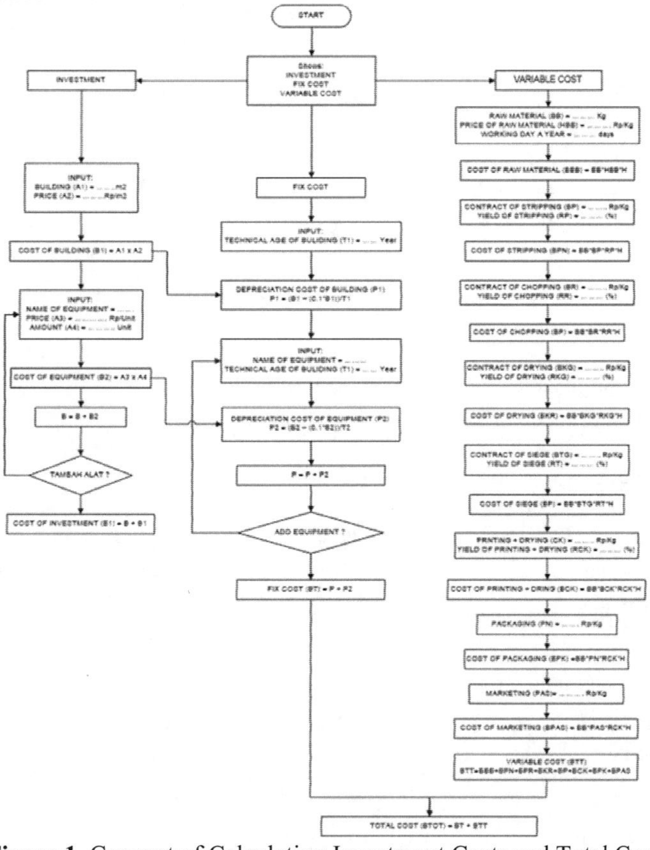

Figure 1. Concept of Calculating Investment Costs and Total Costs

The synthesis stage is to determine the choice of net production chosen based on economic evaluation using ROI criteria, B / C ratio, IRR, and NPV [22]. Environmental evaluations are based on criteria for changes in the use of raw and auxiliary materials, changes in water and energy use, and characteristics of the resulting waste [23-24]. Added value is calculated based on Kartasapoetra, [25] and income analysis calculated based on Elida and Hamidi [26].

Making the DSS-Rasi application stems from an economic calculation concept that is integrated with a database information system including a spatial database information, so that in this application there are 3 large systems that work in it namely, a database system, a base system of economic feasibility calculation and an operational system between advances. Flowchart system calculation of total costs and flowchart calculation of the economic viability of constellations as shown in Figures 1 and 2 below.

The concept of combining database information and spatial database is intended so that users can not only conduct economic feasibility studies but also search for suppliers of raw materials as well as marketing. In this spatial database the user can also add supplier or marketing data, so that the longer there will be more choices of suppliers and markets. Spatial database is created by connecting the DSS-Rasi system with Google-map, so users can access suppliers and markets in real time.

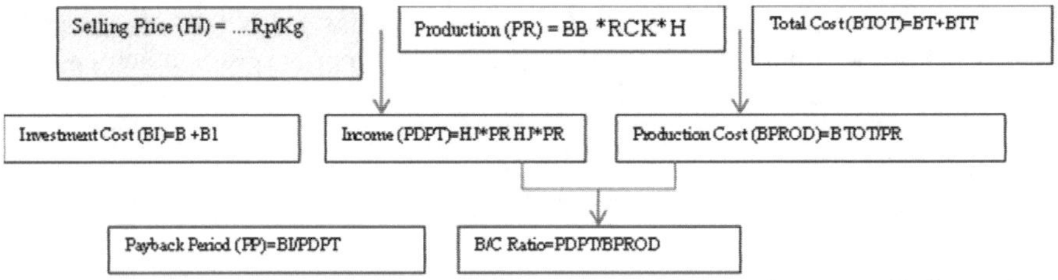

Figure 2. Concept of Calculating Economic Eligibility for Constellations

Based on the results of the analysis in the field, the results were obtained that the software was able to identify the feasibility and optimum capacity of the production business based on the results of input data for several parameters that had been identified based on the system.

3. Result and Discussion

3.1 Economic Techno Analysis of RASI

The raw material used to produce constellations is cassava. The source of raw material supply comes from Cimahi, West Bandung, Cianjur and Sukabumi, which are cassava production centers in West Java. Economic analysis was carried out to obtain information on the feasibility of constellation agro-industry in Cirendeu. The factors taken into account in preparing this analysis are investments which include the rental of land, buildings, equipment, supporting facilities, office equipment, trial costs and initial work costs.

Table 1. Investment Required in RASI Production

No.	Description	Unit	Price per Unit	Cost (Rp)
1.	**Lant Rent**		**10.000.000**	**10.000.000**
2.	**Building**			
	Production Room	13 x 6 m	2.000.000	
	Flour Production Room	2 x 2 m	2.000.000	
	Drying Room	2 x 2 m	2.000.000	
			6.000.000	**6.000.000**
3.	**Equipment**			

	Extrusion	1	63.000.000	
	Polisher	1	14.000.000	
	Dryier	1	20.000.000	
	Mixer	1	9.500.000	
	Siever	1	9.500.000	
	Balancer	1	250.000	
	Packaging	1	750.000	
	Flour Machine	1	6.000.000	
	Slicer	1	4.000.000	
			127.000.000	**127.000.000**
4.	**Supporting Facilities**			
	Steamer	1	700.000	
	Gas Stove	1	400.000	
	Bamboo Trays	10	25.000	
	Filter Cloth	10	15.000	
			1.140.000	**1.140.000**
5.	**Office Tools**			
	Table	1	1.500.000	
	Chair	6	25.000	
	Stationary	1	25.000	
			1.550.000	**1.550.000**
6.	**Experimental Fees**		**3.000.000**	**3.000.000**
7.	**Early Working Fees**		**4.500.000**	**3.000.000**
	TOTAL		**153.190.000**	**153.190.000**

Furthermore, based on information obtained from field observations and interviews with informants, it is assumed that the average daily constellation orders are 50 kg / day and the average weekly order orders are 500 kg / week. From the data collected, information on cassava needs, constellation production costs, constellation prices, gross income and profits are obtained. From the information obtained then analyzed to determine the value of ROI (Rate of Investment), B / C (Net Benefit Cost) ratio, IRR (Incremental Rate of Return) and NPV (Net Present Value).

ROI is the time of return on investment. NPV is analyzed to determine the value of investments by considering changes in currency values. NPV is the difference between the present value of profits and costs. IRR investment level is the prevailing interest rate that shows the present value (NPV) equal to the total investment of the project. B / C ratio is a comparison between total revenue and total costs, which shows the value of revenue obtained from each rupiah issued.

In this economic analysis of racial agro-industry two assumptions are used for raw materials, namely from cassava and cassava flour (Table 2 and Table 3). The price of raw materials is also made two prices, namely Rp. 1500 / kg and Rp. 2000, - / kg to obtain information on the results of the analysis if the price of raw materials changes. From the results of the analysis note that if cassava is used at a price of Rp. 1500, - / kg obtained a B / C ratio of 1093; IRR 123.49 and NPV Rp. 5,529,669.77, - whereas if the price of raw materials is Rp. 2000, - / kg, there will be a loss, if the sale price of the constellation remains Rp. 12,000, -, which can be seen from the B / C ratio <1.

Table 2. Cassava Raw Materials

No.	Component	Price of Cassava (Rp/kg)	
		1500	2000
1.	Cassava Needs (Tons / day)	0.51	0.51
2.	Production Cost (Rp / kg)	11.787.56	12.944.97
3.	Rasi Price (Rp/kg)	12.000	12.000
4.	Bruto Income (Rp/hari)	1.457.143	1.457.143
5.	Profit (Rp/Day)	25796	-114746
6.	ROI (year)	21.48	-
7.	B/C rastio	1.02	0.93
8.	IRR	123.49	229.89
9.	NPV	5.529.669.77	-49.495.799.47

Apart from fresh cassava, raw material for constellations can also be made from cassava flour. The price of raw material made from cassava flour ranges from Rp. 6000, - to Rp. 8,000.0 / kg. From the analysis it is found that if cassava flour is used at a price of Rp. 6,000 up to Rp. 7,500 / kg still provides benefits that can be seen from the B / C ratio of more than 1. The profit value will decrease with increasing raw material prices, and when the price of cassava flour reaches Rp. 8,000 / kg, there will be a loss, if the selling price of the constellation remains Rp. 12,000, -, which can be seen from the B / C ratio <1.

Table 3. Raw Material of Cassava Flour

No	Component	Price of Cassava Flour (Rp / kg)				
		6.000	6.500	7.000	7.500	8.000
1.	Cassava Needs (Tons / day)	0.15	0.15	0.15	0.15	0.15
2.	Production Cost (Rp / kg)	9.764,35	10.389,35	11.014,35	11.639,35	12.264,35
3.	Rasi Price (Rp/kg)	12.000	12.000	12.000	12.000	12.000
4.	Bruto Income (Rp/hari)	1.457.143	1.457.143	1.457.143	1.457.143	1.457.143
5.	Profit (Rp/Day)	271.471	195.579	119.686	43.793	-32.100
6.	ROI (year)	3.59	4.78	7.17	14.34	-
7.	B/C ratio	1.23	1.16	1.09	1.03	0.98
8.	IRR	203,35	200.36	193.55	162.63	273.92
9.	NPV	101.912.508	72.198.755	42.485.001	12.771.248	-16.942.505

The availability of raw materials in the form of fresh cassava and cassava flour greatly influences the calculation of techno-economics, feasibility and continuity of constellation production. The analysis parameters above can be used as a reference for further development of the success rate of a processing unit in the field. The flowcharts in Figures 1 and 2 above were further developed using the Java programming language on an android studio system. The interface design of this application is shown in Figure 3 below. This application is stored in a file with an APK extension, then this application can be directly installed and operated on virtual devices such as smartphones. Some examples of work pages from the DSS-Rasi application are shown in Figure 3 below.

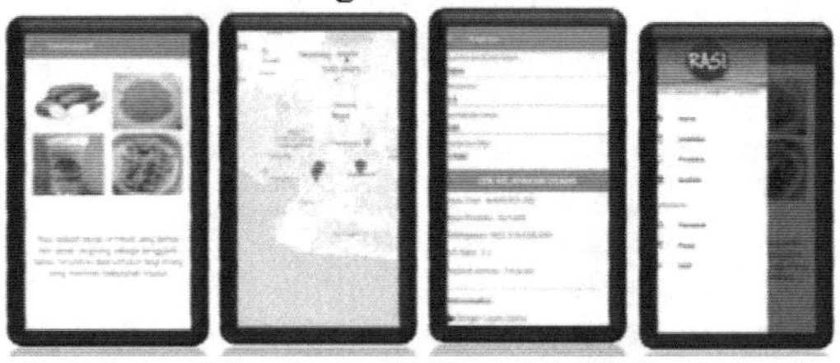

Figure 3. Some of the Android-Based DSS-Rasi Applications

Through the use of the DSS-Rasi application, users can conduct economic feasibility studies on various production capacities and selling prices. In addition, this application provides a facility to search for suppliers of raw materials (cassava) and product marketing. On the top left-hand corner bar there is a menu icon which when checked will display a menu list to conduct an economic feasibility study of the constellation production business. In this application there are many work pages where in Figure 3 above are displayed only a few examples of these work pages, and in Figure 3 above respectively are the main pages of the DSS-Constellation application, menu pages, results pages of economic feasibility studies and pages a map that shows the location of the supplier or the market which if clicked on the balloon location will display supplier / market data complete with mobile phone number and the web so that it can be directly contacted.

4. Conclusion

The results of the analysis note that if cassava is used at a price of Rp. 1500, - / kg obtained a B / C ratio of 1093; IRR 123.49 and NPV Rp. 5,529,669.77, - whereas if the price of raw materials is Rp. 2000, - / kg, there will be a loss, if the sale price of the constellation remains Rp. 12,000, - which can be seen from the B / C ratio <1. The profit value will decrease with the increase in raw material prices, and when the price of cassava flour reaches Rp. 8,000 / kg, there will be a loss, if the selling price of the constellation remains Rp. 12,000, -, which can be seen from the B / C ratio <1. The feasibility of the constellation development efforts developed in Cirendeu Village-Cimahi City can be identified more easily by using android for further implementation at the field level. This technology can be used and replicated as a mobile software-based decision-making system approach.

References

[1] H. Herawati, E. Kamsiati, and Abubakar. Influence of peanut flour addition on rasi process production. 2nd International Conference on Agriculture Postharvest Handling and Processing. IOP Conf. Series: Earth and Environmental Science 309 .2019. 012044. doi:10.1088/1755-1315/309/1/012044.

[2] H. Herawati, A. Arif, K. Oktaviani, and S. Widowati. Karakteristik beras artifisial berbasis ubikayu dan kedelai. Disampaikan pada Seminar Nasional Teknologi Pascapanen. 2011,. Bogor.

[3] H. Herawati, F. Kusnandar, D.R. Adawiyah, and S. Budijanto. Teknologi proses pembentukan butiran beras artifisial instan dengan metode ekstrusi. Pangan. 2013. 22(4): 317–327.

[4] H. Herawati, F. Kusnandar, D.R. Adawiyah, and S. Budijanto. Teknologi proses produksi beras tiruan mendukung diversifikasi pangan J. Litbang Pert. 2014. Vol. 33 No. 3 September 2014: 87-94.

[5] S. Budijanto. Beras analog product vehicle diversifikasi pangan untuk memperkokoh ketahanan pangan nasional. 2012. Ftechnopark. Institut Pertanian Bogor, Bogor.

[6] H. Herawati, and S. Widowati. Karakteristik beras mutiara dari ubi jalar (Ipomoea batatas). Buletin Teknologi Pascapanen Pertanian. 2009. 5(1): 39–48.

[7] A. Subagio, Y. Witono, D. Hermanuadi, A. Nafi, and W.S. Windarti. Pengembangan beras cerdas sebagai pangan pokok alternatif berbahan baku mocaf. Prosiding InsiNas. Jember, 29–30 November 2012. hlm. 157-161.

[8] W. Novia, W.A. Zakaria, and D.A.H. Lestari. Analisis nilai tambah dan kelayakan pengembangan agroindustri beras siger. JIIA. 2013. VOLUME 1 No. 3, JULI 2013. 210-217.

[9] B. Hidayat. Prospek pengembangan dan teknologi pengolahan beras siger (Tiwul dan oyek yang telah dimodernisasi). UP. Politeknik Negeri Lampung, Lampung. 2016. ISBN:978-602-70530-3-8.

[10] W. Amrinola, S. Widowati, and P. Hariyadi. Metode pembuatan sorgum sosoh rendah tanin pada pembuatan sorgum (Sorghum bicolor) instan. ComTech. 2015. Vol. 6 No. 1 Maret 2015: 9-19. \

[11] R.P. Scelia, E. Hegedus, J. Giacone, H.B. Bruins, and E.J. Benjamin. Extruded quick cooking rice like product. European Patent,. 1986. 0226375.

[12] M.L. Wenger, and G.R. Huber. Low shear extrusion process for manufacture of quick cooking rice. US Patent. 1988. 4.769.251.
[13] K. Koide, T. Fukushima, T. Tomita, and T. Kuwata. Fabricated rice. United States Patent. 1999. 5932271.
[14] A. Mishra, H.N. Mishra, and P.S. Rao. Preparation of rice analogues using extrusion technology. International Journal Food Science and Technology. 2012. 1-9.
[15] J.P. Wang, H.Z. An, Z.Y. Jin, Z.J. Xie, H.N. Zhuang HN, and J.M. Kim. Emulsifiers and thickeners on extrusion-cooked instant rice product. Journal of Food Science Technology, published online 28th may 2011.
[16] M.N. Riaz. Selecting the right extruder. Didalam: Guy, R (ed) Extrusion Cooking Technologies and Application. 2001. CRC Press- Boca Raton, USA.
[17] H. Herawati, F. Kusnandar, D.R. Adawiyah,, and S. Budijanto. Thermal characteristics and state diagram of extruded instant artificial rice. Thermochimica Acta. 2014. 593: 50–57. https://doi.org/10.1016/j.tca.2014.08.017.
[18] H. Herawati, F. Kusnandar, D.R. Adawiyah, and S. Budijanto. Optimisation process of twin screw extruder technology for instant artificial rice production. Proceedings International Food Conference. Universitas Katolik Widya Mandala, Surabaya. 2017. ISBN 978-602-61032-1-5.
[19] S. Budijanto and N.D. Yuliana. Development of Rice Analog as a Food Diversification Vehicle in Indonesia Development of Rice Analog as a Food Diversification Vehicle in Indonesia. Journal of Developments in Sustainable Agriculture,. 2015. 10(August), 7–14. https://doi.org/10.11178/jdsa.10.7.
[20] S. Wahjuningsih, D. Praseptiangga and B. Haryanto. A study of sago starch and red bean flour-based analog rice development as functional food. 2016. (August), 23–25.
[21] S. Sumardiono, I. Pudjihastuti, N.A. Handayani and H. Kusumayanti. Physicochemical Characteristics of Artificial Rice from Composite Flour: Modified Cassava Starch, Canavalia ensiformis and Dioscorea esculenta. E3S Web of Conferences. 2018. 31, 6005.
[22] I. Soeharto. Studi kelayakan proyek industri, Penerbit Erlangga. 2002. Jakarta.
[23] G. Alaerts and S.S. Santika. Metode Penelitian Air. Usaha Nasional. 1984. Surabaya. Indonesia.
[24] APHA. Standard Methods for Examination of Water and Wastewater, 18th ed., American Public Health Assosiation,. 1992. Washington.
[25] G., Kartasapoetra. “Masalah Pertanahan di Indonesia”, Jakarta. 1986. Penerbit PT. Bina Aksara.
[26] S. Elida and W. Hamidi W. Analisis pendapatan agroindustri rengginang ubi kayu di Kabupaten Kampar. 2009. Pekanbaru: Fakultas pertanian UIR.

Water pump control system in hydroponic plants using the ebb and flow method

David Setiawan*, Hamzah, Latifa Siswati, Anto Ariyanto, Guntoro

Universitas Lancang Kuning, Pekanbaru, Indonesia

*dsetia@unilak.ac.id

Abstract. Along with the restrictions during the covid-19 pandemic, many new farmers are growing vegetables for their own needs or for sale and there are also many new farmers who are planting ornamental flowers which are currently booming, these new farmers are mothers or people who have free time at home. In growing vegetables using the hydroponic method, it does not require a large area of land and yields faster harvests by paying attention to the accuracy in providing nutrients, the intensity of light and temperature around the plant and the flow of water that is maintained to circulate oxygen and nutrients to the plant roots. In previous research, Hydroponic Plants powered by Solar Panels have been designed and tested but in the middle of the night the motor stops flowing water to the plants due to insufficient battery voltage, so it is necessary to find a solution so that the existing equipment can be optimized again, so this research aims to get a solution so that the existing system can be optimized by using the control system carried out by Arduino so that the pump can continue to flow water (oxygen and nutrients) for 24 hours continuously.

1. Introduction

At this time, hydroponics became a way to grow plants that are on the rise because to grow plants no longer needed extensive soil and land. We can plant anywhere with planting media using water so that we can hone creativity in processing and creating new media for planting. Planting using hydroponic means, the harvest will be faster. However, it should be considered some other aspects, namely, accuracy in the provision of nutrients, the intensity of light and also the temperature around the growing plant. Generally, beginners pay less attention to nutrient water that is really right and can be absorbed perfectly. Plants by hydroponic means require nutrients obtained only through nutrient water only.

There are 2 (two) ways to grow hydroponics, namely using NFT (Nutrient Film Technique) and using WICK. Using the NFT method is the most popular way used by many people in applying it. While using the WICK method is no less famous in the first way, this method is preferred because of its easy manufacture and materials that are easy to obtain and cheap / using used goods. Growing plants using NFT means requires a water pump so that nutrients flowing with water can be streamed to the roots of the plant.

While plants using the WICK method do not require a water pump. Planting using NFT is the most popular way, and the nutrient supply lies in water so it takes water rotation / water flow so that the nutrients and oxygen needed by the roots can be supplied. Imagine if we forget to turn on the pump or there is a rotating blackout / PLN electrical disturbance, then plants that need nutrients and oxygen will song and die or not develop, then a solution is needed so that lovers of hydroponic plants that use NFT methods remain creative without having concerns about the above problems. To design solar panels that are efficient in the use of batteries and the size of solar panels, a control system is needed so that the energy consumption in water pumps is not large at night.

In this study, the author will provide a solution that is the Water Pump Control System in Solar Home System Hydroponic Plants Using Arduino. It is hoped that this research can contribute and benefit especially for lovers of hydroponic plants. Based on the description above, the problem of a small battery that cannot turn on the pumping water until the morning or sunrise the next day becomes able to drain water containing nutrients and oxygen to the roots of the plant without adding batteries or batteries but with other solutions.

2. Literature Review

Hydroponics is a term used to grow crops without using soil as a growing medium. Plants can be planted in pots or other containers using water and or other porous materials, such as gravel, precarious shards, sand, shards of threshold stones, and so on as a planting medium. Hydroponic planting can develop quickly because it has advantages. The main advantage is the success of the plant to grow and produce more guaranteed.

Other advantages are more practical care, the use of fertilizers is more efficient, plants can grow rapidly and not dirty, production is more continuous, and some types of plants can be cultivated outside the season (Lingga, 2005).

Figure 1. Hydroponic Plants

2.1. Electric Motor as Water Pump

An electric motor is an electromagnetic device that converts electrical energy into mechanical energy. This mechanical energy is used to rotate or pump water on hydroponic plants. Electric motors are used also in the home (mixers, electric drills, fans or fans) and in industry.

The working principle of an electric motor is essentially the same for all types of motors in general :

a. The electric current in the magnetic field will give a force if the wire carrying the current is bent into a loop, then both sides of the loop, i.e. at the right angle of the magnetic field, will get the force in the opposite direction.
b. The force pair generates rotary power/torque to rotate the coil.
c. The motors have several loops on their dynamos to provide more uniform rotational power and their magnetic fields are generated by electromagnetic arrays called field coils.

In understanding a motor, it is important to understand what is meant by the load of an electric motor. Load refers to the output of rotary power / torque in accordance with the required speed. Expenses can generally be categorized into three groups (BEE India, 2004) :

a. Constant torque load is a load where the demand for energy output varies with its operating speed, but the torque does not vary. Examples of loads with constant torque are conveyors, rotary kilns, and constant displacement pumps.
b. A load with variable torque is a load with torque that varies with operating speed. Examples of loads with variable torque are centrifugal and fan pumps (torque varies as quadrat speed).
c. A load with constant energy is a load with a changing torque demand and is inversely proportional to speed. An example for a load with constant power is machine tools.

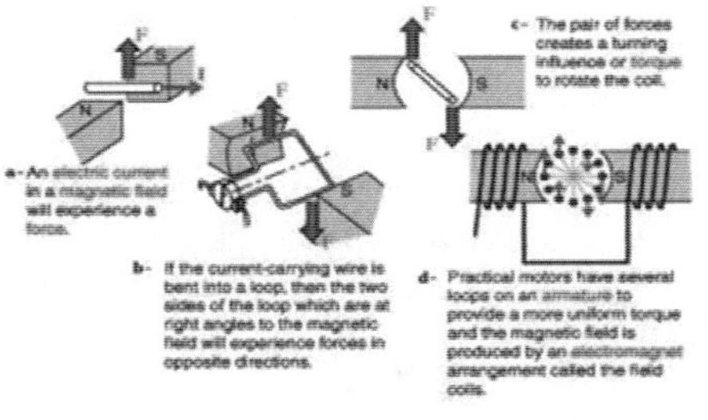

Figure 2. Working principles of electric motors

2.2. Solar Panels

Electricity can be generated using a variety of methods. The photovoltaic effect (wikipedia.org) is the most popular method used to generate electricity. The photovoltaic effect is the phenomenon of converting solar energy into an electric current. Scientists developed solar panels based on the principle of converting light energy into electrical energy through the law of conservation of energy. We need to know, that the notion of solar panels is a device consisting of solar cells that can be used to convert light into electricity.

These solar cells need to be protected from moisture and damage that could occur. This is done so as not to significantly damage the efficiency of solar panels and so as not to decrease their life. Usually, these solar panels have a lifespan of about 20 years. Usually, in that time period the use of solar panels will not experience a significant decrease in efficiency. Today, despite the advancement of advanced technology, most commercial solar panels are only capable of achieving an efficiency of about 15%. Commercial solar panels are very rare that can exceed 20% efficiency.

Do we need to use solar panels? Why? The answer is very simple. Solar panels do not emit harmful greenhouse gas emissions as in the burning of fossil fuels. So, the use of solar panels does not contribute to the effects of climate change. By using solar panels, we can get clean energy from the most abundant energy source on planet Earth, namely the sun. So, why are there still so few people who use solar energy as a source of electricity through these solar panels? Because there are still many people who do not know the advantages and disadvantages of solar panels.

Figure 3. Solar Panels

Advantages of Solar Panels : 1. Solar panels are environmentally friendly because they do not emit harmful greenhouse gas emissions, such as carbon dioxide. Solar panels also do not contribute to climate change. 2. Solar panels utilize solar energy, and the sun is the most abundant energy source on planet Earth. 3. Solar panels are easy to install and also have very low maintenance costs. 4. Many countries in the world offer favorable incentives for homeowners who use solar panels. 5. Solar panels do not lose much efficiency in their life which reaches about 20 years. 6. Because of the long pie period, which reaches 25-30 years. So, solar panels guarantee their users to save energy costs.

Disadvantages of Solar Panels : 1. Currently, solar panels are still relatively expensive. Although solar panels have decreased in price, the price of solar panels still tends to be expensive. 2. Solar panels still need to significantly improve efficiency. The average solar panel currently achieves an efficiency of less than 20%. This is one of the reasons many people do not choose solar panels. 3. Solar panels are made of some materials that are not environmentally friendly. For example, it is made of silicone. 4. If not careful, recycling solar panels can cause environmental damage, because the content of solar panels such as silicon, selenium, and others, of which they are all greenhouse gases, can be found in solar panels. This is dangerous because it can be a source of pollution during the recycling process.

3. Result and Discussion

3.1. Hydroponic design with solar panels

Some trials have been conducted before, namely solar panel hydroponic plant systems without batteries, from the results of experiments obtained the result that dc motors as water pumps can only supply water and nutrients when the solar panel receives enough light from the sun so that when it will be dark and dark, nutrients and water do not flow to hydroponic plants.

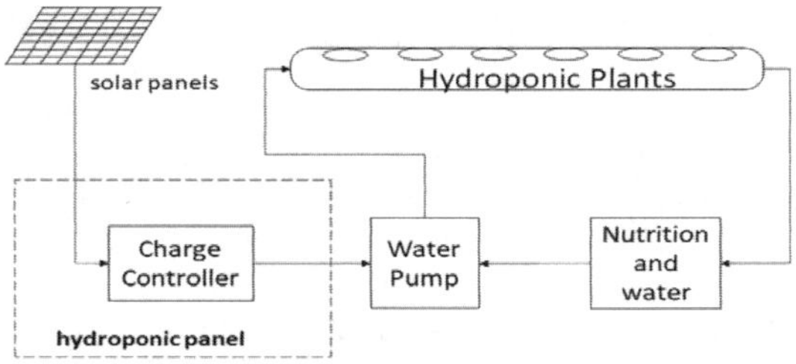

Figure 4. Solar Panel Hydroponic Plant System Without Batteries

Furthermore, a trial of the solar panel system using the battery as a solution to the above problems with the purpose of the battery is used to store electrical energy when the solar panel gets enough light from the sun, but the problem is that it takes a large enough battery to be able to turn on the water pump until morning or sunrise again.

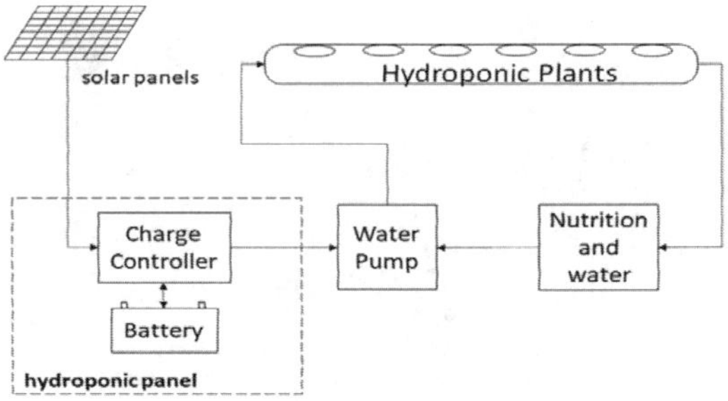

Figure 5. Solar Panel Hydroponic Plant Systems Using Batteries

While the technical specifications of the dipping water pump used are Material: Plastic, Dimension: Approx. 6.5(D) x 8(H) cm, Inlet/Outlet: 1/2" male head, Voltage: 12V DC/24V DC, Maximum Rated Current: 1000MA, Power: 22W, Max Flow Rate: 800 L/H, Max Water Head: 5m and Max circulating water temperature: 100°C

Figure 6. Hydroponic Plant Water Pump

Based on the technical specifications above, information is obtained that the power of the dipping water pump is 22 W which means, if the pump is turned on from 18:00 (afternoon / sunset) to 07:00 (sunrise again) then it takes a total power of 13 hours x 22 Watts = 286 Watts. While the battery used is a dry battery with a capacity of 7.2 A (dry battery, commonly used in UPS) so the battery power capability is 7.2 A x 12 V = 86.4 Watts.

To be able to produce 286 watts, it takes more than 3 batteries or 4 batteries so that the procurement of batteries becomes large, so the idea arises how 1 battery can turn on or supply water and nutrients continuously until the morning before sunrise again. The selection of time intervals is adjusted to the battery owned against the load or power needs on the water pump where the power requirement to turn on when sunlight is less or less is not divided by the ability on battery power stored as the internal Val of time to be used so that the total load power / total battery power = 286 / 86.4 = 3.31 or rounded to 4.

3.2. Improvements made to hydroponics with solar panels

Here are some common hydroponic methods, and the need for circulation pumps :

a. Axis Method (Wick System). No circulation pump is required. With this method, the axis serves to absorb nutrient fluids from the reservoir. The roots of the plant will come into contact with the axis, so it can always absorb nutrients.
b. Deep Flow Technique (DFT) method. A circulation pump takes 12-16 hours. This is because when the pump dies, there are still hydroponic nutrients left around the roots of the plant.
c. Tidal Method (Ebb and Flow). A circulation pump takes 24 hours, but not continuously. Circulation pumps with this hydroponic technique usually turn on and off at certain intervals (for example, 5 minutes on, 5 minutes off)
d. Thin Film Method (NFT, Nutrient Film Technique). The circulation pump takes 24 hours. If the pump dies for a long time, then there are no hydroponic nutrients left in the roots of the plant.

In this study, the authors used the ebb and flow method in which the water pump is simulated to drain water and nutrients for a full 24 hours but not continuously, where the pump is on and off using a specific time interval. Based on the calculations above, it takes intervals of 1 : 3 where life is 5 minutes and turned off 15 minutes so that the total power is obtained by: 5 minutes x 3 = 15 minutes (lifetime) and 15 minutes x 3 = 35 minutes (dead time). So that the lifetime of 15 minutes x 13 is 3.2 hours of life for 13 hours from 18:00 to 07:00 am, meaning the power needed is 3.2 hours x 22 watts = 70.4 watts which means 70.4 watts < 86.4 watts (existing batteries can turn on the water pump for 13 hours with a time interval of 1:3 on and off.

To control the time of life and death according to the above design of 1:3 where life lives 5 minutes and dies 15 minutes, then a system of control or control is needed. In this study, the authors used Arduino as a control system to set the water pumps on and off according to predefined time intervals. The script or command can be stated as follows:

```
int pump = 1;
void setup() {
  use pin 1 as output to the water pump relay
```

```
pinMode (pump, OUTPUT);
}
void loop() {
  living arrangements for 5 minutes or 300secs
digitalWrite (pump, HIGH);
delay(300);
  living arrangements for 5 minutes or 300secs
digitalWrite (pump, LOW);
delay(900);
}
```

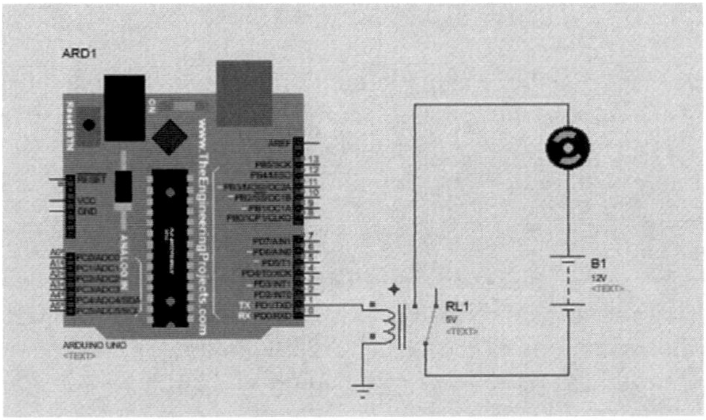

Figure 7. Simulation of tidal methods

After the simulation was done, then the author made a prototype with the following design:

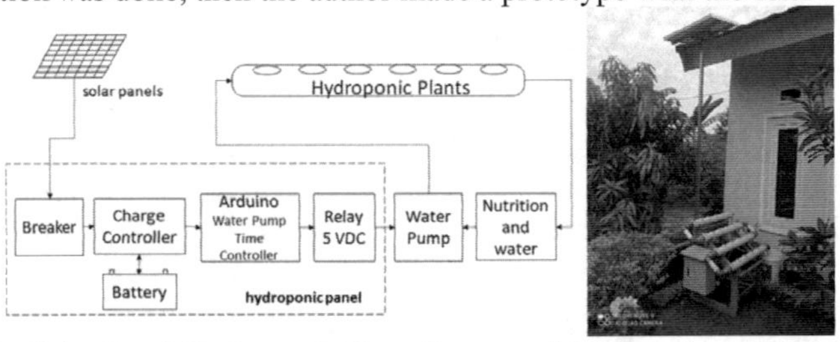

Figure 8. Solar Panel Hydroponic Plant System with Time Control/tidal method

Comparison of water pump conditions on solar panel hydroponic plants using a 7.2 Ampere battery between not using the tidal method and using the tidal method as follows:

Table 1. comparison does not use tidal methods and uses tidal methods

No	Time	Water Pump Status	
		Not using the tidal method	using the tidal method
1	18:00	Working	Working
2	19:00	Working	Working
3	20:00	Working	Working
4	21:00	Working	Working
5	22:00	Working	Working
6	23:00	Working	Working
7	00:00	Not Working	Working
8	01:00	Not Working	Working
9	02:00	Not Working	Working
10	03:00	Not Working	Working

11	04:00	Not Working	Working
12	05:00	Not Working	Working
13	06:00	Not Working	Working
13	07:00	Working	Working

From the results of the comparison can be seen that the tidal method can work at night until the morning while not using the tidal method water pump does not work after consuming battery energy for 5 to 6 hours.

4. Conclusion

Based on the results of experiments conducted, the tidal method can turn on the water pump in hydroponic plants longer because the tidal method does not turn on the water pump continuously but is regulated the time of life and death so that when it dies, electrical energy is not consumed by the water pump but consumed when the water pump is working only. Tidal methods are suitable for being used in hydroponic plants that use solar panels as an energy source to turn on water pumps that serve as an introduction to nutrients and oxygen to plant roots.

References

[1] Abdul Kadir, 2017, Dasar Raspberry Pi, Andi Offset, Yogyakarta, ISBN: 978-979-29-5811-9
[2] Affan Bachri, 2015, Analisis Efisiensi Pemakaian Daya Listrik Di Universitas Islam Lamongan, Jurnal Teknika Vol. 7 No.1, ISSN No. 2085-0859
[3] David Setiawan, Usaha Situmeang, 2018, Evaluasi Sistem Penerangan Lampu Jalan Di Lingkungan Universitas Lancang Kuning.Jurnal Teknik Elektro Fakultas Teknik Universitas Lancang Kuning.
[4] Deni Almanda, Krisdianto, Erwin Dermawan, 2014, Managemen Konsumsi Energi Listrik Dengan Menggunakan Sensor PIR dan LM 35, Jurnal Elektrum, Vol. 14 No. 1, ISSN : 1979-5564, e-ISSN : 2550-0678
[5] Effendi, asnal. Asep Suryana. 2013. “Evaluasi Sistem Pencahayaan Lampu Jalan,” 86–94.
[6] Faizal Reza, Hartono, Siswanto Nurhadiono, 2015, Analisa Deviasi kWh Meter Memanfaatkan Aplikasi Android “App Tole”, Jurnal Institusi Teknologi Dan Seni, Vol.7, No 3, ISSN: 1978-2497
[7] Indrakoesoema Koes, Kiswanto, Muhammad Taufiq, 2013, Kajian Penghematan Energi Listrik Dengan Pemasangan Inverter Pada Motor Fan Menara Pendingin Rgs – Gas, Seminar Nasional VIII SDM Teknologi Nuklir, ISSN : 1978-0176
[8] Irwan Dinata, Wahri Sunanda, 2015, Implementasi Wireless Monitoring Energi Listrik Berbasis Web Database, Jurnal Nasional Teknik Elektro, Vol : 4, No. 1, ISSN: 2302-2949
[9] Itmi Hidayat Kurniawan, Latiful Hayat, 2014, Perancangan Dan Implementasi Alat Ukur Tegangan,Arus Dan Frekuensi Listrik Arus Bolak-Balik Satu Fasa Berbasis Personal Computer, Techno Jurnal Vol.15 No.1, hal. 21-31 ISSN 1410 - 8607
[10] Margioni Abdillah, 2014, Desain dan Instalasi Penerangan jalan Raya, Yayasan Kemajuan Teknik, Pontianak, ISBN: 978-602-72774-6-5
[11] Panggih Sudarmono, Deendarlianto, Adhika Widyaparaga, 2018, Energy efficiency effect on the public street lighting by using LED light replacement and kwh-meter installation at DKI Jakarta Province, Indonesia, Journal of Physics: Conf. Series 1022(2018) 012021 doi:10.1088/1742-6596/1022/1/012021
[12] P.Van.Harten, E.Setiawan, 1985, Instalasi Listrik Arus Kuat 2, Trimitra Mandiri.
[13] Sinaulan M. Olivia, Yaulie D. Y. Rindengan,Brave A. Sugiarso, 2015, Perancangan Alat Ukur Kecepatan Kendaraan Menggunakan ATMega 16, Jurnal Teknik Elektro dan Komputer, Vol 4, No 2, ISSN: 2301-8402
[14] Siti Ahadaiah, Muharnis, Agustiawan, Implementasi Sensor PIR Pada Peralatan Elektronik Berbasis Microcontroller, Jurnal INOVTEK POLBENG, Vol: 01 No,1, ISSN:2088-6225 E-

ISSN: 2580-2798

[15] William Timotius, Muhammad Sofrodin, Suryono, 2014, Efisiensi Penerangan Jalan Umum Menggunakan Sensor Gerak Berbasis Mikrokontroler, Prosiding Seminar Nasional Sains dan Pendidikan Sains UKSW Salatiga, Vol: 5, N0.1, ISSN : 2087-0922

[16] Zubaili Isfarizky, Fardian, Alfatirta Mufti, 2017, Bangun Sistem Kontrol Pemakaian Listrik Secara Multi Channel Berbasis Arduino (Studi Kasus Kantor LBH Banda Aceh), JurnalTeknik Elektro,Fakultas Teknik Universitas Syiah Kualam Vol.02-02 hal: 30-35 e-ISSN: 2252-7036

[17] Zuriman Anthony, 2011, Penggunaan Kontaktor Sebagai Sistem Pengaman Motor 3 Phasa Terhadap Kehilangan 1-Phasa Sistem Tenaga, Jurnal Momentum Vol.11 No.2 ISSN: 1693 – 752X

Resetting the ground fault relay on a 28 MVA power transformer due to grounding method retrofit

Hamzah Eteruddin[1], Arlenny[1], Hadi Siswanto[1], Yanuar Z. Arief[2], Fathimah Hasanti[3]

[1]Universitas Lancang Kuning, Pekanbaru, Indonesia
[2]Universiti Malaysia Sarawak, Sarawak, Malaysia
[3]Universiti Teknologi Malaysia, Johor, Malaysia

*hamzah@unilak.ac.id

Abstract. The goal of the retrofit at Zamrud substation is to transform a 28 MVA power transformer's neutral point grounding system from direct grounding to grounding using a 20 Ohm Neutral Grounding Resistor (NGR). The amount of short-circuit current produced will fluctuate with the installation of NGR. It is necessary to re-setting the Ground Fault Relay (GFR) on Zamrud substation. To estimate the value of the short-circuit current after installing the NGR, the impedance matrix approach is used to calculate it. According to the results, LG's largest short-circuit current is 397.83 A and its smallest is 313.55 A. Meanwhile LLG has a maximum short-circuit current of 5.334.30 A and a minimum short-circuit current of 943.31 A. Therefore, after NGR installation, the value of the relay setting is as follows: GFR pickup current on the transformer grounding GFR relay is 234.29 A, GFR pickup current on the neutral feeder 1 is 19.96 A, GFR pickup current on the neutral feeder 2 is 30.42 A, and GFR pickup current on the neutral feeder 3 is 38.40 A.

1. Introduction

The Joint Operations Agency PT. Bumi Siak Pusako - Pertamina Hulu, hereinafter abbreviated as BOB, is a Contractors of Cooperation Contract ("KKKS") based in the Siak Regency in Riau Province. This company engaged in the oil and gas industry. BOB generates its own electrical energy to support its operations and ensure operational continuity. Short-circuit failures can result in flashovers, damage equipment, and expose danger to employees [1]–[5]. Numerous researches have been conducted to reduce the incidence of disruptions as well as to ensure that the energy distribution to load is maintained [6]–[10].

The grounding method influences the short-circuit fault current to ground [11], [12]. The Neutral Grounding Resistor (NGR) is a technique of grounding [13], [14]. Resistance is installed in series between the transformer neutral and earth in the NGR method. The amount of short-circuit fault current flowing from the neutral side to ground is controlled by NGR [15]–[17].

1.1 Theory and Hypotheses

Ground Fault Relay works when fault current flows through the return ground (ground fault). There are several ground fault protection methods, including: a). Residual current method, b). Neutral circuit to ground method and c). Combined phase fault and neutral fault protection [18]–[20]. Meanwhile, the working duration of the Overcurrent Relay (OCR), GFR, and Recloser is determined by the type of

IOP Conf. Series: Earth and Environmental Science 1041 (2022) 012021 doi:10.1088/1755-1315/1041/1/012021

curve applied, such as: a). IEC overcurrent relay curve, b). IAC (GE) Overcurrent Curve [21]. This will be essential to adjust the coordination settings of the protection system, such as the Ground Fault Relay (GFR), in order to install the NGR [20].

2. Research Methods

The electrical system studied includes three feeders, each of which is fed by a single unit step-down transformer (115 kV to 13.8 kV) as indicated in Figure 1.

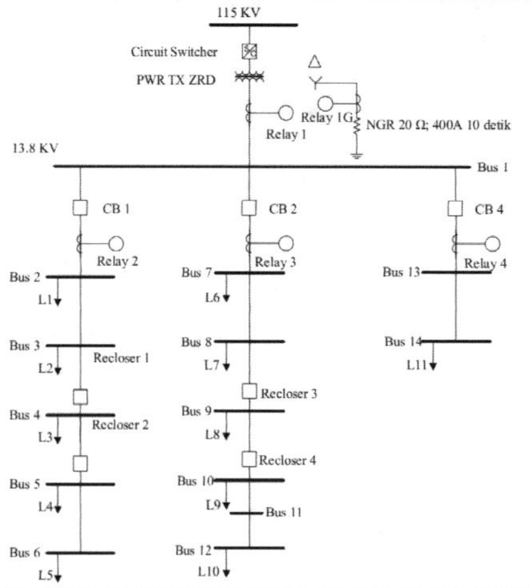

Figure 1. Overview of the Research Single Line Diagrams

The short-circuit current must be calculated first before the relay setting can be determined. The single-phase line to ground (LG) short-circuits current for neutral protection and a double phase to ground (LLG) short-circuits current for phase protection are utilized to calculate the relay setting.

3. Result and Discussion

Short circuit current on each bus is shown in Table 1. From this table, it can be seen that the highest LG short-circuit current is 397.83 A, which corresponds to the fault current situated upstream of the feeder, and the lowest LG short-circuit current is 313.55 A, that correlates to the fault current located downstream of the feeder 2. The greatest LLG is 5,334.30 A, which is the fault current upstream of the feeder, and the smallest short-circuit fault current is 943.31 A, which is the fault current downstream of the feeder 2.

3.1 Phase Protection Settings

The phase protection in the BOB electrical system is provided by four recloser devices: two in each feeder, three OCR relay units at each feeder upstream, and one OCR relay unit on the transformer's outgoing side. The recloser unit is a protection that is located as far downstream as possible from the feeder, allowing the recloser to work as quickly as possible. The quickest working time is 0.3 seconds, with a 0.4-second time gap between protection.

With the following considerations, a timing difference of 0.4 seconds between one relay and the main or back-up relay's may be calculated: a). Relay time error = 0.2 seconds; b). PMT opening time until the spark vanishes = 0.083 to 0.14 seconds; c). Safety factor = 0.05 seconds; d). Overcurrent auxiliary relay delay = 0.05 seconds. Table 2 also displays the results of the phase protection unit settings for each BOB electrical system phase protection unit.

Table 1. LG and LLG fault currents on various buses

Disturbance Location	Line to Ground Fault (A)	Line to Line to Ground (A)
Bus1	397.83	5,334.30
Bus2	397.82	5,330.89
Bus3	391.78	3,936.13
Bus4	390.63	3,792.60
Bus5	360.38	1,922.51
Bus6	313.55	1,074.01
Bus7	397.82	5,330.89
Bus8	385.76	3,148.90
Bus9	374.59	2,436.48
Bus10	360.14	1,882.73
Bus11	315.29	1,086.39
Bus12	300.42	943.31
Bus13	397.82	5,330.89

Table 2. Setting Each Phase Protection Unit

Position	Unit Type	$I_{Prim.}$ (A)	$I_{Sec.}$ (A)	t (s)	TMS
Out-going	Relay 1	1,405.72	4.69	1.50	4.07
Feeder 1	Relay 2	119.76	1.00	1.10	10.25
	Recloser 1	91.68	-	0.70	0.39
	Recloser 2	39.48	-	0.30	0.17
Feeder 2	Relay 3	182.52	1.52	1.10	9.47
	Recloser 3	118.44	-	0.70	0.31
	Recloser 4	27.60	-	0.30	0.19
Feeder 3	Relay 4	230.40	1.92	0.30	2.44

Table 3. Neutral Protections Settings

Position	Unit Type	$I_{Prim.}$ (A)	$I_{Sec.}$ (A)	t (s)	TMS
Out-going	Relay 1	234.29	1.95	1.50	0.74
Feeder 1	Relay 2	19.96	0.17	1.10	8.60
	Recloser 1	15.28	-	0.70	0.33
	Recloser 2	6.58	-	0.30	0.18
Feeder 2	Relay 3	30.42	0.25	1.10	7.47
	Recloser 3	19.74	-	0.70	0.30
	Recloser 4	4.60	-	0.30	0.20
Feeder 3	Relay 4	38.40	0.32	0.30	1.84

3.2 Neutral Protection Setting

Similar to phase protection, the BOB electrical system features neutral protection in the form of four recloser units, two in each feeder, three GFR devices in each feeder upstream, and one GFR unit on the transformer's grounding side. The findings of neutral protection in BOB electrical system are presented in Table 3.

IOP Conf. Series: Earth and Environmental Science 1041 (2022) 012021 doi:10.1088/1755-1315/1041/1/012021

3.3 Protection Coordination Curve

The protection setting is then computed and utilized for ETAP 12.60's OCR and GFR relay and recloser settings. The protection coordination curve can be plotted in ETAP by providing these values. Figure 2 shows the phase protection (Figure 2.a) and neutral protection (Figure 2.b) coordination curves in feeder 1.

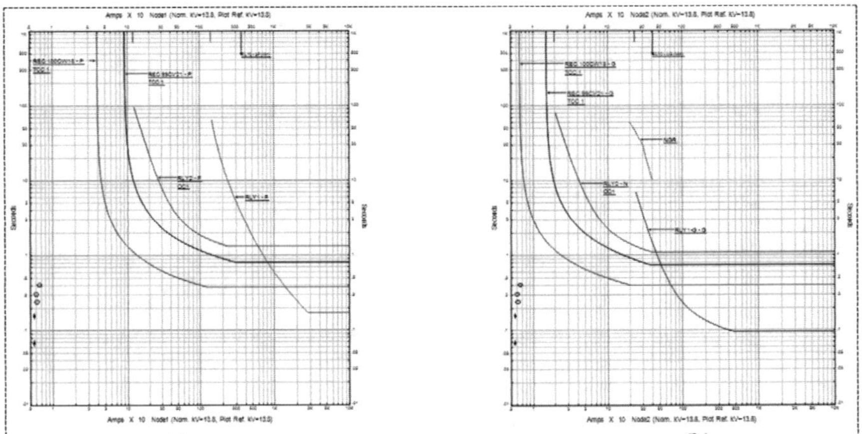

(a) (b)

Figure 2. Coordination of Protection Phase and Neutral Feeder 1 Side

Relay 1, relay 2, recloser 1 (REC 992CV1), and recloser 2 (REC 992CV2) are the four protection components that connect the feeder 1 side to the transformer's outgoing side (REC 100CW18). The phase and neutral protection have worked well, which can be seen in the diagram of coordination curve above, it is shown from the recloser 2 at the very end, which has a smaller pickup current and a faster working time, followed by recloser 1, relay 2, and relay 1 on the outgoing transformers. Figure 3 depicts the phase protection (Figure 3.a) and neutral protection (Figure 3.b) coordination curves in feeder 2.

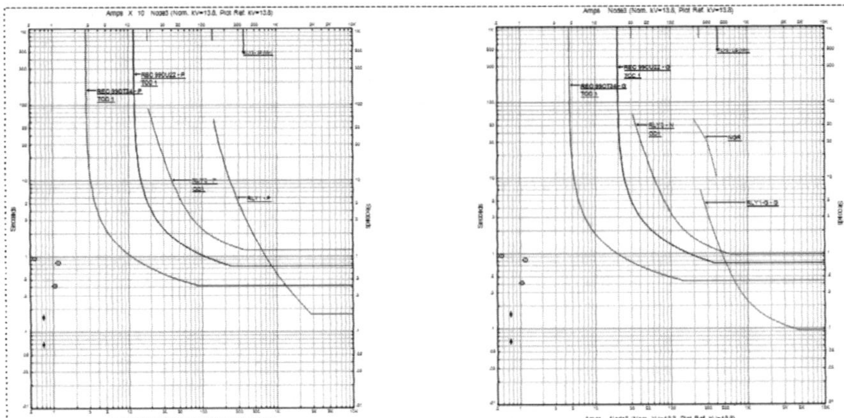

(a) (b)

Figure 3. Coordination of Protection Phase and Neutral Feeder 2

The feeder 2 side to the transformer's outgoing side is made up of four protection units: relay 1, relay 3, recloser 3 (REC 99CU22), and recloser 4 (REC 992CT34). The phase and neutral protection have functioned well, as shown in the diagram of coordination curve above, it can be seen from the recloser

4 at the very end, which has a smaller pickup current and a faster working time, followed by recloser 3, relay 3, and relay 1 on the outgoing transformers. While Figure 4 depicts the phase protection (Figure 4.a) and neutral protection (Figure 4.b) coordination curves on feeder 3.

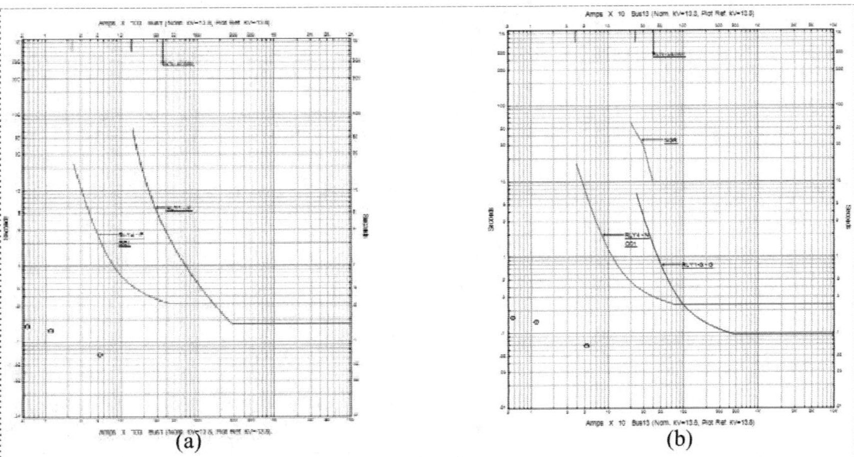

Figure 4. Coordination of protection Phase and Neutral Feeder 3

Only two protection devices, relay 1 and relay 4, connect the feeder 3 side to the outgoing transformers. Phase protection and neutral protection have operated effectively, as seen in the coordination curve above, it can be observed from relay 4, which is positioned at the far end. The opposite end of the transformer has a lower pickup current and a faster running duration, followed by relay 1 on the transformer's outgoing side. When a ground fault occurs in all feeders, the neutral protection coordination curve might open the circuit faster than the NGR time limit to ground fault currents, as shown in Figures 2 – Figure 4.

4. Conclusion

The value of the relay setting after the installation of Neutral Grounding Resistor (NGR) for the pickup current on the GFR relay on the transformer grounding is 234.29 A, while the GFR pickup current on the neutral feeder 1 is 19.96 A, 30.42 A on neutral feeder 2, and 38.40 A on neutral feeder 3. The fault current placed upstream of the feeder has the highest LG short-circuit current of 397.83 A, whereas the fault current located downstream of feeder 2 has the lowest LG short-circuit current of 313.55 A.

References

[1] M. E. El-Hawary, Electrical Energy Systems, 2nd ed. Boca Raton: CRC Press, 2018.
[2] C. Weinschenk, D. Madrzykowski, and P. Courtney, "Impact of Flashover Fire Conditions on Exposed Energized Electrical Cords/Cables," Fire Technology, vol. 56, no. 3, pp. 959–991, 2020.
[3] M. M. Islam, Handbook to IEEE Standard 45: A Guide to Electrical Installations on Shipboard. New York: Wiley, 2004.
[4] T. E. Browne, Circuit Interruption: Theory and Techniques. London: CRC Press, 1984.
[5] "IEEE Guide for Breaker Failure Protection of Power Circuit Breakers," IEEE Std C37.119-2016 (Revision of IEEE Std C37.119-2005). IEEE, New York, pp. 1–73, 2016.
[6] H. Eteruddin, A. A. Mohd Zin, and B. Belyamin, "Line Differential Protection Modeling with Composite Current and Voltage Signal Comparison Method," Telkomnika, vol. 12, no. 1, Mar. 2014.
[7] Y. Z. Arief, A. A. Jalaluddin, M. H. I. Saad, L. Gurusamy, and H. Eteruddin, "Effect of Lightning Surge in AC Power and Telecommunication Lines for Electrical Devices," in 2020 13th

International UNIMAS Engineering Conference (EnCon), 2020, pp. 1–4.

[8] Y. Z. Arief, S. S. A. Sa-Or, N. Mubarakah, M. H. I. Saad, and H. Eteruddin, "Model of Grid-Connected PV System in Sarawak, Malaysia Rural Area," in 2020 4rd International Conference on Electrical, Telecommunication and Computer Engineering (ELTICOM), 2020, pp. 250–259.

[9] H. Eteruddin, A. Atmam, and D. Setiawan, "The Impact of Solar Panel Temperature to Solar Home System (SHS) Output Voltage," IOP Conference Series: Earth and Environmental Science, vol. 469, no. 1, 2020.

[10] H. Eteruddin, D. Setiawan, and A. Atmam, "Web Based Raspberry Monitoring System Solar Energy Power Plant," IOP Conference Series: Earth and Environmental Science, vol. 469, no. 1, 2020.

[11] U. C. Ogbuefi, J. O. Onyishi, and S. E. Oti, "Effectiveness of Neutral Grounding on Power System Application," Asian Journal of Basic Science & Research, vol. 2, no. 2, pp. 44–57, 2020.

[12] N. A. Shalash, A. M. A. Haidar, and K. A. Sattar, "Grounding Locations Assessment of Practical Power System," Energy Power Engineering, vol. 4, no. 1, p. 9, 2012.

[13] J. Blumschein and Y. Yelgin, "New Design of Ground Fault Protection," in 2018 71st Annual Conference for Protective Relay Engineers (CPRE), 2018, pp. 1–7.

[14] A. R. Sultan, M. W. Mustafa, M. Saini, and A. Gaffar, "Identification of Ground – Fault Current in Various Transformer Connection and Generator Grounding using Symmetrical Components Method," in The 5th International Symposium on Material, Mechatronics and Energy, 2019, p. 10.

[15] A. Y. Ahmed and S. I. Al-Mously, "Sensitivity Improvement of the Digital Differential Relay for Internal Ground Fault Protection in the Power Transformer with Tap Changer," in 2001 IEEE Porto Power Tech Proceedings (Cat. No.01EX502), 2001, vol. 4, p. 3 pp. vol.4.

[16] J. C. Das, "Ground Fault Protection for Bus-Connected Generators in an Interconnected 13.8 kV System," in Conference Record of 2006 Annual Pulp and Paper Industry Technical Conference, 2006, pp. 1–10.

[17] ASCO Power Technologies, "Neutral Grounding Resistors for Limiting Fault Current," Cleveland, 2018.

[18] H. Ricardo and T. A. Nogueira, "Ground Fault Protection Methods for Distribution Systems," Neutro-à-Terra, no. 17, pp. 15–20, 2016.

[19] C. Siemens, "Ground Fault Application Guide WL Low Voltage Power Circuit Breakers." 2014.

[20] J. Roberts, H. J. Altuve, and D. Hou, "Review of Ground Fault Protection Methods for Grounded, Ungrounded, and Compensated Distribution Systems," in 28th Annual Western Protective Relay Conference, 2001, p. 40.

[21] GE Multilin, "F35 Multiple Feeder Protection System : UR Series Instruction Manual." GE Industrial Systems, Ontario, p. 590, 2007.

Intelligent ubiquitous technology as a precision agri-food framework: a proposed framework

T A Cahyanto[1,2*], S Wahjuni[2], H Sukoco[2], H Rahmawan[2], S N Neyman[2]

[1]Universitas Muhammadiyah Jember, Jember, Indonesia
[2]IPB University, Bogor, Indonesia

*triawanac@unmuhjember.ac.id

Abstract. Agri-Food is another term for the agriculture-based food sector, which is currently one of Indonesia's main pillars of the national economy. One of the well-known plants for agricultural-based food processing called Agri-Food in Indonesia is Cocoa. Cocoa is one of the plantation commodities that has an essential role in economic activities in Indonesia. Given the critical role of cocoa as a source of the country's economy, it must guarantee productivity and quality of cocoa and sustainability. To ensure cocoa productivity and quality sustainability, agricultural activities in today's era can integrate intelligent ubiquitous technology as a framework for obtaining precise production results. What can measure precision production results from three aspects: increasing agricultural land productivity, reducing production costs, and minimizing the environmental impact. This paper discusses the proposed framework with intelligent ubiquitous technology to support precision agri-food that focuses on context-aware systems. The context-aware system in the proposed framework is expected to become a control center in collecting and processing data in real-time to help farmers make the best decisions regarding weather conditions and diseases caused by plants.

1. Introduction

Agri-Food is another term for the agriculture-based food sector, which is currently one of the main pillars of the national economy in Indonesia, which accounts for more than a third of the country's total gross domestic product (GDP) in 2019 [1]. One of the well-known plants for agricultural-based food processing (Agri-Food) in Indonesia is Cocoa. Cocoa is one of the plantation commodities that has an essential role in economic activities in Indonesia. Indonesia is the third-largest cocoa producer and exporter globally after Ghana and Ivory Coast [2].

Given the critical role of cocoa as a source of the country's economy, it must guarantee the productivity and quality of cocoa and sustainability. To ensure the sustainability of cocoa productivity and quality, agricultural activities in the current era can integrate intelligent ubiquitous technology as a framework to obtain precise Agri-Food production results. What can measure precision Agri-Food production results from three aspects: increasing agricultural land productivity, reducing production costs, and minimizing the environmental impact [3].

What can achieve the success of these three aspects is by applying intelligent ubiquitous technology. What can measure the advantages of ubiquitous, innovative technology based on three components: data acquisition at an appropriate scale, data interpretation and analysis, and implementation of response management at a proper scale and time? Ubiquitous technology often referred to as ubiquitous computing, increases the usability of computers by making many computers available in the physical environment but making them invisible to the user [3].

Ubiquitous technology often referred to as ubiquitous computing, increases the usability of computers by making many computers available in the physical environment but making them invisible to the user [4], [5]. Ubiquitous technology allows hundreds of devices to be connected and computer functionality to be embedded in a device. Users can use the integration of these devices and computers without thinking and realizing, which will be very helpful in carrying out other activities without the user touching or interacting directly with the activity in question. The application of technology as a framework to support Agri-Food precision has been widely carried out [6], [7], [16]–[21], [8]–[15].

Table 1. Desain of Precision Agriculture or Smart Farming

Author	Title	Architecture Systems	Computing Model	Integration with AI
Dholu, M et al., 2018	Internet of Things (IoT) for Precision Agriculture Application.	3 Layers	Cloud	No
Mocnej, J et al., 2021	Quality-enabled Decentralized IoT Architecture with Efficient Resources Utilization.	4 Layers	Cloud	No
Muangprathub, J et al., 2019	IoT and Agriculture Data Analysis for Smart Farm	3 Layers	Cloud	No
Yang, J et al., 2018	Botanical Internet of Things: Toward Smart Indoor Farming by Connecting People, Plant, Data and Clouds	4 Layers	Cloud	No
Li, Z et al., 2017	Design of an Intelligent Management System for Agricultural Greenhouses Based on the Internet of Things	3 Layers	Cloud	Yes
Sahitya, G. et al., 2017	Designing a Wireless Sensor Network for Precision Agriculture Using Zigbee	2 Layers	WSN	No
Shadrin, D. et al., 2020	Enabling Precision Agriculture Through Embedded Sensing With Artificial Intelligence	3 Layers	Cloud	Yes
Shadrin, D. et al., 2019	Designing Future Precision Agriculture: Detection of Seeds Germination Using Artificial Intelligence on a Low-Power Embedded System.	2 Layers	WSN	No
Saha, R. et al., 2021	DLSense: Distributed Learning-Based Smart Virtual Sensing for Precision Agriculture	3 Layers	Cloud	No
Zhang, X. et al., 2017	Monitoring Citrus Soil Moisture and Nutrients Using an IoT Based System	3 Layers	Cloud	No
Colezea, M. et al., 2018	CLUeFARM: Integrated web-service platform for smart farms	3 Layers	Cloud	No
Olivo, B-M. et al., 2018	Rules engine and complex event processor in the context of	3 Layers	Cloud	Yes

	internet of things for precision agriculture			
Gaikwad, S.V. et al., 2021	An innovative IoT based system for precision farming	3 Layers	Cloud	No
Chen, W.-L. et al., 2020	RiceTalk: Rice Blast Detection using Internet of Things and Artificial Intelligence Technologies	3 Layers	Cloud	Yes
Shafi, U. et al., 2020	A Multi-Modal Approach for Crop Health Mapping Using Low Altitude Remote Sensing, Internet of Things (IoT) and Machine Learning	3 Layers	Cloud	Yes
Cahyanto, T.A et al., 2021	Proposed Framework	4 Layers	Edge+Cloud Computing	Yes

The literature study was obtained based on the keywords "Technology on precision agriculture," "edge or fog computing on the smart environment," on the IEEE, ACM, and Elsevier pages in the last ten years. There are more than 400 articles relevant to the topic of this paper based on these keywords. However, after conducting the review process, fifteen articles can represent the uniqueness of the framework that other researchers have done. The categories of frameworks that have been made in the eight articles include design or architecture of precision agriculture technology, computational models, and integration of sensors with artificial intelligence.

Based on literature studies, both articles from conferences and journals, there is still little research on the framework to support precision Agri-Food with edge and cloud computing models integrated with artificial intelligence technology. Table 1. illustrates that the current trend of ubiquitous technology research has begun to incorporate sensors with artificial intelligence. This paper will discuss the design and proposed framework based on intelligent ubiquitous technology to support precision Agri-Food. It is hoped to implement the plan immediately and proposed framework into a natural environment to help accuracy Agri-Food.

2. Methodology

The research produced in this paper is an initial study in obtaining a proposed framework for intelligent ubiquitous technology to support precision Agri-Food.

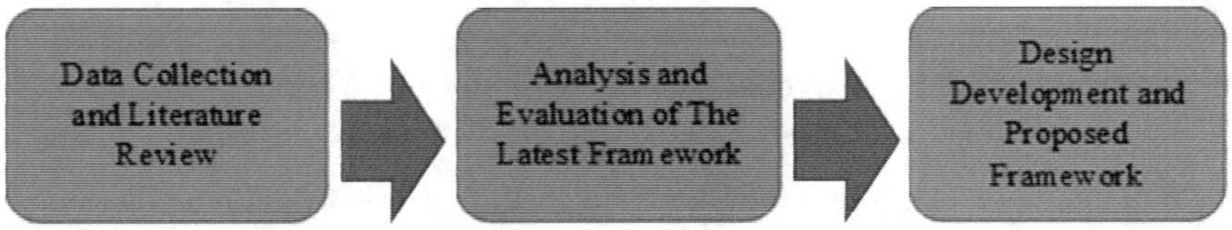

Figure 1. Research Methodology

The stages of research carried out include:
a. Data collection and literature study
Data collection is the initial stage of research that aims to obtain an overview of literature studies from other researchers related to the theme or research topic relevant to this paper.
b. analysis and evaluation of the latest framework

IOP Conf. Series: Earth and Environmental Science 1041 (2022) 012022 doi:10.1088/1755-1315/1041/1/012022

Analysis and assessment of the current framework are carried out to find the shortcomings and non-optimality of the existing framework so that who can propose to develop the framework that What will make in this paper.
c. Design development and proposed framework
The story of the design and proposed framework was made based on the two previous research stages. It could be implemented in real terms in precision Agri-Food, especially for cocoa farmers.

3. Result and Discussion

Technological trends in the Agri-Food sector are starting to develop frameworks integrated with artificial intelligence technologies such as machine learning, deep learning, and decision-making automation. The movement of machine learning technology that can support precision Agri-Food is for classification, prediction, association, and grouping based on data from the Agri-Food sector. It can use deep learning technology trends in helping precision Agri-Food for data processing, both image and non-image data such as multispectral data, etc.

Trends in decision-making systems that can support precision Agri-Food are generally related to the automation of actions, such as fertilization, irrigation, harvesting based on the knowledge provided by sensors. The Internet of Things (IoT)-based framework model proposed in this paper refers to the ITU-T Y.2060 standard [22].

3.1 Device Layer

The proposed framework uses the concept of direct data transmission capability. Directly means that it will send the data obtained from the sensor directly to the base station. The capability of the base station to receive data in this proposed framework is to use a WiFi gateway interface.

Table 2. Sensors used in Precision Agri-Food [23]

	Sensor		Sensor		Sensor
Soil	Temperature	Plant	Photosynthesis	Weather	Temperature
	Moisture		Moisture		Humidity
	Rain/water flow		Hydrogen		Atmospheric pressure
	Water level		Wetness		Wind speed
	Conductivity		CO_2		Wind direction
	Salinity		Temperature		

3.2 Network Layer

The network layer capability in the proposed framework will handle two essential things: mobility management to ensure data transmission runs well and connectivity management between sensor devices and base stations.

3.3 Service Support and Application Support Layer

The proposed framework will have two primary services in terms of support systems, namely data storage and data processing. The primary data store in the framework to be created focuses on cloud services so that farmers can directly manage data. Data processing will be carried out at the local station (edge), hoping it will filter the initial data sent from the sensor to become valid. It will later send this valuable data to cloud services for processing as decision support in Agri-Food Precision.

3.4 Application Layer

The application layer in the proposed framework is the display of the software in the concept of a mobile-based application. Mobile-based applications will be created using the webview application architecture. The output of this application is expected to facilitate farmers in making decisions regarding plant growth data.

3.5 Proposed Framework

According to ITU-T Y. 2060, the following is a block diagram of the proposed framework based on the standard IoT framework.

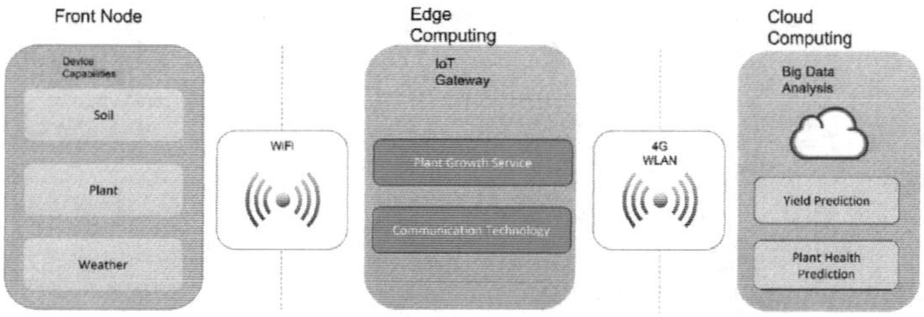

Figure 2. Block Diagram of Proposed Framework

The proposed framework architecture system uses four-layer technology, hoping that the framework that has been made will later support precision Agri-Food. Productivity and quality aspects are undoubtedly important aspects of whether the framework is successful in helping precision Agri-Food. The productivity aspect is measured using the basic needs of plant cultivation, including the need for fertilization, seeds, and plant supervision. Crop yield quality can be measured using tree quality parameters (stems, leaves, roots) and land quality (soil and weather). It will compare these two aspects with the production results using the traditional Agri-Food concept with the Agri-Food concept using the proposed framework.

4. Conclusion

Frameworks with intelligent ubiquitous technology to support precision Agri-Food are still few in actual life applications. This paper tries to propose a framework with innovative, ubiquitous technology, hoping it can later help farmers manage crops. It will measure the success of this framework from two aspects, namely productivity, and quality. The results of the comparison of the traditional Agri-Food concept with the precision Agri-Food concept with the proposed framework will be obtained so that it can show whether this framework can support precision Agri-Food or not. The proposed framework uses a four-layer system architecture and integrates artificial intelligence both on the edge computing side as a base station and on the cloud side. The integration of sensor data with edge architecture and cloud computing is expected to assist farmers in analyzing plant growth data. The plant growth analysis based on this intelligent framework is expected to predict crop yields and crop quality.

References

[1] Antaranews, "Agri food hadapi risiko pemulihan selama COVID-19," 2021. https://www.antaranews.com/berita/2178394/agri-food-hadapi-risiko-pemulihan-selama-covid-19 (accessed Aug. 26, 2021).

[2] B. P. Statistik, Statistik Kakao Indonesia Tahun 2019. 2020.

[3] N. R. Council, Precision Agriculture in the 21st Century: Geospatial and Information Technologies in Crop Management. Washington, DC: The National Academies Press, 1997.

[4] M. Weiser, "Ubiquitous computing," Computer (Long. Beach. Calif)., vol. 26, no. 10, pp. 71–72, 1993, DOI: 10.1109/2.237456.

[5] M. Weiser, "Some computer science issues in ubiquitous computing [Algunas cuestiones

informáticas en la computación ubicua]," Commun. ACM, vol. 36, no. 7, pp. 75–84, 1993.

[6] J. Yang, M. Liu, J. Lu, Y. Miao, M. A. Hossain, and M. F. Alhamid, "Botanical Internet of Things: Toward Smart Indoor Farming by Connecting People, Plant, Data and Clouds," Mob. Networks Appl., vol. 23, no. 2, pp. 188–202, 2018, DOI: 10.1007/s11036-017-0930-x.

[7] J. Muangprathub, N. Boonnam, S. Kajornkasirat, N. Lekbangpong, A. Wanichsombat, and P. Nillaor, "IoT and agriculture data analysis for smart farm," Comput. Electron. Agric., vol. 156, no. December 2018, pp. 467–474, 2019, DOI: 10.1016/j.compag.2018.12.011.

[8] J. Mocnej et al., "Quality-enabled decentralized IoT architecture with efficient resources utilization," Robot. Comput. Integr. Manuf., vol. 67, 2020, DOI: 10.1016/j.rcim.2020.102001.

[9] M. Dholu and K. A. Ghodinde, "Internet of Things (IoT) for Precision Agriculture Application," Proceedings of the 2nd International Conference on Trends in Electronics and Informatics, ICOEI 2018. pp. 339–342, 2018, DOI: 10.1109/ICOEI.2018.8553720.

[10] I. Pedersen, G. Patil, B. Berget, C. Ihlebæk, and M. T. Gonzalez, "Mental health rehabilitation in a care farm context: A descriptive review of Norwegian intervention studies," Work, vol. 53, no. 1. pp. 31–43, 2016, DOI: 10.3233/WOR-152213.

[11] Z. Li, J. Wang, R. Higgs, L. Zhou, and W. Yuan, "Design of an Intelligent Management System for Agricultural Greenhouses Based on the Internet of Things," Proc. - 2017 IEEE Int. Conf. Comput. Sci. Eng. IEEE/IFIP Int. Conf. Embed. Ubiquitous Comput. CSE EUC 2017, vol. 2, pp. 154–160, 2017, DOI: 10.1109/CSE-EUC.2017.212.

[12] S. Roy et al., "IoT, big data science & analytics, cloud computing and mobile app based hybrid system for smart agriculture," 2017 8th Industrial Automation and Electromechanical Engineering Conference, IEMECON 2017. pp. 303–304, 2017, DOI: 10.1109/IEMECON.2017.8079610.

[13] D. Shadrin, A. Menshchikov, D. Ermilov, and A. Somov, "Designing Future Precision Agriculture: Detection of Seeds Germination Using Artificial Intelligence on a Low-Power Embedded System," IEEE Sens. J., vol. 19, no. 23, pp. 11573–11582, 2019, DOI: 10.1109/JSEN.2019.2935812.

[14] D. Shadrin, A. Menshchikov, A. Somov, G. Bornemann, J. Hauslage, and M. Fedorov, "Sensing With Artificial Intelligence," vol. 69, no. 7, pp. 4103–4113, 2020.

[15] R. Saha, A. Chakraborty, S. Misra, S. K. Das, and C. Chatterjee, "DLSense: Distributed Learning-based Smart Virtual Sensing for Precision Agriculture," IEEE Sens., 2021.

[16] X. Zhang, J. Zhang, L. Li, Y. Zhang, and G. Yang, "Monitoring citrus soil moisture and nutrients using an IoT based system," Sensors (Switzerland), vol. 17, no. 3, pp. 1–10, 2017, DOI: 10.3390/s17030447.

[17] M. Colezea, G. Musat, F. Pop, C. Negru, A. Dumitrascu, and M. Mocanu, "CLUeFARM: Integrated web-service platform for smart farms," Comput. Electron. Agric., vol. 154, no. April 2017, pp. 134–154, 2018, doi: 10.1016/j.compag.2018.08.015.

[18] B. Mazon-Olivo, D. Hernández-Rojas, J. Maza-Salinas, and A. Pan, "Rules engine and complex event processor in the context of internet of things for precision agriculture," Comput. Electron. Agric., vol. 154, no. September, pp. 347–360, 2018, doi: 10.1016/j.compag.2018.09.013.

[19] S. V. Gaikwad, A. D. Vibhute, K. V. Kale, and S. C. Mehrotra, "An innovative IoT based system for precision farming," Comput. Electron. Agric., vol. 187, no. June, p. 106291, 2021, DOI: 10.1016/j.compag.2021.106291.

[20] W. L. Chen, Y. B. Lin, F. L. Ng, C. Y. Liu, and Y. W. Lin, "RiceTalk: Rice Blast Detection Using Internet of Things and Artificial Intelligence Technologies," IEEE Internet Things J., vol. 7, no. 2, pp. 1001–1010, 2020, DOI: 10.1109/JIOT.2019.2947624.

[21] U. Shafi et al., "A Multi-Modal Approach for Crop Health Mapping Using Low Altitude Remote Sensing, Internet of Things (IoT) and Machine Learning," IEEE Access, vol. 8, pp. 112708–112724, 2020, DOI: 10.1109/ACCESS.2020.3002948.

[22] ITU-T, "Overview of the Internet of things," Series Y: Global information infrastructure, internet protocol aspects and next-generation networks - Frameworks and functional architecture models,

2012. https://www.itu.int/rec/T-REC-Y.2060-201206-I.
[23] Aqeel-Ur-Rehman, A. Z. Abbasi, N. Islam, and Z. A. Shaikh, "A review of wireless sensors and networks' applications in agriculture," Comput. Stand. Interfaces, vol. 36, no. 2, pp. 263–270, 2014, DOI: 10.1016/j.csi.2011.03.004.

 doi:10.1088/1755-1315/1041/1/012023

Development of internet of things-based monitoring system for application solar energy technology in Bunaken island

M Rumbayan[1], I Pundoko[1], D Ruindungan[1], N V Panjaitan[2]

[1]Sam Ratulangi University, Manado, Indonesia
[2]Tokyo International University, Tokyo, Japan

*meitarumbayan@unsrat.ac.id

Abstract. There are some barriers to monitoring and evaluating power plant systems in a remote location; thus, Internet of Things (IoT) enabled systems can be leveraged remotely. This solution is a part of implementing renewable energy-based technology for coastal communities on the small remote island, such as Bunaken Island. Furthermore, this study aims to develop an IoT-based monitoring system of solar energy utilization as a renewable energy source for off-grid solar home systems in remote areas. The methods of this study refer to the stages of the method research and development (R&D) that consist of designing and testing the product of IoT-based solar PV monitoring system. Moreover, this study indicates that the IoT-reported system can report the performance of the remote solar power systems installed in Bunaken Island.

1. Introduction

Indonesia has unique characteristics of having large populations, diverse cultural backgrounds, and astonishing geographical conditions. There is a plethora of small islands that require special handling due to the issues of energy supply. Dependence on the fossil fuels supply from island to island needs to be minimized because of the high transportation costs and producing greenhouse gases. For this reason, leveraging solar energy as an alternative energy source to support green energy programs in small islands needs to be examined.

Solar energy is one of the renewable energy sources locally available in the Islands, which can be developed and studied. Bunaken Island was chosen as the location of the solar energy application potential observation due to its specificity. It is a small and outermost island, well-known by national and international tourist locations for beaches and marine parks. The map of Bunaken Island can be seen in Figure 1.

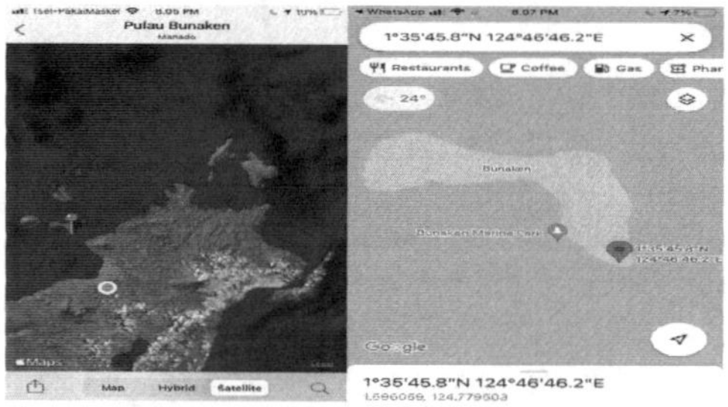

Figure 1. Map of Bunaken Island as the isolated island that located in North Sulawesi Indonesia

The problem of limited electricity supply on the island is due to the dependence on electricity supply from the grid, which operates diesel power plants with high operating costs. One solution to implement an off-grid solar energy system for supplying electricity demand on Bunaken island has been analyzed and presented [1, 2].

It is found that there are some barriers in monitoring the power plant system, considering that the island is a remote location; thus, the Internet of Things can be leveraged as a solution for sustainable energy-based technology implementation for coastal communities on small and islands, such as Bunaken Island.

For the present study, this work is limited only to the current time and voltage monitoring of PV panels to supervise the operation and the performance of small off-grid power systems from a distance. The function of this PV monitoring is to get data in real-time and to support the operation and maintenance of a small off-grid PV system installed on Bunaken island as the pilot plant.

2. Literature Reviews

Many researchers have studied IoT-based PV monitoring systems. Different architectures and topologies of solar PV monitoring systems were highlighted by Shariff et al. [3]. The work addressed several monitoring characteristics such as sample time, data transfer mechanisms, accompanying software, and so on. The architecture and features of several monitoring systems for solar PV power plants were discussed by Rahman et al. [4]. Triki-Lahiani et al. [5] described the various monitoring systems and focused on the key solar PV system failures. Madeti and Singh [6] examined solar PV systems from sensors' standpoint, controllers, and data transfer mechanisms. Shaheer Ansari et al. [7] reviewed monitoring technologies for solar PV systems in terms of categories, specifications, design implementation, software platforms, results, and limits, a full explanation of various data processing modules.

The technology for solar PV system monitoring has been explored and reported by many researchers around the world. Vargas et al. [8] used an Arduino Uno to create a low-cost data logger system for monitoring remote PV systems. Allafi et al. [9] used a low-cost Supervisory Control and Data Acquisition (SCADA) system with an Arduino Uno to create a monitoring system for off-grid PV systems. The current sensor ACS712 and a voltage sensor extracted the data then transferred it to the Arduino Uno microcontroller and connected it to the computer through a USB cable. Jamil et al. [10] presented an Arduino-based performance monitoring system for floating solar PV systems. Harmini and Nurhayati [11] created an independent monitoring system through the ethernet, a PV plant controlled by an ATMEGA8 microcontroller.

The studies about PV monitoring systems may be necessary based on the system's location and under specific circumstances. Therefore, the innovation of a small off-grid PV monitoring system for application in Bunaken island as the remote island in Indonesia is especially significant to be initiated and developed.

3. Methodology

The methods of this study refer to the stages of the method research and development (R&D). The Research and Development research method is a method that could be utilized for producing products and testing their effectiveness before deciding whether or not the products are capable of achieving the previously defined goals. The methods that used in this study are described in Figure 2.

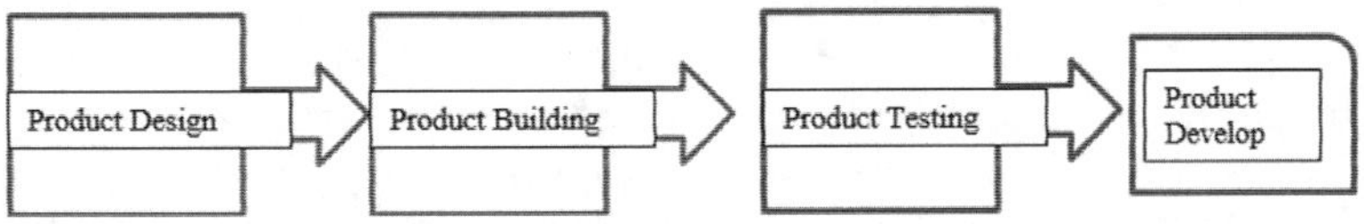

Figure 2. Research Design Stage of Small off-grid PV System Monitoring

doi:10.1088/1755-1315/1041/1/012023

In this method, the main controller was used as Arduino Nano. Electrical parameters such as voltage, current, and output power of solar PV modules with a power output of 100 Wp were also measured and presented in Figure 3. Furthermore, the small off-grid PV system's monitoring was carried out on site of Bunaken island as the remote area.

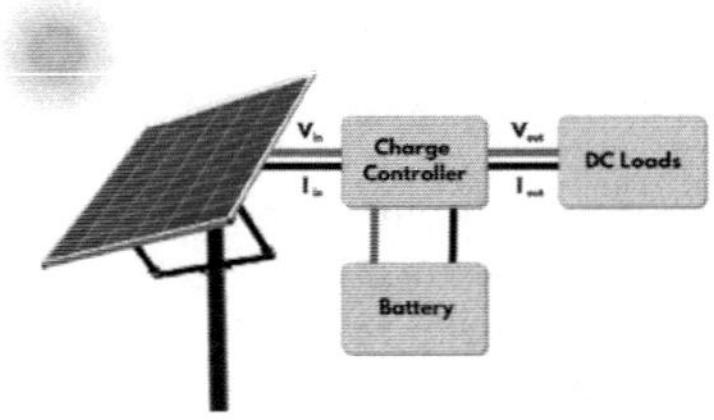

Figure 3. The measurement of electrical parameter of small off-grid PV system

The flowchart of monitoring control system that built-in for this study is shown in Figure 4.

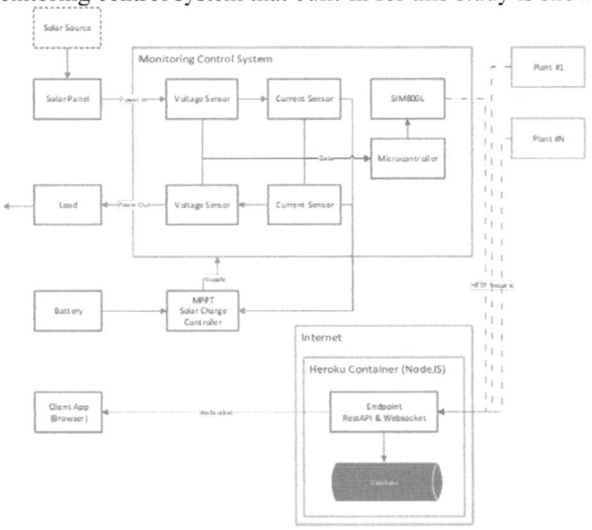

Figure 4. The flowchart of monitoring control system that building for this study

This monitoring system is designed to record data from the integrated solar panels. The purpose of this system is to record flow data and register measured inlet and outlet voltages in the environment in which the solar panels are placed at certain locations. With this monitoring system, the performance of the solar panels can be measured and observed through any media such as laptops, smartphones, or smart TVs. The platform can also be accessed on the internet through a browser.

The data of electrical parameters (i.e, voltage) and current PV panel were obtained through voltage and current sensors. Then, the recorded data were processed in the microcontroller and SIM. The data is sent via 3G, and information is stored in a cloud server. The connection to the internet allows monitoring of the small off-grid PV system from any device or computer.

The connection between the microcontroller and the server is established by a Hypertext Transfer Protocol (HTTP) petition. After setting the connection between the datalogger and the server, the data transferring process starts. The data is transmitted to an application designed to collect and visualize data in real-time in the Heroku platform. The information is stored in a PostgreSQL database hosted on a server.

4. Result and Discussion

The circuit diagram of system's monitoring is shown in Figure 5.

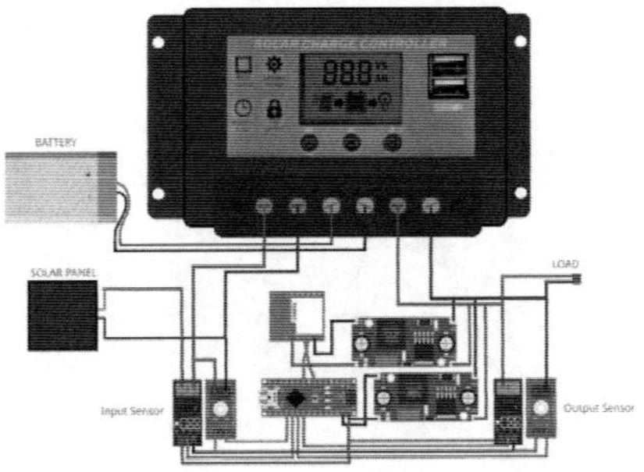

Figure 5. The circuit diagram of system PV monitoring product

The result of system's monitoring product is shown in Figure 6.

Figure 6. The Iot-based system PV monitoring product

The testing of monitoring system based on IoT was conducted at the location in Bunaken island as depicted in Figure 6.

Figure 7. On-site testing of small off-grid monitoring system

The results of monitoring could be accessed in https://solar-power-monitor.herokuapp.com/ was presented in Figure 7.

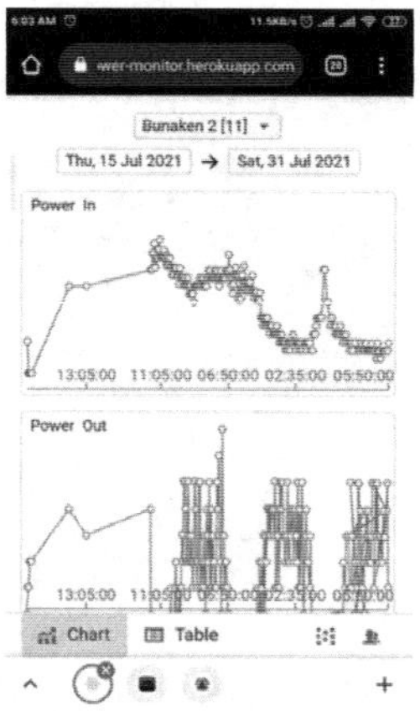

Figure 8. The result of IoT-based of small off-grid PV monitoring system

As the small off-grid PV monitoring system in Bunaken Island works well to visualize data, the system could be installed to solve the problems encountered when observing the PV systems of an isolated island.

5. Conclusion

This monitoring system is designed to record data from solar panels in remote locations. The objective of having an IoT-enabled monitoring system of small off-grid solar power that could report real-time data has been achieved. With this system, solar panels' performance can be measured and observed through any media such as laptops, smartphones, or smart TVs, accessible from an internet browser. Furthermore, it is necessary to develop more data monitoring systems with other parameters of solar power systems (e.g., temperature, environment humidity, solar irradiance). In the future, the PV monitoring systems could also be implemented in multiple locations, particularly in remote areas.

References

[1] Rumbayan M, Sompie R U S, Ruindungan G S D and Panjaitan V N 2021 Model of solar energy utilization in Bunaken Island Communities IOP Conf. Ser.: Earth Environ. Sci. 739 012082

[2] Rumbayan M, Ruindungan D, Sompie S and Sambul A 2019 The concept of the internet of things framework for remote monitoring of solar home system, Journal of Physics: Conference Series, Volume 1402, Issue 4, 1402 044006

[3] Shariff F, Rahim N.A, Hew W.P 2014 Grid-connected photovoltaic system: Monitoring insights. In Proceedings of the 3rd International Conference on Cleaner energy and Technology, Kuching, Malaysia, 24–26 November 2014; pp. 1–5.

[4] Rahman M M, Selvaraj J, Rahim N A, Hasanuzzaman M 2018 Global modern monitoring systems for PV based power generation: A review. Renewable and Sustainable Energy Reviews 2018, 82, 4142–4158.

[5] Triki-Lahiani A, Bennani-Ben Abdelghani A and Slama-Belkhodja I 2018 Fault detection and monitoring systems for photovoltaic installations: A review. Renewable and Sustainable Energy Reviews 2018, 82, 2680–2692.

[6] Madeti, S.R and Singh, S.N. Monitoring system for photovoltaic plants: A review. Renewable and Sustainable Energy Reviews 2017, 67, 1180–1207.

[7] Ansari S, Ayob A, Lipu M S H, Saad M H M, Hussain A 2021 A Review of Monitoring Technologies for Solar PV Systems Using Data Processing Modules and Transmission Protocols: Progress, Challenges and Prospects. Sustainability 2021, 13, 8120. https://doi.org/10.3390/su13158120

[8] Ascensión López-Vargas, Manuel Fuentes and Marta Vivar 2018 IoT application for real-time

monitoring of Solar Home Systems based on ArduinoTM with 3G connectivity, IEEE Sens. J. 2019, 19, 679–691

[9] Allafi I and Iqbal T 2017 Design and implementation of a low cost web server using ESP32 for real-time photovoltaic system monitoring. In Proceedings of the 2017 IEEE Electrical Power and Energy Conference, EPEC 2017, Saskatoon, Canada, 22–25 October 2017; pp. 1–5.

[10] Jamil N A A, Jumaat S A, Salimin S, Abdullah M N, Nor A F M 2020 Performance enhancement of solar powered floating photovoltaic system using arduino approach. Int. J. Power Electron. Drive Syst. 2020, 11, 651–657.

[11] Harmini, H.; Nurhayati, T. Monitoring system of stand alone solar photovoltaic data. In Proceedings of the ICECOS 2017— Proceeding of 2017 International Conference on Electrical Engineering and Computer Science: Sustaining the Cultural Heritage Toward the Smart Environment for Better Future, Palembang, Indonesia, 22–23 August 2017; pp. 254–258.

Technical analysis on household energy consumption: LMDI decomposition index and innovative

Yetty Murni, Eka sudarmaji*, Sri Ambarwati, Ismiriati Nasip

Universitas Pancasila, Jakarta, Indonesia

*esudarmaji@univpancasila.ac.id

Abstract. For decades, Indonesia's Economic Development had achieved great high growth rates by using more labor input, capital investment, physical and human, and fostering competition. Over time, Indonesia's energy intensity, which historically always been high compared to other industrialized economies. This study aimed to analyze decomposition factors of energy consumption by analyzing the link between the other nine factors and household energy use. The Logarithmic Mean Divisia Index were used to identify the determinant variables that drove a household's energy usage. This study used a Time-series VECM model for prediction model analysis. The Logarithmic Mean Divisia Index (LMDI) and decomposition index were used to identify the determining technological, economic, and social factors driving a household's energy consumption evolution. We analyzed the outcomes after applying the LMDI. We provided econometric individual time series VEC models, which we used for three targeting models: social, economic, and technical factors. We discovered that economic variables were the most significant factors that aggravated the change in household energy consumption based on LMDI. Technical factors, on the other hand, were the lowest. We found that the government might reduce its energy use over time by improving technological factors. Energy efficiency had been a cornerstone in managing growing energy use. The declining relevance of fewer labor-force in the previous decade was reflected in the negative indication of social factors at home. We discovered that the models were more stable while using Time Series VECM.

1. Introduction

Indonesia's demand and energy supply forecasts for the years 2019-2050 were based on three scenarios: business as usual (BaU), sustainable development (PB), and low carbon (RK). The three scenarios are based on the same simple GDP growth assumptions and population growth: average GDP growth of 5.6 percent per hour and population growth of 0.7 percent. Energy demand in 2025 is already dominated by the transportation sector, which accounts for about 35 percent of total demand. By 2050, it is dominated by the manufacturing sector, which accounts for 37-42 percent of total demand. Electricity demand in 2025 will increase by 11-12 percent in each case, reaching 576.2 TWh (BaU), 537 TWh (PB), and 520.7 TWh (RK), and by 6-7 percent in 2050, reaching 2,214 TWh (BaU), 1,917.9 TWh (PB), and 1,625.2 TWh (RK) (RK). The household sector dominates electricity demand until 2050 in all scenarios, followed by the manufacturing and commercial sectors.

Total primary energy production, including petroleum, gas, coal, and renewable energy, totaled 411.6 MTOE in 2018—coal and LNG account for 64 percent of total output, or 261.4 MTOE. Indonesia also imports energy, especially crude oil and fuel products worth 43.2 MTOE, and a small amount of high-calorie coal to meet the industrial sector's needs. Transportation accounts for 40% of total final energy consumption (excluding conventional biomass) in 2018, with manufacturing accounting for 36%, households for 16%, industrial and other industries accounting for 6% and 2%, respectively.

By 2025, Kebijakan Energi Nasional (KEN) needs a renewable energy mix of at least 23 percent in the primary energy mix and a petroleum usage of less than 25 percent. Furthermore, energy utilization is expected to decrease by 1% per year to promote energy usage reductions in all industries. BaU, PB, and RK's national final energy demand scenarios will rise at a rate of 5.0 percent, 4.7 percent, and 4.3 percent each year, respectively, before demand reaches 548.8 MTOE, 481.1 MTOE, and 424.2 MTOE in 2050.

In 2050, the final energy demand saving scenario PB against BaU is 12%, while the final energy demand saving scenario RK against BaU is 23%. The manufacturing and transportation sectors will continue to control final energy demand until 2050, as they did in 2018. By 2050, the manufacturing sector would have surpassed all other industries in importance, accounting for 42% in bau scenarios, 40% in PB scenarios, and 37% in RK scenarios. By 2050, energy demand would outnumber supply by 35 percent (BaU), 34 percent (PB), and 33 percent (RK).

On the other hand, the increasing population of Indonesia increased the number of households and urbanization. According to the Handbook of Energy & Economic Statistics of Indonesia in 2017, Indonesia's population stands at 261,891 million with a total household number of 67,173 million see chart.1 below.

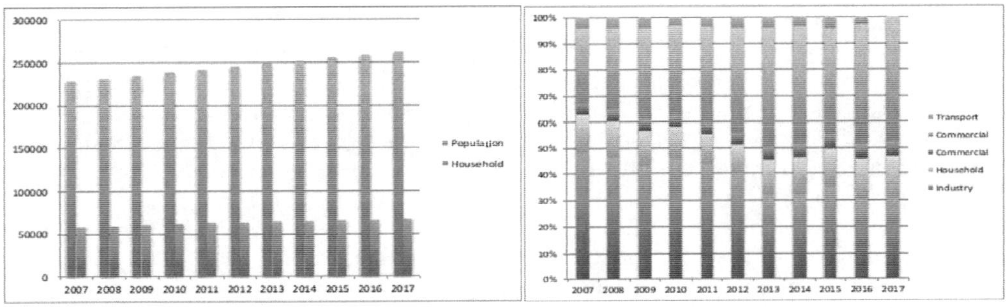

Figure 1. Number of Indonesians and Households & Presentation of energy use by sector (MEMR, 2018)

Based on data from the world bank (2018), more than 55.33% of the total population lives in cities. It made Indonesia have become one of the fastest urbanizations in the world. Increasing households' triggers demand and the drastic use of electronic home appliances in Indonesia. Increasing demand for electric devices increased energy consumption. As a result, that can be seen from Figure.1, the household's sectors have become the second-largest energy consumers in Indonesia (Ministry of Energy and Mineral Resources Republic of Indonesia, 2018). Given the increasing trends, Indonesia must continue to improve energy efficiency standards through labeling and certification systems. The energy consumption source can be divided into four, namely electricity, LPG, Kerosene, and Gas. Those high sources of large fossil cause an increase in CO2 emissions.

The carbon intensity increased the use of coal as fuel in power plants. During 1997-2017, coal consumption increased from 258.19 million tons of BOE to 407.50 million tons, although the percentage of coal usage decreased overall from 19.30% in 2007 to 7.40% in 2017 (Ministry of Energy and Mineral Resources Republic of Indonesia, 2018). Meanwhile, as stated in the National Energy Plan, coal plays a vital role in the Indonesian electrical industry due to its reserves, simplicity of use, and low operation cost.

Reducing the intensity of carbon emissions in electricity is critical to lowering overall emissions. Indonesia needs to diversify its energy mix to include cleaner, more renewable sources such as geothermal energy to minimize carbon emissions. With a high share of coal-fired power plants, Indonesia needs to increase efficiency by adopting the newest fossil fuel technology in power plants. On the other hand, the programs commonly known as Demand Side Management (DSM) need to optimize comprehensively. Demand Side Management (DSM) was first conceived by Gellings (1985). Energy efficiency enables the reduction of energy consumption. Many countries were able to limit the growth of energy demand through DSM. With the ability to limit the increase in energy demand by utilizing renewable energy and maximize policies on energy, the government can reduce CO2 emissions. To reduce emissions and achieve a transition to sustainable development, a shared commitment is needed among all stakeholders, including the public and private sectors.

For decades, Indonesia's Economic Development has achieved great high growth rates by using more labor input, capital investment, physical and human, and fostering competition. Over time, Indonesia's energy intensity, which historically always been high compared to other industrialized economies. This study aims to measure household energy efficiency by examining the relationship between the other eight variables and household energy used.

The Logarithmic Mean Divisia Index decomposition and extended KAYA index were deployed to recognize the determinant factors driving the evolution of Indonesia's energy intensity. This study aims to quantify household energy consumption by studying the link between the nine determinant factors. The LMDI decomposition index was used to identify factors influencing a household's subsidized energy usage.

2. Literature Review

Since the OECD environment minister placed it as an OECD environment policy in 2001, research on the decoupling analysis of energy consumption has been widespread. It became increasingly popular due to studies that have been done by Bacon, Bhattacharya, and Kojima (2009) and de Freitas and Kaneko (2011), which combined decoupling with the index decomposition technique. Some studies used decomposition decoupling analysis with LMDI and econometric techniques such as VECM (Moutinho, Madaleno, & Silva, 2015; Sadikova, Faisal, & Resatoglu, 2017; J. Wu, 2014).

Despite the national scope, some chose to investigate in various sectors (Zhao, Zhang, Li, Shao, & Geng, 2016), believing that it was critical to assess sectoral circumstances at each step. Toba and Seck (2016) categorize all decomposition elements as social, technological, environmental, or economic. They believed that combining energy systems' technological, environmental, and socioeconomic elements might be the key energy policy instrument. Meanwhile, Wang, Liu, Zhang, and Song (2014) pick ten rural home energy use indicators and place them in four dimensions: social, economic, technical, and environmental.

This study offered two significant contributions to the present body of knowledge. Using the same idea as Pui and Othman (2019a), this research classified aggregate decomposition findings into three aspects or categories: economic, technical, and social factors. The goal was to assess the relative importance of these three factors of energy use. Using the LMDI technique, the study divided the household energy consumption into ten components that consider all economic, technological, and social issues. As previously said, we begin by analyzing the change in household energy consumption in terms of economic, technological, and social factors. The VECM was then used to explore the relationship between those three elements of increasing a household's energy consumption. We provided t time series VEC models, which we use for three targeting models: social, economic, and technical factors.

Decomposition analysis is performed by decomposing changes in home energy use into many pre-defined variables using an identical equation. So far, the Laspeyres index and the Divisia index have been the most often used index decomposition analysis (IDA). The Laspeyres index calculates the percentage change in some feature of a set of items over time by employing weights based on values in a base year. The Divisia index, on the other hand, is a weighted sum of logarithmic growth rates, with the weights representing the components' shares of the total value. Because of its adaptability, simplicity of usage, and comparatively minimal data demand. Ang (2015) summarized IDA methods over their advantages and disadvantages, and he advocated the logarithmic mean Divisia index (LMDI) for general use.

IEA is using LMDI then followed by most of the researchers in energy. They used to use the Laspeyres index in early 1990, then AMDI. The Laspeyres index technique calculates the aspects in the same way as described in the scenarios section above, but using the percentage change from base year to year t. This technique was flawed because it did not consider the interplay of the other elements in the decomposition. It meant that changes in the decomposition factors did not always tally up precisely to changes in energy consumption.

3. Methodology

A time series of data were included in this research was household energy consumption. Consequently, the household energy data consisted of consecutive observations of the given variable at a uniform interval over some time. The data were obtained on an annual basis. The time-series data on household energy consumption show a consistent pattern, shocks, and volatility over time. As well as meandering and working with other shows, the action of variables over time is meandering and sharing co-movements with other variables. If the value of a time series continued to return to its long-run average value, it changed did not influence the data series' properties in time alone. In that case, it is said to be stationary.

On the other hand, Nonstationary time series do not appear to return to their long-run average value because their mean, variance, and co-variance shift over time. Household energy consumption has a definite upward trend with no propensity to return to a fixed mean over time. They are thus nonstationary sequences. The time series is said to have a unit root if it is nonstationary. Consequently, in econometrics, the unit root test is used to determine a time sequence's stationarity.

This study presented the Logarithmic Mean Divisia Index (LMDI). The KAYA decomposition index was used to identify the technological, economic, and social factors driving a household's energy consumption evolution. We analysed the outcomes after applying the LMDI. We provided econometric individual time series VEC models, which we used for three targeting models: social, economic, and technical factors.

We discovered that economic variables were the most significant factors that aggravated the change in household energy consumption based on LMDI. KAYA identity was a popular technique for researching large-scale energy and environmental issues, and it may define the link between people, economy, policy, energy, and carbon emissions (Kaya & Yokobori, 1997; Ma & Cai, 2018; O'Mahony, 2013; R. Wu, Mavromatidis, Orehounig, & Carmeliet, 2017). On the other hand, LMDI is being used by IEA, followed by most of the researchers in energy. They used to use the Laspeyres index in early 1990, then AMDI. This approach had a flaw in the interplay of the other elements in the decomposition. It meant that changes in the decomposition factors may not always correspond directly to changes in energy consumption.

VECM was used to examine the innovative structure of home energy use. The authors attempted to verify and discover dynamic connections between endogenous and exogenous factors using this VECM model. It started with the testing phases examined the root unit to see the behaviour of time series economic data. It may be viewed as the first stage in building a time series model, regardless of whether the data utilized is stationary or not, which can be accomplished using the dickey-fuller enhanced test (Dickey & Fuller, 1979).

Second, use Johansen's maximum probability technique to evaluate the cointegration and causality of granger temporal; last, replace VECM and test its exogenous variables in the third stage. The most effective order selection criteria Select the optimal order (p) using four criteria: final prediction error (FPE), Akaike Information Criterion (AIC), Hannan-Quinn Criterion (HQ), and Schwarz Information Criterion (SIC). The vector error correction model was used to evaluate nonstationary multivariate time series data. The VECM model was the VAR model with a linear cointegration relationship.

The dimensions of the and are N x r, where N was the number of variables and r was the degree of cointegration. The degree of cointegration implied a long-term link between the changes Y_t of the model. Therefore, cointegration was the primary criterion for utilizing VECM. The Johansen test was used to determine the degree of cointegration. VAR and VECM models have both short-term and long-term constraints. The short-term limitation arises when one variable could not respond promptly to changes or shocks in another variable. Long-term restriction arises when the variables utilized have a cointegration or long-term connection. IRF and PEVD were examined and reviewed at the end of the modeling process. The final results and the best models, whether stationer or stationer, may be used to anticipate home energy use.

3.1 LMDI Household Energy Consumption Model

Household energy demand has been affected mainly by the rising number of households, reaching 70.6 million in 2025 and 80 million in 2050. Furthermore, the pace of urbanization often encourages increasing energy demand. Based on the BPS forecast, the degree of urbanization in 2035 increased by 67 percent from 2010, just 49.8 percent. For Household energy consumption, the formula become:

$$\text{Household Energy Consumption} = \text{Population} \times \frac{GDP}{Population} \times \frac{Energy\ Primary\ Energy}{GDP} \times \frac{National\ Energy\ Consumption}{Primary\ Energy} \times \frac{Household\ Energy\ Consumption}{National\ Energy\ Consumption}$$

when we Include Renewal Energy, Capital Formation, and Labour force, the formula is as follows:

$$\text{Household Energy Consumption} = \text{Population} \times \frac{GDP}{Population} \times \frac{Energy\ Primary\ Energy}{GDP} \times \frac{Renewal\ Energy\ Consumption}{Primary\ Energy} \times \frac{Fossil{-}Fuel\ Energy\ Consumption}{Renewal\ Energy\ Consumption} \times \frac{Gross\ Fixed\ Capital}{Fossil{-}Fuel\ Energy\ Consumption} \times \frac{Labor\ Force}{Gross\ Fixed\ Capital} \times \frac{Household\ Activity\ Intensity}{Labor\ Force} \times \frac{Household\ Energy\ Consumption}{Household\ Activity\ Intensity}$$

The LMDI formula can be rewritten as follows:

$$HEC^T = Pop^T \times \frac{GDP^T}{Pop^T} \times \frac{EPC^T}{GDP^T} \times \frac{RE^T}{EPC^T} \times \frac{FF^T}{RE^T} \times \frac{GFC^T}{FF^T} \times \frac{LF^T}{GFC^T} \times \frac{HAI^T}{LF^T} \times \frac{HEC^T}{HAI^T}$$

$$HEC^T = Pop^T \times GDP^T \times EPC^T \times RE^T \times FF^T \times GFC^T \times LF^T \times HAI^T \times HEC^T$$

$$HEC^T = Pop^T_{effect} \times GDPeffect \times EPC^T_{effect} \times RE^T_{effect} \times FF^T_{effect} \times GFC^T_{effect} \times LF^T_{effect} \times HAI^T_{effect} \times HEC^T_{effect}$$

Whereas:

POP = Population effect

GDP = Gross Domestic Products effect $\frac{GDP^T}{POP^T}$

EPC = Ratio Energy Primer Consumption effect $\frac{EPC^T}{GDP^T}$

RE = Renewal Energy - Energy Substitution, Ratio Renewal Energy effect $\frac{RE^T}{EPC^T}$

FF = Fossil Fuels - Renewal Energy Substitution, Ratio Fossil Renewal Energy effect $\frac{FF^T}{RE^T}$

GFC = Investment Efficiency - Ratio Gross Capital Stock over Renewal Energy $\frac{GFC^T}{FF^T}$

LF = Capital Labour substitution - Ratio Capital labor $\frac{LF^T}{GFC^T}$

HAI = Ratio Number Household over labor Force $\frac{HAI^T}{LF^T}$

HEC = Ratio Household Consumption per Household $\frac{HEC^T}{HAI^T}$
$\Delta HEC^T = (\Sigma_{t=1}^{K} \Delta HEC)$; if undecompensed
$(\Delta HEC = (\Sigma_{t=1}^{K}(\Delta POP + \Delta GDP + \Delta EPC + \Delta RE + \Delta FF + \Delta GFC + \Delta LF + \Delta HAI + \Delta HEC)$; if decompensed

Household Energy consumption (HEC) can be decomposed using the LMDI Approach as follows:

Social Aspects = $\left(\Sigma L(HEC^t, HEC^0)Ln\left(\frac{POP_{effect^t}}{POP_{effect^0}}\right)\right) + \left(\Sigma L(HEC^t, HEC^0)Ln\left(\frac{LF_{effect^t}}{LF_{effect^0}}\right)\right) + \left(\Sigma L(HEC^t, HEC^0)Ln\left(\frac{HAI_{effect^t}}{HAI_{effect^0}}\right)\right) + \left(\Sigma L(HEC^t, HEC^0)Ln\left(\frac{HI_{effect^t}}{HI_{effect^0}}\right)\right)$

Economic Aspects = $\left(\Sigma L(HEC^t, HEC^0)Ln\left(\frac{GDP_{effect^t}}{GDP_{effect^0}}\right)\right)$

Technical Aspects = $\left(\Sigma L(HEC^t, HEC^0)Ln\left(\frac{RE_{effect^t}}{RE_{effect^0}}\right)\right) + \left(\Sigma L(HEC^t, HEC^0)Ln\left(\frac{FF_{effect^t}}{FF_{effect^0}}\right)\right)$

$\left(\Sigma L(HEC^t, HEC^0)Ln\left(\frac{GFC_{effect^t}}{GFC_{effect^0}}\right)\right) + \left(\Sigma L(HIC^t, HEC^0)Ln\left(\frac{EPC_{effect^t}}{EPC_{effect^0}}\right)\right)$

3.2 Econometric Model

VECM Panel Analysis recognized the technical, economic, and social aspects that drive the evolution of energy intensity in Indonesia. Researchers used a regression method in data analyst techniques through the decomposition approach using the Logarithmic Mean Divisia Index (LMDI) and KAYA index. Regression analysis was a statistical technique to model and investigated the influence between independent variables on one response variable (Dependent *variable).* The regression equation used was as follows:

HEC = α + β₁Social-aspects+ β₂Economic-aspetcs+ β₃Technical-aspects + e

Whereas:
HEC = Household Energy Subsidy
α = Regression constant
β_1 = Regression Coefficient for Social Aspects
β_2 = Regression Coefficient for Economic Aspects
β_3 = Regression Coefficient for Technical
e = *error.*

The dependent variable in this study was household energy consumption, and there were three independents' variables. Authors classified nine variables into three main aspects, i.e, economic, technical, and social. Economic aspects contain variable yang Gross Domestic Product effect $\frac{GDP^T}{POP^T}$. Social aspects contained variable population effect, labor force effects $\frac{LF^T}{GFC^T}$, Household activity intensity over Labor Force $\frac{HAI^T}{LF^T}$, and Ratio Household Consumption per Household $\frac{HEC^T}{HAI^T}$. Meanwhile, Technical aspects contained Renewalenergy - Energy Substitution, Ratio National Renewal Energy effect $\frac{RE^T}{EPC^T}$, Ratio Fossil Renewal Energy effect $\frac{FF^T}{RE^T}$. Gross Fixed Capital over Fossil Fuel substitution $\frac{GFC^T}{FF^T}$. And Ratio Energy Primer Consumption effect $\frac{EPC^T}{GDP^T}$. After applying the LMDI KAYA analysis, the next step used VECM. The long-run model regression model in the study was as follows:

$$\Delta y_t = \alpha\beta^T y_{t-1} + \Gamma_1 \Delta y_{t-1} + \ldots + \Gamma_{p-1}\Delta y_{t-p+1} + U_t\, 0,\ \Gamma_i = -\,(I - A_1 - \ldots A_{t})$$

4. Result and Discussion

There were 567 data observations on energy primer consumption obtained from British Petroleum World Statistics, World Bank, IEA, Minister Sumber Daya Mineral, and Biro Pusat Statistik.

4.1 Decomposition Analysis

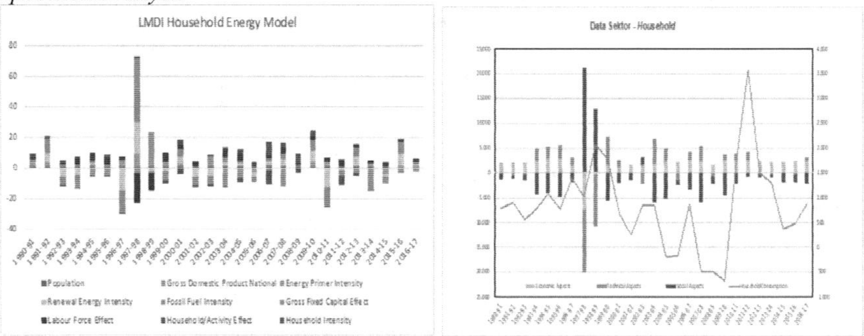

Figure 2. Decomposition Factor Analysis

To determine the relevance of each emission reduction component, as stated in the preceding paragraph. This study used the KAYA identity to decompose the household energy consumption component into the population effect, GDP effect, energy primer intensity effect, renewal energy intensity, fossil fuel intensity, gross fixed capital effect, labor force effect, and household activity, and household intensity effect. The total of all nine of these impacts equaled home energy usage. As seen in Fig 2 and Table 1, the aggregate of all nine aspects increased from 1997 to 1998. The Asian Crisis of 1997-1998 was the only occasion when GDP and energy intensity components fell. In 1998, real GDP fell by 13.1 percent, and the economy hit a low point in mid-1999, with real GDP growth of 0.8 percent.

Based on Figure 2, Economic changes were implemented during the 1980s, including the controlled depreciation of the rupiah to boost export competitiveness and the deregulation of the banking sector. Foreign investment poured into Indonesia, particularly in the export-oriented manufacturing sectors. As a result, between 1990 and 2017, Indonesia saw substantial emissions increase of about 20,48 metric tons per year, driven only somewhat by high economic development and significantly better energy intensity.

According to the IEA, total CO2 emissions in Indonesia from 1971 to 2017 were 941,40 metric tons over 37 years. The Indonesian GDP accounted for more than 7% of global GDP on average between 1989 and 1997. Then, in 2012, Indonesia's real GDP growth rate hit 6%, after progressively declining to 5.1 percent in 2004 and 5.6 percent in 2005. As a result, over nearly 37 years, the economic effect was the only component that had a significant and constant effect on household energy use.

Table 1. Decomposition of Household Energy Consumption

Year	Economic Aspects	Technical Aspects	Social Aspect	Remarks
1990-91	4,993%	-0,366%	-2,733%	
1991-92	4,653%	-1,490%	-1,043%	
1992-93	4,681%	-0,765%	-2,608%	
1993-94	5,726%	5,669%	-9,621%	
1994-95	6,432%	5,261%	-9,274%	
1995-96	6,083%	6,076%	-10,484%	
1996-97	3,187%	3,682%	-3,878%	

1997-98	-15,655%	-26,264%	44,054%	ASEAN Crisis
1998-99	-0,620%	-21,308%	26,132%	
1999-00	3,482%	10,858%	-10,854%	
2000-01	2,225%	2,729%	-3,654%	
2001-02	3,051%	0,187%	-2,791%	
2002-03	3,347%	-4,103%	2,340%	
2003-04	3,594%	8,861%	-10,887%	
2004-05	4,193%	4,790%	-9,326%	
2005-06	4,019%	-2,787%	-1,550%	
2006-07	4,866%	2,785%	-6,053%	
2007-08	4,497%	5,368%	-10,733%	
2008-09	3,182%	-1,279%	-2,783%	Lehman
2009-10	4,671%	2,090%	-7,983%	
2010-11	4,714%	2,542%	-4,005%	
2011-12	4,646%	2,968%	-1,221%	
2012-13	4,118%	-0,526%	-1,079%	
2013-14	3,613%	-0,538%	-0,939%	
2014-15	3,504%	0,127%	-3,047%	
2015-16	3,705%	-0,537%	-2,395%	
2016-17	3,794%	1,036%	-3,443%	
Geomean	**3,201%**	**-0,142%**	**-2,391%**	
Average	**3,282%**	**0,176%**	**-1,866%**	

The social influence defined by population and urbanization exacerbated the growth in home energy demand. The consequence of decomposition was correctly predicted by urbanization. The bulk of variables have led to increased consumption as a result of increased energy consumption by families. Fortunately, the results suggest that Indonesia's energy efficiency has increased during the previous decade due to the government encouraging the business sector to diversify away from the petroleum economy. Most of the energy intensity impact was caused by a rise in technical aspects or was defined by advancements in technological equipment innovation.

From 1971 to 2017, Indonesia's energy intensity minimized CO2 emissions through energy efficiency by increasing the technical element. As a result, energy efficiency had become the cornerstone of controlling energy consumption efficiency. The energy intensity impact was connected to a decrease in home energy usage during the timeframe.

Based on LMDI results, we found that the economic aspects were the most significant factors aggravating energy consumption change for all four sectors. The immense energy consumption was on household sectors with 26,782.83koe. On the other hand, the technical aspects contribute positively to change in energy consumption. It has the lowest aggravates to fuel change with 3,797.67koe, for household sectors. It showed that through improvement in the technical aspects, the government could impede its energy used over the period. Energy efficiency has been the cornerstone in controlling the rising energy used in these sectors. This finding aligned with other similar Malaysian results (Pui & Othman, 2019).

4.2 Time Series VECM

We analyzed the outputs after applying the LMDI KAYA analysis. In the data analyst technique, we realized that there were always advantages and disadvantages to using Panel VCEM. Therefore, we were also applying a time series analysis to seek more precise and reliable results. The method selection was based on the unit root test results that define the variable's stationarity for time series analysis. For the individual time series model, a VECM model was used for the second time. Three equations in all the VECM where every accumulation of LMDI decomposition variables in economic, technical, and social aspects included was the dependent variable in one of the equations.

The empirical framework of the analysis has the following components:

a. Unit Root Tests & Cointegration Tests

b. Optimal Lags Selection
c. VEC Model Estimations
d. Causality Analysis Tests
e. Diagnostic and Stability Tests

4.3 Unit Root Tests & Cointegration Tests

Since the authors used the first difference, all variables were taken (denoted by d( )). Based on the result shown in Table 2, the ADF test results show that the null hypothesis of an existence of a unit root cannot be rejected, as their p-values exceed the selected significance level of 0.01.

Table 2. Unit Root Test

		t-Statistic	Prob.*	t-Statistic	Prob.*	t-Statistic	Prob.*
Augmented Dickey-Fuller test statistic		-5.82167	0.0002	-6.17055	0.0001	-4.4597	0.0025
Test critical values:	1% level	-3.85739		-3.85739		-3.80855	
	5% level	-3.04039		-3.04039		-3.02069	
	10% level	-2.66055		-2.66055		-2.65041	

*MacKinnon (1996) one-sided p-values.

Table 3 shows the output for the Johansen Cointegration test. To clarify, “None” meaning that there was no cointegration among the three variables, “At most 1” and “At most 2,” meaning that there was no cointegration for at most one and most two co-integrated equations. The null hypothesis of no cointegration was rejected by both tests (Trace and Eigenvalue)

Table 3. Co-Integration Test

Unrestricted Cointegration Rank Test (Trace)

Hypothesized No. of CE(s)	**Eigenvalue**	Trace Statistic	0.05 Critical Value	**Prob.****
None *	0.656859	53.48109	29.79707	0.0000
At most 1 *	0.492150	28.88000	15.49471	0.0003
At most 2 *	0.439027	13.29589	3.841466	0.0003

Trace test indicates three co-integrating eqn(s) at the 0.05 level
* denotes rejection of the hypothesis at the 0.05 level
**MacKinnon-Haug-Michelis (1999) p-values

Unrestricted Cointegration Rank Test (Maximum Eigenvalue)

Hypothesized No. of CE(s)	**Eigenvalue**	Max-Eigen Statistic	0.05 Critical Value	**Prob.****
None *	0.656859	24.60109	21.13162	0.0156
At most 1 *	0.492150	15.58411	14.26460	0.0308
At most 2 *	0.439027	13.29589	3.841466	0.0003

Max-eigenvalue test indicates 3 cointegrating eqn(s) at the 0.05 level
* denotes rejection of the hypothesis at the 0.05 level
**MacKinnon-Haug-Michelis (1999) p-values

Max-eigenvalue test indicates 3 cointegrating eqn(s) at the 0.05 level
* denotes rejection of the hypothesis at the 0.05 level

**MacKinnon-Haug-Michelis (1999) p-values

4.4 Optimal Lags Selection
The optimal lag length was determined in Table 4. The Akaike Information Criterion (here, AIC), used for consistency, provides an optimal lag length of 1, denoted by the asterisk next to 55.01154. This optimal lag length holds for the whole model.

Table 4. Optimal Lag

Lag	Lo		lo LR FPO	AIC	SC	HQ
0	-629.9392	NA*	1.60E+20	55,0382	55.18630*	55.07544*
1	-620.6327	15.37602	1.58e+20*	55.01154*	55,60397	55.16054
2	-615.3836	7.303075	2.30E+20	55,33771	56.37446	55.59845

* indicates lag order selected by the criterion
LR: sequential modified LR test statistic (each test at 5%level)
FPE: Final prediction error
AIC: Akaike information criterion
SC: Schwarz information criterion
HQ: Hannan-Quinn information criterion

4.5 VEC Model Estimations
Estimated VECM with Social aspects as target variable, social aspects = 5,3901ectt-1 – 1,6007 ΔEconomic aspects t-1 – 4,3958 ΔTechnical aspects t-1 + 12.7382, and co-integration equation (long-run model); ECTt-1 = 1.0000 Social aspectst-1 + 0,5750 Economic aspectst-1 + 1,3945 Technical aspects t-1 + 12,7382.

Based on the empirical result above – it showed that in the long-run causality, variable economic-aspects had significant at 0.01 level had positive impact on social-aspects with coefficient 0.5750%. it meant that 1% change in economic-aspects increased 0.5750% in social-aspects. On the other hands, variable technical-aspects had significant at 0.01 level has positive impact on social-aspects with coefficient 1.3945%. which means that 1% change in technical-aspects increased 1.3945% in social-aspects.

Table 5. Co-Integration Equation (Long-Run)

Cointegration Equation	Social		
	Coefficient in ICT	Coefficient of ECT (Dependent Var)	Product= Speed of Adjustment
Social(-1)	1	5.390106 [2.60523] **	5.390106
Economic(-1)	0.575067 [5.10244] *	-1.600767 [-1.56324]	-0.920548276
Technical(-1)	1.394537 [21.1960] *	-4.395864 [-3.45900] *	-6.130194995
Coefficient	12.73817		

* accepted Null hypotheses on 0.01 level, ** on 0.05 level

The approximate results of the three models in Table 5 indicated that all three models have long-term causality. The first model showed that the speed of adjustment was 5,3901, which implied that the long-term equilibrium relationship between social, economic, and technical aspects was not valid. It meant that 539.01% of the previous shock duration was not imbalanced. The second economic-aspects model, which had an ECT coefficient of -0,9205, meant that the long-term equilibrium relationship between social, technological, and economic aspects was valid. It meant that the 92.05 percent disparity of previous period shocks reunited favorably into long-run equilibrium in the current period. In other words, social, technological, and economic aspects were not causal in the long term.

Table 6. Short-Run Causality

ECT	5.390106	[2.60523] **	-1.600767	[-1.56324]	-4.395864	[-3.45900] *
Short-Run	△Social		△Economic		△Technical	
△Social(-1))	-2.56957	[-1.31025]	0.57762	[0.59509]	1.85838	[1.54271]
△Economic(-1))	-1.613414	[-1.08927]	-0.04918	[-0.06709]	1.37626	[1.51268]
△Technical(-1))	-2.679418	[-1.17384]	0.511622	[0.45286]	2.098499	[1.49669]
C	-48.87126	[-0.02611]	28.81561	[0.03111]	25.04165	[0.02178]
R-squared	0.507103		0.501927		0.60010	
Adj. R-squared	0.397570		0.391244		0.511234	
F-statistic	4.629693		4.534824		6.752817	

* accepted Null hypothesis on 0.01 level, ** on 0.05level

In the social aspects, reducing the labor force for household sectors contributed to the lowest change to the energy used for both sectors. The negative sign of social aspects in the last decade signifies the decreasing significance of less labor force used denotes that substitution flexibility between machine efficiency and labor-force. This negative sign was contradictive with (Pui and Othman (2019) result. Conversion to energy-saving technologies, intuitively, was a preventing factor in increasing energy consumption.

These findings suggested encouraging indicators of an investment efficiency effect in the manufacturing sector. It indicated that the sector has not been able to allocate sufficient technology improvement in investment funds. It showed that the effect of the technical aspect was successful, embracing a more comprehensive mix of cleaner energy and renewable energy through fossil fuels. Andreoni & Galmarini (2016), Pui & Othman (2019), Shahiduzzaman & Layton (2015), and Wang, Su, Sun, Zhou, & Zhou (2015) were in line with this unique finding.

Similarly, the last technical-aspects model had a speed-of-adjustment of -6,1302, indicating that the long-term balance between social, economic, and technical aspects was valid. It meant that the 613.02% imbalance of shocks reunited in long-term equilibrium over the previous period. It suggested that there was a long-term causality between social and economic aspects in the technological aspects model. There was no short-run causality between variables in the technical-aspects model above.

4.6 Innovative accounting approach

4.6.1 Impulse-Response Function

Based on figure 3 below, △Social response to the variable shock of △Social, △Economic, △Technical began to surprise with opposing trends. Long-term dynamics response on △Social, △Economic, △Technical occurred in the 2nd period. The Social response in the event of a shock to the Economic was adverse effects in the long run. The response of the Social variable to the shocks of the Economic variable reached stability after the 10th period.

Meanwhile, the response of the Social variable to the Technical variable was positive in the long term and reached stability after the 10th period. The results of this IRF indicated that the Economic variable had an impact on suppressing changes in the negative Social variable permanently in the long term. It showed that the response to the Economic variable is inversely proportional to the response to the technical variable.

Meanwhile, in the IRF test, the response to the Economic variable when a shock occurred to the technical variable is negative in the long term, and the Social variable was initially negative close to 4% percent when a shock occurred but then strengthened towards lower negative territory in the long run. It showed that the Economic response had a negative effect on the technical variable and the Social variable. On the other hand, the response of the technical variable to both the Social variable and the Economic variable had a positive effect on the movement of the two variables. The shocks to the technical variables to both social and economic variables almost had the same effect. From the three IRF tests, it could be seen that the response effect on the technical variable to the other variables tested had the same effect. In contrast, the response effect for both the economic and social variables produced the opposite effect.

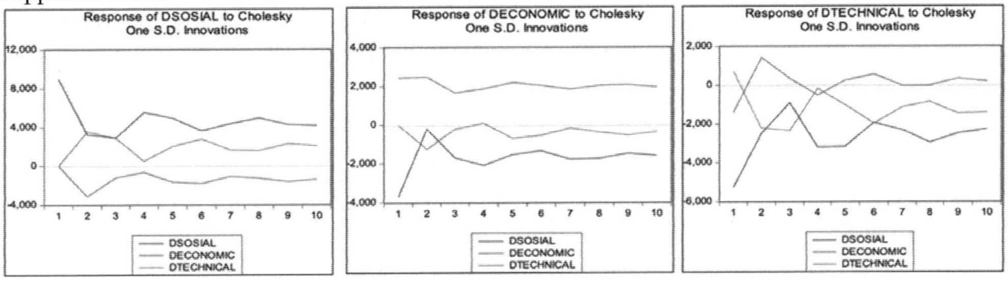

Figure 3. Response's variables of △Social, △Economic and △Technical due to Shocks

4.6.2 Variance Decomposition

Variance decomposition describes how many variance errors were predicted from certain effect variables described by innovations resulting from other effect variables in the system for some time. Variance decomposition was used to describe the variance proportion of errors from different impact factors of △Social, △Economic, and △Technical. It described the relative effect that explained each variable's contribution to the system variable.

Variance decomposition results were presented in Table 7. Subsequently, we highlighted the most critical shocks that changed each effect that was decomposed. The empirical evidence indicated that 77.424 % of △Social was due to innovative shocks, 7.814% due to △Economic and 14.761% due to △Technical. The variable △Economic mainly affected by its innovator shocks △Economic and △Social by 43.76% and 52.627% respectively.

At the same time, 3.614 % due to △Technical with a standard error of 9,067.22 of the inflation rates was explained by one standard deviation shock in its innovative shocks. The △Technical variable component was mainly affected by △Social with 75.911%, △Economic, and △Technical with 4.459% and 19.629%, respectively.

	Variance Decomposition of ΔSocial				Variance Decomposition of ΔEconomic				Variance Decomposition of ΔTechnical			
Period	S.E.	ΔSocial	ΔEconomic	ΔTechnical	S.E.	ΔSocial	ΔEconomic	ΔTechnical	S.E.	ΔSocial	ΔEconomic	ΔTechnical
1	8.969,94	100,000	-	-	4.439,58	69,715	30,285	-	5.509,76	91,787	6,462	1,751
2	10.639,13	80,279	8,732	10,989	5.243,72	50,128	44,148	5,724	6.594,64	78,314	9,229	12,457
3	11.459,14	75,624	8,625	15,750	5.758,40	50,157	44,956	4,887	7.060,85	69,917	8,297	21,786
4	12.758,70	79,936	7,219	12,845	6.407,36	51,098	44,929	3,973	7.772,38	74,698	7,284	18,018
5	13.932,63	79,686	7,444	12,870	6.982,15	47,850	47,802	4,348	8.458,47	77,120	6,239	16,641
6	14.772,88	76,940	8,103	14,958	7.416,23	45,637	49,979	4,384	8.911,32	74,136	6,026	19,838
7	15.532,67	77,554	7,809	14,636	7.848,12	45,833	50,213	3,954	9.274,85	74,659	5,563	19,778
8	16.430,64	78,443	7,569	13,988	8.302,30	45,341	50,930	3,729	9.772,20	76,421	5,011	18,568
9	17.208,19	77,705	7,793	14,502	8.701,74	44,153	52,080	3,768	10.191,70	76,189	4,724	19,087
10	17.883,62	77,424	7,814	14,761	9.067,22	43,760	52,627	3,614	10.536,52	75,911	4,459	19,629

Figure 4. Decomposition of △Social, △Economic, and △Technical

4.7 Validity and Stability Tests

Several diagnostic tests, such as the VEC Residual serial correlation Lagrange Multiplier (LM) test, the Jarque-Bera normality test, and the White heteroskedasticity test, were needed validity of the Vector Error Correction model. We rejected the Null hypothesis that the errors were normally distributed according to the results of Jarque-Bera standard value 35,51023 with PValue,0.05. When we wanted to interpret the estimated econometric equation, the normal distribution of errors was significant. Meanwhile, for the serial correlation test, the two LRE*F-statistic and Rao's F-test statistical tests and the probabilities associated with these tests showed a value of 11,64123 1,369687 respectively for lag-2. The null hypothesis of these two experiments was that the equation errors were not serially correlated when all lags were acknowledged (there was no serial correlation).

CUSUM test was based on the cumulative sum of the equation errors in regression with critical lines of 5%. The equation parameters were considered stable since the whole sum of recursive errors gets inside the two critical lines. Based on the CUSUM test, the whole aspects were considered stable. CUSUM of Squares test was similarly calculated and interpreted as the CUSUM test, with the difference that instead of recursive errors, we use recursive doubled errors. According to this test, the whole aspect was considered stable.

The last validity test was White's heteroskedasticity test. The results of White's heteroskedasticity test. The test's null hypothesis states that there was no heteroskedasticity present. Based on the result, no heteroskedasticity's null hypothesis could not be rejected as the conducted p-value of 0,4027. It exceeded the significance level; hence it was concluded that the specified VECM does not suffer from heteroskedasticity (Fig 5).

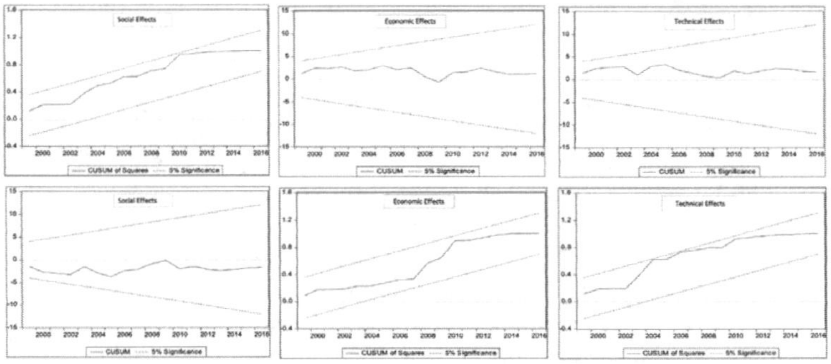

Figure 5. Stability Test CUSUM & CUSUM of Square Test

4.8 Causality Analysis Tests

After running the model's validity, the causal relationships of the variables were to be determined—the Granger causality/ Wald test to measure potential short-run causal aspects between the variables. The p-values of those variables were based on the Chi-square distribution were 3,97881, 4,91595, and 5,603669. The results showed no causal short-run relationship for social, economic, and technical aspects based on the Granger causality/ Wald test. Hence, the null hypothesis of no causality was accepted for social, economic, and technical aspects.

The results show that there was no causal relationship between the variable's social aspects and economic aspects. The null hypothesis for Granger pair social aspects did not Granger Cause economic aspects (p-value 0,5617), and economic aspects did not Granger Cause social aspects (p-value 0,8141). Hence social aspects and economic aspects have No Direct Granger Causality relationship. There was no causal relationship between the variable's social aspects and technical aspects and technical aspects and social aspects, as the null hypothesis for Granger pairwise (p-value 0,5186) and (p-value 0,74). Hence social aspects and economic aspects have No Direct Granger Causality relationship. According to the Pairwise Granger Causality test results, no Granger Causality was implied between social and technical aspects. There was no causal relationship between the variable's economic aspects and technical aspects and technical aspects and economics-aspects, with Granger pairwise p-value 0,5588 and p-value 0,5351.

5. Conclusion

Indonesia's enormous energy demand is driven by increased activity, mainly due to growth in developing infrastructure. Electricity consumption continues to increase by more than 8.5% per year; meanwhile, energy investment is still scarce and never meets consumption demand (PT Perusahaan Listrik Negara (Persero), 2017). Beyond the increasing energy demand and limited supply, there are tremendous untapped opportunities for energy efficiency in various sectors in Indonesia, one of which is in the household sector. Energy efficiency is one of the indicators that consists of both energies use itself and the driving forces behind energy use. Growth in energy use is closely linked to economic growth, goods and services, population buildings, and transport. These energy intensity indicators were put together to analyze a not directly observable development from the underlying statistics. Energy indicators in specific are helpful because they link energy use to relevant activity measures like GDP and Production Value. Energy indicators are a helpful tool for policymakers and can also predict future development in energy use.

The study of decomposition decoupling measured how much energy was used relative to an activity measure (for example, GDP) that was important. The meaning of decomposition does not automatically say anything about how efficient the use of energy was. The elements of decomposition depend on the structure of the economy and the environment of the country. This type of indicator aims to quantify how effective we use energy, how decomposition factors differ across industries and assess the impact of government-initiated energy use control measures. Comparing decomposition factors and energy used in each of the various industries was valuable when decomposing energy consumption.

Therefore, Indonesia should emphasize improving the technical aspect through energy conservation and replacing green inputs to compensate for energy usage and avoid subsidies. Many empirical studies have found that energy efficiency changes were often not adequate to resolve the burden of economic activity. It happens unwittingly because policymakers frequently emphasize economic gains rather than environmental priorities. In the years 1997-1998 and 2007-2008, the Asian and Lehman crises appeared to be successful in minimizing the impact of economic factors through their impact on technical and social aspects. In other words, during the crisis, all sectors responded by reducing energy consumption in the face of an economic recession.

Unfortunately, thanks to government subsidies, it will double even more in the next recovery years. We find that the influence of technical aspects indicates a tendency to become successful in the post-crisis era. We presented individual time series VEC models for three targeting models, namely Social Aspects, Economic Aspects, and Technical Aspects. The robustness of the measurement model during 1971-2017 was evaluated through the Creative Accounting Approach (IAA), Granger Causality, Long-run, and Short-run Co-integration. On the Panel VECM, the optimistic ECT was established in the targeting models of social aspects and technical aspects. Only the model targeting economic aspects had a negative ECT. For the models, the positive ECT symbol was not a good sign. It means that the phase of social aspects and technological aspects has not converged in the long run.

Under Time Series VECM, a few anomalies in the CUSUM square test for technical and economic aspects were found in the industrial sectors. The equation parameters were stable as the entire total of recursive errors gets within the two critical lines. The CUSUM test was stable as the entire sum of recursive errors gets within the two critical lines. Based on the CUSUM measure, the overall outcomes were considered stable. The Squares CUSUM test was calculated and interpreted similarly to the CUSUM test, except that we use recursive duplicate errors instead of recursive errors. According to this test, the general aspects were considered stable. The last validity test was White's Heteroskedasticity Test. The test's null hypothesis states that no existing heteroskedasticity exists. Based on the outcome, the null hypothesis of heteroskedasticity should not be discarded since the p-value carried out approaches the significance stage. Therefore, it is presumed that heteroscedasticity does not impact the specified VECM.

References

[1] Andreoni, V., & Galmarini, S. (2016). Drivers in CO2 emissions variation: A decomposition analysis for 33 world countries. Energy, 103, 27–37.

[2] Ang, B. W. (2015). LMDI decomposition approach: A guide for implementation. Energy Policy, 86, 233–238.

[3] Bacon, R., Bhattacharya, S., & Kojima, M. (2009). Changing Patterns of Household Expenditures on Energy: A case study of Indonesia and Pakistan. World Bank Extractive Industries and Development Series #6, (June).

[4] de Freitas, L. C., & Kaneko, S. (2011). Decomposing the decoupling of CO2 emissions and economic growth in Brazil. Ecological Economics, 70, 1459–1469.

[5] Dickey, D. A., & Fuller, W. A. (1979). Distribution of the Estimators for Autoregressive Time Series With a Unit Root. Journal of the American Statistical Association, 74(366), 427.

[6] Gellings, C. W. (1985). The Concept of Demand-Side Management for Electric Utilities. In The IEEE (pp. 1468–1470).

[7] Kaya, Y., & Yokobori, K. (1997). Environment, energy, and economy : Strategies for sustainability. Tokyo conference on Global Environment, Energy and Economic Development (1993), United Nations Univ. Press, Tokyo, 1997.

[8] Ma, M., & Cai, W. (2018). What drives the carbon mitigation in Chinese commercial building sector? Evidence from decomposing an extended Kaya identity. Science of the Total Environment, 634, 884–899.

[9] Ministry of Energy and Mineral Resources Republic of Indonesia. (2018). Energy & Economic Statistics of Indonesia. Handbook of Energy & Economic Statistics of Indonesia.

[10] Moutinho, V., Madaleno, M., & Silva, P. M. (2015). Which factors drive CO2 emissions in EU-15? Decomposition and innovative accounting. Energy Efficiency, 9(5), 1087–1113.

[11] O'Mahony, T. (2013). Decomposition of Ireland's carbon emissions from 1990 to 2010: An extended Kaya identity. Energy Policy, 59, 573–581.

[12] PT Perusahaan Listrik Negara (Persero). (2017). Keputusan Menteri ESDM No. 1415 K/20/MEM/2017 Tentang Rencana Usaha Penyediaan Tenaga Listrik (RUPTL) PLN 2017-2026.

[13] Pui, K. L., & Othman, J. (2019). The influence of economic, technical, and social aspects on energy-associated CO2 emissions in Malaysia: An extended Kaya identity approach. Energy, 181, 468–493.

[14] Sadikova, M., Faisal, F., & Resatoglu, N. G. (2017). Influence of energy use, foreign direct investment and population growth on unemployment for Russian Federation. Procedia Computer Science, 120, 706–711.

[15] Shahiduzzaman, M. D., & Layton, A. (2015). Changes in CO2 emissions over business cycle recessions and expansions in the United States: A decomposition analysis. Applied Energy, 150, 25–35.

[16] Toba, A. L., & Seck, M. (2016). Modeling Social, Economic, Technical & Environmental Components in an Energy System. Procedia Computer Science, 95, 400–407.

[17] Wang, Q., Su, B., Sun, J., Zhou, P., & Zhou, D. (2015). Measurement and decomposition of energy-saving and emissions reduction performance in Chinese cities. Applied Energy, 151, 85–92.

[18] Wang, W., Liu, X., Zhang, M., & Song, X. (2014). Using a new generalized LMDI (logarithmic mean Divisia index) method to analyze China's energy consumption. Energy, 67, 617–622.

[19] Wu, J. (2014). Urban ecology and sustainability: The state-of-the-science and future directions. Landscape and Urban Planning, 125, 209–221.

[20] Wu, R., Mavromatidis, G., Orehounig, K., & Carmeliet, J. (2017). Multiobjective optimisation of energy systems and building envelope retrofit in a residential community. Applied Energy, 190, 634–649.

[21] Zhao, X., Zhang, X., Li, N., Shao, S., & Geng, Y. (2016). Decoupling economic growth from carbon dioxide emissions in China: A sectoral factor decomposition analysis. Journal of Cleaner Production.

Optimization of electrical energy in the fan using fuzzy logic controllers

Arlenny, David Setiawan*

Universitas Lancang Kuning, Pekanbaru, Indonesia

*dsetia@unilak.ac.id

Abstract. Single-phase induction motors are a type of electric machine that are widely used in people's lives due to it has advantages such as lower prices, simple maintenance and construction. However, the single-phase induction motor also has a weakness, which is difficulty in setting up the rotation. The rotation setting of a single-phase induction motor can be affected by changes in the input frequency and changes in the input voltage. In this study is to design the control of a single-phase induction motor in a commonly used fan. The speed of the single-phase induction motor is regulated based on the number of visitors, using artificial intelligence namely fuzzy logic. The fuzzy logic controller can be group visitors to the room to determine the speed of the fan. The visitors are divided into three groups, namely "many" to indicate the maximum speed, "medium" for medium speed and "few" for low speed. Therefore, this research may contribute as an energy-saving solution using artificial intelligence.

1. Introduction

The single-phase induction motors are alternating current motors that are most commonly used in people's lives such as motors in fans, washing machines, blenders and so on. The single-phase induction motors have advantages such as cheaper prices, simple maintenance and simple construction. But the single-phase induction motor also has the disadvantage such as difficult to adjust the speed. The rotation setting of a single-phase induction motor depends on many Poles and Frequency. So, the action can be taken to adjust the number of poles and their frequency.

The method of changing the number of poles is considered more difficult and requires a lot of consideration to be applied. Therefore, another way that can be done is to adjust the input voltage or input frequency, so that a constant speed can be maintained. Fuzzy logic is one type of artificial intelligence technique that has good performance with a fast response rate in terms of overcoming the performance of control system for changes.

Furthermore, there are several main problems that need to be solved, namely: (1) How to adjust the rotation speed of a single-phase induction motor?, (2) Can the fuzzy logic method control the rotation of a single-phase induction motor?, (3) How is the ATmega328 microcontroller programmed for the fuzzy logic function?, and (4) Is the fuzzy logic controller on the 1 Phase AC motor a solution to energy saving?.

Several studies have been carried out by researchers regarding induction motors using fuzzy logic such as Manan et al. (2012). They conducted study in term of induction motor speed controller based on computer-aided fuzzy logic. According to Manan & Islam (2012), the simulation results for controlling the performance of the induction motor using fuzzy logic are superior to conventional controllers. Rao and A. Rajani (2013) conducted research to investigate the speed control of induction motor, which used the fuzzy logic method.

More, Ahmad et al. (2016) used fuzzy logic as a Proportional Integral (PI) controller to control the speed of a single-phase induction motor. Then, the fuzzy logic system is simulated using MATLAB/Simulink and shows the controller has a high response to load changes to speed variations (Ahmad et al., 2016). Therefore, this paper aim is to design, and analysis of speed control of single-phase induction motors based fuzzy logic controller.

2. Literature Review

2.1 Single Phase Induction Motor

Induction motor is one type of electric motor that works based on electromagnetic induction. The induction motor has a source of electrical energy, namely on the stator side, while the electrical system on the rotor side is induced through the air gap of the stator with electromagnet media. This motor uses Alternating Current (AC). This motor is called an induction motor because in real conditions the motor rotor current does not come from a particular source but is an induced current due to the relative difference between the rotation of the rotating field in the stator coil and the rotation of the rotor itself.

The working principle of a single-phase induction motor is a single phase voltage source connecting to the stator coils, a rotating field will arise with a certain speed. The motion of the rotating field will cut the conductor rod on the rotor, causing electromagnetic induction so that the rotor conductor will cause an induced EMF. Because the rotor coil is a closed circuit, there will be a current (I). The presence of a current in a magnetic field creates a force (F). If the initial coupling is large enough, then the force will rotate the rotor.

To produce an induced voltage, there must be an intersection between the rotating field and the rotor conductors. So, it takes a difference in relative speed between the stator and rotor. Because there is a difference between the rotation of the rotating field on the stator and rotor, this motor is called an asynchronous motor (Zuhal, 1988).

Basically, a single-phase induction motor rotates at a relatively constant speed. But, it affects the rotation of the motor due to a load. In certain systems, a significant change in rotation of a three-phase induction motor is not expected because some losses. The rotation speed of the induction motor is influenced by the frequency and the number of poles on the motor. Sequence, it is difficult to set up the speed regulation. In general, there are two types of single-phase induction motor rotors, namely squirrel cage rotors and winding rotors. Each type of rotor has a different speed regulation method. For example, a single-phase induction motor with a rotor winding can be adjusted by adjusting the amount of resistance that enters the rotor.

This method cannot be applied to motors with squirrel cage rotors. Although it is quite difficult to adjust the rotation, this single-phase induction motor has several advantages, so it is widely used in various needs. Some of the advantages are simple and sturdy construction, relatively cheap price compared to other motors, and easy maintenance.

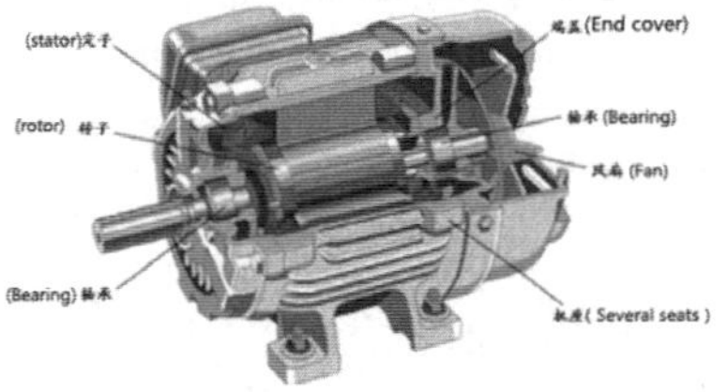

Figure 1. The Basic Elements of Slip Ring Induction Motors

2.2 Single Phase Induction Motor Speed

In general, induction motors is work at a constant speed (Zuhal, 1988). But in real conditions, the effect of loading can affect the rotational speed of the induction motor. Consequence, the action is needed to maintain speed due to loading. The amount of rotation in the stator depends on the magnitude of the frequency and the number of poles, which is stated in the following equation.

$$Ns\frac{120f}{P} \quad (1)$$

Where :
Ns = Stator Rotation Speed (rpm)
f = Frequency (Hz)
P = Number of Poles

The induction motor works when there is a difference between the speed of the stator and the rotor. The "asynchronous" speed is represented in the slip. The greater the load received by the motor, the slip value will also increase.

$$S = \frac{ns-nr}{ns}\, x\ 100\% \quad (2)$$

Where :
Ns = Stator Speed (rpm)
Nr = Rotor Speed (rpm)

2.3 Relationship of Speed, Load and Torque
There are several conditions for a 3-phase induction motor, related to torque and speed, which are as follows (Husnawan, 2008):

- When it starts to light up, it turns out that in this condition the initial ignition current is high with low torque (Pull-Up Torque).
- When it reaches 80% full speed, the torque reaches the highest level (Pull-Out Torque) and the current starts to decrease.
- If it reaches full speed or synchronous speed, the current decreases to zero.

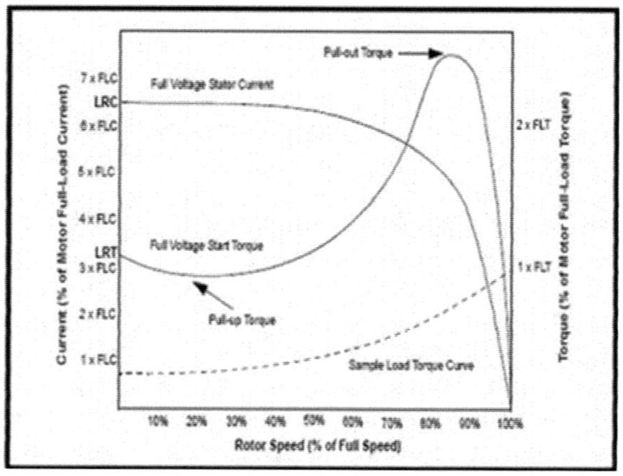

Figure 2. Torque vs Speed Graph (Husnawan, 2008)

2.4. Fuzzy Logic Controller
The theory of fuzzy sets is introduced by Lotfi A. Zadeh in 1965. Some of the reasons for using fuzzy sets are as follows (Klir, 2001; Kahraman et al, 2016):

- Systems that exist in the real world are sometimes difficult to describe exactly and completely. Therefore, an approach is needed to overcome the existing fuzziness.
- Human knowledge and experience are two aspects that have an important role in recognizing system behavior in the real world.

Therefore, a systematic theory is needed to formulate human knowledge along with other information in the form of mathematical models and measurement results.

Fuzzy logic control is one form of intelligent control. The fuzzy logic control can be used to solve problems in systems that have complex behavior. The fuzzy system acts as a bridge between precision language and the language of words so that it is more effective and efficient (Naba, 2009). The basic structure of fuzzy logic control can be described in Figure 3.

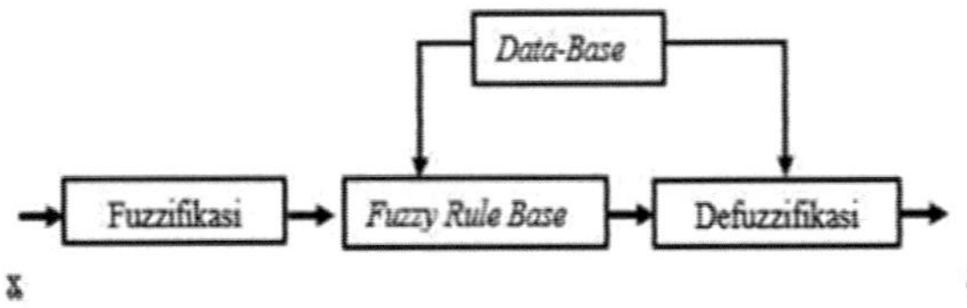

Figure 3. Block Diagram of Fuzzy Logic Controller (Naba, 2009)

The following is a description of the block diagram in Figure 3 (Naba, 2009):

a. Fuzzification
The fuzzification is process crisp variables converting into fuzzy variables through membership function techniques. The transformation process occurs at the fuzzification stage is a mapping of the input space with the help of a scaling factor. There are two kinds of fuzzification methods, namely singleton fuzzification and non-singleton fuzzification (Hasibuan, et al, 2007). In singleton fuzzification, the crisp point is mapped into a fuzzy set whose support is only at one x point. The non-singleton fuzzification, crisp points are mapped into fuzzy sets with more than one support point. The most commonly used type of fuzzification is singleton fuzzification. By applying singleton fuzzification fuzzy set a only has support at one point.

b. Fuzzy Rule Base
A set of fuzzy rules is a set of fuzzy rules that are grouped into a rule base, which is the basis for decision making (inference process) to obtain a control signal output action from an input condition. The output signal generated from the decision-making process is still in the form of fuzzy numbers, namely the degree of membership of the control signal.

c. Membership Function
This function is a function that states the level of fuzzy membership whose value is between zero and one. There are several forms of fuzzy membership functions, including: triangle (Triangular Function), trapezium (Trapezoid Function), sigmoid, gauss, bells and others (Jang, Sun, and Mizutani, 2004). The following is a form of membership function (membership function), namely triangle and trapezoid membership functions.

d. Defuzzification

e. Defuzzification can be defined as the process of changing the fuzzy quantities presented in the form of output fuzzy sets with their membership functions to regain their crisp form. Center of gravity (COG) is the most commonly used deffuzification method with the following equation:

$$y = \frac{\sum_{j=l}^{R} c^{j} \int u_{u} j(u) du}{\sum_{j=l}^{R} \int u_{u} j(u) du} \qquad (3)$$

Where :
y = output value

C^j = middle value of the jth membership function output
$\mu u(u)$ = membership function output
U^j = fuzzy set
R = number of rules

2.5 Arduino Uno
The microcontroller is used the Atmega328, which is embedded in the Arduino-Uno R3 unit. The Arduino Uno R3 is designed to be a ready-to-use microcontroller, programming and I/O are available. So that the Arduino makes a board that is easy to use and the programming uses the C language. Microcontroller chip ATMega8, ATMega168 and ATMega328, in general the function and number of legs of the microcontroller is the same, but the difference is the storage capacity / memory.

Figure 4. Arduino Uno

3. Result and Discussion

3.1 Fuzzy Logic Controller (FLC) Design
To set the fan speed automatically, an input is needed that can read the temperature and the number of visitors in the room so that there are 2 inputs, namely a temperature sensor using an LM35 sensor that can measure temperatures between -550C to 1500C and a photodiode sensor to determine the number of visitors in the room. The method used / FIZ is Mamdani with a fan output.

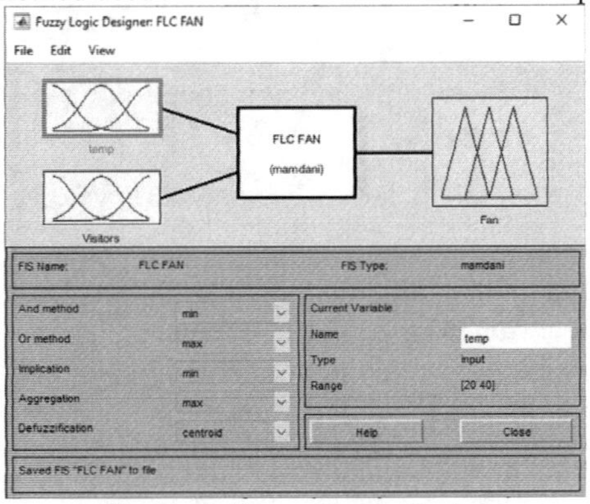

Figure 5. Fuzzy Logic Design

After determining the input, FIS and output, then determine the membership function on both input and output. As for the input in the form of temperature, the membership function is divided into 5 groups, namely cool, slightly cool, medium, slightly hot and hot.

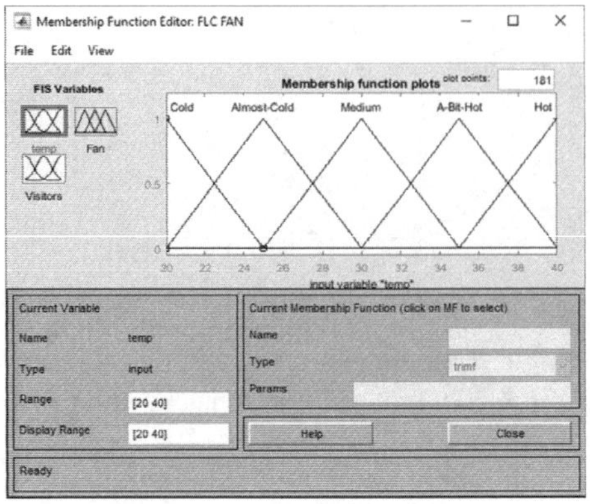

Figure 6. Membership Function Temperature Input

For visitor input, the membership function is also divided into 5, namely a little, almost a little, moderate, almost a lot and a lot. While the membership functions for the output in the form of a fan are slow, medium and fast.

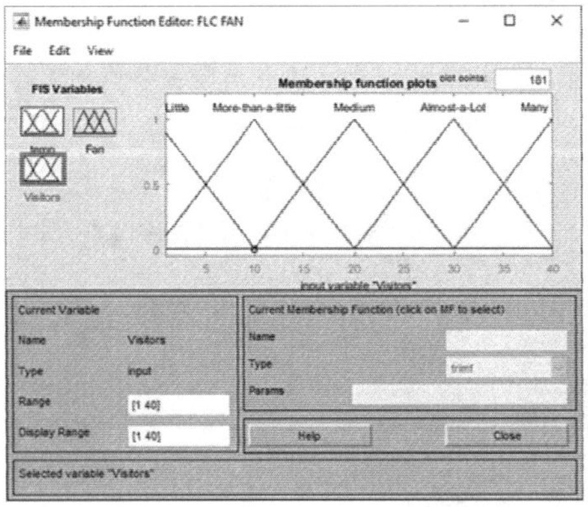

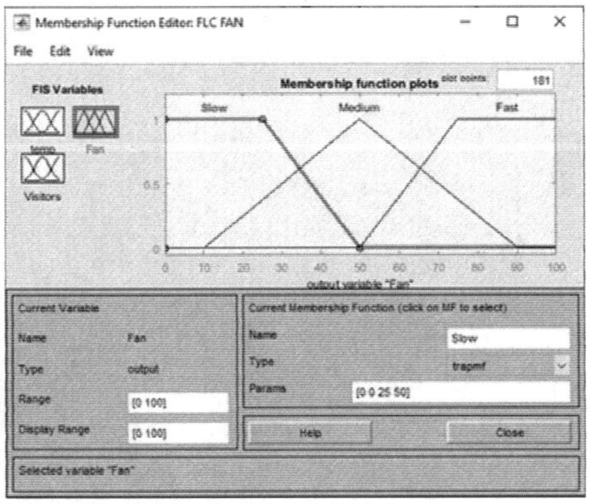

Figure 7. Visitor Input Membership Function **Figure 8.** Membership Function Fan Output

Table 1. Fan rotation settings for temperature and visitors

Temp. / Visitors	**Cold (C)**	**Almost Cold (AC)**	**Medium (M)**	**A Bit Hot (BH)**	**Hot (H)**
Little (L)	Slow	Slow	Medium	Medium	Fast
More than a little (ML)	Slow	Slow	Medium	Medium	Fast
Medium (Md)	Slow	Slow	Medium	Fast	Fast
Almost a Lot (AL)	Slow	Slow	Medium	Fast	Fast
Many (My)	Slow	Medium	Fast	Fast	Fast

After the membership function, then determine the rules of fuzzy logic that was be made. The rules made were then entered into the rule editor to try fuzzy logic simulations that would be applied to the microcontroller late

1. If (temp is Cold) and (Visitors is Little) then (Fan is Slow) (1)
2. If (temp is Cold) and (Visitors is Medium) then (Fan is Slow) (1)
3. If (temp is Cold) and (Visitors is Many) then (Fan is Slow) (1)
4. If (temp is Cold) and (Visitors is More-than-a-little) then (Fan is Slow) (1)
5. If (temp is Cold) and (Visitors is Almost-a-Lot) then (Fan is Slow) (1)
6. If (temp is Almost-Cold) and (Visitors is Little) then (Fan is Slow) (1)
7. If (temp is Almost-Cold) and (Visitors is More-than-a-little) then (Fan is Slow) (1)
8. If (temp is Almost-Cold) and (Visitors is Medium) then (Fan is Slow) (1)
9. If (temp is Almost-Cold) and (Visitors is Almost-a-Lot) then (Fan is Slow) (1)
10. If (temp is Almost-Cold) and (Visitors is Many) then (Fan is Medium) (1)
11. If (temp is Medium) and (Visitors is Little) then (Fan is Medium) (1)
12. If (temp is Medium) and (Visitors is More-than-a-little) then (Fan is Medium) (1)
13. If (temp is Medium) and (Visitors is Medium) then (Fan is Medium) (1)
14. If (temp is Medium) and (Visitors is Almost-a-Lot) then (Fan is Medium) (1)
15. If (temp is Medium) and (Visitors is Many) then (Fan is Fast) (1)
16. If (temp is A-Bit-Hot) and (Visitors is Little) then (Fan is Medium) (1)
17. If (temp is A-Bit-Hot) and (Visitors is More-than-a-little) then (Fan is Medium) (1)
18. If (temp is A-Bit-Hot) and (Visitors is Medium) then (Fan is Fast) (1)
19. If (temp is A-Bit-Hot) and (Visitors is Almost-a-Lot) then (Fan is Fast) (1)
20. If (temp is A-Bit-Hot) and (Visitors is Many) then (Fan is Fast) (1)
21. If (temp is Hot) and (Visitors is Little) then (Fan is Fast) (1)
22. If (temp is Hot) and (Visitors is Medium) then (Fan is Fast) (1)
23. If (temp is Hot) and (Visitors is Many) then (Fan is Fast) (1)
24. If (temp is Hot) and (Visitors is More-than-a-little) then (Fan is Fast) (1)
25. If (temp is Hot) and (Visitors is Almost-a-Lot) then (Fan is Fast) (1)

Figure 9. Fuzzy Logic Controller Rules

After entering the rules into fuzzy logic, then testing the rules that have been made to ensure that the fuzzy logic made can work properly and correctly.

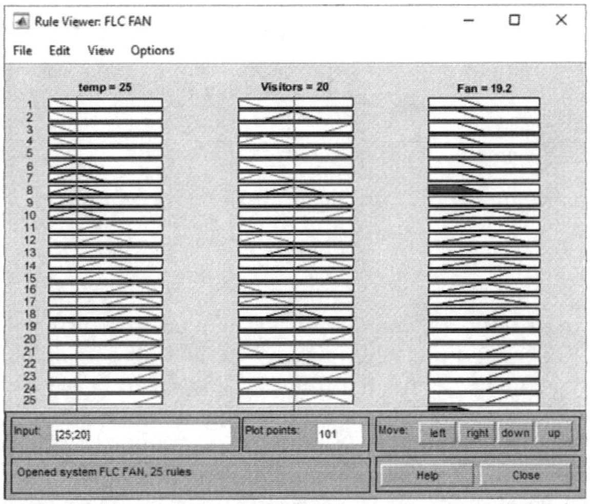

Figure 10. Simulation of 200C Temperature and 20 Visitors

Based on the simulation, if the temperature is 200C and there are 10 visitors, the fan is worth 19.2, then it tries several random experiments to prove that the rules made work well and correctly. From the general view, the fan speed with respect to temperature and the number of visitors can be seen from the shape of the Face Viewer, which is presented in a fuzzy logic designed.

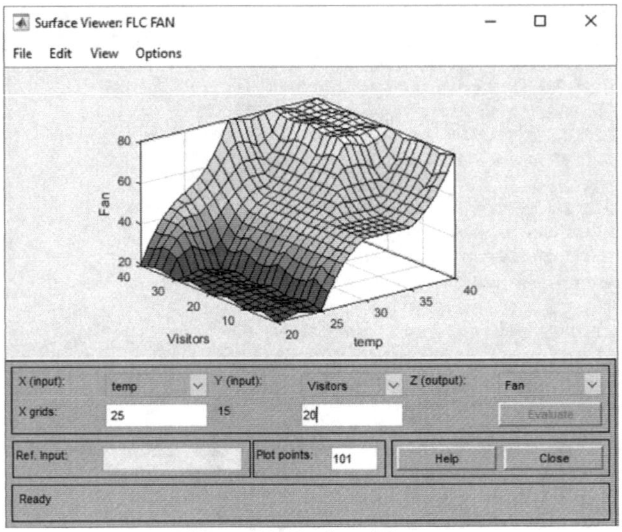

Figure 11. Surface Viewer

3.2. Hardware Design

The hardware or prototype / prototype of the fan controller using the fuzzy logic controller method consists of 2 sensors as input, namely the temperature sensor using the LM35 sensor and the visitor number sensor using a photodiode sensor. Both sensors produce a DC voltage that varies depending on the sensor reading, for LM35 to read a temperature of -550C to 1500C produces an output voltage between 0 to 1.5 volts DC. While the controller uses an Arduino Uno with an Atmega 328 microcontroller and the output is an actuator or final drive to determine the fan speed, whether it is slow, medium or fast.

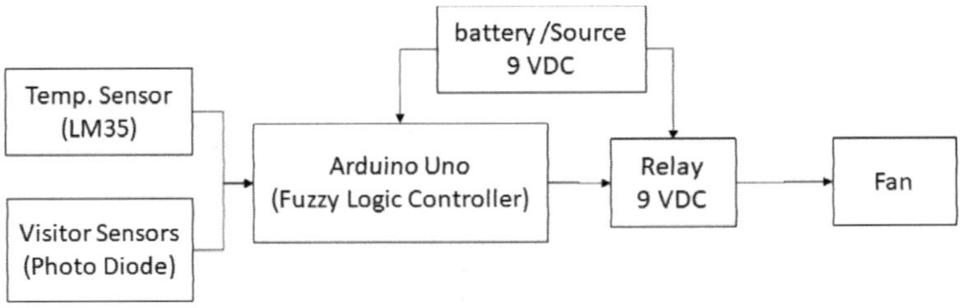

Figure 12. Block Diagram of Fan Control System using FLC

The rules in fuzzy logic in the form of tables, in a microcontroller or arduino are converted into a programming language where the 25 specified rules can be stated in the programming language as follows:

1. if (temperature = C && visitor = L){analogWrite(fan, slow)}
2. if (temperature = C && visitor = ML) {analogWrite(fan, slow)}
3. if (temperature = C && visitor = Md) {analogWrite(fan, slow)}
4. if (temperature = C && visitor = AL) {analogWrite(fan, slow)}
5. if (temperature = C && visitors = My) {analogWrite(fan, slow)}

6. if (temperature = AC && visitor = L) {analogWrite(fan, slow)}
7. if (temperature = AC && visitor = ML) {analogWrite(fan, slow)}
8. if (temperature = AC && visitor = Md) {analogWrite(fan, slow)}
9. if (temperature = AC && visitor = AL) {analogWrite(fan, slow)}
10. if (temperature = AC && visitor = My) {analogWrite(fan, medium)}
11. if (temperature = M && visitor = L) {analogWrite(fan, medium)}
12. if (temperature = M && visitor = ML) {analogWrite(fan, medium)}
13. if (temperature = M && visitor = Md) {analogWrite(fan, medium)}
14. if (temperature = M && visitor = AL) {analogWrite(fan, medium)}
15. if (temperature = M && visitor = My) {analogWrite(fan, loud)}
16. if (temperature = BH && visitor = L) {analogWrite(fan, medium)}
17. if (temperature = BH && visitor = ML) {analogWrite(fan, medium)}
18. if (temperature = BH && visitor = Md) {analogWrite(fan, loud)}
19. if (temperature = BH && visitor = AL) {analogWrite(fan, fast)}
20. if (temperature = BH && visitor = My) {analogWrite(fan, fast)}
21. if (temperature = H && visitor = L) {analogWrite(fan, fast)}
22. if (temperature = H && visitor = ML) {analogWrite(fan, fast)}
23. if (temperature = H && visitor = Md) {analogWrite(fan, fast)}
24. if (temperature = H && visitor = AL) {analogWrite(fan, fast)}
25. if (temperature = H && visitor = My) {analogWrite(fan, loud)}

In programming language, temperature is divided into 5 groups, namely the symbols C for cool, AC for Almost Cold, M for medium, BH for a Bit Hot and H for hot. Meanwhile, the number of visitors was also divided into 5 groups, namely L for a little, ML for More than a Little, Md for moderate/medium, AL for Almost a Lot and My for Many. While the speed setting is determined by the output value of the fuzzy logic controller, namely slow speed if the value is less than 33 (> 33), medium speed if the value is between 33 – 66 and fast speed if the value is more than 66 (< 66).

Figure 13. Fan Speed Control Prototype using FLC

From the experimental results data, it can be seen in Table 2.

Table 2. The results of the experiments carried out.

Experiment	Input		Output	
	Temp. (^{0}C)	Visitors	Simulation	Fan Speed
1	20	10	19,2	Slow
2	20	20	50	Medium
3	20	30	50	Medium
4	25	10	40	Medium

5	25	20	50	Medium
6	25	30	60	Medium
7	30	10	50	Medium
8	30	20	50	Medium
9	30	30	80,8	Fast

Based on the experiment above, the results obtained are the more visitors and the higher the room temperature, the more the fan spinning. conversely, the fewer visitors and low temperatures, the fan slows down.

4. Conclusion

Based on the simulation and prototype, the fuzzy logic controller design, which is made to control the fan speed to regulate the room temperature has been successfully simulated and tested. The simulation uses the MATLAB application with the Mamdani method, while the hardware test uses a prototype with a DC fan output to ensure fan speed. The fan DC can change according to the input given, namely changes in temperature and the number of visitors.

References

[1] Ahmad, A.H, Yasien, F.R. and Abdullah, A.S. (2016). Speed Control of Single Phase Induction Motor Using Fuzzy Logic Controller, 26(4), 17-29.
[2] Barry G. Woollard, H. Kristianto, Practical Electronics, Pradnya Paramita, Jakarta, fifth edition, 2003.
[3] Cengiz Kahraman, Başar Öztayşi, Sezi Çevik Onar 2016. A Comprehensive Literature Review of 50 Years of Fuzzy Set Theory, International Journal of Computational Intelligence Systems, 9(1), 3-24.
[4] Chitra V. 2008. Induction Motor Speed Control Using Fuzzy Logic Controller, International Journal of Electric, Computer, Energetic, Electronic and Communication Engineering Vol. II no. 11.
[5] George J. Klir 2001. Foundations of Fuzzy Set Theory and Fuzzy Logic: A Historical Overview, International Journal of General Systems, 30(2), 91-132.
[6] Hall, Denis T. Practical Knowledge of Ship Electrical. Translated by Sarwito S. Jakarta: IMarE/IMarEST.
[7] Hasibuan, Fahrudin 2007. Control of DC Shunt motor speed with Fuzzy Logic Controller. ITS Electrical Engineering Strata II Program, Surabaya.
[8] Heriyanto, Hery 2008. Planning of PID Control on Induction Motor Based.
[9] Husnawan, Anwar 2008. Speed Regulation of a 3-Phase Induction Motor, Jurnal Electrical Engineering, University of Indonesia, Depok.
[10] Jang, JSR., Sun, CT dan Mizutani, E. 2004. Neuro-Fuzzy and Soft Computing. Singapore: Pearson Education.
[11] Kuriawan, Adi 2015. Induction Motor Speed Control Strategy using PWM Inverter Based on Artificial Neural Networks. ITS Marine Engineering, Surabaya.
[12] M. Nageswara Rao and A. Rajani (2013). Speed Control of Induction Motor Using Fuzzy Logic Approach, International Journal of Engineering Research & Technology, 2(8).
[13] Mannan, M., Islam, A. and Uddin, M.N. (2012). Fuzzy-Logic Based Speed Control of Induction Motor Considering Core Loss into Account, Intelligent Control and Automation, 3, 229-235. 10.4236/ica.2012.33026.
[14] Naba, Agus 2009. Fast Learning Fuzzy Logic using MATLAB. Malang:Andi Offset.
[15] Ogata, Katsuhito 2010. Modern Control Engineering.
[16] Petruzella, Frank 1996. Industrial Electronics. Yogyakarta : ANDI Yogyakarta.
[17] Purwanto, Era 2011. Development of an Induction Motor Model as an Electric Car Propulsion Using Vector Control Method, Scientific Journal of Elite Elektro Vol. II no. 02, Surabaya State

Electronics Polytechnic, Surabaya.
[18] Said, Muhammad Iqbal, 2013. Diesel Electric Propulsion System As An Alternative Propulsion On 10 GT Fishing Vessels. Postgraduate Program in Naval Architecture, Faculty of Engineering, Hasanuddin University, Makassar.
[19] Simulink Matlab. Electrical Engineering, Semarang University, Semarang.
[20] Zuhal, 1988. Basic Electrical Engineering and Power Electronics. Jakarta: Gramedia Utama Pustaka.

The effect of giving chia seeds on the bodyweight of pregnant mice (Mus Musculus L.)

R Uthia*, **I O Rz**

Politeknik Kesehatan Kemenkes Riau, Pekanbaru, Indonesia

*rahimatul1089@gmail.com; ira@pkr.ac.id

Abstract. Chia seeds (Salvia hispanica L.) are becoming a trend for consumption by employees of the Health Ministry Polytechnic of Riau. This support by many studies on the benefits of consuming these seeds for a diet to lose weight or maintain ideal body weight. Several studies also reported that Chia seeds lower and maintain blood cholesterol levels. This study was to see the effect of giving Chia seeds on the body weight of pregnant mice. This research use laboratory experimental methods, which is analogous to the condition of pregnant women on mice. Chia seeds are administrating into three groups of 4 g; 8 g; 16 g and compared to one control group. The administration starts on the 6th to 15th day of pregnancy. Bodyweights measured at the beginning of pregnancy until before parturition. The results showed that giving Chia seeds did not affect the weight gain of pregnant mice (P>0.05). So from this study concluded that giving Chia seeds does not affect body weight in pregnant mice.

1. Introduction

Chia seeds (Salvia hispanica L.) are becoming a trend for consumption by employees of the Health Polytechnic of the Ministry of Health of Riau. Many studies about the benefits of consuming these seeds to lose weight or maintain ideal body weight. Several studies have reported that chia seeds can lower and maintain blood cholesterol levels [1] and have a weight loss effect in obese people [2]. Chia seed can reduce the risk of cardiovascular disease, inflammation, central nervous system disorders, and diabetes [3] thus making people enthusiastic about consuming it to get an ideal body weight. This article is a fraction of the author's research entitled The Effect of Chia Seeds on The Development of The Fetus of Mice (Mus musculus L.).

Chia seeds are natural ingredients with health potential, which use as an alternative for developing functional food products. This plant is native to Central America, specifically Mexico and Guatemala. Chia seeds contain protein (15–25 percent), fat (30–33 percent), carbohydrates (26–41 percent), fiber (18–30 percent), and minerals (4–5 percent) [4]. Chia seeds also contain omega-3 fatty acids (linolenic acid) by 17.83 percent [5]. In 2009 chia seeds were approved as a novel food source by the European Parliament and the Council of Europe. The use of chia seeds as a food ingredient report being safe because it has no side effects or allergenicity [6].

Thus, chia seeds and their derivatives are promising sources for development. Chia seeds contain 17.83% omega-3 fatty acids (-linolenic acid) [5] and are a source of dietary fiber, protein with high biological value, and antioxidants [7]. In addition to containing essential fatty acids, chia seeds also report containing phenolic compounds. These phenolic compounds are bioactive components that contribute to the health benefits of chia seeds. The phenolic components in chia seeds are flavanols and phenolic acids (myricetin, quercetin, kaempferol, caffeic acid [8]. These compounds are primary and synergistic antioxidants that provide a high proportion of antioxidant activity of chia seeds [9].

Caffeic acid and chlorogenic acid found in chia seeds showed protection to cells from free radicals and inhibition of lipid peroxidation. It was signified stronger than vitamin C protection, ferulic acid, and vitamin E. Chia seeds are also high in dietary fiber and protein, rich in many exogenous amino acids. This grain is also a good source of minerals and vitamins as bioactive compounds with high antioxidant activity, especially polyphenols and tocopherols [10].

2. Methodology

This study used an experimental design using mice as test animals consist of four treatment groups (control, dose 4 g, dose 8 g, and dose 16 g). The parameter to be observed is the mother's weight during pregnancy. The study began by characterizing the test preparations, including organoleptic observations, measurement of water content, ash, carbohydrates, total fat, protein, Ca, Fe, P, Zn, crude fiber, Mg, and K. The steps for testing are as follows:

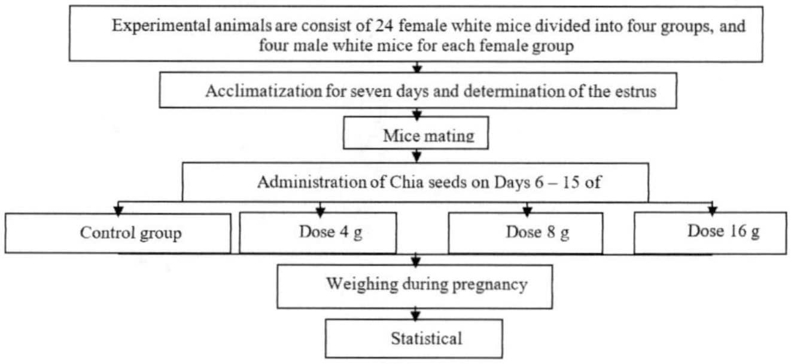

Figure 1. Working Scheme of Observing the Effect of Chia Seed on Pregnant Mice Body Weight

2.1 Dosage Determination

Chia seeds administered to experimental animals were 4, 8, and 16 g per daily given orally.

2.2 Preparation of experimental animals

Animals used in this experiment were female mice aged about three months, never treated with drugs, and body weight ranged from 20-30 grams. Mice were acclimatized for seven days to get the animals used to the experimental environment. The feeds gave as *ad libitum, t*he body weight was measured every day, and the behavior observes daily. The estrus period determines visually by observing the mice's vagina during acclimatization. The mice during estrus marks by the red and gummy vagina of the mice. Animals used are considered healthy if the change in body weight is not more than 10%, visually shows normal behavior, and has an estrus cycle of 4-5 days [11].

2.3 Grouping of experimental animals

Mice divide into four groups, where one group consists of 6 female mice and one male mouse.

2.4 Animal mating

At the next stage during estrus, animals are mated with a ratio of males and females of 1:6. Male white mice were put into female cages at four in the afternoon and separated in the morning. The vaginal plug examine. Vaginal plugs indicate that the mice have copulated and are on the 0th day of gestation. The pregnant mice were separated, unmarried mice were mate again with male mice [12][13].

2.5 Administration of chia seed

The test preparations in three dose variants were given orally for ten consecutive days starting from the 6th to the 15th day of gestation. The weighing takes every day for 18 days of observation. If there is a drastic weight loss and accompanied by bleeding around the vagina, the mice may have suffered a miscarriage or aborted (excluded). It noted that mice were sick due to treatment (also excluded [14].

3. Result and Discussion

Table 1. Organoleptic of Chia

Parameters	Flour Organoleptic	Seeds Organoleptic
Form	Fine powder, coarse fibrous	Small round, slippery
Color	Dark-brown	Dark-brown
Flavor	Tasteless	tasteless
Smell	Smelly	unsmelly

Table 2. Characteristics of Chia flour

Parameters	%	Analysis Method
Water content	3.09	SNI 01-2891-1992,9
Ash Level	4.59	SNI 01-2891-1992,8.2
Carbohydrate	8.47	SNI 01-2891-1992,7.1
Total fat	24.5	SNI 01-2896-1998,5
Protein	23.9	SNI 01-2896-1998,5
Ca	1.57×10^{-1}	AOAC Official method 986, 24
Fe	6.36×10^{-2}	SNI 01-2896-1998,5
P	1.34×10^{-1}	SNI 01-2891-1992,11
Zn	5.59×10^{-3}	SNI 01-2896-1998,5
Coarse Fiber	44.1	SNI 01-2896-1998,5
Mg	3.7×10^{-1}	SNI 01-2891-1992,5.1
K	7.8×10^{-1}	SNI 01-2891-1992,6.1

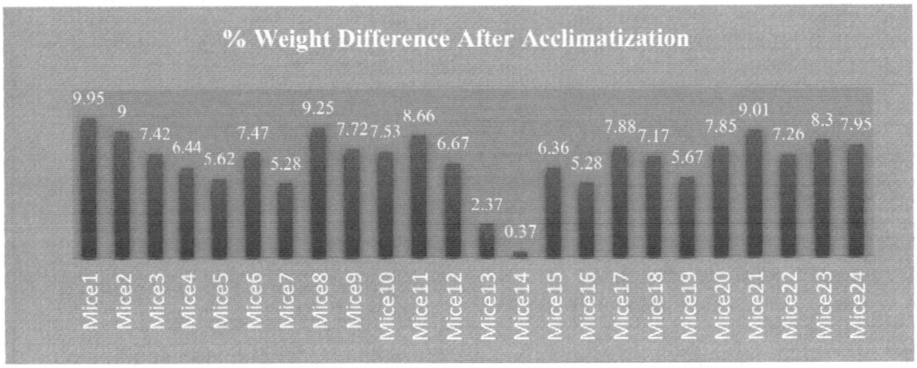

Figure 2. % Weight Difference after Acclimatization

IOP Conf. Series: Earth and Environmental Science 1041 (2022) 012026 doi:10.1088/1755-1315/1041/1/012026

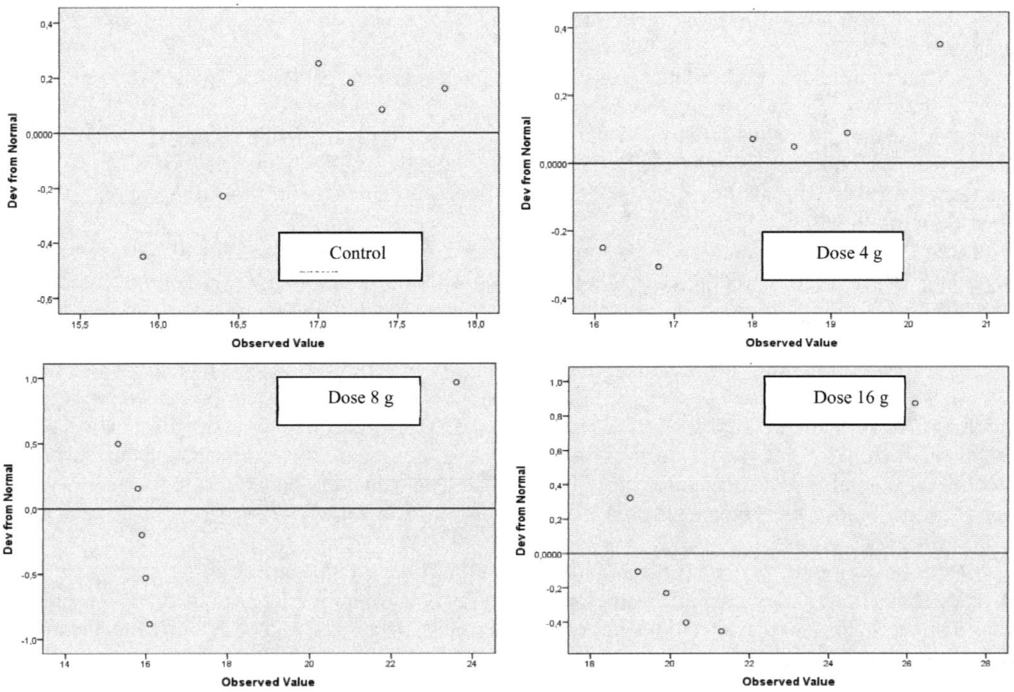

Figure 3. Broodstock Average Weight Gain

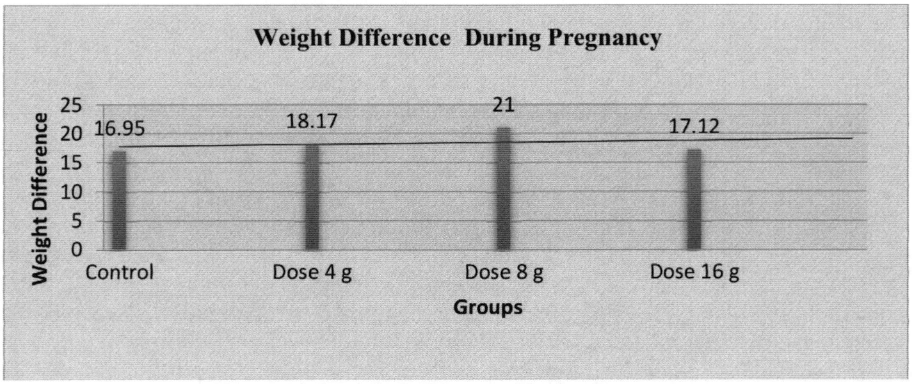

Figure 4. Weight Difference During Pregnancy

There are organoleptic differences in flour and seeds of Chia, namely in the smell. Chia seed flour has a slight odor. One of the distinctive characteristics of chia seeds is their ability to form a gel layer. This gel layer formed after the seed hydrate with water. On contact with water, the outer epidermis of the seed ruptures and releases gum filament absorbs water immediately, and the gel layer looks like a transparent capsule.

The transparent capsule secretes where this gum retains and absorbs water, can even give different viscosities to any given concentration of chia seeds. Chia seed gum absorbs the water up to 12 times its dry weight. Dry gum obtained from the extraction of chia seed gum reports absorbs the water up to 27 times [15]. The smell of the flour may release during flour processing. The organoleptic of Chia show in table 1.

Chia seed gum has a hydration process that is almost the same as xanthan gum. Both require time and a certain way for the swelling process they dispersed in solution homogeneously. However, chia seeds and xanthan gum have different visual appearances of dispersion. In chia seeds, the gum dispersion shows from the chia seeds spread evenly in the solution, while xanthan gum does not exist [16]. This is the reason why the administration route of chia is eaten by the mice directly. And the Characteristics of Chia flour are shown in table 2.

Table 2 shows that Chia seeds contain macro and micronutrients needed by pregnant women, including carbohydrates, protein, fat, calcium, iron, phosphorus, zinc, fiber, magnesium, and potassium. In addition to good nutritional value, chia seeds also have a low level of allergenicity and are also food with a safe category to use. These seeds gave as additional food for pregnant women [6].

No testing animals after acclimatization for seven days experienced weight gain and loss exceeding 10 %, and no testing animals were sick and could be used for research, as shown in Figure 2. Acclimatization is the maintenance of experimental animals to adapt to a new environment [17]. The length of acclimatization varied from 3-14 days, but most researchers acclimatized for seven days. Acclimatization also prevents stress on animals in a new environment [18]. During the acclimatization period, animals adjust to a new environment hoped that animals will no longer stress because of moving from their previous cage.

Basically, during pregnancy, the brood stock will experience weight gain as gestational age increases. A study stated that maternal weight gain during pregnancy is directly related to the baby's weight [19]. Based on the theory, pregnancy causes increased metabolism of energy and other nutrients. Fetal growth and development increased energy and nutrients. The increasing size of the uterine organs changes the composition and metabolism of the mother. The nutritional status of pregnant women determines the weight of babies born; the nutritional adequacy of pregnant women see from their weight gain during pregnancy [20].

In this study, the body weight of pregnant mice measures during pregnancy. The average weight gain of mice before parturition can be seen in Figure 3. It shows that in each dose group of chia seeds, the bloodstock experienced weight gain during pregnancy, including the control group. This indicates that the pregnancy is healthily taking place and hoped that the foetus is also in top condition. Weight gain occurred in all treatment groups, indicating that the provision of Chia seeds did not affect the weight of pregnant women ($p>0.05$). The same thing is also clarified in figure 4, the difference in maternal weight during pregnancy.

Figure 4 sees that the difference in maternal weight gain in each treatment group includes the control group. If a straight line draws, it shows that the weight gain of the parent during pregnancy appears linear, or there is no difference in the weight gain of the parent when given chia seeds or not given chia seeds. The data processing using ANOVA also showed no effect of giving chia seeds to the bodyweight of the parent ($p>0.05$). Nieman's hypothesis states that high dietary fiber and ALA content in chia seeds can reduce body weight, risk of heart disease, and obesity.

However, the results of clinical trials of chia seeds in humans found that increasing plasma ALA levels did not cause significant weight loss [21]. It proves from this research that giving chia seeds to pregnant mice did not have a weight-lowering effect. The bodyweight of the bloodstock in all treatment groups increased.

4. Conclusion

Giving chia seeds during pregnancy (in the study to mice) did not affect body weight, meaning that the mice continued to gain weight which was no different from the control group.

Acknowledgment

Thank you to the Health Ministry Polytechnic of Riau for research funding so that this research can run smoothly.

References

[1] Ayerza R, Coates W. 2007. Effect of dietary a-linolenic fatty acid derived from chia when fed as ground seed, whole seed an oil on lipid content and fatty acid composition of rat plasma. Ann Nutr Metab. 51 (1): 27-34
[2] Brissette CE, Jenkis AL, Choleva L, Vuksan V. 2013. Abstract. The effect of Salvia hispanica L seeds on weight lost in everweight anda obese individuals with type 2 diabetes mellitus. Can J Diabetes. 37(4): 64
[3] Vuksan V, Whitman D, Siepenviver J, Jenkis A, Rogovik A, Bazinet R, Vidgen E, Hanna A. 2007. Supplementation of conventional therapy with the novel grain Salba (Salvia hispanica L) improves major and emerging cardiovascular risk factors in type 2 diabetes. Diabetes care 30(11): 2804-2810
[4] Ixtaina, V.Y., S.M. Nolasco, dan M.C. Tom. 2008. Physical Properties of Chia (Salvia hispanica L.) Seeds. Industrial Crops and Product. Vol. 28(3): 286–293.
[5] [USDA] United States Department of Agriculture. 2011. National Nutrient Database for standard reference. Release 24. Nutrient data Laboratory Home Page, U.S. Dpeartment of Agriculture Research Service
[6] EFSA. 2009. Scientific Opinion of the Panel on Dietetic Products Nutrition and Allergies on a Request From The European Commission on The Safety of 'Chia seed (Salvia hispanica) and Ground Whole Chia Seed' as a Food Ingredient.The EFS Journal.Vol. 996: 1-2.
[7] Craig R. 2004. Application for approval of whole chia (Salvia hispanica L) seed and ground while seed as novel food ingredient. UK:ACNFP
[8] Ali, N.M., S.K. Yeap, W.Y. Ho, B.K. Beh, S.W. Tan, dan S.G. Tan. 2012. The promising future of chia, Salvia hispanica L. Journal of Biomedicine and Biotechnology. doi: 10.1155/2012/171956
[9] Fernandez I, Ayerza R, Coates W, Vidueros SM, Slobodianik N, Pallaro AN. 2006. Nutritional characteristis of chia. Actualizacion en Nutricion. 7: 23-25
[10] Reyes CE, Tecante A, Valdivia Mal. 2008. Dietary fiber content and antioxidant activity of phenolic compund present in Mexicam chia (Salvia hispanica L) seed, Food Chemc. 107(2): 656-663
[11] Almahdy, A. 2012. Teratologi Eksperimental. Padang: Andalas UniversityPress.
[12] Almahdy, A. 2011. Uji aktivitas vitamin A terhadap efek teratogen warfarin pada fetus mencit putih. Medan: USU Press.
[13] Kauffman, M. H. 1992. The atlas of Mouse Development. London: Academic Press Limited
[14] Lu, F.C. 1995. Toksikologi Dasar (Edisi kedua). Penerjemaah: E. Nugroho. Jakarta: Universitas Indonesia Press.
[15] Hernandez, L.M. 2012. Gum Form Chia Seeds (Salvia hispanica): Microstructure, Physico-Chemical Characterization and Application in Food Industry. PhD Thesis of Pontificia Universidad Catolica de Chile, 120h.
[16] Campos, M.R., N.C. Solis, G.R. Rubio, L.C. Guerrero, dan D.B. Ancona. 2014. Chemical and Functional Properties of Chia Seed (Salvia hispanica L.) Gum. Mexico. International Journal of Food Science.
[17] Hasanah A. 2017. Efek jus bawang bombay (Allium cepa Linn.) terhadap motilitas spermatozoa mencit yang diinduksi Streptozotocin (STZ). Saintika Medika 11(2): 92-101.
[18] Purwantono P, Kusrini MD, Masy'ud B. 2016. Manajemen penangkaran empat jenis kura-kura peliharaan dan konsumsi di Indonesia. Jurnal Penelitian Hutan dan Konservasi Alam 13(2): 119-135.
[19] Khoiriah, F., Angraini, D. I., Carolina. N., & Sukohar, A. 2015. Hubungan Pertambahan Berat Badan Ibu Selama Hamil Dengan Berat Bayi Lahir Rendah. J Majority. Vol.4(3):52-57
[20] Nadesul, H. 2008. Makanan Sehat Untuk Ibu Hamil. Puspa Swara. Jakarta.
[21] Nieman, D.C., E.J. Cayea, M.D. Austin, D.A. Henson, S.R. McAnult, dan F. Jin. 2009. Chia seed does not promote weight loss or alter disease risk factors in overweight adults. Nutrition Research.

Vol. 29(6): 414–418.

Supplementing mineral selenium and vitamin e in diets on *in vivo digestibility*, blood glucose, and urea levels of cows

D. Pamungkas*, A.S. Putri, F.Firdaus, R.Widiyawati, D.M.Dikman

Indonesian Agency for Agriculture Research and Development Ministry of Agriculture, Pasuruan, Indonesia

*dpamungkas2000@yahoo.com

Abstract. In such condition as of mating, pregnant and delivering calves, cows were often having deficiency in micro minerals, causing their reproductive functions to be less than optimal. The aim of this study was to determine the effect of Selenium and Vitamin E (SeVE) addition in diets on in vivo digestibility, blood glucose and urea levels. A total of 30 Ongole Crossbred cows (I3-I4 Aged, Body Weight 423.8 75.0 kg, and Body Condition Score 3.1 0.2) were used in this experiment. The study was conducted for 13 weeks. Animals were prepared to be inseminated by the Belgian Blue straws. They were kept in individual cages and were divided into two treatments, namely (P0) = basal diets and (P1) = basal diets + SeVE. The diets given were elephant grass and concentrate (3% of body weight). The results showed that addition SeVE increased digestibility of organic matter and crude protein, increased blood Se and urea levels, but did not show significant effect on blood glucose levels. In subsequent observations, SeVE addition turned out to have a positive effect on the incidence of pregnancy.

1. Introduction

The reproductive status of cow must be a major concern in terms of obtaining optimal pregnancy and birth rates. Several cases of reproductive failure and disturbance that occurred in the field were related to nutritional shortages, including macro and micro minerals and or vitamin. Selenium is an essential component of selenoproteins playing an important role in many biological functions, such as antioxidant defense, formation of thyroid hormones, DNA synthesis, fertility and reproduction. Selenium can be converted in the organism into various metabolites. Some, like methylselenol, play a role in cancer prevention [1].

However, selenium supplementation of feedstuffs appears to have no significant effect in chemical composition milk production as of lactose fat and protein [2], [3]. The last months of gestation cows represent the critical period for the selenium availability. The transfer of selenium to the fetus and subsequently to the calf occurs even when cows are deficient in selenium. Cows sacrifice selenium available to them to ensure adequate intake for newborns [5]. The right selenium concentration can reduce the incidence of metritis and ovarian cysts in the postpartum period, conversely, chronic selenosis decreases fertility by supporting the prolonged growth of ovarian cysts and anesthesia [6].

In addition, [7] and [8] reported deficiency of selenium leads to disorders in the perinatal period altering milk quality of cows. Selenium consumption is affected by dietary of Vitamin. It should be noted that selenium and vitamin E antioxidant functions are interdependent. A diet low in vitamin E may increase the amounts of selenium necessary for the prevention of certain anomalies [9].

The results of previous studies showed that the administration of selenium and vitamin E injections at a dose of 1.5 mg / mL of selenium and 50 mg / mL of vitamin E five times during gestation at seven months to the 14th day after childbirth encourages the successful involution of the uteri in dairy cows. Cows will soon return to estrus [10]. The condition of the cows after giving birth is associated with immune pressure which results in increased susceptibility to infection. This is mostly related to endocrine changes and decreased nutritional intake which have implications for the performance and health status of cows.

Selenium and vitamin E (SeVE) are closely involved with biological functions and antioxidant activities. Increased energy diets together with SeVE reduce oxidative stress in cows, improve immune function and improve reproductive performance of cows after calving [11]. In another study [12], antioxidant supplementation (SeVE) increases insulin sensitivity thereby reducing the effects of oxidative stress which can lead to insulin resistance. Zigo et al. [13] reported that parenteral administration of SeVE in dairy cows showed a positive effect on increasing the concentration of alpha-tocopherol and Se in blood plasma.

Based on the statements above, this study examined the supplementation of mineral Selenium combining Vitamin E in diets which initiate the changes on in vivo digestibility, blood glucose and urea levels and encourage the performance of cows specially to maintain reproductive states of the cows.

2. Methodology

A total of 30 Ongole Crossbred cows (I3-I4 Aged, Body Weight 423.8 ± 75.0 kg, and Body Condition Score 3.1 ± 0.2) were used in this experiment. The study which conducted for 13 weeks was located in Beef Cattle Research Institute (BCRI) barns at Grati, Pasuruan Indonesia. Animals were prepared to be inseminated by the Belgian Blue straws. They were kept in individual cages and were divided into two treatments, namely (P0) = basal diets and (P1) = basal diets + SeVE. The diets given were elephant grass and concentrate (3% of body weight).

The chemical composition of feed was shown in Table 1. The mixture of Selenium and vitamin E were given in powder form as of 9 g/kg concentrate. Parameter observed were: intake and the feed digestibility (dry matter, organic matter, and crude protein), and the level of Glucose, Uric, and Selenium in the blood. They were analyzed using Spectrophotometry. Meanwhile the data collected were analyzed by using T test.

Table 1. Chemical composition of feed (dry matter basis)

Note: CP = crude protein; EE = ether extract; CF = crude fibre; TDN = total digestible nutrient

Feed	Ash	CP	EE	CF	Starch	TDN[1]	Selenium
			(%)				(ppm)
Basal diets	11.6	17.3	6.3	12.5	52.3	72.7	0.70
Concentrate + SeVe	13.8	17.1	6.0	13.0	50.1	69.9	12.5
Grasses	12.2	10.9	1.8	34.3	40.8	48.0	0.2

3. Result and Discussion

3.1. Dietary selenium

The selenium content appeared to be higher in the P1 diet, as a logical consequence of supplementation in concentrate feed. Se content within grasses at P0 and P1 is relatively the same (1.4 vs 1.4 ppm). As shown in Fig1. there was a significant increase in the Selenium content in the total feed, namely an increase of 59.7 ppm (93.0%). This means that there is a significant contribution of Se content in the concentrate. This is in line with the report of Setyowati [16] that organic selenium supplementation (0.03 ppm) in concentrate increased the selenium content of treated feed.

Our findings showed the crude protein intake was increased (Table 2). Although research has been conducted to evaluate the effect of organic and inorganic selenium supplementation in beef cattle, it was found that different form of selenium did not affect average daily intake, average daily gain, and feed efficiency [17]. Dietary selenium intake can be affected by different form of selenium because of its difference in bioavailability and biopotency [18]. Both form of selenium supplementation in diet could increase blood selenium concentration although organic selenium has the highly impact [19], [20].

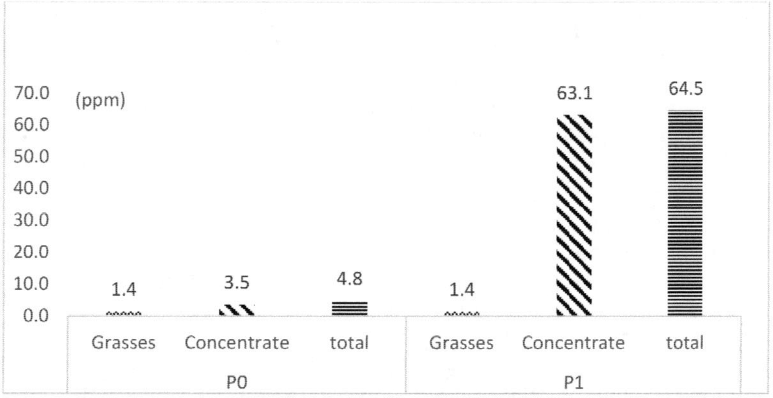

Figure 1. Dietary Selenium

Selenium levels in plants are influenced by selenium levels in the soil, plant species, plant age, plant production levels and climate [21]. Selenium levels are not evenly distributed in the soil resulting in different levels in plants. In addition, plants have different abilities to absorb selenium.

3.2. Intake and digestibility

Supplementation SeVE in the diet was not different among the intake of dry mater, organic matter and crude protein. This condition reflected that the diets was preferred by animals, and this is a way to explore this supplement in regards increasing feed digestibility and cow's performance. The average of daily DMI in this research (11.2 kg/head) was higher compared to earlier experiment (0.92 kg/head) [22]. Another experiment reported that animal supplemented with different form of selenium (organic vs inorganic) has similar daily feed intake [17].

Although not significantly different, data showed that average daily OMI and CPI obtained in this research was in agreement with earlier report [23]. There is an interesting result that although the dry matter digestibility of P1 is lower than P0, the digestibility crude protein of P1 is higher than that of P0. This shows positive effect resulting from SeVE supplementation in the diet in regard to digestibility. *In vivo* dry matter digestibility (DMD) and crude protein digestibility (CPD) were affected ($P<0.05$) by SeVe supplementation, while organic matter digestibility (OMD) was not ($P>0.05$) (Table 2).

Inorganic selenium supplementation in this research decreased DMD, similar with earlier experiment although animal received organic selenium supplementation [22], [24]. Previous study conducted by Kumar et al. [23] showed that OMD was similar between control and supplemented group. In agreement with this research, higher CPD was found in earlier experiment with animal supplemented lower to medium selenium yeast [24] while other experiment reported lower CPD affected by selenium supplementation [22].

Table 2. Feed intake and *in vivo* digestibility

Treatment	Daily intake (kg/head)			Digestibility (%)		
	DM	OM	CP	DM	OM	CP
P0	11.2 ± 1.0	9.7 ± 0.1	1.3 ± 0.1	72.1 ± 0.5[a]	74.5 ± 0.4	65.1 ± 0.7[b]
P1	11.2 ± 1.1	9.5 ± 0.0	1.3 ± 0.1	68.3 ± 0.9[b]	74.2 ± 0.2	66.7 ± 1.0[a]

Note: DM : dry matter; OM: organic matter; CP: crude protein; Means followed with different letters in same column differ significantly ($P<0.05$).

The apparent digestibility reflects process in the body to ensure the energy available and to estimate the protein microbial synthesis in the rumen. The report of [25] that the supplementation of high-degradation feed ingredients has increased digestibility and nutrient uptake, especially in the use of ingredients which from agricultural residues as basal feed.

According to the National Research Council [26], selenium can be legally supplemented in beef cattle diets to provide 3 mg/head daily or 0.3 mg/kg in complete diets, although the selenium requirement of beef cattle can be met by 0.1 mg/kg of diet. From this research, it was found that without SeVE supplementation, the diets contain enough selenium for animal according to NRC recommendation. Selenium supplementation can increase fertility and reduce embryonic death in first moth of gestation [6] and has an antioxidant role during belated follicular development as reported by previous research [27].

3.3. Blood glucose, urea, and selenium

Blood glucose levels did not show a significant difference even though blood glucose at P1was higher than that of P0. This result was reasonable due to the energy content of the feed with mineral Se + vitamin E supplementation and the control treatment feed relatively the same. Blood glucose in the results of the study ranged from 69.5 to 73.8 mg/dl.

The glucose level is still in the normal ranges. [28] reported that normal bovine blood glucose levels were in the range of 43-100 mg / dl. Blood glucose concentration of the heifers was 57 ± 2 mg/dL in the low group and 78 ± 2 mg/dL in the high group. Heifers that had low blood glucose at arrival consumed more feed (P=0.02), tended to have increased final bodyweight and rate of gain (P<0.10), had increased backfat thickness (P<0.05), and tended to have heavier hot carcass weights and fewer standard carcasses (P<0.10) compared with heifers that had high blood glucose at arrival.

Meanwhile, blood urea levels showed a significant difference. Blood urea levels at P1 was higher than those of P0. The increasing blood urea level was reasonable due to simultaneously effects of supplementation. Blood urea levels in dairy cows in the study of Juniper et al. [20] with supplementation of 0.40 mg / kg selenium yeast in the feed was 8.4 mmol / L. The blood urea levels in the study ranged from 39-44.8 mg / dl. The blood urea levels were still in the normal range.

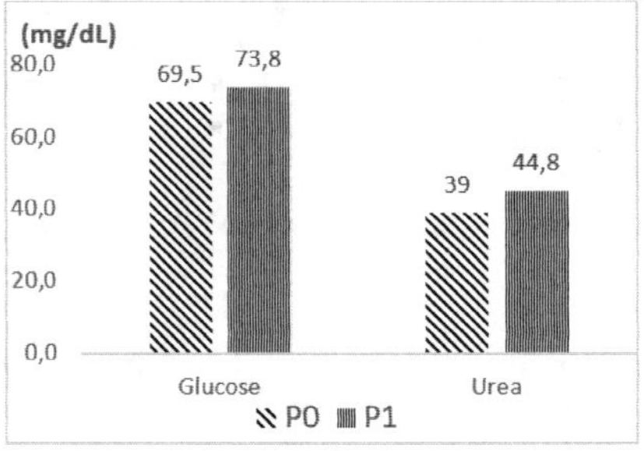

Figure 2. The level of glucose and urea in the blood

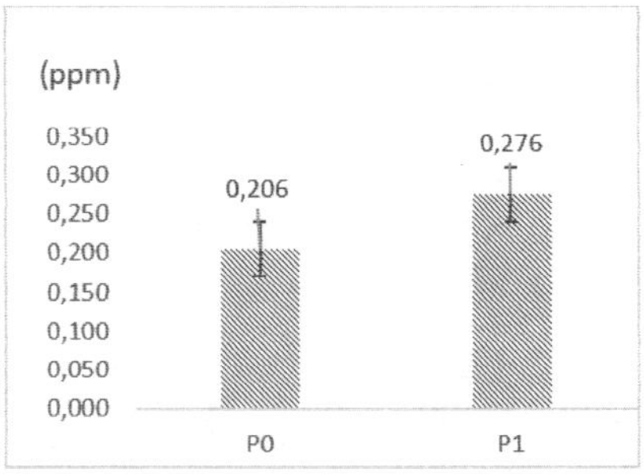

Figure 3. The level of selenium in the blood

In blood, the selenium is bound to the α and β-globulins, LDL (low density lipoprotein) and VLDL (very low density lipoprotein) and albumin [29]. Ten selenium atoms as selenocysteine are included in the selenoprotein-P which is a selenoprotein supplier of selenium to body tissues [30]. According to [31], the rate of adequate selenium in the blood of cattle is between 0.08 and 0.16 mg/L. In plasma, the selenium is mainly found in the albumin fraction.

According to Villar et al. [32], the appropriate level in plasma selenium is between 51 and 85 μg/L. To achieve this adequate rate, additional dietary intake of 0.5 mg/kg DM may be enough [9]. An increase in Hungate [33] stated that normal blood urea levels in cattle range from 26.6 to 56.7 mg / dl. In this study blood selenium levels showed a significant difference. Blood selenium in the supplementation treatment was higher than the control treatment. This is because the selenium supplements in feed are absorbed to bloodstream. Juniper et al. [20] reported levels of selenium in the blood of dairy cows in the selenium yeast supplementation treatment to the feed of 0.40 mg / kg was 251 μg / L. The treatment without supplementation of selenium was 211 μg /L.

In subsequent observations (after a period of 13 weeks), we found that the cattle in P1 (four out of 15 animals) showed pregnancy, after insemination with *Belgian Blue* straw as donors, whereas in P0 still did not show signs pregnancy of animals.

4. Conclusion
Supplementation of SeVE to the basal diet gives positive impact of cows in regard to crude protein digestibility and the level of glucose, urea, and mineral selenium in the blood.

Acknowledgment
This project was the part of collaborative research, assessment, and agricultural development program (KP4S) conducted in 2018 by Indonesian Centre of Animal Research and Development (ICARD) and Livestock Embryo Centre, Indonesia Ministry of Agriculture with regards of developing Indonesian Belgian Blue Cattle Breeding Programs.

References
[1] Mehdi Y, Hornick J-L, Istasse L, et al. Selenium in the environment, metabolism and involvement in body functions. Molecules 2013; 18: 3292–3311.
[2] Meyer U, Heerdegen K, Schenkel H, et al. Influence of various selenium sources on selenium concentration in the milk of dairy cows. J für Verbraucherschutz und Leb 2014; 9: 101–109.
[3] Salman S, Dinse D, Khol-Parisini A, et al. Colostrum and milk selenium, antioxidative capacity and immune status of dairy cows fed sodium selenite or selenium yeast. Arch Anim Nutr 2013; 67: 48–61.
[5] Hefnawy AEG, Tórtora-Pérez JL. The importance of selenium and the effects of its deficiency in animal health. Small Rumin Res 2010; 89: 185–192.
[6] Mehdi Y, Dufrasne I. Selenium in Cattle: A Review. Molecules 2016; 21: 545.
[7] Horkỳ P. Effect of selenium on its content in milk and performance of dairy cows in ecological farming. Potravinarstvo 2015; 9: 324–329.
[8] Ran L, Wu X, Shen X, et al. Effects of selenium form on blood and milk selenium concentrations, milk component and milk fatty acid composition in dairy cows. J Sci Food Agric 2010; 90: 2214–2219.
[9] Council NR. Nutrient Requirements of Beef Cattle: Seventh Revised Edition: Update 2000. Washington, DC: The National Academies Press. Epub ahead of print 2000. DOI: 10.17226/9791.
[10] Prasdini WA, Rahayu S, Djati MS. Determination of the success uterine involution in Friesian Holstein dairy cow based estrogen levels after multiple injection of selenium-vitamin E. J Vet 2015; 16: 351–356.
[11] Xiong X, Lan D, Li J, et al. Selenium supplementation during in vitro maturation enhances meiosis and developmental capacity of yak oocytes. Anim Sci J 2018; 89: 298–306.
[12] Abuelo A, Alves-Nores V, Hernandez J, et al. Effect of Parenteral Antioxidant Supplementation During the Dry Period on Postpartum Glucose Tolerance in Dairy Cows. J Vet Intern Med 2016; 30: 892–898.

[13] Zigo F, Vasiľ M, Farkašová Z, et al. Influence of selenium and vitamin e supplementation during pregnancy on udder health and milk quality in dairy cows at parturition. Potravin Slovak J Food Sci 2017; 11: 58–64.
[16] Setyowati EY, Tanuwiria UH, Santoso M. The organic selenium identification on the cattle feedstuffs in Sumedang region. In: Proceedings of national seminar neutron activation analysis. Indonesia: National Nuclear Energy Agency, p. 351.
[17] Liao SF, Brown KR, Stromberg AJ, et al. Dietary Supplementation of Selenium in Inorganic and Organic Forms Differentially and Commonly Alters Blood and Liver Selenium Concentrations and Liver Gene Expression Profiles of Growing Beef Heifers. Biol Trace Elem Res 2011; 140: 151–169.
[18] Ceballos A, Sánchez J, Stryhn H, et al. Meta-analysis of the effect of oral selenium supplementation on milk selenium concentration in cattle. J Dairy Sci 2009; 92: 324–342.
[19] Gunter SA, Beck PA, Phillips JM. Effects of supplementary selenium source on the performance and blood measurements in beef cows and their calves. J Anim Sci 2003; 81: 856–864.
[20] Juniper DT, Phipps RH, Givens DI, et al. Tolerance of ruminant animals to high dose in-feed administration of a selenium-enriched yeast. J Anim Sci 2008; 86: 197–204.
[21] Dilaga SH. Nutrisi Mineral pada Ternak-Kajian Khusus Unsur Selenium. Akad Presindo Jakarta.
[22] Vinu MN, Gangadevi P, Mercy AD, et al. Effect of Dietary supplementation of organic selenium on growth performance and nutrient utilization in crossbred calves. J Anim Vet Sci 2012; 43: 36–40.
[23] Kumar N, Garg AK, Mudgal V, et al. Effect of different levels of selenium supplementation on growth rate, nutrient utilization, blood metabolic profile, and immune response in lambs. Biol Trace Elem Res 2008; 126: 44–56.
[24] Wang C, Liu Q, Yang WZ, et al. Effects of selenium yeast on rumen fermentation, lactation performance and feed digestibilities in lactating dairy cows. Livest Sci 2009; 126: 239–244.
[25] Pamungkas D, Utomo R, Ngadiyono N, et al. Use of Corn Pericarps and Coffee Pods as Basal Diet Supplementing with Energy and Protein Source at Different Rates of Degradability. JITV; 19.
[26] Council N-NR. Nutrient requirements of beef cattle.
[27] Ceko MJ, Hummitzsch K, Hatzirodos N, et al. X-Ray fluorescence imaging and other analyses identify selenium and GPX1 as important in female reproductive function. Metallomics 2015; 7: 71–82.
[28] Mitruka BM, Rawnsley HM. Clinical biochemical and hematological reference values in normal experimental animals. Clin Biochem Hematol Ref values Norm Exp Anim.
[29] Schrauzer GN. Selenomethionine: a review of its nutritional significance, metabolism and toxicity. J Nutr 2000; 130: 1653–1656.
[30] Burk RF, Hill KE. Selenoprotein P-expression, functions, and roles in mammals. Biochim Biophys Acta 2009; 1790: 1441–1447.
[31] Dargatz DA, Ross PF. Blood selenium concentrations in cows and heifers on 253 cow-calf operations in 18 states. J Anim Sci 1996; 74: 2891–2895.
[32] Villar D, Arthur JR, Gonzalez JM, et al. Selenium status in cattle. Bov Pract 2002; 73–80.
[33] Hungate RE. The rumen and its microbes. Elsevier, 2013.

Effect of Saccharomyces cerevisiae on Probiotic Properties of Goat Milk Kefir

L E Radiati[1*], D L Hati[1], D Fardiaz[1] and L R H Sari[2]

[1]Universitas Brawijaya, Malang, Indonesia
[2]Universitas Airlangga, Surabaya, Indonesia

*lilik.eka@ub.ac.id

Abstract. Goat milk kefir is a fermented milk product with kefir grain, as a probiotic agent and contains bioactive compounds. Kefir grains consist of bacteria and yeast. One of these yeasts are Saccharomyces cerevisiae, its role in the kefir system can be enhanced by restructure of kefir grain with increased Saccharomyces cerevisiae population. The objectives of this research were to analysis effect of the Saccharomyces cerevisiae concentration levels on kefir, as antibacterial against Escherichia coli, Salmonella typhi. Klebsiella pneumoniae, antioxidant activity, β-galactosidase synthesis. The research method was true experiment by Completely Randomized Design consist of 5 kind Saccharomyces cerevisiae concentration levels (S0: control; S1: 0.10%; S2: 0,25%; S3: 0.5% and S4: 1.0% v/w) of kefir grain. The results showed that kefir inhibited Escherichia coli, Salmonella typhi and Klebsiella pneumoniae, able to produce β-galactosidase and antioxidant activity. Kefir is relevant as food functional development.

1. Introduction

Kefir from goat's milk is beneficial for health and has a unique aroma. The protein hydrolysis process in milk produces active peptides which are responsible for the functional value of goat milk kefir [1] [2]. Kefir is fermented milk that claimed as functional dairy products with probiotics are living microbial [3], when consumed with recommendation diet, provide significant effect to increase human health [4]. Probiotics are needed to balance the intestinal microflora; it can competitively suppress the growth of pathogenic microbes [2] The development of probiotic microbes in the intestinal tract is influenced by the symbiotic matrix as protection and nutrient for growth [5].

Kefir is fermented milk product with kefir grain that rich beneficial microbes containing bacteria and yeasts [6]. Microbes of kefir proliferate, produce enzyme and other bioactive compound causing biochemical reaction applied to medical benefits. Lactobacillus and Bifidobacterium are bacteria have been often reported as probiotics, but reduction exposed to gastrointestinal tract fluids [7]. While Bacillus strain as probiotic showed stability in food products [8].

Currently, Yeasts is developed as probiotic, rather than as Single Cell protein. The development knowledge promise yeast as probiotic because they are resistant to antibiotic agents, and this property is used for combination therapies of probiotics with antibiotic for gastrointestinal disease [9]. Yeast have been used in traditional food processing [10], and produce metabolite compounds that play role as immunomodulator, cholesterol and toxin binding, antimicrobial and antioxidant [11].

Saccharomyces boulardii has been studied and other researcher suggested to try other species to develop research in vitro and in vivo assays [11]. Various analytical methods have been used to evaluate probiotics [12]. For this reason, relevant study was needed to develop required probiotic properties such as anti-enteropathogenic, antioxidant and produce β-galactosidase. The addition of Saccharomyces cerevisiae to kefir preparation can increase the functional value of kefir as a probiotic agent and functional food.

Saccharomyces cerevisiae in fermentation produce compound as preservative agent and improve the nutritional [13]. The studies of mixed probiotics are limited [14], kefir is common example as multi culture probiotics consist of bacteria and yeasts. One of yeast are Saccharomyces cerevisiae. This yeast produces ethanol and simple compounds such as esters and peptide compound through the fermentation of substrates [15].

In addition, has been mentioned above, many research results increase insight on the understanding of the antioxidant activity of probiotics. The mechanism of antioxidation by probiotics namely probiotics produce antioxidant metabolites that regulate antioxidant activities, chelating ion agent and regulates intestinal microbiota [16]. Goat milk kefir showed the antioxidant and antimicrobial activities [17]. The addition of Saccharomyces cerevisiae to kefir preparation increases antimicrobial and antioxidant [18]. One of the mechanisms of antimicrobial activity are discussed in the context of food microbial interactions, bio preservation and the role of GAPDH (Glyceraldehyde 3-phosphate dehydrogenase) in the defense system of Saccharomyces cerevisiae [19].

Therefore, research on other Saccharomyces is needed to improve data supporting the role of Saccharomyces cerevisiae possess probiotic properties. Thus, the objectives of the research were to determine effect of Saccharomyces cerevisiae on probiotic properties of goat milk kefir.

2. Methodology

2.1 Research methods

The research method was true experiment, consist of four treatments *Saccharomyces cerevisiae* ATCC 36858 (10^7CFU/ml) (0; 0.10; 0, 25; 0.50; 1.00% v/w) levels of kefir grain and four replications. Statistical analysis was used SPSS for windows 16 then continued with Duncan's multiple ranged tests.

2.2 Kefir preparation

Kefir grains were obtained from animal product technology Laboratory, Animal Science Faculty, Universitas Brawijaya. Malang Indonesia. Kefir was produced by adding 3% (w/v) kefir grain into goat milk and was incubated at 25 ±2° C for 48 h. The kefir was separated from the grains by filtration through a plastic sieve and stored before the next analysis.

2.3 Antimicrobial activity assay

The antimicrobial activity of kefir was established on the supernatant sample by preparing the sample as follows: Kefir was centrifugated at 4°C, 4000 rpm for 15 min. The antimicrobial activity was determined by the spot-on-paper dish method, according to Tulini et al. [20].

In these steps was used the bacteria test namely *Escherichia coli* ATCC 25922, *Salmonella typhi* ATCC 14028 and *Klebsiella pneumoniae* ATCC 29665 were obtained from Medicine Faculty Laboratory. Universitas Brawijaya. Twenty microliters of kefir supernatant were spotted on paper disc (ϕ 6 mm) on Sabouraud dextrose agar plate and incubated at 37°C for 72 h under aerobic condition. After incubation, the plates were overlay with 7 ml of soft BHI and seed with bacteria test (10^5CFU/ml), then incubated at 37°C for 24h. Observation on the inhibition with indicator clear zone or halos (mm).

2.4 β-galactosidase activity assay

β-galactosidase activity was determined by converted chromogenic substrate o-nitrophenyl-β-D-galactopyranoside (ONPG) [9] and modified by Radiati et al. [21]. Prepared 4.5 ml of 0.2 phosphate buffer was added into 5 g of kefir. The sample was diluted 1:10 with 0.1 M phosphate buffer (pH 7.0; 0.001 M MgSO4 M β-mercaptoetanol) at 4°C, added two drop of chloroform and one drop of 0.1% SDS.

The mixed sample was vortex for 10 s and incubated for 15-20 min, adjusted with pH 11, was centrifuged at 16,000 x g for 10 min. The OD was observed at 420 nm (Spect-UV-Zimadzu). β-galactosidase activity obtained by curve standard of o-nitrophenol obtained in one unit of enzyme equal wit 1.0 μmole of o-nitrophenol released per min.

$$\beta\text{-galactosidase activity (units/mL)} = 1000 \times A_{420}/tv)$$

Where: t: time of reaction (min), v: sample volume, A_{420}: Abs. at 420 nm

2.5 Antioxidant activity assay
Principal of antioxidant activity is radical 1,1-diphenyl-2-picryhydrazyl ($DPPH^+$) scavenger [22] Briefly. 1 mL of kefir was centrifugated at 4000 x g for 5 min, at 4°C. Kefir supernatant was collected. The reagent and solution are mixed sequentially by 1 mL of $DPPH^+$ (0,2 mmol/L in methanol) was added with 800 μL kefir supernatant, vortexed and incubated at room temperature in darkness for 30 min, were centrifuged at 4°C, 15.000 x g for 5 min, 300 μL of supernatant were transferred into 24 well plates. Measure the absorbance at η 517 nm. The percentage of reduction of $DPPH^+$ was calculated according to Equation:

$$\% AA = \{[Ai - (As - Ab)]/Ai\} \times 100$$

Where: Ai: Abs. methanolic and $DPPH^+$,
As: Abs. of $DPPH^+$ and sample
Ab: Abs of blank sample.

The result are presented in accordance with the classification of Gil-Rodriguez et al [22]: low activity (20-30%), good activity (30-40%), very good activity (40- 50%) and excellent activity (> 50%)

2.6 Statistic analysis
The data were obtained as mean of four replication and were evaluated with analysis of variance (ANOVA) software MINITAB 16.0 and Tukey's multiple range test to determine the different means between treatment. The differences were considered significant at $P < 0.05$.

3. Result and Discussion

3.1 Antimicrobial activities of goat milk kefir
Table 1 presents the antimicrobial activities of the goat milk kefir by addition various *Saccharomyces cerevisiae* level in the grain studied. All the *supernatant of* kefir inhibited the *Escherichia coli* ATCC 25922 *Salmonella typhi* ATCC 14028, *Klebsiella pneumoniae* ATCC 29665 by 56.00, 64, 28 and 68.18% respectively. While the highest susceptible was found for the strain *Salmonella typhi* which produced a halo of inhibition of 28.00 ± 2.0 mm.

Table 1. Antimicrobial Activity of Goat Milk Kefir

S. cerevisiae (%)	*E. coli*	*S. typhi*	*K. pneumoniae*
Control	$11^a \pm 1.0$	$10^a \pm 1.6$	$7^a \pm 2.8$
0.1	$11^a \pm 2.6$	$10^a \pm 1.4$	$10^b \pm 1.5$
0.25	$19^b \pm 2.1$	$15^b \pm 0.5$	$12^c \pm 1.0$
0.50	$20^b \pm 0.0$	$25^c \pm 1.8$	$18^d \pm 2.2$
1.00	$25^c \pm 2.7$	$28^d \pm 2.0$	$22^e \pm 2.3$

Note: The different superscript in the same column shows a highly significant ($P \leq 0.01$)

The grain kefir and Saccharomyces cerevisiae form matrix complex increased probiotic properties. One of primary properties of yeasts is their antimicrobial activity against pathogenic bacteria. Speranza et al [23] [24] the antimicrobial activity of yeasts was caused yeast produce biofilm compound and other metabolites. Yeasts isolated from cultures of cereal foods processing showed no antimicrobial activities against Escherichia coli O157:H7 and Staphylococcus aureus growth.

The antimicrobial activity and the capacity of grain kefir to produce antimicrobial compounds are considered critical characteristics and depend on yeast species and their growth condition for effective competitive with pathogens. Therefore, the grain kefir design can be modified according to the target component and benefits for health.

3.2 Antioxidant activity

Goat milk kefir after adding Saccharomyces cerevisiae increased the antioxidant activity, it is interesting to discuss of the functional characteristics. The evaluation scavenger or percentage of reduction of the 1,1 diphenyl-2-picrylhydrazyl (DPPH+) radical is present Table 2. The addition 1% Saccharomyces cerevisiae.

Table 2. Effect of Saccharomyces cerevisiae on β-D Galactosidase (U/g protein) Antioksidan Activities (% DPPH inhibition)

S. Cerevisiae (%)	β-D Galactosidase (U/g protein)	Antioxidant Activities (% DPPH inhibition)
Control	$2.88^a \pm 0.76$	$45^a \pm 1{,}50$
0.1	$6.10^b \pm 0.28$	$45^a \pm 1.70$
0.25	$13.08^c \pm 0.11$	$50^b \pm 2{,}00$
0.50	$80.16^d \pm 0.28$	$60^c \pm 1.20$
1.00	$120.33^e \pm 0.57$	$72^d \pm 2.30$

In the kefir grain showed the highest antioxidant activity. Several studies have reported the yeast cells as natural antioxidant [22]. Other research determined that intact cells had higher antioxidant capacity than the extract, which may be due mainly to the high content of β-D-glucan and β-glucans found in their cell wall rather than in cellular compounds. In the present research, kefir supernatant content yeast, bacteria in intact cells and other metabolites that support antioxidant activity. According to Gil-Rodríguez et al. [22] the activity of supernatant kefir was classified into the excellent antioxidant activity (> 50%). The *Saccharomyces cerevisiae* has important role to develop food functional.

3.3 β-galactosidase ctivity

β-galactosidase is the enzyme that convert lactose to galactose and glucose in milk, relieve the diary symptoms of lactose intolerance [25], therefore, β-galactosidase play role in functional food. Table 2.The higher *Saccharomyces cerevisiae* concentration in grain kefir was followed by the increasing β-galactosidase. *Saccharomyces cerevisiae* produce β-galactosidase in various media [26]. Besides that, this enzyme convert milk fat into short-chain fatty acids that are useful material for the host [27].

The β-galactosidase activities of *L. rhamnosus* and *L. plantarum* were induced with 1% lactose. All of lactic acid bacteria showed higher β-galactosidase enzyme activity in medium containing 1% of lactose than without 1% of lactose. This result indicated that the addition of 1% lactose could increase β-galactosidase enzymatic activity [27]. Similar with this research result that *Saccharomyces cerevisiae* in milk showed the high β-galactosidase concentration in goat milk kefir.

4. Conclusion

Saccharomyces cerevisiae in grain kefir complex improved bioactive compound of goat milk kefir, simultaneously support the increasing antimicrobial activity, antioxidant and β-galactosidase as improvement of probiotic properties of Goat milk kefir.

Acknowledgement

This research was supported by Professor Research Grant Animal Science Faculty, Universitas Brawijaya. Thank to our research team and Laboratorium Sentral Ilmu Hayati staff to support to who support for research activity.

References

[1] Amorim FG, Coitinho LB, Dias AT, Friques AGF, Monteiro BL, Rezende LCD de, et al. Identification of new bioactive peptides from Kefir milk through proteopeptidomics: Bioprospection of antihypertensive molecules. Food Chem. 2019 Jun 1;282:109–19.

[2] Lin YC, Chen YT, Hsieh HH, Chen MJ. Effect of Lactobacillus mali APS1 and L. kefiranofaciens M1 on obesity and glucose homeostasis in diet-induced obese mice. J Funct Foods. 2016 May 1;23:580–9.

[3] Kim DH, Jeong D, Kang IB, Lim HW, Cho YJ, Seo KH. Modulation of the intestinal microbiota of dogs by kefir as a functional dairy product. J Dairy Sci [Internet]. 2019;102(5):3903–11. Available from: http://dx.doi.org/10.3168/jds.2018-15639

[4] Gholizadeh P, Mahallei M, Pormohammad A, Varshochi M, Ganbarov K, Zeinalzadeh E, et al. Microbial balance in the intestinal microbiota and its association with diabetes, obesity and allergic disease. Vol. 127, Microbial Pathogenesis. Academic Press; 2019. p. 48–55.

[5] Hemarajata P, Versalovic J. Effects of probiotics on gut microbiota: Mechanisms of intestinal immunomodulation and neuromodulation. Vol. 6, Therapeutic Advances in Gastroenterology. 2013. p. 39–51.

[6] Thohari I, Purnomo H, Radiati LE, Fanani Z. A developmental strategy for consumer buying - choices of goat milk kefir in East Java. Livest Res Rural Dev. 2012;24(3).

[7] Arslan S. A review: Chemical, microbiological and nutritional characteristics of kefir. Vol. 13, CYTA - Journal of Food. Taylor and Francis Ltd.; 2015. p. 340–5.

[8] Soares MB, Martinez RCR, Pereira EPR, Balthazar CF, Cruz AG, Ranadheera CS, et al. The resistance of Bacillus, Bifidobacterium, and Lactobacillus strains with claimed probiotic properties in different food matrices exposed to simulated gastrointestinal tract conditions. Food Res Int. 2019 Nov 1;125.

[9] Lima M dos SF de, Souza KMS de, Albuquerque WWC, Teixeira JAC, Cavalcanti MTH, Porto ALF. Saccharomyces cerevisiae from Brazilian kefir-fermented milk: An in vitro evaluation of probiotic properties. Microb Pathog. 2017 Sep 1;110:670–7.

[10] Vickers CE, Bydder SF, Zhou Y, Nielsen LK. Dual gene expression cassette vectors with antibiotic selection markers for engineering in Saccharomyces cerevisiae. Microb Cell Fact. 2013;12(1).

[11] Gut AM, Vasiljevic T, Yeager T, Donkor ON. Characterization of yeasts isolated from traditional kefir grains for potential probiotic properties. J Funct Foods. 2019 Jul 1;58:56–66.
[12] Hong JY, Lee NK, Yi SH, Hong SP, Paik HD. Short communication: Physicochemical features and microbial community of milk kefir using a potential probiotic Saccharomyces cerevisiae KU200284. J Dairy Sci. 2019 Dec 1;102(12):10845–9.
[13] Shi H, Zhang M, Wang W, Devahastin S. Solid-state fermentation with probiotics and mixed yeast on properties of okara. Food Biosci [Internet]. 2020;36(July 2019):100610. Available from: https://doi.org/10.1016/j.fbio.2020.100610
[14] Hikmetoglu M, Sogut E, Sogut O, Gokirmakli C, Guzel-Seydim ZB. Changes in carbohydrate profile in kefir fermentation. Bioact Carbohydrates Diet Fibre. 2020 Jul 1;23.
[15] Arslan-Tontul S, Erbas M, Gorgulu A. The Use of Probiotic-Loaded Single- and Double-Layered Microcapsules in Cake Production. Probiotics Antimicrob Proteins. 2019;11(3):840–9.
[16] Wang Y, Wu Y, Wang Y, Xu H, Mei X, Yu D, et al. Antioxidant properties of probiotic bacteria. Vol. 9, Nutrients. MDPI AG; 2017.
[17] Kim DH, Jeong D, Kim H, Kang IB, Chon JW, Song KY, et al. Antimicrobial activity of kefir against various food pathogens and spoilage bacteria. Korean J Food Sci Anim Resour. 2016;36(6):787–90.
[18] Fakruddin M, Hossain MN, Ahmed MM. Antimicrobial and antioxidant activities of Saccharomyces cerevisiae IFST062013, a potential probiotic. BMC Complement Altern Med [Internet]. 2017;17(1):1–11. Available from: http://dx.doi.org/10.1186/s12906-017-1591-9
[19] Branco P, Francisco D, Chambon C, Hébraud M, Arneborg N, Almeida MG, et al. Identification of novel GAPDH-derived antimicrobial peptides secreted by Saccharomyces cerevisiae and involved in wine microbial interactions. Appl Microbiol Biotechnol. 2014;98(2):843–53.
[20] Tulini FL, Winkelströter LK, De Martinis ECP. Identification and evaluation of the probiotic potential of Lactobacillus paraplantarum FT259, a bacteriocinogenic strain isolated from Brazilian semi-hard artisanal cheese. Anaerobe [Internet]. 2013;22:57–63. Available from: http://dx.doi.org/10.1016/j.anaerobe.2013.06.006
[21] Eka Radiati L, Jaya F, Oktavia H. Effect of Carrot-Juice on Exopolisaccharides and β-D Galactosidase Activity in Yogurt. Anim Prod. 2016;18(3):173–9.
[22] Gil-Rodríguez AM, Carrascosa A V., Requena T. Yeasts in foods and beverages: In vitro characterisation of probiotic traits. LWT - Food Sci Technol [Internet]. 2015;64(2):1156–62. Available from: http://dx.doi.org/10.1016/j.lwt.2015.07.042
[23] Speranza B, Corbo MR, Campaniello D, Altieri C, Sinigaglia M, Bevilacqua A. Biofilm formation by potentially probiotic Saccharomyces cerevisiae strains. Food Microbiol [Internet]. 2020;87(June 2019):103393. Available from: https://doi.org/10.1016/j.fm.2019.103393
[24] Silva DR, Sardi J de CO, Pitangui N de S, Roque SM, Silva ACB da, Rosalen PL. Probiotics as an alternative antimicrobial therapy: Current reality and future directions. Vol. 73, Journal of Functional Foods. Elsevier Ltd; 2020.
[25] Lee KW, Park JY, Sa HD, Jeong JH, Jin DE, Heo HJ, et al. Probiotic properties of Pediococcus strains isolated from jeotgals, salted and fermented Korean sea-food. Anaerobe [Internet]. 2014;28:199–206. Available from: http://dx.doi.org/10.1016/j.anaerobe.2014.06.013
[26] Grosová Z, Rosenberg M, Gdovin M, Sláviková L, Rebroš M. Production of d-galactose using β-galactosidase and Saccharomyces cerevisiae entrapped in poly(vinylalcohol) hydrogel. Food Chem [Internet]. 2009;116(1):96–100. Available from: http://dx.doi.org/10.1016/j.foodchem.2009.02.011
[27] Son SH, Jeon HL, Jeon EB, Lee NK, Park YS, Kang DK, et al. Potential probiotic Lactobacillus plantarum Ln4 from kimchi: Evaluation of β-galactosidase and antioxidant activities. LWT - Food Sci Technol [Internet]. 2017;85:181–6. Available from: http://dx.doi.org/10.1016/j.lwt.2017.07.018

Cattle production system in semi-arid area of Timor Island

Y Ngongo*, S Ratnawaty and PR Matitaputty

Researcher at Assessment Institute for Agriculture Technology East Nusa Tenggara (AIAT-ENT), Kupang, Indonesia

*yohanisngongo@gmail.com

Abstract. Since Cattle introduced in Semi Area of East Nusa Tenggara (ENT), Bali Cattle in Timor Island is considered as the most productive cattle in terms of population. In the last 10 years Bali cattle in West Timor (Indonesia) is around 550,000 – 980,000 head, however the quality/performance of Bali cattle is considered degraded and also native vegetation continuing depleted. Until recently, forage is the main limitation for cattle production in semi-arid region of Timor. This paper discusses cattle farming development in semi-arid region of ENT, how to address the forage shortage and farmers' response to the introduced cattle-forage innovations. The data and information were mainly taken from the collaboration research program of ACIAR and AIAT-NTT, particularly in Leucaena (Leucaena leucocephala Var. Tarramba) development. The study revealed that traditional cattle farming relies on grazing land and abandoned/fallow land during dry season for forage. Fattening systems practiced mainly in the area or farmers that have enough forage, particularly Leucaena. Farmer's plant Leucaena as part of local agro-forestry and or integrated forage into maize farming as main strategy to overcome forage shortage problem. Farmers considered innovation related Leucaena as the best innovation/ approach to improve cattle farming in semi-arid area. It is concluded that introduced forage legume, particularly Leucaena in semi-arid areas have increased carrying capacity of pastureland, improved soil fertility and decreased shifting cultivation or encourage more permanent crop land.

1. Introduction

Cattle are not a native animal in East Nusa Tenggara (ENT). It was introduced by the Dutch colony in the beginning of 20th C. According to Ormeling [1], Dutch introduced Ongole cattle to Sumba Island in 1909 – 1915 from India, Bali cattle introduced to Timor Island in 1912, and Madura cattle to Flores Island.

Although introduced later, Bali cattle population in Timor exceed far higher than Ongole cattle in Sumba. Up to recently (2019), population Bali cattle in Timor closed to 700,000 head, while Ongole cattle was only closed to 65,000 head or just 6% of the total population of cattle in ENT. On the other side, Water buffalo population is the lowest in Timor Island (Table 1).

Table 1. Cattle and water buffalo population 2018

District	Cattle	Buffalo
North Central Timor	125,165	743
Belu	69,930	849
Kupang	225,697	1,836
South Central Timor	194,401	324
Malaka	78,298	478
Kupang Municipality	6,323	27
Timor Island	699,814	4,257
%	68.124596	2.719242

West Sumba		1,596		12,768
East Sumba		51,886		38,910
Middle Sumba		8,055		9,254
Southwest Sumba		3,257		15,121
Sumba Island	64,794		76,053	
%		6.30748324		48.58033
East Flores		2,478		13
Sikka		14,409		1,976
Ende		36,172		2,655
Ngada		40,977		14,134
Manggarai		25,307		7,822
West Manggarai		14,846		18,614
Nagekeo		33,884		7,253
East Manggarai		15,356		1,594
Flores Island	183,429		54,061	
%		17.8562111		34.53252
Alor		4,970		-
Lembata		4,850		-
Rote Ndao		63,902		11,781
Sabu Raijua		5,497		10,399
Other	79,219		22,180	
%	7.71		14.17	
East Nusa Tenggara	1,027,256		156,551	
%	100.00		100.00	

Source: ENT Statistical Bureau, 2019

Bali cattle farming in Timor have been able revolutionize farming system and improving income portfolio of people in Timor Island. Bali cattle were slowly replacing water buffalo' roles in socio-cultural of Timor which led sharp declining of buffalo population in Timor. The roles of Bali cattle highlight early, and successful innovation brought to Timor in order to improve well-being of local people and in the same time regional income through export and inter-island cattle trade [2]. Although there was little attempt by the Dutch colony improve living conditions of people in semi-arid areas of Timor [1, 3], introduction of Bali cattle have changes in farming practices, broadening people's livelihood strategies and therefore promoted agricultural development in general.

Livestock, especially cattle, play important roles for upland communities both for economic and socio-cultural purposes. In this semi-arid area, forage quantity and quality is a major problem during the long dry seasonal and when cattle population increased while raising system (extensive) remains unchanged it seems that there is a conflict of interest between those growing food crops and those raising livestock. Interestingly, in some (small) areas, local communities have been able to overcome the forage problem and therefore enable them improve household living conditions (food production, income and ecological sustainability).

Beside its positive impact to the farmers' livelihood, introduction of Bali cattle in semi-arid of Timor have also brought serious problem to the environment, especially when the cattle population have growing very fast [1, 4]. Increasing population of Bali cattle, swidden cultivation and increasing demand of Sandalwood have also contributes to the declining environment quality in West Timor [1, 5, 6]. Without proper management of common pastureland and traditional cattle growing (extensive system), it has been devastated more the environment [7, 8].

Attempt to solve the problems related to the development of Bali cattle in Timor have been done for the long time. Forage deficit during the long dry season is still the main problem in raising cattle in semi-arid areas. Upland farmers besides embracing innovations being introduced, they have also developed their own strategies to overcome the problems faced in raising cattle. This paper aims to describe Bali cattle production system in Timor and how farmers mediating, and response innovations related to Bali cattle development, particularly forage related innovation.

The paper organized into four main sections. After this Introduction, the next section discussed methodology shortly. The third section discussed the results of the research. The last section (Fourth) is a conclusion drawn from previous sections.

2. Materials and Methods

The data and information were mainly taken from the collaboration research program of ACIAR and AIAT-ENT, particularly in Leucaena (Leucaena leucocephala var. Taramba) development. Data from related cattle development program have also used to enrich the narratives. Population of the research is all households that have upland and raising cattle in West Timor. Purposive sampling employed to determine villages for research sites based on the involvement in forage (Leucaena) and cattle development, and therefore respondents for in-depth interview were mainly taken from the ACIAR-AIAT' research sites in two villages: Camplong II and Oebola Dalam of Fatuleu Sub-district - Kupang districts.

Secondary data gathered from Statistical Bureau and other relevant/related sources. Some relevant forage research reports and related studies were also used. Primary data and information gathered through multiple visiting during the program implemented. Data and information were processed in tabulation and analyzed descriptively.

3. Result and Discussion

3.1. General Characteristics of Cattle Raising Environment in Timor

Timor is one of the three main Islands in East Nusa Tenggara province. West Timor is part Republic of Indonesia (RI) and Eastern part of Timor and Oecusse enclave in West Timor is part of Republic Democratic Timor Leste (RDTL). Topography of the Island of Timor dominated with mountainous and has various different agro-ecosystem zones (AEZ). The region of Timor Island is considered as semi-arid area, rain period of 3 – 4 months and dry period/season of 8 – 9 months. Within this environment and various AEZ, Bali cattle have well adaptable.

Due to low, erratic and unpredictable rainfall pattern, harvest failure or low productivity of food crops are common this semi-arid environment. The rainfall pattern also affects production and quality of livestock product, however it lesser than crops production impact. As the cropland related activity in general is limited by the rainfall, livestock related activity is long years round and therefore raising livestock is part of economic and socio-cultural activity of people in semi-arid of Timor. Semi-arid climate and dominated savannah ecosystem in Timor has suitable for Bali cattle development. It can be seen from the fast increased population since it was introduced.

The main problems are remains in the forage availability as it close related with rainfall pattern. Cattle productivity in the traditional extensive system fluctuates follow seasonal pattern. During the rainy season, forage available abundantly and the cattle gain weight while during the dry season forage availability is not enough both in quantity and quality – its crude protein and mineral of decrease while crude fiber increases that lead cattle lose weight and high mortality rate for calves.

Agricultural land in ENT dominated with the upland or dryland. Arable land constitutes around 1.65 million ha or close to 35% of total land area. Of the 1.65 million ha, only 127.000 ha or close 2,7% is lowland rice and the rest 1.5 million ha or close to 32.7% is drylands which includes 832.228 ha as grazing lands [9]. Due to population pressure and cropland expanded, this grazing land tends to be decreased every year which have direct impact to the traditional cattle raising.

3.2. Bali Cattle profile

It was believed that around 90% famers in ENT raising livestock in common land as the cheapest way for small farmers. Almost no investment allocated to protect or improve the quality of pastureland in common grazing land. The quality of livestock, particularly cattle raising in traditional system associated with the seasonal pattern. During the rainy season, cattle gains weight significantly, however cattle get loose weight 20 – 25% during long dry season (Bamualim, 1994).

IOP Conf. Series: Earth and Environmental Science 1041 (2022) 012029 doi:10.1088/1755-1315/1041/1/012029

Beside lose weight, calf mortality rate for Bali Cattle in Timor is quite high mainly due to lack of forage/low milk of cows that mostly calved during dry season. Dry season is normally started on April which grass/forage in communal grazing land contain very low crude protein and continued until rainy season started on November/December (depend on the region/altitude). Bali Calved normally calved down on during cooler weather during June/July which is also just the mid of dry season.

Around 20 – 25 % of these calves died before the rainy season starts or before the mother cow regain weight or have produce enough milk to feed calves [10]. Compared to other big ruminants, cattle population increases steadily that exceed to one million head since 2017 (Table 2). This figure showed that among big ruminants, cattle still play important roles for livelihoods of people in semi-arid area of Timor and for provincial/regional economy.

Table 2. Livestock population in ENT, 2014 – 2018

No	Livestock	Year (head)				
		2014	2015	2016	2017	2018
1	Cattle	865.731	899.577	984.508	1.007.608	1.027.256
2	Water buffalo	134.567	141.075	156.927	162.658	165.551
3	Horse	112.948	111.047	112.557	112.589	114.514
4	Goat/lamb	674.012	691.852	704.853	741.806	761.965
6	Pig	1.755.058	1.812.449	1.845.408	2.073.466	2.141.246
	Total	3.542.316	3.656.000	3.804.253	4.098.127	4.210.532

Cattle raising in Timor is actually integrated part of the farming system in semi-arid areas. Almost no external input used for cattle farming in traditional system. Cattle have considered can maximize use by-products left from the farmland and in the same time provide manure for the farmlands when cattle roaming around in the dry season. Although, there are some variations in cattle raising in some degree, cattle raising in Timor can be divided into three main systems such as:

a. Intensive system which mostly for bull fattening. Within this system, cattle penned permanently all the time before sold or traded. There are different types of intensive systems that mostly dictated by the availability of forage. In Amarasi region of Kupang district, most cattle penned in Local agro forestry so-called Mamar and rely on Leucaena as dominant source of forage. In other places, cattle penned around, or part of house garden practicing and the forages can be generated from the crop's residues, from the forest and bush/grassland. The bulls being raised in intensive system is normally for inter-island trade.
b. Semi-intensive which cattle tethered during the day and fenced during the night. Most small holder's farmers have practiced this system which still has enough land or communal land. The cattle rely on native grasses during rainy season and sometimes feed with by product from farmland during the dry season. Banana trunk and from Corypha gebanga trunk (local term: putak) are the most common feed during the long dry season.
c. Extensive system which cattle in the gazing land all the time and the cattle fenced only once or twice a month. This system practices for farmers own around 5 or more cattle that cannot be managed with cut-and carry system. A group of cattle in this system can be owned by single household or some farmers/household joint together in look-after it. Cattle mostly rely on native grass/vegetations in common pastureland or forest during rainy season and during the dry season cattle have also depend on fallowed/abandoned agricultural lands.

Cattle distributed relatively equal across districts in West Timor, however the quality of the cattle are varying which mostly dictated by forage availability. In the region dominated with Leucaena like in Amarasi – Kupang District, most farmers do intensive cattle raising/fettering, while region with less forage particularly Leucaena, most farmers do breed or traditional extensive cattle raising. Calves from breading region normally traded in the local market and ended for fattened in the region that have enough forage.

Cattle are the main export commodity from ENT. Bali cattle from Timor had been the main export commodity to Hongkong up to the 1960s era. However, when domestic demand for beef increased, Bali cattle have been supplied only for domestic market, mostly to supply beef in Jakarta. During 1980s – 1990s, ENT does inter-island cattle trades around 50,000 head yearly. In the beginning of monetary crisis 1997/1998, inter-island cattle trade was increased double as the cattle price increased sharply. Until then, inter-island cattle trade tends to be decrease below 50,000· head/year as direct impact of sharp increased during Monetary crisis before back to normal in the mid of 2000s [11].

Besides as main source of cash income and saving for rural farmers, cattle are also as important source of animal protein for people in ENT. The number of cattle for inter-island trade from ENT in the last 10 years almost similar for slaughtered in registered abattoir for domestic (provincial) market (Table 3) and obviously even higher if including for unregistered slaughtered in rural areas.

Table 3. Cattle population, Inter-Island trade and local consumption in ENT 2007 – 2017

Year	Number of cattle (Head)		
	Population	Inter-island trade	Local Consumption
2007	555,383	63,036	49,153
2008	566,464	61,889	50,166
2009	577,551	58,392	54,051
2010	588,814	49,876	50,674
2011	778,633	59,670	69,921
2012	825,972	63,449	66,078
2013	803,450	60,528	61,473
2014	865,731	49,658	62,562
2015	899,577	55,000	81,990
2016	984.551	63,429	82.944
2017	1,007,653	66,554	68,252

Source: 2007 – 2012: Buku Investasi Peternakan NTT
2013 – 2014: Dirjen PKH 2015 and 2018
2015 – 2016: Statistik Pertanian NTT 2016 [12]
2017 : Statistik Daerah Propinsi NTT 2018

3.3 Forage Development

Aware that forage as the main constraint for cattle development in Timor and semi-arid areas of ENT in general, various forage development programs have been introduced. One of the important programs is introduced forage legume, particularly Lamtoro (Leucaena leucocephala). This legume tree is now becoming important part of traditional farming system and local agro forestry (so-called *Mamar*) in Timor.

The most practical ways for forage availability improvement are through introducing forage legume (herbaceous and tress) both local and improved/introduced varieties into local farming system. These forages can provide good quality forage both for energy and protein sources. By feeding with the leguminosae forage, it can be provided enough crude protein for cattle with the main diet of grass that have low nutrients.

Forage trees legume in the upland, particularly in the slope's areas, besides for forage it can be functioning to improve soil quality and soil and water conservation. Depend on the cropping pattern and cutting management, forage trees legumes could provide forage all year round for small intensive cattle raising. For small holders farmers who spend more time looking forages during dry season, by planting forages could saves labour and time.

Herbaceous legume could maintain and even increased cattle weight by 0,37-0,45 kg/head/day [13] compare with pure king grass and other native grass. This figure indicated that herbaceous legume can be developed and used for fattening in the highland that have longer rainfall pattern. It can be planted in during rainy season or managed in order to be harvested during the dry season where native grass quality decreases. Bamualim *et al* [14] reported that crude protein of native grass in Timor decreased in June – November up to below 7% which is a minimum basic requirement for cattle.

The research of Ratnawati et al. [15] showed that there was significant increase of soil fertility in the land planted with herbaceous legumes of *C.ternatea, D. pernambucanus* and *C. Pascuorum*) which contributed NO_3 as much 95.2 kg/ha, 42.05 kg/ha and 21.14 kg/ha respectively. Integration legumes into maize farming system can also improve maize productivity. *Nulik et al, (2013 p.76)* in their research in Timor showed that: "*In the situations where maize was sown after Clitoria ternatea, Centrosema pascuorum and Macroptilium bracteatum and all legume biomass remained in the field (similar to a green manure crop without the incorporation of the material), maize grain yield increased by 1,800–2,700, 1,400 and 660–1,900 kg/ha, respectively, compared with an area with no legume*".

One of the important forage tree legumes in ENT is Leucaena leucocephala. As the Leucaena (local) attached by psyllic in the mid-1980s, ACIAR with collaboration with AIAT-ENT have developed psyllic resistance Leucaena. One of the promising new introduced *Leucaena* is *Leucaena leucocephala* var. Taramba. According to Nulik et al. [16], the superiority of *L. leucocephala* var. taramba is preferred by cattle over local leucaena; was less affected by psyllids; provided better dry season growth; and produced poles suitable for construction.

Two villages such as Camplong II and Oebola Dalam are among the five villages in Kupang district where *L. leucocephala* var Taramba introduced to the farmers for further developed in their farming (Table 4). Since introduced in 2011, most farmers in two villages have developing *L. leucocephala* var Taramba by their own initiatives both in pastureland or integrated into maize farming land. Almost two third of dominated marginal land in two villages have covered with Leucaena and enable them to do cattle fattening.

Table 4. General figures of camplong ii and oebola dalam villages that plant leucaena

Item	Village	
	Camplong II	Oebola Dalam
Land area (Km^2)	48.63	40.85
Population (people)	3,557	1,180
No. of Household (HH)	880	251
No. of HH farmers plant Leucaena	660	150
Total Land covered with Leucaena var. Taramba (Ha)	550	145
Cattle ownership (% HH):		
➢ Own	50	60
➢ Profit sharing	50	40
Cattle per household	1 – 5	1 – 10
Raising method (% HH):		
➢ Fattening/Cut and carry	70	50
➢ Semi-intensive	20	10
➢ Extensive system	10	40

Source: Fatuleu in Figures 2020 and FGD

The availability of good quality forage long years round in two villages has enabled more farmers' changes from extensive system to intensive/fattening system. They have increased their capacity from only fattening 1 head per year before no Leucaena/forage to become 2 – 5 head/year/hh. They also attract investor to invest in cattle fattening by joint with local farmers in profit sharing arrangements.

Waldron et al. [17] who conducted an economic analysis of a leucaena-fattening system in a village in West Timor showed that under all measures of profitability, leucaena-fattening systems in representative households are profitable in the wet season, and it considerably lower in the dry season and, as would be expected, are sensitive to relative prices of feeder and finished cattle. Further,

Waldron et al. [17] concluded that importantly, 'returns to person days' are higher than off-farm incomes.

Feeding Bali cattle with Leucaena for fattening have improved live weight gain significantly. Study conducted by Soares et al. [18] showed that bull feed with 100% Leucaena could gain weight up to 0.5 kg/day compared with only 0.2 kg/d for growing bulls fed grass only. Study conducted by Bamualim [19] showed that feeding lactating cows with Leucaena as supplement can increase milk for lactating cows. The study of Bamualim [19] concluded that the use of supplement during the lactating period would improve the performance of cows and calves in the area, which in turn would increase the farmers' income.

Although forage development programs have been implemented for at least since 1930s when Leucaena introduced in Timor [1], the impact of the programs varies among regions in Timor. The most successful forage development, particularly Leucaena is limited in Amarasi region of Kupang Districts, while other regions have been considered less successful. Some constraints encountered in the regions that considered less successful are:

a. Free grazing. By this system it is quite difficult to manage the livestock for not destroying the forage up to the level for unrecovering for the next rainy season. Forage programs have generally more success in the region or farmers which practicing intensive and permanent farming.
b. Time-lag and limited forage innovation. Unlike innovation relate to food crops, Innovation related to forage is quite limited and takes longer time to reach farmers.
c. Limited seed availability. Forage seeds normally unavailable in the market. So far there is no commercial forage seed provider.
d. Farmers' motive/behavior. Small farmers normally have good response to plant or develop forage that has direct economic impact for the household. In case of L. leucocephala var. Taramba, besides for feeding the cattle, farmers also can sell Leucaena forage and Leucaena seed with good price.
e. Less passionate of extension worker. Extension workers prefers to disseminate food crops related innovations rather than forage crops as extension worker' performance counted based on the successful in food crops development.

4. Conclusion

In the semi-arid area of Timor Island, forage shortage both in quantity and quality during the long dry season becomes a major constrains in cattle development. Extensive cattle raising system have been practiced by dominant cattle farmers as the simple and economical strategy. Extensive system is also practiced when farmers still have access to the common grazing lands, forest and abandoned/fallowed farming land. However, some farmers have practicing intensive cattle raising mainly in the area or farmers that have enough forage, particularly Leucaena.

Forage development program, particularly introduction of Leucaena and herbaceous legumes in recent years enable farmers at least in the program/research sites to be shifted into intensive/fattening cattle system. Besides, by planting L. leucocephala var. Taramba and herbaceous legume, It have shortening fattening period, improving cattle quality and improving soil quality that allow most maize farming to get better yield. Most farmers in the areas that planting forages, particularly L. leucocephala var. Taramba, have considered as the best strategy/innovation to overcome forage deficit during the long dry season. By planting Leucaena, farmers have able to convert unproductive or dominated rocky land to become productive land.

Acknowledgement
The authors would like to thanks to Dr. Jacob Nulik and Ir. Debora Kanahau, MSi. as Country representative of Forage Research for allowing us to have access for the data being used in this paper. Thanks also for all farmers group in Camplong II and Oebola Dalam villages for providing the Tim with data and information during the research program. We also thank you to our colleague Noldy R.

Kotta, MSi for editing and managing the References of this paper. Any remaining errors are authors' responsibility, and they have equal contribution in writing and publishing this paper.

References

[1] Ormeling FJ. 1957. The Timor problem: a geographical interpretation of an underdeveloped island. Groningen: J. B. Wolters. 284p : ill ; 24cm.
[2] Ngongo Y. 2011. The Political Ecology of Agricultural Development in West Timor, Indonesia, in School of Agriculture and Food Sciences The University of Queensland: Brisbane. p. 269.
[3] Pelzer KJ. 1945 Pioneer Settlement in Asiatic Tropics. Studies in land and agricultural colonization in southeastern Asia New York.: American Geographical Society.
[4] Ataupah H. 1992. Ekologi, Persebaran Penduduk, dan Pengelompokan Orang Meto di Timor Barat. Unpublished PhD thesis, Universitas Indonesia, Indonesia: Jakarta. p. 513.
[5] Fox JJ. 1977. Harvest of the Palm Cambridge: Harvard University Press.
[6] Yoder LSM. 2011. Political ecologies of wood and wax: Sandalwood and beeswax as symbols and shapers of customary authority in the Oecusse enclave, Timor. Political Ecology. 18(1): p. 11-24.
[7] Fynn RWS and O'connor TG. 2000. Effect of stocking rate and rainfall on rangeland dynamics and cattle performance in a semi-arid savanna, South Africa. Journal of Applied Ecology. 37(3): p. 491-507.
[8] Kgosikoma O, Mojeremane Wand Harvie BA. 2012. Pastoralists' perception and ecological knowledge on savanna ecosystem dynamics in semi-arid Botswana. Ecology and Society. 17(4).
[9] BPS NTT. 2019. NTT Dalam Angka 2019, Starictics, Editor BPS NTT: Kupang.
[10] Banks D. 1986. Analysis of livestock survey data, in NTTLDP Completion Report. II. Livestock. ACIL.: Melbourne.
[11] Dinas Peternakan Propinsi Ntt. 2016. Data Pembangunan Peternakan di NTT 2004; 2005; 2006; 2016., Peternakan., D, Editor NTT Provincial Livestock Office.: Kupang.
[12] Bps Ntt. 2017. Statistik Pertanian Provinsi Nusa Tenggara Timur 2017, Statistics, Editor BPS NTT: Kupang. p. 56.
[13] Ratnawaty S, P.Th.Fernandez, Y.L.Sera, D.A.Budiantoand E.Budisantoso. 2009. Introduksi leguminosa herba melalui sistem ley farming untuk perbaikan kualitas pakan ternak sapi Bali di dataran tinggi Nusa Tenggara Timur. , in Competitive Collaborative Research Grant (CCRG)-ACIAR SARDI,2008/2009, Penelitian, LH, Editor.
[14] Bamualim A, Saleh A, Liem Cand Fernandez PT. 1994. Produksi dan kualitas hijauan rumput alam di Nusa Tenggara. Dalam: Chaps Book A. held at the Disease Investigation Centre. Denpasar-Bali, May 15-17, 1994. Mataram: , in Final Seminar of the Cattle Health and Production Survey (CHAPS) P.202., EIVSP, Editor: Mataram.
[15] Ratnawaty S, Ratnada M, Yusufand J.Nulik. 2004. Pengelolaan pakan ternak di lahan kering Nusa Tenggara Timur in Sistem dan Kelembagaan Usahatani Tanaman-Ternak: Badan Penelitian dan Pengembangan Pertanian, departemen Pertanian.
[16] Nulik J, Hau DK, Halliday MJand Shelton HM. 2019. Tarramba leucaena: A success story for smallholder bull fattening in eastern Indonesia. Tropical Grasslands-Forrajes Tropicales. 7(4): p. 410-414.
[17] Waldron S, Ngongo J, Kusuma Putri Utami S, Halliday MJ, Panjaitan T, et al. 2019. Economic analysis of cattle fattening systems based on forage tree legume diets in eastern Indonesia. Tropical Grasslands-Forrajes Tropicales. 7(4): p. 437-444.
[18] Soares WO, Gunartha IGE, Mullik ML, Sutaryono YAand Dahlanuddin. 2018. Feed intake, feed digestibility and live weight gain of male bali cattle fed different combinations of Leucaena leucocephala and maize stover under farm conditions in Timor Leste. Livestock Research for Rural Development. 30(7).
[19] Bamualim A, Fernandez PT, Liem Cand Wirdahayati RB. 2000. Effect of feeding regimes on the performance of Bali and Ongole cows in dry tropic region in Nusa Tenggara, Indonesia. Asian-Australasian J*ournal of Animal Sciences*. **13**(SUPPL. 1): p. 207-209.

Sustainable indigenous cattle production in Brebes Regency, Indonesia: opportunities and threats during pandemic

M Sugiarto*, Y. N Wakhidati, O. E. Djatmiko, S. Nur and A Einstein

Universitas Jenderal Soedirman, Purwokerto, Indonesia

*mochamad.sugiarto@unsoed.ac.id

Abstract. This study assesses the sustainability of indigenous cattle production (Jabres Cattle) during the COVID 19 pandemic in Brebes Regency, Indonesia. The sustainability of Jabres cattle production is considered from the aspects of production, socio-economic, and environmental aspects. The study was conducted on 64 Jabres farmers in 5 sub-districts (Ketanggungan, Bantarkawung, Banjarharjo, Larangan, and Salem). Respondents were taken using a simple random sampling method and data was taken using online questionnaires. The collected data were analyzed using descriptive statistics. The indigenous cattle farmers in Brebes Regency have a long experience as beef cattle farmers (10.7 years on average) with an average ownership scale of 7 heads. During the pandemic, there was an increase of 54.7 percent of farmers who felt it was difficult to get cattle breeds compared to before the pandemic. The number of farmers who found it difficult to get forage also showed an increase of 36 percent compared to before the pandemic. Likewise, an increase in the number of farmers (75 percent) felt that it was increasingly difficult to sell beef cattle, which resulted in not making it easy for profits. However, the COVID 19 pandemic can encourage all farmers to pay more attention to the cleanliness of cowshed and the cage environment. Most of the indigenous cattle farmers (89 percent) remain committed to continuing their business despite difficulties in obtaining calves, feed, and selling cattle. Farmers still have confidence that the difficult situation will pass, and demand slaughter cattle will recover.

1. Introduction

Brebes Regency is an agriculture-based area with an area filled with mountainous plains. The people carry out the business of raising cattle and other agricultural activities not only because of profit but it is a way of life for rural communities. Jawa Brebes beef cattle known as Jabres are local beef cattle raised to produce calves. Jabres beef cattle is an indigenous breed that needs to be preserved, guarded, and protected following the mandate of the Decree of the Minister of Agriculture Number 2842 / Kpts / LB.430 / 8/2012. In 2019, the number of Jabres cows reached 27,000 heads [1] and spread across 5 sub-districts in the highlands of Brebes Regency (Ketanggungan, Bantarkawung, Banjarharjo, Larangan, Salem). The sustainability of Jabres cattle farming is an interesting issue and a challenge that must be formulated in the long term.

The issue of food security is very important in rural development which currently facing natural and biological disasters because it is related to the availability and ability of the community to obtain agricultural products at affordable prices. Food availability is an important dimension in creating people and families who have physical, social, and economic access to food [2]. Food availability is closely related to the supply of livestock products to the market/community, and this depends on livestock production. Jabres cattle production has become very strategic in helping the government achieve community food security. The attention of the Brebes Regency government with its various development programs is aimed at increasing the Jabres cattle population so that it can supply more meat products to the community and market. Small-scale farming has been able to contribute to the availability of food and nutrition for farmers households at local, regional, and national levels [3].

However, efforts to increase sustainable production are not easy because they face internal and external challenges. Environmental challenges that are always changing due to natural and biological disasters as well as economic challenges are things that must be considered in increasing livestock production. The sustainability of livestock production is a continuous effort to increase livestock production by taking into 3 dimensions, namely environmental, economic, and social [4]. The presence of the Covid 19 pandemic is one of the environmental and biological changes that must be assessed for its impact on the production of Jabres cattle farming. This study aims to describe the characteristics of Jabres cattle farmers, the challenges and opportunities faced by Jabres cattle farmers in sustaining their farming during the Covid 19 Pandemic.

2. Materials and Methods

The study was conducted using a survey method of 177 respondents of Jabres cattle farmers who were selected using the multistage sampling method. In the first stage, 5 sub-districts were selected purposively based on these areas as Jabres cattle production centers (Ketanggungan, Bantarkawung, Banjarharjo, Larangan, Salem). In the second stage, from the 5 sub-districts, 15 groups of Jabres cattle farmers were selected randomly, 20 percent of the total farmers' groups. Then the respondents were selected 20 percent of the total group members in each selected group.

The data collected is primary data from farmer interviews using a structured questionnaire. The data collected was then analyzed using descriptive statistics to describe characteristics Jabres cattle farmers and farmers 'perceptions of the impact of Covid 19 on the easiness of doing Jabres cattle production as well as farmers' commitment to the continuation of Jabres cattle farming.

3. Result and Discussion

3.1 Respondents' characteristic

Jabres cattle farmers in Brebes Regency have an average age of 49.57 years with a low level of education (mean 7.31 years or equivalent to passing elementary school). Most of the farmers (41.8 percent) had graduated from elementary schools. Jabres cattle farmers have a small family of 4 people per family. Raising cattle is not only an economic requirement, but also a way of life for people. Jabres cattle farmers have experience raising beef cattle with an average of 17.35 years and also with an average involvement of 7.35 years in farmer groups. Jabres cattle farmers have an average number of livestock ownership of 5 cows. Detailed profiles of Jabres cattle farmers can be seen in Table 1.

Age of farmers, the farming experience of farmers, and farm size are important assets to maintain Jabres cattle farming and also increase production. Livestock production is not only related to the technical ability of farming but also to the motivation to raise livestock. Farmers with high motivation are believed to be able to sustain their farming. Farmer motivation is influenced by the age of the farmer, farming experience, and farm size [5].

Table 1. Profile of Jabres beef cattle farmers in Brebes Regency

No	Variables	Mean	Standard of Deviation
1	Age of farmers (year)	49.57	8.91
2	Farmers' education (year)	7.31	4.22
3	Family members (person)	3.93	1.45
4	Farming experience (year)	17.35	11.57
6	Experience in farmers group (year)	7.35	3.51
7	Farm size (heads)	5.51	5.04

3.2 Respondents' perception on impact of pandemic covid 19

The Covid 19 pandemic has social implications with lockdown and social distancing, and this atmosphere has an impact on several Jabres cattle farming activities such as the ease of getting calves, looking for animal feed, selling cattle, and getting farming profits.

Table 2. Perception of Jabres beef cattle farmers on the easiness of doing business before and during the pandemic

No	Variables	Before Pandemic (percentage)	During Pandemic (percentage)
1	Getting a calf	97.8	77.4
2	Find and getting animal feed	87	64.4
3	Selling cattle	97.1	65.4
4	Getting farming profits	93.8	41.8

Based on Table 2, it can be seen that farmers were greatly affected by Covid 19 in the aspect of ease of obtaining business profits. During the Covid 19 pandemic, only 41.8 percent of Jabres cattle farmers found it easier to get business profits than in the pre-Covid 19 periods (93.8 percent). One of the reasons for the decrease in the ease of getting a profit is because farmers have difficulty selling their cows during the Covid 19 pandemic. Before Covid 19, as many as 97.1 percent of farmers found it easy to sell their cows.

However, at the time of Covid 19, farmers felt it was difficult and only 65.4 percent felt it was easy to sell their cows. Social distancing implemented by the government and information about the dangers of Covid 19, caused farmers to find it difficult to get calves and animal feed. Farmers with small scale farming and extensive farming orientation are under more pressure due to the Covid 19 pandemic than farmers who use a business approach. Families with limited farming income are more vulnerable than families with high-income categories [6]. The lockdown that was implemented when Covid 19 occurred has caused food consumption to decline, prices for agricultural commodities to decline, and resulting in decreased demand for agricultural products [7] [8].

3.3 Challenges and opportunities on production sustainability of Jabres beef cattle farming

The sustainability of the Jabres beef cattle farming is a systematic effort that takes into account the social, economic, and environmental aspects so that the Jabres cattle farming continues to operate. Various challenges have come and may be able to hamper Jabres beef cattle production. The obvious challenges are natural disasters and biological disasters in the form of the Covid 19 pandemic. However, several opportunities are still available and must be further explored to bring benefits to the management of Jabres cattle farming in Indonesia.

Based on the data collected, it is found that 89 percent of Jabres cattle farmers remain committed to continuing their farming despite difficulties in obtaining calves, feed, and selling cattle. This illustrates that farmers generally are not felt affected by the difficulties faced during the Covid 19 pandemic. They believe that these problems will occur only in the short term and are believed to have a better recovery soon. The belief in getting benefits like before the conditions of Covid 19 makes farmers committed to the sustainability of their farming. Adequate economic performance from the livestock business is one of the reasons for the sustainability of agricultural business [9]. However, the challenge that must be resolved immediately is the availability of adequate calves so that Jabres cattle production can continue to operate.

Besides, the availability of animal feed and the improvement of the market situation are expected to be resolved soon. The availability of quality animal feed (grass) can contribute to environmental improvement by improving gas emissions from livestock business [10]. The role of various parties such as the government and the private sector is expected to help smallholder farmers in facing the challenges of the Jabres beef cattle production during the Covid 19 Pandemic.

4. Conclusion

Based on the description above, it can be concluded that the Jabres beef cattle farming in Indonesia which was carried out on a small scale was quite affected by Covid 19, especially in the difficulty of selling cows, obtaining production inputs which included calves and forage during the Covid 19 period, and in obtaining farming income compared to the situation before Covid 19. Even though they are faced with quite difficult business challenges, Jabres beef cattle farmers are still committed to continuing their cow's business with the belief that these difficult conditions will soon pass, and they will return to doing business as a way of life that is not easy to abandon.

References

[1] BPS Kabupaten Brebes 2020 Kabupaten Brebes dalam Angka Tahun 2019 (Kabupaten Brebes)
[2] Peng W and Berry E M 2018 The concept of food security vol 2 (Elsevier)
[3] Galli F, Grando S, Adamsone-Fiskovica A, Bjørkhaug H, Czekaj M, Duckett D G, Almaas H, Karanikolas P, Moreno-Pérez O M, Ortiz-Miranda D, Pinto-Correia T, Prosperi P, Redman M, Rivera M, Toma I, Sánchez-Zamora P, Šūmane S, Żmija K, Żmija D and Brunori G 2020 How do small farms contribute to food and nutrition security? Linking European small farms, strategies and outcomes in territorial food systems Glob. Food Sec. 26
[4] Van der Linden A, de Olde E M, Mostert P F and de Boer I J M 2020 A review of European models to assess the sustainability performance of livestock production systems Agric. Syst. 182 102842
[5] Andarwati S, Haryadi T, Guntoro B, Sulastri E, Putra R A R S and Gunawan G 2018 Relationship Between Farmer's Characteristics with the Motivation of Goat Milking in the Girikerto Village Turi District Sleman Regency Bul. Peternak. 42 256–61
[6] Kansiime M K, Tambo J A, Mugambi I, Bundi M, Kara A and Owuor C 2021 COVID-19 implications on household income and food security in Kenya and Uganda: Findings from a rapid assessment World Dev. 137
[7] Elleby C, Domínguez I P, Adenauer M and Genovese G 2020 Impacts of the COVID-19 Pandemic on the Global Agricultural Markets Environ. Resour. Econ. 76 1067–79
[8] Ceballos F, Kannan S, and Kramer B 2020 Impacts of a national lockdown on smallholder farmers' income and food security: Empirical evidence from two states in India World Dev. 136
[9] Michaličková M, Krupová Z, Krupa E, and Zavadilová L 2016 Economic weights as a tool for sustainable livestock farming 137–43
[10] Pullar D, Allen N, and Sloyan M 2011 Challenges and opportunities for sustainable livestock production in the UK Nutr. Bull. 36 432–7

IOP Conf. Series: Earth and Environmental Science 1041 (2022) 012031 doi:10.1088/1755-1315/1041/1/012031

The role of philanthropy in reducing impact of disaster on economies of ASEAN-9

M S A Majid*, R Nurdin, T Azhar, S Sartiyah

Universitas Syiah Kuala, Banda Aceh, Indonesia

*mshabri@unsyiah.ac.id

Abstract. This study aims to determine the mediating effect of economic growth on the relationship between disaster and unemployment in the ASEAN-9 economies during the 2000-2020 period. It also attempts to examine the moderating role of religious philanthropy in reducing the impacts of disasters on economic growth and unemployment in the ASEAN-9. By using a combination of path analysis and panel moderated regression model, the study found that disasters have adversely affected economic growth and caused an increase in the employment rate. In addition, philanthropy is found to have a negative moderating role in the effects of disaster on economic growth and unemployment. The findings showed a significant role of philanthropy in reducing the impacts of disasters on economic growth and unemployment in ASEAN-9. Our findings provide an important benchmark for the formulation of government policies to mitigate disaster risks on economic problems through the enhancement of religious philanthropic institutions.

1. Introduction

Economic development is a process in which the government and all components of the society manage existing resources and then form a partnership pattern between local governments and the private sector to create new jobs and stimulate the development of economic activities in the region. Economic development, basically, aims to improve people's welfare. Sustainable economic growth causes the economy to further develop. High economic growth is accompanied by equitable distribution of income and the reduction of the unemployment rate.

However, in reality, non-inclusive economic growth causes poverty and unemployment to remain high. Unstable economic growth is related to poverty and unemployment. Usually, people who are categorized as poor do not have work (unemployed). Similar to other countries, ASEAN countries are also vulnerable to economic problems, such as poverty, inequality, and unemployment. Many efforts were taken to reduce poverty and unemployment through sustainable economic growth. However, the target of economic growth is not always smooth, there are several obstacles both from within and from outside. One of the internal factors that cause the economic growth target to be slow or even fail is a disaster.

Natural disasters often cause damage to various dimensions of the community's economy [1] both in the short- and long-term. In the short-term, disasters have the potential to cause inflation due to disruption of economic and market activities, in the form of production, distribution, marketing, and consumption [2]. In the post-disaster period, the demand for building materials, food, energy, and water increased. On the other hand, damage to infrastructure in agriculture, industry, transportation, and

IOP Conf. Series: Earth and Environmental Science 1041 (2022) 012031 doi:10.1088/1755-1315/1041/1/012031

marketing and communication facilities has disrupted the circulation of goods and services and exacerbated the rate of inflation. In the long-run, disasters have led to a larger decline in output in developing countries than in developed countries. Disasters have caused reduced economic growth, increased poverty, unemployment, and economic inequality [3].

Furthermore, disasters can disrupt or destroy various functions, and many institutions at the same time can bring about a wider or systemic community crisis [4]. Natural and humanitarian disasters do not only destroy physical infrastructure, but they have the potential for a greater systemic impact on regional and national economies [5].

In summary, natural disasters cause various socio-economic problems, such as poverty, economic instability, income inequality, and unemployment. The impact of disasters on the decline in output and economic problems in various sectors can be explained by supply-side economics or also known as trickle-down theory. Disasters have disrupted production activities, causing a deficit in the supply of goods and services to meet public demand. Consequently, poverty, unemployment, and income disparities will occur due to slowing economic growth due to natural disasters.

However, previous research on the role of philanthropy in reducing the impact of disasters on economic instability and unemployment in ASEAN countries was not found in scientific publications. This is what motivated this research to be carried out. There are at least three main differences between this study and previous research.

First, previous research only looked at the impact of disasters on economic growth, while our study explores a more comprehensive impact of disasters on economic growth and unemployment. Second, previous studies only investigated the effects of disasters on economic problems but none of those studies assessed the role of philanthropy as a moderating variable in reducing the impact of the disaster on economic growth and unemployment.

Finally, this study combines path analysis and panel moderated regression model to assess the investigated issue, while previous studies only used a single approach.

The results of this study are expected to be a reference for making policies related to planning, handling, and mitigating disasters in a more comprehensive manner. The results of the study help the policymakers to formulate a strategy to strengthen religious philanthropy as an important solution in minimizing the impact of disasters on economic instability and unemployment in ASEAN countries.

2. Methodology

2.1 Site and time

This study aims to empirically examine the impact of disasters on the economy and the role of philanthropy in reducing the impact of disasters on economic growth and unemployment. The focus of this research is 9 ASEAN countries, namely Indonesia, Malaysia, Philippines, Singapore, Thailand, Cambodia, Laos, Myanmar, and Vietnam during the period 2000-2020.

2.2 Sampling and measurement

The data used for this research study is annual secondary data. The entire data is in the form of panel data, which is a combination of time series data for the period 2000-2020 and cross-section data covering 9 ASEAN countries. Data were collected from reports containing information on the variables studied, such as disaster data sourced from EM-DAT (Emergency Events Database), population, economic growth, unemployment sourced from International Financial Statistics (IFS) – International Monetary Fund (IMF), and philanthropic funds from the World Giving Index. Disasters are measured by the number of losses due to disasters in USD, economic growth is measured by changes in GDP per capita, unemployment is measured by the percentage of unemployment per capita, and philanthropy is measured by the number of donations for humanity calculated in USD.

2.3. Data and analysis

This study examines the mediating effect of economic growth on the effects of disaster on unemployment and the moderating effect of philanthropy in reducing the impact of the disaster on economic growth and unemployment in ASEAN-9 countries. This study uses a combination of path analysis models and moderated panel regression model. The path analysis model was used to test the effect of mediation, while the moderated panel regression analysis model was used to measure the effect of moderation.

This research model is able to analyze the pattern of relationships between causal variables (exogenous variables) and effect variables (endogenous variables) to test and measure the effect partially and simultaneously [6]. Mathematically, the path analysis model used in this study examines the mediating effect of economic growth on the effect of the disaster on unemployment in ASEAN-9 countries as follows:

$$\text{Econ growth} = \alpha_{01} + \alpha_{11}\text{Disaster} + \varepsilon_1 \quad (1)$$
$$\text{Unemployment} = \alpha_{02} + \alpha_{21}\text{Disaster} + \varepsilon_2 \quad (2)$$
$$\text{Unemployment} = \alpha_{03} + \alpha_{31}\text{Disaster} + \alpha_{32}\text{economic growth} + \varepsilon_3 \quad (3)$$

Furthermore, to examine the moderating role of philanthropy in reducing the impact of disasters on economic growth and unemployment in ASEAN-9 countries, the following Moderated Regression Analysis (MRA) equations will be estimated:

$$\text{Economic growth} = \alpha_{04}\text{Disaster} + \alpha_{41}\text{Philanthropy} + \alpha_{42}\text{Disaster*Philanthropy} + \varepsilon_4 \quad (4)$$
$$\text{Unemployment} = \alpha_{05}\text{Disaster} + \alpha_{51}\text{Philanthropy} + \alpha_{52}\text{Disaster*Philanthropy} + \varepsilon_5 \quad (5)$$

where γ_{0i} is the constant term; γ_{ii} is the estimated regression coefficients, the symbol * indicates the interaction between disaster and philanthropy variables to measure the moderation effect, and ε_i is the structural *error term*.

To determine which panel regression model is the most suitable for use in this study, of the three possible panel regression models (common effect model, fixed-effect model, and random effect model), the suitability of the panel model to be used will be tested first. The panel regression model was selected using the Chow test (to choose either the common effect model or the fixed effect model), the Hausman test (to choose between the random effect model and fixed effect model), and the Lagrange test (to select between the random effect model or common effect model).

3. Result and Discussion

Table 1 reports the findings of the panel regression model selection based on the Chow, Hausman, and Lagrange tests. As observed from Table 1, the significant value of the cross-section of the Chow test indicated the fixed effect model is more suitable to be used in our study than the common effect model.

Based on the Hausman test, the insignificance value of the random cross-section showed the suitability of the random effect model to be adopted in the study as compared to the fixed-effect model. Finally, the Langrage test findings of the significant value of the random cross-section ascertained that the random effect model is more appropriate to be utilized in our study as compared to the common effect. Based on these tests, thus the study used the random effect model to provide the answers for our research objectives.

Table 1. Testing a suitable panel regression model

	Effects Test	Statistic value	P-value
Chow test	Cross-section F	3.546	0.000
Hausman test	Cross-section Chi-square	52.860	0.000
	Cross-section random	0.426	0.467
Breusch-Pagan Lagrange test	Cross-section	88.643	0.000

Source: Secondary data processed using the Eviews program (2021).

3.1. Mediating effect of economic growth on disaster-unemployment relationships

Table 2 reports the findings of the direct effects of disaster on economic growth and unemployment in the ASEAN-9 economies. As illustrated in Table 2, the study found a significant impact of the disaster on economic growth and unemployment. The presence of disaster adversely impacted economic growth and caused a higher unemployment rate in the ASEAN-9. The disaster caused a greater impact on unemployment rather than on economic growth.

Table 2. Effect of disaster on economic growth and unemployment

Dependent variable	Constant	Independent variable		Diagnostic test
		Disaster	Economic growth	
Economic growth	3.618^{***} (6.897)	-0.133^{***} (4.733)	-	$R^2 = 0.107$, F-stats = 22.410 F-value = 0.000
Unemployment	11.645^{***} (7.329)	0.159^{***} (2.733)	-	$R^2 = 0.009$, F-stats = 1.777 F-value = 0.001
Unemployment	21.056^{***} (8.854)	-	-0.121* (1.623)	$R^2 = 0.009$, F-stats = 1.751, F-value = 0.002
Unemployment	11.246^{***} (7.743)	0.177** (1.981)	0.962** (1.874)	$R^2 = 0.027$, Adj-$R^2 = 0.017$, F-stats = 2.657, F-value =0.002

Source: Secondary data processed using the Eviews software (2021).

Our findings of the negative significant effects of disaster on economic growth are supported by many previous studies. For example, a previous study documented that at high disaster rates, the losses are enormous (as Hurricane Katrina occurred in the United States), but its impact on the economy is proportionately smaller. This reflects increased assistance in mitigation and preparedness measures and better environmental management.

Disasters have caused various direct damages to humans, crops, houses, buildings, and other infrastructures [7]. Humanitarian, climate, and other types of natural disasters adversely impacted economic growth in the short term and its effect is more severe in poor countries where the development of the financial sector is relatively weak [8] and lack of disaster mitigation preparedness [9] and [10].

Furthermore, our evidence of the significant positive effect of disasters on unemployment is in line with many previous studies. When a disaster occurs, many projects are stopped or forced to stop due to the paralysis of supporting infrastructure. These incidents cause workers to quit and were unemployed. This condition further endangered the economy to become shakier and more fragile due to an increase in unemployed people during the post-disaster period [11]. To further identify the mediating effect of economic growth on the disaster-unemployment relationships, the findings from Table 2 are represented in Figure 1 and tested using the Sobel test [12] within Baron and Kenney's mediating framework [13].

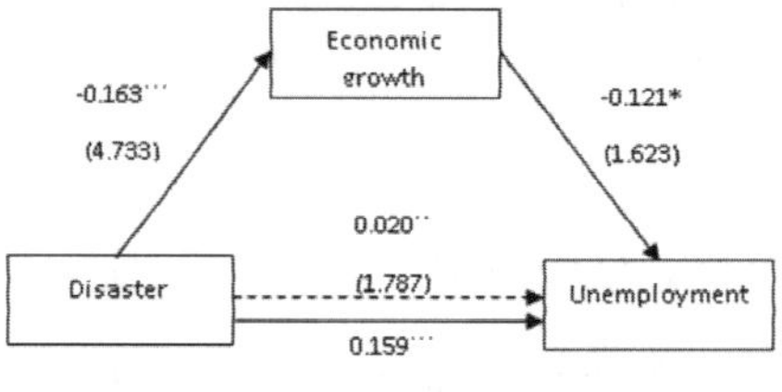

Source: Secondary data processed using the Eviews software (2021).

Figure 1. Mediating effect of economic growth on disaster- unemployment relationships

As observed from Figure 1, the study found a significant mediating effect of the disaster on unemployment through the economic growth in the ASEAN-9 (indicated by the dotted line). This finding is not surprising since the disaster has directly harmed the economy and the economy, in turn, has directly impacted unemployment, thus this showed the existence of the indirect significant effect of the disaster on unemployment through economic growth. Thus, the nature of this mediating effect showed that economic growth acted as a partial mediator. These findings showed the importance of mitigating the disaster impact on economic growth as it consequently would reduce the disaster impact on unemployment.

Disruption of production activities in the real sector due to declining demand during the disaster caused companies to reduce production capacity to save their business and automatically reduce the workforce. The increasing unemployment rate will bring problems in the economy because it will burden the family and reduce income and consumption which will eventually increase. Disruption of economic growth results in rising unemployment rates when disasters occur. Therefore, reconstruction efforts in post-disaster areas need to consider expanding employment opportunities.

In particular, in the local earthquake-hit areas, the demand for goods and services such as demand for health care and especially construction services rose. In such cases, job opportunities for people who were not affected by the earthquake could be improved. Although the initial effects of a disaster such as an earthquake will be devastating in the affected area, in the long term both local and aggregate labor markets improve [14].

3.2. Moderating role of philanthropy in reducing the impact of the disaster on economic growth and unemployment.

Table 3 reports the results of the estimation of the moderating effect of philanthropy in reducing the impact of disasters on economic growth and unemployment in ASEAN-9 during the period 2000-2020. The results showed that philanthropy moderated the impact of disasters on economic growth. This finding shows that philanthropy can restore economic conditions when a disaster occurs.

Our finding is following previous research which states that zakat (Islamic philanthropy) could be used to build community resilience to disasters in two stages. The first phase is the use of zakat in disaster emergency response to meet the basic needs of the community, including food, water, sanitation, shelter, and health care. The second phase is in reducing risk. Zakat does not only address the short-term needs of people affected by disasters. When an emergency occurs, it is difficult for vulnerable people to work and support their families. This puts them at greater risk in the future, making them less able to cope with [15].

Furthermore, zakat can be used in the community in the long-term by helping to rebuild lives and livelihoods, and by implementing livelihood programs, for example, distribution of livestock, seeds, tools, etc. Zakat can also be used to reduce exposure to emergencies by supporting communities with funds to import technology and expertise that will help them improve disaster prediction, preparedness, response, and vulnerability to similar events in the future.

Table 3. Moderating effect of philanthropy on disaster' impacts to economic growth and unemployment

Dependent variable	Constant	Independent variable			Diagnostic test
		Disaster	Philanthropy	Disaster*Philanthropy	
Economic growth	3.986^{***} (3.854)	0.256^{***} (2.072)	0.005^{*} (1.779)	-0.337^{***} (2.963)	$R^2 = 0.259$; Adj-$R^2 = 0.247$; F-stats = 21.580; F-sig.= 0.000
Unemployment	8.289^{***} (5.765)	5.448^{**} (1.947)	0.570^{*} (1.408)	-2.618^{***} (-7.589)	$R^2 = 0.020$; Adj-$R^2 = 0.004$ F-stats = 1.271; F-sig. = 0.028

Source: Secondary data processed using the Eviews software (2021).

Table 3 also provides evidence of the significant moderating role of philanthropy in reducing the disasters' impact on unemployment in ASEAN-9 economies. This shows that philanthropy can weaken the impact of disasters on unemployment. A previous study documented the tropical storms effect on the supply of labor. Damage to infrastructure due to hurricanes has limited the supply of tourism services.

This vulnerability reduces the demand for labor in the aftermath of a natural disaster. Financial helps in the form of philanthropy given during a disaster can raise the spirits of victims [16]. In the United States and elsewhere, government training programs have the best records. Some have suggested that the unemployed are encouraged to start their own businesses, and entrepreneurship is a valid option for some.

With the presence of many religious-based social and philanthropic institutions globally, it is not surprising that these institutions play an active role in helping various elements of society, both individually and collectively when natural disasters and social, economic, and political instabilities occur. Philanthropy is documented as an effective way to reduce the impact of disasters on slowing economic growth and reducing unemployment.

Overall, our findings show the importance of philanthropic institutional support during the episodes of disastrous periods to reduce and mitigate the disaster risks on economic growth and unemployment. During times of disaster, philanthropic organizations are always at the forefront to help communities meet their basic needs to ensure survival and rebuild their future lives better. Their presence has contributed to ensuring economic stability and reduced unemployment during the disastrous periods.

4. Conclusion

This study empirically measured and analyzed the mediating effect of economic growth on disaster-unemployment relationships. It also attempted to measure and analyze the moderating role of philanthropy in reducing the impact of disasters on economic growth and unemployment in ASEAN-9 during the period 2000-2020. Using a combination of path analysis and panel moderated regression analysis; the study found an adverse impact of disasters on ASEAN-9 economies and caused a higher unemployment rate in the region. In addition, the study documents the significant moderating role of philanthropy in reducing the impact of disasters on economic growth and unemployment.

Our findings demonstrate the important role of religious philanthropic institutions during disaster periods to strengthen government efforts to mitigate disaster risks to the ASEAN-9 economies. Therefore, it is very important for policymakers and the public to support the existence of religious philanthropic institutions locally, regionally, and globally. During times of disaster, philanthropic organizations are always at the forefront to help communities meet their basic needs to ensure survival and rebuild their future lives better.

This study only limited its focus on the mediating effect of economic growth on disaster-unemployment relationships and the moderating role of philanthropy in reducing the impact of disasters on economic growth and unemployment in ASEAN-9. To enrich the existing literature and empirical evidence on this issue, further studies are suggested to evaluate the holistic impact of disasters on overall economic performance as well as the role of philanthropy to mitigate the disaster impact on entire macroeconomic performance. Adding more countries into the analysis across the region would also enrich the existing empirical evidence on this topic.

Acknowledgment

This study is part of the research grant funded by the Institute of Research and Community Services, Universitas Syiah Kuala (USK), Indonesia, under the scheme of Professor Grant No. 42/UN11.2.1/PT.01.03/PNBP/2021.

References

[1] Hochrainer, S. (2009). Assessing the Macroeconomic Impacts of Natural Disasters: Are There Any?. USA: The World Bank
[2] Noy, I. (2015). Natural Disaster and Economic Policy for ASEAN and the Pacific Rim: A Proposal for a Disaster Risk Reduction 'Seal of Approval' Fund. In Resilience and Recovery in Asian Disasters (pp. 219-237). Springer, Tokyo.
[3] Degasperi, N. C., & Mainardes, E. W. (2017). What motivates money donation? A study on external motivators. Revista de Administração (São Paulo), 52(4), 363-373.

[4] Wilhelm, M.O., 2006. New data on charitable giving in the PSID. Economics Letters 92, 26–31.
[5] Parida, Y., Agarwal Goel, P., Roy Chowdhury, J., Sahoo, P. K., & Nayak, T. (2020). Do economic development and disaster adaptation measures reduce the impact of natural disasters? A district-level analysis, Odisha, India. Environment, Development and Sustainability, 1-33.
[6] Kuncoro, R., & Achmad, E. (2008). Cara Menggunakan dan Memaknai Analisis Jalur (Path Analysis). Bandung: Publisher Alfabeta
[7] Cavallo, Eduardo A., & Ilan Noy. 2010. The economics of natural disasters: A survey. IDB Working Paper, No. 35, Inter-American Development Bank, Washington, D.C.
[8] Noy, I. (2009) The macroeconomic consequences of disasters. Journal of Development Economics, 88, 221–31
[9] Loayza, N.V., Olaberria, E., Rigolini, J., & Christiaensen, L. (2009) Natural disasters and growth—going beyond the averages, Policy Research Working Paper Series 4980. World Bank, Washington, DC
[10] Raddatz, C. (2007) Are external shocks responsible for the instability of output in low-income countries? Journal of Development Economics, 84, 155–87
[11] Hsiang, S. M., & Jina, A. S. (2014). The causal effect of environmental catastrophe on long-run economic growth: Evidence from 6,700 cyclones. Working Paper (No. w20352). National Bureau of Economic Research.
[12] Preacher, K. J., & Leonardelli, G. J. (2001). Calculation for the Sobel test. Retrieved: August 20, 2021, from: http://www.people.ku.edu/~ preacher/sobel/sobel.htm.
[13] Baron, R. M., & Kenny, D. A. (1986). The moderator-mediator variable distinction in social psychological research: Conceptual, strategic, and statistical considerations. Journal of Personality and Social Psychology, 51(6), 1173.
[14] Ewing, B. T., Kruse, J. B., & Thompson, M. A. (2009). Twister! Employment responses to the 3 May 1999 Oklahoma city tornado. Applied Economics, 41(6), 691-702
[15] Kasri, R. A. (2013). Giving behaviors in Indonesia: motives and marketing implications for Islamic charities. Journal of Islamic Marketing, 4(3), 306-324.
[16] Lee, C. Y., & Tang, C. F. (2019). How do natural disasters influence the rate of poverty?. Journal of Poverty, 23(6), 478-486

Community participation in disaster management in Gondoriyo village administration, Ngaliyan sub-district, Semarang city

Supratiwi*, Yuwanto, Kushandajani

Universitas Diponegoro, Semarang, Indonesia

*tiwik75@gmail.com

Abstract. Indonesia as a disaster-prone country demands the active role of the community in disaster management. So, the perspective that focuses on handling during and after a developing disaster must be changed. In this matter, the government establishes the Desa Tangguh Bencana (Destana)/Disaster Resilient Village Policy. With various activities with the community involvement, the risk and impact of disasters can be eliminated because the community has been trained to prepare for disasters. The research aimed at determining the types and forms of community participation in disaster management in Gondoriyo Village Administration, Ngaliyan Sub-District, Semarang City. Gondoriyo Disaster Resilient Village is one of the pioneers in Semarang City, which started from the volunteers of the Disaster Preparedness Group. The research results show that the type of community participation is direct; either in a kind of money, support or their involvement in decision making. Their participation is in the form of thoughts, energy, skills, and property as well. It indicates that the awareness of Gondoriyo community to actively participate in prevention, preparedness, disaster risk reduction and capacity building efforts for the post-disaster recovery. It is proven later that Gondoriyo Village Administration has received the title as the Intermediate Disaster Resilient Village. However, what must be improved is how the disaster management programs are integrated into village administration development plans; particularly in relation to assistance to increase capacity and funding as well. Therefore, its existence can be maintained and sustainable.

1. Introduction

Indonesia is a disaster-prone country that lies at the intersection between three tectonic plates, i.e. the Eurasian, Indo-Australian, and Pacific Ocean plate. This condition has made Indonesia known as a disaster supermarket (1). According to BNPB (Badan Nasional Penanggulangan Bencana/National Disaster Management Agency), the most common disasters are volcanic eruptions, earthquakes, tsunamis, floods and landslides (2). Not a few victims fell in a kind of loss of life and property. The number of victims depends on the perspectives of disaster management whether is more concentrated during and after the disaster and less pay attention to the pre-disaster stage.

The enactment of Law No. 24 of 2007 concerning Disaster Management has changed the paradigm of disaster management from the perspective of responsive emergency response to preventive disaster risk reduction. The follow-up is carried out, among others, with the policy of PRB (Pengurangan Resiko Bencana/Disaster Risk Reduction) directed at the pre-disaster stage through efforts to increasing community capacity for disasters. The government through the BNPB has developed a system and strategy through some policies and so, the PRB are community-based efforts. The strategy that the government has made is to establish Destana; as stated in the Regulation of Head of the National Disaster Management Agency Number 1 of 2012.

The Destana is villages/village administrations with the status of the independent ability to adapt and deal with disaster threats as well as recover quickly from the adverse impacts of disasters, if a disaster hit the villages/village administration. Destana has the capacity to recognizing potential disasters, reducing potential disasters and being able to organize all elements of the society to participate in disaster management.

The objectives of establishing Destana are as follows: a. Protecting communities located at hazard-prone areas from the adverse impacts of disasters; b. Increasing community participation, particularly for vulnerable groups in resource management and so it can reduce disaster risks; c. Improving the capacity of community institutions in resource management and the maintenance of local wisdom for the PRB; d. Increasing the government capacity in providing resource and technical support for the PRB; e. Increasing cooperation between stakeholders in the DRR, local government, business institutions, universities, Non-Governmental Organizations (NGOs), community organizations, and other concerned groups (3).

To achieve these goals, the presence of community participation is very important as shown in the research of Kalangkahan (4), Oktari (5), Widyasari (6), and Julius (7). According to Isbandi (8), the community participation is the community involvement in the process of identifying problems and the existing potentials in the society, selecting and making decisions about alternative solutions to deal with the problems, implementing efforts to overcome the problems, and the community involvement in the process of evaluating the existing changes.

Therefore, this research aims at determining the level and form of community participation in Gondoriyo Village Administration in the village-level disaster management. Gondoriyo Village Administration is one of the village administrations in Ngaliyan Sub-District, Semarang city which is prone to floods, landslides, hurricanes, and land fires described in the Disaster Hazard Map compiled in 2017. Since 2012 Gondoriyo Village has had a disaster management volunteer organization, i.e. KSB (Kelurahan Siaga Bencana/Disaster Preparedness Village); which is one of the pioneers of KSB in Semarang City. The success of its program makes the status upgraded to a Disaster Resilient Village / Village Administration in 2017. The involvement of various community elements is increasing whether they are youth and women.

2. Methodology

The research design used descriptive qualitative research method. The research approach will explore data and information. The main research questions were formulated to study and answer the formulation of the research problem. Primary data were obtained by interviewing the related informants, i.e. Head of Gondoriyo Disaster Resilient Village, Head of Gondoriyo Village Administration, head of BPBD in Semarang City, an NGO that observes environmental issues in Semarang City. To fulfill the data that the research need, it used secondary data in a kind of documents and the related library sources.

3. Result and Discussion

3.1. Community Participation in Disaster Resilient Village Program (Destana)

Disaster is an event or a series of events that threaten and disrupt the life and livelihood of society caused by both natural and/or non-natural factors and human factors resulting in the occurrence of human casualties, environmental damage, property losses and psychological impacts (9). Therefore, it is must to reduce the impact; as stated by Mc Entire (10), that to reduce losses caused by disasters, it is necessary to create disaster mitigation plans, plan evacuation routes, and have capacity development in emergency response management and guide the community in steps of reducing losses due to the disasters.

The introduction of the Destana is expected to increase the community preparedness capacity in dealing with disasters, with the active role of the community by applying the Community Base Disaster Management methods, i.e. placing the community as the subject and object of the disaster management.

With the community participation, disaster risks is reduced, and so, victims and losses can be reduced as well. Participation is very important in the sustainability of a program in the community; therefore, the program can run according to the goals having been set previously. As stated by Zhang (11), public participation can contribute positively to the government decision-making by increasing community participation in decision-making and by increasing the likelihood of successful implementation of the decisions. The community involvement is expected in all components of Destana including a. Legislation; b. Planning; c. Institutional; d. Funding; e. Capacity building; and f. Implementation of Disaster Management (PB).

There are several types of participation; these include (12):

a. Direct participation in physical and face-to-face activities;
b. Participation in a kind of money and goods. It is better if the participation in a kind of money and goods comes from within the community itself. If it is forced to carry out it can come from outside of the community. However, it only occurs temporary and it is only as bait;
c. Participation in a kind of support;
d. Participation in the decision-making process;
e. Representative participation, i.e. by giving confidence to community representatives who sit in organizations or committees.

Moreover, the forms of community participation according to Abu Huraerah (13), include: a. Ideas, the community participates in contributing ideas in the planning process; b. Energy, involving the community actively in the implementation and participation of the intended activity; c. Skills, the community is involved in activities by emphasizing the skill elements; d. Property, a form of community participation by contributing money and objects to achieve something desired in the community.

The research results show that; the community participation of Gondoriyo Village Administration in disaster management is manifested in the Destana components and these include as follows:

a. Legislation

Gondoriyo Village Administration already has various disaster management documents/stipulations, consisting of the 2017-2022 Disaster Management Plan for Disaster Risk Reduction. These documents contain an explanation on the choice of actions and the mechanism of action when a disaster occurs, starting from pre-disaster, during disaster to post-disaster. It also contains details regarding the allocation and roles of activity actors, program priorities, community action plans, and monitoring and evaluation.

Another document that Gondoriyo Village Administration has is the Disaster Risk Assessment. It contains an explanation of the most serious disaster threats in Gondoriyo Village Administration, i.e. floods, landslides, hurricanes, and land fires. Floods that occur can be caused by high and continuous rainfall, water transfers from higher areas (Wates and BSB Village Administration), unregulated waterways, deforestation, and river sedimentation. Gondoriyo Village Administration has also made the document of Early Warning System and so, the community can be prepared if a disaster strikes at any time.

For the handling if a flood disaster comes; Gondoriyo Village Administration already has a flood disaster contingency guideline. It contains the Village Administration Vulnerability Study and the Gondoriyo Capacity Study as well as the risk reduction action plan. All these regulations are contained in the Decision of Head of Gondoriyo Village Administration, which was changed in 2017. Finally, Gondoriyo Village Administration has a disaster preparedness map, i.e. the form of mapping areas that have a disaster impact and how the flow charts of disaster management. This map contains an explanation that in Gondoriyo Village Administration there are 12 RWs (Citizens Association) with different potential disaster. When a disaster occurs, they will be directed to temporary shelters. However, if it is needed, they can be directed to the final evacuation sites.

b. Planning

Gondoriyo Village Administration has prepared the disaster management planning documents such as The Disaster Management Plans, PRB Action Plans or Contingency Plans that are efficiently arranged together with the formation of a Disaster Management task force. However, these documents have not been integrated into the Village Administration Development Plans.

c. Institutionalization

Gondoriyo Village Administration already has FPRB (Forum Pengurangan Resiko Bencana/Disaster Risk Reduction Forum, which the composition of the administration structure consists of community leaders, heads of youth organizations and women's groups (PKK). The forum has been actively functioning with well-planned and implemented the PRB programs since 2017. The number of active members amounts to 30 people.

The forerunner of the Disaster Resilient Village is the Disaster Preparedness Village (KSB) which has existed since 2012; where previously named the Disaster Preparedness Group. This volunteer group was established since the flash flood hit Gondoriyo Village Administration in 2010. The volunteer team has held several training activities, simulations, and standby meetings for its members and the community. The implementation is carried out in collaboration with the BPBD of, National Search and Rescue Agency (Basarnas), Indonesian Red Cross (PMI), DLH, and the others. It also makes the collaboration with some universities, such as Diponegoro University and State University of Semarang (Unnes) and some NGOs, such as Bintari Foundation.

d. Funding

For funding, Gondoriyo Village Administration does not have a special budget yet. So far, it definitely relies on donations from the members and the community; as well as non-routine assistance from the Gondoriyo Village Administration. The proposal for budgeting into the Village Administration Development Plans has actually been approved in the Village Administration Development Plans; However, this has not been realized yet due to the COVID-19 pandemic.

e. The Capacity Development of Gondoriyo Village Administration

For the disaster management, there have been efforts to carry out disaster training for Village Administration officials as well as for personnel working as disaster preparedness volunteers in the level of village administration. It aims to carry out the efforts of the PRB, emergency response operations, and post-disaster recovery. The form of the disaster training is such simulation activities, evacuation, and emergency response operations for men and women, including youths.

f. Implementation of Disaster Management

Based on Law No. 24 of 2007 concerning Disaster Management, in efforts to implementing mitigation and reducing disaster risks in the level of villages, mapping and analyzing village threats, vulnerabilities and capacities have been carried out. This is carried out by making various disaster documents, disaster trainings, and the mutual cooperation of cleaning up rivers and waterfalls.

3.2. Types and Forms of Community Participation in Gondoriyo Village Administration

Based on the data at above, it can be analyzed that the types of participation of the Gondoriyo community, according to Huraerah, include the first to the fourth participations. If the participation is direct; it means that citizens are directly involved in various activities physically and face to face. This can be seen from their participation in various educational programs, trainings, and the mutual cooperation of cleaning up the village areas.

Finally, the participation is also realized in a kind of donating money and financing the activities and operations of the organization. Thyerefore, the community provides support for the activities, both material and non-material; including being involved in the decision-making process, as shown in their involvement in the administration structure of the FPRB.

Concerning the forms of community participation in Gondoriyo Village Administration, these include:

a. Ideas, the community participates in contributing ideas in the planning process.

This is carried out in the meetings of Destana or FPRB. The community provides suggestions and responses to disaster management activities.

b. Energy, involving the community actively in carrying out and participating in an activity. It is carried out in a kinf of volunteering and participating in various activities held by Destana.
c. Skills, the community is involved in such activities by emphasizing the skill elements.

This is carried out in various trainings and disaster simulations, both pre-disasters, during and post-disaster, such as the building of emergency shelters, public kitchens, aid for disaster victims.

d. Property, a form of community participation by donating money and objects to achieve something desired in the community.

This participation is carried out in the form of giving voluntary donations in various disaster activities organized by Destana.

The active role of Gondoriyo Village Administration in disaster management has successfully achieved the predicate of the Intermediate Disaster Resilient Village. Gondoriyo Village Administration has achieved the ranking because it has the following indicators: a) There is the PRB policy developed in the level of village administration; b). The existence of the disaster management planning documents that have been compiled but have not been integrated into the village planning instruments; c). The existence of the PRB Forum consisting of representatives from the community, including women's groups and vulnerable groups, but they are not yet fully functional and active; d). The existence of Village Level Disaster Management Volunteers Team involved in capacity building, knowledge and disaster education activities for its members and the community in general; however, is not routine and not very active; e). There are efforts to conduct risk assessment, risk management and vulnerability reduction, including alternative productive economic activities to reduce the vulnerability, but it not yet tested; f). There are efforts to increase disaster preparedness and response capacity that have not yet been tested and made systematically.

4. Conclusion

The community participation in Gondoriyo Village Administration in the Kelurahan Tangguh Bencana/Disaster Resilient Village program which aims to reduce disaster risks is definitely quite good. Gondoriyo Village Administration as the Destana has been actively conducting various trainings as an effort to develop capacity in disaster management. For the type of their direct participation, these include the contribution of money, support, or involvement in decision making, and for the type of indirect participation their contribution is in a kind of thoughts, energy, skills and property.

However, what the government must improve is how the disaster management programs are integrated into the village administration development plans, particularly in relation to assistance to capacity development and routine funding. Therefore, the existence of the Destana program can be maintained and sustainable when carrying out its role in reducing disaster risks based on community participation.

References

[1] Haryadi, W. Gempa Tektonik di Pulau Sumbawa dan Dampaknya Terhadap Bangunan Sipil (Suatu Kajian Geologis). 2012. 6:13.
[2] BNBP. go.id . “Potensi Bencana Indonesia.” 2017
[3] BNPB. Peraturan Kepala Badan Nasional Penanggulangan Bencana No. 1 Tahun 2012 tentang Pedoman Umum Desa/ Kelurahan Tangguh Bencana.2012
[4] Kalangkahan PH, Areros WA, Sampe S. Kajian Partisipasi Masyarat Dalam Program Pengurangan Risiko Bencana Berbasis Komunitas Di Kota Bitung. Jurnal Ilmiah Society. 2017 Aug 1;1(24):9.
[5] Oktari RS. Peningkatan kapasitas desa tangguh bencana. Jurnal Pengabdian kepada Masyarakat (Indonesian Journal of Community Engagement). 2019;4(2):189-97.
[6] Widyasari N. Pendidikan Kebencanaan dan Partisipasi Masyarakat dalam Program Kelurahan Tangguh Bencana di Kelurahan Kebondalem Kecamatan Kota Kendal Kabupaten Kendal (Doctoral dissertation, UNNES).
[7] Julius AM, Widyaningrum N, Najib A, Aminullah AA, Syarifah H, Pratikno H, Fadlurrahman I,

Adri K, Suroso T, Ramadhani RM, Widana ID. Implementation of the Resilient Village at Gunung Geulis Village, Sukaraja Sub-regency, Bogor, West Java. Swabumi. 2020 Mar 19;8(1):1-0.

[8] Adi, Isbandi Rukminto. Perencanan Partisipatoris Berbasis Aset Komunitas: Dari Pemikiran Menuju Penerapan. Depok: FISIP UI Press.2007

[9] Peraturan Indonesia. UU No 24 Tahun 2007 tentang Penanggulangan Bencana.Lembaran Negara No.66. Sekretariat Negara. Jakarta

[10] McEntire, David. Understanding and Reducing Vulnerability: From The Approach of Liabilities and Capabilities. Journal of Disaster Prevention and Management. 2011. Volume 20. Number 3

[11] Zhang, Shanqi. Public Participation in the Geoweb era: Defining A Typology For Geo-Participation In Local Goverments. Journal of Cities. Volume 85. 2019. Pages 38-50.

[12] Huraerah, Abu. Pengorganisasian dan Pengembangan Masyarakat : Model dan Strategi Pembangunan Berbasis Kerakyatan. Bandung : Humaniora. 2011

[13] Pasaribu, I., & Simandjuntak, B. Sosiologi Pembangunan. Bandung: Tarsito.1986

IOP Conf. Series: Earth and Environmental Science 1041 (2022) 012033 doi:10.1088/1755-1315/1041/1/012033

Assessing and preparedness for earthquake disaster in Salaf-Khalaf Islamic Boarding School

Fatiya Rosyida*, Neni Wahyuningtyas, Ardyanto Tanjung, Abdul Kodir

Universitas Negeri Malang, Malang Indonesia.

*fatiya.rosyida.fis@um.ac.id

Abstract. Trenggalek is an earthquake-vulnerable area because it is located in the subduction zone of the Indo-Australian and Eurasian plates. Therefore, it is necessary to have the preparedness of the people who live there in order to reduce disaster risk. The Islamic boarding school that was there was also affected by the disaster. Santri have a high risk of experiencing a disaster because they live in the Islamic boarding school. There have been many studies related to the preparedness of the community and school students, but not many studies related to santri's preparedness. Therefore, the purpose of the study was to analyze the level of santri's preparedness against the earthquake disaster in Trenggalek. The research method is a quantitative description using questionnaires and interviews. The sample amounted to 30 and was obtained by stratified random sampling. Data analysis uses the earthquake preparedness index. The results showed that 10% were classified as extremely prepared, 35% were classified as prepared, 43% were classified as unprepared, and 17% were classified as extremely unprepared. This happens because of the lack of knowledge of students about the vulnerability of disasters that exist around the Islamic boarding school and the disaster mitigation efforts.

1. Introduction

Natural disasters are common in Indonesia. The occurrence of this natural disaster is due to Indonesia's geographical location, which is in a disaster-prone zone. Natural disaster-related problems in Indonesia become worse. The number of disaster events is very high, with around 3,000 events per year, according to data presented by the National Agency for Disaster Management Indonesia in table 1 from 2018 to 2020. Every year, the number of victims who are suffering and displaced grows. Hydrometeorological disasters are the most common type of disaster, followed by geological disasters [1] [2]. Climate change is to blame for the rise in hydrometeorological disasters [1]. Climate change causes an increase in hydrometeorological disasters like floods, landslides, and cyclones. However, geological disasters have the greatest impact [3] [4].

Geological disasters, in this case earthquakes, have a significant impact on people's lives and property [5].The cost loss is also significant, both economically and socially [6] [7]. 74 earthquakes occurred between 2018 and 2020. (Table 1). Recently, earthquakes have been more frequent than in previous years, with 429 earthquakes recorded from 1900 to 2015 [2] with an average of 3 earthquakes per year and the last three years with an average of 24 earthquakes per year. Since 1900, the death toll has been recorded. More than 250,000 people have been affected, with a population of more than 30 million [2] [8].

IOP Conf. Series: Earth and Environmental Science 1041 (2022) 012033 doi:10.1088/1755-1315/1041/1/012033

Table 1. Disasters and their Impacts in Indonesia 2018-2020

Years	**2018**	**2019**	**2020**
The number of disasters	3.397	3.814	2.925
Type of disasters			
Tornado	1.113	1.387	873
Flood	871	784	1.065
Forest fires	527	746	326
Landslide	615	719	572
Drought	152	123	29
Earthquake	28	30	16
TidalWave/Abrasion	33	18	36
Volcanic eruption	58	7	7
Covid 19	0	0	1
Disaster Impact			
victim died	3.874	478	370
missing victim		111	39
Suffering and Refuge	563.000	6,1 million	3,8 million
Injuries	2.171	3.421	536
Home Damage	370.153	73.723	29.964
Damage to Public Facilities	4.774	2.024	1.317

(Source: National Agency for Disaster Management Indonesia 2018, 2019, 2020)

The history of disasters in Indonesia demonstrates that the number of incidents and disaster victims is increasing and becoming more complex [9]. In human history, disaster is inextricably linked [10] [11]. Natural disasters are the result of a series of natural events. Earthquakes, tsunamis, volcanic eruptions, floods, droughts, hurricanes, and landslides are examples of natural disasters. The Indonesian region is, by definition, extremely vulnerable. This is due to the seismic path that the earthquake took through the Indonesian islands of Sumatra, Java, Bali, Nusa Tenggara, Sulawesi, and Maluku. Other possible disasters include soil movement, tropical climate, and peat bogs. The basis for major disasters that occur is vulnerability to risk and the impact of disasters.

Earthquakes of various magnitudes occur frequently in Indonesia, ranging from large to minor. The number of earthquakes that occur in Indonesia is inextricably linked to the country's highly active tectonic conditions. Many destructive earthquakes have struck Indonesia, and they are frequently followed by powerful tsunami waves. The Indonesian islands are at high risk of earthquakes, with 80 percent of the country's territory located in high-risk seismic areas [12]. The southern part of Java Island is close to the tectonic plate subduction zone (Eurasia and Indo-Australia), making it prone to earthquakes [13].

Trenggalek is located on Java's southern coast and is prone to earthquakes. Several earthquakes were recorded in Trenggalek by the Meteorological, Climatological, and Geophysical Agency in 2020. The most recent earthquake occurred on Friday, July 24, 2020, at 21:41 WIB (local time), with a magnitude of 4.0 on the Richter Scale. The earthquake occurred at 8.49 south latitude and 111.59 east longitude. The earthquake struck the sea 46 kilometers southwest of Trenggalek. At 01:28:06 WIB (local time) on Monday, August 31, 2020, an earthquake occurred in the Trenggalek area and its surroundings. According to BMKG data, the earthquake had a Richter scale magnitude of 4.7. The earthquake's epicenter was located at 8.80 latitude and 111.221 east longitude. The location is 70 kilometers southeast of Pacitan in the Sea. The earthquake's magnitude was determined using the MMI Scale, II Trenggalek, II Tulungagung, II Karangkates, and II Bantul.

Trenggalek is one of the earthquake-prone areas. As a result, there is a need for community-owned disaster preparedness. Awareness and concern for the importance of community preparedness in disaster management only became important after the massive disaster in Aceh that shocked the entire world [14]. The experience teaches that the community takes part in disaster preparation activities. Community involvement is critical in disaster management [14] [15] [16]. Communities in disaster-prone areas, along with authorities, are responsible for anticipating disaster risks in the surrounding environment.

IOP Conf. Series: Earth and Environmental Science 1041 (2022) 012033 doi:10.1088/1755-1315/1041/1/012033

Participation in a limited capacity takes the form of preparation for themselves and their respective families. Community and community group preparedness evolves into broader participation.

One of the pesantren affected by the earthquake is Al Ittihad Darunnajah Islamic Boarding School. Santri are particularly vulnerable to disaster effects because they attend Islamic boarding schools. According to the interviews, the students experienced earthquakes several times. There are students who panic in the face of disasters, those who are unconcerned, and only a few who understand the steps that must be taken when an earthquake occurs. The level of preparedness of the students must be determined as the basis for determining the next steps in dealing with the earthquake disaster. The importance of disaster preparedness and mitigation as a means of reducing disaster risk is stated in the Hyogo Protocol. The protocol emphasizes the importance of strengthening disaster resilience in countries and communities [17].

2. Methodology

A quantitative survey method is used in the research. Data was gathered through interviews and questionnaires. The stratified random sampling technique was used in the study, which included male and female students from Madrasah Tsanawiyah (junior high school equivalent) grades VII, VIII, and IX. The preparedness study is based on four parameters: 1) knowledge and attitude/KAP, 2) Emergency Planning (EP), 4) Warning System (WS), and 4) Resource Mobilization Capacity (RMC) [21]. The Sturgess Formula K = 1 + 3.3 Log 100 is used to assess readiness based on the index. Table 3 displays the prepередness Index.

Table 2. Disaster Preparedness Index for Earthquakes

Index	Score
Extremely prepared	**76-100**
Prepared	**51-75**
Unprepared	**25-50**
Extremely unprepared	**<25**

3. Result and Discussion

In general, disasters are caused by the interaction of a threat (hazard) and a vulnerability (vulnerability) [21] [22] [23]. Vulnerability to the impact or risk of disaster is a condition or feature of a biological, geographical, social, economic, political, cultural, and technological community in a specific area for a set period of time that reduces the community's ability to prevent, mitigate, achieve preparedness, and respond to the impact of certain hazards [24]. The increased risk of earthquake disasters will be exacerbated by high vulnerability. Community resilience in the face of disasters reduces disaster risk by increasing community capacity [25].

Increasing preparedness can help to improve disaster resilience. Preparedness is a set of activities undertaken to anticipate disasters through organization and the implementation of appropriate and efficient measures. [7][16]. This activity is performed to reduce the negative effects of a hazard. Actions that can be taken include effective disaster prevention, rehabilitation, and recovery to ensure the timely and effective arrangement and delivery of aid and assistance following a disaster. [24][25]. Preparedness is defined as all efforts and activities undertaken prior to an earthquake in order to respond quickly and effectively to circumstances/situations during and immediately following a disaster. The community requires this effort to mitigate the effects of the earthquake, including loss of life, property damage, and environmental damage.

Santri (students studying in Islamic boarding schools) must also carry out disaster preparedness. They are highly vulnerable because they still require the assistance of adults to care for and protect them in the event of a disaster [26]. Their level of preparedness is low because more than half of them are not prepared for an earthquake. As many as 40% are prepared to be ready in the event of a disaster (Diagram 1). This condition is evidenced by the fact that few people understand the dangers that can occur in an

Islamic boarding school environment. They also have no idea what to do in the event of an earthquake. They have no plans to increase preparedness through simulation activities, training, seminars, and other means. This is due to the fact that the activities carried out by students have been determined by the Islamic boarding school, so they are unable to choose their own activities. Furthermore, the activities are very dense because they deepen knowledge, religion, and general science all at once [27].

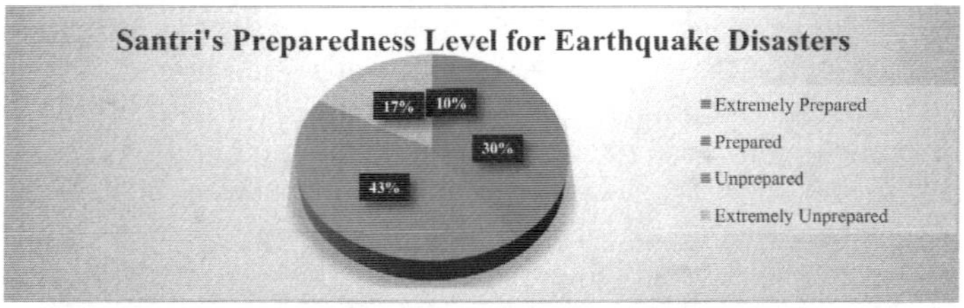

Figure 1. Santri's Preparedness Level for Earthquake Disasters

Students have the highest level of preparedness in the parameters of knowledge and attitude (KAP) and emergency planning, according to the parameters of preparedness (EP). The lowest levels are Resource mobilization Capacity/RMC and Warning System/WS, in that order (Diagram 2). The first parameter is understanding of earthquake phenomena and readiness to deal with such disasters. Most of the students who have a high level of knowledge obtained through school learning and experience. Students in Class IX have more knowledge as a result of their longer residence time.

They have been subjected to a number of earthquakes. Interaction with the community surrounding the Islamic boarding school also provides additional information about the possibility of an earthquake. Because of their short stay experience, seventh grade students who have just entered the Islamic boarding school have less knowledge than those who have been there for a longer period of time. Students' levels of knowledge are influenced by factors such as experience and length of stay [28]. Knowledge can usually influence students' attitudes and concerns in order for them to be ready and alert in the event of a disaster [29].

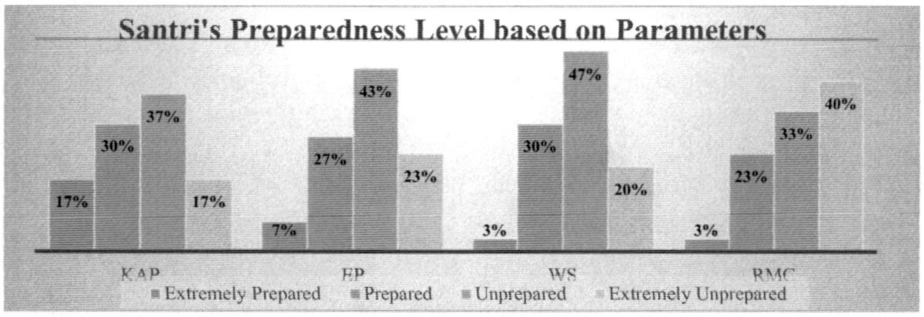

Figure 2. Santri's Preparedness Level based on Paramaters

The second parameter is the response strategy in the event of an earthquake. This plan is an important part of disaster preparedness, particularly for evacuation, first aid, and rescue, in order to reduce the number of disaster victims [30]. This effort is critical, particularly during a disaster and in the days following the disaster, particularly before outside aid arrives [16] [31]. Santri, who is well-prepared, will dash to the field in the boarding school complex.

When an earthquake strikes, Santri who are unprepared remain silent and even panic. The majority of them who panicked were female. There is no one to secure important items and documents, as well as carry items necessary for survival. This occurred because they assumed the earthquake was brief and did not destroy the boarding school building. The third parameter is concerned with the disaster warning system. This system includes impending disaster warning signs and information distribution [31] [30]. They can take appropriate action, especially to reduce casualties, with disaster warnings.

When there was an earthquake, information was given by students who were the first to feel it and then passed it on to their friends. This information is especially useful for students who live a long distance away from the Islamic boarding school and have never experienced an earthquake. They usually remain silent and wait for guidance from other students. Due to a lack of access to information, they are unaware of the possibility of aftershocks.

They usually wait up to 10-15 minutes after the earthquake to re-enter the room. The disaster warning system is very low because there is no source of information on disaster warnings, and no one has participated in regular earthquake disaster training and simulations. As a result, training and simulation are required to teach them what to do when they hear a warning, where and how to save themselves at a specific time, based on where they are at the time of the warning [31] [32].

The fourth parameter is the ability to mobilize available resources, including human resources, funding, and critical infrastructure [14]. This ability has the potential to either help or hinder preparedness for earthquake disasters [20]. As a result, resource mobilization is critical. When an earthquake strikes, no one has prepared them in a systematic manner to save themselves. When students panic and scramble to find an evacuation location, this becomes dangerous. The low number of students participating in disaster simulations and training demonstrates low mobilization. Santri who participate in scouting activities are better prepared because they have been trained to set up tents and make stretchers in emergency situations. They are also given disaster-related information. It's just that the skills and knowledge aren't shared with other friends.

4. Conclusion

Students at the Darunnajah Islamic boarding school have a low level of preparedness because 60% of them are classified as unprepared or very unprepared, while the rest are classified as ready or very ready. The most important parameter is knowledge and attitude, with as many as 47 percent classified as ready or very ready. The lowest parameter is Resource mobilization Capacity, where up to 73% are classified as unprepared or very unprepared. As a result, training and simulations are required to improve student preparedness.

References

[1] Rosyida, A. and Nurmasari, R., 2019. A Comparison of the Impact of Hydrometeorological and Geological Disasters in Indonesia as Measured by the Number of Victims (Study: Indonesian Disaster Occurrence Data 2018). Jurnal Dialog dan Penanggulangan Bencana, 10(1), pp.12-21.

[2] Yulianto, S., Apriyadi, R.K., Aprilyanto, A., Winugroho, T., Ponangsera, I.S. and Wilopo, W., 2021. The History of Disasters and Their Management in Indonesia from a National Security Perspective. PENDIPA Journal of Science Education, 5(2), pp.180-187.

[3] Djalante, R. and Garschagen, M., 2017. A review of disaster trend and disaster risk governance in Indonesia: 1900–2015. Disaster risk reduction in Indonesia, pp.21-56.

[4] Chang, S.E., 2000. Disasters and transport systems: loss, recovery and competition at the Port of Kobe after the 1995 earthquake. Journal of transport geography, 8(1), pp.53-65.

[5] Li, M., Lv, J., Chen, X. and Jiang, N., 2015. Provincial evaluation of vulnerability to geological disaster in China and its influencing factors: a three-stage DEA-based analysis. Natural Hazards, 79(3), pp.1649-1662.

[6] Hallegatte, S. and Przyluski, V., 2010. The economics of natural disasters: concepts and methods. World Bank Policy Research Working Paper, (5507).

[7] Dore, M. and Etkin, D., 2000. The importance of measuring the social costs of natural disasters at a time of climate change. Australian Journal of Emergency Management, The, 15(3), pp.46-51.

[8] Parwanto, N.B. and Oyama, T., 2014. A statistical analysis and comparison of historical earthquake and tsunami disasters in Japan and Indonesia. International Journal of Disaster Risk Reduction, 7, pp.122-141.

[9] Jati, W.R., 2015. Demographic Bonus as an Engine of Economic Growth: Window of Opportunity or Window of Disaster in Indonesia. Populasi, 23(1), pp.1-19.
[10] Anies. 2017. A Million Disasters in a Million Countries: Identifying, Analyzing, and Overcoming Disasters through Disaster Management. Yogyarakarta: Ar-Ruz Media.
[11] Jameson, F., 2003. The dialectics of disaster. In Dissent from the Homeland (pp. 55-62). Duke University Press.
[12] Naryanto, H.S., 2012. Analysis of the Risk of Earthquakes and Tsunamis on Lampung's West Coast in Relation to Mitigation and Regional Planning. Jurnal Sains dan Teknologi Indonesia, 10(2).
[13] Pasari, S., Simanjuntak, A.V., Mehta, A. and Sharma, Y., 2021. The Current State of Earthquake Potential on Java Island, Indonesia. Pure and Applied Geophysics, pp.1-18.
[14] Hidayati, D., 2008. Community preparedness: A new paradigm of natural disaster management, 3(1), pp.69-84.
[15] Shaw, R., Kobayashi, K.S.H. and Kobayashi, M., 2004. Linking experience, education, perception and earthquake preparedness. Disaster Prevention and Management: An International Journal.
[16] Becker, J.S., Paton, D., Johnston, D.M. and Ronan, K.R., 2012. A model of household preparedness for earthquakes: how individuals make meaning of earthquake information and how this influences preparedness. Natural hazards, 64(1), pp.107-137.Nations, U., 2006. International Strategy for Disaster Reduction ISDR (2006)." words Into Action: Implementing the Hyogo Framework for Action". documents for consolation. draft.
[17] Halim, W., 2017. The Role of Islamic Boarding Schools in Discourse and Empowerment of Civil Society. AKADEMIKA: Jurnal Pemikiran Islam, 22(2), pp.191-218.
[18] Haningsih, S., 2008. The strategic role of Islamic boarding schools, madrasah and islamic schools in Indonesia. El Tarbawi, 1(1), pp.27-39.
[19] Thahir, M., 2014. The role and function of Islamic boarding school: An Indonesian context. TAWARIKH, 5(2).
[20] LIPI, T., 2006. Study of Community Preparedness in Anticipating Earthquake and Tsunami Disasters in Indonesia. Bandung: LIPI.
[21] Flanagan, B.E., Gregory, E.W., Hallisey, E.J., Heitgerd, J.L. and Lewis, B., 2011. A social vulnerability index for disaster management. Journal of homeland security and emergency management, 8(1).
[22] Birkmann, J., 2006. Measuring vulnerability to promote disaster-resilient societies: Conceptual frameworks and definitions. Measuring vulnerability to natural hazards: Towards disaster resilient societies, 1, pp.9-54.
[23] Faas, A.J., 2016. Disaster vulnerability in anthropological perspective. Annals of Anthropological Practice, 40(1), pp.14-27.
[24] Maharani, N., 2020. Tingkat Pengetahuan Siswa Tentang Kesiapsiagaan Bencana Gempa Bumi Di SMPN 3 Kuta Selatan Badung Provinsi Bali. PENDIPA Journal of Science Education, 4(3), pp.32-38.
[25] Tekeli-Yeşil, S., Dedeoğlu, N., Tanner, M., Braun-Fahrlaender, C. and Obrist, B., 2010. Individual preparedness and mitigation actions for a predicted earthquake in Istanbul. Disasters, 34(4), pp.910-930.
[26] Rosyida, F. and Adi, K.R., 2017. An exploratory study of knowledge and attitudes towards flood disaster preparedness at Pilanggede Elementary School, Balen District, Bojonegoro Regency. Jurnal Teori dan Praksis Pembelajaran IPS, 2(1), pp.1-5.
[27] Ilyasin, M., 2020. Transformation of Learning Management: Integrative Study of Islamic Boarding School Curriculum. Dinamika Ilmu, 20(1), pp.13-22.
[28] Thorup-Binger, C. and Charania, N.A., 2019. Vulnerability and capacities of international students in the face of disasters in Auckland, New Zealand: a qualitative descriptive study. International journal of disaster risk reduction, 39, p.101136.

[29] Odai, E.D., Azodo, C.C. and Chhabra, K.G., 2019. Disaster management: Knowledge, attitude, behavior, willingness, and preparedness among Nigerian dentists. Prehospital and disaster medicine, 34(2), pp.132-136.
[30] Hosseini, M. and Izadkhah, Y.O., 2006. Earthquake disaster risk management planning in schools. Disaster Prevention and Management: An International Journal.
[31] Irvine, J., Buttimore, A., Eastwood, D. and Kendrick-Jones, J., 2014. The C hristchurch earthquake: Dialysis experience and emergency planning. Nephrology, 19(5), pp.296-303.
[32] Wang, J.J., 2016. Study on the context of school-based disaster management. International Journal of Disaster Risk Reduction, 19, pp.224-234.

Disaster education in disaster-prone schools: a systematic review

R Yusuf[1*], Razali[1], Sanusi[1], Maimun[1], I Fajri[2,3], S A Gani[1]

[1]Universitas Syiah Kuala, Aceh, Indonesia
[2]Universitas Pendidikan Indonesia, Bandung, Indonesia
[3]Universitas Islam Negeri Ar-Raniry, Aceh, Indonesia

*rusliyusuft@unsyiah.ac.id

Abstract. This paper aims to analyze disaster education in schools in disaster-prone areas, focusing on the resulting policies, supporting programs, and challenges in implementing disaster education in schools. The literature review in this paper is carried out systematically and analyzes the content as a methodology in investigating policies, programs, and challenges in implementing disaster education in disaster-prone schools in the Scopus database. The results of the analysis of the literature review highlight several important points related to disaster education in disaster-prone schools, namely (1) In supporting increased understanding of environmental management and resilience in dealing with disasters in disaster-prone areas, governments in various countries make policies by including environmental content in school materials and policies for environmental conservation. (2) School cooperation with the community and government institutions related to the environment becomes a program in increasing understanding of environmental management and disaster resilience. (3) The literature review also reveals that the material's content, the approach/learning model used, and the teacher's ability to understand the environment are challenges for implementing disaster education in disaster-prone schools. The problems in the discussion of this literature review are usable as implications for further research considerations. The findings show that disaster education in disaster-prone schools can effectively increase environmental understanding and resilience if existing policies, programs, and challenges are implemented optimally.

1. Introduction

A significant event occurred in December 2004. An earthquake measuring 9.0 on the Richter scale that rocked the west coast of Aceh, Indonesia is one of the natural phenomena that triggered not only a tsunami in that area but also in other countries such as Thailand, Sri Lanka, Malaysia, Myanmar, Somalia, Bangladesh, India, and Maldives [1]. Other similar events or disasters directly affect significant numbers of people, such as physical infrastructure, death, and psychological disturbances in large numbers of people [2]. These conditions affect the function of life in the community due to disasters that occur [3], so it is necessary to reduce disaster risk in the community in various ways by the government to overcome various existing problems.

A study revealed that children affected by Hurricane Katrina suffered from the impacts about three to four years after the disaster. About 28% more have psychological problems or mental disorders, so they are taken to the hospital [4]. In-depth, the author could not provide further explanation of what happened to these children. However, another study provides information that a disaster event has the most significant impact on children, especially those who live in disaster-prone areas [5]. So it is necessary to carry out a series of special activities in areas prone to disasters, especially educational institutions in the area.

As educational institutions, schools have a great responsibility in providing education that creates a conducive learning atmosphere and process [6]. One of the responsibilities of schools is to provide post-disaster responses as a way to reduce problems that have occurred in the local community [7]–[9]. In detail, schools have a role as a place of life for children in disaster-prone areas. Furthermore, schools are responsible for gathering points for people affected by disasters with an adjusted atmosphere. On the other hand, schools are an essential part of the local community, and in some areas, schools are used as the first and long-term early emergency response [10], [11].

Based on what has been described previously, the impact of disasters on schools and their personnel becomes a critical issue to consider directly and in the long term from disasters in the community, such as organizing an educational process-oriented to disaster knowledge. As a result, disaster education is an alternative in providing disaster knowledge to students as basic knowledge in dealing with future disasters.

The argument against the importance of teaching disaster education in schools which is increasing day by day is caused by several reasons as described by experts, namely as follows: (1) children are one of the most vulnerable groups during a disaster, (2) the impact of Disaster education obtained in schools can be information for parents, families, and communities, (3) children as the generation who will inherit the future and (4) schools as locations for community meetings and planned group activities [12].

From this explanation, the school community is critical when considering the impact of disasters on children. Consistent with the results of research that state that teachers can be a valuable source in terms of the first assessment of the impact of disasters that occur on their students, school-based psychological interventions for students' psychosocial health and school-based psychological interventions for students' psychosocial health [13]. Teachers also offered knowledge of disaster management in schools which were mentioned as key in managing these disasters [14], other studies reveal the importance of implementing disaster education in schools, especially in schools of the areas prone to natural disasters [15]. The programs and policies created must be oriented to increasing knowledge of disasters in the school environment.

This literature review research intends to look at the experience of disaster education in disaster-prone schools. This experience will be explored from the results of research that has been carried out so far related to disaster education in schools located in disaster-prone areas. The exploration focused on policies, programs, and challenges in implementing disaster education in disaster-prone schools. The systematic literature review consists of studies published in the Scopus database, then analyzed using thematic and content. Based on the focus of this research, several research questions have been adapted to the research needs:

a. RQ 1: What policies have been implemented to support disaster education in disaster-prone schools?
b. RQ 2: What are the programs that support the implementation of disaster education in disaster-prone schools?
c. RQ 3: What are the challenges faced in implementing disaster education in disaster-prone schools?

2. Research Method

This literature review paper uses a research method with a content analysis approach with 30 articles analyzed on disaster research in disaster-prone schools. The selection of the analysis is used to achieve the expected objectives of the study. Content analysis is a research technique often used in the social sciences and humanities [16]. Furthermore, Creswell, provides a detailed explanation of the content analysis of written texts, recordings, images, and artifacts [17]. In addition, the systematic methodological approach uses a combination of content analysis and thematic analysis to facilitate the author in synthesizing all data from the selected literature for analysis according to the purpose of this paper.

To help the writer finish this paper, the writer utilizes the Scopus database search facility, which investigates disaster education in disaster-prone schools. The researcher accessed all articles from May 1, 2021, to August 15, 2021, with the terms used (TITLE-ABS-KEY (disaster AND education) AND TITLE-ABS-KEY (disaster-prone AND schools). In the Scopus database, the author tries to limit the articles considered most related to writing this article, significantly articles indexed by Scopus, to maintain the quality of the articles produced. Martin et al, mention that Scopus as a database is considered incorrect [18]. One of the world's most trusted citation index platforms for quality scientific information.

3. Result and Discussion

This study began by looking for articles related to disaster education in schools in disaster-prone areas. The search process was carried out through the Scopus database. All articles were accessed from May 1, 2021, to August 15, 2021, with the terms used (TITLE-ABS-KEY (disaster AND education) AND TITLE-ABS-KEY (disaster-prone AND schools). The discussion will be explained in detail following the research objectives, starting from disaster education policies in schools of disaster-prone areas, programs that support disaster education, and the challenges of implementing disaster education in schools of disaster-prone areas.

3.1 Disaster education policy in disaster-prone schools

In realizing the safety of school communities in disaster-prone areas, schools must be safely built. The process applied must also be continuous and dynamic. In realizing this, its management must involve the school community, such as teachers, students, parents, students, and residents who live in the school environment.

In addition, the policies produced by the government to deal with disasters in schools in disaster-prone areas must be comprehensive. It means that the resulting policies must be oriented towards increasing the disaster knowledge of school residents. For example, several regions have developed geographic information systems for mapping disaster-prone areas to facilitate the identification of schools that require special attention, such as schools located in volcanic areas [19].

Disaster management policies produced in schools must include several indicators such as hazard assessment, vulnerability, resource capacity, planning and implementation of physical risks, maintenance of safe facilities, standard operating procedures (SOP), and training policies for disaster response for the school community [20]. Bisri and Sakurai, note that in Indonesia, national policies and regulations can develop to potential regional regulations (Perda) or governor/mayor instructions that enable schools to carry out disaster education or safe school activities [1]. In addition, several countries also make policies regarding disaster education that are important to be taught in schools, such as Japan making disaster education a unique material that is taught in schools [21]–[24].

Another policy on increasing disaster knowledge is exemplified by Pakistan, a country with particular attention to disaster risk management. It is evidenced by a national framework for disaster risk management developed in 2007 to direct the entire system to build capacity, policy, and planning in each field, including in educational institutions [25], [26]. In more detail, the policies of each country in implementing environmental education through the content of specially taught materials [5], [27], [28], in the integration of other subjects [15], [29]–[31] and even employing environmental conservation policies located in disaster-prone areas [15], [32], [33].

3.2 Programs that support disaster education in disaster-prone schools

Schools have essential roles and responsibilities in building community resilience in conveying knowledge related to the environment and disaster and disaster risk mitigation education. As a result, implementing disaster education through various programs is becoming increasingly important in various existing schools, without exception, schools located in disaster-prone areas.

For example, Yogyakarta Province in Sleman district, a school located in the danger zone living near Mount Merapi, participates in a disaster risk reduction (DRR) program in collaboration with the local government and is also involved with several other institutions such as universities, non-governmental organizations, and other governmental institutions [34].

In addition, other programs that support the improvement of disaster knowledge and disaster risk reduction are through the Disaster Preparedness School Program (SSB). The SSB program is a form of activity aimed at increasing disaster awareness for school community members with the principles of preparedness, reducing the impact and risk of disasters, increasing disaster resilience and knowledge about disasters [35].

Furthermore, several areas, such as Aceh province, are one of the areas that are vulnerable to tsunamis and earthquakes. So that the Aceh region, after the tsunami disaster in 2004, became the center of research studies on world disasters. Some researchers have even conducted research related to disaster education in schools affected by the tsunami [5]. Some of these concerned researchers, such as Oktari et al, illustrate the importance of collaborative networks of schools and communities in increasing the resilience of coastal communities in Banda Aceh, Indonesia [36].

Sakurai reinforced it by stating that around 56% of public elementary schools in Banda Aceh City are exposed to high tsunami risk, and some disaster preparedness activities in these schools are driven by external factors which schools now discontinue due to lack of funding, regulatory institutions and lack of knowledge [5]. Other programs are followed by educational programs taught in the form of materials in the learning process, such as the Adiwiyata school, local wisdom, and other government programs [37].

On the other hand, several countries also carry out various programs that support improving disaster knowledge and knowledge of disaster risk mitigation in schools. Japan is a country with particular attention to disaster risk mitigation, as it has a high vulnerability to disasters. According to Uchida et al, the Japanese government builds and develops a disaster program in ICT-based schools designed to provide disaster prevention and mitigation to increase awareness of disaster mitigation among young people [38]. In addition, Japan also conducts public education programs on tsunami disaster mitigation and rehabilitation in elementary schools in Japan [39]. Based on the previous description, programs that support disaster education are essential to be implemented, particularly in schools that are prone to disasters.

3.3 Challenges faced in implementing disaster education in disaster-prone schools

In achieving comprehensive school safety, disaster preparedness programs should not be entirely carried out by external parties. However, they must be carried out jointly and sustainably by the school and the local community so that the process of implementing disaster education can be carried out correctly and adequately.

Based on the literature, several challenges may hinder the implementation of disaster education in disaster-prone schools, such as the lack of teacher and student capacity, limited learning resources, and the model/approach used. Similar research also states that schools in implementing preparedness have challenges regarding their resources and limitations on the facilities they have [1], [30], [40]. Other studies also state that teachers' knowledge of disaster education needs to be improved to successfully implement disaster education in disaster-prone schools [2], [41].

In addition, the challenges faced also lie in the material or curriculum regarding disaster education that has not been correctly implemented, so that material about disaster becomes increasingly important to provide knowledge to students at school and the community. The curriculum used to integrate disaster risk reduction in schools can effectively reduce the impact of disasters [2]. In addition, other challenges also come from the methods or approaches used in the implementation of disaster education for disaster risk reduction. However, the method becomes an essential medium for teachers to teach materials related to the disaster. Aghaei et al, expressed that effective disaster risk mitigation strategies will positively impact disaster education implementation [42].

4. Conclusion

Based on the literature review analysis on disaster education in disaster-prone schools, emphasis on several vital points is needed. Most countries in the world have launched disaster education activities both in disaster-prone schools and not in the context of disaster risk mitigation. However, the actions taken are not enough, and there are some gaps between expectation and reality. The gap starts from the not optimal policies, including disaster education in the curriculum and other policies. In addition, the gap in the form of programs produced is sometimes only limited to programs supporting disaster education, but in reality, it is not as expected to support disaster education in disaster risk mitigation.

The results of the literature test reveal that the school's collaboration with the community and related institutions is a program in increasing understanding about a disaster. In addition, the literature results also state that schools faced several challenges in implementing disaster education programs in schools, such as the abilities of teachers and students, the material taught is still inadequate, and the approach/strategy used is still not optimal. So that the conclusion is that policies, programs, and challenges faced must be completed comprehensively for the implementation of disaster education in schools to the fullest.

References

[1] M. B. F. Bisri and A. Sakurai, "Disaster Education and School Safety Governance after the 2004 Indian Ocean Tsunami in Indonesia: From National Policy to Local Implementation," in Disaster Risk Reduction in Indonesia, Springer, Cham, 2017, pp. 189–212.

[2] N. Nurdin, I. Rafliana, S. Hidayati, R. S. Oktari, and R. Djalante, "Integrating Disaster Risk Reduction and Climate Change Adaptation into School Curricula: From National Policy to Local Implementation," in Disaster Risk Reduction in Indonesia, 2013, pp. 213–234.

[3] T. Wicke and R. C. Silver, "A community responds to collective trauma: An ecological analysis of the James Byrd murder in Jasper, Texas," Am. J. Community Psychol., vol. 44, no. 3, pp. 233–248, 2009.

[4] A. Olteanu, R. Arnberger, R. Grant, C. Davis, D. Abramson, and J. Asola, "Persistence of mental health needs among children affected by hurricane Katrina in New Orleans," Prehosp. Disaster Med., vol. 26, no. 1, pp. 3–6, 2011.

[5] A. Sakurai et al., "Exploring minimum essentials for sustainable school disaster preparedness: A case of elementary schools in Banda Aceh City, Indonesia," Int. J. Disaster Risk Reduct., vol. 29, no. August 2017, pp. 73–83, 2018.

[6] R. Yusuf, M. Yunus, Maimun, and I. Fajri, "Environmental Education: A Correlational Study among Environmental Literacy, Disaster Knowledge, Environmental Sensitivity, and Clean-Living Behavior of Post Tsunami Disaster in Aceh Communities, Indonesia," Polish J. Environ. Stud., vol. 31, no. 1, 2022.

[7] S. Clettenberg, J. Gentry, M. Held, and L. A. Mock, "Traumatic loss and natural disaster: A case study of a school-based response to hurricanes Katrina and Rita," Sch. Psychol. Int., vol. 32, no. 5, pp. 553–566, 2011.

[8] M. A. Heath, A. B. Nickerson, N. Annandale, A. Kemple, and B. Dean, "Strengthening cultural sensitivity in children's disaster mental health services," Sch. Psychol. Int., vol. 30, no. 4, pp. 347–373, 2009.

[9] L. H. Ying, X. C. Wu, and C. De Lin, "Longitudinal linkages between depressive and posttraumatic stress symptoms in adolescent survivors following the Wenchuan earthquake in China: A three-wave, cross-lagged study," Sch. Psychol. Int., vol. 33, no. 4, pp. 416–432, 2012.

[10] K. Evans, D. Dougherty, M. Pollack, and S. Rauch, "Using neuroimaging to predict treatment response in mood and anxiety disorders," Ann. Clin. Psychiatry, vol. 18, no. 1, pp. 33–42, 2006.

[11] L. Wolmer, N. Loar, and Y. Yazgan, "School Reactivation Programs after disaster: Could teachers serve as clinical mediators?," Child Adolesc. Psychiatr. Clin. N. Am., vol. 10, no. 3, pp. 363–381, 2001.

[12] R. Shaw, Y. Takeuchi, Q. Ru Gwee, and K. Shiwaku, Disaster education: An introduction, vol. 7. Emerald Group Publishing Ltd, 2011.

[13] D. C. Seyle, C. S. Widyatmoko, and R. C. Silver, "Coping with natural disasters in Yogyakarta, Indonesia: A study of elementary school teachers," Sch. Psychol. Int., vol. 34, no. 4, pp. 387–404, 2013.

[14] A. F. Jorm, B. A. Kitchener, M. G. Sawyer, H. Scales, and S. Cvetkovski, "Mental health first aid training for high school teachers: A cluster randomized trial," BMC Psychiatry, vol. 10, 2010.

[15] D. N. Pascapurnama, A. Murakami, H. Chagan-Yasutan, T. Hattori, H. Sasaki, and S. Egawa, "Integrated health education in disaster risk reduction: Lesson learned from disease outbreak following natural disasters in Indonesia," Int. J. Disaster Risk Reduct., vol. 29, no. July 2017, pp. 94–102, 2018.
[16] J. H. McMillan, Educational Research: Fundamentals for the Consume. Boston: Pearson, 2012.
[17] J. W. Creswell, A concise introduction to mixed methods research. SAGE Publications, 2014.
[18] A. Martín, M. Mike, T. Enrique, and O. Malea, Web of Science , and OpenCitations ' COCI : a multidisciplinary comparison of coverage via citations, vol. 126, no. 1. Springer International Publishing, 2021.
[19] Y. Ariyanti, R. A. Yuana, and A. Budianto, "Web-based geographic information system for school mapping and disaster mitigation," 2018 Int. Conf. Inf. Commun. Technol. ICOIACT 2018, vol. 2018-Janua, pp. 136–140, 2018.
[20] M. D. Rina Susanti, Sri Adelila Sari, Sri Milfayetty, "Hubungan Kebijakan, Sarana dan Prasarana dengan Kesiapsiagaan Komunitas Sekolah Siaga Bencana Banda Aceh," J. Ilmu Kebencanaan Pascasarj. Univ. Syiah Kuala, vol. 1, no. 1, pp. 42–49, 2014.
[21] K. Kitagawa, "Continuity and change in disaster education in Japan," Hist. Educ., vol. 44, no. 3, pp. 371–390, 2015.
[22] Y. Tanaka, "Disaster Policy and Education Changes over 15 Years in Japan," J. Comp. Policy Anal. Res. Pract., vol. 14, no. 3, pp. 245–253, 2012.
[23] A. Sakurai, T. Sato, and Y. Murayama, "Impact evaluation of a school-based disaster education program in a city affected by the 2011 great East Japan earthquake and tsunami disaster," Int. J. Disaster Risk Reduct., vol. 47, no. May, p. 101632, 2020.
[24] J. Á. Del Brío, E. Fernández, and B. Junquera, Management and employee involvement in achieving an environmental action-based competitive advanrage: An empirical study, vol. 18, no. 4. 2007.
[25] A. A. Shah, Z. Gong, M. Ali, R. Sun, S. A. A. Naqvi, and M. Arif, "Looking through the Lens of schools: Children perception, knowledge, and preparedness of flood disaster risk management in Pakistan," Int. J. Disaster Risk Reduct., vol. 50, no. September, p. 101907, 2020.
[26] A. A. Shah, Z. Gong, I. Pal, R. Sun, W. Ullah, and G. F. Wani, "Disaster risk management insight on school emergency preparedness – A case study of Khyber Pakhtunkhwa, Pakistan," Int. J. Disaster Risk Reduct., vol. 51, no. July, p. 101805, 2020.
[27] R. P. Dewi, "Fieldwork and Research Impact on Learning of Disaster Risk Reduction," Univers. J. Educ. Res., vol. 8, no. 8, pp. 3718–3724, 2020.
[28] P. A. Kamil, S. Utaya, Sumarmi, and D. H. Utomo, "Improving disaster knowledge within high school students through geographic literacy," Int. J. Disaster Risk Reduct., vol. 43, no. October 2019, p. 101411, 2020.
[29] M. B. de Mendonca, T. da Silva Rosa, and A. R. Bello, "Transversal integration of geohydrological risks in an elementary school in Brazil: A disaster education experiment," Int. J. Disaster Risk Reduct., vol. 39, no. June, p. 101213, 2019.
[30] V. A. Johnson, K. R. Ronan, D. M. Johnston, and R. Peace, "Implementing disaster preparedness education in New Zealand primary schools," Disaster Prev. Manag., vol. 23, no. 4, pp. 370–380, 2014.
[31] R. Shaw, A. Sakurai, and Y. Oikawa, "New Realization of Disaster Risk Reduction Education in the Context of a Global Pandemic: Lessons from Japan," Int. J. Disaster Risk Sci., vol. 12, no. 4, pp. 568–580, 2021.
[32] A. L. Burrow and P. L. Hill, "Purpose by design or disaster: Preserving a sense of purpose amid environmental uncertainty," J. Environ. Psychol., vol. 69, no. March, 2020.
[33] P. T. Apronti, S. Osamu, K. Otsuki, and G. Kranjac-Berisavljevic, "Education for disaster risk reduction (DRR): Linking theory with practice in Ghana's basic schools," Sustain., vol. 7, no. 7, pp. 9160–9186, 2015.
[34] E. T. W. Mei et al., "Building volcanic disaster resilience community through school and

education," IOP Conf. Ser. Earth Environ. Sci., vol. 451, no. 1, 2020.
[35] R. Yusuf, Maimun, Sanusi, J. Saputra, and S. A. Gani, "Identifying the Green Supply Chain for Environmental Sanitation Management in Banda Aceh , Indonesia," Int. J. Supply Chain Manag., vol. 9, no. 3, pp. 1175–1185, 2020.
[36] R. S. Oktari, K. Shiwaku, K. Munadi, Syamsidik, and R. Shaw, "A conceptual model of a school-community collaborative network in enhancing coastal community resilience in Banda Aceh, Indonesia," Int. J. Disaster Risk Reduct., vol. 12, pp. 300–310, 2015.
[37] R. Yusuf, Maimun, Sanusi, I. Fajri, and J. Saputra, "A comparison of student environmental literacy: Public and Islamic schools in Banda Aceh, Indonesia," Int. J. Innov. Creat. Chang., vol. 12, no. 9, pp. 222–239, 2020.
[38] O. Uchida, S. Tajima, Y. Kajita, K. Utsu, Y. Murakami, and S. Yamada, "Development and Implementation of an ICT-based Disaster Prevention and Mitigation Education Program for the Young Generation," Inf. Syst. Front., pp. 1–11, 2020.
[39] H. Ohta, Susumu lai, S. Morioka, and A. Iizuka, "Tsunami Induced by 2011 Tohoku-Pacific Ocean Earthquake and a Possible Renewal Plan," in Geotechnical Predictions and Practice in Dealing with Geohazards, vol. 25, 2013, pp. 33–46.
[40] C. L. Elrod, J. L. Hamblen, and F. H. Norris, "Challenges in Implementing Disaster Mental Health Programs: State Program Directors' Perspectives," Ann. Am. Acad. Pol. Soc. Sci., vol. 604, no. 1, pp. 152–170, 2006.
[41] S. I. A. Dwiningrum, "Developing school resilience for disaster mitigation: A confirmatory factor analysis," Disaster Prev. Manag. An Int. J., vol. 10, no. 18, pp. 270–277, 2004.
[42] N. Aghaei, H. Seyedin, and H. Sanaeinasab, "Strategies for disaster risk reduction education: A systematic review," J. Educ. Health Promot., vol. 7, no. January, pp. 1–6, 2018.

IOP Conf. Series: Earth and Environmental Science 1041 (2022) 012035 doi:10.1088/1755-1315/1041/1/012035

LCA methodology for detecting environmental impacts on natural gas drilling process

Masayu Rosyidah[1], Andiyan Andiyan[2*], Haniek Listyorini[3], Pranoto Hadi Prayitno[4], Yuswardi Yuswardi[5], Yuhanah Yuhanah[6]

[1]Universitas Muhammadiyah Palembang, Palembang, Indonesia
[2]Universitas Faletehan, Bandung, Indonesia
[34]Sekolah Tinggi Ilmu Ekonomi Pariwisata Indonesia, Semarang, Indonesia
[5]Universitas Jabal Ghafur, Aceh, Indonesia
[6]Universitas Sembilan Belas Nopember Kolaka, Kolaka, Indonesia

*andiyanarch@gmail.com

Abstract. This study uses LCA analysis to monitor potential damage to eco-efficiency. This approach uses an environmental life cycle assessment (LCA) to detect ecological impacts in natural gas production systems. The study results found that the lowest environmental impact on the production process at PT was seen. X is located on the Ozone layer and Resp. Organics are 4.71E-06 DALY and 3.22E-05 DALY. The highest impact lies in Resp. Inorganics, Fossil fuels, and Climate change. This shows that the drilling system hurts the environment, especially in the climatic conditions of soil, plants, and fossil fuels which continue to decrease. In addition, because the drilling process is carried out using generator power, it causes air pollution that can interfere with the respiratory system (respiratory inorganics).

1. Introduction

Natural gas usage in Indonesia has grown significantly since 1974, when PERTAMINA started delivering natural gas through pipes from the Prabumulih natural gas resource in South Sumatra to the Palembang fertilizer facilities Pusri II, Pusri III, and Pusri IV. Simultaneously, in 1974, PERTAMINA provided natural gas to fertilizer plants and medium and heavy industries in West Java and Cilegon Banten through gas pipes from natural gas reserves off the Java Sea and Cirebon regions. The natural gas pipeline that extends from Cirebon to Cilegon, Banten, provides natural gas to a variety of industries, including cement plants, fertilizer plants, ceramic plants, steel plants, and gas and steam power plants.

In addition to local use, natural gas is exported in the form of LNG in Indonesia (Liquefied Natural Gas)[1]. One of the companies that process natural gas is PT. XYZ. The company is located in Muara Enim Regency. The natural gas exploration process uses only simple technology. Gas waste at PT. The combustion process removes XYZ[2]. The results of this combustion are predicted to have an impact on the surrounding environment.

Therefore, it is very important to analyze the potential environmental effects in the raw gas production system. The problem in this research is how much potential ecological impact is produced in the crude gas production process using the Life Cycle Assessment (LCA) method. The aim is to find out and identify what potential will impact the environment as a result of the production process that occurs[3].

2. Literature Review

Raw natural gas comes from several adjacent wells and is collected, and the first treatment process that occurs is removing water content and natural gas condensate. Condensation usually flows to the oil refinery, and the water is disposed of as wastewater. The raw natural gas is then sent to a processing plant, where the initial purification usually removes the acid content (H2S and CO2). The process used in general is Amine Treating which is commonly called Amine Plant. The next process is to remove moisture using the adsorption process in liquid trimethylene glycol (TEG). The next phase is to convert into liquefied natural gas (NGL), which is the most complex process and uses modern gas processing plants, shown in figure 1[4].

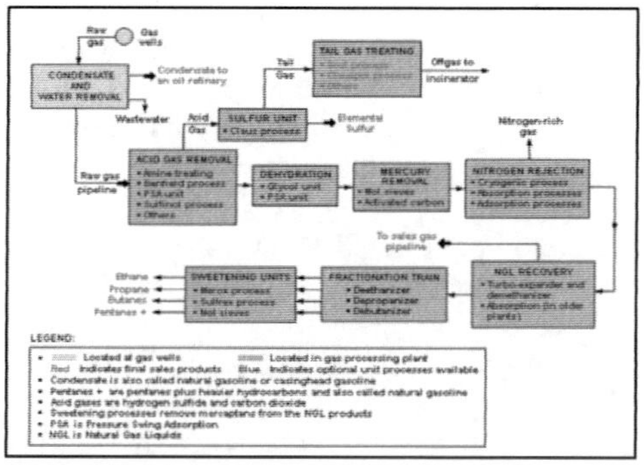

Figure 1. Natural Gas Production Process [4]

According to Marques and Silva, 2019, natural gas is a non-renewable energy commodity that plays an important role in meeting the world's primary energy needs. In 2017, natural gas demand worldwide reached 23.4% of the total global, immediate energy demand. Natural gas is one of the energies needed for every country, wherein developing a nation. It takes a certain amount of energy which will grow more and more every year.

Natural gas has a very important role as the third most widely used primary energy in the country after oil and coal, supporting people's lives and driving the Indonesian economy. In everyday life, natural gas is used as fuel for cooking, heating water, welding, driving turbines and generators to produce fertilizers and electricity, fuel gas for transportation (BBG), petrochemical raw materials, steel smelting, glass industry, ceramics. Additives and so on [5].

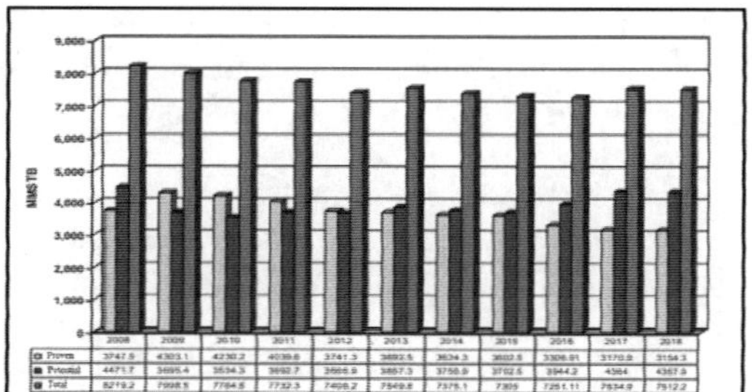

Figure 2. Graph of Indonesia's Petroleum Reserves for the Last 10 Years Based on MMSTB

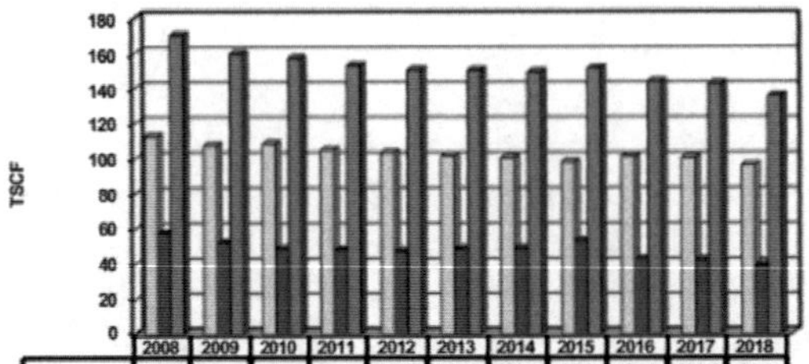

	2008	2009	2010	2011	2012	2013	2014	2015	2016	2017	2018
Proven	112.5	107.34	108.4	104.7	103.3	101.5	100.3	98.0	101.2	100.4	96.1
Potential	57.6	52.29	48.74	48.2	47.4	48.9	49.0	53.3	42.8	42.4	39.4
Total	170.1	159.63	157.14	152.9	150.7	150.4	149.3	151.3	144.0	142.7	135.5

Figure 3. Graph of Indonesia's Petroleum Reserves for the Last 10 Years Based on TSCF

Although some areas produce just natural gas, oil and natural gas are often discovered in the same field. In general, Indonesia's oil and gas reserves have been declining during the past decade. From 8.21 billion barrels in 2008, oil reserves dropped to a range of 7.5 billion barrels in 2018.[6].Reserve to Production (as a percentage of Proven Reserves) is in the range of ten to eleven years. The extension to 12 years occurred in 2014 as a result of substantial additions to prove oil reserves, particularly from the Banyu Urip Cepu Field.

Following that, the fall in global oil prices in 2015, which is still being felt today, is cited as one of the reasons for the low rate of new resource finding. In Indonesia, KKKS are more cautious and have a smaller footprint when it comes to oil and gas exploration and production. Indonesia's natural gas reserves are also diminishing. Natural gas reserves peaked at 170 TSCF in 2008 and have since declined to a range of 136 TSCF in 2018.

However, Indonesia's Reserve to Production ratio of natural gas (as a percentage of Proven Reserves) has been relatively constant during a 34-year period from 2009 to 2018, after falling substantially from 41 years in 2008. This is partially because proved reserves can be maintained while production continues at its current pace[7]. One method that can be used to analyze environmental impacts is the Life Cycle Assessment (LCA) method. Life Cycle Assessment (LCA) is a mechanism to analyze and calculate the total environmental impact at each stage of its life cycle. Life Cycle Assessment (LCA) uses an overall "cradle to grave" approach that starts with taking raw materials from the earth to make products and ends at the point where all materials return to the world [8]. There must be an in-depth study to identify the best environmental management scenario, particularly to reduce potential environmental impacts [9].

Life Cycle Assessment (LCA) is a process for analyzing and quantifying a product's overall environmental effect throughout its life cycle. Beginning with the preparation of raw materials, manufacturing processes, sales and transportation, and product disposal. The idea of Life Cycle Assessment (LCA) is often referred to as the "cradle to grave" concept. The Existence Cycle Assessment (LCA) method employs a quantitative approach to assessing environmental effects by quantifying all exchange fluxes between the system and the environment at each stage of the system's life. LCA operates from four locations: Definition of the objective and scope of work, life cycle inventory, assessment of the life cycle impact, and interpretation[10].

With many phases of study, life cycle assessment (LCA) is a tool or technique for identifying environmental effects. Life cycle assessment (LCA) is a technique or tool for analyzing the environmental effects associated with the manufacturing process of a product. The benefit of LCA is that it is exhaustive since it can evaluate the possible environmental consequences of the life cycle activities.[11].

With LCA, it is possible to observe how resources are used (input) and materials are created (output) during a process. LCAs are often focused on contributions to regional and global-scale effects, which include resource use. Life Cycle Analysis (LCA), or more commonly referred to as Life Cycle Assessment, is a cradle to grave method (an analysis of the entire cycle from raw materials to waste treatment) that is used to determine the amount of energy, costs, and environmental impacts associated with the various stages of a product's life cycle—from the time raw materials are obtained to the time consumers use the finished product.

Each stage of the LCA process is specified in international standards (ISO 14040, ISO 14041). This phase is repeated, with varying degrees of detail and effort in accordance with the study goals (World Business Council for Sustainable Development, 2002). The stages are as follows: (1) establishing goals and scope, (2) doing an inventory analysis, (3) conducting an impact analysis/assessment, and (4) conducting an interpretation (ISO 14040, 2006)[10].
According to Pujadi (2013), Life Cycle Assessment (LCA) has the following principles:

- We see the life cycle as a perspective, in other words, considering the entire physical life cycle of a product (or service), starting from the extraction of raw materials, the use of energy and production materials, the production process, the use of the product, to the end of the product's life. Another perspective is to look at the life cycle of a particular method currently being carried out as research.
- Cover all environmental aspects into one general assessment so that environmental impacts can be identified.
- Provide transparency to ensure the correct interpretation of the results obtained by calculations.
- Narrative in nature consists of four stages, namely determining the purpose and scope of the research, Life Cycle Inventory (LCI), Life Cycle Impact Assessment (LCIA), and interpretation.
- Focus on the environment by studying the environmental aspects of the production system and leaving economic and social elements aside from research.
- It is a science-based method, even though scientific circumstances are always changing. LCA provides an overview of a certain state at a certain time.

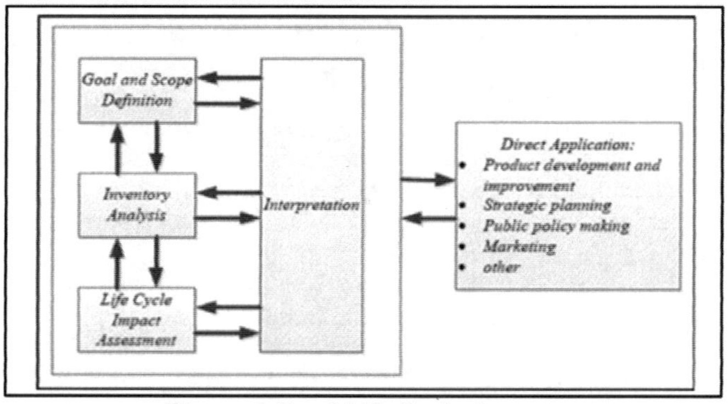

Figure 4. Fasel Life Cycle Assessment

3. Methods

The research took place from March to July 2020. The study was conducted on the natural gas production process in the Muara Enim district[12]. Data were collected from primary data obtained from interviews and observations. The research approach uses quantitative and qualitative methods (mixed method)[13]. Integrated Methods Research (MMR) is a research method applied when researchers have questions that need to be tested in terms of results and processes and involves a combination of quantitative and qualitative methods in one study[14].

Because it focuses on developments and techniques, MMR designs are commonly used in program evaluation research. The stages of the LCA method used are LCA procedures according to ISO 14040, which consist of four steps, namely goal and scope definition, life cycle inventory, life cycle impact assessment, and interpretation, assisted by the Simapro program [10].

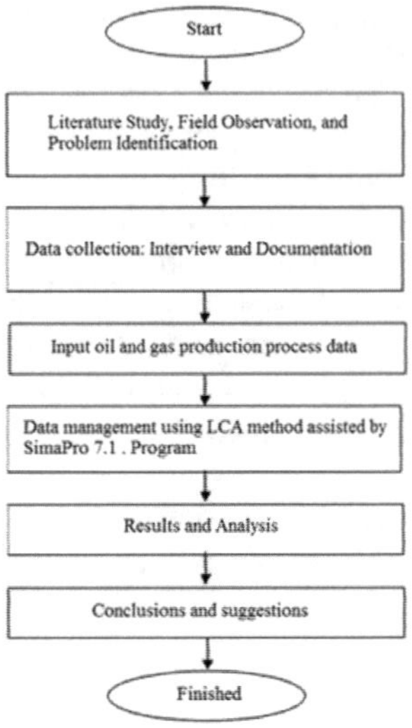

Figure 5. Research Flowchart

4. Result

PT. XYZ produces oil and gas. The gas produced is condensate, a liquid hydrocarbon obtained from gas wells or oil wells mixed with gas. Production of crude gas and oil from wells to collection stations is carried out by natural flow (not aspirated using machines). After entering the collection station, filtering is carried out by flowing it into the high separator, then flowing back into the low partition. After that, the dry gas is filtered back into the scrubber, then the gas is sent to the main Pertamina to be produced further, and the waste from the Production is flowed to the disposal by burning. From the production process, there are no chemicals used by PT. XYZ[15].

However, in the production process, there is waste from the rest of the disposal and the production process. Producing crude gas and oil begins with determining the point of the well to be drilled. After that, from the borehole directly into the collection station, filtering is carried out by flowing it into a high separator. Next, it has rushed back to the low partition[16]. The next stage is to filter into the scrubber. The results from the filtration are sent to the main Pertamina to be produced further, and the waste from the Production is sent to disposal by burning.

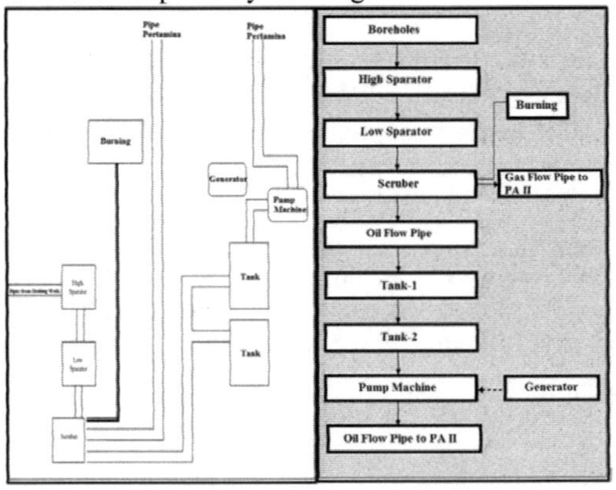

Figure 6. Production Process Flow of PT. XYZ

IOP Conf. Series: Earth and Environmental Science 1041 (2022) 012035 doi:10.1088/1755-1315/1041/1/012035

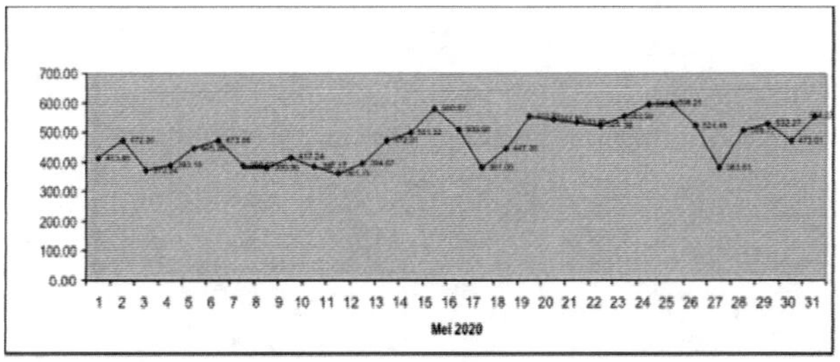

Figure 7. Oil output histogram of PT. XYZ Period May 2020 can be seen in the image below

Table 1. Total KWh Engine Generator Usage PT. XYZ (Period May 2020)

No.	K
1	43.07
2	46.19
3	36.28
4	37.12
5	53.42
6	55.29
7	42.12
8	36.27
9	48.56
10	52.72
11	32.18
12	34.46
13	62.43
14	78.21
15	81.05
16	63.34
17	57.18
18	68.45
19	70.07
20	80.13
21	76.24
22	78.61
23	82.3
24	88.27
25	92.14
26	76.08
27	62.14
28	74.08
29	83.92
30	75.42
31	84.17
Amount	1951.91
Average	62.96

The total output of PT. XYZ in the May 2020 Period was 1951.91 KWh/day, with an average of 62.96 KWh/day.

Table 2. Total Fluid Combustion PT. XYZ (Period May 2020)

No.	Barrel
1	4.71
2	5.36
3	3.59
4	3.28

5	**3.65**
6	**4.98**
7	**3.19**
8	**3.65**
9	**4.32**
10	**3.54**
11	**3.19**
12	**3.25**
13	**3.25**
14	**5.36**
15	**6.25**
16	**5.24**
17	**3.28**
18	**4.25**
19	**5.98**
20	**5.65**
21	**5.32**
22	**4.39**
23	**5.14**
24	**6.32**
25	**6.95**
26	**6.54**
27	**3.25**
28	**3.95**
29	**5.08**
30	**5.32**
31	**6.32**
Amount	**144.55**
Average	**4.66**

Table 3. Input and Output Data PT. Bama Bima Sentosa Pagar Dewa Village, Lubai District (May 2020 Period)

Type	***Output***	
	Amount per month	**Average per day**
Raw Gas (CH4)	538.54 MMSCFD	**17.37 MMSCFD**
Crude Oil	14674.42 Barrel	**473.37 Barrel**
Hydrogen fluoride	144.55 Barrel	**4.66 Barrel**
Natural Gas	**1951.91 KWh**	**62.96 KWh**

Description :
1 kWh = 1.000 watt
1 barrel = 158.99 liters
MMSCFD = Million Standard Cubic Feet per Day (gas)

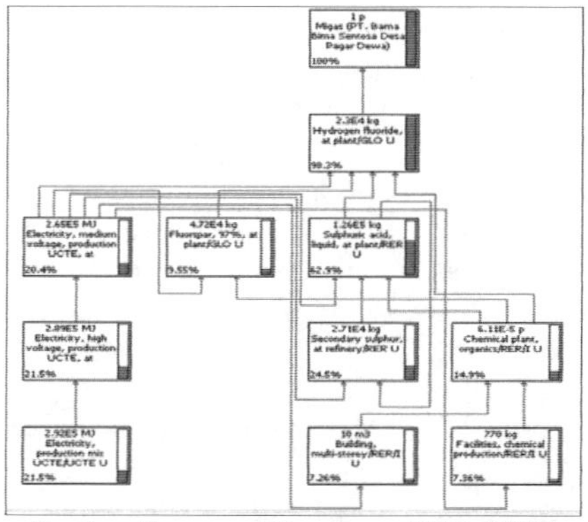

Figure 8. Production Process Diagram of PT. XYZ with Simapro

Table 4. Environmental Impact Based on Damage Assessment

Impact category	Unit	Total
Carcinogens	DALY	**0.006415**
Resp. organics	DALY	**3.22E-05**
Resp. inorganics	DALY	**0.171476**
Climate change	DALY	**0.013872**
Radiation	DALY	**0.000704**
Ozone layer	DALY	**4.71E-06**
Ecotoxicity	PDF*m2yr	**2909.298**
Acidification/ Eutrophication	PDF*m2yr	**3651.186**
Land use	PDF*m2yr	**1074.951**
Minerals	MJ surplus	**7646.472**
Fossil fuels	**MJ surplus**	72521.72

(Source: SimaPro 7.1.8 Data Processing)

Based on the table above, it can be seen that the lowest environmental impact on the production process at PT. XYZ is the Ozone layer and Resp. Organics are 4.71E-06 and 3.22E-05[17]. This is because of the production process at PT. XYZ does not damage the ozone layer and organic response. The Ozone layer has almost no impact. That is the use of materials in the production process by PT. XYZ does not damage the ozone layer[18].

Furthermore, it can be seen that the highest environmental impact on the production process at PT. XYZ is in Resp. Inorganics and climate change. This is due to the drilling carried out by PT. XYZ hurts the environment, especially in the climatic conditions of the soil and plants[19].

Furthermore, the environmental impact based on a single score can be seen in the table below.

Table 5. Environmental Impact Based on Single Score

Impact category	Unit	Total
Carcinogens	Pt	167.0443
Resp. organics	Pt	0.837839
Resp. inorganics	Pt	4465.243
Climate change	Pt	361.2335
Radiation	Pt	18.34125
Ozone layer	Pt	0.12263
Ecotoxicity	Pt	226.9252
Acidification/ Eutrophication	Pt	284.7925
Land use	Pt	83.8462
Minerals	Pt	181.986
Fossil fuels	Pt	1726.017

The table above shows that the highest impact lies in Resp. Inorganics, Fossil Fuels, and Climate Change[20]. This indicates that PT carried out the drilling. XYZ hurts the environment, especially on the climatic conditions of soil, plants, and fossil fuels that continue to decrease[21].

5. Conclusion

a. Based on the study results, it was found that the lowest environmental impact on the production process at PT. XYZ is the ozone layer and Resp. Organics are 4.71E-06 and 3.22E-05. This is because the production process at PT. XYZ does not damage the ozone layer and organic response. The ozone layer has almost no impact. That is the use of materials in the production process by PT. XYZ does not damage the ozone layer.

b. Based on the study results, it was found that the highest impact was on respiratory inorganics of 4,465,243, fossil fuels of 361,234, and climate change of 1,726,017. This shows that PT carried out the drilling. XYZ hurts the environment, especially on the climatic conditions of soil, plants, and fossil fuels that continue to decrease. In addition, drilling carried out using a generator can pollute the air to interfere with breathing (respiratory inorganics).

References

[1] "Gas alam," 2021. .

[2] B. Rugani, "Towards integrating the ecosystem services cascade framework within the Life Cycle Assessment (LCA) cause-effect methodology," Sci. Total Environ., vol. 690, pp. 1284–1298, 2019, doi: 10.1016/j.scitotenv.2019.07.023.

[3] D. Chisalita, "Assessing the environmental impact of an integrated steel mill with post-combustion CO<inf>2</inf> capture and storage using the LCA methodology," J. Clean. Prod., vol. 211, pp. 1015–1025, 2019, doi: 10.1016/j.jclepro.2018.11.256.

[4] I. Bastian, "Optimasi Nilai Gas Alam," 2014.

[5] Andriawan, R. Budiman, and D. Febriansyah, "Pemanfaatan Pengembangan Jaringan Gas Bumi Sebagai Pengganti Lpg Rumah Tangga Di Kabupaten Bojonegoro, Jawa Timur," J. ASIIMETRIK J. Ilm. Rekayasa Inov., vol. 2, no. 1, pp. 1–7, 2020, doi: 10.35814/asiimetrik.v2i1.1030.

[6] S. García-Pérez, "Environmental assessment at the urban level combining LCA-GIS methodologies: A case study of energy retrofits in the Barcelona metropolitan area," Build. Environ., vol. 134, pp. 191–204, 2018, doi: 10.1016/j.buildenv.2018.01.041.

[7] ESDM, "Annual Archivement of Development Report: Utilization of Natural Gas," Kementeri. Energi dan Sumber Daya Miner., p. 120, 2018.

[8] Pujadi and M. Yola, "Analisis Sustainability Packaging dengan Metode Life Cycle Assessment (LCA)," UIN Sultan Syarif Kasim Riau, vol. 1, pp. 1–127, 2013.

[9] W. D. N. Desinta Sawitri Giandadewi, Pertiwi Andarani, "Potensi Dampak Lingkungan Dalam Sistem Produksi Minyak Kelapa Sawit Mentah (Crude Palm Oli-CPO) Dengan Menggunakan Metode Life Cycle Assesment (Eco-Indicator 99) (Studi Kasus PT. Sinar Mas Agro Resources And Technology Tbk)," J. Tek. Lingkung., vol. 327, no. 6, pp. 1–10, 2017.

[10] M. Rosyidah, A. Masruri, and R. A. Putra, "Assessment (LCA) Method On Tofu Production," vol. 1, no. 4, pp. 428–435, 2020, doi: https://doi.org/10.46729/ijstm.v1i4.73.

[11] J. Laso, "Assessing energy and environmental efficiency of the Spanish agri-food system using the LCA/DEA methodology," Energies, vol. 11, no. 12, 2018, doi: 10.3390/en11123395.

[12] Sugiyono, Metode Penelitian Kuantitatif Kualitatif dan R&B. Bandung: CV. Alfabeta, 2012.

[13] L. J. Moleong, Metodologi Penelitian Kualitatif. Yogyakarta: Gadjah Mada University Press, 2007.

[14] S. Sugiyono, "Metode penelitian kuantitatif dan kualitatif dan R&D," Alf. Bandung, 2010.

[15] J. J. López-Andrés, "Environmental impact assessment of chicken meat production via an integrated methodology based on LCA, simulation and genetic algorithms," J. Clean. Prod., vol. 174, pp. 477–491, 2018, doi: 10.1016/j.jclepro.2017.10.307.

[16] P. Kerdlap, "M3-IS-LCA: A Methodology for Multi-level Life Cycle Environmental Performance Evaluation of Industrial Symbiosis Networks," Resour. Conserv. Recycl., vol. 161, 2020, doi: 10.1016/j.resconrec.2020.104963.

[17] C. Di Noi, "Can S-LCA methodology support responsible sourcing of raw materials in EU policy context?," Int. J. Life Cycle Assess., vol. 25, no. 2, pp. 332–349, 2020, doi: 10.1007/s11367-019-01678-8.

[18] M. Schulz, "Circular economy considerations in choices of LCA methodology: How to handle EV battery repurposing?," Procedia CIRP, vol. 90. pp. 182–186, 2020, doi: 10.1016/j.procir.2020.01.134.

[19] E. Campiglia, "Hemp seed production: Environmental impacts of Cannabis sativa L. Agronomic practices by life cycle assessment (LCA) and carbon footprint methodologies," Sustain., vol. 12, no. 16, 2020, doi: 10.3390/su12166570.

[20] F. Wu, "Assessing the environmental impact and payback of carbon nanotube supported CO<inf>2</inf> capture technologies using LCA methodology," J. Clean. Prod., vol. 270, 2020, doi: 10.1016/j.jclepro.2020.122465.

[21] D. E. G. Bizarro, "Potential carbon footprint reduction for reclaimed asphalt pavement innovations: Lca methodology, best available technology, and near-future reduction potential," Sustain., vol. 13, no. 3, pp. 1–20, 2021, doi: 10.3390/su13031382.

Community based ecotourism management to strengthen environmental ethics and supports sustainable development in Pelalawan district, Indonesia

Nawari*, Thamrin, Nofrizal, Almasdi Syahza, Juandi Muhammad, Nur Islami

Universitas Riau, Pekanbaru, Indonesia

*n4wari@gmail.com.

Abstract. Pelalawan District has several natural attractiveness and indigenous knowledge potentially to be developed as ecotourism. The community-based ecotourism management (CBEM) by utilize indigenous knowledge as a part of tourist attractions is one of ways to improve environmental ethics and achieve sustainable development goals (SDGs). This paper was prepared to examine the potential of indigenous CBEM in Pelalawan District. Secondary data was used compiled from relevant literatures of previous research or provided by authorized agency. The data analysis method carryout by comparative descriptive interpretative qualitative analysis. It identified the gaps between the existing condition of CBEM versus standards. A conceptual framework and CBEM strategy were arranged to eliminate current management gaps. The analysis result shows that current indigenous CBEM yet optimal. It shows by the lack of management dimension aspect, includes: tenurial license, local ecotourism norm, land use planning, marketing, supporting regulation, environmental education, management capacity, infrastructures development, and business incentives. The strategies to eliminates CBEM gaps, as follows: (1) Integrated planning; (2) Feasibility analysis of village own enterprise (BMUDes); (3) Capacity development of staffs and community; (4) Enhances marketing system; (5) Accountability of management; (6) Integrated environmental education curriculum; (7) Continuous improvement; and (8) Supporting regulation. The research result was recommended as one of references for policy makers to develop indigenous CBEM planning for strengthen environmental ethics and SDGs achievement as integrated manner. Authors recommend to future researchers would be expanded the scopes of the literatures and collect primary data for its accuracy improvements.

1. Introduction

Sustainable development is become priority to be implemented in Pelalawan District. Rapid economic growth still relies on the utilization of natural resources. World agroforestry explained that there are two mainland-based sectors have been driving the economy in this district which agriculture and industrial processing. The agricultural sector was dominated by rubber and palm oil plantations [1]. Extensification of plantations has caused forest fires [2], reduced biodiversity and social conflicts [3], and changes the rural socio-ecological behaviors [4]. Significant behavioral changes are starting to fade the local wisdom of community in managing natural resources and environment. High economic inequality will further diminish the values of community environmental awareness because indigenous peoples who hold the local wisdom values in the majority have lower economic stability than people who behave exploitatively. Handrian and Andry in their research results explain that the poverty rate in Pelalawan District reaches 11.15%, the second rank in Riau Province after Meranti Islands District [5].

On the other side, Pelalawan District has natural resources potential that could be developed as ecotourism destination. Several tourist destinations are Malay Cultural Sites, where the peoples still adhere to local wisdom in managing natural resources and the environment. Ecotourism development is one of ways to achieves sustainable development. It's can improve the local economic, create environmental awareness of the community and tourists and encourage the achievement of natural resource conservation [6].

The participation of local communities at all stages of management, both at the planning, implementation and evaluation will be increased the achievement of ecotourism sustainability [7]. Community participation accommodated in a community-based ecotourism management (CBEM) model, by aligning sustainable development goals on a bottom-up basis [8]. Community-based ecotourism packaging by emphasizing aspects of local wisdom in the natural resources management would increases the local community's income as well supporting the preservation of natural and cultural resources [9].

The development CBEM in Pelalawan District a way to preserve local community wisdom while increasing environmental awareness of local communities and tourists. Alania and Mzarelua in their research explained that the development of ecotourism has the potential to shape the ethics of environmental conservation by taking into account several aspects, namely: (1) Identification of tourism potential and markets; (2) Environmental suitability and carrying capacity analysis; (3) Strengthening local economy and culture; (4) Tourist services; and (5) An effective marketing strategy [10].

This paper was prepared to examine the potential of CBEM by optimizing natural resources and local wisdom as one of ecotourism attraction in Pelalawan District. Ecotourism management was expected would become a media to strengthening the community environmental ethics, which to achieved public environment awareness by applying aspects of natural resources conservation into sustainable development.

2. Methodology

Research was conducted by qualitative descriptive-interpretative [11] and [12]. Identify CBEM existing condition in Pelalawan District, Riau Province, Indonesia compared best practices or standard. The data used in this study is secondary data which collected from various relevant literatures based on the results of previous studies and authorized agencies. Gap analysis [13] was used to identify CBEM gap (Figure 1). The results of the comparative analysis are used as the basis for the preparation of the CBEM conceptual framework in Pelalawan District to eliminate management gap and ensure future sustainability achievement [14].

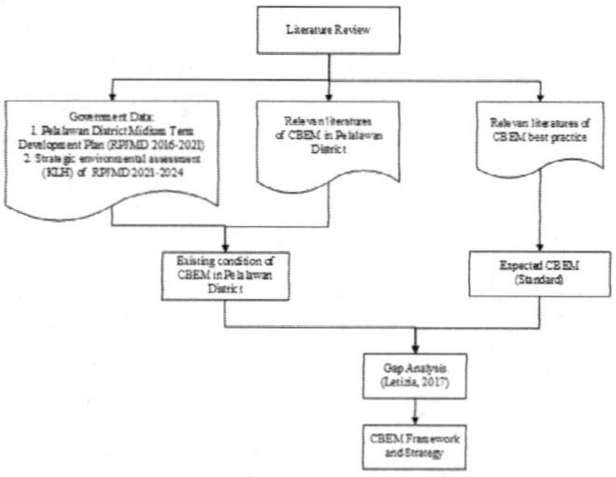

Figure 1. Research method

3. Result and Discussion

3.1 Existing condition of CBEM development in Pelalawan District

Several local wisdoms in natural resources management that can be packaged into objects of ecotourism attraction in Pelalawan District include: (1) Sialang Kepungan management in Betung Village, Pangkalan Kuras Sub-district; (2) Conservation of tree biodiversity in Tajwid Lake in Langgam Village,

Langgam Sub-district; and (3) Sialang Honey Management around Teso Nilo National Park in Lubuk Kembang Bunga Village, Ukui Sub-district.

In the three villages have natural attractiveness and are supported by the indigenous community. Sari et al in their research results explain that the development of an economic business based on local wisdom in environmentally friendly of Sialang Kepungan management has provided sufficient economic value and technical convenience for the community than oil palm plantations [15]. Some aspects of local wisdom in environmental management that can be used as ecotourism attractions include: (1) traditional fishing gear in fishery; (2) Trees maintenance; (3) Biodiversity conservation; (4) Organic agriculture, medicines and crafts [16]. The development of tourist attractions by utilizing indigenous knowledge potentially increase tourist interest as knowledge for environmental awareness than only for picnic [17].

Based on Pelalawan District Medium-Term Development Plan (RPJMD) 2016-2021 review, local government has setup tourism development as an alternative resource [18]. However, based on the strategic environmental assessment (KLHS) document of 2021-2024's RPJMD review shows that the future tourism development yet supported by specific strategy to optimize this industry [19]. This condition was caused the acceleration of ecotourism development did not optimal. Which indicated by the lack of management capacity both government staffs and the community. It worse by unsettled institutions, inadequate development of infrastructure, transportation, supporting facilities and weakness of current CBEM marketing [20]. The weak of management dimensions both in terms of natural resource management, human resources and financial resources can affect the sustainability of ecotourism management in this district.

It's shows by perceptions of tourists and the public who provide feedback on requirement to improve the infrastructures. The sufficient of tourism facilities and utilities potentially increase tourist attractions [21] and [22]. In addition of facility management aspect, there are three key factors should be considered in order to achieve sustainable ecotourism management, such as: the local workers employments, increasing environmental knowledge and the conditions of traders around tourist sites [23]. Management aspects that affect the satisfaction and frequency of tourist visits are tourism products, prices, location arrangements, promotion systems, community friendliness and transaction processes [24].

3.2 CBEM Best Practice in the Other Location as Standard

Several tourist destinations that have developed CBEM by utilize local wisdom in Indonesia as a reference for developing CBEM in Pelalawan District, include: (1) Tri Hita Karana-based ecotourism in Bali Province; and (2) Ecotourism based on local wisdom development in Lopati Village, D.I. Yogyakarta Province. Hakim et al stated that ecotourism management based on local wisdom has improved the image and authenticity of Bali Island and has become a significant tourism attraction [25].

Meanwhile, the development of ecotourism in Lopati Village, apart from being able to improve the economy, has also increased environmental awareness of the community and tourists. The key factors in the success of CBEM development in this area is community participation in all management cycle and building motivation for continuous improvement together [26].

Based on a review of indigenous CBEM in Asia, Africa and Latin America, there are several prerequisites for developing indigenous CBEM, includes: (1) A journey to remote areas that have natural attractions and customs; (2) Minimizing negative impacts on the environment and culture; (3) Building environmental awareness; (4) Provide sustainable economic value; (5) Improving the economy and empowerment of local communities; (6) Respect local culture; and (7) Supporting people's human rights. The key factors driving the sustainability indigenous CBEM, includes: (1) Certainty of land management; (2) financial support from various parties; and (3) Strengthening the network to the private tourism industry [27]. Sustainable CBEM has several dimensions, namely: cultural heritage, social, ecological, economic, marketing, policy, local knowledge, and spiritual/religious dimensions [28].

The efforts to increase environmental awareness through ecotourism packages development by collaborate with formal or informal educational institutions to organize environmental education at the destination area. Saat in his research explained that environmental education in ecotourism areas can increase the awareness of tourists towards natural resources and environment conservation [29]. Ecotourism packages to improve environmental ethics could be arranged with edu-ecotourism concept which provides various themes of environmental conservation for tourists and the community [30].

In the formal education curriculum, educational goals are directed at forming characters who have a set of environmental values for elementary to senior education [31]. In the learning process, the environment must be supported by field practicums that bring students closer to the real world [32]. The development of environmental education in formal institutions is determined by management commitment, participation of all components in schools to increase environmental awareness, as well as stakeholder support outside schools for environmental education enrichment programs [33].

One of the potential learning methods for environmental education is an inquiry learning model which guide the participants to think critically and care about the environment [34]. Environmental education programs that are integrated with ecotourism have a positive relationship in increasing public and tourist awareness to the environmental conservation [35].

3.3 Gap Analysis of Existing CBEM Compared the Standards

Based on the comparative's descriptive-interpretative qualitative analysis of CBEM management explained that the current indigenous CBEM in Pelalawan District is still not optimal compared to the expected conditions both in terms of aspects of tenure certainty, policies, planning, marketing, regulation, environmental education programs, staff development, infrastructure development and business incentives (Table 1).

It's caused the development of CBEM with local wisdom in this district was slow growth. Has not been able to provide economic improvement for local communities and significantly improve environmental ethics. This indication can be seen from the existence of environmentally unfriendly behaviors in the moderate category which has caused the degradation of the peatland ecosystem in Pelalawan District [36]. Environmental awareness behaviors that support sustainable development from the government and the community is reflected in prevention efforts against greater exploitative behaviors towards natural resources [37].

The development of local wisdom-based ecotourism businesses to increase environmental awareness is determined by several factors, includes: the capacity of local communities as managers, ecotourism market conditions, political stability, marketing technology, conservation financial support, job creation, the local community adaptation capabilities, cultural integrity and the ecological status of the conservation area. Implementing economic instruments which include payment environmental services (PES), price regulation and certification potentially increase the benefits of developing indigenous ecotourism [38].

Table 1. The gap analysis between existing CBEM condition versus standards

CBEM's determinant factors	Existing condition of CBEM in Pelalawan District	Expected condition (CBEM Standards)
1. The certainty of management unit area	Communities still has limited access to land. This is due to the lack of capacity, facilitation, and financial resources of the community to obtain CBEM permits	There are certainty of management unit area permits and thus provides flexibility for the community to develop sustainable ecotourism
2. Ecotourism development policy	The government program of tourism development has not been followed by a more specific achievement strategy and derivative regulations causing the direction of ecotourism development has not expressed	Local governments encourage the ecotourism development based on local wisdom as one of efforts to preserve and strengthen environmental ethics as well as to increase regional and national income
3. Land use planning	land suitability and environmental carrying capacity analysis is not yet available which causes ecotourism development planning not yet integrated	There are an analysis of the suitability and environmental carrying capacity for sustainable ecotourism development

CBEM's determinant factors	Existing condition of CBEM in Pelalawan District	Expected condition (CBEM Standards)
	with other sectors and with regional spatial plans	
4. Ecotourism marketing	Ecotourism marketing is yet optimal. Tourists come to destination locations are dominated by local people who know the existence of tourist sites personally	There is cooperation between managers, NGOs and market networks in ecotourism marketing has been able to increase tourist visits, both domestic and foreign tourists
5. The equity to access economic benefits of ecotourism	Ecotourism development has not been supported by local regulation to arrange the distribution of benefits from ecotourism services	There are regional regulations that support tourism in terms of lodging regulations and profit sharing of ecotourism services
6. Education and interpretation of nature for strengthening environmental ethics	Ecotourism activities have not been packaged as part of environmental education and still not yet able to increase environmental awareness behavior both to local communities and to tourists	Environmental education and interpretation are part of the ecotourism package so as to increase environmental awareness both in local communities and tourists
7. Staff's development	Staff development is still not optimal, indicates by weak capacity and commitment to sustainable ecotourism development	There is regular staff development that is able to increase the capacity and commitment of ecotourism managers in supporting sustainable development
8. Ecotourism infrastructures development	Infrastructure development has not been optimal, which is characterized by inadequate road conditions for transportation, and the lack of facilities and utilities to support ecotourism activities	Adequate ecotourism facilities and utilities development to support all ecotourism activities
9. Financial management and business incentives	Business incentives are not optimal, where the current tourism development with local wisdom is an initiative of the local community, some of which are the older generation and have not been packaged into an ecotourism	Business strengthening through payment environmental service's (PES) program, pricing and certification. Business incentives for the success of ecotourism management are carried out for green investments and increase future ecotourism development

3.4 Conceptual Framework and Strategy of CBEM Development on Pelalawan District

Based review of existing conditions of CBEM in Pelalawan District compared with CBEM best practice in other locations and considering its success story, a conceptual framework for CBEM management for improving environmental ethics is drawn up as shown in Figure 2. The ecotourism management dimension is carried out by managing natural resources focusing on the use of ecosystem services for ecotourism development.

Natural resource management is supported by strengthening the capacity of managers with good leadership skills oriented to active the community participation, collaborative institutions and getting support from local governments and private companies. The support could be a policy, capacity building facilitation, provision of investment funds and CSR. Both aspects must have reliable financial resource management in a village-owned enterprise (BUMDes) that is able to create accountability and transparency in the budgeting process, profit sharing, and the provision of social safety funds.

In the financial aspect, management must create an effective marketing system and implement LEAN in order to be able to maintain positive cash flow at an adequate level of liquidity. This is in line with Kampetch and Jitpakdee whom state that aspects of leadership management, community participation, and strengthening of ecotourism networks increases the success of CBEM [39].

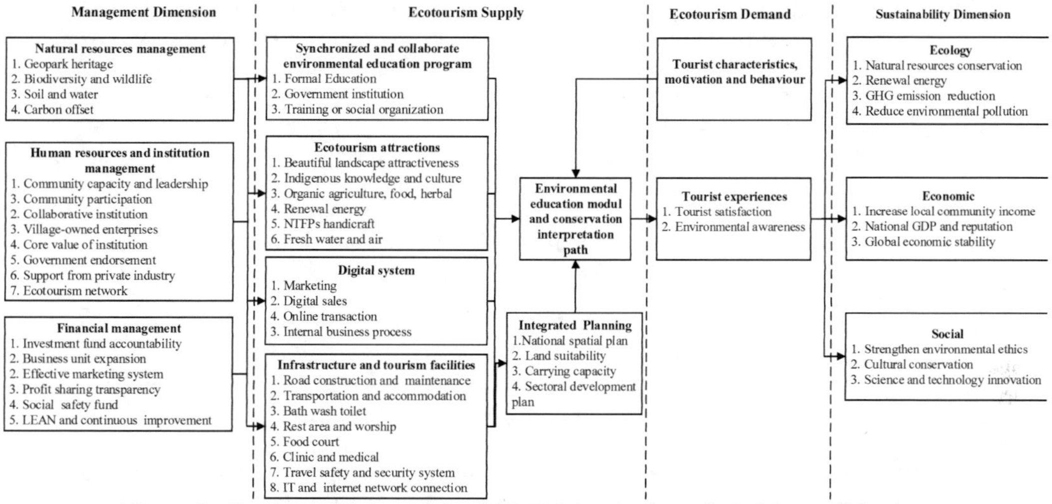

Figure 2. Conceptual framework of CBEM development in Pelalawan District

The reliable management dimension will be able to increases ecotourism supply that are competitive, which being able to provide good attractions, infrastructure, facilities and utilities, digital systems and environmental education modules. Which are prepared based on integrated planning and capture market opportunities, based on the pattern recognition of tourist characteristics, motivation and its behavior. So that a tour package was integrated with environmental education will be able to increase tourists' stultification and their interest in environment.

In the end of the results or sustainability dimension of ecotourism development will be realized, which being able to achieve environmental conservation, economic improvement, cultural preservation and improvement of environmental ethics accompanied by the development of science and technology. Bhuiyan et al based on their research results explained that the basic principle of ecotourism development is to achieve the sustainability of economic, ecological and social functions, also increases environmental awareness [40]. This is in line with the main goal of establishing environmental ethical principles, namely minimizing damage [41]

To achieve the Indigenous CBEM model which available to increase environmental ethics and contribute to sustainable development, the following management strategies can be carried out:

a. Develop an integrated plan of indigenous CBEM by taking into account aspects of suitability, environmental carrying capacity and integrated economic valuation of cross-sectoral and structural development (village, district, provincial and national governments development program) as well as the application of collaborative management accompanied to strengthening of collaborative institutions. This can be integrated with village development program which give chances to develop village-owned enterprises (BUMD) as a business unit in the development of ecotourism villages.
b. Using the integrated planning as the basis for the feasibility of developing an indigenous CBEM business in Pelalawan District which involves the participation of the community and policy makers to obtain an allocation of investment funds that will be used for ecotourism facilities and utilities construction to increase the attractiveness and competitiveness of ecotourism products
c. Improve the staff and communities' capabilities for indigenous CBEM innovation with engage universities, research institutions and private companies in Riau Province

d. Implementing a digital-based ecotourism product marketing system by considering ecotourism demand trends
e. Improving the accountability system for indigenous CBEM as integrated program with village development and building a reputation for the ecotourism market
f. Establish an environmental education curriculum in the development of ecotourism based on local wisdom that is integrated with formal education (elementary to higher education, and government or professional staff training). To improve the environmental ethics of the wider community in Pelalawan District. Curriculum development is integrated system educational program. Accompanied best practices information on its implementation in the field.
g. Implement continuous improvement by developing ecotourism development technology that is able to contribute on achievement of green society and sustainable development.
h. Local governments would establish the supporting regulations of indigenous CBEM for strengthening environmental ethics accompanied by law enforcement to increase the spirit of environmentally sound development

4. Conclusion

Based on the analysis of the existing condition indigenous CBEM for strengthening environmental ethics in Pelalawan District was yet optimal compared with standards. It indicates by the weak of management aspects, includes tenure license, indigenous policy, land use planning, marketing, supporting regulation, education and environmental interpretation program, staff and community capacity, infrastructures development and business incentives. Due to the weak aspects of ecotourism management, the development Indigenous CBEM yet able to become one of the alternative development resources that can improve the local economy as well as improve environmental ethics in Pelalawan District.

The strategies to eliminate the gap as follows: (1) Formulating an integrated plan taking into account natural resources supply and ecotourism market demand; (2) Analyzing the feasibility of developing an ecotourism village business entity; (3) Capacity building of management staff and community; (4) Enhances marketing system; (5) Improved management accountability system; (6) Implementation of educational curriculum and environmental interpretation programs; (7) Continuous improvement; and (8) Establish supporting regulation.

The research result is expected to be one of references in the indigenous CBEM development and as an effort to improve environmental ethics for the community in supporting sustainable development in Pelalawan District and the other location. The enhancement of this research was opened, especially by add more literatures and primary data which not covered by this research

References

[1] Worldagroforestry. 2019. Land Use Plan for Pelalawan: Targeted Scenario Analysis Results. South East Asia Regional Office. Bogor. https://goodgrowthpartnership.com/wp-content/uploads/FINAL-Deliverable2-Sustainable-Land-Use-Scenario-of-Pelalawan_TSA-Results-1.pdf. Accessed at March 4th, 2021.
[2] Tata, H.L., Narendra, B.H & Mawazin. 2018. Forest and Land Fires In Pelalawan District, Riau, Indonesia: Drivers, Pressures, Impacts and Responses. Biodiversitas, Volume 19(2), 494-501. DOI: https://doi.org/10.13057/biodiv/d190224.
[3] Suryadi, Dharmawan, A.H, & Barus, B. 2020. Ekspansi Perkebunan Kelapa Sawit : Persoalan Sosial, Ekonomi dan Lingkungan Hidup (Studi Kasus Kab. Pelalawan, Riau). Jurnal Ilmu Lingkungan, Volume 18(2), 367-374.
[4] Hidayah, N., Dharmawan, A.H., & Barus, B. 2020. The Expansion of Palm Oil Plantation and Changes of Rural Social Ecology. Jurnal Sosiologi Pedesaan, Volume 4(3), 249-256. DOI: https://doi.org/10.22500/sodality.v4i3.14434
[5] Handrian, E., & Andry, H. 2020. Sustainable Development Goals: Tinjauan Percepatan Pencapaian di Provinsi Riau. PUBLIKA: Jurnal Ilmu Administrasi Publik, Volume 6(1), 77-87.

[6] Nawari, Isjoni, & Zulkarnaini. 2021. Environmental suitability and carry capacity for Ecotourism Development in the North Mareje Mountain Forest Area of Central Lombok. Dinamika Lingkungan Indonesia, Volume 8(1), 17-28. DOI: https://doi.org/10.31258/dli.8.1.p.17-28

[7] Giriwati, O.S.S., Hawa, L.C., Pamungkas, S.T., Iyati, W., &. Hidayat, A.R.T. 2019. Local Community Participation in Ecotourism Development: The Case of Sumberwangi Hamlet Destination, East Java, Indonesia. People: International Journal of Social Sciences, 5(1), 81-98. DOI: https://dx.doi.org/10.20319/pijss.2019.51.8198

[8] Müllera S., Hucka, L. , & Markovab, J. 2020. Sustainable Community-Based Tourism in Cambodia and Tourists' Willingness to Pay. Austrian Journal of South-East Asian Studies, Volume 13(1), 81-101. DOI: https://doi.org /10.4324/9780429264191-12

[9] Mendoza, A., Ramos & Prideaux, B. 2017. Assessing ecotourism in an Indigenous community: using, testing and proving the wheel of empowerment framework as a measurement tool. Journal of Sustainable Tourism, 277-291. DOI: http://dx.doi.org/10.1080/09669582.2017.134717

[10] Alania, A., & Mzarelua, L. 2020. Pr Activities As An Effective Way to Promote Ecotourism. Ovateur Publications Journalnx-A Multidisciplinary Peer Reviewed Journal Volume 6(10), 253-258.

[11] Marjan, M. 2017. A Comparative Analysis of Two Qualitative Methods: Deciding Between Grounded Theory and Phenomenology for Your Research. Vocational Training: Research and Realities, Volume 28 (1), 23-40. DOI: https://doi.org/10.2478/vtrr-2018-0003

[12] Kim, H., Sefcik,J.S., & Bradway, C.2017. Characteristics of Qualitative Descriptive Studies: A Systematic Review. es Nurs Health, Volume 40(1), 23–42. DOI: https://dx.doi.org/10.1002/nur.21768

[13] Letizia, A.J. 2017. Using Strategic Planning to Create the Public Good for Higher Education in Volatile Times. International Journal of Progressive Education, Volume 13 Number 2. 144-164

[14] Adom, D., Hussein, E.K, & Agyem, J.A. 2018. Theoretical and Conceptual Framework: Mandatory Ingredients of a Quality Research. International Journal of Scientific Research, Volume 7(1), 438-441.

[15] Sari, G., Yoza, D & Mardiansyah. (2014). Pola Pengelolaan Pohon Sialang Oleh Masyarakat Desa Betung Kecamatan Pangkalan Kuras Kabupaten Pelalawan. Jom Faperta Vol.1 No. 2 Oktober 2014.

[16] Sutrisno, E. & Wahyudi, A. (2015). Keragaman hayati dan pola pemanfaatan Danau Tajwid di Kabupaten Pelalawan, Riau. Pros Sem Nas Masy Biodiv Indon Volume 1(3), Nomor 3, 635-641, DOI: https://doi.org/10.13057/psnmbi/m010342

[17] Widyaningsih, D. (2014). Atraksi Wisata Menumbai Di Taman Nasional Tesso Nilo Desa Lubuk Kembang Bunga Kecamatan Ukui Kabupaten Pelalawan. Jom FISIP Volume 2 (2), 1-15.

[18] Bappeda Pelalawan. 2021. Rencana Pembangunan Jangka Menengah Daerah Kabupaten Pelalawan Tahun 2016-2021. Kantor Badan Perencana Pembangunan Daerah Kabupaten Pelalawan. Pangkalan Kerinci. https://bappeda.pelalawankab.go.id/download-dokumen/dokumen-perencanaan/rpjmd-20162021/. Accessed at March 6th, 2021.

[19] Bappeda Pelalawan. 2021. Kajian Lingkungan Hidup Strategis (KLHS) Rencana Pembangunan Jangka Menengah Daerah Kabupaten Pelalawan Tahun 2021-2024. Kantor Badan Perencana Pembangunan Daerah Kabupaten Pelalawan. Pangkalan Kerinci. https://bappeda.pelalawankab.go.id/download-dokumen/dokumen perencanaan/kajian-lingkungan-hidup-strategis-klhs-rpjmd-pelalawan-tahun-20212024/. Accessed at March 7th, 2021.

[20] Syahputra. 2014. Peran Pemerintah Daerah dalam Pengembangan Obyek Wisata Istana Sayap di Kabupaten Pelalawan tahun 2009 - 2011. Jom Fisip, Volume 1(2), 1-15.

[21] Purwanto, Sadjati, E., & Insusanty, E. 2019. Visitor's Perception Of Tourist Attraction Danau Tajwid Kecamatan Langgam Kabupaten Pelalawan. Wahana Forestra: Jurnal Kehutanan, Volume 14(1), 46-62. DOI: https://doi.org/10.31849/forestra

[22] Melani, N.P., Maemunaty, T., & Jas, J. 2021.Persepsi Masyarakat terhadap Pengembangan

Objek Wisata Taman Kreatif Publik Kecamatan Pangkalan Kerinci Kabupaten Pelalawan Provinsi Riau. Jom Fkip – Ur Volume 8 (2), 1-8.
[23] Thamrin, T., Putra, R.M., Nofrizal, N., & Karnila, R. 2020. Sustainable Management of Guruh Gemurai Waterfall Ecotourism in Kuantan Singingi. ECOTONE, Volume 1 (2), 93-104. DOI: http:// doi.org/10.31258/ecotone.1.2.p.93-105
[24] Febrina, L.K., Kartikowati, S., & Jahrizal3. 2021. Effect of marketing mix on tourist satisfaction and loyalty in tourist area of bono Teluk Meranti Pelalawan Regency. Jurnal Manajemen dan Bisnis,Volume 10(1), 27- 35. DOI: https://doi.org/10.34006/jmbi.v10i1.254
[25] Hakim, L., Kim, J.E & Hong, S.K. 2009. Cultural Landscape and Ecotourism in Bali Island. Journal of Ecology and Environment 32(1), 1-8. DOI: https://doi.org/10.5141/JEFB.2009.32.1.001
[26] Vitasurya, V.R. (2016). Local Wisdom for Sustainable Development of Rural Tourism, Case on Kalibiru and Lopati Village, Province of Daerah Istimewa Yogyakarta. Procedia Social and Behavioral Sciences Volume 216, 97 – 108. DOI: https://doi.org/10.1016/j.sbspro.2015.12.014
[27] Zappel, H.D. 2006. Indigenous Ecotourism Sustainable Development and Management. CABI. Wallingford.
[28] Theingthae, S. 2017. Sustainability of Community Based Ecotourism Development After The Impact of Tsunami Disasters: Comparison Between Buddhism Community and Muslim Community In Puket Province, Thailand. Journal Of Tourism Research and Hospitality. 6(4), 100-175. DOI: https://doi.org/10.4172/2324-8807.1000175
[29] Saat, M.H.M., Thamrin, Sudjianto, & Yoswati, D. 2015. Ecotourism Development: Educational Media of Environmental Care. Al-Ta'lim Journal, Volume 22(2),129-141. DOI: http://dx.doi.org/10.15548/jt.v22i2.131
[30] Sutisno, A.N., & Afendi, A.H. 2017. The Implementation of Edu-Ecotourism Concept As A Media Of Character Education Environment Based. Ecolab, Volume 12(1), 1 – 52.
[31] Prihantoro, C.R. 2015. The perspective of curriculum in Indonesia on environmental education. International Journal of research studies in Education, Volume 4(1), 77-83.
[32] Dewi, N.L.P.R, Suastra, I.W., & Pujani, N.M. 2018. Pengembangan Modul Praktikum IPA SMP Kontekstual pada Materi Pencemaran Lingkungan untuk Meningkatkan Keterampilan Proses Sains dan Karakter Peduli Lingkungan. IVCEJ, Volume 1(2), 57-67.
[33] Nurhafni, Syahza, A., Auzar, Nofrizal. 2019. The Strategy of Environmental School through the Program of National Adiwiyata School in Pekanbaru (High School Level). Interdisciplinary Journal of Environmental and Science Education, Volume 15(1), e02204. DOI: https://doi.org/10.29333/ijese/6289
[34] Azriani, N., Islami, N., Hermita, N., Nor, M., Syaodih, E., Handayani, H., Zulirfan, Z., Suhandi, A., Malik, A., Mahbubah, K., & Samsudin, A. 2019. Implementing Inquiry Learning Model to Improve Primary School Students Critical Thinking on Earth and Universe Concept. IOP Conf. Series: Journal of Physics: Conf. Series 1227 (2019) 012033. DOI: https://doi.org/10.1088/1742-6596/1227/1/012033
[35] Safitri, D., Yunaz, H., Umasih, Marini, A., & Wahyudi, A. 2020. Effect of Environmental Education on Ecotourism: Evidence from Jakarta. International Journal of Control and Automation, Volume 12(4), 37-48.
[36] Syaufina, L., & Hamzah, A.A. 2021.Changes of tree species diversity in peatland impacted by moderate fire severity at Teluk Meranti, Pelalawan, Riau Province, Indonesia. Biodiversitas, Volume 22(5), 2899-2908. DOI: https://doi.org/10.13057/biodiv/d220555
[37] Juandi, M. 2016. Sustainability Model for Unconfined Aquifers. International Journal of Science and Applied Technology, IJSAT, Volume 1(1), 8-14.
[38] Coriaa, J. & Calfucura, E. 2021. Ecotourism and the Development of Indigenous Communities: The Good, the Bad, and the Ugly. Ecological Economics 73, 47-55. DOI: https://doi.org/10.1016/j.ecolecon.2011.10.024
[39] Kampetch, P., and Jitpakdee, R. 2019. The Potential for Key Success Of Community-Based

Tourism Sustainability: Case Study Baan Rim Klong Homestay, Samut Songkram, Thailand. BAC Journal, Volume 39(4), 111-122.

[40] Bhuiyan M.A.H., Siwar, C., Ismail, S.M., Islam R. 2011. The role of home stay for ecotourism development in East Coast Economic Region. Am J Appl Sci, Volume 8(6), 540–546. DOI: https://doi.org/10.3844/ajassp.2011.540.546.

[41] Aciksoz, S., Bollukcu, P., & Celik, D. 2016. Ecotourism and Ethics In Protected Areas: Bartin-Sogutlu Village. Oxidation Communication, Volume 39(4), 3621–3636.

Community forestry for environmental sustainability and ecotourism: the context and problems in Indonesia

R Sulistiowati[1], S Wahyuni[2], M K Yunanto[3], M Elsera[2], W E Yudiatmaja[2*], T Samnuzulsari[2]

[1]Universitas Lampung, Bandar Lampung, Indonesia
[2]Universitas Maritim Raja Ali Haji, Kepulauan Riau, Indonesia
[3]Sekolah Tinggi Ilmu Administrasi "AAN", Yogyakarta, Indonesia

*wayuguci@umrah.ac.id

Abstract. Social forestry is one of the programs delivered by the Indonesian government to distribute forest ownership to society. The program consisted of village forest, customary forest, community forest, a people forest plantation, and partnership for forestry. This study merely focused on the community forest in Indonesia. The purpose of this research is to analyze the context and problems during the implementation process of permit delivery for community forests in Gunungkidul, Yogyakarta, Indonesia. A qualitative method was used in this research to understand the case of community forests in the study site. Data were collected using an interview, non-participant observation, and secondary document. We find that many groups of farmers have not received the permit for forestry management because of regulatory constraints. The problems are induced by many policies in the forestry sector regulating various forms of forest management permits contradicting the principle of the decentralized forest management policy. We recommend to the government to simplify the process of forest permit management, especially for the community forest.

1. Introduction

Environmental issues are interesting topics frequently discussed along with the development process. It occurs because most of the development process affects the natural and social environment [1]. The influence on the social environment can be acknowledged by an increase in the socio-economic conditions of the community, while the influence of development on the natural environment can be indicated by the utilization of the results of resources, both energy, water, air, land, or forests that contribute to the implementation of the development process. One of the developments intensively carried out by all countries in the world, including in Indonesia, is development in the forestry sector [2].

Forest development is not only relied on to trees but also the people. Trees are important as long as their existence is beneficial to the public interest because the community participates in utilizing their production. This is mainly because forests are one of the resources that support the life of human life to prevent damage to natural ecosystems. Indonesia still had the original forest condition until the late 1960s with an area of 143 million hectares of the forest [3]. The forest still exists because the community views the forest as a home, being part of their culture and life. There is an important message that destroying the forest means destroying life [4].

However, forest exploitation activities, especially state forests, began to flare up, precisely during the New Order administration. At that time forest management was controlled by the central government through private parties provided the authority in the form of forest concession rights for forest areas outside Java and appointed Perum Perhutani to manage forest areas in the form of forest concession rights in Java Island [5].

In practice, the forest concession rights regularly abuse their authority to enrich themselves or for other illegal practices endangering forestry conditions in Indonesia. Not to mention the addition of centralized nature of government in the use of forest products, which makes people around the forest unable to enjoy one of the natural wealth of forest products. In addition, timber concessions carried out by the government through forest concession rights tend to be timber extraction, or it can be said to use timber forest products excessively through illegal logging, smuggling, logging, and theft of timber triggering deforestation.

Yogyakarta province has the characteristics of forest management which is slightly different from other regions because the status of Yogyakarta's land is the place belonging to the Sultanate and Pakualaman ground [6]. An area that has a forest area of 14,895.5000 ha out of a total area of 1,485.36 km^2 is not necessarily free from the occurrence of forest area damage. It was recorded that more than 50% of the damage occurred in state forests in 1997 [7].

From the management of the existing forest land area, it is state forests that are often damaged. Along with the increasing forest destruction, the government is increasingly aggressive by starting to issue and promote a series of programs related to community-based rehabilitation of state forest areas. The issuance of a series of programs is none other than a form of policy in the era of decentralization, namely Law No. 22 of 1999 concerning regional autonomy in the forestry sector. From many programs created, it is the community forest program that has its unique characteristics because community forestry has been born since 1995.

In fact, there are obstacles in the implementation of community forestry, especially related to the issue of the permit to utilize the forest for the society which has an impact on the not yet issued community forestry permit. Departing from the emergence of various problems and exposure to facts in the form of data analysis from the results of the initial review that has been carried out, the researchers are interested in discussing the problems of implementing a community forest permit. Considering that the permit is the actualization of the implementation of community forestry policies supporting the implementation of forestry decentralization by prioritizing community participation in the implementation process, as well as increasing the efficiency of the central government. It can be achieved through sharing of authority to the regions to manage policies related to the implementation of forestry decentralization.

This research will focus on the problematization of the implementation of community forest permits in Gunungkidul Regency, Yogyakarta because by focusing this research on this, it provides a deeper understanding of the main problems surrounding the community forest permit managed by several farmer groups in Gunungkidul. The problems of this research can be formulated as follow: How is the process of implementing community forest permit and what is the problems in community forest management for environmental sustainability and ecotourism? This study aims to understand the process and mechanism of the implementation of community forestry and the ultimate problems in community forest governance in Gunungkidul.

2. Methodology

To answer the research problem comprehensively, we applied the qualitative research method. As explained by Miles and Huberman qualitative research methods are defined as an investigative process in which researchers slowly interpret a social phenomenon by identifying the object of research [8]. They also explained that the application of qualitative methods in research aims to understand certain situations, events, groups, or social interactions. By using this research method, it is hoped that we can precisely investigate the problems in the process of implementing community forest permits through various data collection systems, to obtain comprehensive study results from the results of data interpretation and analysis.

Data in this research were collected by using several methods. First, we made observations to see the natural conditions of a community experience in utilizing forest for community interests in Gunungkidul. The observation was conducted to obtain et an overview and find out the extent of the process of community forestry permit. Second, we interviewed numerous informants using unstructured interview techniques. It was used so that informants feel free to answer the questions asked and can further develop their opinions according to their capacity without any limitation on answers reflected in a structured interview.

Several parties and figures interviewed were the Watershed Management Development Center, the Forestry and Plantation Service of Yogyakarta Province, several farmer groups of community forest, non-government organizations (NGOs), and academics concerning forestry issues [9]. Third, we performed documentation techniques by collecting official documents from parties related to regulations and the implementation process of community forestry carried out by the government, community, or other related institutions.

After completing the data, the next step was to analyze the results of these results to obtain valid and credible findings. Data analysis took place along with the process of data collection. In other words, this process indirectly took place during the initial review process previously carried out by researchers through the results of observations, interviews, and documentation. The data analysis was conducted in several steps, namely processing and preparing data for analysis, reading all the data obtained from various data sources, and interpreting the data [10].

3. Result and Discussion

3.1 Current situation of forest area and community forestry in Gunungkidul

The development of community forestry has long been initiated in Gunungkidul. At a peak in 2005, the government of Gunungkidul and related parties formed a working group for sustainable community forest to realize sustainable community forest management encouraging sustainable community forest certification. The purpose of this cooperation was to increase community and regional income. The development of community forests continues until this day.

It is noted that Gunungkidul has the largest forest area, reaching 87% or 14,895,5000 ha of 18,715,0640 Ha of the total forest area spread in Yogyakarta. Forests in Gunungkidul are divided into several functions, namely production forest covering an area of 12,810,1000 Ha, protected forest covering an area of 1,016,7000 Ha, and conservation forest covering an area of 1,068,7000 Ha as shown in the following table.

Table 1. Number of Forest Based on Its Function in Gunung Kidul

Function	Amount (Ha)
Production Forest	12,810
Protected Forest	1,016
Conservation Forest	1,068

Source: Processed from [11]

One of the forest functions in Gunungkidul is production forest. Production forest located in Gunungkidul is under the management system of the Forestry and Plantation Service of Yogyakarta Provincial Government, with teak and eucalyptus as the main commodities. Production forest can be classified into AB (*Afkiren Boschs*) forest and non-AB forest. AB forest is a forest area that is very small and separated. In the previous period, AB forest was slightly neglected due to inefficient operational costs for its management.

Even though AB forest has status as state land, its management is delegated to the community through the village government with the institutional concept of people's plantation forest. In people's

IOP Conf. Series: Earth and Environmental Science 1041 (2022) 012037 doi:10.1088/1755-1315/1041/1/012037

plantation forests, the community can work to manage their forest-on-forest land with assistance from the Forestry and Plantation Service.

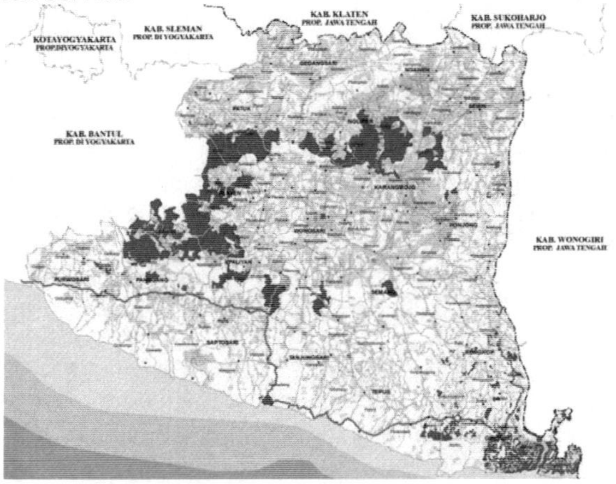

Figure 1. Forest Area in Gunungkidul

Meanwhile, a non-AB forest is a forest area that has a large area and is not separated. The non-AB forest area is grouped into many plots with a forest chairman in charge of several forest plots. The forest chairman leads several forest foremen. There are three forest foremen with different tasks, including forest security foreman, forest harvest foreman, and forest planting foreman. The forest chairman is directly responsible to the head of the forest area. In Gunung Kidul Regency, there are four forest areas, namely Paliyan covering an area of 3,872,500 Ha, Playen covering 3,711,000 Ha, Panggang covering 2,143,600 Ha, Karangmojo covering 3,324,900 Ha. Several non-AB forest lands in Gunungkidul are managed using the community forest model [12].

Community forestry is a state forest whose main use is intended to empower local communities. To facilitate the management of community forestry, the government has divided four forest areas spread over four sub-districts in Gunungkidul. From around 13,052.0 Ha of state forest area, only a small portion of an area of 1,089.40 Ha of land has been declared and granted a permit for forest management with a community forestry model [13].

3.2 The problems of community forestry for sustainability and ecotourism in Gunungkidul

In general, the process of community forestry has anxiety at several farmer groups because the profit-sharing system that had been determined was not commensurate with the results of their hard work so far. The profit-sharing system following the governor regulation arranges that the proportion of profit-sharing from the use of forest and non-timber forest products is 60:40 percent model, after deducting 10 percent of forest tax. The proportion of 60 percent is for farmer groups working on community forestry land, and 40 percent is for the Yogyakarta Provincial Government.

Although the community gains 60 percent of the results from the use of forest utilization, the results seem to be far from the expectations of the people of several farmer groups [14]. The lack of understanding of the majority of the community regarding the pattern of community forestry in forest development is the primary of the community's problems. The community has an understanding that the forest has existed since they were there, so they understand that the forest is theirs.

The community forestry pattern after the regional autonomy law is considered as one of the government's efforts to invite the community to jointly develop and maintain forest sustainability. The government considers this policy to be just a pattern used by the government to be able to maintain forest sustainability, without handing over a full forest land ownership system to the community [15]. In the name of decentralized forest management, the government has succeeded in producing various forest development programs, one of which is community forestry.

The community is provided a contract for a certain period to be able to work in the forest area under applicable regulations. Indirectly, the community is only mobilized by the government to maintain forest sustainability without adequate rewards. This can be seen from the non-issue of the permit of community forest utilization at several farmer groups of community forestry in Gunungkidul. The community forestry permit is the final legality required by the community to take forest products in the form of wood, environmental sustainability, and ecotourism.

On the other hand, the district government is only authorized to provide counseling regarding the technical implementation of community forestry, program socialization, distribution of teak seeds, and general monitoring [16]. Limited funds and human resources are the main considerations in this regard. Local governments consider that the central government only gives them an obligation without accompanying them with adequate resources to carry out these obligations. This condition further complicates the implementation of community forestry in Gunungkidul

4. Conclusion

The problems of implementation of community forest permit yielding in the never-ending issuance community forestry. In other words, the process can be simplified as a half-hearted decentralization that the government has rolled out to the community. It occurs because there are various irregularities in the community forestry program marked by three things. First, there are many sectoral policy points in the forestry sector regulating various forms of forest management permits contradicted with the essence of the decentralized forest management policy.

It reflects in the community forestry program considered as an effort to decentralize forest management. Second, the regulations governing the implementation of community forestry are periodically changed, resulting in confusion between the local government and the community in its implementation. Third, the lack of harmony in the relationship existing within the organizational structure of the ministry of forestry between the river management agency and forest business management agency.

References

[1] Yudiatmaja W E et al., 2021 Reducing solid waste through waste banks: an empirical study in Kepulauan Riau, Indonesia IOP Conf. Ser. Earth Environ. Sci. 755, 1 p. 012076.

[2] Nogueira C, 2019 Contradictions in the concept of sustainable development: An analysis in social, economic, and political contexts Environ. Dev. 30 p. 129–135.

[3] Dai L et al., 2013 The progress and challenges in sustainable forestry development in China Int. J. Sustain. Dev. World Ecol. 20, 5 p. 394–403.

[4] Akbar D Mariani M Yudiatmaja W E and Edison E, Jul. 2021 Governance of mangrove restoration and conservation to climate change resilience in Bintan Island IOP Conf. Ser. Earth Environ. Sci. 824, 1 p. 012048.

[5] van Noordwijk M Agus F Dewi S and Purnomo H, Sep. 2014 Reducing emissions from land use in Indonesia: Motivation, policy instruments and expected funding streams Mitig. Adapt. Strateg. Glob. Chang. 19, 6 p. 677–692.

[6] Carey P, Feb. 1986 Waiting for the "just king": The agrarian world of south-central Java from Giyanti (1755) to the Java war (1825–30) Mod. Asian Stud. 20, 1 p. 59–137.

[7] Sandholz S, 2017 Urban centres in Asia and Latin America: Heritage and identities in changing urban landscapes Springer.

[8] Miles M B Huberman A M and Saldana J, 2014 Qualitative data analysis: A methods sourcebook 3rd Ed. Los Angeles: Sage.

[9] Samnuzulsari T et al., 2021 The strategic role of civil society organisations in handling climate change: A case of Riau in Indonesia IOP Conf. Ser. Earth Environ. Sci. 824, 1 p. 012104.

[10] Creswell J W, 2007 Qualitative inquiry and research design: Choosing among five approaches 2nd Ed. Thousand Oaks, CA: Sage.

[11] Environment and Forestry Office, 2019 Buku statistik kehutanan [Forestry statistics book]

Yogyakarta: Provincial Government of Yogyakarta.
[12] Sadono R Pujiono E and Lestari L, Jul. 2020 Land cover changes and carbon storage before and after community forestry program in Bleberan village, Gunungkidul, Indonesia, 1999–2018 Forest Sci. Technol. 16, 3 p. 134–144.
[13] Ota M, 2011 Implementation of the Community Forest (Hutan Kemasyarakatan) scheme and its effects on rural households in Gunungkidul district, Java, Indonesia: an exploration of the local agrarian context Tropics 19, 3 p. 123–133.
[14] Maryudi A et al., Oct. 2015 Complex regulatory frameworks governing private smallholder tree plantations in Gunungkidul District, Indonesia For. Policy Econ. 59 p. 1–6.
[15] Yudithia Y Edison E Kristanti D Samnuzulsari T Suyito S and Yudiatmaja W E, 2020 Evaluating the Impact of Technological Adoption Policy for Rural Coastal Communities E3S Web Conf. 202 p. 1–7.
[16] Yudiatmaja W E et al., 2021 Social policy on the rural coastal communities: Why the implementation fails? E3S Web Conf. 232 p. 1–7.

Study on the commitment of oil palm companies to achieve sustainable agriculture in Riau Province from the perspective of pesticide use

S Syafrani[1], I Purnama[1,4], A Mutamima[2,4], W N Dewi[3]

[1]Universitas Lancang Kuning, Pekanbaru, Indonesia
[2]Universitas Riau, Pekanbaru, Indonesia
[3]Monash University, Victoria, Australia
[4]Research Center of Sustainable Indonesia, Pekanbaru, Indonesia

*indra.purnama@unilak.ac.id

Abstract. Oil palm has become the main plantation commodity in Riau Province, Indonesia. As the province with the largest oil palm plantation at around 2.7 million hectares, Riau Province has 19% of Indonesia's total oil palm plantation area. In the midst of environmental issues such as forest fires or using hazardous chemicals, companies or individuals who own oil palm plantations are trying to obtain sustainability certification from the Roundtable of Sustainable Palm Oil (RSPO). However, many companies have not obeyed the correct principles and criteria for sustainable palm oil production, especially in the perspective of using pesticides in oil palm plantations. Based on those problems, this study was to carry out an in-depth examination of the data available from the Plantation Office and Environment-Forestry Office of Riau Province. As a result, we discovered that 70% of the pesticides used were classified as hazardous pesticides, according to the World Health Organization (WHO). In addition, it was also found that oil palm companies used glyphosate and paraquat as the active ingredient in pesticides, which have been banned in many countries due to environmental issues. By this finding, the commitment of oil palm companies to achieve sustainable agriculture in using appropriate and environmentally friendly pesticides needs to be questioned..

1. Introduction

Blessed with its fertile land, the tropical country of Indonesia has many various plantations. One of them is oil palm plantation which accounted for around 12 million hectares in 2017. Besides being a versatile crop as it can be used for food and energy, palm oil has many advantages compared with other vegetable oil. For example, it has the highest yield, producing ten times higher oil than soybean, rapeseed, and canola oils [1]. In addition, it also has the lowest production cost and minimum land area to produce more oil [2].

The plantation of palm oil plays a vital role in the economy of Indonesia. The plantation sector in Indonesia accounts for 2.31 % of the national Gross Domestic Product (GDP) in 2000 prices after the food crops by 6.96% [3]. It also has employed more than 6 million people [4]. Being gradually increased from 9.5 million hectares in 2012 to around 12 million hectares in 2017 of oil palm plantation area, crude oil palm (CPO) production in Indonesia rose to around 42 million tons [3]. Indonesia still becomes the largest palm oil producer with this production amount and controls 48% of the world market share [5].

IOP Conf. Series: Earth and Environmental Science 1041 (2022) 012038 doi:10.1088/1755-1315/1041/1/012038

On the other hand, Indonesia's oil palm plantation industry also faces a challenge that must be determined carefully. Due to the expansion of the plantation area, oil palm plantation has an impact on damaging natural resources and environmental sustainability. To overcome these problems, the Indonesian government makes efforts to ensure the sustainability of the development of the palm oil industry by creating a sustainability standard called the Indonesian Sustainable Palm Oil (ISPO). Meanwhile, another oil palm certifications scheme from nongovernmental and consumer organizations is accepted globally, namely the Roundtable on Sustainable Palm Oil (RSPO).

There are eight principles for the oil palm plantation industry that must be done to get the certificate from RSPO. Those principles include: commitment to transparency; environmental responsibility and conservation of resources and biodiversity; responsible consideration of workers as well as individuals and communities affected by the activities of the palm oil industry; responsible development of new plantings; commitment to continuous improvement for main activities; compliance with relevant laws and regulations; commitment to long-term financial and economic viability; application of best practices by smallholders and palm oil mills [6].

Having the largest area of oil palm plantation [7], Riau has 364 plantation companies which, according to the Riau Plantation Office, only more than a half of those companies had the licenses issued by Plantation Service [3]. In addition to dealing with licenses, oil palm plantation companies in Riau are also questioned whether they have implemented the RSPO standard properly. Therefore, this study was conducted by an in-depth examination of the companies' commitment to dealing with RSPO standards, especially in using pesticides.

2. Methodology

This research had been carried out in oil palm plantations of Tapung, Tapung Hulu, and Tapung Hilir districts. In addition, a literature study was conducted from the environmental impact analysis (ANDAL) documents of plantation companies provided by the Environment-Forestry Office (DLHK) and the Plantation Office of Riau Province (Disbun). Informants in this study were asked to provide information about the situation and condition of the targeted object. Informants were determined using the purposive sampling technique, namely sampling deliberately in accordance with the requirements set by the research team and by choosing sources who know the conditions and situations of oil palm plantations in the targeted areas.

The informants in this study were staff companies or individuals who manage oil palm plantations and the supervisors from the Disbun and DLHK of Riau Province. The data in this study were analyzed using descriptive analysis, where the purpose of this analysis was to describe systematically, factually, and accurately the facts and the relationship between the phenomena investigated. The analysis was carried out after the data were collected.

3. Result and Discussion

3.1 Oil Palm Plantation Company

Based on the data from the Disbun and the DLHK of Riau Province as the agencies that monitor environmental conditions for oil palm companies, 17 oil palm companies actively planted oil palm in Tapung, Tapung Hilir, and Tapung Hulu District. As a result, the total area of oil palm plantations managed by these companies reached 83,175.65 hectares. Table 1 shows the oil palm companies, locations, and the area each company manages for the three districts.

Table 1. Oil palm plantation companies in district of Tapung, Tapung Hilir, and Tapung Hulu

No	Company Name	Location	Area (Ha)
1	PT. Bumi Mentari Karya	Tapung	36.15
2	PT. Egasuti Nasakti	Tapung	3,608.44
3	PT. Padasa Enam Utama	Tapung	4,626.00
4	PT. Ramajaya Pramukti	Tapung	760.00

5	PT. Sewangi Sawit Sejahtera	Tapung	14.98
6	PT. Wira Karya Pramitra	Tapung	22.90
7	PT. Bumi Sawit Perkasa	Tapung Hulu and Tapung Hilir	11,826.68
8	PT. Inti Kamparindo Sejahtera	Tapung Hulu	7,341.36
9	PT. Lindai Jaya Lestari	Tapung Hulu	1,171.00
10	PT. Multi Agrindo Sawit Sejahtera	Tapung Hulu	457.00
11	PT. Multi Agro Sentosa	Tapung Hulu	10.00
12	PT. Multi Sarana Agro Lestari	Tapung Hilir	785.00
13	PT. Rajawali Khatulistiwa Abadi	Tapung Hulu and Tapung Hilir	3,335.00
14	PT. Sarana Inti Pratama	Tapung Hulu	228.00
13	PT. Sekar Bumi Alam Lestari	Tapung Hilir	6,200.00
14	PT. Subur Arum Makmur II	Tapung Hulu	4,800.00
15	PT. Sugih Indah Sejati	Tapung Hulu	6,00.00
16	PT. Perkebunan Nusantara V	Tapung Hulu, Tapung and Tandun	33,024.14
17	PT Bina Pitri Jaya	Tapung Hilir	4,329.00
Total Area			83,175.65

Sources: Disbun and DLHK of Riau Province

The total area in Table 1 accounted for 3% of the total plantation area referred to as the Riau plantation area by 2018. It showed that three districts in Tapung, Tapung Hulu, and Tapung Hilir had oil palm plantations managed by various oil palm companies. The largest plantation area was managed by PT Perkebunan Nusantara V (State-owned enterprises), while PT Multi Agro Santoso handled the smallest area of 10 hectares. For the types of pesticides used, three companies were selected as samples, with the total area was almost 39,000 ha.

3.2 Types of Pesticides Used

The World Health Organization has classified pesticides based on hazardous levels [8,9]. This classification distinguishes pesticides formed on the active ingredient compounds which cause the effect to humans. To be more precise, every sign has a meaning with the following order: Ia = extremely hazardous; Ib = highly hazardous; II = moderately hazardous; III = slightly hazardous; U = unlikely to pose an acute hazard under normal use. The active pesticide ingredients used by the selected sample companies can be seen in Table 2, which was compared to the WHO classification in 2005 and 2020

Table 2. Pesticides used by palm oil companies in Tapung area

Pesticides	Classification	
	WHO (2005)	WHO (2020)
Glyphosate	U	III
Paraquat	II	II
Glufosinate	III	II
Metsulfuron-methyl	U	U
Triclopyr	III	II
Aminopyralid potassium	NA	U

IOP Conf. Series: Earth and Environmental Science 1041 (2022) 012038 doi:10.1088/1755-1315/1041/1/012038

Kao® adjuvant A-134 (surfactant)	NA	NA
Deltamethrin	II	II
Alkyl aryl polyglycol ether	NA	NA
Cypermethrin	II	II
Lambda-cyhalothrin	II	II
Carbaryl	II	II
Carbosulfan	II	II
Acephate	III	II
Mancozeb	U	U
Triadimenol	III	II

Ia = extremely hazardous; Ib = highly hazardous; II = moderately hazardous; III = slightly hazardous; U = unlikely to pose an acute hazard under normal use; NA = Not mentioned on WHO's list.

Table 2 showed that almost all pesticides used by palm oil companies were classified as dangerous and negatively impacted the environment and non-targeted organisms; only metsulfuron-methyl, aminopyralid potassium, and mancozeb were classified as safe chemicals for use under normal conditions. In addition, changes in the classification of several active ingredients from 2005 data to 2020 data indicated that each pesticide's hazardous level continues to change from time to time, depending on the testing results by WHO or other studies [10,11].

For example, in 2005, glyphosate was considered harmless in regular use, but later discovered by the international cancer association, glyphosate had been considered one of the factors that cause pancreatic cancer [12,13]. Meanwhile, paraquat, which has clearly been stated in the principles of sustainable palm oil which should not be used [6], was still being used by several oil palm plantations in the Tapung area.

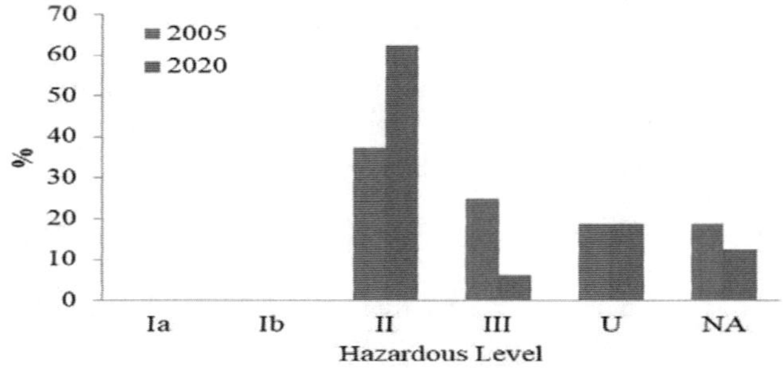

Figure 1. The comparison of total pesticides classified by WHO in 2005 and 2020

In addition, there were two types of pesticides that had not been included in the WHO's list as hazardous pesticides, i.e., Kao® adjuvant A-134 (surfactant) and alkyl aryl polyglycol ether. Had been innovated by Japan's Kao Corporation, Kao Adjuvant A-134 (A-134) is claimed to be a new eco-friendly botanical adjuvant. However, this pesticide was used in tandem with other herbicides in the field, such as glyphosate [14,15], classified as a slightly hazardous chemical. At the same time, alkyl aryl polyglycol ethers are not widely known and have no information related to this active ingredient.

The bar chart in Figure 1 illustrates the percentage of total pesticides used by selected companies based on WHO classification of pesticide types in 2005 and 2020 (see Table 2). If referring to the WHO document in 2005, it was clear that the most used pesticides are type II and type III, account for 37.5% and 25%, respectively. In that year, the use of pesticides in the Tapung area classified as undangerous was 18.75%. On the other hand, the percentage of pesticide use that was not hazardous with normal use using the WHO classification in 2020 was the same as in 2005; thus, almost 70% of pesticide used was classified hazardous for non-targeted organisms, including humans.

4. Conclusion

This study has demonstrated the commitment of oil palm plantation companies to sustainable agriculture from the perspective of pesticide use. It was assumed that the use of pesticides was not too much of a concern for many oil palm plantations in realizing sustainable palm oil. Almost 50% of the total land in three districts (Tapung, Tapung Hulu, and Tapung Hilir) managed by three selected companies was contaminated with hazardous pesticides. Furthermore, only 18.75% of the total pesticide used was classified as no hazardous level by WHO with normal use, and the rest remains dangerous.

Acknowledgment

We much appreciate the funding from the Annual Activity Plan and Budget (RKAT) of Universitas Lancang Kuning Pekanbaru under contract no: 043/LPPM/Pn/2021.

References

[1] Dermawan A and Hospes O 2018 Achieving Sustainable Cultivation of Oil Palm 1 33-48
[2] Corley R H V and Tinker P B 2015 Oil palm. 5th ed. London: Wiley-Blackwell
[3] Apriyanto M, Rujiah and Marlina 2020 Indonesian Sustainable Palm Oil (ISPO) Standards in Management of Palm Oil Preprints doi:10.31219/osf.io/b97v4.
[4] Apriyanto M, Chairul M and Umanailo B 2019 Decrease Polyphenols, Ethanol, Lactic Acid, and Acetic Acid during Fermentation with Addition of Cocoa Beans Inoculum International Journal of Scientific & Technology Research 8 461–465
[5] Anwar R, Sitorus S R, Fauzi A M, Widiatmaka N and Machfud N 2018 Achievement of Indonesian sustainable palm oil standards of palm oil plantation management in East Borneo Indonesia Journal of Indonesia Research Plantation 22 11
[6] Roundtable on Sustainable Palm Oil (RSPO) 2013 Principles and criteria for sustainable palm oil production Kuala Lumpur 19-21
[7] Direktorat Jenderal Perkebunan (2019) Statistik perkebunan Indonesia komoditas kelapa sawit tahun 2018 – 2020.
[8] WHO 2005 WHO recommended classification of pesticides by hazard and guidelines to classification 2004. In World Health Organization.
[9] WHO 2020 The Who recommended classification of pesticides by hazard and guidelines to classification 2019. In World Health Organization.
[10] Damalas C A and Eleftherohorinos I G 2011 Pesticide exposure, safety issues, and risk assessment indicators Int J Environ Res Public Health 8 1402–1419
[11] Boedeker W, Watts M, Clausing P and Marquez E 2020 The global distribution of acute unintentional pesticide poisoning: estimations based on a systematic review BMC Public Health 20 1-19.
[12] WHO 2015 Evaluation of five organophosphate insecticides and herbicides. In IARC Monographs.
[13] WHO/FAO 2004 Glyphosate. Pesticides residues in food 2004 Joint FAO/WHO Meeting on Pesticides Residues Part II Toxicological IPCS/WHO 95–162.
[14] Hao Y, Zhang Y, Xu W, Gao J and Tao L 2019 Journal of Pesticide Science 44 1–6
[15] Mustikawati M, Sembodo D R J, Sanjaya P, and Pujisiswanto H 2020 Effect of surfactant and time of rainfall after application on the efficacy of glyphosate herbicide J. Agrotek Tropika 8 461-470

Assessing the feasibility of migration policy from LPG stoves to induction stoves in Indonesia

T Anggono[1], I Ruslan[2], C Anditya[3], D G Cendrawati[4], M I A Irsyad[5]

[1,2,4,5]Research and Development Agency of Energy and Mineral Resources, Ministry of Energy and Mineral Resources, Jakarta, Indonesia
[3]Directorate General of New and Renewable Energy, and Energy Conservation, Ministry of Energy and Mineral Resources, Jakarta, Indonesia

Email: tri.anggono@esdm.go.id[1], iyung.ruslan@esdm.go.id[2], chrisnawan.anditya@esdm.go.id[3], dian.cendrawati@esdm.go.id[4], m.indra.al.irsyad@brin.go.id[5]

Abstract. Assessing the progress of lectures ensures the involvement of lectures in the exercise of science dissemination by research, education, and community service. That job affects the performance of lecturers at universities throughout the country. The performance of lecturers is evaluated regularly because of the vast number of evaluation attributes. Manual assessment can take time and cannot be carried out. In the evaluation of lecturer's outcomes, a decision support model is needed to facilitate decision making. This study applies the preference ranking organization method for enrichment assessment (PROMETHEE) as a decision model in Multi-Criteria Decision Making (MCDM) and Geometric Analysis for Interactive Aid (GAIA) as a geometric analysis in determining lecturer performance. This analysis resulted in the maximal value of an alternative of 2 with Phi 0.2333, followed by 1 and 3 alternatives. These results suggest that alternative 2 is the best way to assess the performance of lecturers as the main indicator. Alternative 3 is the least recommendable alternative because of its lowest precision on the basis of visualization results.

1. Introduction

Switching kerosene to LPG is also a common policy by many developing countries because of its benefits, such as time saving for cooking preparation, health-related benefits, and fuels saving [1]. Indonesia started the switching program from kerosene to LPG in 2007 by initiating the Conversion Program from Kerosene to LPG or also known as the Zero Kerosene Program [2, 3]. As results, LPG consumption increased from 979 thousand ton in 2007 to 1,592 thousand ton in 2008 and 7,449 thousand ton in 2019 [4-6]. Meanwhile, kerosene consumption declined from 8,474 thousand kilo liter (kl) in 2007 to 6,765 thousand kl in 2008 and 484 thousand kl in 2019. The program is also successful to reduce kerosene subsidy, which had increased from US$ 1.29 billion in 2001 to US$ 3.78 billion in 2006 [2]. Budya and Arofat [2] estimated that accumulative kerosene subsidy saving from the program in 2007 to 2010 was US$ 2.94 billion.

However, the increasing LPG consumption brought back the energy subsidy problem since the government aims to keep the subsidy of a 3 kg LPG cylinder. Meanwhile, the LPG distribution uses an open system, meaning that anyone can buy subsidized 3 kg LPG cylinders. Consequently, households initially using a non-subsidized 12 kg LPG cylinder now also use the 3 kg LPG cylinder [7]. Moreover, illegal practices refilling 12 or 50 kg LPG cylinders using LPG in 3 kg cylinders also contribute to the LPG subsidy increase. The LPG subsidy was US$ 1.1 billion in 2006 increased to US$ 3.9 billion in 2018 [2, 3, 7, 8]. Consequently, the LPG import is also increasing from 138 thousand ton in 2007 to

5,715 thousand ton. Therefore, the government now intends to reduce LPG usages by promoting the usages of induction stoves.

Studies analyzing costs and benefits of LPG stoves and electricity-based cooker emerge in various developing countries [7, 9-14]. However, those studies focused on households' perspectives only and ignored macroeconomic impacts from government perspective. This study aims to fill this gap by using a holistic approach integrating micro analysis (i.e., energy equivalence measurement, and LPG consumption survey) and macro analysis (i.e., input-output analysis, and feasibility analysis on both government and household perspective).

2. Literature Review

Indonesia is not the first country aiming LPG consumption reduction through induction stove program. Unlike in developed countries, electric stove is uncommon appliance used by households lived in developing countries [11, 12, 15, 16]. Cooking in developing countries mostly relies on biomass, kerosene, or LPG [9]. More common electrical appliances found on kitchens in those countries are rice cooker and water dispenser/ kettle. Only 0,76% of total households in Indonesia used electricity as the main energy source for cooking [17].

Ecuador has been implementing the induction stove program since 2014 through the National efficient cooking plan (NEFC) [18-20]. After a pilot project for 3.433 stoves, Ecuador sets 3 million electric stoves, but the induction stoves sold were only 740 thousand units by 2018 [7]. Studies assessing the feasibility of similar programs for other developing countries are emerging, such as in Nepal [13], India [10], and East Africa countries [11]. However, implementing the program should tackle various barriers such as less comfortable, electricity bill concerns, and suitability with food cooking methods [7].

In India, Banerjee, Prasad, Rehman and Gill [10] assessed the cooking behaviors of 1,020 rural households who were using a discounted induction stove with free cookware for one year. The induction stoves successfully replace LPG stoves on 84% of households using LPG as a secondary cooking fuel. Main factors contributing to the successful result are reliable electricity supply and affordable electricity price. Households not shifting to induction stoves still worry on electricity bill impact. Another study in India is Dhevika, Latasri and Kalidhasan [21] who conducted a satisfactory survey on 50 households using induction stoves in India. Main factors influencing induction stove usages are fast cooking and free cookware given. Dhevika, Latasri and Kalidhasan [21] also did not find electricity consumed by induction stoves as a barrier. Most households using induction stoves concerned on coil failures (20%), poor customer services (18%), and shut off too quickly (16%).

Batchelor, Brown, Scott and Sumanik-Leary [11] conducted field experiments on 60 households in Kenya, Tanzania, and Zambia to cook daily food using electric resistance stoves, induction stoves, and EPC. The conclusion is that EPC is more desirable. Respondents can cook almost everyday food with EPC with minimal training. Moreover, EPC is the most cost-effective cooking technology followed by kerosene, LPG, hotplate stove, and charcoal. Higher efficiency occurs when cooking heavy foods such as beans, and meat. Batchelor, Brown, Scott and Sumanik-Leary [11] also supported that electricity price and grid stability are the successful keys for the electric-based stove program.

Indonesia has a unique electricity retail market by dividing electricity customers based on connection capacity, i.e., 450 volt-ampere (VA), 900 VA, unsubsidized 900 VA or 900 VA/M, 1300 VA, 2200 VA, 3500 to 5500 VA, and higher than 6600 VA for residential customers. The capacity connection is simply determined by different mini circuit breaker (MCB) capacity installed on the meter, but it effectively limits the consumption of subsidized electricity enjoyed by 450 VA and 900 VA connections. Therefore, the induction stove migration policy must reform the connection capacity system because capacity lower than 2200 VA is inadequate to use the stove. However, shifting LPG stoves to induction stoves allows the government to control energy subsidy for poorer households.

3. Methodology

Our study estimated the costs and benefits of the proposed induction stove migration policy by using multiple analysis approaches, i.e., an online LPG consumption survey, energy equivalence measurement, I-O analysis to estimate the macroeconomic impacts of the proposed policy, and the cost and benefit analysis. Energy consumed by induction stoves and EPC was measured by boiling 2,500-gram water from initial temperature $\pm$ 20 °C to $\pm$ 90 °C. The EPC was set at the soup mode. We calculated the heat calories ($\boldsymbol{Q_{water}}$) required to boil the water by using following formula:

$$Q_{water} = m.c_{water}.\Delta T \quad (1)$$

where Q_{water} is heat calories (kCal), m is water mass (kg), c_{water} is water specific heat (kCal/kg), and ΔT is temperature increases (°C).

$$Q_{water} = 2.5 * 1 * 70 = 175 \quad (2)$$

The efficiency η of the induction stove is:

$$\eta = \frac{Q_{water}}{Q_{test}} \times 100\% \quad (3)$$

where Q_{test} is the measured heat calories (kCal) converted from E_{test}, the measured electricity consumption (kWh), by using following formula:

$$Q_{test} = 860.421 \times E_{test} \quad (4)$$

The measurement of energy consumed by the LPG stove has a similar procedure. In addition, we measure LPG mass before and after boiling the water. We then used the mass difference and boiling time data to calculate M_n, gas flowrate (kg/hour). Heat input Q_n (kWh) is calculated by using following formula:

$$Q_n = \frac{(1000 \times Mn \times Hs)}{3600} \quad (5)$$

where H_s is gas caloric value, i.e., 49.14 MJ/kg. We then used Equations 3 and 4 to calculate the efficiency of the LPG stove.

We conducted an online survey in December 2020 to collect data on shares of LPG cylinder sizes used by 1,006 households and their average monthly LPG consumption. We used the data for estimating changes on LPG and electricity consumptions. We admitted that the sample size is very small compared to total actual households, but to the best of our knowledge, this survey involved more respondents than other studies. For an example, Pranadji, Djamaludin and Kiftiah [22] only had 80 respondents in a city to analyze LPG consumption behaviors in households. Larger survey is the National Socioeconomic Survey conducted annually by the Central Bureau of Statistics (BPS). The survey asks the main energy source for cooking, but it does not ask about LPG consumptions [17].

The input-output (I-O) analysis used the I-O table 2010 [23] to estimate the economic output multiplier due to final demands changes on LPG, electricity, cookware, induction stoves, and electrical installation service. The assumptions for LPG prices, LPG subsidy, LPG import price, stove price, cookware price, installation cost, and commissioning test cost are in Table 1. Customers with capacity lower than 1300 VA have lower tariffs as in Table 2 and therefore, higher electricity subsidy. One cost component omitted is the cost for increasing the connection capacity or MCB replacement cost. We assumed that the government would ask PLN to not charge the cost to support the electric stove migration policy.

Table 1. Price and cost assumptions

Items	Unit	Prices
3 kg LPG price	US$ cent/kg	49.3
3 kg LPG subsidy	US$ cent/kg	40.1
12 kg LPG price	US$ cent/kg	90.8
LPG import price	US$/metric ton	438.0
2000 W induction stove price – commercial	US$/unit	121.1
2000 W induction stove price – government project	US$/unit	42.3

Induction cookware prices – commercial	US$/3 set	79.2
Induction cookware prices – government project	US$/3 set	45.8
[a] Electrical installation service cost	US$	79.0

[a] Note: The service is to install a new cable that meets the safety standard for using 2 kW induction stove. Most of cables in poor households are not suitable for the stove. The used exchange rate is 1 US$ for IDR 14,200.

Table 2. Electricity tariffs and subsidy

Items	Unit	450 VA	900 VA	900 VA/M	1300 VA
Electricity tariff	US$ cent/kWh	2.9	4.3	9.5	10.2
Electricity subsidy	US$ cent/kWh	4.1	2.7	0.0	0.0

4. Result and Discussion

Table 3 shows the results of energy consumption measurements to boil 2,500-gram water using four induction stoves, and an LPG stove. In this light, our analysis will focus on 2,000 W induction stoves that have better performances than the 1000 W induction stove. Therefore, we calculated the conversion factor from LPG consumption to electricity consumption by using measurement data of the LPG stove and the best induction stove (i.e., the induction stove - 2) by using following formula:

$$Conversion\ factor = \frac{0.250\ kWh}{34.73\ gram\ LPG} = 7.2\ Wh/gram\ LPG \quad (6)$$

Table 3. The results of energy equivalence measurements

Cooker	Burner Setting	Boiling Time	Energy Consumption In original unit	In kcal	Efficiency (%)
Induction stove - 1	1,000 W	20'22"	0.319 kWh	274	64
Induction stove - 2	2,000 W	8'12"	0.250 kWh	215	81
Induction stove - 3	2,000 W	10'43"	0.279 kWh	240	73
Induction stove - 4	2,000 W	8'40"	0.265 kWh	228	77
LPG stove	Maximum	10'58"	34.728 gram	391	45

The LPG Consumption Survey concluded the subsidized 3 kg LPG cylinder as the primary option for electricity customers with installed capacity 1300 VA or lower as shown in Table 4. Yet, 3 kg LPG usages in customers with installed capacity 2200 VA to 5500 VA are also high between 31.6 % to 46.7 % because of its lowest price. Customers with installed capacity 450 VA consumed LPG for 8.6 kg/ month while customers with higher installed capacity consumed higher LPG for 10 kg/ month (900 VA and 900 VA/M), 11 kg/ month (1300 VA), 12.7 kg/ month (2200 VA), 16.1 kg/ month (3500 to 5500 VA), and 32.7 kg/ month (higher than 6600 VA).

Table 4. The results of LPG consumption survey

Customer category (VA)	450	900	900/M	1300	2200	3500 - 5500	> 6600
n	160	131	290	289	107	19	10
Size of LPG cylinders							
• LPG 3 kg (%)	97.5	86.3	80.3	70.2	46.7	31.6	10.0
• LPG 5,5 kg (%)	0.6	2.3	4.1	5.9	6.5	-	-
• LPG 9 kg (%)	-	-	0.7	-	-	-	-
• LPG 12 kg (%)	1.9	11.5	14.8	23.9	46.7	68.4	50.0
• LPG 14 kg (%)	-	-	-	-	-	-	40.0
Average LPG consumption (kg/ month)	8.6	10.5	10.0	11.0	12.7	16.1	32.7

Table 5 shows the impacts of the migration policy from LPG stoves to induction stoves on energy costs and energy subsidy for cooking. The residential customers with electricity connecting capacity 450 VA dan 900 VA will have lower energy costs for cooking because they have highly subsidized electricity tariff. Consequently, electricity subsidy for those customers (i.e., US$ 4.9 to 5.9 per month) will be higher than LPG subsidy (i.e., US$ 4.6 per month). Other residential customers will have higher energy cost when cooking with the 2000 W induction stove compared cooking costs with 3 kg LPG cylinder. Net energy subsidy for this customer category will reduce from US$ 4.6 per month (i.e., LPG subsidy) to maximum US$ 0.5 per month (i.e., electricity subsidy).

The induction stove program will affect several economic sectors. First of all, it will affect the oil and gas product sector through setbacks on residential LPG consumptions, LPG import, and LPG subsidy. We calculated the setbacks by multiplying price assumptions in Table 1 with LPG consumption data in Table 4. For simplicity, we assumed all customers with connection 1300 VA and lower used 3 kg LPG cylinders and the rests used 12 kg LPG cylinders. Second, the induction stove program will create new demands of ferromagnetic cookware, induction stove, electrical installation service, and commissioning test with values as in Table 1. Third, the program will increase electricity consumptions.

Based on those changes on final demands, we calculate the output multiplier by using the I-O table 2010 [23] with results in Table 6. Residential customers with electricity connecting capacity from 450 VA dan 1300 VA have similar final demand changes due to they all are assumed to previously use 3 kg LPG cylinders and to be entitle for free induction stove and free ferromagnetic cookware. In contrast, residential customers with electricity connecting capacity 2.200 VA or higher are assumed to use non-subsidized 3 kg LPG cylinders.

We also assume that they will use a commercial stove and luxury cookware since they are not entitled for free stoves and cookware. Therefore, they have higher final demand changes on five related sectors in Table 6. Consequently, the output multiplier will be US$ 690.8 per residential customer 2.200 VA or higher; and US$ 551.2 per residential customer 1,300 VA or lower.

Table 5. Costs and benefits of stove usages on each residential customer category

Parameter	LPG stove		Induction stove			
	3 kg	12 kg	450 VA	900 VA	900 VA/M	1300 VA & higher
LPG consumption (kg/month)	11.4	11.4	-	-	-	-
Electricity consumption (kWh/month)	-	-	82	82	82	82
Energy cost for cooking (US$/month)	5.6	10.4	2.4	3.5	7.8	8.3
Energy subsidy (US$/month)	4.6	-	5.9	4.9	0.5	-

Table 6. Output multiplier for each household due to the stove migration policy (in US$)

Commodity	Sector	450 to 1300 VA	≥ 2,200 VA
Changes on final demands			
LPG	Oil and gas refinery products	-61.1	-63.1
Cookware	Metal kitchen, carpentry, home and office furniture	45.8	79.2
Induction stove	Electronic goods, communications and equipment	42.3	121.1
Electricity	Electricity	100.1	100.1
Electrical installation service	Professional, scientific and technical services	79.0	0
Output multiplier		551.2	690.8

5. Conclusion

Our analysis shows that the proposed migration policy from LPG stoves to induction stoves is economically feasible. The policy will reduce energy costs for cooking in relatively poor households (i.e., residential customers 450 VA and 900 VA); and substitute the imported LPG with locally produced electricity. The government may apply electricity subsidy block to control electricity subsidy on those household group. Relatively higher income households (i.e., residential customers 900 VA/M and higher) will have higher energy costs for cooking when using the 2,000 W induction stove because they still use the subsidized LPG 3 kg cylinder. Compared to LPG 12 kg cylinder, the induction stove has lower energy costs due to higher efficiency and lower energy prices.

Yet, technical and social challenges may hamper the implementation of the migration policy. The connection capacity grouping entails electricity retail market transformation. The government have two options either to cease the grouping or to install a dedicated electricity line for the induction stove on a house that has connecting capacity lower than 2200 VA. Regarding the social challenges, the policy surely draws rejections by households who is unfamiliar with the induction stove, and by LPG retailers and distributors who will lose their business.

For that, we firstly recommend a massive campaign of the induction stove to encourage people voluntarily using the induction stove. The second recommendation is to withdraw the subsidy for the LPG 3 kg cylinder, and the same time, to offer free induction stove, free cookware, and free special electricity line for people who want to use the induction stove. In this light, people still have choices whether to use unsubsidized LPG 3 kg cylinder or subsidized electricity tariff.

Acknowledgments
All authors acknowledge funding supports from the Research and Development Center for Electricity, New-Renewable Energy, and Energy Conservation Technology - Ministry of Energy Mineral and Resources.

References
[1] WLPGA 2018 The Socioeconomic Impact of Switching to LPG for Cooking. (France: WLPGA)
[2] Budya H and Arofat M Y 2011 Providing cleaner energy access in Indonesia through the megaproject of kerosene conversion to LPG Energy Policy 39 7575-86
[3] Thoday K, Benjamin P, Gan M and Puzzolo E 2018 The Mega Conversion Program from kerosene to LPG in Indonesia: Lessons learned and recommendations for future clean cooking energy expansion Energy for Sustainable Development 46 71-81
[4] MEMR 2020 Handbook of Energy & Economic Statistics of Indonesia 2019. ed M o E a M R (MEMR) (Jakarta
[5] MEMR 2016 Handbook of Energy & Economic Statistics of Indonesia 2016. ed M o E a M R (MEMR) (Jakarta
[6] Hidayati J and Panama J 2019 Tinjauan Permintaan Gas Global dan Distribusi LPG di Indonesia: Studi Pustaka TALENTA Conference Series: Energy & Engineering (EE) 2 386-96
[7] IEA 2020 LPG-to-electric cooking transition. (Paris, France: International Energy Agency (IEA)) p 41
[8] Suharsono A, Lontoh L and Maulidia M 2021 Indonesia's Energy Policy Briefing. (Jakarta, Indonesia: The Global Subsidies Initiative (GSI), the International Institute for Sustainable Development (IISD))
[9] Aemro Y B, Moura P and de Almeida A T 2021 Experimental evaluation of electric clean cooking options for rural areas of developing countries Sustainable Energy Technologies and Assessments 43 100954
[10] Banerjee M, Prasad R, Rehman I H and Gill B 2016 Induction stoves as an option for clean cooking in rural India Energy Policy 88 159-67
[11] Batchelor S, Brown E, Scott N and Sumanik-Leary J 2019 Experiences of electric pressure cookers in East Africa. In: 10th International Conference on Energy Efficiency in Domestic Appliances and Lighting (EEDAL'19), (Jinan, China: Modern Energy Cooking Services (MECS)) p 29
[12] Couture T D and Jacobs D 2019 Beyond Fire: How to achieve electric cooking. (Hague,

IOP Conf. Series: Earth and Environmental Science 1041 (2022) 012039 doi:10.1088/1755-1315/1041/1/012039

Netherland: Hivos and World Future Council)

[13] Dahal G R 2020 Cost-benefit analysis of replacing LPG stoves with induction stoves in rural households of Kavre district Nepal. In: Environmental Systems: Energy, Technology and Policy, (California, US: Humboldt State University) p 119

[14] Jeuland M, Soo J-S T and Shindell D 2018 The need for policies to reduce the costs of cleaner cooking in low income settings: Implications from systematic analysis of costs and benefits Energy policy 121 275-85

[15] Quinn A K, Bruce N, Puzzolo E, Dickinson K, Sturke R, Jack D W, Mehta S, Shankar A, Sherr K and J P R 2018 An analysis of efforts to scale up clean household energy for cooking around the world Energy for Sustainable Development 46 1-10

[16] Griffin M, Ramsson T and Gibson G 2012 Cooking Appliances In: Technology Brief R06, (Paris, France: International Energy Agency (IEA) Energy Technology Systems Analysis Programme (ETSAP)) p 8

[17] BPS 2020 Statistik Kesejahteraan Rakyat (Welfare Statistics) 2020. (Jakarta, Indonesia: Central Bureau of Statistics (BPS))

[18] Gould C F, Schlesinger S B, Toasa A O, Thurber M, Waters W F, Graham J P and Jack D W 2018 Government Policy, Clean Fuel Access, and Persistent Fuel Stacking in Ecuador Energy for Sustainable Development 46 111-22

[19] Martínez-Gómez J, Ibarra D, Villacis S, Cuji P and Cruz P 2016 Analysis of LPG, electric and induction cookers during cooking typical Ecuadorian dishes into the national efficient cooking program Food Policy 59 88-102

[20] Gould C F, Schlesinger S B, Molina E, Bejarano M L, Valarezo A and Jack D W 2020 Household fuel mixes in peri-urban and rural Ecuador: Explaining the context of LPG, patterns of continued firewood use, and the challenges of induction cooking Energy policy 136 111053

[21] Dhevika V, Latasri O and Kalidhasan C 2014 A Study on Customer Satifcation and Problems of Induction Stove User International Journal of Research Instinct 1 233 - 9

[22] Pranadji D K, Djamaludin M D and Kiftiah N 2010 Analisis Perilaku Penggunaan LPG Pada Rumah Tangga di Kota Bogor Jurnal Ilmu Keluarga & Konsumen 3 172-83

[23] BPS 2015 Tabel Input Output Indonesia 2010. (Jakarta, Indonesia: Central Bureau of Statistics (BPS))

The innovation of ecological fiscal transfers policy at Siak Regency

Muhammad Ichsan Kabullah*

Universitas Andalas, Padang, Indonesia

*muhammadichsankabullah@soc.unand.ac.id

Abstract. For several decades, the problem of forest fires has been recurring almost every year in Siak regency, Riau Province. This phenomenon affected a huge portion of damage for public. However, the previous policy has been failed to solve the forest fires problem. However, some local government has attempting to address the problem of forest fires with policy innovation. This article focuses on innovation by the Siak local government through ecological transfer policy. Although the idea of ecological transfer policy is not new, there are few Indonesian local governments has initiate in the policy. Siak is one step ahead to initiate the ecological transfer policy during Siak Hijau vision. This article analyses the success story of Siak in formulating the ecological transfer policy in local budget. Through a qualitative research method, we found that the ecological transfer policy can occur due to combination of local elite support and advocacy of civil society organizations. Both actors played an active role in the ecological transfer policy. However, the women tended to be alienated in the formulation of the ecological transfer policy due to cultural and political barriers. In that sense, it is important to create inclusiveness for policy formulation..

1. Introduction

The problem of forest fires in Indonesia is hard problem to solve by the government. The forest fires caused more than 100,000 people deaths in Indonesia, Malaysia, and Singapore due to smoke exposure in 2015 (World Bank, 2016). Another study conducted by Glauber and Gunawan (2016) describes the 2015 forest and land fires in Indonesia caused economic losses of US$16.1 billion. At the same time, the forest fires also lead negative impact for public education especially in Sumatera, Kalimantan, and Papua (Purnomo et al, 2017). These implications required a serious response from the government to reducing forest fires problem.

Attention can be directed to the Siak local government, which is trying to distribute a share of fiscal transfer based on ecological indicators in local budget (Anggaran Pendapatan Belanja Daerah/ APBD). This step is in line with the prevent forest fires agenda. It is based on a common practice in many countries, where programs for maintain forest were carried out through a special financing mechanism or better known as the National Forest Funds (NFF).

Research by Matta (2015) shows that several countries have allocated priority budget for forestry sector such as the United States through the Knutson-Vandenberg Fund in 1930 and Spain through the Patrimonio Forestal del Estado in 1939. This step became widespread in many countries such as Africa, Asia, the Pacific, and Europe (Rosenbaum & Lindsay, 2001). However, the Indonesian government efforts to tackle forest fires have been limited through NFF scheme.

Previous research identifies the main problem of local governments to solve forest fires due to the lack of government capability (Meiwanda, 2016), and the occurrence of budget policy alienation (Kabullah et al, 2020). Other scholars explore the forest fires prevention could success through actor networks (Purnomo et al, 2017), institutional effectiveness (Putra et al, 2019), and collaborative governance (Wicaksono, 2019). Despite the richness of this research's, there has limited discussion relate the ecological fiscal transfer by the local government.

Relatively few studies have considered the ecological transfer policy in the Indonesia as research subject. A study by Haryanto (2015) finds that the ecological transfers could adapt to the region from the General Allocation Fund by inserting a form of comprehensive ecological indicators of forest closure. Another study by (Mumbunan et al, 2012) argued that the ecological transfer policy is essential for environmental conservation mission because the mission requires large funds from the government. Although the possibility of ecological transfer policy has discussed often by scholars, only four Indonesian regions have adopted a special financing mechanism for dealing with forest fires problems in the local budget policy and one of these regions is Siak.

In 2021, the Siak local government took the initiative to formulate more pro-environmental budgeting policy called the ecological transfer policy (Transfer Anggaran Kabupaten berbasis Ekologi/ TAKE). In short, the ecological transfer policy is a local government-funding scheme by carrying out ecological indicator in the fiscal sharing with a village government (kampung). In Siak Regency, the ecological transfer policy is carried out through reformulate of the fiscal transfer scheme from three variables namely basic, proportional, and sakinah allocations into four basic allocation variables, proportional, sakinah, and village performance. The inclusion of village performance as the basis for determining the amount of Village Fund Allocation in Siak regency makes every village must fulfill goal that state in the Green Village Index.

The objective of this article is to reveal how the Siak local governments formulate the ecological transfer policy. I show that the difference actor were involved to formulate the ecological transfer policy. At the same time, by portraying the ecological transfer policy at Siak will stimulate the ecological transfer policy in the other regions in Indonesia.

2. Methodology

Research for this paper conducted from May to August 2021 in the Siak regency. The analysis is carried out through qualitative research methods with data collection techniques, namely in-depth interviews, observation, and documentation. In-depth interviews were conducted with people who have the capacity and are involved in the ecological transfer policy. The informants in this study came from executive office, local parliament, and the lowest government apparatus, namely the village.

For triangulation, in-depth interviews were also conducted with scholars and activists from several civil society organizations. The other data collection technique is observation by focusing on the ecological transfer policy formulation process, which initiate by Regional Development Planning Agency into one forum called forum OPD. The other data collection technique, namely documentation, is carried out through collecting data related to the ecological transfer policy in the form of budget documents, regulations and other relevant data that improve the analysis.

3. Result and Discussion

Siak is a region in the west part of Riau Province with a recurrent case of forest fires. Around 53.22% of its area is peat land that is easily burned in the summer season (Riau Daily, 2017). According to the local government performance report of Siak in 2014, of the 855,603 hectares of forest area, only 9.23% was natural. With the small availability of natural forests, the threat of forest and land fires always appears every year. This is shown by the Meteorology, Climatology and Geophysics Agency (BMKG) report in October 2016, which indicates that 70% of hotspots in Riau Province came from Siak regency. Within a landscape that resist with forest fires, the Siak local government has been attempting the forest fires through the ecological transfer policy.

The commitment of Siak local government to solve forest fires problem has showed in 2016 when the Regent of Siak, Syamsuar, launched Siak Regency as Green Regency (*Siak Hijau*). This vision has accommodated in the local regulation number 12 in 2016 about Regional Medium-Term Development Plan (*Rancangan Pembangunan Jangka Menengah Daerah/* RPJMD) of Siak. In that regulation, one of the missions is infrastructure development based on environmentally including prevention of forest fires.

As a result, the Siak local government has decided to allocate funds for environmental activities like graph 1.

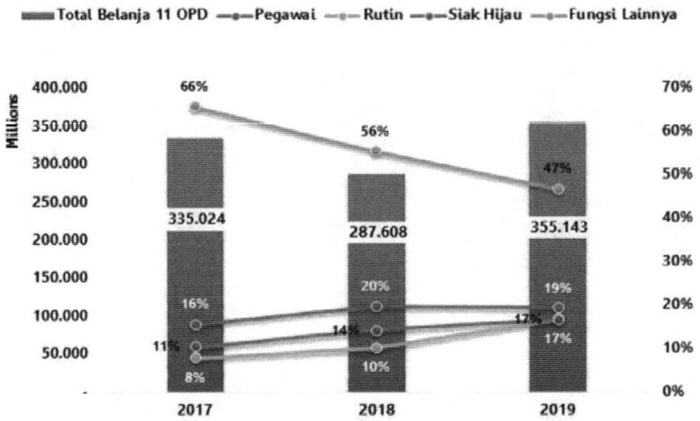

Figure 1. The number of Siak Hijau budget from 2017-2019

As indicated in the graph 1, *Siak Hijau* has supported by 11 agencies (*dinas/ badan/ kantor*) in the local government of Siak such as the Office of Community and Village Development (*Dinas Pemberdayaan Masyarakat Kampung*), the Environmental Office (*Dinas Lingkungan Hidup*) and Regional Disaster Management Authority (*Badan Penanggulangan Bencana Daerah*). The collaboration among 11 agencies creates opportunities for the acceleration of forest fires prevention. This can be seen from the increasing number of *Siak Hijau* budget on 11 agencies from 2017 until 2019. The budget of Siak Hijau has continuously increased from Rp. 35.5 billion in 2017 to Rp. 41.5 billion in 2018, and in Rp. 59.6 billion in 2019.

It means that the proportion of *Siak Hijau* budget has approximately 11 until 17 % from the total spending of 11 agencies. However, the amount of *Siak Hijau* budget is still not enough when compared to the performance indicators that must be achieved. Whereas in the Regent's Decree No. 650 of 2019 concerning the *Siak Hijau* Roadmap, there are three objectives, five targets, fourteen indicators and forty-one programs. With this picture, *Siak Hijau* vision was not enough to ensure success of forest fire prevention in Siak.

Therefore, the Siak local government has formulate the ecological transfer policy. The policy has launched since 2021 and contains a mission to ensure forest fires programs in Siak. This policy also has purpose as a strategy for improving environment quality by integrating regional development policies in the fields of natural resource management, economic improvement and social welfare. The determination of the forest fires prevention as a priority agenda is the right step considering that the fiscal capacity of the Siak local government is very limited. Interestingly, there are big attentions from political elites and Civil Society Organizations (CSOs) activists with the ecological transfer policy.

In Siak, the involvement of political elites and CSOs activists in initiating the ecological transfer policy were very strong. Syamsuar, as Regent of Siak, has to develop an effective strategy to rally support from as many voters as possible. As one of Governor candidates at that moment, he tries to construct a positive image in the purpose that this image will attract voter support in the Governor election. With the Regent of environmental mission, he wants public know that he have vision for forest fires problem. With the *Siak Hijau* and the ecological transfer policy, these terms 'image' have influenced significantly and bring a positive image for public consumption.

The *Siak Hijau* and the ecological transfer policy can be defined as an effective political campaign of Syamsuar to gaining public sympathy before the Governor election. His political image is strong in the mind of voter as candidate who has a commitment for environment. At the same time, other candidates do not have a clear vision for forest fires prevention. As a result, Syamsuar won the Riau Gubernatorial election 2019-2024 because his political images presence his strength among other candidates. In the electoral context, a political image is essential to bridging a mind of the voter to the candidate popularity (Tapsell 2015, 41-43). In this context, Riau voters know that the leader should solve a forest fires problem quickly.

At the same time, some CSOs activists are motivated to encourage the ecological transfer policy. CSOs activists likely to supervise the ecological transfer policy because they will get incentives to participate. For instance, Winrock, as one of leading international CSOs, has involved in the ecological transfer policy. Some Winrock local staffs explained:

"Siak has selected as pilot project due to characteristic of peatlands areas and palm oil production that affected greenhouse gas emissions" (interview S, 29/6/2021).

Though S shows her organizations has big interest in participating because they get benefit from foreign aid. Interestingly, some activists from other CSOs might engage at different stages of the ecological transfer policy such as Fitra Riau in the budget advocacy. Thus, collaboration of CSOs has become understood to be a central element to ensure the forest fires problems solve with sufficient budget by the Siak local government.

Although the ecological transfer policy is associated with inclusive participation between local elites and CSOs activists, the women participation is still weak in this policy. This can be seen that only two woman could directly involve in the ecological transfer policy. Gustimar is successful local parliament member because she is only female of 40 members. However, Gustimar's performance in the local parliament still has not succeeded to against the patriarchal culture in Siak Regency. She also does not have a strong vision related to gender rights in the ecological transfer policy in Siak.

Another female actor is Febriyeni as secretary the Office of Community and Village Development. Although most of her work experience is involved in village affairs, she has not yet had the opportunity to get the position of head of office. It means that she became inferior because she doesn't get support by elite that tends to be masculine. Suryakusuma's research (2011) clearly describes how the ruling regime succeeded in marginalizing the role of women in political life. If some women have successful in sitting in the power structure, it will not guarantee her competitively (Kabullah & Fajri, 2021).

The challenge of women in the ecological transfer policy is also caused by the patriarchal culture that grows in the community, including in political life. It is common due to the weak gender perspective in the political process (Kittilson, 2019). Women defined as supporting actor in the family with a duty to take care of household. Suryakusuma (2011) clearly describes how the ruling regime succeeded in defining the role of women only as wives to serve the family and mothers to take care of the children. At the same time, women do not have chance to involve in the public agenda. As a result, the ecological transfer policy should evaluate based on inclusion view in the future.

4. Conclusion

This article presents that the innovation of the ecological transfer policy at Siak is not enough to prevent of forest fire. The Siak local government needs to promote an idea of inclusive for all participants. Wider inclusion will attract the policy achievement. In that sense, the formulation of every policy should focus to social context, which focuses on social relations among actors and their networks in which they are embedded.

Acknowledgments

This article is supported by research grant of the Faculty of Social and Political Sciences, Andalas University in 2021. We especially thank all informants for this research. We also express our deepest gratitude to the lecturers and students at the Faculty of Social and Political Sciences, Andalas University who have supported this research. We also give appreciation to the International Conference on Environmental, Energy, and Earth Science (ICEEES) organizer.

IOP Conf. Series: Earth and Environmental Science 1041 (2022) 012040 doi:10.1088/1755-1315/1041/1/012040

References

[1] Glauber, A.J., Gunawan, I., (2016). The Cost of Fire: An Economic Analysis of Indonesia's 2015 Fire Crisis. The World Bank http://pubdocs.worldbank.org/en/643781465442350600/Indonesia-forest-fire-notes.pdf.

[2] Kabullah, M. I, & Fajri, M. N. (2021). Neo-Ibuism in Indonesian Politics: Election Campaigns of Wives of Regional Heads in West Sumatra in 2019. *Journal of Current Southeast Asian Affairs*, *40*(1), 136-155.

[3] Kabullah, M. I., Koeswara, H., & Rahmadi, D. (2020). Alienasi Kebijakan Anggaran Provinsi Riau dalam Pencegahan dan Pengendalian Kebakaran Hutan dan Lahan. *Jurnal Administrasi Dan Kebijakan Publik*. *5*(2), 114-133.

[4] Kittilson M. C. (2019) Gender and Electoral Behavior. In Franceschet, S., Krook, M. L., & Tan, N. (Eds.). The Palgrave Handbook of Women's Political Rights. London: Palgrave Macmillan.

[5] Matta, R. (2015). *Towards Effective National Forest Fund*. Rome: FAO Foresty Paper 174.

[6] Meiwanda, G. (2016). Kapabilitas Pemerintah Daerah Provinsi Riau: Hambatan dan Tantangan Pengendalian Kebakaran Hutan dan Lahan. *Jurnal ilmu sosial dan ilmu politik*, *19*(3), 251-263.

[7] Mumbunan, S., Ring, I., & Lenk, T. (2012). *Ecological fiscal transfers at the provincial level in Indonesia* (No. 06/2012). UFZ Discussion Paper.

[8] Nugroho, R. (2017). *Kebijakan Publik: Formulasi, Implementasi, dan Evaluasi*. Jakarta: Elex Media Komputindo.

[9] Purnomo, E. P., Ramdani, R., Setyadiharja, R., & Muzwardi, A. (2018). Collaborative Governance Tata Kelola Hutan Berbasis Masyarakat. Yogyakarta: Lembaga Penelitian, Publikasi dan Pengabdian Masyarakat (LP3M) Universitas Muhammadiyah Yogyakarta.

[10] Putra, I. K., Saharjo, B. H., & Wasis, B. (2019). Tantangan kelembagaan pengendalian kebakaran hutan dan lahan pada tingkat tapak. *Jurnal Ilmu Pertanian Indonesia*, *24*(2), 151-159.

[11] Rosenbaum, K.L. & Lindsay, J.M. (2001). *An overview of national forest funds: current approaches and future opportunities*. Rome: FAO Legal Papers Online 15.

[12] Suryakusuma, J. (2011), *State Ibuism: The Social Construction of Womanhood in New Order Indonesia*. Depok: Komunitas Bambu.

[13] Tapsell, R. (2015), Indonesia's Media Oligarchy and the "Jokowi Phenomenon", *Indonesia*, 99, 29-50.

[14] Wicaksono, A. (2019). Kolaborasi Multi Aktor dalam Program Restorasi Gambut di Provinsi Riau. *Jurnal Administrasi dan Kebijakan Publik*, *4*(2), 99-113.

[15] World Bank. (2016). *The Cost of Fire: An Economy Analysis of Indonesia's 2015 Fire Crisis*. Jakarta: World Bank.

IOP Conf. Series: Earth and Environmental Science 1041 (2022) 012041 doi:10.1088/1755-1315/1041/1/012041

Farmers' perceptions of the performance of agricultural instructors in mangosteen farmers' groups in Segati Village, Langgam District, Pelalawan Regency

Asgami Putri, Latifa Siswati, Hamdan Yasid

Universitas Lancang Kuning, Pekanbaru, Indonesia

*asgami@unila.ac.id, latifasiswati@unilak.ac.id, hamdanyasid@unilak.ac.id

Abstract. Implementation of agricultural extension operations in the form of Agricultural Extension Center (BPP) as an extension institution that fosters mangosteen farmer groups, especially in Segati Village with 1 (one) extension worker and 12 fostered Mangosteen Farmer Groups that have been formed by the government. things that greatly affect the welfare of existing farmer groups. This study aims to: determine farmers' perceptions of the performance of extension workers. This research was conducted in Segati Village, Langgam District, Kab. Pelalawan. The samples taken were farmers from the mangosteen farmer group. The analysis used in this research is factor analysis. The results obtained from this study are from several indicators that are included which affect the performance of the instructor based on the perception of the farmer is to provide guidance (X18).

1. Introduction

Extension activities in agricultural development act as a bridge that connects the practices carried out by farmers with the knowledge and technology of farmers who are always developing into the needs of these farmers. In order for farmers to carry out practices that support farming, farmers need information on innovation in agriculture. This information can be obtained by farmers, among others, from PPL (Field Agricultural Extension) through the implementation of agricultural extension activities. Extension can be an effective policy tool to encourage agricultural development in situations where farmers are unable to achieve their goals due to limited knowledge and insight. As a means of extension policy, only if it is in line with the interests of the government or the organization that funds extension services to achieve the farmer's goals.

The joining of farmers in farmer group forums can help explore potential, solve farming problems for their members to be more effective, make it easier to access information, markets, technology, capital and other resources [1]. deliberately grown through farmer group dynamics. Group dynamics are defined as the motion or forces that exist within the group, which determine or influence the behavior of the group and its members in achieving goals [2].

One of the prospective horticultural commodities to be developed in Indonesia is mangosteen. The government has designated mangosteen as a national superior commodity in the RUSNAS Buah (National Strategic Excellence Research for Fruits) since 2000. This is because mangosteen has the potential to continue to be developed. The area of the mangosteen garden in Segati Village is 130 hectares, as a village producing mangosteen production in Pelalawan, of course it must also have a

IOP Conf. Series: Earth and Environmental Science 1041 (2022) 012041 doi:10.1088/1755-1315/1041/1/012041

growing and independent farmer institution so that farmers can more optimally carry out their farming activities to increase their production, one of which is through Gapoktan.

2. Methodology

This research was conducted in Segati Village, Langgam District, Pelalawan Regency. Determination of this location is done intentionally (purposive). This research location was chosen because in Segati Village there are Mangosteen farmers who are members of a farmer group. The research time is carried out for 3 (three) months, from May to July 2021. The population is the entire Segati Village farmer group. With a total of 12 farmer groups and 130 farmer group members and has 1 instructor. For Sampling in this study using the Purposive Random Sampling technique, with the criteria for taking samples of the management consisting of the chairman, secretary and treasurer and 3 members of farmer groups who are active in each farmer group and extension workers are used as samples.

Factor Analysis

Factor analysis is a multivariate analysis method based on correlations between variables. Factor analysis is one of the statistical techniques that can be used to provide a relatively simple description by reducing the number of variables called factors. Factor analysis is used to reduce data or summarize, from many old variables that have been changed to a few new variables called factors, and still contain most of the information contained in the original variables.

[3]Factor analysis in multivariate analysis is classified as an interdependence analysis (interdependence technique) where the entire set of interdependent relationships is examined. Variables in one group will have a high correlation while variables in different groups will have a low correlation. In Supranto's [3] writing, it is said that factor analysis is used to reduce data/variables.

Factor analysis is used in the following conditions:

a. Recognizing or identifying the underlying dimensions or factors, which explain the correlation between a set of variables.
b. Recognizing or identifying a new set of variables that are not correlated (independent) which are less in number for
c. replace an original set of correlated variables in subsequent multivariate analyses.
d. Recognizing or identifying an important set of variables from a larger set of variables to be used in subsequent multivariate analysis.

There are two ways that can be used in conducting factor analysis, especially factor score coefficients, namely Principal component analysis (PCA) and Common factor analysis (CFA).

a. Principal component

The amount of variance present in the data set must be considered. The diagonal of the correlation matrix consists of numbers one and the full variance is carried in the factor matrix.[4] The principal component is recommended when it is essential to specify that the number of factors must be minimum taking into account the maximum variance in the data for use in further multivariate analysis.

b. Common factor analysis

The estimated factors are only based on the common variance and communalities which are included in the correlation matrix. [5] This method is considered very appropriate if the main goal is to identify/identify the underlying dimensions and common variances that attract attention.

3. Result and Discussion

3.1 Farmers' Perceptions of Extension Performance

From the results of data processing obtained from filling out questionnaires that have been distributed to both farmers and extension workers, the following results are obtained: data is processed using factor analysis from the SPSS program:

3.1.1 KMO dan Bartlett's Test

The results of the values from the KMO and Bartlett's Test, the first data entered into the application shows the number 48.4% can be seen in table 1 and this figure is not in accordance with the provisions where the analysis will continue if the KMO and Bartlett's Test values are more than 50%. So that further follow-up is needed, namely by removing several indicators whose MSA value is less than 50%. After this is done, namely the disposal of the missing indicator, the KMO and Bartlett's Test values are obtained to be 54.1% and this is in accordance with the provisions so that the analysis can be continued. The results can be seen in the table 1

Table 1. KMO dan Bartlett's Test 1

KMO and Bartlett's Test

Kaiser-Meyer-Olkin Measure of Sampling Adequacy.		.484
Bartlett's Test of Sphericity	Approx. Chi-Square	966.156
	df	253
	Sig.	.000

Table 2. KMO dan Bartlett's Test 2

KMO and Bartlett's Test

Kaiser-Meyer-Olkin Measure of Sampling Adequacy.		.541
Bartlett's Test of Sphericity	Approx. Chi-Square	828.123
	df	190
	Sig.	.000

3.1.2 Measures of Sampling Adequency (MSA)

The results obtained from this analysis show that the Anti Image Correlation value which is less than 50% will be removed and by eliminating the indicator whose anti-image correlation value is less than 50% will increase the KMO and Bartlett's Test values. From this result, the previously entered indicators amounted to 23 indicators, after deleting indicators whose value was less than 50%, there were 20 indicators. [6] From this calculation the missing indicators include providing the latest information in group development efforts (X7), providing agricultural tools with their own business (X9), facilitating farmer group members in accessing information from various sources (X10).

3.1.3 Communalities

The results obtained in this table show that the indicator with the highest value is the indicator based on the farmer's perception which largely determines the good performance of the instructor. From these results, the highest value is the indicator Providing guidance (X18) with a value of 90.5. [7]This means that farmers expect the extension workers to always provide guidance from the extension workers both in using the latest innovations, providing guidance in terms of solving problems faced by farmers and other things needed by farmers.

3.1.4 Total Variance Explained

The results obtained from Total Variance Explained are factors formed from the indicators that we enter into the analysis. From these results, it was found that the factors formed were 6 factors. [8]This can be seen from the Extraction Sum of Squared Loading value which is more than one. You can see the results in the table 3.

Total Variance Explained

Component	Initial Eigenvalues			Extraction Sums of Squared Loadings			Rotation Sums of Squared Loadings		
	Total	% of Variance	Cumulative %	Total	% of Variance	Cumulative %	Total	% of Variance	Cumulative %
1	3.992	19.958	19.958	3.992	19.958	19.958	3.829	19.147	19.147
2	3.485	17.427	37.385	3.485	17.427	37.385	2.808	14.041	33.188
3	2.209	11.046	48.431	2.209	11.046	48.431	2.335	11.675	44.864
4	1.782	8.908	57.339	1.782	8.908	57.339	2.266	11.331	56.194
5	1.425	7.127	64.466	1.425	7.127	64.466	1.516	7.578	63.772
6	1.167	5.833	70.299	1.167	5.833	70.299	1.305	6.527	70.299

Extraction Method: Principal Component Analysis.

Figure 1. Result of Total Variance Explained

3.1.5 Rotated Component Matrix

The results obtained in the rotated component matrix analysis are that each indicator that we include in this study includes any factor [9]. It can be seen that each indicator has been placed according to its respective factors.

Included in a) factor 1 are: Helping to solve problems (X13); assisting farmers in developing cooperation (X16); activate the roles of administrators and members of farmer groups (X17); apply the method of doing business (X19); provide input on how to get capital (X22); b) Factor 2 are : providing information and technological innovation (X5); Help market access (X11); play a role in assisting farmers to develop groups (X12); c) Factor 3 are : provide training (X2); Giving farmers time to conduct consultations (X15); carry out the review (X20); provide input and advice in increasing the economic added value of the product (X21); arouse the potential and capabilities that exist in farmers (X23); d) Factor 4 is : explaining the calculations in determining something in farming; e) factor 5 are : increase farmers' knowledge (X1); improve skills (X4); provide information on the price of production and production output (X8); Provide facilities and infrastructure to solve problems (X14); provide guidance (X18);f) factor 6 is: active instructor (X3).

4. Conclusion

Farmers' perceptions of the performance of the extension workers who are very influential are providing guidance (X18).

References

[1] Peraturan Menteri Pertanian Nomor 82 Tahun 2013. Tentang Pedoman Pembinaan Kelompok Tani dan Gabungan Kelompok Tani. BKP5K Kabupaten Bogor.

[2] Van Den Ban, A W and Hawkins H S. 2012 Penyuluh Pertanian Cetakan Ketiga Belas Yogyakarta (ID): Kanisius

[3] Poluan Juwendi. 2017. Dinamika Kelompok Tani Maesaan Waya Di Desa Manembo Kecamatan Langowan Selatan. Jurnal Agri sosial ekonomi. Vol 13

[4] Sukino, 2013. Membangun Pertanian dengan Pemberdayaan Masyarakat Tani. Pustaka Baru Press. Yogyakarta

[5] Aslamia Mardin. Awaluddin Hamzah (2017) "Peran Penyuluh Pertanian Dalam Pengembangan Kelompok Tani Di Kelurahan Mata Bubu Kecamatan Poasia Kota Kendari" Jurnal ilmiah membangun desa dan pertanian-Vol.2 (1):6-9 ISSN:2527-2748.

[6] Ginting, Nina Maksimiliana. dan Gardis Andari (2020) "Peran Penyuluh Pertanian Terhadap Pengembangan Usahatani Padi" Journal Agricola-Vol 10 (1) Maret : 2020. Hal 19 - 24 e-ISSN: 2354 – 77311

[7] Handayani, Furry. Sutadji. A. Margono (2015) "Analisis Persepsi Petani Terhadap Kompetensi Penyuluh Pertanian Lapangan Dalam Pembangunan Pertanian Di Kabupaten Kutai Timur" Journal Administrative Reform- (2015) 3 (2) : 276-285

[8] Lini, La. Awaluddin Hamzah. Sukmawati Abdullah (2018) "Peranan Penyuluh Pertanian Dalam Pengembangan Kelompok Tani Di Kelurahan Benua Nirae Kecamatan Abeli Kota Kendari" Jurnal Ilmiah Membangun Desa dan Pertanian Vol:3 (5) : 128-132

[9] Inten, Sekar. Dewi Elviana dan Budi Rosen Nover (2017) "Peranan Penyuluh Pertanian Dalam Peningkatan Pendapatan Petani Komoditas Padi Di Kecamatan Tanjungselor Kabupaten Bulungan Kalimantan Utara" Jurnal Agrifor-Volume XVI Nomor 1 Maret : 2017

IOP Conf. Series: Earth and Environmental Science 1041 (2022) 012042 doi:10.1088/1755-1315/1041/1/012042

Environment Conservation Responsibility in Community Development

N Sutisna[1], H Herdiansyah[2]*, and P Sanjatmiko[3]

University of Indonesia, Depok, Indonesia

Email: nana.sutisna@ui.ac.id[1], herdis@ui.ac.id[2], rihandoko09@ui.ac.id[3]

Abstract. Business sustainability is a synergy of three pillars, namely; economic, social and environmental. Implementation is reflected in the performance of corporate social responsibility, so that CSR programs that have been characterized of being reactionary and charitable can be transformed into community development activities that encourage participatory planning with the community by providing program assistance for sustainability. This study aims to analyze and describe the extent of business sustainability through the implementation of Corporate Social Responsibility (CSR) programs that transform caricature programs into community development programs by oil and gas companies in Indonesia. This research uses a qualitative approach and is carried out with descriptive analysis methods using literature study of secondary data in the form of documents based on relevant references. The results of this study provided a CSR model with community development through mentoring and capacity building as a form of transformation from charity programs while encouraging sustainable business practices. This study is expected to be able to contribute to sustainability studies, particularly in the context of corporations with CSR models in the extractive industry.

1. Introduction

The success of economic growth during the New Order government was supported by the extractive industry or the mining sector, especially the oil and gas sector which experienced the peak of the highest price increase known as the ‘oil boom’ which eventually became an additional foreign exchange for the country as development capital. In this decade, the extractive industry as one of the driving forces of development, on the economic side provided additional state income, but also brought social and ecological impacts and problems [1].The issue of the extractive industry that has came into the public spotlight is environmental issues as a result of development and industrialization policies during the New Order era until now. The potential for environmental damage caused by extractive industry activities, as well as the oil and gas industry, for example, includes air pollution, oil spills, industrial waste and disruption of biodiversity [2].

Another problem that arises from the extractive industry is social issues. Prayogo (2010) identified an increasing conflict between communities and corporations. Conflicts between companies or corporations with local communities are prone to occur in the extractive industry. The high tendency for conflicts between companies and local communities in the extractive, mining and oil and gas industries, as mentioned above is because it is related to the understanding and perception of local resources, which will eventually trigger problems. -Other issues include local workers, local entrepreneurs, as well as corporate social responsibility (CSR) activities [3].

Implementation of Corporate Social Responsibility is often understood as an activity or action that is charity, only takes the form of direct assistance or merely a reaction to community pressure " bush fire" [4]. Based on these problems, the paradigm and implementation of the company's CSR in an effort to business sustainability can be transformed into programs with a community development approach. Community development programs are expected to be able to contribute to the company's sustainability, thus success is no longer seen from the funds spent, but the magnitude of the contribution in solving social, economic and environmental problems [5]. Previous studies related to community development were more focused on community strategy [6], model reconstruction in the context of empowerment [7] and discourse analysis of business sustainability at the company level [8]. The study of ideas and practices of corporate social responsibility becomes a new spectrum in studying sustainability in relation to companies and business practices in the extractive industry.

This study aims to analyze and describe the concept of business sustainability through the corporate social responsibility (CSR) model in community development programs in oil and gas companies in Indonesia. The contribution of the literature on the concept of business sustainability, corporate social responsibility, community development is essential in an effort to understand sustainability through economic, social and environmental pillars in the context of business practices.

2. Method

This research uses a qualitative approach and is carried out with descriptive analysis method using literature study of secondary data in the form of documents based on relevant references in the context of knowledge and implementation of corporate social responsibility. Qualitative research is research conducted on the process and meaning of social reality, the thing being done is more focused on explaining social phenomena to be understood and interpreted as a reality [9].The analysis stages of this research were carried out by collecting data, compiling data and analyzing data. Analysis of the study begins by analyzing the concept of sustainable business and corporate social responsibility, then analyzing the implementation and assistance of community development programs.

The subject of this research is JOB Pertamina Medco E&P Tomori, an oil and gas company that has implemented a corporate social responsibility program that has received a Gold Proper award from the Ministry of Environment and Forestry..

3. Results and discussion

3.1. Business Sustainability: Economic, Social and Environmental Synergy

Efforts to achieve sustainability in a business context have various considerations and benefits, including high share value, cost savings, increased competitiveness, company image and reputation [10]. Various experts have defined business sustainability in various variants. Academics and practitioners define it with environmental conservation to create a better environment for future generations by focusing on internal corporate governance, efficiency, and sustainable collar organizational practices and fulfillment of corporate social responsibility (CSR). in meeting community needs that are beyond the company's obligations [11].

Table 1. Stages of corporate sustainability. [10]

	Compliance	**Business-Centered**	**Systemic**	**Regenerative**	**Co-evolutionary**
Sustainability spectrum position	Very Weak	Weak	Intermediate	Strong	Very Strong
Orientation	Economic Science-oriented. Business-oriented.	Economic Science-oriented. Business-oriented,	Economic Science-oriented. Business-oriented.	Ecological science-oriented. Ecology-oriented.	Ecological science-oriented. Ecology-oriented.
Understanding of sustainability	Meet compliance requirement internal firm-centric view.	"Do less bad" Internal firm-centric view.	"Do more good" Begins to look externally in defining sustainability. Business is part of a larger industry and community working together toward systemic change.	Repair damage to system.	Human and all earth's beings are in a mutually enhancing and beneficial relationships.
Relationships to natural world	To be managed and controlled; anthropocentric resources exploitation.	To be managed and controlled; anthropocentric resources exploitation. Eco-Efficiency.	To be managed and controlled; anthropocentric resources exploitation. Eco-Efficiency.	Part of natural world Operate within planetary boundaries. Manage and repair.	Self-management as part of the natural world. Participate in cooperative symbolic relationship with the natural world.
Economic growth	Pursuit of production, consumption, and growth.	Pursuit of production, consumption, and growth.	Pursuit of production, consumption, and growth.	Qualitative development without production, consumption and growth. Steady-state growth.	No growth in production or consumption. Qualitative improvement.
Sustainability concerns	Externally enforce or regulated activities. Defensive action with regard to economic,	"Business case" is the motivation and measure of success. Adoption and internal enforcement of activities	Integrates three realms of sustainability (economic, environmental, social). Work with other human system.	Integrates three realms of sustainability (economic, environmental , social).	Work in balance with other system. Contribute to flourishing of other systems.

	environmental, social concerns.	incremental improvements to business-as-usual. Many focus on one or more realms of sustainability (economic, environmental, social).		Work with other human system.	

The increasing number of companies implementing sustainable business, turns out to be a paradox. It often occurs that while many companies adopt sustainability, the environmental quality declines instead [12]. Based on the Landrum study [10], the model provides insight and understanding about sustainability in the corporate stage, so that the corporate sustainability paradigm will be more widespread. Business sustainability is a necessary component in planning and management cycles that are integrated into emerging risk aspects, so business sustainability also helps organizations to develop and balance an integrated approach to meet their economic, environmental and social aspects and their responsibilities to stakeholders [13].

Bose and Mahajan [14] in their work titled 'Business Sustainability: Exploring the Meaning and Significance' explain that the concept of business sustainability is influenced by the perspective of Elkinton (1997) with the Triple Bottom line approach which focuses on the synergy of Profit, People, Planet that in the third slice will promote sustainability. Sustainable business is explained as the process of managing an organization by considering three different aspects, namely, economic, social and environmental [14].

3.2. Corporate Social Responsibility: Efforts Towards Business Sustainability

The basic principles of corporate social responsibility [15] are distinguished between the charity principle or charity and the stewardship principle. These differences are described in the following table.

Table 2. Comparison of the basic principles of corporate social responsibility. [15]

	Charity Principles	**Stewardships Principles**
Definition	1. Business should give voluntary aid to society need person or groups	1. Business acting as a public trustee, should consider the interest of all who are affected by business decision and policies
Type of activity	1. Corporate philanthropic 2. Voluntary action to promote social good	1. Acknowledging business and society interdependence 2. Balancing the interest and need of many divers group in society
Examples	1. Corporate philanthropic foundation 2. Private initiative to solve social problem 3. Social Partnerships and needy group	1. Enlightened self interest 2. Meeting legal requirement 3. Stakeholder approach to corporate strategic planning

The basic principles of CSR cannot be separated from the definition of CSR proposed by A. Carrol (1979). An integrated definition of "the social responsibility of business encompasses the economic, legal, ethical, and discretionary expectation that society has of organization at a given point in time" . This definition ultimately gives way into the four corporate responsibilities known as the Carroll Pyramid. These responsibilities are economic, legal, ethical and philanthropic [16].

3.3. Community Development: Program Implementation and Assistance

The perspective of Jim Ife [17] sees that the concept of empowerment is closely related to the concepts of power (strength) and disadvantage (inequality). The concept of empowerment can be studied with four perspectives namely; pluralist, elitist, structuralist and post structuralist. Referring to these perspectives, it is necessary to understand the conceptual components of community development, namely the community component as a social unit that interacts with each other and the elements that exist in society such as human resources and physical elements and economic resources within the framework of livelihoods [18]. Community development is directed at building support communities,; a community structure based on the development and fair distribution of resources and the existence of social interaction, participation and mutual efforts to encourage one another [19].

Referring to this basic understanding and principle, the transformation of corporate social responsibility in the community development program is carried out by JOB Pertamina Medco E&P Tomori as an effort to build synergy between the company, the community and the surrounding environment. The transformation of the CSR program is directed at empowering the community with programs based on the potential and problems surrounding the community. The program planning carried out is only top-down, but is carried out by prioritizing community participation based on the results of social mapping.

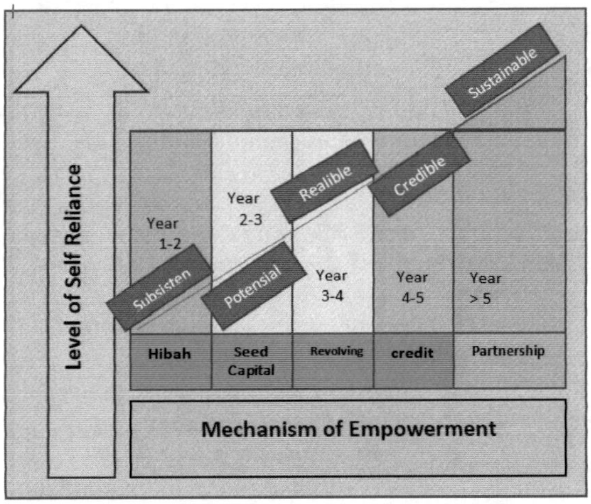

Figure 1. Empowerment mechanism. [4]

The implementation of community development programs requires a mentoring strategy so that the program beneficiaries are able to increase their capacity, both individually and institutionally. Community mentoring or assistance strategy will be able to increase the level of self-defense, be it groups or individuals, so that the empowerment mechanism will continue to lead from subsistence to sustainable. This community evelopment is long-term and continuous, so efforts to encourage subsistence in communities towards an independent and sustainable society are carried out through interrelated activities between mentoring and capacity building.

4. Conclusion

Sustainable business as a praxis is implemented in corporate social responsibility. However, the framework of knowledge and implementation tends to be understood as a charitable activity so that programs are often discontinuous and do not contribute to solving environmental and social problems. The transformation of corporate social responsibility model in community development programs is a relevant thing to do with reference to program sustainability and synergizing economic, social and environmental aspects.

IOP Conf. Series: Earth and Environmental Science 1041 (2022) 012042 doi:10.1088/1755-1315/1041/1/012042

The implementation of the CSR model in the community development program encourages stakeholder involvement and community participation in preparing a more participatory program plan. This was done by JOB Pertamina Medco E&P Tomori Sulawesi as an effort to synergize. Community development programs that are implemented as a program transformation from charitable activities to empowerment must be carried out with mentoring strategies to increase individual and institutional capacities aimed at independence and sustainability.

In the context of the extractive industry, the CSR model through community development is relevant to be carried out for the purpose of promoting social harmony, environmental preservation and economic improvement within the framework of business sustainability.

References

[1] Widyanta A 2013 Sampyuh: Genealogi Konflik Industri Ekstraktif di Lanskap Masyarakat Agraris *J. Pemikir. Sosiol.* **2** 87–106
[2] Sulistyono S 2015 Kegiatan Usaha Industri Migas Hubungannya dengan Dampak dan Tanggung Jawab Kelestarian Lingkungan Hidup *Swara Patra* **5**
[3] Prayogo D 2008 K*onflik Antara Korporasi dengan Komunitas Lokal Studi Empirik pada Industri Geotermal di Jawa Barat.* (Depok: FISIP UI Press)
[4] Sumardjo, Firmansyah A, Dharmawan L and Wulandari Y P 2014 *Implementasi Program Pengembangan Masyarakat* (Bogor: Care IPB)
[5] Nurani M D 2019 *Menuju Bisnis Berkelanjutan Petunjuk Praktis Pelaksanaan CSR* (Jakarta: Balai Pustaka)
[6] Setiowati S E 2006 *Analisis pemberdayaan masyarakat pesisir dalam pemanfaatan sumberdaya alam Kecamatan Tepus dan Tanjungsari Kabupaten Gunung Kidul* (Universitas Indonesia)
[7] Pratomo N 2015 *Rekonstruksi Model Pemberdayaan Masyarakat dalam Rangka Kerangka Pembangunan Berkelanjutan Berbasis Lingkungan Hidup* (Universitas Indonesia)
[8] Suryani G I 2016 *Visi Retorikal Korporasi Tentang Keberlanjutan Bisnis* (Universitas Indonesia)
[9] Cresswell J W and Clark V L P 2018 *Mendesain dan Melaksankan Mixed Method Research* (Yogyakarta: Pustaka Pelajar)
[10] Landrum N E 2017 Stages of Corporate Sustainability: Integrating the Strong Sustainability Worldview *Organ. Environ.* **31** 287–313
[11] Rezaee Z, Tsui J, Cheng P and Zhou G 2019 *Business Sustainability in Asia: Compliance, Performance and Integrated Reporting and Assurance* (New Jersey: John Wiley & Sons Inc)
[12] Dyllick T and Muff K 2015 Clarifying the Meaning of Sustainable Business: Introducing a Typology From Business-as-Usual to True Business Sustainability *Organ. Environ.* **29** 156–74
[13] Pojasek R B 2007 A framework for business sustainability *Environ. Qual. Manag.* **17** 81–
[14] Mahajan R and Bose M 2018 Business Sustainability: Exploring the Meaning and Significance *IMIKonnect* Vol 7 (2).
[15] Post J E, Lawrence A T and Weber J 2002 *Business and society : corporate strategy, public policy, ethics* (New York: McGraw-Hill)
[16] Carroll A B 2016 Carroll's pyramid of CSR: taking another look *Int. J. Corp. Soc. Responsib.* **1** 3
[17] Ife, J and Tesoriero F 2008 *Community Development; Alternative Pengembangan Masyarakat di era Globalisasi* (Yogyakarta: Pustaka Pelajar)
[18] Matarrita-Cascante D and Brennan M A 2012 Conceptualizing community development in the twenty-first century *Community Dev.* **43** 293–305
[19] Zubaeidi 2013 *Pemberdayaan Masyarakat Wacana dan Praktik* (Jakarta: Prenada Media)

IOP Conf. Series: Earth and Environmental Science 1041 (2022) 012043 doi:10.1088/1755-1315/1041/1/012043

Environmental Conservation in Ciletuh Geopark, West Java

R A Zuvara[1], H Herdiansyah[2*] and D Asteria[3]

University of Indonesia, Central Jakarta, Indonesia

Email: ressy.amalia01@ui.ac.id[1], herdis@ui.ac.id[2], donna@ui.ac.id[3]

Abstract. The implementation of sustainability is a keys concept for many industries. The tourism industry is no exception. Geopark as a conservation area that pays attention to sustainability by utilizing tourism activities as the socio-economic development of the surrounding community still has obstacles in its implementation. The low level of public understanding about the importance of tourism development in the Geopark area is still an obstacle. By using qualitative descriptive analysis methods through literature studies, this study aims to analyze the condition of the community and the role of stakeholders in realizing sustainable tourism. The substantial actor of stakeholders is required as a forum for coordination and communication for the development of tourist areas. It is significant because it determines community participation in the management of Geopark tourism areas.

1. Introduction

The implementation of sustainability is a leading concept for many industries, likewise with the tourism industry. The adoption of sustainability in tourism is a form of maintaining and maintaining the quality of the environment from overexploitation of resources [1]. Sustainable tourism is a guideline and best practice in tourism development that can run in a balanced and long-term property, regarding environmental, economic, and socio-cultural aspects [2][3].

Geopark is a protection concept that has a focus on Geological Heritage (Geoheritage), Geological Diversity (Geodiversity), and landscapes that are integrated with the concepts of conservation, education, and sustainable development through a bottom-up approach [4]. Geopark also has advantages in terms of utilization, which is an area that focuses on local communities for socio-economic improvement by utilizing Geological resources as a tourist area [5]. According to UNWTO [2], sustainable tourism development requires the informed participation of all relevant stakeholders and strong political leadership to ensure broad participation and reach a mutual agreement.

The Ciletuh Geopark area is an area that has unique geological potential (mineral rocks and landscapes) and developed as a tourist destination with comparative and competitive advantages and very varied natural potential [6]. Thus, the Geopark area is very likely to provide opportunities for job creation for the local community in terms of obtaining economic benefits through sustainable tourism [7].

With the recognition of Geopark Ciletuh as a Global Geopark in 2018, the government and the private sector still have many tasks ahead to maintain this status. Low public awareness about the importance of tourism development because most of the population is more interested in doing mineral rock mining activities as a livelihood. Thus, socialization and community empowerment must be carried out–to increase public awareness regarding Ciletuh Geopark tourism [8]. According to Dinata, [9] the use of community social capital in developing tourism potential in Geoparks has not been fully supported by capacity building in the management of tourist attractions, so the management is not yet professional. Whereas community involvement in tourism management will make tourism objects sustainable because of ecological conservation, which is carried out directly by the community (there is a sense of belonging so that people maintain existing tourism assets because it can improve the welfare). Ecological conservation and tourism management also supported by Raharjana [10] that the environment and tourism synergize and support each other. Nature can be maintained, which is followed by an increase in the progress of the community. Nevertheless, this assumption can only be implemented if the stakeholders are aware that environmental quality is the basis of selling value for tourists. This is also related to the environmental quality of the Ciletuh Geopark, where the maintenance of the natural environment is a priority.

Ortis et al. [11] stated that the community as a critical factor in the management and development of Geopark tourism areas is a future challenge in realizing sustainable tourism. Sustainable tourism uses natural and cultural resources and has an integral part in conservation efforts, empowering local economies and encouraging tremendous respect for cultural or cultural differences. Thus, stakeholders have an essential role in the implementation of sustainability in tourism [12]. The community, government, and active management bodies needed to carry our tourism activities to develop the Geopark area sustainably.

Through the concept of sustainable tourism approach, this study aims to analyze the potential and condition of the sustainability of the Geopark tourism area holistically including, natural, social, and economic resources, especially the involvement of stakeholders in managing the Geopark area. Thus, this can be used as material for evaluation and input for the government to implement future policies.

2. Method

The research was conducted with a qualitative descriptive analysis method, using a literature study from several relevant references. This research was carried out with the analysis stages, including data collection activities, data preparation, and data analysis. The analysis phase begins with analyzing the concept of sustainable tourism in the Geopark area, community readiness, and the role of stakeholders. Then the economic analysis is carried out by describing the tourism potential of the Geopark area based on literature data, as an effort caused by Geopark tourism to change the community's economy.

3. Results and Discussion

3.1. Sustainable Tourism in Geopark Area

Sustainable tourism is emerging as a concept that takes complete account of the current and future economic, social and environmental impacts to address visitors' needs, the tourism industry, the environment, and host communities. [13].

Geopark is a protection concept that focuses on Geological Heritage and landscapes, integrated with conservation, education and sustainable development [4]. Geopark also includes a geological heritage conservation management system and a sustainable development management system that aims to improve the welfare of the people living in the geological site area [14], Sustainability development is closely related to community development [15], which also requires the involvement of stakeholders in the implementation of sustainability [12].

These criteria are fundamental requirements for the tourism sector, which ensure that it refers to balanced environmental, economic, and socio-cultural aspects to ensure long-term sustainability based on Table.1 [11]. From the table above, it becomes the basis for implementing sustainable tourism development in a Geopark area. Because it is undeniable in the mechanism, every construction will affect the region or the surrounding area.

Thus, tourism development will develop, where planners must ensure harmony with the local environment, with efficacious planning and clear guidelines [16]. However, this still leaves obstacles in interpreting a balanced use (the distribution of resources and the ability of tourism) to involve and involve all stakeholders simultaneously [17].

Table 1. Substantive Criteria of Sustainable Tourism in Geopark. [2]

Citeria	Description
Ecologically sustainable	Synergizing geological diversity (minerals rock and landscape), biodiversity, and culture for conservation of natural resources and the environment
Socio-culturally acceptable	• Local communities can absorb tourism activities without causing social resistance and refer to traditional values. • Geopark management must come from the community or local authorities with a solid commitment to develop and implement management. • Involving the private sector and government to coordinate sustainable tourism development by having an effective organization, group, or committee.
Economically profitable	Distributed the job and opportunities for the host community to generate income for poverty alleviation.

The sustainable tourism approach is designed to encourage the positive impacts of tourism development on local social and cultural values and to recognize and manage any negative impacts. Problems may also arise if local people have to fight for their resources and are alienated from the comforts enjoyed by tourists. At the same time, sustainable tourism development offers a better environment for the community. It creates mutually beneficial relationships by placing local culture as part or even being the core of tourism products [11]. Thus, Geoparks can be a holistic way to integrate conservation initiatives consider elements of sustainable resource use, infrastructure provision, and local socio-economic development. Community empowerment through education, communication, strengthening tolerance and respect, cultural exchanges, and practical cooperation is much needed [18].

3.2. The Role of Stakeholders in Community Involvement

The process of community involvement participation is essential to ensure that community involvement continues to increase over time [19]. Local communities are essential stakeholders who must work closely with protected area managers. More conservation benefits need to be communicated to local communities so they can benefit from protecting a park. Thus, the use of the park impact on local communities in sustainable income, such as the use of tourism in protected areas. So, they also tend not to exploit it in a less sustainable and excessive [20].

Geopark is one of the concepts that contribute to sustainability, which requires community participation in its development and management, in addition to contributions from the government and the private sector [4]. Furthermore, Rosyidie said that the sustainable development of Geoparks is also to improve the socio-economic development of local communities through their participation [21]. So, it is not surprising that people in relatively poor local communities aspire to be involved in tourism [20].

According to Tao dan Læssøe [22][23], a viable model for a sustainable society should build strong adaptive capacities and encourage consistent participation from all stakeholders. Participation and empowerment of stakeholders is an essential element to support sustainable practices and improve the ability of communities to manage and respond to planned and unpredictable changes.

In addition, an institution also plays a role in sustainable tourism. An agency in a tourist area plays a valuable role as they provide a forum for discussion and offer a means of communication for many of interested individuals [3]. However, this strategy encounters several obstacles: resistance from the surrounding community due to the lack of public understanding of Geoparks [24], causing people not to care and not be fully involved [25]. Furthermore, according to Ortis [11], distortions in tourist areas may occur because residents may not get a share in the economic activities that grow from tourism.

Practically and theoretically, public acceptance of tourism is an important instrument. The open attitude of the community to be involved in tourism activities will form pro-tourism actions, thus triggering the development of a tourist area [24]. So, it is hoped that the community can take advantage of tourism as a livelihood. However, on the other hand, participation is also a dynamic process with its limitations. The growth and development of community participation are determined by three elements:

the opportunities provided, the willingness of the community, and the community's ability to participate [26].

3.3. Geotourism as an Economic Development

Geotourism emerged as a type of tourism that refers to environmental innovation, which is also being developed worldwide, especially as a sustainable development tool for the development of local and regional communities. Understanding the identity or character of an area or region becomes vital in developing Geotourism in the Geopark area. Geotourism was founded on the concept of the promotion of geo conservation that encourages appreciation and understanding of the earth's heritage through proper interpretation of the economic benefits generated through Geotourism [27].

Geotourism is one of the Geopark activities that have the potential to improve the welfare of its people with various activities such Geo-site Sightseeing (going around to see geological sites), Geo-tours (geological tours), Geo-sports (sports rides in the geopark area), Geo-study (educational vehicle in geopark area), Geo-conservation (nature conservation vehicle in geopark area) and Geo-festival [28]. Moreover, according to Dowling, Geotourism is also used as a vehicle to encourage sustainable tourism in some regions around the world [27].

Through Geotourism, livelihood diversification becomes a structural process by which sources of employment and income for villagers are developed and expanded [29] through tour guides and homestay owners, which integrates strategies for sustainable regional economic development [30]. In addition, geotourism provides a better understanding and utilization of the tourism market segmentation contributing to the development and marketing of more specific geotourism products (traditional products) and providing competitive global insight on the advantages of a Geopark [27].

However, Geoparks is not always seen as a valuable thing and puts to good use by the community. In terms of economy, geological environment, and traditionally, Geoparks are considered a potential source of minerals, thus making local communities tend to exploit to improve the economy [31][32]. Geotourism activities in the Geopark area have provided economic benefits for the community as business actors and workers, contributing to the local community's economy, for instance, in the form of absorption of workers who spread out as management units for Geotourism objects and tourism business units.

4. Conclusion

The existence of the Ciletuh Geopark as a form of sustainable tourism still leaves obstacles in its implementation, especially in terms of community involvement. The contribution of stakeholders in managing of the Geopark tourist area is essential as a forum for coordination and communication for the development of the tourist area. The challenge becomes complex when faced with the unpreparedness of the community in terms of public understanding of Geoparks to participate in tourism activities. So, this triggers the emergence of resistance to tourism development in the Geopark itself. Geotourism can be a potential for the growth of livelihood diversification for local communities to improve the economy.

The involvement of stakeholders in the Geopark is strongly influencing the community. It is essential because it will determine community participation in the management of the Geopark tourist area. Thus, strong collaboration and communication are needed between the community and the Regional Government as policymakers to realize sustainable tourism in Ciletuh Geopark.

References

[1] Pranita D 2017 Membangun Kapabilitas Dan Strategi Keberlanjutan Untuk Meningkatkan Keunggulan Bersaing Pariwisata Bahari Indonesia. Jurnal Vokasi Indonesia, **4,2**

[2] UNWTO 2020 Sustainable Development *unwto.org*

[3] Costas C , Hillel O, Matus S and Sweeting J 2003 Sustainable Tourism Concepts. Conservation International and UNEP, 1–30.

[4] UNESCO 2014 *United Nations Educational, Scientific and Cultural Organization: Guidelines*

and Criteria for National Geoparks seeking UNESCO's Assistance to Join the Global Geoparks Network (GGN)

[5] Henriques M H and Brilha J 2017 *UNESCO Global Geoparks: A strategy towards global understanding and sustainability*. IUGS, **40(4),** 349–355

[6] Yanuar Y, Anna Z, Rosana M F, Rizal A, Sudrajat A and Zakaria Z 2018 Keberlanjutan Pengembangan Geopark Nasional Ciletuh-Palabuhan Ratu Dalam Perspektif Infrastruktur Sustainable Development of Geopark National Ciletuh-Palabuhanratu in the Infrastructure Perspective. *J. sos. ekon*, **10.1** 64–76

[7] Hasriadi R 2020 Pengaruh Pengembangan Potensi Geowisata Geopark Nasional Karangsambung-Karangbolong Menjadi Geopark UNESCO

[8] Raseuki F and Choiriyati W 2019 Aktifitas Branding Geopark Ciletuh Oleh Dinas Pariwisata Dan Kebudayaan Dalam Perspektif Penta Helix *Communiverse* **3(2)** 165–175

[9] Dinata I P and Mussadun M 2015 Pengaruh Pengembangan Kawasan Wisata Geopark Merangin Terhadap Kesejahteraan Masyarakat Desa Air Batu *J. Pembang. Wil. Kota* **11(3)** 328

[10] Raharjo S T, Apsari N C, Santoso M B, Wibhawa B and Humaedi S 2019 Ekowisata Berbasis Masyarakat (Ebm): Menggagas Desa Wisata Di Kawasan Geopark Ciletuh-Sukabumi. *Soc Work* **8(2)** 158

[11] Ortis O and Gunawan M 2012. *Rencana Strategis Pariwisata Berkelanjutan dan Green Jobs Untuk Indonesia*. Kementerian Pariwisata dan Ekonomi Kreatif & ILO (International Labour Organization ii)

[12] Budeanu A, Miller G, Moscardo G and Ooi C S 2016 Sustainable tourism, progress, challenges and opportunities: An introduction. *J. Clean. Prod.* **111** 285–294

[13] UNWTO. (2013). Sustainable Tourism for Developlment Guidebook (1st ed.) *library1.nida.ac.th*

[14] Keever M P J and Zouros N 2005 Geoparks: Celebrating earth heritage, sustaining local communities. *Episodes* **28(4)** 274–278

[15] Zubaedi D 2013 *Pengembangan Masyarakat: Wacana dan Praktik* (Jakarta: KencanaPrenada Media Group)

[16] Alam A S 2008. Tantangan dan Peluang Indonesia Dalam Pemberdayaan Maysarakat Lokal. Ilmu Pemerintahan, Vol. 1.

[17] Hall C M, Gössling S and Scott D 2015 The Routledge Handbook of Tourism and Sustainability (London:Routledge)

[18] Halim S, and Komoo I 2011. The Geopark as a Potential Tool for Alleviating Community Marginality *Shima* **5(1)** 94–113

[19] Halim S A and Ishak N A 2017 Examining Community Engagement In Heritage Conservation Through Geopark Experiences From The ASIA Pacific Region Kajian Malaysia **35** 11–38

[20] Goodwin H 2002 Local community involvement in tourism around National Parks: Opportunities and constraints. *Curr. Issues Tour.* **5(3–4)** 338–360.

[21] Rosyidie A, Sagala S, Syahbid M M and Sasongko M A 2018 The Current Observation and Challenges of Tourism Development in Batur Global Geopark Area, Bali Province, Indonesia. *IOP Conf. Ser. Earth Environ. Sci.* **158(1)**

[22] Tao T C H and Wall G 2009 Tourism as a sustainable livelihood strategy. *Tour. Manag* **30(1)** 90–98

[23] Læssøe J 2010. Education for sustainable development, participation and socio-cultural change. *Environ. Educ. Res.* **16(1)** 39–57

[24] Zaenal A, Sagala G H and Armayanti N 2016 Analisis Partisipasi Masyarakat Dalam Peningkatan Daya Saing Pariwisata Di Kawasan Geopark Danau Toba. *J.l Bisnis & Akuntansi* **16(1)** 1

[25] Yuliawati A K, Pribadi K N and Hadian M S D 2016 Geotourism Resources as Part of Sustainable Development in Geopark Indonesia **15** 962–965

[26] Mardikanto T and Soebiato P 2013 *Pemberdayaan Masyarakat (Dalam Perspektif Kebijakan Publik)* (Bandung: Alfabeta)

IOP Conf. Series: Earth and Environmental Science 1041 (2022) 012043 doi:10.1088/1755-1315/1041/1/012043

[27] Dowling R K 2014 Global Geotourism – An Emerging Form of Sustainable Tourism. *Czech J. Tour.* **2(2)** 59–79
[28] Putri S M , Deliarnoor N A and Nurasa H 2020 Implementasi Kebijakan Pemerintah Provinsi Jawa Barat Tentang Pengembangan Kawasan Geopark (Studi Pada Pemerintah Kabupaten Sukabumi Dalam Pengembangan Kawasan Pariwisata Geopark Ciletuh 2019) *MODERAT.* **6** 171–187
[29] Scoones I 2015 Prelims - Sustainable Livelihoods and Rural Development. Sustainable Livelihoods and Rural Development, 23(Figure 1), iv–xvi.
[30] Jones C 2008 History of Geoparks. *Geol. Soc. Spec. Publ.* **300** 273–277
[31] Carvalho C N D and Rodrigues J 2010 Building a Geopark for Fostering Socio-Economic Development and To Burst Cultural Pride : the Naturtejo European Geopark (Portugal) 467–480.
[32] Ruban D A 2017 Geodiversity as a Precious National Resource: A Note on the Role of Geoparks. *Resour. Policy.* **53** 103–108

Effectiveness of silica gel from palm kernel shell ash as a moisture absorber of bottle packaging medicine

Lidya Novita[1*], Iswadi Idris[2]

[1]Poltekkes Kemenkes Riau, Riau, Indonesia
[2]Sucofindo Company, Riau, Indonesia

*lidya@pkr.ac.id

Abstract. Silica gel is widely used in industrial and pharmaceuticals, especially as a moisture absorber. Silica gel can be synthesized from a material containing high SiO2. One of natural material SiO2 resource is palm kernel shell ash (PKSA). PKSA contains 55.2% SiO2. The purpose of this study was to determine the effectiveness of silica gel from PKSA as a moisture content absorber for bottle packaging medicine. This study used experimental method with laboratory analysis. The steps in this study were initiated with the preparation of PKSA, synthesis of silica gel from PKSA, characterization of the synthesized silica gel with X-Ray Fluorescence (XRF) and X-Ray Diffraction (XRD) instruments, water content analysis to determine the optimum absorption time of silica gel as a moisture absorber for bottle packaging medicine followed by testing the effectiveness silica gel at the optimum absorption time. The results showed that obtained silica gel from PKSA with a yield of 34%. XRF characterization shows content of SiO2 by 45.13 %. XRD characterization shows the nature of the solid silica gel in the form of crystalline phase. The optimum absorption time is 2 weeks for silica gel from PKSA. The effectiveness of the silica gel from the PKSA is 85.65 % as bottle packaging medicine moisture content absorber. Further analysis is needed on how to minimize the level of impurities in silica gel of PKSA to increase the effectiveness as a bottle packaging medicine moisture absorber.

1. Introduction

Palm Kernel Shell (PKS) is one of the palm oil processing wastes, which is quite large, which reaches 60% of oil production. Combustion of PKS generated PKSA containing high silica (SiO2). Zarina [1], studied characterization of PKSA and obtained levels of SiO2 51.18%, Al2O3 4.61%, Fe2O3 3.42%, CaO 6.93% and MgO 4.02%. High silica content in PKSA indicated that this material can be used as adsorbent and as SiO2 source to produce natural silica gel. SiO2 is a porous solid compound consisting of the elements Silicon and Oxygen bonded covalently. SiO2 has a large surface area. In addition to composition and polarity, pore structure is also an important factor to consider. The pore structure is related to the surface area, the smaller the pores, the larger the surface area [2].

Silica gel as an adsorbent has been widely known because of its amorphous (unordered) polymer structure, the combination of Si-OH aggregates that form Si-O-Si directed polymers, making this compound has a fairly good stability. Si atoms in silica gel are covalently bonded to four O atoms in a tetrahedral arrangement. Each O atom is covalently bonded to another Si atom to form siloxane (-Si-O-Si-) and silanol (-Si-OH) functional groups. In general, the Si-O bond length is ±0.16nm and the Si-O-Si bond angle is around 1480 [3]. Heating does not change the structure of the silica gel, even when the silica gel has been saturated, heating is often carried out to remove or break the H2O bonds from the surface of the silica gel.

Silica gel has wide surface and high adsorption rate to water, so it is widely used in the manufacture of diapers and moisture absorbing agents in the packaging of various shoe products, medicines, and clothing to prevent mold and damage. Silica gel is a highly porous form of glass-like granule, silica gel is made synthetically from sodium silicate. Silica gel is a natural mineral that is purified and processed into one form of granules or beads. Silica gel has an average pore size of 2.4 nm and has a strong affinity for water molecules [4]. Silica gel can be synthesized from natural material that contains high SiO2. A study by Fahmi [5], on the analysis of the absorption of silica gel based on rice husk ash, showed that silica gel from rice husk ash had a greater absorption of water content than synthetic silica gel with a difference in absorption of 0.285%.

Research by Kalapathy [6], on a simple method for the manufacture of silica gel from rice husk ash. The silica gel was then characterized by XRF (X-Ray Fluorescence) to determine the type and amount of SiO2 and FTIR (Fourier Transform Infra-Red) to determine the functional groups in the silica gel. The silica gel produced has a purity of 93% SiO2 and 2.6% water. Research by Novita [7], regarding the use of PKSA to improve the quality of used cooking oil, it was found that the SiO2 content in PKSA was 55.2% from the XRF characterization. Silica gel is widely used, including in the pharmaceutical industry as a moisture content absorber. Therefore, this study aims to utilize PKSA for synthesis silica gel and analysis its absorption effectiveness on bottle packaging medicine.

2. Methodology

2.1. Preparation of PKSA

PKSA is obtained from palm oil processing plant in Siak Regency. PKSA burned in a furnace at temperature of 600 C for 8 hours to remove organic material. After cooling, PKSA then soaked with 1.2 M HCl for 24 hours to dissolve the metal content. After that PKSA then neutralized with NaOH, washed with distilled water, dried at 120°C for 5 hours until a constant weight was obtained [8].

2.2. Synthesis of Silica Gel from PKSA

10 g of PKSA was adjusted to pH 5, stirred for 2 hours and filtered with Whatman 41 filter paper. The residue was washed with distilled water and 60 ml of NaOH was added. Then boil with stir for 1 hour and filter with Whatman 41 paper. Wash the residue. All the filtrates were combined, 1N HCl was added to pH 7. Allow the filtrate to form to form a gel. Centrifuge at 2500 rpm for 15 minutes. Separate the supernatant and wash again. The gel was dried at 80 C.

2.3. Silica Gel Packaging

Silica gel is packed with a weight of 1 g into semi-permeable paper and then tested for its effectiveness in absorbing the moisture content of bottle packaging medicine.

2.4. Silica Gel Characterization

XRD characterization is a crystal structure determination of a solid with X-ray radiation. X-ray scattering is produced when a metal electrode is fired with electrons at high speed in a vacuum tube. A crystal can be used to diffract X-ray beams because the order of the X-ray wavelength is almost the same as or smaller than the order of the distances between atoms in a crystal [9]. XRF (X-Ray Fluorescence) or X-ray fluorescence is an analytical technique that can analyze the elements that build up a material. This technique can also be used to determine elemental concentrations based on the wavelength and amount of X-rays that are re-emitted after a material is bombarded with high-energy X-rays [10].

2.5. Moisture Content Analysis to Determine Optimum Absorption Time of Silica Gel

Silica gel is weighed before packaging. The packaged silica gel was put into a brown bottled medicine with a weight of 150 g medicine. The absorption process is left for 1 week, 2 weeks, 3 weeks and 4 weeks. Silica gel after the absorption process is weighed again. The moisture content of absorption of silica gel in bottle packaging medicine is determined as follows:

$$\textbf{\textit{Moisture Content}} = \frac{(\text{g silica gel after absorption}) - (\text{g silica gel before absorption})}{\text{g silica gel before absorption}} \; x\ 100$$

2.6. Determination of Silica Gel Effectiveness at Optimum Absorption Time
The silica gel obtained with the optimum absorption time was heated to a constant weight. Then the process of absorbing the moisture content of the bottle packaging medicine was repeated until 2 times repetition.

$$\%R = \frac{(\text{moisture content after repetition})}{(\text{moisture content before repetition})} \; x\ 100$$

%R = Recovery Percentage

3. Result and Discussion
The results and discussion are based on the methods of this study.

3.1. Synthesis of Silica Gel from PKSA
The first step PKSA is heated in furnace at 600 °C for 8 hours to remove the organic fraction so that only the inorganic fraction remains and to increase the quantity of SiO2 in the PKSA. The ash produced in this process has a light gray color. PKSA then washing with Hydrochloric Acid (HCl) 1.2 M solution, this process is to eliminate the levels of impurities such as metal oxides. The impurities will form salts and water molecules when reacted with HCl. The resulting salt has a high solubility in water, so the impurities in the form of salt will be lost during the washing process [6].

The next step is to extract the silica contained in the PKSA using 1N sodium hydroxide (NaOH) solution. The extraction process was carried out using a hot plate with constant stirring with a magnetic stirrer for 1 hour at a temperature of ±120 °C. The sodium silicate resulting from the extraction process is cooled and filtered to separate the insoluble residue or precipitate from the filtrate in the form of a sodium silicate solution. Then the residue was washed with 100 mL of distilled water to optimize the dissolution of sodium silicate. The resulting sodium silicate solution is cloudy white. The mechanism formed during the formation of the sodium silicate is shown in Figure 1 below :

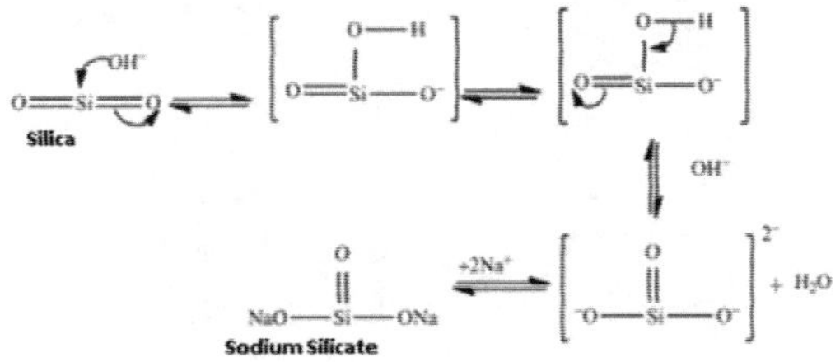

Figure 1. Reaction Mechanism of Sodium Silicate Formation [11]

Based on the reaction, NaOH dissociates to form sodium ions (Na+) and hydroxide ions (OH-). The hydroxide ion is able to attack the Si atom in SiO2 to form a silicate. The Si atom in the presence of an empty d orbital is able to accept an electron pair from O to OH-. In addition to SiO2, the high electronegativity of O atoms causes Si to be more electropositive and an unstable intermediate [SiO2OH]- is formed, resulting in dehydrogenation and the second hydroxyl ion will bind to hydrogen to form a water molecule. Sodium ions will balance the negative charge formed and bind to SiO32- ions to form sodium silicate [12].

The formation of silica gel is carried out through the sol-gel process, by adding an acid solution to the sodium silicate solution produced in the previous stage. The acid used in this study was 1N HCl. The addition of 1N HCl in the process of forming silica gel will cause a condensation reaction to the silicate ion. The picture is the reaction mechanism for the formation of siloxane bonds in the xerogel formation process.

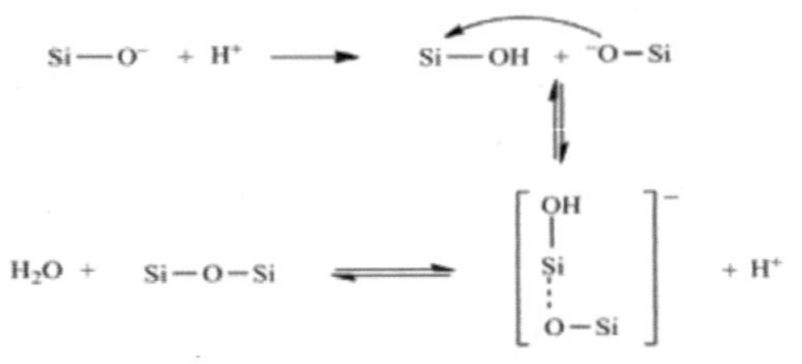

Figure 2. Reaction Mechanism of Siloxane Bond Formation in Xerogel Forming Process [11]

The sol-gel process will produce xerogel which is then dried to form dry white silica. The resulting Xerogel can be seen in the figure 3.

Figure 3. Xerogel

Xerogel was dried at 80°C until it dries to form dry silica gel in the form of a white and odorless solid. The yield of silica gel from PKSA is 34%. Figure 4 is silica gel from PKSA.

Figure 4. Silica Gel from Palm Kernel Shell Ash

The dry silica gel formed was then characterized using XRF (X-Ray Fluorescence) and XRD (X-Ray Diffraction) instruments.

3.2 XRF Characterization of Silica Gel from PKSA

The results of XRF characterization of silica gel from PKSA are shown in table 1 below :

Table 1. XRF Characterization of Silica Gel

Element	Conc	Unit	Compound	Conc	Unit
Mg	1,29	%	MgO	1,98	%

Al	0,98	%	Al_2O_3	1,67	%
Si	23,56	%	SiO_2	45,13	%
P	1,83	%	P_2O_5	3,49	%
Cl	69,99	%	NaCl	45,55	%
K	1,27	%	K_2O	1,11	%
Ca	1,03	%	CaO	1,03	%
Fe	0,02	%	Fe_2O_3	0,02	%
Zn	0,01	%	Zn	0,00	%

From table 1 can be seen that the content of SiO2 in the silica gel from PKSA is 45,13 %. Other compounds which also dominates is NaCl at 45,55 %, because in the process of synthesis of silica gel from PKSA, use of NaOH and HCl as a reagent to form a byproduct in the form of NaCl. The presence of oxides and other elements in silica gel indicates that the process of immersing PKSA using 1.2M HCl is not optimal as well as when washing the soaking PKSA with distilled water. PKSA naturally contains various elements and oxides. Novita [7] characterized PKSA with XRF obtained SiO2 as 55,2 %, meaning that there is an impurity compound, or another oxide contained in the PKSA.

3.3. XRD Characterization of Silica Gel from PKSA

The results of the characterization of silica gel from PKSA are shown in Figure 5 below:

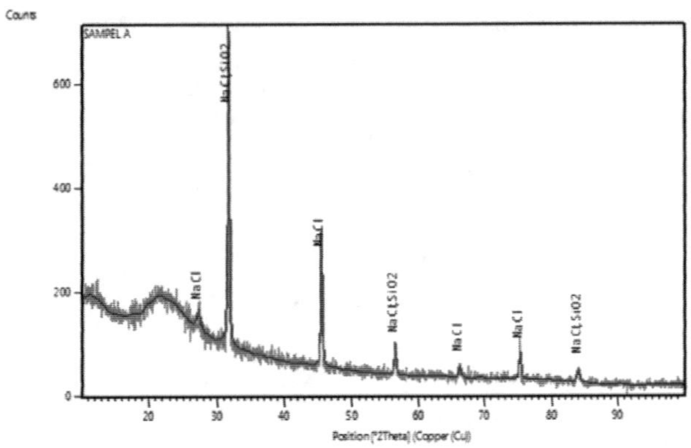

Figure 5. Diffractogram of Silica Gel from PKSA

Figure 5 shows that silica gel solids from PKSA are crystalline solids with the mineral name Cristobalite. In addition, other minerals with solid properties which are also crystalline are also detected, namely solid NaCl. The presence of NaCl in silica gel is a by-product of the silica gel synthesis process using NaOH and HCl as reagents. The purity of the silica gel produced has not been optimized, there are the presence of peaks from other compounds, allegedly caused by a less than optimal ash washing process so that there are other mineral residues left behind [11][13].

Based on its molecular structure, silica is divided into two, namely amorphous and crystalline silica. Amorphous silica is an irregular molecular arrangement silica, in other words, it does not have an arrangement pattern. Crystalline silica is a certain pattern molecular arrangement silica. Crystalline silica is also called polymorph [14,15]. Silica gel with amorphous solid properties has a higher surface area than silica gel with crystalline solid properties. From the XRD analysis, silica gel from PKSA is crystalline so it has a lower surface area. Dipowardhani [16] synthesized crystalline silica using sodium silicate and TEOS as silica sources. The result of silica characterization using XRD analysis method is that the synthesized crystalline silica material is a type of impure stishovit, quartz and cristobalite.

3.4. Moisture Content Analysis for Determination of Optimum Absorption Time of Silica Gel

The optimum absorption time of silica gel from PKSA can be seen in Figure 6 below :

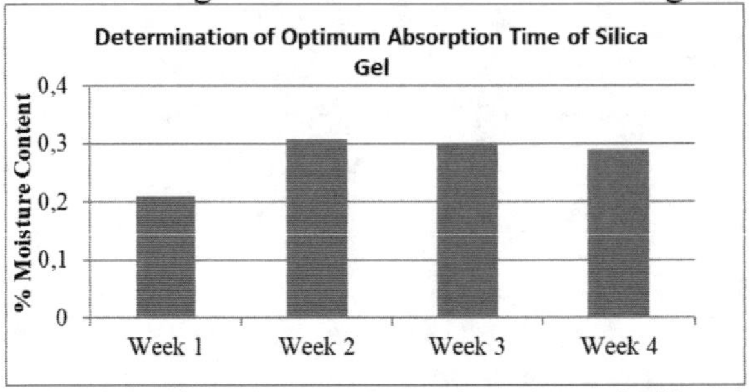

Figure 6. Determination of Optimum Absorption Time of Silica Gel from PKSA

Figure 6 shows the absorption of moisture content in the bottle packaging medicine by silica gel increased from the first week to the second week. Furthermore, there was a decrease in the absorption of silica gel in the third and fourth weeks. In the first week, the moisture content absorbed by silica gel was only 0.2096%. Silica gel had more capacity to absorb moisture content because not all siloxane groups (Si-O-Si) bonded with water molecules to form silanol groups (Si-OH). The optimum absorption of silica gel occurred in the second week of 0.3094%.

The tendency of absorption of silica gel from PKSA after 2 weeks, decreased with increasing absorption time. Based on the XRF characterization data, silica gel from PKSA has impurities in the form of other oxides. The absorption of silica gel PKSA in the fourth week tends to decrease, presumably due to the presence of impurities in the silica gel so that the absorption of silica gel decreases overtime. The mechanism of absorption of water (H2O) by SiO2 in silica gel is the formation of hydrogen bonds between water molecules and the oxygen of the silica gel [11].

3.5 Silica Gel Effectiveness As a Moisture Content Absorber At Optimum Absorption Time

The effectiveness of silica gel from PKSA, can be seen in Figure 8 below :

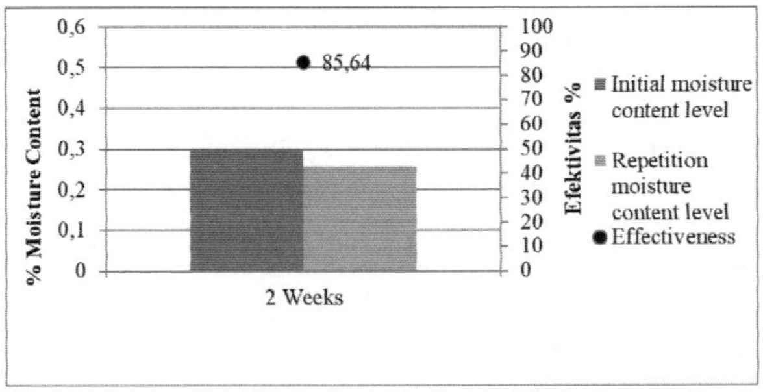

Figure 7. Silica Gel Effectiveness as a Moisture Content Absorber at Optimum Absorption Time

From Figure 7, the effectiveness of silica gel from PKSA is 85.64%. If the silica gel has more silica content, then the silica will bind more –OH and O which comes from the water vapor that absorbs the silica gel. This causes an increase in the absorption of silica gel [5]. From XRD analysis, the solid nature of silica gel of PKSA is crystalline. Amorphous solids have a higher porosity which makes the surface area larger, and the absorption capacity of H2O also increases. Another advantage of this silica gel is that it comes from natural ingredients, namely PKSA, so it is safer to use. Silica gel in daily applications is used as an adsorbent in food, because it has the ability to absorb moisture so as to prevent food spoilage during storage [17]. Silica gel is also needed to store laboratory equipment and electronic equipment so that they are not damp. Silica gel, which is known as a drying agent, is in great demand by consumers [18].

4. Conclusion

Silica gel from PKSA with a yield of 34%. XRF characterization shows content of SiO2 by 45.13 %. XRD characterization shows the nature of the solid silica gel in the form of crystalline phase. The optimum absorption time is 2 weeks for silica gel from PKSA. The effectiveness of the silica gel from the PKSA is 85.65 % as bottle packaging medicine moisture content absorber. Further analysis is needed on how to minimize the level of impurities in silica gel of PKSA to increase the effectiveness as a bottle packaging medicine moisture absorber.

References

[1] Zarina Y, Mustafa Al Bakri AM, Kamarudin H, Nizar IK, Rafiza AR. Review on the various ash from palm oil waste as geopolymer material. Rev Adv Mater Sci. 2013;34:37–43.
[2] Meisrilestari Y, Khomaini R, Wijayanti H. Pembuatan Arang Aktif Dari Cangkang Kelapa Sawit Dengan Aktivasi Secara Fisika, Kimia Dan Fisika-Kimia. Konversi. 2013;2(1):46–51.
[3] An D, Guo Y, Zhu Y, Wang Z. A green route to preparation of silica powders with rice husk ash and waste gas. Chem Eng J [Internet]. 2010;162(2):509–14. Available from: http://dx.doi.org/10.1016/j.cej.2010.05.052
[4] Lima SPB de. Production of silica gel from rice husk ash. Quim Nov. 2011;34(1):71–5.
[5] Fahmi H, Nurfalah AL. Analisa daya serap silika gel berbahan dasar abu sekam padi. J Iptek Terap. 2016;10(3):176–82.
[6] Kalapathy U, Proctor A, Shultz J. A simple method for production of pure silica from rice hull ash. Bioresour Technol. 2000;73:257–62.
[7] Novita L, Asih ER, Arsil Y. Utilization of palm kernel shell ash to improve used palm cooking oil quality. Atl Press. 2020;22(ISHR 2019):255–60.
[8] Ma X, Zhou B, Gao W, Qu Y, Wang L, Wang Z, et al. A recyclable method for production of pure silica from rice hull ash. Powder Technol [Internet]. 2012;217:497–501. Available from: http://dx.doi.org/10.1016/j.powtec.2011.11.009
[9] Hadi I, Arsa M, Sudiarta IW. Sintesis silika gel dari abu sekam padi dan abu limbah pembakaran batu bata dengan metode presipitasi. J Kim. 2013;7(1):31–8.
[10] Pausa Y, Malino MB, Arman Y. Optimasi tingkat kemurnian silika SiO2 dari abu cangkang sawit berdasarkan konsentrasi pengasaman. Prism Fis. 2015;3(1):01–4.
[11] Yusuf M, Suhendar D, Hadisantoso EP. Studi karakterisasi silika gel hasil sintesis dari abu ampas tebu dengan variasi konsentrasi asam klorida. 2014;VIII(1):16–28.
[12] Suprihatin E, Anita Zaharah T, Wahyuni N, Hadari Nawawi JH. Pembuatan Membran Silika Dari Fly Ash Dan Aplikasinya Untuk Menurunkan Kadar COD Dan BOD Limbah Cair Kelapa Sawit. JKK. 2015;4(3):48–53.
[13] Ramadani K. Sintesis dan karakterisasi silika gel dari limbah kaca untuk menurunkan kesadahan air. J Saintifik. 2018;4(2):179–85.
[14] Zaki A, Saputra E, Fadli A. Pembuatan silika high grade dari fly ash sawit dengan proses ekstraksi dan cation exchange. Jom FTEKNIK. 2017;4(2):1–4.
[15] Tadjarodi A, Haghverdi M, Mohammadi V. Preparation and characterization of nano-porous silica aerogel from rice husk ash by drying at atmospheric pressure. Mater Res Bull [Internet]. 2012;47:2584–9. Available from: http://dx.doi.org/10.1016/j.materresbull.2012.04.143
[16] Dipowardhani BT, Sriatun, Taslimah. Sintesis Silika Kristalin Menggunakan Surfaktan Cetiltrimetilamonium Bromida (CTAB) dan Trimetilamonium. J Kim Sains dan Apl. 2008;11(1):20–8.
[17] Nopianingsih N, Sudiarta I, Sulihingtyas W. Sintesis silika gel terimobilisasi difenilkarbazon dari abu sekam padi melalui teknik sol gel. J Kim. 2015;9(2):226–34.
[18] Yusrin A, Susatyo E, Mahatmanti F. Perbandingan kemampuan silika gel dari abu sabut kelapa dan abu sekam padi untuk menurunkan kadar logam Cd. J MIPA Unnes. 2014;37(2):154–62.

Comparative study of village community participation in peat restoration in the Peat Swamp Forest Management Unit (PSFMU) Tebing Tinggi Island, Indonesia

H. Malik, Ig. L. S. Purnama, Sudarmadji, L. W. Santosa, U. Widyanarko, M. Naufal

Universitas Gadjah Mada, Yogyakarta, Indonesia

*hapriadi.malik@mail.ugm.ac.id

Abstract. Community participation greatly determines the success of a peatland restoration program because people are the core component of its implementation. This research was intended to compare levels of participation in the restoration program between communities in the villages where a governmental intervention program for a good village and peatland governance termed Desa Peduli Gambut had been introduced (DPG villages) and those in non-DPG villages. A questionnaire survey was conducted to collect data directly in the field using several relevant questions related to peatland restoration, then the responses to these questions were quantified to measure community participation. The results showed that the community participation in DPG villages had higher mean ranks than in non-DPG villages, indicating higher public awareness of program participation in the former than the latter. Further, the difference in participation level between the two villages was of high significance, which means that the government's intervention through the DPG program is effective in raising public awareness of peatland restoration measures.

1. Introduction

The Government of Indonesia facilitates better management of tropical peatlands by issuing and enforcing a number of policies and laws. Despite these efforts, peatlands continue to degrade at an alarming rate [1]. In 2016, the government established the Peat Restoration Agency (Badan Restorasi Gambut or BRG) to manage peat hydrological units (PHUs), vegetation, and socioeconomic carrying capacity of the community in degraded peat swamp forest management units (PSFMUs) [2]. BRG has introduced several courses of action to achieve the target of involving and assisting 109 villages in the Desa Peduli Gambut (DPG) program—i.e., a governmental intervention to establish a good village and peatland governance—through collaboration with institutional partners.

This DPG mentoring pattern has five target aspects: participatory planning, participatory mapping (tenurial certainty), peatland restoration, economic revitalization, and cultural revitalization and local wisdom. DPG is expected to take part in aligning development programs in villages located in peatlands, especially within and around restoration areas.

The basis in determining DPG villages includes those located in peatlands that show indications of environmental degradation. Some examples include frequent occurrences of land fires and the construction of canals on peatlands [3] [4]. These can be found in several villages in PSFMU Tebing Tinggi Island where degradation persists. Therefore, BRG designated several DPG villages in the unit,

IOP Conf. Series: Earth and Environmental Science 1041 (2022) 012045 doi:10.1088/1755-1315/1041/1/012045

namely Lukun, Sungai Tohor, Nipah Sendanu, and Sendanu Darul Ihsan. However, despite the program implementation, some of these villages still show a sign of degradation, i.e., annual land fires from 2016 until 2019 [5].

Several studies of peatland management agree that active participation of the community is the key to sustainable peatland management. Thus, it is imperative that the level of community participation in peatland restoration measures be assessed to trace the root of the problem. Furthermore, comparing levels of community participation in DPG and non-DPG villages is expected to be able to address the root causes of persistent peatland degradation.

Based on the above problems, the research focuses on comparing the levels of participation of the people living in villages with and without the DPG concession. Local communities in DPG villages have been engaged in various governmental intervention programs relevant to sustainable peatland restoration strategies. This is not the case for those in non-DPG villages.

2. Methodology

The study area was PSFMU Tebing Tinggi Island, part of the Tebing Tinggi Timur District. The samples were Sungai Tohor and Nipah Densanu for DPG villages and Tanjung and Teluk Buntal for non-DPG villages (Figure 1). This research used a quantitative method and collected data through a questionnaire survey.

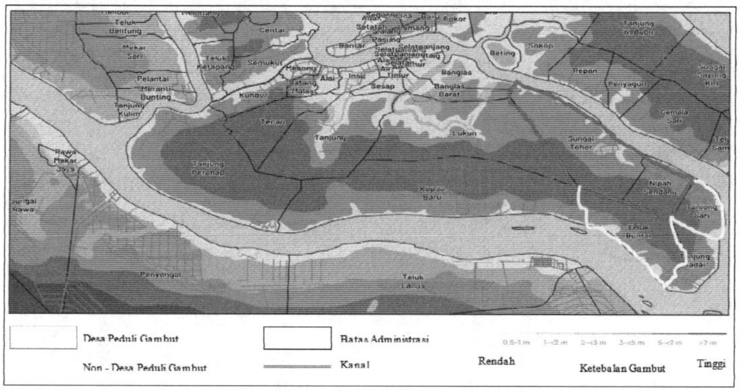

Figure 1. Map of the Research Locations: DPG and Non-DPG villages

Purposive sampling was applied to collect data from 729 respondents (Table 1). The targeted respondents differed in age, sex, education level, livelihood, income, and length of stay in the area. Each answer was given a weight value and tested with the Mann-Whitney test to determine the level of community participation from DPG and Non-DPG villages. These results can then be used as a basis to follow up the established government programs, especially related to effectiveness in increasing public awareness of peatland restoration.

Table 1. Questionnaire Survey Respondents in DPG and Non-DPG Villages

No	Village Name	Population Size (people)	Head of household (household)	Village criteria	Sample (people)	Duration
1	Sungai Tohor	1.252	347	DPG	188	August 25 until September 25, 2020
2	Nipah Sendanu	1.232	341	DPG	168	
3	Teluk Buntal	1.252	347	Non-DPG	188	
4	Tanjung Sari	1.122	324	Non-DPG	185	
		Total			729	

3. Result and Discussion

There are nine participatory actions asked in the questionnaire. Responses given during the interviews were qualitatively analyzed then quantitatively processed with the Mann-Whitney test (mean ranks). The graph below (Figure 2) shows the test results for communities in DPG villages.

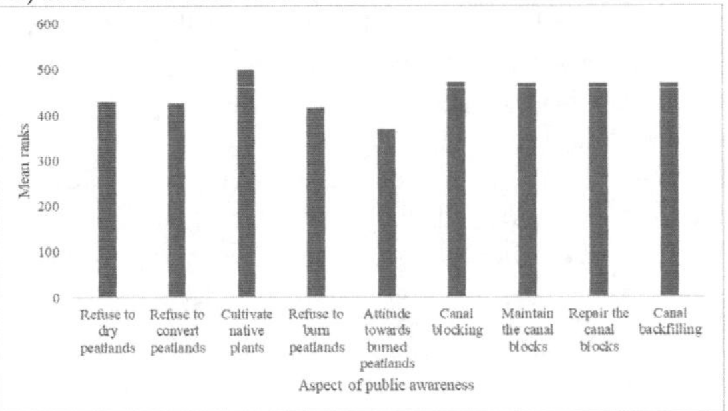

Figure 2. Community Participation in DPG Villages

The results showed that cultivating native plants had the highest mean rank among the forms of participation, i.e., 498.97. Meanwhile, the attitude towards burned peatlands had the lowest mean rank, 367.86. These results mean that any interventions established to raise public awareness of land burning in peatlands in the DPG villages must be improved.

Unlike in the DPG villages, the attitude towards burned peatlands had the highest mean rank, 362.27, among the participatory actions in non-DPG villages (Figure 3). Meanwhile, cultivating native plants had the lowest mean rank, 237.14 (Figure 3). Therefore, strategies to increase public awareness and knowledge of all participatory actions in non-DPG villages need to pay more attention to native plant cultivation.

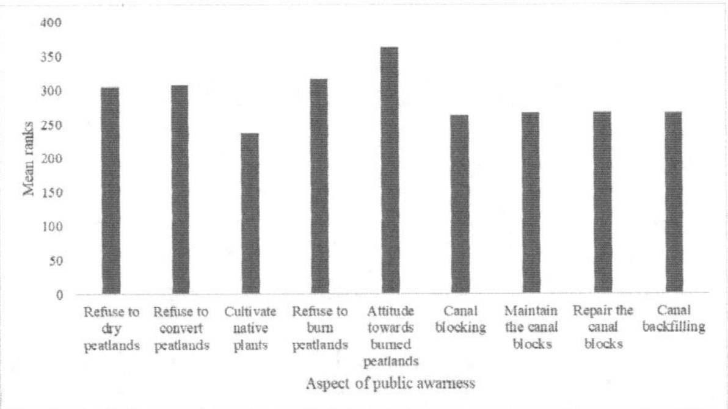

Figure 3. Community Participation in Non-DPG Villages

The mean ranks of community participation in both DPG and non-DPG villages are presented in a graph (Figure 4). The results showed that the people living in the DPG villages showed a higher interest in peatland restoration than those in the non-DPG villages. However, further analysis revealed that public attitudes about burned peatlands in both villages had similarly low mean ranks, which indicates that public education about land fires and their impact, especially in DPG villages, is less optimal.

However, these low mean ranks can indicate the contrary, i.e., the public education is optimal but some people in the community still choose to burn land for peatland management. Thus, public education and interventions need to pay specific attention to raising awareness of peatland fires and their impacts to prevent further practices.

Moreover, there should also be more strategies for improving public knowledge and interest in peatland restoration targeting people in non-DPG villages who live around peatlands. As previously mentioned, in addition to raising public knowledge of all aspects of awareness, future interventions need to emphasize native plant cultivation to avoid massive land conversion. This condition is based on the mean rank of this aspect, which was the lowest compared with other aspects in non-DPG villages.

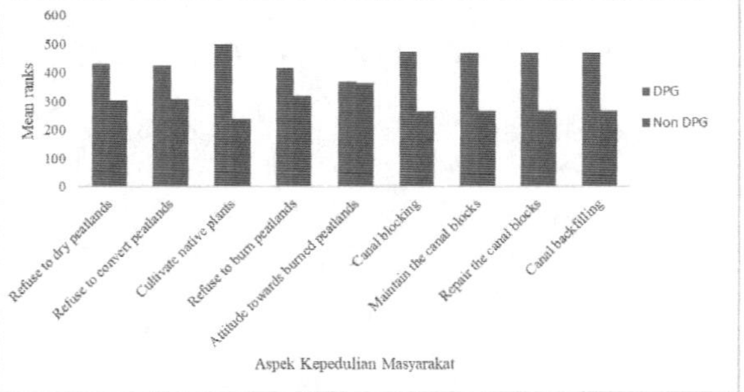

Figure 4. Bar Chart Comparing Levels of Community Participation in DPG and non-DPG Villages

The Mann-Whitney test was also used to measure to what extent the levels of public awareness in two villages differed. The results showed that the public awareness in DPG had a statistically significant interval, as evidenced by the critical value of 17 > 0 (Table 2).

Therefore, it can be concluded that the levels of public awareness in DPG and non-DPG villages are significantly different at a 5% significance value. The test results can provide a basis or reference for the government to decide on follow-up actions in both villages, and in this case, current interventions prove quite effective in raising public awareness of the peatland restoration programs.

Table 2. The Results of the Mann-Whitney Test of Public Awareness in DPG and non-DPG Villages

Aspects	**Scores**
N1 (DPG)	9
N2 (Non-DPG)	9
Total Rank (DPG)	126
Total Rank (Non - DPG)	45
U1 (DPG)	0
U2 (Non-DPG)	81
Statistical Value	**0**
Critical Value (5% significance level)	**17**
Conclusion	**Statistically significant**

4. Conclusion

The overall level of community participation in DPG villages is higher than in non-DPG villages, as evident from the mean ranks. Also, the Mann-Whitney test results have confirmed a statistically significant difference between these levels at a 5% significance level. In conclusion, current governmental interventions in DPG villages have been effective in raising public awareness of peatland restoration measures.

Acknowledgment

The authors would like to thank the Government of Riau Province for supporting and funding this research through the Doctoral Dissertation Research grant scheme in the fiscal year 2020.

References

[1] Uda, S., K. Hein, L., Atmoko, D. (2018). Assessing The Health Impacts of Peatland Fires: A Case Study for Central Kalimantan, Indonesia. doi.org/10.1007/s11356-019-06264-x.

[2] BRG. (2017). Siaran Pers Badan Restorasi Gambut Republik Indonesia No: Sipres/Brg/19/12/2017. Retrieved from https://brg.go.id/wp-content/uploads/2017/12/281217_SIPRES-AKHIR-TAHUN-BRG.pdf

[3] Wahyunto, Ritung, S., Nugroho, K., Sulaeman, Y., Tafakresnanto, H. C., Suparto, & Sukarman. (2013). Atlas Lahan Gambut Terdegradasi Pulau Sumatera. Jakarta.
[4] Masganti, Wahyunto, Dariah, A., Nurhayati, & Yusuf, R. (2014). Karakteristik dan Potensi Pemanfaatan Lahan Gambut Terdegradasi di Provinsi Riau. Jurnal Sumberdaya Lahan, 8(1), 59–66.
[5] Khotimah, G. K., Sutikno, S., Yusa M., Wijatmiko, I. (2020). Analisis Pengaruh Penyekatan Kanal Untuk Pembasahan Lahan Gambut Tropis. Jurnal Rekayasa Sipil 4(2), 129-135

Swot analysis of utilization of palm oil mill effluent to improve soil quality and crop productivity (case study at PT X, Lampung, Indonesia)

Dela Febriana, Sylvia Madusari*, Vira Irma Sari

Politeknik Kelapa Sawit Citra Widya Edukasi, Bekasi, West Java, Indonesia

*smadusari@gmail.com

Abstract. Crude Palm Oil (CPO) manufacturing in palm oil processing plants produces by- products known as palm oil mill effluent (POME). POME can improve the physical and chemical properties of the soil because it contains nutrients N, P, K, Mg, and Ca. One of the companies that have processed and utilized POME is PT X, Lampung Province, Indonesia. This study was conducted to analyze the strengths, weaknesses, opportunities, and threats (SWOT) of the effect of using POME on oil palm plantations for waste quality, physical soil conditions, soil chemical conditions and productivity of oil palm plants. The methodology of the research was using descriptive statistical methods of the secondary data and SWOT analysis. The results showed that the palm oil effluent applied in the PT X complied with the national quality utilization and waste management standards. The POME application has a beneficial effect on soil conditions. The involved area had higher pH, N, P, K, Ca, and Mg than control, but C- Organic and CEC in the application area were lower. Furthermore, the application of palm oil effluent tends to increase crop productivity in a range of 7.57 - 63.23%. The primary strategy obtained from the SWOT analysis is that it is necessary to conduct an intensive assessment and to implement appropriate and proper management and application. This recommended strategy is to get the value of the application of palm oil liquid waste suitable for a long-term sustainable waste application for soil and plants to improve the physical and chemical properties of the soil optimally.

1. Introduction

Indonesia is one of the largest crude palm oil or CPO producing countries in the world. Production CPO in Indonesia reached 42.88 million tons/year [1]. The high production rate causes an increase in the amount of waste produced, one of the wastes produced is palm oil mill effluent (POME). POME contains Biological Oxygen Demand (BOD) of 20,000-30,000 mg/l and Chemical Oxygen Demand (COD) of 40,000-60,000 mg/l [2].

Palm oil liquid waste cannot be directly discharged into the waters because it can cause a decrease in dissolved oxygen (DO) content in the air, causing the environment, so special treatment is needed to deal with these problems. PT X is one of the companies that utilize and utilize POME to be applied on land. POME is a brownish viscous liquid consisting of 95–96% water, along with 0.6–0.7% oil and 2–4% suspended solids [3]. The brown color comes from carotene (8 mg/L), pectin (3400 mg/L), tannins, phenolic (5800 mg/L), and lignin (4700 mg/L), resulting in a nutrient-rich substance [3]. POME has the potential to increase soil fertility because it contains nutrients N, P, K, Mg, and Ca [4]. This study was conducted to analyse the strengths, weaknesses, opportunities and threats of the influence of the use of POME at PT X on soil physical conditions, soil chemical conditions and oil palm productivity.

2. Methodology

This research was conducted in April – July 2021 at PT X Central Lampung Regency, Lampung Province. The method used in this research is descriptive method with data analysis in the form SWOT analysis. The data used is secondary data obtained by observation, interviews, literature studies, and documentation. Observations are carried out by observing the object of research directly in the field to ensure the truth of the data obtained. Observations were made to obtain data on the application process of POME in oil palm plantations. Interviews were conducted by questioning and answering the officers of the POME management, laboratory officers and environmental staff at PT X to obtain the required information.

Literature study is a method that is carried out by collecting library data, reading, recording and managing research data obtained. Documentation is a method of collecting data in the form of POME content, physical and chemical properties of the soil on the observation area and plant productivity. Documentation is also carried out to collect data in the form of images to strengthen the research results. The secondary data obtained in the form of data on the content of POME after processing, technical data on the application of POME in oil palm plantations, data on the analysis of physical and chemical properties of the soil, and data on productivity of oil palm plants.

3. Result and Discussion

3.1 POME Content

The results of the analysis of the data obtained from PT X showed that the parameters of the palm oil liquid waste were in accordance with the quality standards of the Decree of the State Minister of the Environment Number 28 of 2003 and the Regulation of the Minister of Agriculture Number 70 of 2011. Comparison of the results of the analysis of POME with waste quality standards can be seen in Table 1.

Table 1. Comparison of the results of the analysis of POME with waste quality standards

Parameter	Unit	Quality standards	POME content
pH	-	6-9	7,83
BOD	mg/l	3500-5000	566,83
COD	mg/l	-	1.656,58
Fat Oil	mg/l	-	21,58
Cu	mg/l	5000	0,14
Pb	mg/l	50	0,09
Zn	mg/l	5000	0,72
Cd	mg/l	2	0,005

POME at PT X has a pH of 7.89, this value is still below the quality standard value so that it can be applied to oil palm plantations. BOD in POME is relatively low at 566.83 mg/l compared to the quality standard ranging from 3,500-5,000 mg/l. The low BOD in palm oil effluent is caused by waste processing that is too long so that BOD continues to be degraded due to bacterial activity [5]. The fat oil contained in POME is 21.58 mg/l, the presence of the fat oil content is caused by losses in the processing of fresh fruit bunches into CPO. Palm oil liquid waste to be applied on land has a heavy metal content value below the quality standard, so it can be applied to oil palm plantations.

3.2 Soil Chemical Properties

The application of POME can increase soil pH, the soil pH on the application land is 4.12-4.87 while in the control area 3.43-3.90. Oil palm has an optimum pH of 5.0-5.5 [6]. The application of palm oil effluent is able to increase soil pH to approach the optimum value. The application of POME can increase soil pH even though it is still classified as acidic [7]. POME can increase soil pH because palm oil effluent contains organic material that is able to absorb amphoteric metals, causing soil pH to increase

[8]. In addition to the pH, the application of palm oil effluent can increase the P value, but decrease the C-Organic and N-Total. Soil chemical properties at land application and control land can be seen in

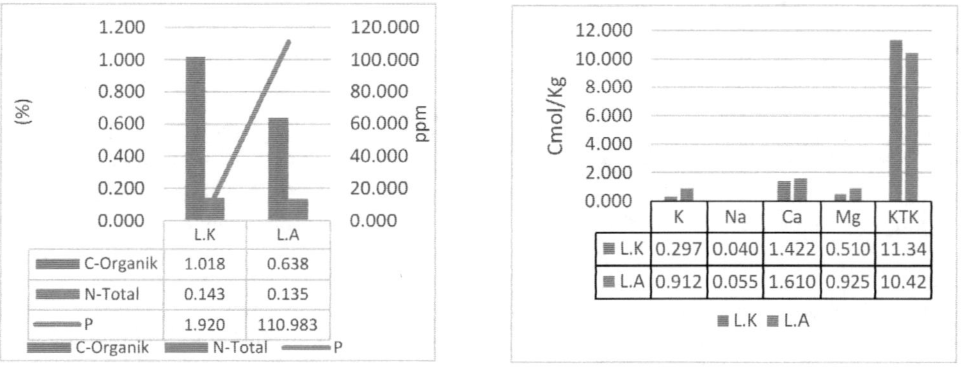

Figure 1. Soil chemical properties at land application and land control

C-Organic in the POME application land is lower than the control land. This is because the initial organic matter present on the land as well as through the application of palm oil liquid waste has decomposed, and some of it has been leached by water to the layer below it. Decomposition process of organic matter produces simpler compounds, CO2 and H2O [9]. The decomposition process can reduce the organic matter content of the soil.

In addition to this, it is suspected that the C-Organic content originating in the control land has a higher value because in the control land there are resident cows being grazed and cow dung is thought to be the cause of the increase in C-Organic in the control land. Phosphorus (P) is the nutrient that has the highest increase with an average increase from 1,920 ppm in control land to 110,983 ppm in application land. Given palm oil effluent contains P element, and in the decomposition process it can liberate P into the soil solution [10]. Other parameters analysed were elements of K, Na, Ca and Mg and CEC.

The elements of K, Na, Ca and Mg did not experience a significant increase because during waste treatment, the content contained therein became very low so that the elements contained also decreased. Insignificant increase in nutrients is because the applied liquid waste has a low BOD value (<1,000 mg/l) so that the nutrient content in the applied liquid waste is also low [11]. CEC in the liquid waste application land has a smaller value than the control land because the availability of CEC in the soil is influenced by the availability of C-Organic.

3.3 Soil Texture

The application of palm oil effluent at PT X can reduce the sand fraction and increase the dust fraction at all depths. Analysis of soil texture in the application land and control land of PT X can be seen in Table 2.

Table 2. Analysis of soil texture in the application land and control land of PT X

Sample	Texture	----------------Depth (cm)----------------					
		0-20	20-40	40-60	60-80	80-100	100-120
Control land	Sand	13,42	10,88	8,89	7,01	7,47	5,88
	Silt	8,59	7,06	9,39	6,19	8,70	9,64
	Clay	77,99	82,06	81,72	86,80	83,84	84,48
	Class	Clay	Clay	Clay	Clay	Clay	Clay
	Sand	7,36	6,03	6,10	4,69	4,84	4,91

Land application	Silt	12,23	18,55	10,05	15,16	12,70	14,14
	Clay	80,41	75,42	83,85	80,15	82,46	80,95
	Class	Clay	Clay	Clay	Clay	Clay	Clay

Source: PT X Central Office (2021); Lab soil analysis results. IPB, 2020

The comparison of soil fractions (sand, silt, and clay) at both locations is classified as clay. From the results of the analysis, it can be seen that at all depths there is a decrease in the sand fraction in the application area and an increase in the dust fraction. Soil that has a high value of dust fraction is indicated as fertile soil [12]. The increase in the dust fraction in the application area indicates further weathering from the sand to silt fraction and then to the clay fraction which has smaller soil particles

3.4 Oil Palm Productivity

The productivity of oil palm plantations at PT X in the POME application land has a higher productivity than the control land. Productivity in the application area is greater than the control area with a production difference of 7.57% to 63.23%. This is in line with the statement by Dirgantoro and Robiatul [13] that the application of palm oil effluent can increase FFB production by approximately 16%-60%.

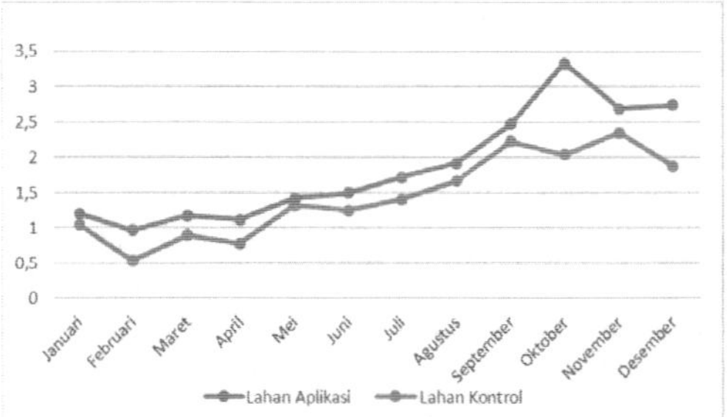

Figure 2. Differences in the productivity of PT X oil palm plantations

The increase in productivity is caused by the addition of nutrients in the application area and POME can increase the availability of water in the application area. The application of palm oil effluent can increase the productivity of oil palm plants because palm oil effluent is able to provide water continuously [14].

The availability of water in oil palm plantations greatly affects crop productivity. Lack of water in oil palm plants can cause plant productivity to decrease due to a decrease in the rate of photosynthesis so that the distribution of assimilate is disrupted and a decrease in plant productivity also occurs because the number of male flowers produced is more than the number of female flowers at the time oil palm plantations lack water [15].

The addition of liquid waste to the application area can increase the percentage of female flowers so that productivity is higher than the control land. Although palm oil effluent can increase the productivity of oil palm plants, the nutrient needs have not been able to be fulfilled by palm oil effluent. For this reason, land that is applied to palm oil liquid waste at PT X still needs to be given inorganic fertilizer to meet the nutrient needs of oil palm plants.

3.5 SWOT Analyst

External / Internal	**Opportunity**	**Threat**
External **Internal**	**Opportunity** 1. Maintain good relationship with stakeholders 2. Committed to protecting the environment by not polluting 3. Received the Green Industry Award given by the Indonesian Ministry of Industry 4. The requirements for the application of POME on land are easier to implement than flowing into rivers	**Threat** 1. There is a change in government regulations regarding the POME 2. Causes an odor that makes people uncomfortable 3. The occurrence of overflow of waste in the rainy season so that the waste flows into water bodies and causes environmental pollution 4. There is no quality standard made by the government regarding the complete content of POME which is applied to oil palm plantations (the only quality standards are BOD and pH)
Strength 1. Criteria for palm oil liquid waste after processing at PT X has a content below the quality standard value (fit for application on land) 2. Able to increase crop productivity 3. Assessment of palm oil effluent at PT X is carried out periodically and structurally	**Strength and Opportunity Strategy** 1. Periodically reviewing palm oil effluent to obtain a suitable value for palm oil effluent content for soil and oil palm plantations at PT X in order to optimally improve the physical and chemical properties of the soil.	**Strength and Threat Strategy** 1. Using competent POME management personnel so that when there are changes related to POME management it is easy to adjust 2. Waste application is carried out in blocks far from community settlements 3. Make a backup ditch along the control road to anticipate overflow that flows into the river
Weakness 1. The content of palm oil liquid waste has a value that is too small after processing so that its ability to improve the physical and chemical properties of the soil is not optimal 2. The existence of a ditch at the	**Weakness and Opportunity Strategy** 1. Testing the content of palm oil liquid waste regularly and adjusting to the processing time so that the content of palm oil liquid waste is suitable for application in oil palm plantations.	**Weakness and Threat Strategy** 1. Adjusting the management and assessment of palm oil effluent in accordance with government regulations. 2. Evaluate regularly so that any problems

conservation gate makes it difficult for harvesters to harvest because FFB often enters the ditch	2. Make a modification tool so that the harvested FFB does not fall directly into the ditch.	and deficiencies in waste management are immediately resolved.

4. Conclusion

The Promethee-based decision support framework assists decision-makers in determining the performance of lecturers based on the guidelines presented. The PROMETHEE and GAIA approaches are used in this analysis to evaluate the performance of lecturers based on five criteria. This study yielded an alternative of 2 with phi 0.2333 as the maximum value, accompanied by alternatives of 1 and 3. These findings indicate that alternative 2 is the better option for assessing lecturer performance as the primary indicator. Alternative 3 is the least recommended choice based on the visualization results since it has the lowest accuracy.

References

[1] [DITJENBUN] Direktorat Jendral Perkebunan Indonesia. 2019. Statistik kelapa sawit indonesia. [internet]. [diunduh 2020 Nov 15]. Tersedia pada www.ditjenbun.pertanian.go.id.
[2] Rambe, S.M., Iriany., Irvan.2014. Pengaruh waktu tinggal terhadap reaksi hidrolisis pada pra-pembuatan biogas dari limbah cair pabrik kelapa sawit. Jurnal Dinamika Penelitian Industri.25(1):23-30.
[3] Mohammed, R.R.; Chong, M.F. 2014. Treatment and decolourization of biologically treated Palm Oil Mill Effluent (POME) using banana peel as novel bio sorbent. J. Environ. Manag. 132, 237–249.
[4] Widhiastuti, R., Suryanto., Mukhlis., Wahyuningsih, H., 2006. Pengaruh pemanfaatan limbah cair pabrik pengolahan kelapa sawit sebagai pupuk terhadap biodiversitas tanah. Jurnal Ilmiah Pertanian Kultura. 41(1).
[5] Budianta, Dedik. 2004. Pengaruh pemberian limbah cair pabrik kelapa sawit untuk pupuk cair terhadap kualitas air. Pengelola Lingkungan dan Sumber Daya Alam. 2(3):147-154.
[6] Pahan, I. 2015. Panduan teknis budidaya kelapa sawit untuk praktisi perkebunan. Cibubur (ID) : Penebar Swadaya. 411 hal.
[7] Banuwa, I.S., Pulung, M.A. 2008. Pengaruh land application limbah cair kelapa sawit terhadap ketersediaan unsur hara dalam tanah dan kandungannya pada tanaman kelapa sawit. Jurnal Tanah Tropika. 13(1): 35-40.
[8] Siregar, A.F., Dedik, B., Napoleon, A., 2015. Pengaruh aplikasi limbah cair kelapa sawit terhadap beberapa sifat kimia tanah. Jurnal Sriwijaya.
[9] Hakim, N., Nyakpa, M.Y., Lubis, A.M., Nugroho, S.G., Diha, M.A., Hong, G.B., Bailey, H.H., 1986. Dasar-dasar ilmu tanah. Lampung (ID): Universitass Lampung. 488 hal.
[10] Zulkarnain. 2014. Perubahan beberapa sifat kimia tanah akibat pemberian limbah cair industri kelapa sawit dengan metode land application. Jurnal Agrifor. 13(1): 125–130.
[11] Silalahi, B.M., Supijatno. 2017. Pengelolaan limbah kelapa sawit (Elaeis guineensis Jacq.) di angsana estate, kalimantan selatan. Jurnal Agrohorti. 5(3): 373-383.
[12] Hanafiah, K.A., 2005. Dasar-dasar ilmu tanah. Jakarta (ID) : Raja Grafindo Persada. 359 hal.
[13] Dirgantoro, M.A., Robiatul, A., 2018. Nilai ekonomi pemanfaatan limbah kelapa sawit menuju zero waste production. Biowallaca. 5(2): 825-837.
[14] Pujono, H.R., Kukuh, S., Erizal, R., Afandi., Rahmat., A. 2021. The effect of POME application on production and yield components of oil palm in Lampung, Indonesia.
[15] Darmosarkoro, W., Harahap, I.Y., Syamsudin, E. 2001. Pengaruh kekeringan pada tanaman kelapa sawit dan upaya penanggulangannya. Warta PPKS. 9(3): 83-96.

Groundwater level fluctuations in peatlands as responses to canal block construction in Sungai Tohor, Tebing Tinggi Island, Indonesia

H. Malik, Ig. L. S. Purnama, Sudarmadji, L. W. Santosa, M. Naufal

Universitas Gadjah Mada, Yogyakarta, Indonesia

*hapriadi.malik@mail.ugm.ac.id

Abstract. Peatlands play an essential role in an ecosystem and generally have three main functions, e.g., a buffer zone in a water system. However, they are vulnerable and susceptible to environmental changes. In response to the increasing demand for habitable space, more peatlands have been converted to productive agricultural land by canal blocking to transform an originally anaerobic environment into an aerobic one. This condition also occurs in peatlands in Sungai Tohor, Tebing Tinggi Island, Kepulauan Meranti Regency. This study sought to determine the groundwater level fluctuations in peatlands after the construction of canal blocks by the Peatland Restoration Agency (BRG). A water level logger was installed at the monitoring well, and the daily groundwater levels were observed for one year. The results showed that the groundwater levels fluctuated from 8.78 m (the lowest) to 9.72 m (the highest), with an average of 9.21 meters. This means that canal blocking might have a sufficient wetting impact, i.e., maintaining the presence of water in the surrounding peatlands.

1. Introduction

Peatlands play an ecologically vital part in the environment. In general, this landform has three primary functions, including a hydrological function that is as a buffer area for water systems [1][2][3]. However, peatlands are vulnerable and susceptible to environmental changes [4][2]. In Indonesia, peatlands cover 14.9 million hectares of the land surface [5][2][6], which makes them the most extensive tropical peatland worldwide. Further, it constitutes 45% of the total area of the global tropical peatlands [7][8][9]. Moreover, peatlands are distributed in many regions in the country, including along the east coast of Sumatra [10] [11].

As the population rate grows, the need for land in various sectors constantly increases, especially for agricultural fields and settlements. The implications of this condition are, for instance, intensive conversion of peatlands to accommodate agricultural economic activities, such as food crop production, plantations, agroforestry, fisheries, and livestock farming [2][3]. Peatland conversion involves regulating water management by constructing canals or drainage channels.

In this context, canals transform originally anaerobic peatlands into aerobic environments [12][13]. As a result, the peatlands over drainage, thus accelerating peat decomposition [14][15], destroying the spongy characteristic of peat, and eventually reducing its hydrological ability to reserve water [12][14][16]. More than one million hectares of peatland in Riau Province are degraded and used for food crop production, horticulture farming, and plantations [17]. Degraded peatlands are responsible for fires in several regencies in the province, which occur most extensively in Kepulauan Meranti (928 hectares).

IOP Conf. Series: Earth and Environmental Science 1041 (2022) 012047 doi:10.1088/1755-1315/1041/1/012047

Therefore, the Peatland Restoration Agency (BRG) and the Riau Provincial Government introduced a peatland rewetting program that involves canal blocking. It is expected to decelerate surface runoff and increase water storage in the canal structure and its surroundings to prevent and suppress peatland fires. Based on these formulated problems, this research aimed to determine the groundwater level fluctuations in peatlands as responses to the canal block construction, especially in Sungai Tohor Village on Tebing Tinggi Island, Kepulauan Meranti Regency.

Kepulauan Meranti Regency is a group of islands on the east coast of Sumatra Island and comprises one peat hydrological unit (PHU). One of the islands is Tebing Tinggi, whose PHU has two peat domes in the west and east and non-peat domes in the remaining parts of the unit. This research focuses on the eastern peat dome that is hypothetically degraded, particularly in Sungai Tohor Village (Figure 1).

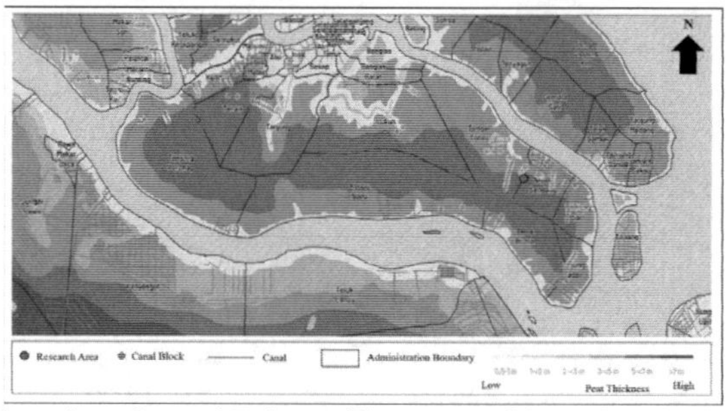

Figure 1. The research location: Sungai Tohor Village on Tebing Tinggi Island, Kepulauan Meranti Regency, Indonesia

2. Methodology

The research employed a field survey to obtain primary data: groundwater levels and rainfall. The groundwater level data were generated by processing the hydrostatic pressure and air pressure data derived from the water level logger at the monitoring well in the research location (Figure 2a). The groundwater level data were combined with one-year periodic rainfall from the automatic rain gauge recorder installed in the area where the monitoring well was located (Figure 2b).

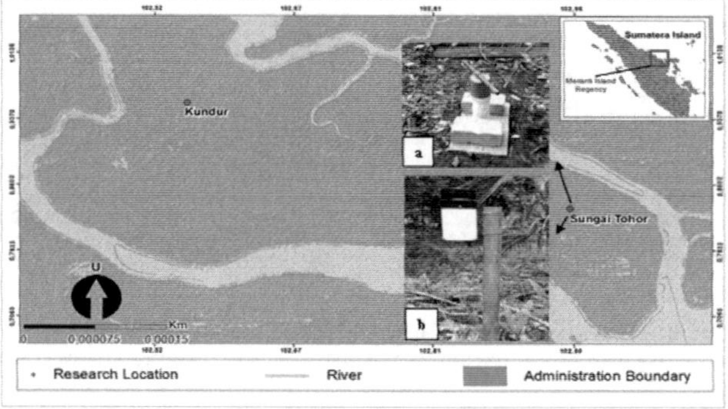

Figure 2. Research location and data measurement stations: (a) Rain weather station; (b) Monitoring well

3. Result and Discussion

The analysis of the data derived from the monitoring well revealed that the lowest and highest groundwater levels in peatlands in Sungai Tohor were 8.78 m and 9.72 m (Figure 3), with an average of 9.21 m. Based on the shape of the hydrograph, the groundwater levels showed a relatively significant fluctuating pattern. The condition of the peatlands in Sungai Tohor is believed to contribute to this pattern substantially.

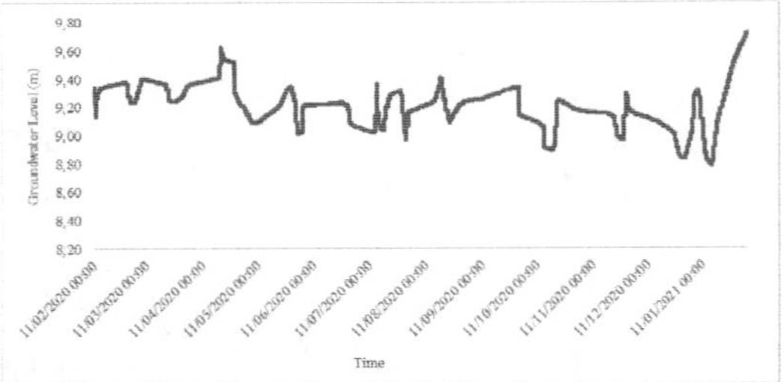

Figure 3. Daily groundwater levels in peatlands in Sungai Tohor

The groundwater level data were then paired with rainfall records, and the resultant graph shows that rainfall affected the increase or decrease in the groundwater level (Figure 4). The rain gauge station recorded that the highest rainfall in Sungai Tohor was 330.48 mm. Further, the rainfall data indicate that the research location has an equatorial rainfall pattern, which tends to be wet and is characterized by high rainfall throughout the year.

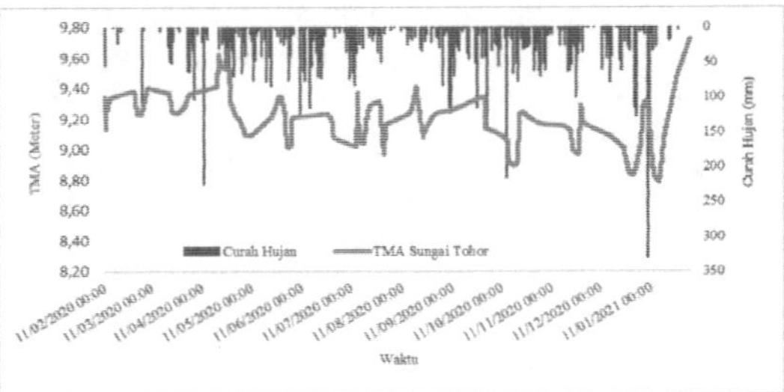

Figure 4. Integration of groundwater level and rainfall data

In addition to rainfall, canal blocking also influences the dynamics of the groundwater level. This effect is apparent after comparing the groundwater level analysis results in Sungai Tohor with those of the less degraded peatlands in Kundur. The groundwater levels in the former were consistently higher than in the latter (Figure 5). This condition is closely related to the water absorption function of Sungai Tohor's peatlands that is less optimal than Kundur. In other words, surface runoffs will infiltrate the peatland more quickly in Kundur.

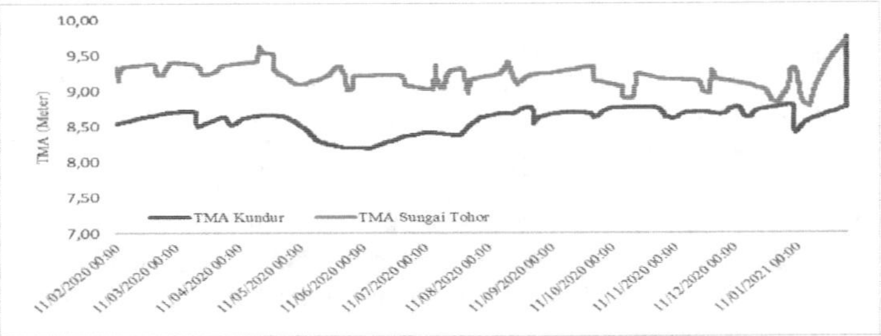

Figure 5. Daily groundwater levels (GWL) in Sungai Tohor and Kundur

Moreover, the groundwater levels in Sungai Tohor, which were always higher than those in Kundur, are most likely influenced by the relatively massive canal blocking around its peatlands. Sungai Tohor had 28 canal blocks (Figure 6), while Kundur only had 16. Although canal blocks are commonly constructed to help maintain the presence of water (wetting) in degraded peatlands, they cannot substitute the peatland's natural function as a water absorber (sponge). Nevertheless, canal blocking proves effective in retaining water and keeping the peatlands in Sungai Tohor wet.

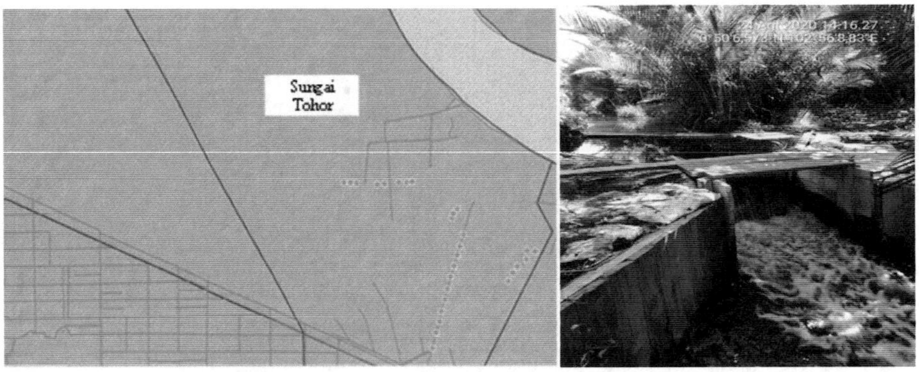

Figure 6. The spatial distribution (a) and a sample of canal blocks (b) in Sungai Tohor Village

4. Conclusion

The groundwater levels of the peatlands in Sungai Tohor vary between 8.78 m and 9.72 m, with an average of 9.21 m. Based on the data comparison, they are always higher than the groundwater levels of the peatlands in Kundur. This means that canal blocking has a wetting impact sufficient to continuously retain water in Sungai Tohor's peatlands.

Acknowledgment

The authors would like to thank the Government of Riau Province for supporting and funding this research through the Doctoral Dissertation Research grant scheme in the fiscal year of 2020.

References

[1] Sabiham, S., & Sukarman. (2012). Peatland Management for Oil Palm Development in Indonesia, 55–66.

[2] Masganti, Wahyunto, Dariah, A., Nurhayati, & Yusuf, R. (2014). Karakteristik dan Potensi Pemanfaatan Lahan Gambut Terdegradasi di Provinsi Riau. Jurnal Sumberdaya Lahan, 8(1), 59–66.

[3] Agus, F., Anda, M., Jamil, A., & Masganti. (2014). Lahan Gambut Indonesia, (August 2016), 250.

[4] Maswar, & Dariah, A. (2013). Teknologi Pengelolaan Lahan Gambut dan Sumbangannya Terhadap Mitigasi Gas Rumah Kaca. In Bunga Rampai III Konservasi Tanah dan Air (p. 249). Jakarta: Masyarakat Konservasi Tanah dan Air Indonesia.

[5] Mulyani, A., Susanti, E., Dariah, A., Maswar, Wahyunto, & Agus, F. (2012). Basisdata Karakteristik Tanah Gambut di indonesia. In Husen dkk (Ed.), Seminar Nasional Pengelolaan Lahan Gambut Berkelanjutan (pp. 143–154). Bogor: Balai Penelitian Tanah Kementerian Pertanian.

[6] Wahyunto, Nugroho, K., Ritung, S., & Sulaeman, Y. (2014b). Indonesian Peatland Map: Method, Certainty, and Uses Peta. In Pengelolaan Berkelanjutan Lahan Gambut Terdegradasi untuk Mitigasi Emisi GRK dan Peningkatan Nilai Ekonomi (pp. 81–96). Retrieved from http://balittanah.litbang.pertanian.go.id/ind/dokumentasi/prosiding/prosiding gambut icctf/05 Wahyunto Indon peatland map-draft1-20juli2014_HN_1-EditWt11Agt-tika.pdf

[7] Page, S. E., Banks, C. ., & Rieley, J. O. (2007). Tropical Peatlands: Distribution, Extent and Carbon Storage – Uncertainties and Knowledge Gaps. Peatlands International, 2(2), 26–27.

[8] Dariah, A., & Maswar. (2015). Isu Lingkungan Gambut Tropika Indonesia, (12), 101–129.

[9] Noor, M., Masganti, & Agus, F. (2016). Pembentukan dan Karakteristik Gambut Tropika Indonesia. In Lahan Gambut Indonesia (pp. 7–32). Jakarta: IAARD Press Badan Penelitian dan Pengembangan Pertanian Kementerian Pertanian.

[10] Sukarman. (2001). Pembentukan, Sebaran dan Kesesuaian Lahan Pembentukan Tanah Gambut. Balai Besar Penelitian Dan Pengembangan Sumberdaya Lahan Pertanian, (12), 2–15.
[11] Noor, M., Noor, M., Nursyamsi, D., Alwi, M., & Fahmi, A. (2014). Prospek Pertanian Berkelanjutan di Lahan Gambut: dari Petani ke Peneliti dan Peneliti ke Petani. Jurnal Sumberdaya Lahan, 8(2), 69–79. https://doi.org/10.2018/jsdl.v8i2.6469.
[12] Andriesse, J. P. (1988). Nature and Management of Tropical Peat Soils. Rome: Food and Agriculture Organization of the United Nations.
[13] Noor, M., Masganti, & Agus, F. (2016). Pembentukan dan Karakteristik Gambut Tropika Indonesia. In Lahan Gambut Indonesia (pp. 7–32). Jakarta: IAARD Press Badan Penelitian dan Pengembangan Pertanian Kementerian Pertanian.
[14] Subiksa, I. G. M., Agus, F., Wahyunto, & Ananto, E. . (2007). Mitigasi Degradasi Lahan Gambut, 113–139.
[15] Nusantara, R. W., Sudarmadji, Djohan, T. S., & Haryono, E. (2012). Karakteristik Fisik Lahan Akibat Alih Fungsi Lahan Hutan Rawa Gambut Physical Characteristic of Peatlands As Effect Land Use Change of Peat Swam Forest. Jurnal Perkebunan & Lahan Tropika, 2.
[16] Noor, M., Noor, M., Nursyamsi, D., Alwi, M., & Fahmi, A. (2014). Prospek Pertanian Berkelanjutan di Lahan Gambut: dari Petani ke Peneliti dan Peneliti ke Petani. Jurnal Sumberdaya Lahan, 8(2), 69–79. https://doi.org/10.2018/jsdl.v8i2.6469.
[17] Wahyunto, Ritung, S., Nugroho, K., Sulaeman, Y., Tafakresnanto, H. C., Suparto, & Sukarman. (2013). Atlas Lahan Gambut Terdegradasi Pulau Sumatera. Jakarta.

IOP Conf. Series: Earth and Environmental Science 1041 (2022) 012048 doi:10.1088/1755-1315/1041/1/012048

Identification of land use and land cover using the image Landsat 8 in upstream Lematang sub-watershed by support vector machine and random trees methods

A Dinata*, F Dhiniati, L E Diansari

College of Technology Pagaralam, Pagar Alam City - South Sumatra, Indonesia

*alhariadinata@gmail.com

Abstract. Accuracy of data and information on various types of land use and land cover (LULC) is crucial in watershed management planning. The development of remote sensing technology in predicting LULC has been widely used, such as the Support Vector Machine (SVM) and Random Trees (RT) methods. Both methods are part of the model of machine learning and belong to the category of supervised learning. This method can predict, study, and recognize data patterns from the results of training data. The aim of this research is the use of SVM and RT methods for the identification of LULC. Based on the SVM method analysis results, the forest area is 211.61 km2, coffee plantations are 32.92 km2, shrubs are 16.85 km2, built-up/bare land is 3.40 km2, the water body is 3.17 km2, and cloud/shadow is 6.79 km2. While the RT method obtained a forest area of 215.51 km2, coffee plantations covering an area of 29.32 km2, shrubs are covering an area of 16.32 km2, built-up/bare land surrounding an area of 7.22 km2, water body covering an area of 2.62 km2, and cloud/shadow covering an area of 3.75 km2. Overall, the percentage accuracy of the SVM method of 87%, and the RT of 86%. The success rate of modeling SVM (AUC = 0.73) and RT (AUC = 0.75) based on the results of data validation using the area under the receiver operating characteristic curve (AUC) was considered fair good.

1. Introduction

Remote sensing technology plays a crucial role in monitoring the earth's surface, especially regarding land use and land cover (LULC) [1]. Specifications of remote sensing platforms such as ground vehicles, airplanes, satellites, and handheld gadgets are utilized to mount the sensor. The data obtained from these sensors depend on four resolutions, spatial, temporal, radiometric, and spectral [2]. Landsat 8 imagery has Onboard Operational Land Imager (OLI) and Thermal Infrared Sensor (TIRS) sensors with a total of 11 channels, consisting of 9 channels (bands 1 - 9) located in OLI and TIRS has two channels, namely bands 10 and 11 [3].

The data is widely used for various purposes, such as watershed management planning. In this regard, selecting an analytical method using machine learning is the right thing because this method can process data from large data sets in a non-linear framework from the results of the remote sensing process. This method is one of the most suitable methods and is very effective from an economic point of view in analyzing LULC [3–5]. Various machine learning algorithms, such as support vector machine (SVM) and random trees (RT), have been widely used to obtain information from remote sensing data because the results obtained have high accuracy [6,7]. Thus, the main objective of this study is to determine the level of accuracy of the SVM and RT methods in identifying LULC using Landsat 8 imagery.

2. Methodology

IOP Conf. Series: Earth and Environmental Science 1041 (2022) 012048 doi:10.1088/1755-1315/1041/1/012048

2.1 Study area

The study location is in the upstream Lematang watershed with an area of 274.74 km2. A watershed is a unitary unit of the Musi river area. Location of study area position as seen in (Figure 1).

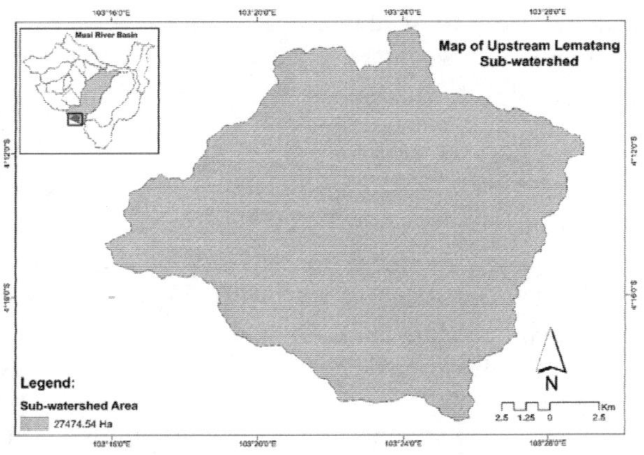

Figure 1. Location of study

2.2 Classification LULC

Classification of the LULC thematic maps using the Landsat 8 OLI/TIRS 2020 image interpretation downloaded via https://earthexplorer.usgs.gov (path 125/row 60). Data analysis stages include pre-processing (image cropping and radiometric correction), data processing stages (this stage consists of color composites, sharpening, training data, and multispectral classification of Landsat 8 OLI/TIRS images using SVM and RT methods), and the stages of assessing the accuracy of the modeling results using the kappa index and area undercover (AUC). All stages of analysis using ArcGIS 10.4 application tools.

2.3. Support vector machine

In 1995 Cortez and Vapnik introduced the first SVM method. This method is a machine learning algorithm rooted in statistical learning theory with better results [8]. The application of SVM modeling in the LULC study demands a systematic exploration of appropriate data sampling, attribute selection, and parameter selection that is not yet fully known. The SVM model built to predict land use classes based on spatial data attributes, more efficient for non-linear problems [9]. Empirically, SVM maximizes the boundary between data points to be separated to find the most optimal hyperplane. So, this indicates that only the essential supporting vectors can be maximized into a linear classifier, while the rest of the training data can be ignored [10].

2.4. Random trees

Random trees model (RT) is one of the predictive models that is better than other classification models in terms of computational accuracy, data management, obtaining more data information by eliminating less data, better data extraction, and can work with many data [11].

2.5. Validation

The data used for validation uses high-resolution imagery (google earth pro) in 2020 by determining the number of sample points using the stratified random sampling method generated from the LULC classification map. These methods work to determine the number of sample points randomly distributed into each class, where each class has several sample points proportional to its relative area [12–14]. So that the determination of the sampling location occurs evenly by minimizing the mean of the shortest distances and by adopting spatial simulated annealing, the most optimal solution was obtained by reducing the fitness function [15].

Furthermore, the accuracy assessment is determined based on the calculation of the overall accuracy (OA) and the kappa coefficient based on the confusion matrices of different classification techniques. Both models were also tested for sensitivity/specificity by plotting the data into the receiver operating characteristic curve (ROC) to see the success rate in predicting LULC. The value of the area under the curve (AUC) resulting from the ROC curve ranges from 0-1. The AUC close to 1 indicates a high level of model performance.

3. Result and Discussion

3.1. LULC classification

After going through the pre-processing data stages (image cropping and radiometric correction) and data processing (color composites and image sharpening), which facilitates the training data process, the LULC prediction process is carried out using the SVM and RT methods. The results of the analysis by classifying seven land-use obtained information that there was a slight difference in the percentage of forest distribution of 77.02% (SVM) and 78.44% (RT), built land/vacant land 1.24% (SVM) and RT with a percentage of 2.63%, plantations mixed 11.98 % (SVM) and RT with a rate of 10.67%, 1.7% shadow (SVM) and 0.62% (RT). While other land uses have a difference of >1% (Table 1) and the LULC map as seen in (Figure 2).

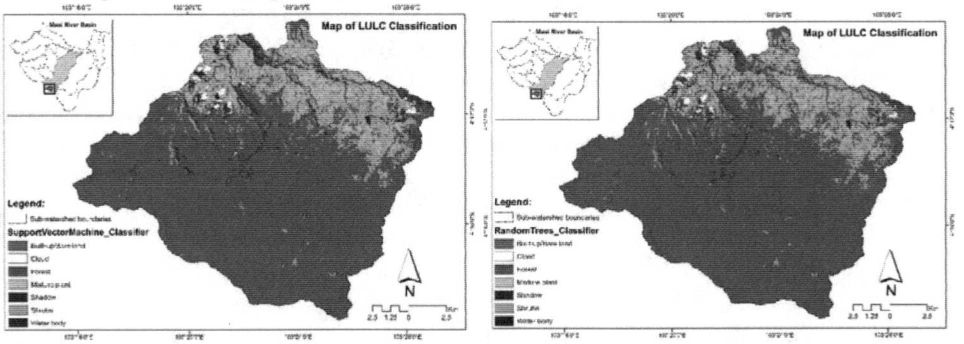

Figure 2. LULC map produced by using support vector machines classifier (left), and LULC map produced by using random trees classifier (right)

Table 1. LULC Classification

Land Use Land Cover	Method			
	SVM (Km²)	SVM (%)	RT (Km²)	RT (%)
Forest	211.61	77.02	215.51	78.44
Mixture plant	32.92	11.98	29.32	10.67
Shrubs	16.85	6.13	16.32	5.94
Built-up/Bare land	3.40	1.24	7.22	2.63
Water body	3.17	1.15	2.62	0.96
Cloud	2.12	0.77	2.04	0.74
Shadow	4.67	1.70	1.71	0.62
Total	274.74	100.00	274.74	100.00

3.2. Validation

Assessment of the accuracy of the LULC classification results using a sample point of 230 points proportional to the relative area in each class (Figure 3) and the modeling accuracy assessment results as shown in (Table 2).

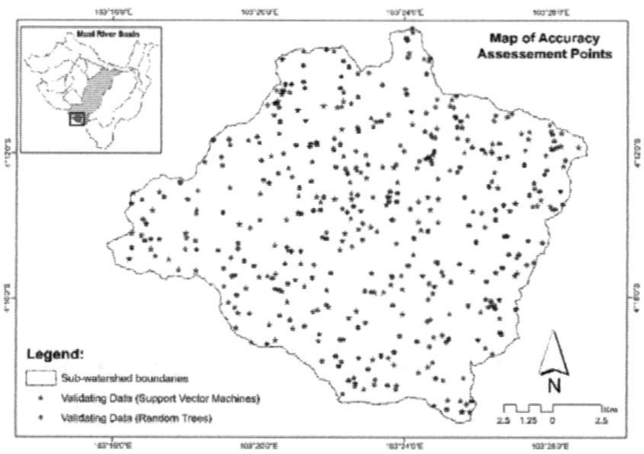

Figure 3. Map of flash flood hazard zones in upper Lematang sub-watersheds

Table 2. Overall accuracy level of the classifiers used SVM and RT

Methods	OA	Kappa
SVM	0.9	0.75
RT	0.9	0.72

From table 2 above, it is known that OA has good accuracy with a value of 0.87 (SVM) and 0.86 (RT) and a kappa index of 0.75 (SVM) and 0.72 (RT). While the results of the evaluation of the model using the AUC curve still show a fair good performance, where the success rate of the SVM model is 0.73, and the RT model is 0.75 (Figure 4). The percentage value of overall accuracy above 0.85 and with a success rate above 0.70 shows that the results of the LULC classification of the two methods can be used in making LULC thematic maps.

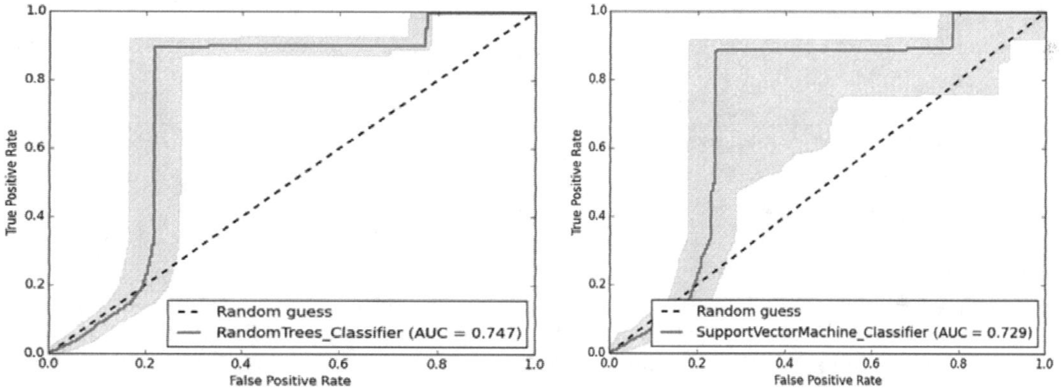

Figure 4. AUC curve analysis of SVM (left) and RT (right) for the LULC prediction model using a test validation dataset

4. Conclusion

The analysis results show that the overall accuracy of the SVM method is 0.87 and 0.86 (RT). While the results of the evaluation of the model's performance using AUC, the success rate of the SVM method is 0.73, and the RT is 0.75. Overall, both methods can produce LULC maps with reasonable accuracy. Finally, LULC using the SVM method results in a forest area of 211.61 km2, a mixture plant covering an area of 32.92 km2, shrubs covering an area of 16.85 km2, built-up/bare land covering an area of 3.40 km2, a water body covering an area of 3.17 km2, and cloud/shadow covering an area of 6.79 km2. Furthermore, LULC using the RT method resulted in a forest area of 215.51 km2, mixture plant covering an area of 29.32 km2, shrubs covering an area of 16.32 km2, built-up/bare land covering an area of 7.22

km2, water body covering an area of 2.62 km2, and cloud/shadow covering an area of 3.75 km2. In conclusion of this study is both methods can predict LULC using Landsat 8 imagery.

Acknowledgments
Researchers higher appreciate the chairman of the College of Technology Pagar Alam for all the support and facilities provided to complete this research.

References

[1] Hütt C, Koppe W, Miao Y and Bareth G 2016 Best accuracy land use/land cover (LULC) classification to derive crop types using multitemporal, multisensor, and multi-polarization SAR satellite images Remote Sensing 8
[2] Talukdar S, Singha P, Mahato S, Shahfahad, Pal S, Liou Y A and Rahman A 2020 Land-use land-cover classification by machine learning classifiers for satellite observations-A review Remote Sensing 12
[3] Indrayani, Buchari E, Putranto D D A and Saleh E 2017 Analysis of land use in the Banyuasin district using the image Landsat 8 by NDVI method AIP Conference Proceedings 1903
[4] Karakus C B, Cerit O and Kavak K S 2015 Determination of Land Use/Cover Changes and Land Use Potentials of Sivas City and its Surroundings Using Geographical Information Systems (GIS) and Remote Sensing (RS) Procedia Earth and Planetary Science 15 454–61
[5] Wondrade N, Dick Ø B and Tveite H 2014 GIS based mapping of land cover changes utilizing multi-temporal remotely sensed image data in Lake Hawassa Watershed, Ethiopia Environmental Monitoring and Assessment 186 1765–80
[6] Singh G and Pandey A 2021 Evaluation of classification algorithms for land use land cover mapping in the snow-fed Alaknanda River Basin of the Northwest Himalayan Region Applied Geomatics
[7] Virnodkar S S, Pachghare V K, Patil V C and Jha S K 2020 Remote sensing and machine learning for crop water stress determination in various crops: a critical review vol 21 (Springer US)
[8] Lee S, Hong S M and Jung H S 2017 A support vector machine for landslide susceptibility mapping in Gangwon Province, Korea Sustainability (Switzerland) 9 15–9
[9] Samardžić-Petrović M, Dragićević S, Kovačević M and Bajat B 2016 Modeling Urban Land Use Changes Using Support Vector Machines Transactions in GIS 20 718–34
[10] Kalantar B, Ueda N, Idrees M O, Janizadeh S, Ahmadi K and Shabani F 2020 Forest fire susceptibility prediction based on machine learning models with resampling algorithms on remote sensing data Remote Sensing 12 1–24
[11] Joloudari J H, Joloudari E H, Saadatfar H, Ghasemigol M, Razavi S M, Mosavi A, Nabipour N, Shamshirband S and Nadai L 2020 Coronary artery disease diagnosis; ranking the significant features using a random trees model International Journal of Environmental Research and Public Health 17
[12] Ramezan C A, Warner T A and Maxwell A E 2019 Evaluation of sampling and cross-validation tuning strategies for regional-scale machine learning classification Remote Sensing 11
[13] Heung B, Bulmer C E and Schmidt M G 2014 Predictive soil parent material mapping at a regional-scale: A Random Forest approach Geoderma 214–215 141–54
[14] Dong S, Chen Z, Gao B, Guo H, Sun D and Pan Y 2020 Stratified even sampling method for accuracy assessment of land use/land cover classification: a case study of Beijing, China International Journal of Remote Sensing 41 6427–43
[15] Gao B, Liu Y, Pan Y, Gao Y B, Chen Z, Li X and Zhou Y 2017 Error index for additional sampling to map soil contaminant grades Ecological Indicators 77 129–38

Conservation and utilization on Sawahlunto ex mining area for cultural landscape

Rika Cheris[1*], Amanda Rosetia[2]

[1]Universitas Lancang Kuning, Pekanbaru, Indonesia
[2]Universitas Internasional Batam, Batam, Indonesia

*rika.cherish@unilak.ac.id

Abstract. As a UNESCO World Heritage, Sawahlunto Ex Coal Mining has the profound and lasting impact of the industrialization and social changing relations which has built the history remarkable as per today. To facilitate public on having mindful memories of the place, an interpretation of evidence and information need to emerge with a suitable flow and approach. Discover perspectives and improving interpretation on conservation area which has changing gradually because the community impact has always been a degrading issue. We often ignore the neighborhoods needs to commercial purpose. The historic district assessed by analysis and qualitative research method from the past and present documentation through site inventory and relevant studies of the living culture. The aim of this paper is to bring better conservation strategies that can enhance the overall cultural image of Sawahlunto. Collective memory, habit changing, site management, and material security supposed to be co-habitant through times. In order to keep the familiarization and to be a place of original community can be secure, content, and proud of.

1. Introduction

The declaration of Sawahlunto City and 6 other city regencies in West Sumatera as the World Heritage Site Ombilin Mining Heritage Site in 2019 (UNESCO, 2019) has brought tourism in West Sumatera Province to the international level. This pride is a valuable moment for each region in the 7 regencies and cities that include in the World Heritage Site. Especially in the core area, namely the City of Sawahlunto drives the force for the creation of this prestige recognition.

One of the outstanding universal values (OUV) of the existence of the Ombilin Coal Mining Heritage Site is the series of a mining industry that is integrated with the natural resources around the site. This means that the visual architecture remains of the building and the surrounding environment since the early 19th century to the 20th century are still well maintained and utilize by the local community for their daily needs and consent.

1.1 Cultural Landscape Paradigm

Helen Armstrong on Interpreting Cultural Landscape interpreted worldviews on Queensland landscapes into Scientific, Romantic, Colonial, National/Postcolonial, Capitalist, Socialist, Ecological, and Traditional Aboriginal Vision. Where development ethos, national identity, and resources exploitation are mainly mentioned on the categorize. Significantly, the scientific, colonial and capitalist worldview are read predominantly associated to contemporary cultural landscape evolvement (The & Geographies, n.d.). Interpreting the landscape is fundamental to an understanding of existing meaning and values of familiar and everyday places, as well as iconographic interpretations. Landscape were repositories of signs and symbols, also vehicle to reveal the socio-political signifiers embedded in representation of place.

Landscape design evolvement especially in contemporary cultural theme, associate to the identity of the area. Sawahlunto has embedded with the image as an industrial area for the unique theme on heritage culture. They configuration within design, focus on the character of space to be recognize as the character of place (Rito, 2019). Recently, people response to some concern on place characteristic especially one that has the tourism purpose. They concern on the ecosystem and community lived within that specific neighborhoods.

Some evident of the history traces purposely "polished" and commercialized to raise people interest to visit the place, to the extent, exploitation. While the management should engage on the regulation, somehow there are inappropriate action taken by the community lives there. Which cause some "violent" to the originality sense of place and shifting the cultural landscape design approach. Besides, the management does not consider this is important for the continuity of the local culture. Urgency on this matter somehow sees differently. So that it does not immediately create a rule that leads to sustainable maintenance. Sometimes, political influence also plays a passive role to restrain the enforcement of these rules due to individual and party interests.

1.2 Familiarity and Sense of Belonging

Throughout years and generations, things change very fast. The transmission of memory from the past into the future can differ perceptions on newcomers and native. How individuals interact with, perceive, or understand their culture produces some feeling of a mankind that are reflected in the landscape. Whether they are natural or cultural, they become that data from which we make inferences about cultural processes (Fleming, 1998).

Nurturing community, like how Chinese temple are continuously renovated, upgraded, and even reconstructed as sign of respect and as a reflection of the community prosperity from the past well into the present (Widodo, n.d.), shows the effort on how conservation necessarily should be done, despite demolitions and re-developments as a common practice. This called the maturity process that allow time, community, and environment to reconnect.

In the other hand, different perceptions on cultural relativism and phenomenal absolutism gauge the cultural continuum on heritage conservation and utilization. Cultural relativism has their own values and concept of reality that are culturally conditioned, while phenomenal absolutism is where the world is exactly as how the observer sees it (Fleming, 1998). Living witness of historical process play the most important part there. They are the most honest observer of how the cultural landscape change throughout the time being and their opinion on how the site should be preserved is as much as necessary than regulations. Though in some cases, perception may not be observable and will remain a matter of conjecture.

2. Methodology

The historic district assessed by analysis and qualitative research method from the past and present documentation through site inventory and relevant studies of the living culture. Relevant studies on cultural landscape are exploratory carried with the data, based on respective site and built environment, ethnography, and history. Visiting the area and learning theoretical framework on cultural landscape vision build a different perspective to develop the heritage site. Conservation areas wish to be sustainable to both community and environment where landscape object can be utilize to an optimum benefit. The location is strategically conveying all the needs for this research.

3. Result and Discussion

Sawahlunto City has four sub-districts with an area of 27,345 Hectares (Widiati, 2017). The distribution center area for industrial area development located at Lembah Segar sub-district with an area of 52,58 Hectares. Meanwhile, the distribution of cultural heritage building centered to the old town of Sawahlunto with an area of ±25 Hectares. The center of this industrial area bordered by four (4) hills, namely Cemara hill, Soegar hill, Poland hill, and Kampung Teleng hill. Right at the valley, there are two (2) river (respectively called "batang"; in local Minangkabau language) which divide the valley plain into three (3) sections. They are the Batang Lunto and Batang Sumpahan. The river meets in the city center and become one stream towards the downstream.

This very favorable geographical condition became a blessing and a good opportunity for Dutch engineers to start designing coal mining areas. Starting from the search for the first mining pit, which is currently a tourist attraction, known as Mbak Seoro. Furthermore, they build a house for mining worker there, then they construct a prison for the chain worker or mining worker which later famous by the name "orang rantai", and lastly the construction for mining installations and a rail transportation route that ended at Teluk Bayur (Emmahaven), Padang City which was the largest port at that particular time. The map below shows all the infrastructure provided by the Dutch Colonies which has bring the city promoted to GEMENTEE from the Queen of Netherlands in 1930.

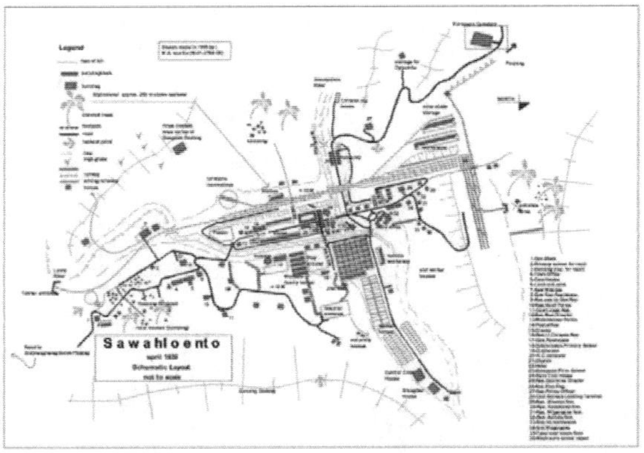

Figure 1. Documentary map from Mr. Van EE, a person who still alive and he drawled a picture for his memories in 2009 before he is path away at 2011. Personal Documentary

The urge and interest of area of expertise of wanting to take advantage of the existing geographical conditions have made city planners or urban planners able to design this small valley to meet the needs of a coal mining industry which coherent with facilitated needs. The arrangement of buildings on hillsides with those on the banks of the river seems to have a calculation of natural phenomena such as the possibility of earthquakes, landslides and floods that occasionally occur in the rivers there. Engineers carry their measurement to the extent the construction has to provide the optimum strength to the land.

Addition, building location seems to be done by considering the contours and geographical conditions of the land condition. Aesthetic part of the building façade based on the human scale considering when they are looking at the buildings in the middle of the city and surroundings. The fact that this city has very well developed, the design is integrated with nature, is a priority to keep this sustain and maintain. Furthermore, maintaining the harmony of development with the local nature should be a continuous work in every development ethos that will be carried out on this world heritage *site.*

3.1 Symbolic Nature of Objects in The Landscape

The basic act of architecture is therefore to understand the 'vocation' of the place. In this way we protect the earth and become ourselves part of a comprehensive totality. What is advocated here is not some of environment determinism. We only recognize that man is an integrated part of environment (Abel, 2000). Humans as actors in being creative to create a comfortable place to live in their environment should really preserve the environment such as avoiding development on unstable land, making rivers a source of life, preserving forests as biodiversity and participating in revitalizing if there is damage to the environment. Structuring the built environment which is often assigned to planners and architects is a professional responsibility that is passed on to the future generations.

Figure 2. Sawahlunto Morphology in two decade which is shown the river sumpahan (a)&(c), and (b) lunto river in the center of the city.

To make the site keep on being alive, development process should create by marking landscape which able to socialize. Which in turn has an effect on community behavior. Topography, natural vegetation and the presence of water called a detail landscape which each of these details described certain dominant element (The & Geographies, n.d.). Scenic quality factors such as naturalness, patterns, and diversity & variety combined to create panoramic yet detail character in cultural landscape element. Engineer driven project processed the understanding on existing landscape function in particular, water bodies.

Ecological approach should be considered involved in this improvement to keep the nature sustain from time to time (Rosetia, 2021). This identification process as part of the data base to start on cultural heritage should have been begun earlier.

Figure 3. Batang Sumpahan (the river) located near to the relic of the former prison for the chain people

This city developed at the beginning of the early 20th century. Their architecture typology and the form of the buildings are Indish (Cheris, 2021). Places in which carries significance enrich people's lives. A deep and inspirational sense of connection to community and landscape advocate the definition of sensibility of design (Rito, 2019).

In Sawahlunto Ex Mining Area case study, the surround landscape where hills are magnificently scenic as the focal point and the river flow just right behind the neighbourhood enhance the experience of transcend memory of the past, are the symbolic nature that every heritage site needs in landscape design. In order to utilize them wisely, environmental sensibility needed in hand, does not rule out the possibility on conserving the site hermeneutically.

4. Conclusion

According to Burra Charter and the Australian Heritage Commission (AHC), criteria of relevance to cultural landscape from the highest relevant to the least relevant respectively are the pattern in history to significant people (which linking landscapes with people) (The & Geographies, n.d.). Pattern in history holds the highest relevance to cultural landscape design that shows study on preserving the flow- and whichever materials come along with them, predominant in utilize how the heritage site should be developed.

Sawahlunto Ex mining area illustrated as an industrial heritage site, but not many recognized the potential of its natural objects. Notwithstanding, many cultural World Heritage Sites benefit from strong supporting geoheritage interests (Gordon, 2018), Sawahlunto properties have the possibilities to be more expanded and developed. To illustrate the site better, the utilization on cultural landscape for this heritage site must represent the influence of the physical constraints and/or opportunities laid. Community engagement is highly important to be considered in the development ethos.

Acknowledgment
This writing contributes to more expand and broader vision on preservation and utilization on Sawahlunto Ex Mining Are as a UNESCO World Heritage Site. Another approach on suitable requirement mentioned in the writing in terms of cultural landscape. The writers wished to keep this valuable cultural heritage well developed and cherish by the local community. Handling the site with content to preserve the history made is why this write done with the mind of cultural revolution within times. Acknowledgment is made to researchers and broad members of regulation maker who contribute to this heritage gem.

References
[1] Abel, C. (2000). Architecture & Identity (2nd ed.). Architectural Press.
[2] Cheris, R. & imbardi. (2021). Elemen Arsitektur Pembentuk Karakter Bangunan Pada Tapak Warisan Dunia Kota Sawahlunto Sumatera Barat, Indonesia. Arsitektura, 19(21 April 2021), 18. https://doi.org/https://doi.org/10.20961/arst.v19i1.4796
[3] Fleming, K. (1998). Cultural Landscape: A Theoretical Perspective. Proceedings of the Society for California Archaeology, 11, 112–117.
[4] Gordon, J. E. (2018). Geoheritage, geotourism and the cultural landscape: Enhancing the visitor experience and promoting geoconservation. Geosciences (Switzerland), 8(4). https://doi.org/10.3390/geosciences8040136
[5] Rito, B. B. R. (2019). Landscape Design Approach in Heritage Context (Case study: Emmahaven Port Coal Storage Facilities Sawahlunto City, West Sumatra). Journal of Architectural Research and Design Studies, 3(1). https://doi.org/10.20885/jars.vol3.iss1.art5
[6] Rosetia, A. (2021). A COMPARATIVE STUDY OF CONTAMINATED RIPARIAN ZONE FOR ECO LANDSCAPE DEVELOPMENT STRATEGY. Vitruvian Jurnal Arsitektur, Bangunan, & Lingkungan, 10(3), 212–218. dx.doi.org/ 10.22441/vitruvian.2021.v10i3.005
[7] The, H. A., & Geographies, C. C. (n.d.). A NEW MODEL FOR CULTURAL LANDSCAPE INTERPRETATION.
[8] UNESCO. (2019). Ombilin Coal Mining Heritage of Sawahlunto.
[9] Widiati. (2017). LAPORAN KAJIAN ZONASI KAWASAN CAGAR BUDAYA KOTA LAMA TAMBANG BATUBARA SAWAHLUNTO.
[10] Widodo, J. (n.d.). Preserving the Memory of Place: Case for Support for Palmer Road Area Conservation in Singapore Preserving the memory of place. 1–9.

Measuring cross-sector collaboration an effective solution for solving unlicensed mining problems in Kuantan Singingi, Indonesia

MY Tiyas Tinov[1], Aras Mulyadi[2], Sujianto[3], Zulkarnain[4], Ardi Putra[5], Tika Mutia[6] and Tito Handoko[7]

[1,2,3,4,7]Universitas Riau, Pekanbaru, Indonesia
[5]Universitas Maritim Raja Ali Haji, Tanjung Pinang, Indonesia
[6]Universitas Islam Negeri Sultan Sarif Khasim, Pekanbaru-Indonesia

Email: mtiyastinov@lecturer.unri.ac.id[1], aras.mulyadi@lecturer.unri.ac.id[2], sujianto@lecturer.unri.ac.id[3], zulkarnain@lecturer.unri.ac.id[4], ardiputra@umrah.ac.id[5], tikamutia@uin-suska.ac.id[6], tito.handoko@lecturer.unri.ac.id[7]

Abstract. The research question of this paper is to explore success in eradicating unlicensed mining activities with a cross-sector collaboration approach. Mining workers without permits are of concern in many sectors because they refer to state assets. The mining sector requires an effective policy governance approach, through well-integrated planning actions between government, security or police agencies, NGOs, local communities and the role of the private sector. To explore possible implementation problems by following factors that influence government policy in dealing with unlicensed mining. The method used in this paper is a qualitative descriptive approach. Data sources are secondary and premier data. Premiere data were obtained from the results of in-depth interviews with relevant informants during the research conducted. The paper locus of this research is Kuantan Singingi Regency, Riau Province, Indonesia. The analysis used refers to the systematic literature review approach to the concept of cross-sector collaboration. The paper concludes with a discussion of the important factors underlying the emergence of unlicensed mining, the fundamental obstacles in eradicating them, how the government's plan of action to create a cross-sector collaboration as an effective solution to resolve the unlicensed mining problems.

1. Introduction

Economic pressure on low-income communities become a problem in each country [1][2]. The efforts of low-income communities are doing everything that works for their economic well-being. Communities in developing countries have survived in an uncertain and dynamic economy [3][4]. There are so many aspects of economic growth that influence people's behavior. Such as social inequality, uneven income, scarcity of employment, poor community environment, and the political influence of economic turbulence [5][6]. The role of governance is the most important in controlling and overcoming these problems.

The economic, social and environmental aspects can be assumed to be the antecedents of the sustainable development in every country, and that thinking is supported by several research findings [7]. The economy is growing, the role of the corporate, or private sector is very important. The private sector contributes to employment and plays a role in regional development [8]. Valuable mining sector not licensed by the corporate, through corporate social responsibility [9]. They are obliged to channel a portion of the company's profitability to the community around the company environment. Besides, the corporate role in terms of employment. Then, the linkages between government, private sector or corporate and local community, must be strong. Then, the sustainability of a country will be developed [10].

However, by projecting into the future with the guidance of the important work, tries to provide a future context for ongoing collaboration between government and the private sector [11][12]. In this context, the focus is not on path dependence, incrementalism, and other manifestations of the institutional status quo. In contrast, in the era of scarcity of natural resources [13], the context of sustainability and collaboration shifted from apocalyptic warnings to reality where a perfect storm of global resource scarcity, population growth, poverty, and economic stagnation were existential challenges [14]. In this context, the dominant strategic focus is on cross-sector collaboration between government, local community, private sector, NGOs, and others.

Based on the literature review, in the last 15 years, there has been an exponential increase in the use of cross-sector collaboration to solve any problems [15][16], such as sustainable development [17], environmental damage [18] social and economic inequality of the community to the problem of illegal mining that often occurs in Indonesia [19]. These collaborations come from any sector, ranging from alliances between a private sector or non-governmental organizations (NGOs) to networks of small rural in Indonesia working with micro-financiers, Government's efforts to manage natural resources or design policies in managing resources, for the body's industrial level, norm-setting is designed instead of government intervention [20].

Organizations have many motivations for linking with others to address sustainability challenges. For private sector, the desire to enhance their reputations for corporate social responsibility is critical, along with finding leaner ways to produce and reducing their environmental footprint [21]. For governments, partnerships are becoming the governance structure of the future. For NGOs, the challenge involves taking up the role that, in many cases, governments are no longer able to fulfill because of shrinking resources or weakened social mandates [22]. For local community, their most important desire is to avoid poverty, eliminate unemployment, ensure sustainable natural resources and improve their overall health and well-being.

These cross-sector collaborations draw on diverse competencies of partners from many sectors to tackle problems that individual organizations (or even whole sectors) cannot solve or cannot solve as well, working independently. Because of their potential to combine resources, skills and knowledge from a wide range of stakeholders to address the challenges of creating a sustainable planet, these multi-sector partnerships have been called the "collaboration paradigm of the 21st century, [23], the "new organizational zeitgeist in dealing with societal issues" and a "stunning evolutionary change in institutional forms of governance"[24].

Each country has its own problems [25][26]. In Indonesia, many natural resources are disastrous [27]. Unlicensed mining problems have arisen in Indonesia, one of which is in Kuantan Singingi Regency. Kuantan Singingi Regency has illegal mining areas managed by local communities, the majority of which are not licensed. Unlicensed mining becomes a threat to all parties, both from the Kuantan Singingi Regency government and the Riau Provincial government in general and the communities directly affected by illegal mining in the form of environmental damage [28].

Based on the description and empirical gap above, this article tried to analyse the role government, local community, private sector and NGOs to solve unlicensed mining problem in Kuantan Singingi Regency, Riau Province in Indonesia through cross-sector collaboration as an effective solution action. Cross-sector collaboration (CSC) is becoming a more familiar fixture on the governance landscape. Alliances involving government, business, and non-profit organizations are now addressing issues that were once viewed as the prerogative of government.

The main objective of this research is to identify and analysed the how an activity of unlicensed mining always keep exits at Kuantan Singingi Regency, Riau Province in Indonesia. However, these activities cause damage to the environment and have an impact on the sustainability of natural resources. Therefore, there must be a problem solving for the unlicensed mining.

2. Methodology

In this article, we provide a systematic review of two related but distinct literature that addresses the implementation of cross-sector collaboration as an effective solution for solving unlicensed mining problems. We used a descriptive qualitative approach, and data is obtained by conducting deep-interviews with the parties concerned, then the interpretation of the findings in the field is associated with theories relating to the Principles of Cross-Sector Collaboration [29].

The result of this research is Cross-Sector Collaboration Framework on Unlicensed Mining as a primary work of the Indonesia's Peoples. Cross-Sector Collaboration Framework will become a reference for the collaboration strategy between private sector, government, NGOs, and Communities as a worker of unlicensed mining, however, that's concept become advantage to create a sustainable collaboration in the future [30][31].

Qualitative research is a study that requires scientific procedures so that the conclusions obtained are truly objective and precise. To test the validity of the data obtained to measure the validity of the results of this study, it was done by increasing perseverance in research, they are making more clearly, careful and continuous observations using triangulation techniques [32][33]. The triangulation referred to in this study is source triangulation, namely comparing data obtained through time and different tools or techniques, by comparing the results of observations with interview data, then comparing the results of interviews with documentation data.

Then, the initial sources or data to the final data are expected to be more sustainable and in accordance with the facts in the field. Finally, in making conclusions the results of this paper research are more systematic and targeted as the data described in the text is in accordance with the actual reality [34] (Neuman, 1994).

3. Result and Discussion

In this century, society will face obstacles with unmatched complexity. Critical issues such as climate change and water scarcity [35], widening disparities between rich and poor, declining educational outcomes, and cultural conflicts require our attention at the right time. Globalization and technology exponentially increase the scope, speed, and interrelation of our challenges.

As speed increases, more challenges arise, so the capacity to quickly switch and adapt will become more important [36]. The irreplaceable element in determining whether we overcome our obstacles and take advantage of opportunities at these times remains the same as has happened throughout history: leadership [37]. As complex, interdependent and emerging challenges arise, effective change to secure a brighter future will require transformative, collaborative leaders who can effectively lead cross-sector collaboration.

In order to apply mining patterns in Indonesia that are environmentally sound, so that natural resource management can be effective and effective with an orientation to the interests of ecosystem sustainability, the Indonesian government deems it necessary special efforts to curb the business of unlicensed mining whose effectiveness is seen as causing concern, disturbing the community and damaging the environment. The efforts of the Kuantan Singingi Regional Government to curb the unlicensed mining business is by issuing a policy through a Decree in the Local Area concerning the Formation of an Integrated Unlawful Mining Control Team that is spread throughout the districts in Kuantan Singingi Regency with the hope that the unlicensed Mining problem can be handled well in the future [38].

Based on interviews with the government in the efforts of Kuantan Singingi Regency in carrying out measures to control illegal or unlicensed mining activities that have damaged the joints of life through the mechanism of the Kuantan Singingi Regent Decree Number 13 of 2013 concerning the Formation of an Integrated Team for the Issuance of Mining without Permits in Kuantan Singingi Regency whose team members come from various elements from government, law enforcement, and society [39][40]. After the government decree was issued since 2016 until now many actions have been carried out either in the form of direct actions in the form of controlling unlicensed mining, or in the form of persuasive actions carried out by carrying out social and cultural approaches.

Monitoring of activities and developments in unlicensed mining can be seen from the location of illegal mining by the community in almost all Sub-Districts in Kuantan Singingi District [41]. The distribution of unlicensed mining locations was identified after field monitoring activities carried out by various government agencies, law enforcers, and the community who were part of an integrated team formed by the Regent [42].

In 2016 there were 56 locations of unlicensed mining spread carried out by the community in various locations (rivers, plantation areas, and vacant land) spread throughout the Districts in Kuantan Singingi Regency. However, after efforts to implement the policy up to the end of 2018, there were 40 locations with mining activities without licenses. The Integrated Team conducts field monitoring that has previously coordinated with the entire Integrated Team. Thus, the monitoring team will continue to coordinate and monitor the activities of unlicensed Mining as initial information to formulate and arrange future actions to curb mining without a license in Kuantan Singingi Regency, Indonesia.

3.1 Factors Affecting Government Policy on Unlicensed Mining

a. Cross-Sector Communication

The environmental damage control policy resulting from Unlicensed Mining requires intense communication between the operators that have been determined, so that the control activities and the socialization of the prohibition against unlicensed mining can be carried out properly in accordance with the mandate contained in the Kuantan Singingi Regent Decree No.13 of 2013. The communication that exists between the government and the community through socialization and appeals to the community by involving leaders in the community regarding the prohibition of illegal mining.

In addition, intense communication between fellow different sectors by coordinating and working together to unite the vision in controlling illegal mining continues to be established in order to achieve common goals going forward. The clarity of unlicensed mining and enforcement in Kuantan Singingi Regency is based on the structure in the Regent's Decree by establishing a Cross-Sector Collaborative Integrated Team consisting of various elements of government, law enforcement, NGOs, and the community by establishing communication, coordination, cooperation, and team alignment, so that all Integrated Team activities are coordinated well [43].

b. Human Resources

Human resources are an important factor that can carry out and implementing policies. The quality and quantity of human resources possessed can be seen from the composition of the Integrated Team issued by the Kuantan Singingi Regent in Bupati Decree No. 13 of 2013 varies greatly from government officials, law enforcement and security (Police, Attorney General's Office, Republic of Indonesia Army, Civil Service Police Unit), and community leaders [44].

The composition of the existing integrated team structure certainly has very good quality and quantity of human resources, so that it is not impossible if the goal to be achieved, namely controlling unlicensed mining will be realized if all that have been set are able to work together and make strategies together and provide a way out for unlicensed mining actors. All unlicensed mining control activity field officers require various forms of facilitation to support these activities such as sinso (wood cutting machines) and other logistics. With this facility, it will be easier to control unlicensed mining locations and to destroy unlicensed mining equipment.

c. Actors involved in Unlicensed Mining

Unlicensed mining activities have powerful actors who control them, making it difficult to touch and these actors have a strong network of both information and capital and can influence the community, government, and law enforcement officials. The actors or persons who sometimes take part have other interests that are not according to the rules in controlling unlicensed mining.

Most actors as financiers come from inside and outside Kuantan singing and have a widespread network in Kuantan singing. Then, it needs intensive deepening in solving the problems of actors who participate in providing capital to the community. Blood ties, cooperation with the private sector as financiers, who make bribes to the government in the form of money as collateral for mining operations often become a hindering factor in carrying out control of unlicensed mining. Where the government needs to review the teams directly involved in the exterminate. In addition, creating cross-sector collaboration will establish a more effective system for solving unlicensed mining problems.

d. Local Community
The local community is giving a big hand in solving the problem of unlicensed mining, community attitudes and participation are needed so that the activities of controlling unlicensed mining can be completed and the environment becomes livable. Lack of community awareness of the environment is a huge obstacle for the integrated team in controlling unlicensed mining in Kuantan District.

Awareness for environmental hazards due to the act of taking natural resources by unlicensed mining, will have a very big impact on the lives of local communities, but the community's unawareness of the problem has led to the emergence of new unlicensed mining actors, and continues to exist today.

e. Sosio-Economic
Socio-economic inequality is the most common cause so that there are still many unlicensed mining actors in Kuantan Singingi Regency. It is undeniable that the economic condition of local people will make them act arbitrarily if their household needs are not prosperous. Unlicensed mining activities will multiply if local communities experience difficulties in the household economy. Another problem is the declining price of rubber trees, which forces them to carry out unlicensed mining activities. That is because of the urgent economic needs of the household. Consequences of economic pressure of the community, a strong impetus is when they must work to get money, so they are looking for easy ways to get money [45].

However, mining in Kuantan Singingi district has been around for a long time and is a job for poor households. Difficulties in employment make other incentives to become illegal miners. Another thing is also caused by the influence of relatives who had previously worked as illegal miners. Limited employment opportunities make us lower class people are forced to carry out illegal activities and violate the law for the needs of families at home can be met at least.

3.2. Cross-Sector Collaboration for Unlicensed Mining Control Action Plans
Integrated Teams that collaborate between different sectors coordinate continuously to obtain the right pattern or formulation in efforts to control unlicensed mining. The formulation that is made is implemented by direct action to the field, so that the actions carried out become more effective and on target. As it is known that during 2016-2018 there were a lot of unlicensed mining that was destroyed by the Sector Collaboration Team.

Thus, the formulation that has been made has met the right target point, so the actions taken will give a deterrent effect to anyone who wishes to carry out unlicensed illegal mining. Every time action and enforcement will be held, the Integrated Team has always coordinated and held meetings or meetings in the context of preparing technical and non-technical action plans and will make a written report to be submitted to the Regent on the activities of enforcement and enforcement carried out.

After the formulation and determination of the unlicensed mining control policy carried out by the Integrated Team, the control measures are carried out. Unlicensed mining control activities in 2016 were carried out in 2 (two) stages. The first phase is in April and the second phase is in May. This activity aims to provide awareness to illegal mining actors of the impacts caused by illegal mining, providing an understanding that momentary profits are not a reason to meet basic needs, but requires careful policies which benefit the community itself or the profits of the investor group. Kuantan Singingi Government wants the community not to take part in providing facilities to unlicensed miners.

Unlicensed mining control activities in 2017 were carried out in the context of Siak Operations conducted in 8 (eight) villages in 7 (seven) sub-districts with obtained unlicensed mining equipment which was destroyed as many as 437 rafts. The curbing in 2017 is quite successful in reducing the number of unlicensed mining actors in Kuantan Singingi Regency and destroying unlicensed mining equipment found in the field by means of annihilation by burning.

Figure 1. Unlicensed mining activity in Kuantan Singingi Regency, Indonesia

From the fact of the implementation carried out by the Integrated Team in the 2016-2018 period, it can be seen that the activities of the control and enforcement of unlicensed mining have been well coordinated and well planned, so that when the control and enforcement in the field all elements in the Integrated Team already understand their duties and functions each of which results in a large number of extermination actions that the Integrated Team can take on unlicensed mining. Strict enforcement and enforcement carried out by the Integrated Team in the field becomes a deterrent effect for unlicensed mining actors who have not been subjected to control, so that with this action it is expected that unlicensed mining actors can stop by themselves and no longer damage the environment just for livelihood.

Efforts to control unlicensed mining involve many parties including the Government, NGOs, Communities, Private sector. The government has an important role in controlling, because the mining actors are not licensed to carry out their activities in rural areas and the community also comes from the village concerned, so the role of the government, especially the village head in controlling mining is not licensed, appeal to the community and socialize prohibition of illegal mining activities. That is the spearhead of illegal mining control activities that damage the environment.

3.3. Cross-Sector Collaboration as an Effective Solution to Resolve The Unlicensed Mining Problem

The general effectiveness shows at the level of achievement of results, in simple language it can be explained that; the effectiveness of the local government is if the objectives of the local government can be achieved in accordance with planned needs [46][47][48]. One of the objectives of the Kuantan Singingi Regency government is to restore the damaged environment by curbing unlicensed illegal mining.

During the period of 2016 - 2018, the control of unlicensed mining was carried out, every action was not yet effective because it was rarely encountered by illegal mining actors at the mining site, which was only found with exterminate teams. This allows for uncertainty between operators, so that information on enforcement is often leaked or known to the illegal mining actors. Limited funds are one of the obstacles of controlling unlicensed mining [49], so that the enforcement is less effective because it is not implemented comprehensively and in a large scale. Another thing that causes ineffective enforcement is because mining actors are not licensed in carrying out their activities often moving places, so the location information obtained is not uncommon that has been left by unlicensed mining actors.

The impact of unlicensed illegal mining activities is very negative, especially on the environment, social and community welfare. Where the environmental impact caused using chemicals and the dredging of the ground which makes large holes in the river and in the area become a factor triggering the bad environment. Rivers are no longer clear and the water smells and is dangerous to humans and the ecosystems in them, while the holes that will one day bring disaster resulting from environmental destruction.

4. Conclusion

Based on the results of the research and discussion in this paper, it can be concluded that the implementation of the Kuantan Singingi Regency government policy in controlling environmental damage due to unlicensed mining has been implemented through the Decree of the Regent No. 13 of 2013 concerning the Establishment of an Integrated Licensing Cross-Sector Collaborative Collaboration Team which in the policy is outlined 4 (four) main points of the formation of an integrated cross-sector collaboration team to coordinate and identify problems and developments of unlicensed mining, formulate and prepare unlicensed mining control action plans, carry out control measures in accordance with applicable procedures and report developments and the results of the eradication of unlicensed mining.

The effectiveness of the implementation of the policy is very influential in efforts to save environmental damage, which results from unlicensed mining activities that damage the environment. However, with the ineffectiveness of the performance of the integrated cross-sector collaboration team, it has not achieved success. Then, unlicensed mining activities are still running until now. The impact of environmental damage is getting worse from year to year. The policy implementation process is influenced by several factors which will determine the success or failure of the implementation of the unlicensed mining control policy. These factors are communication across sectors, human resources, Actors involved in Unlicensed Mining, local communities, and socioeconomics.

The author hopes that in the future the government will re-discuss the procedures for controlling that are confidential and systematic so that effective solutions from implementing unlicensed mining control policies can be successful. In addition, the government has an effort to save the environment so that land that is currently damaged can be reused and become a driver for creating prosperous households and economic well-being.

References

[1] Febrian, A. F., Maulina, E., & Purnomo, M. (2018). The influence of social capital and financial capability on sustainable competitive advantage through entrepreneurial orientation: Empirical evidence from Small and Medium Industries in Indonesia using PLS-SEM. Advances in Social Sciences Research Journal, 5(12).

[2] Towe, V. L., Leviton, L., Chandra, A., Sloan, J. C., Tait, M., & Orleans, T. (2016). Cross-sector collaborations and partnerships: essential ingredients to help shape health and well-being. Health Affairs, 35(11), 1964-1969.

[3] Day, R. H. (1984). Disequilibrium economic dynamics: A post-Schumpeterian contribution. Journal of Economic Behavior & Organization, 5(1), 57-76.

[4] Naghshpour, S., & Sergi, B. S. (2018). The impact of commercial banking development on Russian economic growth. In Exploring the Future of Russia's Economy and Markets. Emerald Publishing Limited.

[5] Kraus, S., Rigtering, J. C., Hughes, M., & Hosman, V. (2012). Entrepreneurial orientation and the business performance of SMEs: a quantitative study from the Netherlands. Review of Managerial Science, 6(2), 161-182.

[6] Purnomo, M., Permana, E., Qosasi, A., Febrian, A. F., & Miftahuddin, A. (2019). Entrepreneurial Orientation in Micro and Small Enterprises of Traditional Food Centers in Bandung, West Java. Binus Business Review, 10(3), 167-174.

[7] Gupta, J., & Vegelin, C. (2016). Sustainable development goals and inclusive development. International environmental agreements: Politics, law and economics, 16(3), 433-448.

[8] McDonald, S., & Young, S. (2012). Cross-sector collaboration shaping corporate social responsibility best practice within the mining industry. Journal of Cleaner Production, 37, 54-67.

[9] Spiegel, S. J. (2012). Governance institutions, resource rights regimes, and the informal mining sector: Regulatory complexities in Indonesia. World development, 40(1), 189-205.

[10] Crosby, B. C., & Bryson, J. M. (2005). A leadership framework for cross-sector collaboration. Public Management Review, 7(2), 177-201.

[11] Erakovich, R., & Anderson, T. (2013). Cross-sector collaboration: management decision and change model. International Journal of Public Sector Management.

[12] Page, S. B., Stone, M. M., Bryson, J. M., & Crosby, B. C. (2015). Public value creation by cross-sector collaborations: A framework and challenges of assessment. Public administration, 93(3), 715-732.

[13] Bachriadi, D. (2007). Mining in a state forest? This is Indonesia! Inside Indonesia.

[14] Miller, P. M., Scanlan, M. K., & Phillippo, K. (2017). Rural cross-sector collaboration: A social frontier analysis. American Educational Research Journal, 54(1_suppl), 193S-215S.

[15] Hamann, R., & April, K. (2013). On the role and capabilities of collaborative intermediary organisations in urban sustainability transitions. Journal of Cleaner Production.

[16] Hamann, R., Giamporcaro, S., Johnston, D., & Yachkaschi, S. (2011). The role of business and cross-sector collaboration in addressing the "wicked problem" of food insecurity. Development Southern Africa.

[17] Herren, H. R. (2018). The Sustainable Development Goals. In Farming, Food and Nature.

[18] Teece, D. D., Pisano, G., & Shuen, A. (1997). Dynamic capabilities and strategic management. Strategic Management Journal, 18(7), 509–533.

[19] Herren, H. R. (2018). The Sustainable Development Goals. In Farming, Food and Nature

[20] Bryson, B. J. M., Crosby, B. C., Stone, M. M., & Saunoi-sandgren, E. O. (2009). Designing and managing cross-sector collaboration: A case study in reducing traffic congestion. In IBM Center for The Business of Goverment.

[21] Baumgartner, R. J., & Ebner, D. (2010). Corporate sustainability strategies: Sustainability profiles and maturity levels. Sustainable Development, 18(2), 76–89.

[22] Forrer, J. F., Kee, J. E., & Boyer, E. (2014). Governing Cross-Sector Collaboration. In Jossey-Bass.

[23] Dienhart, J. W. (2010). Sustainability, Cross-Sector Collaboration, Institutions, and Governance. Business Ethics Quarterly.

[24] Yakovleva, N., Kotilainen, J., & Toivakka, M. (2017). Reflections on the opportunities for mining companies to contribute to the United Nations Sustainable Development Goals in sub – Saharan Africa. Extractive Industries and Society.

[25] Lehtinen, M. J. (2015). Industrial Minerals and Rocks. In Mineral Deposits of Finland.

[26] Spiegel, S. J. (2012). Governance Institutions, Resource Rights Regimes, and the Informal Mining Sector: Regulatory Complexities in Indonesia. World Development.

[27] Saam, Z., Amiruddin, R., & Afandi, D. (2015). The Influence of Illegal Gold Mining Activities toward Health of Workers in Kuantan Singingi , Indonesia. International Journal of Science and Research.

[28] Saam, Z., Amiruddin, R., & Afandi, D. (2015). The Influence of Illegal Gold Mining Activities toward Health of Workers in Kuantan Singingi , Indonesia. International Journal of Science and Research.

[29] Halvorsen, K., Almklov, P. G., & Gjøsund, G. (2017). Fire safety for vulnerable groups: The challenges of cross-sector collaboration in Norwegian municipalities. Fire Safety Journal.

[30] Neuman, W. L. (1994). Social Research Methods: Qualitative and Quantitative Approaches. Boston, MA:Allyn and Bacon.

[31] Teddlie, C., & Yu, F. (2007). Mixed methods sampling: A typology with examples. Journal of Mixed Methods Research, 1(1), 77–100.

[32] Ghozali, I. (2014). SEM Metode Alternatif dengan menggunakan Partial Least Squares (PLS). Semarang: Badan Penerbit Universitas Diponegoro.

[33] Neuman, W. L. (1994). Social Research Methods: Qualitative and Quantitative Approaches. Boston, MA:Allyn and Bacon.

[34] Wildemuth, B. M. (Ed.). (2016). Applications of social research methods to questions in information and library science. ABC-CLIO.

[35] Yakovleva, N., Kotilainen, J., & Toivakka, M. (2017). Reflections on the opportunities for

mining companies to contribute to the United Nations Sustainable Development Goals in sub – Saharan Africa. Extractive Industries and Society.
[36] Bryson, B. J. M., Crosby, B. C., Stone, M. M., & Saunoi-sandgren, E. O. (2009). Designing and managing cross-sector collaboration: A case study in reducing traffic congestion. In IBM Center for The Business of Goverment.
[37] Crosby, B. C., & Bryson, J. M. (2005). A leadership framework for cross-sector collaboration. Public Management Review.
[38] Saam, Z., Amiruddin, R., & Afandi, D. (2015). The Influence of Illegal Gold Mining Activities toward Health of Workers in Kuantan Singingi , Indonesia. International Journal of Science and Research.
[39] Garr, J. S., & Badu, E. (2021). Assessing the social impacts of illegal Gold mining activities at Dunkwa-On-Offin (Doctoral dissertation).
[40] Spiegel, S. J. (2012). Governance Institutions, Resource Rights Regimes, and the Informal Mining Sector: Regulatory Complexities in Indonesia. World Development.
[41] Amoako, C., Adarkwa, K. K., & Koranteng, K. A. (2021). The politics of artisanal small-scale gold mining (ASM) in the Akyem Abuakwa Traditional Area of Ghana. Journal of Contemporary African Studies, 1-16.
[42] Yamindago, R. I. A. (2015). Restoring coastal ecosystems - a case study Malang and Gresik regency, Indonesia. Journal of Coastal Conservation.
[43] Saam, Z., Amiruddin, R., & Afandi, D. (2015). The Influence of Illegal Gold Mining Activities toward Health of Workers in Kuantan Singingi , Indonesia. International Journal of Science and Research.
[44] Bachriadi, D. (2007). Mining in a state forest? This is Indonesia! Inside Indonesia.
[45] Kurnianingsih, F., Zulkarnain, I., & Mahadiansar, M. (2021). How Socio-Economic Impact Tourism Development in Pandemic COVID-19? Study of Bintan Regency, Indonesia. International Journal of Social Science and Religion (IJSSR), 71-85.
[46] Dorisman, A., Muhammad, A. S., & Setiawan, R. (2021). Stakeholder Collaboration in Traffic Accident Management. Jurnal Magister Administrasi Publik, 1(1), 64-75.
[47] Ratten, V., Marques, C. S., & Braga, V. (2018). Knowledge, learning and innovation: research into cross-sector collaboration. In Contributions to Management Science.
[48] Zhang, N., Lu, Z., & Shou, Y. (2017). The dominant role of governing structure in cross-sector collaboration in developing China: two case studies of information integration in local government one-stop services. Information Technology for Development.
[49] Bachriadi, D. (2007). Mining in a state forest? This is Indonesia! Inside Indonesia.
n Pemeringkatan Siswa," *J. MEDIA Inform. BUDIDARMA*, vol. 3, no. 4, p. 381, Oct. 2019, doi: 10.30865/mib.v3i4.1546.

IOP Conf. Series: Earth and Environmental Science 1041 (2022) 012051 doi:10.1088/1755-1315/1041/1/012051

Helminthiasis of rabbits on the upland and lowland areas and the risk factors

D Indrasanti*, M Indradji, Sufiriyanto, M Samsi and E Yuwono

Universitas Jenderal Soedirman, Purwokerto, Indonesia

*diana.indrasanti@unsoed.ac.id

Abstract. Rabbits are a potential livestock commodity to supply quality meat with higher protein and lower fat and cholesterol content than other meat. Helminthiasis is a parasitic worm infection that has imposed economic loss involving some risk factors. This study aimed to investigate the helminthiasis cases in both upland and lowland areas, as well as their risk factors on helminthiasis infection of rabbit farming. A total of 192 rabbit samples was derived from Baturraden and Kalibagor, Banyumas Regency, Central Java Province, Indonesia. The research engaged a survey method with a purposive sampling. Data on the correlation between the areas and helminthiasis incidence were subjected to a chi square analysis. The risk factors were computed in the 2x2 contingency table and analyzed with Odds Ratio. The result showed a significant correlation between the altitude of an area and the helminthiasis incidence ($P<0.05$). Regarding the host, male rabbit is 1.15 times as risky as the female counterparts to get helminthiasis infection. Meanwhile, the environmental factors, such as the metal floor of the rabbit cage contribute 7.42 times as much risk of helminthiasis infection as that of wooden floor.

1. Introduction

Rabbits are small mammals of the Leporidae family with great potential and excellent, low-cholesterol meat [1], [2]. Helminthiasis is one of the diseases that require further investigation because, like most parasitic worm infections, it is without clinical symptoms. When present, the clinical symptoms are usually mild and negligible [3]. Helminthiasis is a worm-parasitic disease which contributes to the inhibited rabbit's maintenance. Some species of parasites in rabbits could expose humans to disease (zoonosis), such as Hymenolepis, Giardia spp and Toxocara sp [4], [5], [6]. Additionally, some risk factors for gastrointestinal parasites include sex, age, and seasons [7].

Banyumas Regency in Central Java Province, Indonesia has a wet tropical climate and fertile agricultural land because it is located near an active volcano, Mount Slamet. Therefore, this area is potentially developed into livestock farming, including rabbits [8]. Banyumas covers an upland (at the foot of Mount Slamet) and lowland areas that potentially affect the condition of livestock maintenance. The high rainfall and humidity are the ideal condition for parasitic diseases, such as helminthiasis.

Epidemiological studies on the incidence and risk factors of diseases have been conducted in many countries [9], [4], [10], [7]. However, data on helminthiasis cases in Indonesia are limited. Meanwhile, rabbit meat is the potential substitute for other livestock meat because of its nutrition value [2]. Supplying ASUH meat (Safe, Healthy, Whole and Halal) to guarantee quality meat for public consumption and securing the health of the veterinary public is one of the main objectives of this study. Accordingly, this study aims to investigate the helminthiasis incidence in the upland and lowland areas of Banyumas Regency, as well as the risk factors of the causes to helminthiasis of rabbits.

2. Materials and Methods

This study conducted an analytical survey with two types of data: laboratory analysis of feces samples and a questioner interview with rabbit farmers. The researcher used the purposive sampling technique to collect rabbit's feces from rabbits farming in the upland (Baturraden) and lowland (Kalibagor) areas in the Banyumas Regency because both areas (Baturraden and Kalibagor) have a relatively large population of rabbits. The total samples were collected using the Slovin Formula: n=N/ 1+Ne2 where n is the total minimum samples, N is the population, and e is the error margins, such as 5% [11]. Therefore, the total samples collected was 192.

Feces were examined to identify worm eggs in the samples of rabbit feces using the native method and float test. Total feces samples were calculated using the Mc Master method as in the previous study [12], [13]. The rabbit farmers filled out the questioner when the researchers conducted the field survey. The dependent variable in this study was helminthiasis incidence. The independent variables included lands, sex, body condition, age, body weight, type of cage, total rabbits per-cage, cage cleanliness, cage density, cage cleaning frequency, type of feed, cage lighting, total drinking water, and water source. Data on the correlation between the lands (upland and lowland) and the helminthiasis incidence were subject to chi-square analysis. The risk factors data was computed into a 2x2 contingency table and analyzed with Odds Ratio [14].

Odds Ratio (OR) is the measure of association between the exposure (risk factors) and the disease incidence, which is calculated from the number of disease incidence in the risk group (exposed to risk factors) compared to the total disease incidence in the no-risk groups (not exposed to risk factors). If OR = 0 (zero), the variables are not the risk factors [15], [16]. The Odds Ratio in this study measures the extent of the risk factors/variables to the helminthiasis infection of rabbits and how many times rabbits exposed to risk factors are more likely to get infected than the not exposed ones. The Odds Ratio formula is ad/bc, where: "a" is cell a, "b" is cell b, "c" is cell c and "d" is cell d. A detailed information is presented in the Table 1.

Table 1. Table of 2x2 contingency of Odds Ratio

Factor	Disease	
	No	Yes
No	a	b
Yes	c	d

3. Result and Discussion

3.1 Helminthiasis in upland and lowland areas

Batturaden upland is located on the slope of Mount Slamet, so it has fresh air. Meanwhile, Kalibagor is a lowland with a temperature of 25°C [8]. Both areas are located in Banyumas Regency, Central Java Province, Indonesia. The geographic differences, such as altitude above sea level (asl) will result in different weather and microclimatic condition in those areas, especially temperature and humidity [17]. Helminthiasis distribution in both areas is presented in Table 2.

The result shows a significant correlation between land altitude and helminthiasis incidence (P<0,05). Different rabbits' age and sex, as well as the climate and season of an area, contribute to the helminthiasis incidence [9], [18], [19].

Table 2. Distribution of rabbit samples on helminthiasis in the upland and lowland areas

Lands	Total samples	Helminthiasis incidence		Percentage
		Positive	Negative	
Upland area	99	7	92	7.07%
Lowland area	93	27	66	29.03%

There was a lower helminthiasis incidence in the upland than that in the lowland area (Table 2). The prevalence of infection that causes helminthiasis in the upland (Baturraden) and lowland (Kalibagor)

areas was 7.07% and 29.03%, respectively. The contributing factors may include the different geographical conditions in which the upland area has a cooler weather and lower rate of rainfall than those of the lowlands; as a result, the rabbits are more comfortable living in the uplands. Furthermore, the maintenance system also contributes to the helminthiasis incidence in both areas. Rabbits are animals that enjoy living in ambiance temperature around 9-19°C [20].

Furthermore, Skriivanova and Nuriyasa [21], [22] reported that rabbits are animals that find it difficult to balance their body hit due to their limited sweat gland. During heat stroke, rabbits would undergo an accelerated respiratory rate and heartbeat. Regarding parasitic worms, the prevalent worm egg in Baturraden includes Strongyle, Trichuris sp, Cittotaenia and Toxocara sp while in Kalibagor are Strongyle, Trichuris sp, Cittotaenia sp, Trychostrongylus sp and Passalurus sp. Siegmund [23] stated that worms that are common in rabbits are Taenia pisiformis, Taenia serialis from Cestoda and Passalurus ambiguus from the Nematode class.

3.2 The risk factors of helminthiasis in rabbits

The transmission models of infectious disease are generally affected by 3 (three) factors, namely an infection agent (etiology), the host, and the environment. In the transmission of helminthiasis disease, the infectious agent and host characteristics are relatively constant; however, the environmental factor is more dynamic and influential to the infectious pattern of helminthiasis.

The risk factors for the parasitic gastrointestinal disease include sex, age, seasons, and climate [24], [7]. In this study, we attempted to dissect the environmental factors to the infection pattern of helminthiasis. We found four factors related to the host (group A) and 18 factors regarding the maintenance system that includes cage (group B), maintenance (group C), and feed and others (D).

Table 3. Odd Ratio (OR) value of the contributing factors to helminthiasis of rabbits

No	Risk factors	Helminthiasis incidence				Total samples	OR
		Infection		No infection			
		Total samples	%	Total samples	%		
A							
1.	Sex						
	a. Male	18	18.75	78	81.25	96	1.15
	b. Female	16	16.66	80	83.33	96	
2.	Body condition						
	a. Medium	27	17.53	127	82.47	154	0.94
	b. Fat	7	18.42	31	81.58	38	
3.	Age						
	a. Offspring	32	17.68	149	82.32	181	0.97
	b. Parent	2	18.18	9	81.81	11	
4.	Body weight						
	a. Below 1 kg	5	8.62	53	91.38	58	0.34
	b. Over 1 kg	29	21.64	105	78.36	134	
B							
1.	Types of cage						
	a. Dump	0	0.00	0	0.00	0	0.00
	b. Stage	34	17.71	158	82.29	192	
2.	Cage material						
	a. Wood	10	17.24	48	82.76	58	0.95
	b. Non Wood	24	17.91	110	82.09	134	
3.	Shape of cage floor						
	a. No slate	3	16.67	15	83.33	18	0.92
	b. Slate	31	17.82	143	82.18	174	
4.	Materials on cage floor	33	20.37	129	20.37	162	7.42
	a. Wood	1	3.33	29	3.33	30	

No	Risk factors	Helminthiasis incidence				Total samples	OR
		Infection		No infection			
		Total samples	%	Total samples	%		
	b. Metal						
C							
1.	Cage cleaning method a. Sweep b. Sweep and wipe with cleanser	11 23	21.57 16.31	40 118	78.43 83.68	51 141	1.41
2.	Separate feses and urin disposal a. No b. Yes	0 34	0.00 18.58	9 149	100.00 81.42	9 183	0.00
3.	Feces and urine deposits a. Around the house b. A dedicated place	1 33	16.67 21.57	5 153	83.33 85,26	6 186	0.93
4	Cage cleaning frequency a. Occasionally b. Everyday	10 24	14.93 19.20	57 101	85.07 80.80	67 125	0.74
5.	Frequency of cleaning feces disposal a. Occasionally b. Regularly	0 34	0.00 18.58	9 149	100.00 81.43	9 183	0.00
6.	Daily residual feed is wasted a. No/accumulated b. Yes	17 17	20.00 15.89	68 90	80 84.11	85 107	1.32
7.	Feeding frequency a. 2 times a day b. > 2 times a day	34 0	18.28 0.00	152 6	81.72 100.00	186 6	0.00
8.	Frequency of feed waste disposal a. Irregular c. Every day	4 30	16.67 17.86	20 138	83.33 82.14	24 168	0.92
9.	Frequency of drinking water supply a. Ad libitum b. Irregular	13 21	18.57 17.21	57 101	81.43 82.79	70 122	1.10
10.	Total rabbits in one cage a. More than one b. One	7 27	12.28 20.00	50 108	87.72 80.00	57 135	0.56
11.	Cage density a. = 60x40x40 per rabbit b. >60x40x40 per rabbit	2 32	7.14 19.51	26 132	92.86 80.49	28 164	0.32
D							
1.	Type of dominant feed a. Grass and pellet	4 30	9.30 15.63	39 119	90.70 79.87	43 149	0.41

No	Risk factors	Helminthiasis incidence				Total samples	OR
		Infection		No infection			
		Total samples	%	Total samples	%		
	b. Pellet						
2.	Water source						
	a. Non tap water	33	18.64	144	81.36	177	3.21
	b. Tap water	1	6.67	14	93.33	15	
3.	Cage lighting						
	a. Bright	30	20.41	117	79.59	147	2.63
	b. Dark	4	8.89	41	91.11	45	

Regarding rabbits' characteristics, the contributing factors to helminthiasis infection include sex (1,15), age (0,97), body condition (0,94) and body weight (0,34). Different sex is the highest factor with the OR value of 1,15. Accordingly, male rabbits are 1,15 times as risky as the female to get infected. Female animals are more resistant to parasitic infection than their male counterparts because of their hormonal performance.

Similar with previous study [25], that result, perhaps attributed to the aggressive nature of male animals when feeding, may pick up eggs of parasite. Previous studies by [26] and [7], this study found that young rabbits are more susceptible to infection than the adult. Additionally, fat rabbits are 0.34 times as risky as those under 1kg to get helminthiasis infection. Table 3 presents the detailed OR values in this study.

In the caging system, the highest risk of helminthiasis exposure is on the material of the cage floor. The wooden cage floor has a 7,42 OR, so it exposes rabbits to helminthiasis infection 7,42 times as much as that with a metal floor. The wooden cage floor may have higher humidity and take a longer time to dry, so it allows worm eggs or larvae to live longer and infect the rabbits. The farmers did not use grounded rabbit housing because it makes the rabbits step on their faces. Rabbits are animals that enjoy clean space, so when the cage is dirty, they prefer lying down in the cleanest spot. When all places are dirty, the rabbits will not lie down, and they show signs of discomfort.

Table 3 presents the detailed OR value of the caging system. The cage cleaning method obtains the highest OR value (1,41). Some farmers only swept the cage while others wiped it, occasionally using a chemical agent to make the cage cleaner. In the first method, the rabbits were 1,41 times exposed to helminthiasis infection than those in a cleaner cage. The technique of feed residue disposal also contributes to helminthiasis infection, with a 1,32 OR values. Farmers who did not discard the residual feed in the feed container, but only added more feed to the old one had made the rabbits 1,32 times as risky to get infected with helminthiasis.

Table 2 presents the detailed OR value of the maintenance system. Furthermore, rabbits' cage with a brighter lighting has the OR value of 2,63, so the rabbits are 2,63 times as risky to get infected with helminthiasis than those in the dark cage. A bright cage is assumed to make the rabbits more active and alert, so they need more energy than the rabbits that stay idle. Rabbits that feed more are more likely to get exposed to infectious parasites in the feed. Table 3 presents the detailed OR value of rabbit feed and other factors. In the cattle, prevalence of helminthiasis is multifactorial in different breeds. That is influenced by type of management system, frequency and type anthelmintic used, physiology and nutrition of the cattle [25].

4. Conclusion

Land altitude is significantly correlated with helminthiasis incidence ($P<0,05$). The infection prevalence of the causes of helminthiasis in the upland and lowland area is 7.07% and 29.03%, respectively. Regarding the host, the male rabbit is 1.15 times as risky as the female counterparts to get helminthiasis infection. Meanwhile, the environmental factors, such as the metal floor of the rabbit cage contribute 7.42 times as much risk of helminthiasis infection like that of the wooden floor.

Acknowledgement
The author would like to thank to University of Jenderal Soedirman because the research funds were obtained from Unsoed BLU funds. The author also thanks the various parties who have contributed to this research. Hopefully, this research is useful for the development of the world of animal husbandry and veterinary medicine and society.

References
[1] Zotte AD and Z Szendro. 2011. The role of rabbit meat as functional food. Meat Science. 88(3):319-331.
[2] Nistor E, VA Bampidis, N Pacala, M Pentea, J Tozer and Prundeanu. 2013. Nutrient content of rabbit meat as compared to chicken, beef and pork meat. Journal of Animal Production Advances. 3(4):172-176.
[3] Anosike JC, V Zaccheaus, CM Adeiyongo, OC Abanobi, EO Dada, IR Keke, JC Uwaezuoke, OU Amajuoyi, CE Obiukwu, DC Nwosu and FI Ogbusu. 2006. Studies on the intestinal worm (helminthiasis) infestation in a Central Nigerian rural community. Journal of Applied Sciences Environmental Management. 10(2):61-66.
[4] Rewatkar SG, SS Deshmukh, PG Kumar, DK Maske and GN Bhangale. 2013. Occurrence of gastrointestinal helminths in rabbits with special reference to importance of Giardia spp. as parasitic zoonoses. Science, Technology and Arts Research Journal. 2(3):142-143.
[5] D'ovidio D, E Noviello, P Pepe, L Del-Prete, G Cringoli and L Rinaldi. 2015. Survey of Hymenolepis spp in pet rodents in Italy. Parasitology Research. 114:1-4.
[6] Ziegler MA and CNL Macpherson. 2019. Toxocara and its species. CAB Reviews. 14(53):1-27.
[7] David OSF, K Hussain, M Rabiu and IA Ganiyu. 2018. Parasitic conditions of domestic owned rabbits in Osun State, Southwestern Nigeria: Retrospective evaluation, risk factors and co-infestations. International Journal of Veterinary Science and Medicine. 6:208–212.
[8] Simda-Green Economy. 2019. Geographical Location. Official Portal of Banyumas Regency. Banyumas Regency Local Government. https://www.banyumaskab.go.id.
[9] Odoi A, JM Gathuma, CK Gachuiri and A Omore. 2007. Risk factors of gastrointestinal nematode parasite infections in small ruminants kept in smallholder mixed farms in Kenya. BioMed Central Veterinary Research. 3:6.
[10] Makitaipale J, I Karvinen, A Virtala and A Nareaho. 2017. Prevalence of intestinal parasites and risk factor analysis for Eimeria infections in Finnish pet rabbits. Veterinary Parasitology Regional Studies Reports. 9:34-40.
[11] Tejada JJ and JRB Punzalan. 2012. On the misuse of Slovin's Formula. The Philippine Statiscian. 61(1):129-136.
[12] Indrasanti D, M Indradji, S Hastuti, E Aprilliyani, Fatikha and KA Rosyadi. 2017. The administration of garlic extract on Eimeria stiedai oocysts and the hematological profile of the coccidia infected rabbits. Media Peternakan (Tropical Animal Science Journal). 40(3): 158-164.
[13] Indrasanti D, M Indradji, E Yuwono, M Samsi, PV Sundari, MN Ichwan, ES Anengseh, MN Hatmadifia and TN Hidayat. 2019. Treatment of rabbit coccidiosis with combination of herbal extract II toward oocysts excretion and hematology parameters. The 1st Animal Science and Food Technology Conference (AnSTC) 2019 IOP Conf. Series: Earth and Environmental Science. IOP Publishing. 372:012008.
[14] Steel RGD and JH Torrie. 1980. Principles and Procedures of Statistics. McGraw-Hill Inc.
[15] Indradji M and E Yuwono. 2004. Grassing as potential factor in gastrointestinal nematodiasis in goat in Banyumas District. Animal Production. 6(2):95-100.
[16] Indradji M and Sufiriyato. 2013. Evaluation of the effectiveness of avian influenza vaccines in native chicken using two by two contingency table. Animal Production. 15(1):62-68.
[17] Andrian S and M Purba. 2014. The effect of elevation and slope on rubber (Hevea brasiliensis Muell. Arg.) production in PTPN III Hapesong Farm of South Tapanuli. Jurnal Online Agroteknologi. 3(2):981-989.

[18] Rahman AKMA, N Begumr, M Nooruddin, MDS Rahman, MA Hossain, HJ Song. 2009. Prevalence and risk factors of helminth infections in cattle of Bangladesh. Korean Journal of Veterinary Service. 32(3):265-273.
[19] Elshahawy I and A Elgoniemy. 2018. An epidemiological study on endoparasites of domestic rabbits (Oryctolagus cuniculus) in Egypt with special reference to their health impact. Sains Malaysiana. 47(1):9-18.
[20] Mc Nitt JI, NM Nephi, SD Lukefahr and PR Cheeke. 1996. Rabbit Production. Interstate Publishers Inc. 78-109.
[21] Skriivanova V, M Marounek, M Skriivan and J Knizek. 2011. Effect of temperature on growth, feed efficiency and mortality of rabbits. Research Instutute of Animal Production. Parague-Unrineves.
[22] Nuriyasa IM, NGK Roni, E Puspani, DPMA Candrawati, IW Wirawan and AW Puger. 2014. The physiological responses of local rabbit fed diets using tofu by-product supplemented with yeast housed in two cage system. Majalah Ilmiah Peternakan. 17(2):2656-8373.
[23] Siegmund OH. 2015. Merck Manual, Veterinary Manual, Parasitic Disease of Rabbits. www.merckvetmanual.com.
[24] Morgan ER and JVan Dijk. 2012. Climate and the epidemiology of gastrointestinal nematode infections of sheep in Europe. Veterinary Parasitology. 189(1):8-14.
[25] David OFS, IA Ganiyu, M Rabiu, H Karimat. 2020. Helminth infections of great concern among cattle in Nigeria: Insight to its prevalence, species diversity, patterns of infections and risk factors. Veterinary World. 13(2):338-344.
[26] Mwangi GW. 2015. Prevalence and intensity of coccidiosis in adult and weaning domestic rabbits under intensive system. BVM Thesis. Nairobi, Kenya: University of Nairobi.

Water Quality Analysis Based on Land Use in Sekampung River, Lampung, Indonesia

K Fitriani[1], T K Nufutomo[1,2], and R Putra[1,2*]

[1]Institut Teknologi Sumatera, Lampung 35365, Indonesia
[2]Research and Innovation Center for Environmental Life and Sanitation, Institut Teknologi Sumatera, Lampung 35365, Indonesia

*Correspondence author: raden.putra@tl.itera.ac.id

Abstract. Currently, the status of water quality in the river area tends to decline where one of the causes is the land use along the Sekampung River, and the land use is dominated by agriculture. Mapping of water quality in rivers aims to provide information in the form of distribution of water quality status in rivers. The purpose of this study is to identify the type of land use in the Sekampung River, Sumur Kucing Village, analyze the water quality of the Sekampung River, Sumur Kucing Village based on physical parameters (Temperature and TDS) and chemical (pH, Nitrate, Nitrite, and Ammonia) and interpreted the results of river water quality status using the STORET method spatially thematically. Determination of sample points was carried out using purposive sampling method with 4 sample points which were representative of each land use. Sampling of river water was carried out by referring to SNI 03-7016-2004. The results show that the land use in the Sekampung River is in the form of agricultural areas, plantations, settlements, ponds, and rice fields. The results of the analysis of water quality are based on physical parameters and chemical parameters, parameters that exceed the quality standards include TDS, Ammonia, and Nitrite. The results of the water quality distribution map by determining the status of water quality using the STORET method show point 1 and point 4 are classified as lightly polluted and point 3 and point 2 are classified as moderately polluted.

Keywords: Sekampung River, Land Use, Water Quality, Sumur Kucing Village.

1. Introduction

Sekampung river is one of the rivers that has an important role for the community, especially for the people of Sumur Kucing Village. The Sekampung River is a river that passes through 8 autonomous regions or regencies/cities starting from the upstream area in the Tanggamus Regency, the Central Region in Central Lampung Regency and the downstream area in East Lampung. The current increase in population has also resulted in an increase in people's water consumption needs [1]. The current status of water quality in the river area has decreased, one of the causes is land use in the Sekampung River watershed. Land use can affect water quality because it comes from activities such as the application of fertilizers, pesticides, feeding, and household activities that produce domestic waste [2]. Plantation and agricultural waste that uses artificial fertilizers and excessive manure is the main source of anthropogenic nitrogen in the waters [3]. Pollution mapping and water quality analysis are needed to convey information about the condition of these waters, so that they can assist in determining the classification of river water quality and determining the sources of river pollutants. The river is a body of water which is an effective means of transportation for the passage of various pollutants from upstream to downstream and can function as a pollutant absorber [4]. So, a water quality distribution map is needed to make it easier to classify pollutant sources as well as river water quality information.

2. Materials and methods

2.1 *Study area*

The research implementation time is planned from June 2021 to July 2021.

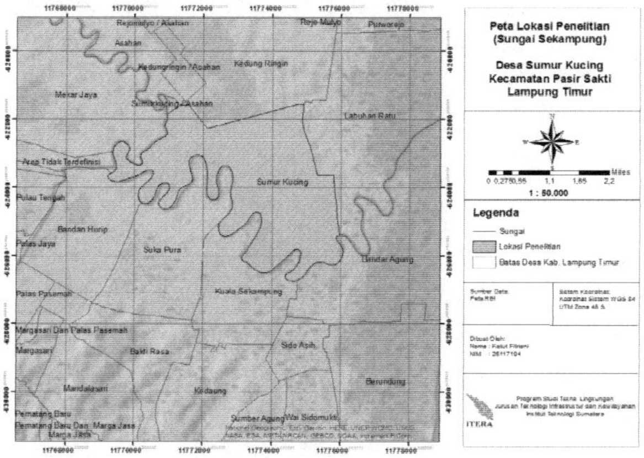

Figure 1. Research site map. The Sekampung River in Sumur Kucing Village, which is located in Pasir Sakti District, East Lampung Regency, is a downstream Way Sekampung watershed in the form of a food and fishery production center by utilizing pond ecosystems and coastal areas. The Sumur Cat Village is the border between East Lampung Regency and South Lampung Regency which borders the Sekampung River

2.2 *Data preparation*

The arrangement of the data used in this study is Thematic Geospatial Information (IGT), which is geospatial information that describes one or more themes. The IGT in this study was sourced from my homeland with a scale of 1:25,000,000, 2020. The data for this research used erath google satellite imagery in 2020. The water quality data was obtained by means of observation and lab tests with parameters in the form of physics and chemistry.

2.3 *Land use*

Identification of land use in the Sekampung River was carried out using thematic Geospatial Information sources in Sumur Kucing Village. Thematic Geospatial Information is used as a base map in making land use maps. The land use map was created using google earth overlay satellite imagery with thematic geospatial information for the research location, namely the administrative shapefile of Sumur Kucing Village. in the digitization process The method used is a visual interpretation method. Sampling points were obtained from analysis using previously made land use maps and observations made in terms of accessibility, time and cost. Determination of river water sample points is done by using purposive sampling method. Purposive sampling is one of the procedures for taking water sample points based on several considerations made by researchers. These considerations are in the form of land use around the Sekampung River in the form of accessibility, time and cost. Determination of sampling points based on representatives of land use around the river in the form of rice fields, gardens, shrimp ponds and settlements.

2.4 *Sampling*

Sample points were obtained from analysis using land use maps. Determination of river water sample points is done by purposive sampling method. . These considerations are in the form of land use around the Sekampung River in the form of accessibility, time and cost. The water quality test is divided into:

land use around the Sekampung River in terms of accessibility, time and cost. Water quality test is divided into:

a. In situ water quality testing includes temperature, pH, and TDS which can change rapidly and cannot be maintained.
b. Ex-situ water quality tests in the form of nitrate, nitrite and ammonia were brought to the laboratory.

Sampling of river water is carried out using an integrated sample (combined sample of place) which is a mixture of instantaneous samples taken from various points/locations at the same time based on SNI 03-7016-2004. The sampling point refers to the Indonesian National Standard (SNI) 6989.57:2008. Water samples were taken every week for 3 weeks. After taking samples and tested in situ and ex-situ, then analyzed using the STORET method to determine the condition of the Sekampung River water.

3. Result and discussion

3.1 Land use

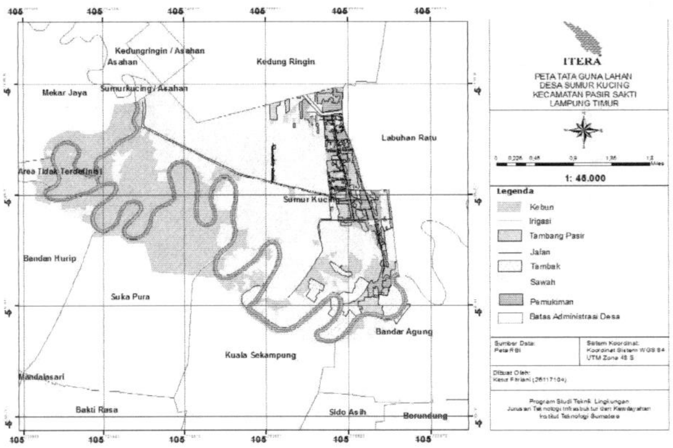

Figure 2. Land use map

Most of the land use in Sumur Kucing Village is agriculture in the form of rice fields. Other land uses include gardens, ponds, settlements and ponds. The results of ground checks carried out at 22 points include 6-point gardens, 4-point settlements, 5-point rice fields, 4-point ponds, and 3-point ponds. A total of 2 points do not match the analysis, namely points 18 and 22. This can be caused by errors in the interpretation results. Based on the ground check that has been carried out, obtained 90% accuracy. The calculation of the accuracy test data obtained is 90% while the accuracy requirement is > 80%, then the accuracy results in Table 3 have met the specified requirements. The land use area of Sumur Kucing Village based on secondary data digitization is 1592.29 Ha. The land use of Sumur Kucing Village consists of 59% agricultural area, 26% gardens, 6% settlements, 5% shrimp ponds and 4% former sand mining.

Table 1. Land use area

No	Types of land use	Area (Ha)
1	Ricefield	938,3
2	Plantation	413,11
3	Settlement	96,26
4	shrimp	74,09
5	former sand mine	70,57
	Amount	**1592,29**

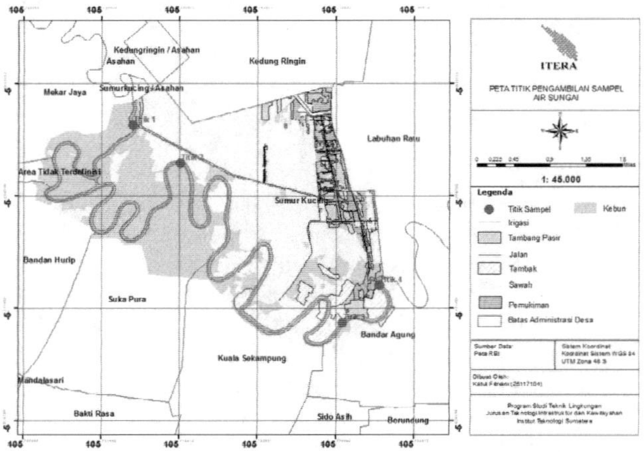

Figure 3. River water sampling point map

a map of the Sekampung River water sampling points containing 4 points in the form of representatives of each existing land use. Sampling points are determined based on accessibility, time and cost taken into account.

Table 2. Water sampling point

Point	Information		Keterangan
	Latitude	Longitude	
1	5°34'53.63"S	105°44'25.12"T	Areas with plantation
2	5°35'23.55"S	105°45'1.24"T	Areas with rice fields
3	5°36'59.30"S	105°46'55.63"T	Areas with shrimp ponds
4	5°36'38.73"S	105°47'4.63"T	Areas with residential

3.2 Water quality

Conditions of land use that are turned into agricultural land, plantations, settlements, and shrimp ponds are indicated to affect the condition of river water quality [5]. The water sampling point is based on the river water discharge obtained from each sampling location point. The average water discharge at point 1 is 24.54 m3/s. The average water discharge at point 2 is 33.67 m3/s. The average water discharge obtained at point 3 is 37.77 m3/s. The average water discharge at point 3 is 37.77 m3/s. The width of the river at point 3 is the widest among the other 3 points, because at point 3 there is a former sand mine so that it has the potential to cause widening of the river body. The average discharge obtained at point 4 is 37.32 m3/s. The average water discharge at the four points is in the range of 5 m3/s – 150 m3/s, then water samples are taken at 2 points with a distance of 1/3 and 2/3 at a depth of 0.5 times the depth of the river water surface and then mixed.

3.2.1 Temperature

the average temperature measurement in 3 weeks of measurement obtained the highest temperature at point 2 which is 29.9 and 3 which is 30.3 then the temperature drops again at point 4 which is 29.7. The water temperature at the research site is influenced by the environmental hue. The high temperature in the river can be caused by the high intensity of sunlight that enters because the weather is sunny and is an open place so that light enters directly into the body of water [6]. Temperature affects the process of photosynthesis in water, oxygen content in water, and metabolic rate in aquatic organisms. The high temperature measurement results were caused when the sample was measured during the day and the weather was hot so that the intensity of sunlight obtained was large [1]. Temperature can be the influence of biochemical processes such as nitrification and ammonification in water [7]. High temperatures will also cause an increase in the solubility of toxic compounds.

3.2.2 Total dissolved solid (TDS)
The results of the average measurement at 3 weeks show that the average TDS value is decreasing, the value is increasing. at points 1 and point 2 the TDS measurement results are below the quality standard while at points 3 and 4 the TDS value has exceeded the quality standard. The downstream TDS measurement results will increase, this is indicated because it is the result of accumulation of contaminants through waste runoff from river bodies [8]. This can happen because the research area along the river has land uses in the form of rice fields, gardens, ponds and settlements that have the potential to become waste contributors. Increasing the concentration of TDS can be dangerous because it will cause changes in salinity. Changes in salinity will disrupt the balance of aquatic biota, biodiversity, will cause less tolerant species to be disturbed and can cause high toxicity [9].

3.2.3 pH
Each point shows the pH value below the predetermined quality standard, the quality standard in class III is 6-9. The highest pH value is at point 3, namely shrimp ponds of 5.7 while the lowest pH value is at point 2, namely rice fields of 5.3 while the average pH value at points 1 and 4 is 5.6. The pH conditions in very acidic or very alkaline waters can endanger the survival of organisms because they can interfere with respiration and metabolism processes [5]. Measurement of the average pH value at each point shows the pH results are below the predetermined quality standard and are acidic.

3.2.4 Ammonia
The lowest ammonia value is at point 1, which is 0.13 mg/L, then it increases at point 2, which is 0.43 mg/L, then the ammonia value tends to increase at point 3 with a value of 0.5 mg/L. However, the ammonia value then dropped to point 4, which was 0.2 mg/L. Ammonia comes from the main metabolic waste products produced by fish. High levels of ammonia indicate pollution in the form of organic matter from runoff from agricultural fertilizers and shrimp ponds [10]. Ammonia comes from the breakdown of protein sourced from leftover feed and dead plankton. The concentration of ammonia in river water can be affected by the concentration of pH and water temperature [11]. Increasing the pH concentration will also increase the toxicity of ammonia to aquatic organisms. High levels of ammonia can be toxic to commercially farmed fish .

3.2.5 Nitrate
Nitrate is the more stable form, nitrite is the transitional form between nitrate and ammonia. Chemical fertilizers used in agricultural systems can result in an increase in the amount of nitrate in river water [12]. During 3 weeks of measurement at 4 sampling points the results showed the nitrate value was below the predetermined quality standard. Existing land uses around river bodies such as shrimp ponds and settlements have the potential to contribute their waste to the waters. The high nitrate content needs to be watched out for because it has the potential to cause eutrophication in the waters [13].

3.2.6 Nitrite
Nitrite can cause eutrophication in river bodies because eutrophication occurs due to the entry of nitrite and nitrate from fertilizers (which are agricultural waste products) or untreated waste. The results of the measurement of the average value of nitrite at point 1 is 0.04 mg/L, at points 2 and 3 the average value of nitrite is 0.34 mg/L and the average value of nitrite at point 4 is 0.04 mg/L. The nitrite value at points 2 and 3 can indicate that the pollutant comes from agricultural waste, as well as shrimp pond waste [14].

High nitrite values can also be indicated because in the 3rd week of measurement there was drizzling rain which could cause nitrite to be washed into the waters [15][10][5].

3.2.7 river water quality status
Point 1 obtained river water quality results of -3 with mildly polluted status, point 2 of -11 with moderately polluted status, point 3 of -12 with moderately polluted status and point 4 of -5 with mildly polluted status. The results showed that at point 1 to point 3 the quality of river water decreased. However, at point 4 there is an increase in water quality.

4. Conclusion
Based on the results of the identification of land use around Sumur Kucing Village, there are 938.30 hectares of rice fields, 413.11 hectares of gardens, 96.26 hectares of settlements, 74.09 hectares of ponds and 70.57 hectares of ponds. The results of the accuracy test on the land use map that have been obtained have an accuracy of 90%. The test measurement results for 6 parameters, namely physics and chemistry, the values of TDS, nitrite and ammonia in the Sekampung River have exceeded the quality standard. The results of the interpretation of river water quality using the STORET method are spatially thematic, the results are obtained at points 1 and 4 with lightly polluted status, points 2 and 3 with moderately polluted status.

References
[1] A. Suriadikusumah, O. Mulyani, R. Sudirja, E. T. Sofyan, M. H. R. Maulana, and A. Mulyono, "Analysis of the water quality at Cipeusing river, Indonesia using the pollution index method," *Shengtai Xuebao/ Acta Ecol. Sin.*, vol. 41, no. 3, pp. 177–182, 2021, doi: 10.1016/J.CHNAES.2020.08.001.
[2] H. Somura, S. B. Yuwono, H. Ismono, B. Arifin, F. Fitriani, and R. Kada, "Relationship between water quality variations and land use in the Batutegi Dam Watershed, Sekampung, Indonesia," *Lakes Reserv. Res. Manag.*, vol. 24, no. 1, pp. 93–101, 2019, doi: 10.1111/lre.12221.
[3] R. Kudubun, D. Rahardjo, J. Jlagran, and J. Dongkelan, "Pengaruh Tata Guna Lahan , Tipe Vegetasi Riparian , dan Sumber Pencemar Terhadap Kualitas Air Sungai Winongo di Daerah Istimewa Yogyakarta yang mengalir secara kontinu dari hulu menyebabkan pencemaran pada air permukaan ," no. September, pp. 392–400, 2020.
[4] J. Y. Zhang, W. M. Ni, Y. M. Zhu, and Y. D. Pan, "Effects of different nitrogen species on sensitivity and photosynthetic stress of three common freshwater diatoms," *Aquat. Ecol.*, vol. 47, no. 1, pp. 25–35, 2013, doi: 10.1007/s10452-012-9422-z.
[5] S. Gyawali, K. Techato, S. Monprapussorn, and C. Yuangyai, "Integrating Land Use and Water Quality for Environmental based Land Use Planning for U-tapao River Basin, Thailand," *Procedia - Soc. Behav. Sci.*, vol. 91, pp. 556–563, 2013, doi: 10.1016/j.sbspro.2013.08.454.
[6] A. E. Briciu, D. Mihaila, A. Graur, D. I. Oprea, A. Prisacariu, and P. I. Bistricean, "Changes in the water temperature of rivers impacted by the urban heat island: Case study of Suceava city," *Water (Switzerland)*, vol. 12, no. 5, 2020, doi: 10.3390/W12051343.
[7] J. Huang *et al.*, "Characterizing the river water quality in China: Recent progress and on-going challenges," *Water Res.*, vol. 201, no. April, p. 117309, 2021, doi: 10.1016/j.watres.2021.117309.
[8] M. G. Y. L. Mahagamage and P. M. Manage, "Mapping Spatial Distribution of Water Quality Parameters Using Gis in Groundwater of the Kelani River Basin, Sri Lanka," *Proc. Acad. World*, vol. 12, no. December, pp. 978–93, 2015.
[9] Rinawati, D. Hidayat, R. Suprianto, and P. Dewi, "Penentuan Kandungan Zat Padat (Total Dissolve Solid Dan Total Suspended Solid) Di Perairan Teluk Lampung," *Anal. Environ. Chem.*, vol. 1, no. 01, pp. 36–45, 2016, [Online]. Available: http://repository.lppm.unila.ac.id/2831/1/Volume 1_Hal 36-45-Rina.pdf.
[10] P. Mahabeer and M. Tekere, "Anthropogenic pollution influences on the physical and chemical

IOP Conf. Series: Earth and Environmental Science 1041 (2022) 012052 doi:10.1088/1755-1315/1041/1/012052

quality of water and sediments of the umdloti river system, Kwazulu-Natal," *Phys. Chem. Earth*, vol. 123, no. January 2020, p. 103030, 2021, doi: 10.1016/j.pce.2021.103030.

[11] T. T. Ding *et al.*, "Water quality criteria and ecological risk assessment for ammonia in the Shaying River Basin, China," *Ecotoxicol. Environ. Saf.*, vol. 215, no. March, p. 112141, 2021, doi: 10.1016/j.ecoenv.2021.112141.

[12] X. Zhang, Y. Zhang, P. Shi, Z. Bi, Z. Shan, and L. Ren, "The deep challenge of nitrate pollution in river water of China," *Sci. Total Environ.*, vol. 770, p. 144674, 2021, doi: 10.1016/j.scitotenv.2020.144674.

[13] M. A. Kusumaningtyas, R. Bramawanto, A. Daulat, and W. S. Pranowo, "Kualitas perairan Natuna pada musim transisi," *Depik*, vol. 3, no. 1, pp. 10–20, 2014, doi: 10.13170/depik.3.1.1277.

[14] Q. Gu *et al.*, "Characterizing the spatial variations of the relationship between land use and surface water quality using self-organizing map approach," *Ecol. Indic.*, vol. 102, no. March, pp. 633–643, 2019, doi: 10.1016/j.ecolind.2019.03.017.

[15] H. Y. Pak, C. J. Chuah, E. L. Yong, and S. A. Snyder, "Effects of land use configuration, seasonality and point source on water quality in a tropical watershed: A case study of the Johor River Basin," *Sci. Total Environ.*, vol. 780, p. 146661, 2021, doi: 10.1016/j.scitotenv.2021.146661.

IOP Conf. Series: Earth and Environmental Science 1041 (2022) 012053 doi:10.1088/1755-1315/1041/1/012053

Application of oil palm empty bunches as organic mulch in oil palm plantation (*Elaeis guineensis Jacq.*): an evaluation and SWOT analysis

Lala Novita Sari, Sylvia Madusari*, Vira Irma Sari

Politeknik Kelapa Sawit Citra Widya Edukasi, Bekasi, West Java, Indonesia

*smadusari@gmail.com

Abstract. Palm oil waste management must be done as a form of implementing zero waste palm oil industry. Oil palm Empty Fruit Bunches (EFB) are one of the solid wastes produced from processing Fresh Fruit Bunches (FFB). PT ABC manages EFB waste by utilizing it as organic fertilizer with the mulching method. EFB is organic material containing nutrients and can improve soil physical properties when applied to plant cultivation areas. This study was conducted to determine the impact of EFB application on plantation land on the level of oil palm production and the potential for pests that can be caused and to analyze the implementation of EFB application at PT ABC, Bengkulu. The study method used is descriptive, using primary data obtained through observation, interviews, literature studies and documentation, and secondary data obtained by identifying organisms at the spot of EFB application and observing oil palm plants for symptoms of horn beetle attack. Then, data analysis was carried out using descriptive statistical methods and SWOT analysis methods. The results showed that the application of EFB with a single layer mulching method at PT ABC could reduce the risk of the appearance of horn beetle pests and could potentially increase the productivity of oil palm plants, although not too high. PT ABC has the opportunity to carry out a comprehensive EFB application to have a good impact on the company by considering the recommended SWOT analysis strategy.

1. Introduction

Palm oil is one of the plantation commodities owned by Indonesia and plays an important role in contributing to the country's foreign exchange because the export value continues to increase. The area of land for oil palm plantations in Indonesia continues to increase, from 12 million ha in 2017 to 14 million ha in 2018 [1]. The increase in land area is certainly accompanied by an increase in the production of fresh fruit bunches (FFB) and Crude Palm Oil (CPO). So, this will also have an impact on increasing the amount of palm oil waste produced. As a form of implementing sustainable palm oil, the waste generated from the palm oil industry is sought to be utilized so that oil palm becomes an industry with zero waste.

Oil palm empty fruit bunches (EFB) are solid waste produced from the processing of fresh fruit bunches (FFB) with the volume of EFB produced reaching 21-23% of the total FFB processed, the large volume encourages several companies to utilize EFB . One form of utilization of EFB is to use it as organic fertilizer in oil palm plantation areas. EFB is considered to be able to improve the physical and chemical properties of the soil because it contains nutrients. [2] EFB contains nutrients N, P, K, Mg and Ca so EFB can give a benefit by applying in the oil palm areas.

PT ABC is one of the companies that has applied EFB as organic fertilizer in oil palm plantation areas, both in mature and immature plants. EFB application at PT ABC uses a mulching system with the aim of improving soil physical properties and as a substitute for inorganic fertilizers. Therefore, this study was conducted to analyze the system and the impact of the management of EFB application as organic fertilizer carried out by PT ABC, so that it can be seen the influence and impact of EFB application on plant production and the potential for pests caused.

In addition, to find out the weaknesses and strengths it has, as well as the opportunities and threats faced by PT ABC in the sustainability of the EFB application. The scope of this study is application of EFB waste can be an alternative to organic fertilizer for oil palm plants and has a positive impact on productivity levels.

2. Methodology

This study was conducted at the PT ABC Oil Palm Plantation, Bengkulu, Indonesia. The study method used is descriptive using primary and secondary data. Primary data was obtained by conducting identification of the type of organism at the spot of EFB application and census of symptoms of Oryctes rhinoceros attack on plants that were applied with EFB.

Secondary data obtained through observation, interviews, literature studies and documentation. Data analysis was carried out using descriptive statistical methods and SWOT (Strengths Opportunities Weaknesses and Threats) analysis. The SWOT analysis is carried out by evaluating the strengths and opportunities in the utilization of oil palm EFB waste, so as to minimize weaknesses and threats. SWOT analysis is carried out to find new strategies that can be applied by the company, so that it can take full advantage of opportunities in addition to existing threats.

3. Result and Discussion

EFB waste applied in PT ABC was taken from the PT ABC's palm oil mill. The EFB application method at PT ABC uses a single layer mulching method. Mulching is an activity to cover the soil surface with litter or plant remains that has the ability to prevent erosion, because it protects the soil from the overwriting power of raindrops and the eroding power of surface water flow [3]. The application of the mulching method on EFB as an inorganic fertilizer was chosen in order to obtain several other benefits such as maintaining soil moisture, suppressing weed growth and reducing the risk of erosion.

EFB was applied at the space of trees each other with dose of 30 tons/ha/year with a rotation once a year (Figure 1). There are 139 trees/ha, the dose of application is 220 kg/spot and EFB weight average is 5 kg, so the EFB application can be arranged in one layer. This can speed up the decomposition process and reduce the risk of pests arising.

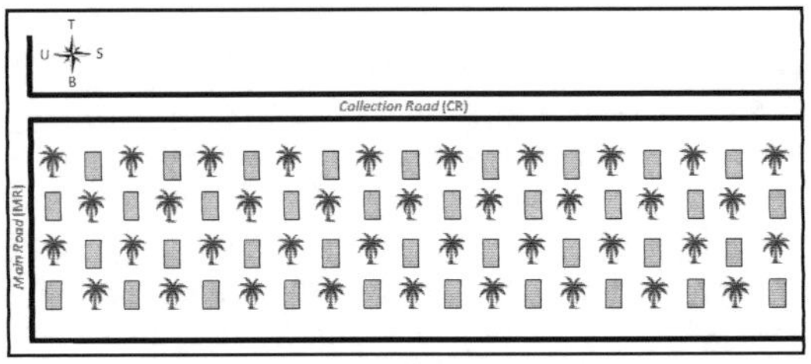

Figure 1. Application Layout of Empty Palm Oil Bunches

3.1 Pest Potential

The application of EFB as mulch in oil palm plantations is considered one of the sources of the emergence of pests. This is considered because EFB is an organic material that has been decomposed for a long time so that it can become a nest for pests, especially the *Oryctes rhinoceros*. However, from the results of the identification of organisms at the application spot of single layer EFB at PT ABC, no

Oryctes rhinoceros' larvae were found and only termites and ants were found. The results of the identification of organisms at the EFB application spot can be seen in Table 1.

Table 1. Identification of Organisms at the EFB Application Spot at PT ABC

Sample	*Isoptera*	*Formicidae*	*Oryctes rhinoceros* Larva
Application spot 1	Unavailable	Available	Unavailable
Application spot 2	Unavailable	Available	Unavailable
Application spot 3	Unavailable	Available	Unavailable
Application spot 4	Unavailable	Available	Unavailable
Application spot 5	Unavailable	Available	Unavailable
Application spot 6	Available	Available	Unavailable
Application spot 7	Available	Unavailable	Unavailable
Application spot 8	Unavailable	Available	Unavailable
Application spot 9	Unavailable	Available	Unavailable
Application spot 10	Unavailable	Unavailable	Unavailable

Oryctes rhinoceros larvae were not found at the spot of EFB application, presumably because the EFB application applied at PT ABC was one layer. According to [4] the application of a single layer of EFB can minimize the risk of to *Oryctes rhinoceros* lay eggs. In addition, the temperature in one layer of EFB is not too humid and the odor generated by the EFB is not too sharp and this can reduce the level of interest of *Oryctes rhinoceros* to the EFB that is applied. The types of ants found at the EFB application spot were *Oecophylla smaragdina*, *Lasius niger* and *Monomorium pharaonis*. Among the three types of ants, *Oecophylla smaragdina* are predatory ants. *O. smaragdina* use for biological control in oil palm plantations as a predatory is consume pupae over larvae (of all instars) and predatory of *P. pendula* [5]. Ants can also play a role in improving the physical properties of the soil. According to [6], ants have similar functional roles to earthworms, which is increasing soil water infiltration due to their tunnels and improved soil nitrogen.

The presence of termites at the EFB application spot is caused because the application spot is fairly moist and contains termite food material. In [7] termites are organisms that feed on cellulose-containing materials such as wood or dead plants. The existence of termites in an area has two roles, namely as decomposers and pests. The emergence of termites at the spot of EFB application is considered to have a positive impact because it helps in the decomposition process. This statement is supported by the literature which states that termites are one of the organisms that play a role in the decomposition of EFB [8].

The type of termites found was *Captotermes gilvus* and the result of observations at oil palm plants that were close to the termite house is, there were no signs of attack and no termites were found on oil palm plants near the termite house. It is because the distance between the termite house and the EFB application was also close, so the EFB is the main food source for termites. In line with [9] which states that *Captotermes gilvus* is a termite that nests in the soil and is close to organic materials such as litter and humus. This statement is also supported by the discovery of termites in the EFB pile.

3.2 Palm Oil Productivity

The percentage of EFB application progress in PT ABC's is calculated by means of the volume of EFB that has been applied (tonnes) divided by the volume of EFB required per block according to a dose of 30 tons/ha/year. The progress of the TKKS application and the per-block productivity value can be seen in Table 2.

Table 2. Palm Oil Productivity in 2018-2020

Nama Blok	Tahun Tanam	Progres Aplikasi TKKS (%)			Produktivitas (ton/ha)		
		2016	2017	2018	2017	2018	2019
D05	1996	263	96	112	11.98	12.10	8.22
D07	1996	239	31	100	11.85	11.46	7.73
D08	1996	256	125	127	12.95	11.87	8.00
D10	1996	114	80	77	11.89	11.94	7.68
D14	1996	129	321	312	11.57	12.88	8.39

The production value is not too difference between each block. This can happen because the EFB application is not carried out as a whole in blocks, but only in areas that are close to roads or piles of EFB. Application activities that are not carried out thoroughly occur due to uneven road contours, namely waves and hills, making it difficult for workers in the process of stacking EFB. The decline in production occurred in all blocks in 2019 because that year the plants were already 23 years old. In [10], oil palm plants will experience a decrease in production when they are more than 22 years old, because the age of the plant has passed the maximum productivity age of oil palm.

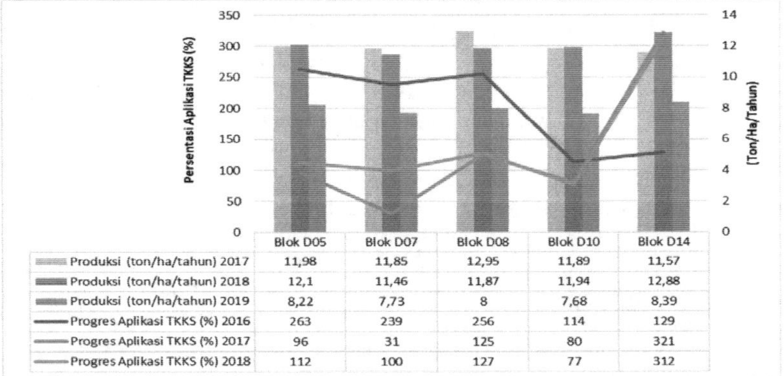

Figure 2. Graphical image of the relationship between the percentage of EFB applications and oil palm productivity

Block D14 had the highest production value and in 2017-2018, the block was also having the highest percentage of EFB applications, namely 321% and 312%. The EFB application can add value to oil palm production within one year after application, although the value added to production is not too different from blocks with a lower percentage value of EFB application. This happens because EFB is an organic material that contains nutrients and also high lignin, so it takes a long time to decompose, and the nutrients contained will also be slowly available to plants. The time required for EFB to decompose without the help of a decomposer can reach 6 months [11]. Due to the slow availability of nutrients to plants, the application of EFB as mulch does not provide high value-added productivity. This is also in line [12] which states that the application of EFB as mulch can only increase FFB production ranging from 10 – 34%.

The application of EFB application at the space of trees each other using the mulching method can also increase the potential for absorption of nutrients in EFB because the area is a place for young roots that are still active in absorbing nutrients. In [12] stated that the highest production was obtained by plants with EFB application on the inters trees. The application of the mulching method is also an indication of the potential for a stable increase in production in oil palm plants, because the application of a mulching system will reduce the risk of nutrients being washed out in the rainy season and can reduce the rate of soil evaporation during the dry season. EFB application as mulch is considered effective because it can maintain soil moisture and can reduce the unfavorable impact on growth and production in the dry season [13]. In [14] EFB mulching was increasing soil chemical properties (pH, organic C, total N, available P, K, Ca and Mg), oil palm nutrition levels (N, P, K, Ca, and Mg) and better for soil conservation.

3.3 SWOT analysis

Table 3. SWOT Analysis of EFB at PT ABC

INTERNAL FACTORS / EXTERNAL FACTORS	STRENGTHS	WEAKNESSES
	- Soil and water conservation efforts. - Expanding the root system of oil palm plants. - A single layer application can reduce the potential for pests to appear and a preventive weed control - The company's commitment to utilizing oil palm EFB waste as a form of implementing environmentally friendly oil palm plantations. - No potential to cause soil and water pollution. - Retaining/limiting losses of fertilizing nutrients. - Maximizing the absorption of nutrients by the roots because the EFB application can meet the requirements of soil conditions when the roots absorb nutrients.	- Lack of human resources as supervisors who are able to supervise and ensure the distribution of EFB to the plantation is carried out properly. - Road conditions are an obstacle to the application of EFB in blocks. - EFB is not completely distributed in the field - The risk of EFB being scattered on the road
OPPORTUNITIES - Application of the zero-waste concept in realizing a sustainable palm oil industry (Reducing methane gas emissions). - Implementation of government regulations to utilize palm oil solid waste. - Does not take a long time to apply because there is no special treatment (composting). - Potential to increase production in long term applications.	**SO, STRATEGY** - Complete application while still applying single layer pulsing application. - Preparation of standard operating procedures (SOP) for EFB application in the oil palm plantation area of PTABC. - Conducting research on the content of EFB produced from PT ABC to determine the appropriate and effective dose to be applied.	**WO STRATEGY** - Improvement of road access to the block so that distribution can be carried out smoothly. - Improved supervisory supervision and performance through work motivation. - Quotation of EFB that is on the road and returned to piles on the side of the road.
THREATS - The duration of EFB decomposition in the field is determined by climatic and weather conditions. - There is no written regulation from the government specifically	**ST STRATEGY** - Giving treatment to EFB so that it decomposes quickly.	**WT STRATEGY** - Coordinate through a persuasive approach with the relevant agencies to conduct a study on the utilization of EFB so that it is carried out optimally. - Implementation of training activities for

regarding the effective and efficient use of EFB.		employees to improving the competence of the management and strategy of using EFB as organic mulch.

4. Conclusion

The EFB application at PT ABC uses a single layer mulching method and it make does not have the potential to cause organisms that can act as plant pest organisms for oil palm plants. The pattern of EFB application, although not too high, has the potential to increase oil palm productivity. Application of single layer EFB at PT ABC can be continued by considering the recommended strategy.

References

[1] [DITJENBUN] Direktorat Jendral Perkebunan. 2019. Statistik Perkebunan Indonesia 2018–2020. Jakarta (ID): Sekretariat Direktorat Jendral Perkebunan. 86 hal.

[2] Bakar RA, Siti ZD, S. Kulaseharan, N. Jamaluddin. 2011. Effects of ten year application of empty fruit bunches in an oil palm plantation on soil chemical properties. Nutrient cycling in agroecosysem. 89: 341-349

[3] Karyati dan Sri, Sarminah. 2018. Teknoogi Konservasi Tanah dan Air. Samarinda (ID): Mulawarman University Press. 103 Hal

[4] Fauzana, Hafiz., Wardati., Al Ikhsan, Amri. 2018. Ketebalan Mulsa Tandan Kosong Kelapa Sawit Pengaruhnya terhadap Oryctes rhinoceros dan Peningkatan Hara Tanah pada Ekosistem Kelapa Sawit. Jurnal Agriculture and Food Security. Vol 1: 78-83

[5] Pierre EM and Azarae HI. 2013. Studies on the predatory activities of Oecophylla smaragdina (Hymenoptera: Formicidae) on Pteroma pendula (Lepidoptera: Psychidae) in oil palm plantations in Teluk Intan, Perak (Malaysia). 5: 163-176

[6] Evans TA, Tracy ZD, Philip RW, Nathan L. 2011. Ants and termites increase crop yield in a dry climate. Nature communications. 2(262): 1-7

[7] Safitri, Maulinda. 2019. Intensitas serangan rayap pada kebun kelapa sawit di PT Bakrie pasaman Plantation Sungai Aur Kabupaten Pasaman Barat [Skripsi]. Padang (ID): Universitas Muhammadiyah Sumatera Barat

[8] Safitri, Maulinda. 2019. Intensitas serangan rayap pada kebun kelapa sawit di PT Bakrie pasaman Plantation Sungai Aur Kabupaten Pasaman Barat [Skripsi]. Padang (ID): Universitas Muhammadiyah Sumatera Barat

[9] Rafli M A, Sylvia M, Jojon S. 2020. Komparasi efektivitas metode pengendalian rayap Macrotermes gilvus di perkebunan kelapa sawit. Jurnal Agrosains dan Teknologi. 5(2): 78-86

[10] Lubis, M. F., Iskandar, Lubis. 2018. Analisis Produksi Kelapa Sawit (Elaeis guineensis Jacq.) Di Kebun Buatan, Kabupaten Pelalawan, Riau. Jurnal Bul Agrohorti. 6(2): 281-286

[11] Purnamayani, Rima., Hendri., E., Salvia, D.S., Gusfarina. 2011. Potensi Tandan Kosong Kelapa Sawit sebagai Pupuk Organik dengan Berbagai Dekomposer. Kementrian Pertanian Balai Pengkajian Teknologi Pertanian Jambi

[12] Arryanto, Dian. 2020. Status Basa Basa Tukar Tanah K Ca Mg Pada Tanah Spodosol Sub Group Typic Plachortod Yang Diaplikasikan Tandan Kosong Kelapa Sawit Selama Lima Tahun Pada Lahan Budi Daya Kelapa Sawit (Elaeis Guineensis Jaqc.) Di Kabupaten Barito Timur Propinsi Kalimantan Tengah [Tesis]. Medan (ID). Universitas Sumatera Utara

[13] Octaviany, W.R., Hariyadi. 2016. Manajemen Pemupukan Tanaman Kelapa Sawit (Elaeis guineensis Jacq.) pada Area Marjinal di Kabupaten Kotawaringin Timur, Kalimantan Tengah. Jurnal Bul Agrohorti. 4(3): 321-326

[14] Moradi A, Christopher TBS, Goh KJ, Ahmad HMH, Che FI. 2012. Evaluation of four soil conservation practices in a non-terraced oil palm plantation. Agronomy journal. 104(6): 1727-1740

Oil palm trunk replanting as brown sugar raw materials

Latifa Siswati*, Enny Insusanty, Neng Susi, Nopryanti

Universitas Lancang Kuning, Pekanbaru, Indonesia

*latifasiswati@unilak.ac.id

Abstract. The trunks of unproductive oil palms that are rejuvenated can be used to make brown sugar as an innovation and additional income for farmers. The goal of this research is to know the utilization of palm trunks as brown sugar raw materials. Additionally, to know the break event point (BEP) and to analyze the fitness of return cost of ratio from agroindustry of oil palm's brown sugar in Kampung Maredan Barat Kecamatan Tualang Kabupaten Siak. The method of data collection is done through interviews with sellers using questionnaires. The result showed that oil palm trunks can be utilized with roughly 45 trunks out of 120 trunks in total per ha (37%). This business can gain 3,811,650 IDR with 310,265.6 IDR as BEP. The ability to produce on average is 900kg per month with BEP score is 22.4kg. The return cost of ratio is (1.44 > 1) and as it is bigger than one, so this business is fitting.

1. Preface

Oil palm that has aged over 25 years have to be rejuvenated. If there are 120 trees in one hectare, there will be 48,000 trunks in 400 hectares that would be waste. Replanting is divided into two phases, which are land preparation and planting phase. The activities in land preparation phase are basic census, grass length, overthrowing, mincing, trees cultivating, hole closing, and path contouring. After this phase is done, the next steps will be point planting, ground cover planting, hole planting, and oil palm planting (Junaedi, 2017).

Riau is one of the provinces in Indonesia where the predominant area is for oil palm plantation. According to Dithenbun in 2020, the area of oil palm in Riau reaches 14,996,010 hectares and its business can employ 622,315 people, making it play the most important part of its economy.

During its cultivation, oil palms have to go through some steps, such us land provision, seed selection, plant care, and rejuvenation. Its productive age is until 25 years until it needs to be replaced to the new ones. It wouldn't produce anything until it reaches three years old, and this is the time where farmers would do other replanting and farming in between (Siswati, 2018).

According to the farmers, during its oil palm rejuvenation, they are losing big parts of their income, making them try to find other way to gain some income back. It piques authors' interest where these farmers and their creative innovation could utilize oil palm trunks as raw materials of brown sugar.

What people of Kampung Meredan Barat do in order to utilize this is by tapping its shoot edge to get the liquid from the trunk. This liquid is the main ingredients for the brown sugar. It's still been done in household scale; traditionally, however this has been going on for quite a while now.

Based on the problem identification, the research goal is (1) To know how the utilization of oil palms are done as the raw materials for brown sugar, (2) To analyze the break event point (BEP) and return cost of ration (RC) of agroindustry of brown sugar in Kampung Maredan Barat Kecamatan Tualang Kabupaten Siak.

2. Research Methods

2.1. Place and Time of Research

This research is done in Kampung Maredan Barat Kecamatan Tualang Kabupaten Siak. The location is decided intentionally since Kampung Maredan Barat is where exactly the brown sugar from oil palm trunks are done.

2.2 Population and Sample

Population is translated as the area consisted of objects or subjects whose certain characteristic determined by the researchers to study and later to conclude. Sample, on the other hand, is part of its population (Sugiono, 2010). Sampling method used is censusing, where the whole population was the sample (Sugiono, 2012). In this case, the population of oil palm farmers in Kampung Maredan Barat is three people.

2.3 Data Collecting Method

Data collected in this research consists of primary data and secondary data. Primary data is obtained through interviews, which is by asking the farmers directly, where the sample is three people using questionnaires prepared in advance. Secondary data is obtained from the head office of Kampung Maredan Barat Kecamatan Tualang Kabupaten Siak.

2.4 Data Analysis Method

$$TC = TFC + TVC \quad (1)$$

Note:
TC = Total Cost (IDR/year)
TFC = TotalFixedCost(IDR/year)
TVC = Total Variable Cost (IDR/year)

Palm sugar farming revenue is calculated using the formula:

$$TR = P.\ Q \quad (2)$$

Note:
TR = Total Revenue (IDR)
P = Price (IDR/kg)
Q = Quantity(kg).

Revenue is obtained from the revenue minus the total cost, with the following formula:

$$I = TR - TC \quad (3)$$

Note:
I = Income(IDR),
TR = Total Revenue(IDR)
TC = Total Cost (IDR)
To determine the feasibility of farming, it can be seen from how much each IDR sacrificed is able to provide income. The feasibility of farming in this case is shown through the Hermanto (2009) formula as follows:

RCR = ... (4)

Where:
RCR = Revenue Cost Ratio
R = Total Revenue
C = Total Cost

With below criteria:
RCR >1 : Palm sugar farming is profitable
RCR < 1 : Palm sugar farming is not profitable
RCR=1 : Palm sugar farming is at an impasse
..................................(5)

....(6)

To find out a business reaches the break-even point where the costs incurred by the company and the income received are used the Break Event Point (BEP) formula (Sastraadmaja, 2009).

3. Result and Discussion

3.1 Cost

3.1.1 Oil Palm Utilization

The replanted palm trunks can be used as raw material for palm oil palm sugar. In 1 hectare of oil palm plantations, an average of 45 trees can be utilized, of which 120 stems per hectare per hectare (37%) can be utilized because not all farmers cut their oil palm trunks. In one oil palm trunk, the juice can be taken for one month. One day can be obtained 175 liters of sap which is converted into palm palm sugar. In 35 saps produce 5 kg of palm sugar so that in one day 25 kg of sugar can be produced.

Nira is a liquid that has a sweet taste because it contains sucrose. Nira can be obtained from palm plants such as oil palm, coconut, and sugar palm. Nira can also be called baged, the term baged is used for sap that is being cooked but has not become sugar. The way to obtain sap water is by tapping (Suwandi, 1993). According to (Apriwinda, 2013) the sap content of each type of plant has a different composition of sap and generally consists of water, sucrose, reducing sugar, other organic materials. Sucrose is the largest part of solids ranging from 12.30%-17.40%. Reducing sugar is between 0.50-1.00% and the rest is an organic compound.

Palm sugar stems are of good quality with a fine texture and are widely liked by people who already consume them at a price of 15,000 IDR per kg so that consumers can get brown sugar at affordable prices, but the availability of sugar is limited because replanted palm stems are not cut down at the same time because some farmers still expect fruit yields. Utilization of oil palm trunks to make palm sugar can only be done in smallholder oil palm plantations and oil palm plantations managed by cooperatives. Until now, oil palm stems from the company's plantations have not been used for brown sugar because the company has its own way of using palm trunks from replanting activities.

3.1.2 Brown Sugar Business Analysis

This brown sugar agro-industry activity has two types of costs, fixed costs and variable costs.
a. Fixed Cost
Fixed costs are costs incurred for the purchase of production assets which are calculated on a depreciation basis for business continuity if the assets have passed their useful life.

Table 1. Fixed Costs of Palm Oil Palm Sugar Agroindustry in Kampung Maredan Barat

Farmer's name	Depreciation Cost/Year (IDR)	Depreciation Cost/Month (IDR)
Susanto	992,488.1	82,710
Aidil Fitri	1,188,083.3	99,000
Anto	1,288,083.3	107,340

Table 1 shows that the amount of equipment costs or depreciation costs in palm sap processing is 82.710 IDR to 107.340 IDR. The difference in equipment costs is due to the different amount of assets used and the various purchase prices between entrepreneurs.

b. Variable costs

Variable costs are costs incurred in one production. Variable costs in processing brown sugar from palm sap such as the cost of purchasing equipment, and labor costs.

- Equipment

Table 2. Total Cost of Palm Sugar Agroindustry Business Equipment in Kampung Maredan Barat

Entrepreneur's name	Total cost (IDR/month)
Susanto	3,026,500
Aidil Fitri	4,189,000
Anto	4,392,500

In table 2 the total costs incurred to purchase equipment for processing palm sugar are different because the time and place of purchase of equipment are not the same with different prices.

- Labor Wages

In the management of replaying oil palm trunks into brown sugar in Kampung Maredan Barat, entrepreneurs need workers outside the family to facilitate and speed up the process of felling and stripping the palm fronds, so that the oil that has been uprooted can be extracted as quickly as possible. For the process of processing sap into sugar, business actors do not use labor outside the family. The following is the total number of working days in a month:

Table 3. Total Wages of Palm Sugar Agroindustry Workers in Palm Oil in Kampung Maredan Barat

Name	Total Wages of Labor (IDR)
Susanto	4.140.000
Aidil Fitri	4.978.000
Anto	4.825.000

The wages of these workers are adjusted to the unit of male working days (HKP) although in its implementation it also uses female workers, and the results are multiplied by the wages of unskilled laborers in the Kampung Maredan Barat area with a standard of 100,000 IDR/working day. For the fee for felling, a fee is charged per unit of oil palm trunk, which is 50,000 IDR/stem. So, for that, if the number of oil palm trunks that are uprooted is 40, the wage issued is 2,000,000 IDR and if the number of oil palm trunks is uprooted as much as 50 stems, the wage issued is 2,500,000 IDR.

- Total Production Cost

Total costs are the result of the sum of fixed costs and variable costs. This analysis is used to determine the total costs incurred by palm oil palm sugar processors in the Kampung Maredan Barat. For more details about the total cost of palm oil palm sugar business can be seen in table 4.

Table 4. Total Cost of Production of Palm Oil Palm Sugar Agroindustry in Kampung Maredan Barat

Entrepreneur's name	Total Cost (IDR)
Susanto	7,249,210
Aidil Fitri	9,266,000
Anto	9,324,840

The costs incurred to make oil palm are wages for cutting trees, transportation fees, equipment costs and the cost of buying firewood to cook sap into palm sugar. Costs vary according to the number of stems harvested and the distance from the plantation to the location where palm sugar is produced.

3.2 Production

Palm sugar production in Kampung Maredan Barat is influenced by the oil palm trunks that are uprooted, both in terms of age and in terms of the number of uprooted oil palm trunks. The older the oil palm stems, the better the quality of the juice. Brown sugar production in Kampung Maredan Barat is calculated from 750 Kg to 1050 Kg in one month. Production is an activity that is carried out to fulfill the use value of an object or create new objects so that it is more useful in meeting needs. Production is not only limited to manufacture, but also storage, distribution, transportation, retail and repackaging or others (Miller & Meiners, 2000).

Table 5. Total Production of Palm Oil Palm Sugar Agroindustry in Kampung Maredan Barat

Enterpreneur's name	Rustle (trunks)	Production/month (Kg)
Susanto	40	750
Aidil Fitri	50	900
Anto	50	1050
On average		900

The average value of production per month is 900 kg with the number of stems ranging from 40 to 50 with oil palm trunks. This condition is caused by not all oil palm trunks in one hectare of plantations can be processed into palm sugar because some of the tree trunks, namely 37%, are used as oil palm sap. The average value of production per month is 900 kg with the number of stems ranging from 40 to 50 with oil palm trunks. This condition is caused by not all oil palm trunks in one hectare of plantations can be processed into palm sugar because some of the tree trunks, namely 37%, are used as oil palm sap.

3.3 Revenues

Revenue is the result of multiplying the amount of brown sugar production with the selling price. The selling price is strongly influenced by the trading channel used by business actors, if producers sell brown sugar products directly to consumers, they use the market price of brown sugar as appropriate, around Rp. 15,000 to Rp. 17,000. However, this is an obstacle because producers cannot fully sell brown sugar directly to consumers and they choose to distribute it first to wholesalers so that their products sell well in the market.

Table 6. Total Revenue and Revenue of Palm Oil Palm Sugar Agroindustry in Kampung Maredan Barat

Entrepreneur's name	Production/ Month (Kg)	Price (IDR/Kg)	Total Revenue (IDR)	Total Cost (IDR)	Net(IDR)
Susanto	750	14.000	10.500. 000	7.249.210	3.250.790
Aidil Fitri	900	14.000	12.600.000	9.266.000	3.334.000
Anto	1050	13.500	14.175.000	9.324.840	4.850.160
Rerata	900		12.425.000	8.613.350	3.811.650

The income earned is 3,811,650 IDR which is a reduction from the total revenue with production costs. According to (Tuwo, 2011), a business is said to be successful if the income that meets the requirements of the business, namely farming must be able to generate enough income to pay for all purchases from production facilities, enough to pay interest on the capital invested, enough to pay wages. Paid labor or other forms of wages, there are savings for investment in business development, and there are sufficient funds to pay for family education and development taxes.

3.4 Break Event Point (BEP)

Break Event Point or breakeven point is a condition where business actors do not get profits and do not experience losses, namely when:

Table 7. BEP Palm Sugar Palm Oil Agroindustry Business in Kampung Maredan Barat

Enterpreneur's name	Production/month (Kg)	Revenue (IDR)
Susanto	18,6	260.523,5
Aidil Fitri	26	363.355,7
Anto	22,7	306.917,7
Average		310.265,63

The average BEP value of brown sugar business is IDR 310,265.63. In addition, the analytical tool that will be used is the Break Event Point (BEP) which is an analytical tool used to determine the limit of production value or production volume of a business reaching the break-even point, namely no loss and no profit. The Break Event Point (BEP) to be used is the Price Break Event Point (BEP) and the Production Volume Break Event Point (BEP) (Soekartawi, 2002).

3.5 Business Feasibility with RCR

RCR is an analysis that is used to determine whether the business being run is feasible or not, it can be calculated by comparing the total revenue with the total cost. From the results of the above calculation that the three businesses are feasible to run because the Return Cost of Ratio figures from the three businesses show that there is no smaller number than one, namely:

Table 8. Feasibility

Entrepreneur's name	RCR (Return Cost of Ratio)
Susanto	1,45
Aidil Fitri	1,36
Anto	1,53
Rerata	1,44

Figure 1. Palm sugar pictures and the making process

4. Conclusion

a. Oil palm trunks can be used as raw material for brown sugar. An average of 45 trees from palm stalks per hectare, 120 stems (37%) from smallholder oil palms.
b. Farmers' income is IDR 3,811,650 per business with a product BEP of IDR 310,265.6,- the average production capacity of palm oil palm sugar agroindustry is 900 Kg per month with an average product BEP value of 22.4 Kg per month. The palm oil brown sugar business in Maredan Barat Village, Tualang District is feasible to run because the RCR value is > 1, namely the RCR value = 1.44.

Acknowledgement

Thank you to the Ministry of Education and Culture who has funded PTUPT's research and also thanks to the Faculty of Agriculture, Lancang Kuning University and the Lancang Kuning University LPPM for facilitating this activity.

References

[1] Junaedi, A. (2017). Peremajaan Kelapa Sawit (ELAIS Guineensis J.). Peremajaan di Seruyan Estate, Minamas Plantation Group, Seruyan, Kalimantan Tengah.
[2] Ditjenbun. (20). Statistik Perkebunan Indonesia-Kelapa Sawit Palm Oil. Jakarta: Direktorat Jendral Perkebunan, Kementrian Pertanian.
[3] Siswati, L. (2018). Optimalisasi Penggunaan Lahan Saat Kelapa Sawit Replanting. Pekanbaru.
[4] Sugiono. (2010). Memahami Penelitian Kualitatif. Bandung: Alfabeta. (2012). Metode Penelitian Bisnis. Bandung: Alfabeta.
[6] Apriwinda. (2013). Studi Fermentasi Nira Batang Sorgum Manis (Sorghum bicolor L.) Untuk Produksi Etanol. Makassar: UNHAS Press.
[7] Miller, R. L., & Meiners, E. R. (2000). Teori Mikroekonomi Intermediate, Terjemahan Haris Munandar. Jakarta: PT. Raja Grafindo Persada.
[8] Tuwo, M. A. (2011). Ilmu Usahatani Teori dan Aplikasi Menuju Sukses. Kendari.
[9] Soekartawi. (2002). Analisis Usahatani. Jakarta: UI Press.

IOP Conf. Series: Earth and Environmental Science 1041 (2022) 012055 doi:10.1088/1755-1315/1041/1/012055

Prediction of palm oil production in Riau Province using the single exponential smoothing method

Guntoro, Lisnawita, Zamzami, David Setiawan*

Universitas Lancang Kuning, Pekanbaru, Indonesia

*dsetia@unilak.ac.id

Abstract. Riau Province is one of the provinces in Sumatra with the most oil palm plantations. Riau Province was once recognized as Indonesia's greatest oil-producing province, but with the fall in output owing to decreasing oil reserves, Riau has become a greater palm oil producer. Not unexpectedly, palm oil has surpassed petroleum in terms of foreign exchange in Indonesia since 2010. There are numerous stages in this research, namely problem analysis, data collection, data preprocessing, single exponential smoothing prediction, data modeling, and calculation precision. Based on the findings of the testing, the Single Exponential Smoothing technique with alpha 0.9 may be used to forecast the requirement for prior time series data in the future, particularly in the case of oil palm output in Riau Province. Oil palm production is expected to grow somewhat over the next five years, according to the forecast.

1. Introduction

Indonesia is a country with agricultural advantages, with the largest plantations and agriculture in ASEAN. Riau Province is one of the provinces in Sumatra with the most oil palm plantations. Riau Province was once recognized as Indonesia's greatest oil-producing province, but with the fall in output owing to decreasing oil reserves, Riau has become a greater palm oil producer. Not unexpectedly, palm oil has surpassed petroleum in terms of foreign exchange in Indonesia since 2010. Palm oil production, on the other hand, does not remain consistent month after month, fluctuating according to a variety of factors such as climate, rainfall, soil fertility, selling price, and so on [1].

Observing the pattern of increasing or decreasing crop yields and predicting the current month's crop yields for the next few months is one way to determine the best times to carry out these activities. There are several methods for prediction and forecasting, ranging from traditional approaches (such as autoregression, exponential smoothing, moving averages, and their variations) to machine learning-based methods. However, when there is a limited and sparse amount of data available, the traditional time series-based method is the best option for prediction.

Exponential Smoothing is a kind of moving averages that is frequently used to tackle difficulties involving time series data. Forecasting using exponential smoothing is accomplished by repeated calculations utilizing the most recent data. Exponential Smoothing is classified into three types: single, double, and triple.

Research [2] Prediction of Oil Palm Fruit Production Level with Single Moving Average Method. Research [3] Application of Triple Exponential Smoothing Method in Predicting Oil Palm Seed Planting at PT . Palmanco Palm Kernel. Research [4] Application of Backpropagation Algorithm in Predicting Paddy Rice Production by Regency/City in North Sumatra. Research [1] Comparison of Moving Average Methods for Predicting Palm Oil Production Results. Research [5] Forecasting Of Palm Oil

IOP Conf. Series: Earth and Environmental Science 1041 (2022) 012055 doi:10.1088/1755-1315/1041/1/012055

Production Using Winter's Exponential Smoothing And Pegels Exponential Smoothing Method (Case Study : PT. Rezeki Kencana Kec. Teluk Pakedai Kab. Kubu Raya). The goal of this study is to use the Single Exponential Smoothing method to predict the amount of palm oil production in Riau Province based on the issues raised above.

2. Methodology

There are numerous stages in this research, namely problem analysis, data collection, data pre-processing, single exponential smoothing prediction, data modelling, and Calculation Precision. Figure 1 illustrates the steps [6].

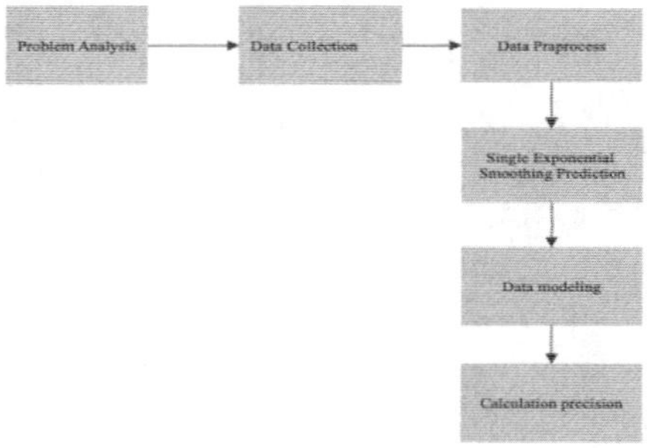

Figure 1. Research Method

2.1 Problem Analysis

This is the stage at which issues relating to the amount of palm oil produced in Riau Province are analyzed, including whether it has increased or decreased. Further study will be conducted on the problems identified to determine how to resolve them.

2.2 Data Collection

The quantity of palm oil produced in Riau Province was collected from the Central Bureau of Statistics of Riau Province via the BPS Riau official website. The information utilized ranges from 2017 to 20121.

2.3 Data Preprocess

At this point, the data will be converted into a more user-friendly format so that an accurate value may be obtained. Data will be verified for oil palm production in Riau Province at the data processing step. At this point, an investigation is performed to see whether any data is missing, unimportant, or needs to be deleted. The quantity of palm oil produced in Riau Province was collected from the Central Bureau of Statistics of Riau Province via the BPS Riau official website. The information utilized ranges from 2017 to 20121.

2.4 Single Exponential Smoothing Prediction

At this step, two phases will be performed: prediction using the Single Exponential Smoothing technique and prediction using the Multiexponential Smoothing approach. The goal at this point is to determine the optimum alpha value and the lowest mistake rate. The data utilized is data from Riau Province on oil palm production.

2.5 Data Modeling

Data modeling is performed at this step based on the prediction results obtained using the Single Exponential Smoothing techniques in the form of error presentations. Data financiers are responsible for presenting data in an understandable manner.

IOP Conf. Series: Earth and Environmental Science 1041 (2022) 012055 doi:10.1088/1755-1315/1041/1/012055

2.6 Calculation Precision

The accuracy of the estimate of the amount of palm oil output in Riau Province will be calculated at this stage. This precision is calculated using the Mean Absolute Precetage Error (MAPE). MAPE is used to compute the error by comparing the actual data presented to the projected data.

3. Result and Discussion

The data used in this study is data on the amount of oil palm production originating in BPS Riau Province from 2017 to 2021. Fig. 2 shows a description of the province of oil palm production in Riau Province. As illustrated in Fig 2, the amount of oil palm in Riau Province has increased every year.

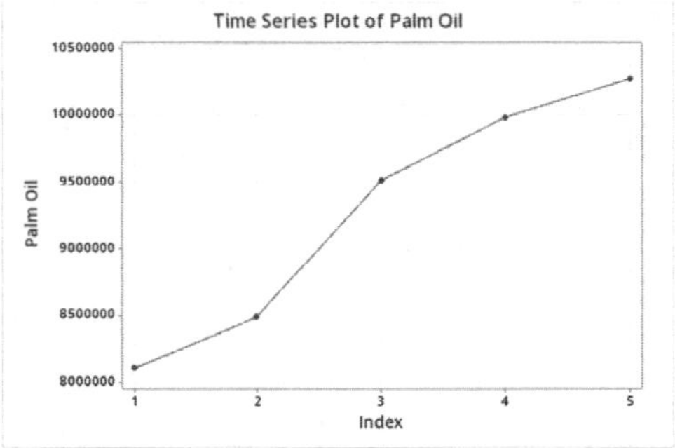

Figure 2. Time Series Plot of Palm Oil

Table 1. Forecasting results of Palm Oil Production with Single Exponential Smoothing

Period	Forecast	Lower	Upper
2026	10235811	8574586	11897036
2027	10235811	8574586	11897036
2028	10235811	8574586	11897036
2029	10235811	8574586	11897036
2030	10235811	8574586	11897036

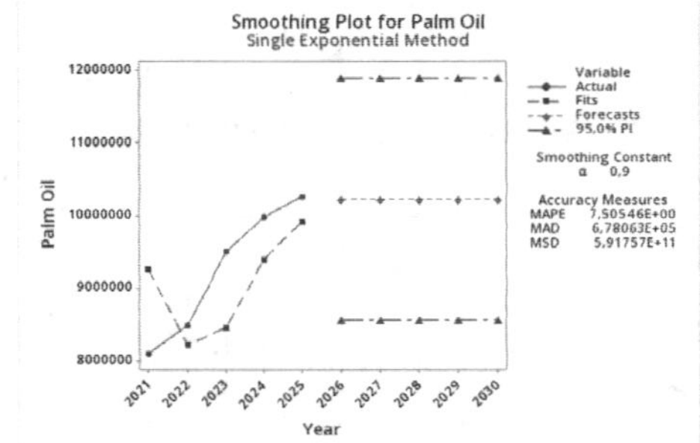

Figure 3. Palm Oil Production Prediction Using Single Exponential Smoothing

According to Table 1 and Figure 2, oil palm production in Riau Province is expected to increase over the next 5 years using the Single Exponential Smoothing method with Alpha 0.9, but it is possible to see a decline, which could be due to the fact that a lot of oil palm land has started to be replanted.

IOP Conf. Series: Earth and Environmental Science 1041 (2022) 012055 doi:10.1088/1755-1315/1041/1/012055

4. Conclusion

Based on the findings of the testing, the Single Exponential Smoothing technique with alpha 0.9 may be used to forecast the requirement for prior time series data in the future, particularly in the case of oil palm output in Riau Province. Oil palm production is expected to grow somewhat over the next five years, according to the forecast.

References

[1] S. Agustian and H. Wibowo, "Perbandingan Metode Moving Average untuk Prediksi Hasil Produksi Kelapa Sawit," Semin. Nas. Teknol. Informasi, Komun. dan Ind., vol. 3, no. 2, pp. 156–162, 2019.

[2] F. Irawan, S. Sumijan, and Y. Yuhandri, "Prediksi Tingkat Produksi Buah Kelapa Sawit dengan Metode Single Moving Average," J. Inf. dan Teknol., vol. 3, pp. 251–256, 2021, doi: 10.37034/jidt.v3i4.162.

[3] D. Santika et al., "Penerapan Metode Triple Exponential Smoothing Dalam Prediksi Penanaman Bibit Kelapa Sawit Pada PT . Palmanco Inti Sawit," vol. 5, no. 1, pp. 12–24, 2020.

[4] H. Aini, H. Haviluddin, E. Budiman, M. Wati, and N. Puspitasari, "Prediksi Produksi Minyak Kelapa Sawit Menggunakan Metode Backpropagation Neural Network," Sains, Apl. Komputasi dan Teknol. Inf., vol. 1, no. 1, p. 24, 2019, doi: 10.30872/jsakti.v1i1.2261.

[5] M. N. Rahmadhani, E. Sulistianingsih, and H. Perdana, "Peramalan produksi kelapa sawit dengan metode winter's exponential smoothing dan pegels exponential smoothing (Studi Kasus : PT. Rezeki Kencana Kec. Teluk Pakedai Kab. Kubu Raya)," Bul. Ilm. Mat. Stat. dan Ter., vol. 07, no. 2, pp. 1–8, 2018.

[6] L. Costaner, G. Guntoro, and S. Sutejo, "E-Learning Feasibility Analysis: Dwi Sejahtera Vocational High School Pekanbaru Context," IOP Conf. Ser. Earth Environ. Sci., vol. 469, no. 1, 2020, doi: 10.1088/1755-1315/469/1/012038.

Study of biogas utilization as waste-to-energy plant and transport modelling of iron (Fe), lead (Pb) and copper (CU) in leachate at Muara Fajar landfill Pekanbaru

H Yuvendius, E Zondra, Zainuri, V I Sari

Universitas Lancang Kuning, Pekanbaru, Indonesia

Email: hazra_yuvendius@unilak.ac.id[1], elviraz@unilak.ac.id[2], zainuri@unilak.ac.id[3], vonny@unilak.ac.id[4]

Abstract. Daily activities such as household, restaurants, markets and others especially in Pekanbaru city will produce waste. Waste produced is both organic and inorganic. A biogas utilization study has been conducted at the Muara Fajar landfill and 888.6 kW potential electricity has been obtained. Each landfill will produce leachate containing metals such as iron (Fe), lead (Pb), copper (Cu) and others. The level of metal toxicity will increasingly pollute the environment. Using the PHREEQC Interactive version 3.0.6.7757 modeling of these contaminants, it can be predicted the transport of the elements and at the same time reduce the risk of soil and water pollution. Modeling with parameter condition of the presence of dolomite show that concentration of Ca, Mg and Na metal decreased, while Cu metal showed a significant increase in the 10th to 50th years and stable at Cu concentration 1.4 mg/L. Lead (Pb) also increased from the 20th to 60th year and stable at Pb concentration 0.25 mg/L. In conditions without dolomite, concentration of iron (Fe) increase in the 15th year namely 0.175 mg/L and constant for the following year while Ca, Na and K have decreased to reach zero concentration.

1. Introduction

Electrical energy consumption, especially in Pekanbaru city, is increasing. However, the increase in electricity demand is not comparable with the electricity supply and the price is tend to increase. Therefore, it is necessary to find alternative electrical generator, such as waste to energy plant. Renewable energy sources can be generated through the use of biogas from organic waste processing at landfills (dump site). Such organic waste is able to generate electricity and will reduce environmental problems by reducing the volume of municipal waste. Organic waste, also known as biomass, refer to all waste originating from living things such as plants and animals. In a landfill, organic waste generally comes from food scraps, wood, paper, animal waste, and others.

Biomass becomes an unlimited energy resource due to the presence of carbon chain cycle that plays a role in photosynthesis reactions. Conversely, fossil resources is limited in nature and increasingly diminishing. Besides that, the emissions of carbon dioxide (CO_2) gas into the air will have a negative effect on global climate change [1].

Table 1. Parameters to estimate biomass waste production [1]

Biomass Specie	**Waste production Ratio (t/y)**	**Energy Conversion Coefficient (GJ/y)**
Paddy	14	16,3
Wheat	1,3	17,5
Corn	1,0	17,7
Roots and Tuber	0,4	6,0
Sugarcane residue	0,28	17,33
Cow	1,10 (t/y/animal)[a]	15,0
Pig	1,10 (t/y/animal)[b]	17,0
Poultry	0,037 (t/y/animal)[c]	13,5
Horse	0,55 (t/y/animal)[d]	14,9
Buffalo and camel	1,46 (t/y/animal)[e]	14,9
Sheep and goats	0,18 (t/y/animal)[f]	17,8
Industrial Woods	1,17	16,0
Fuel Woods	0,67	16,0
Wood waste	0,784	16,0

[a,b,c,d,e,f] manure production rate, dry ton bases

In Pakistan, studies have been conducted on renewable electricity generated from food waste through anaerobic digestion. Under ideal conditions, assumed that 1000 kg of food waste produces 200 m^3 of biogas, then in 2017 and 2030 will produce 1.96 million m^3 and 2.96 million m^3 biogas respectively. With the calculation of 600 kWh of electricity generation from 180 m^3 of biogas, Pakistan generates around 6.5 GWh and 9.96 GWh of electricity in 2017 and 2030 [2]. Khan M.S.M shows that biodegradable process of waste into biogas make Saudi Arabian as potential market and biogas production is a better solution than dumping waste into landfills. Biogas production can be an important part in Saudi Arabia to reduce its dependence on fossil fuels and thus reduce the environmental impact [3].

A modelling can show simulation of organic matter biodegradation which is hydrolyzed to produce biogas and heat. Such modelling is based on the relationship between physical, chemical, thermodynamic and microbial processes that occur in urban landfills. Degradation process of urban waste landfill by microbes is simulated in hydrolysis reaction of solid organic matter to form glucose and acetate as an intermediate carbon substrate and produce CH_4 and CO_2. So, the overall decomposition of organic materials undergoes biochemical processes namely hydrolysis, acidogenesis, acetogenesis and methanogenesis and influenced by environmental factors, namely changes in pH, temperature and hydrogen [4].

Organic waste that processed through biochemical and thermochemical methods to produce energy, is able to generate electrical energy from wood waste biomass around 143,317 GWh [5]. Using the LandGem program, the largest amount of methane production from the Muara Fajar landfill decomposition in 2019 was 1,331,487 m^3/year [6]. Rubbish heap if left over for a long period will produce leachate. Leachate is black water that can damage soil structure around landfill. Leachate contains harmful heavy metals for the environment.

Speciation of heavy metals in leachate was carried out using physical and chemical fractionation methods as well as computer-based thermodynamic models. Free metal ions are found to be less than 30% of the total metal concentration [7]. Calcite ($CaCO_3$) deposition was observed in 25 pollutant samples in landfill leachate. Precipitation experiments of calcite formation of the elements Fe, Sr, Ba, Mn and other trace elements for 6 months show that concentration of elements in the solution decreased significantly. The PHREEQC-2 geochemical model shows a over-saturated code of leachate which was originally polluted with calcite. Mineralogical chemistry studies show that the newly formed calcite contains a number of metals and metalloids which are replaced from the solution. Replacement caused by washing from rainwater [8].

Transport metals that will be studied are iron (Fe), lead (Pb) and copper (Cu). Leachate quality testing at Muara Fajar landfill found that the value of iron (Fe) was 9.83 mg/l, lead (Pb) was 0.06 mg/l and copper (Cu) was 0.08 mg/l [9]. If there is a source of clean water around Muara Fajar landfill, it must have a reference according to the Minister of Health Regulation No. 492/MENKES/PER/IV/2010. Maximum allowable values is as follow: Fe 0.3 mg/l, Pb 0.01 mg/l and Cu 2.0 mg/l [10]. The value of Fe and Pb has exceeding the maximum value and risky to the environment. By using the PHREEQC Interactive modeling version 3.0.6.7757, it can be studied the transport of metals from garbage heap leachate moving into the environment.

2. Methodology

2.1 Calculation of electrical energy generated

Calculation of the volume of methane gas produced is 1,331,487 m^3/year [6]. Then, potential of electrical energy generated through the conversion of methane gas will be studied.

2.2 Metal transport modeling with PHREEQC Interactive version 3.0.6.7757

Leachate produced from Muara Fajar landfill contains iron (Fe) 9.83 mg/l, lead (Pb) 0.06 mg/l and copper (Cu) 0.08 mg/l [9]. From this data, PHREEQC interactive modelling program version 3.0.6.7757 can be used to estimate leachate concentrations in the next 10 to 100 years in Muara Fajar landfill. PHREEQC-3 is used to display chemical reactions, species and transfers of chemical materials in soil, water and pollutants. PHREEQC-3 modelling can calculate the specification of chemical material in aqueous solution, which shows the saturation of minerals, absorption and precipitation of minerals, and transport of reactive compounds. PHREEQC-3 is a reference for geochemical modeling because provides calculation of ions activity and concentration, complexes and equilibrium [11]. The purpose of this study is to simulate the species of Iron (Fe), Lead (Pb) and Copper (Cu) in leachate at Muara Fajar landfill, Pekanbaru.

3. Result and Discussion

3.1 Calculation of electrical energy produced

The data source is from previous researchers [6]. These data show that the volume of methane gas is 1,331,487 m^3/year = 3,647.9 m^3/day.
The calculation is as follows:

Biogas calorific value = 5.000 kcal/m^3 = 5,8 kWh/m^3
Thus, energy recovery potential = 3.647,9 m^3/day x 5,8 = 21.158 kWh.
Electric power potential (kW) = 21.158 kWh/24 = 881,6 kW

3.2 Metal transport modeling with PHREEQC Interactive version 3.0.6.7757

3.2.1 Speciation

In these 0 (zero) solutions, the majority of metal structure in the solution can be determined for each element. An important parameter is toxicity. Thus, the reactivity of an element cannot be determined only by its chemical form. For its elements, it has:

Table 2. Speciation of metal elements

Element	Majority compounds	Concentration (mol)	Total (%)
Cu(2)		1.259e-6	
	Cu^{+2}	4.075e-7	32
	$Cu(OH)_2$	7.913e-7	63,28
Other compounds			*4,72*
Fe(2)		1.760e-4	

	Fe^{+2}	1.220e-4	69,3
	$FeOH^{+}$	3.652e-7	0,20
Other compounds			*30,5*
Pb(2)		2.896e-7	
	Pb^{+2}	2.443e-7	84,2
	$PbOH^{+}$	4.508e-8	15,5
Other compounds			*0,3*

For Fe and Pb, the majority is in the form of ions, while for Cu the majority is a complex compound.

3.2.2 Transport

Model with Dolomite

Decreasing of metal concentration in leachate-contaminated water can be observed from calcite precipitation. Calculations using PHREEQC-3 show that the presence of metals in solution is in the form of ions (Fe^{2+}, Pb^{2+}, Cu^{2+}), carbonate complex ($CaCO_3$, $CaMg(CO_3)_2$, $PbCO_3$, $FeCO_3$) and hydroxyl Cu $(OH)_2$. Saturation Indices values in Table 2 show that Cerrusite ($PbCO_3$) and Siderite ($FeCO_3$) have negative values (undersaturated solution).

Table 3. Saturation Indices (SI) values for carbonate solid species calculated using PHREEQC.

Phase	Minimum	Maximum
Calcite $CaCO_3$	-3,41	0,13
Dolomite $CaMg(CO_3)_2$	-7,7	0,00
Cerrusite $PbCO_3$	-2,09	-0,69
Siderite $FeCO_3$	-8,18	-8,19

Note : If *SI = 0, then the solution is saturated*
SI < 0, then the solution is under saturated
SI > 0, then the solution is over saturated

After modeling in the PHREEQC-3, it is possible to obtain a graph showing the elements in the soil at concentration unit of mol/kg of water on the function of year. At the same time, the graphs are generated in an Excel file which describe the changes of each metal.

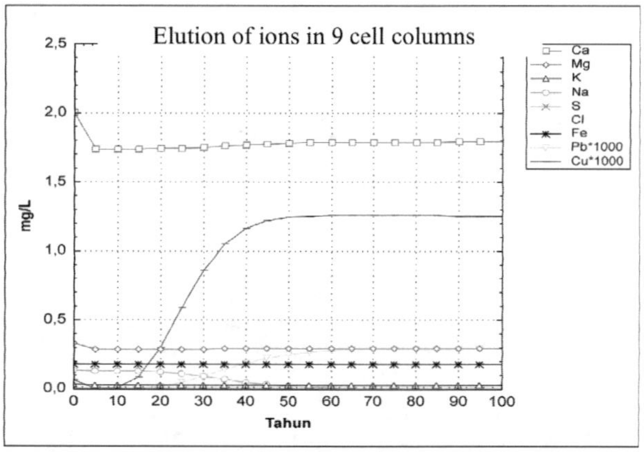

Figure 1. Elution of Ions with contaminants over time.

Figure 1 shows elution of metals over the depth of the soil. Concentration of Ca and Mg decreases in the second year but constant in the following year. Likewise, concentration of Na in the 20th year is decreases to zero. Cu showed a significant increase in the 10th to 50th years and stable at concentration of 1.4 mg/l. Pb is also increased from 20th year to 60th year and stable at concentration of 0.25 mg/l. The concentration values of Cu and Pb have exceeded the maximum value as stipulated in the Minister of Health Regulation No. 492 / MENKES / PER / IV / 2010. Other metals namely Fe, Cl, K and S did not show significant changes.

Model without Dolomite

The Saturation Indices values in Table 3 show that Cerussite ($PbCO_3$) and Siderite ($FeCO_3$) have negative values (unsaturated solution).

Table 4. Saturation Indices (SI) values for carbonate solid species calculated using PHREEQC.

Phase	Minimum	Maximum
Calcite CaCO3	-9,74	-3,41
Dolomite Cam (CO3)2	-20,13	-7,47
Cerussite $PbCO_3$	-10,75	
Siderite $FeCO_3$	-10,82	-10,79

Note : If SI = 0, then the solution is saturated
SI < 0, then the solution is under saturated
SI > 0, then the solution is over saturated

After modeling in the PHREEQC-3, it is possible to obtain a graph showing the elements in the soil at concentration unit of mol/kg of water on the function of year. At the same time, the graphs are generated in an Excel file which describe the changes of each metal.

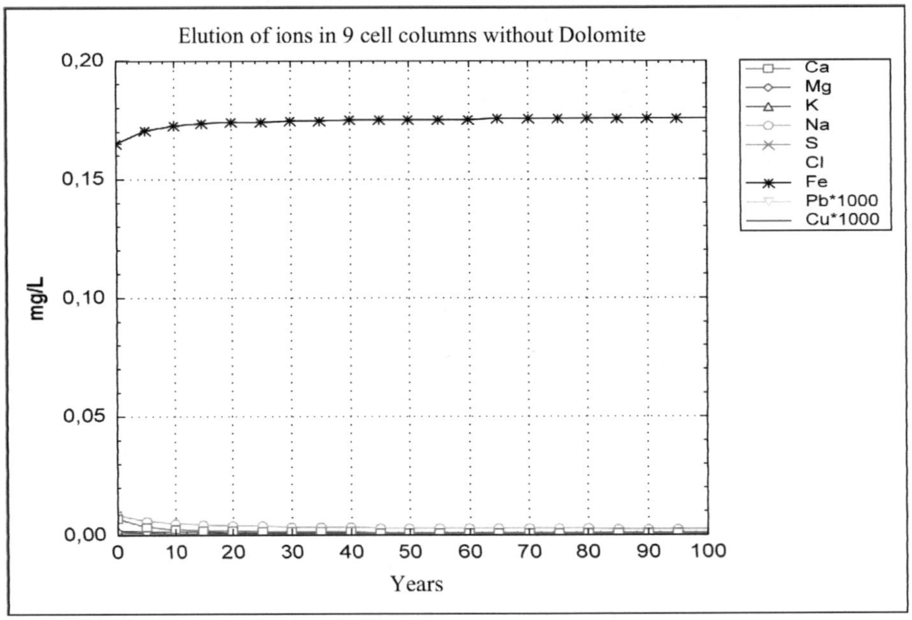

Figure 2. Elution of Ions with contaminants over time

Figure 2 shows the elution of metals over the depth of the soil without dolomite. Fe experiencing concentration increase at 15th year, reaching 0.175 mg/l and constant for the following year. Metals such as Ca, Na and K have decreased concentrations to reach zero. Other metals did not show significant changes.

IOP Conf. Series: Earth and Environmental Science 1041 (2022) 012056 doi:10.1088/1755-1315/1041/1/012056

4. Conclusion

Electric potential at the Muara Fajar Landfill is 881.6 kW. Metal transport modeling using PHREEQC-3 with parameter condition of the presence of dolomite show that concentration of Ca, Mg and Na metal decreased, while Cu metal showed a significant increase in the 10th to 50th years and stable at Cu concentration 1.4 mg/L. Lead (Pb) also increased from the 20th to 60th year and stable at Pb concentration 0.25 mg/L. In conditions without dolomite, concentration of iron (Fe) increase in the 15th year namely 0.175 mg/L and constant for the following year while Ca, Na and K have decreased to reach zero concentration.

References

[1] Yokohama S., 2008, Asian Biomass Handbook. The Japan Institute of Energy

[2] Ali A., Rasool Bux Mahar R.A., and Sheerazi S.T., 2018., Renewable Electricity Generation from Food Waste Through Anaerobic Digestion in Pakistan: A Mini Review, Earth Systems and Environment

[3] Khan M.S.M., and Kaneesamkandi Z., 2013
Biodegradable waste to biogas: Renewable energy option for the Kingdom of Saudi Arabia, International Journal of Innovation and Applied Studies, Vol. 4 No. 1 Sep. 2013, pp. 101-113

[4] Haarstrick A., and Hempel D. C., 2001, Modelling of the biodegradation of organic matter in municipal landfills, Waste Management and Research, Vol 19, pp. 320–331

[5] Skytte K., Meibom P., and Henriksen T.P., 2006, Electricity from Biomass in the European Union-With or Without Biomass Import, Biomass and Energy, Vol. 30, pp. 385 - 392

[6] Sasmita A., 2016, Potensi Produksi Gas Metana Dari Kegitan Lanfilling di TPA Muara Fajar, Pekanbaru, Teknologi Oleo Petro Kimia Indonesia, pp. 169 – 175

[7] Baun D.L., and Christensen T.H., 2009, Speciation of heavy metals in landfill leachate: a review, Waste Management and Research, Vol. 22, pp. 3 – 23.

[8] Ettler V., et al, 2006, Removal of trace elements from landfill leachate by calcite precipitation, Journal of Geochemical Exploration, Vol. 88, pp. 28– 31

[9] Darmayanti L., Yusa M., dan Esther RA., 2011, Identifikasi Tanah Tercemar Lindi Tempat Pembuangan AKhir Sampah Muara Fajar dengan Metoda Geolistrik, Bumi Lestari, Vol. 11, pp. 371 - 378

[10] Permenkes RI NOMOR 492/MENKES/PER/IV/ 2010 Tentang Persyaratan Kualitas Air Minum.

[11] Apello, C.A.J. et Postma, 2005, Geochemistry, groundwater and pollution.

Utilization of oil palm waste as animal feed for Bali cattle in Margomulyo Village, Central Bengkulu

E Ramon, Z Efendi, H Kusnadi, S Yuliasari, A Ishak, WA Wulandari*

Assessment Institute Agricultural Technology of Bengkulu, Bengkulu, Indonesia

*wahyuniwulandari88@yahoo.co.id

Abstract. Oil palm is the leading plantation commodities in Bengkulu province which is able to increase original local government revenue. Besides producing palm oil, this plant also produces solid waste, fronds, and leaves. The purpose of this study was to determine the response of Bali cattle fed by oil palm waste. The research was conducted from August to November 2018 in Margo Mulyo village, Pondok Kubang subdistrict, Central Bengkulu regency, using 2 treatments (introduced and existing) with 13 replications. The composition of introduced feed is forage (70% field grass + 30% fermented fronds and palm leaves) + fermented solid as concentrate. Existing feed composition is field grass (forage) + rice bran (concentrate). Forage is given 10% and concentrate 1% of body weight. The parameters observed were body weight gain and Body Condition Score (BCS) and analyzed by T-test. The results showed a significant effect of treatment on body wight gain. The introduction treatment was able to produce body weight gain by 0.33 kg/ head/day, while the existing treatment produce body weight gain of 0.17 kg/head/day only. BCS score of Bali cattle in both treatments not significantly different at the beginning was 2-3 and in the end 3-4, categorized in medium conditions.

1. Introduction

Oil palm (Elaeis guinensis Jacq.) is a leading agricultural commodity because it has higher vegetable oil productivity compared to other vegetable oil sources [1]. This plant is capable of producing oil of around 7 tonnes/ha with low labor costs of only 0.2 people per hectare [2]. Oil palm began to be developed since the 1970s in Sumatera and Kalimantan [3]. Oil palm has become one of the leading plantation commodities in Bengkulu Province with an area of 208,627 hectares or 50.36% of the total plantation area [4].

The development of oil palm plantations causes an increase in the production of waste originating from plantation businesses and palm oil processing factories which can cause environmental problems [5] because they have not been optimally utilized [6]. Production of biomass produced from oil palm plantations can accommodate 25,623 beef cattle iin Central Bengkulu regency [7] for 8 beef cattle per hectare [8] and has the opportunity in developing beef cattle farms [9]. In addition, oil palm fronds and leaves contain 5.3-8.9% crude protein so that they can replace forage or superior grass [7,10].

One of the obstacles in developing beef cattle is the ability of breeders to provide feed. This is due to the increasingly limited agricultural land which is a source of forage, narrowing grazing areas, and the difficulty of land availability to plant superior grass as cattle feed. Even though the weight gain of cattle is highly dependent on the quality of feed [11]. This requires the use of alternative resources that are capable of developing ruminant livestock in a sustainable manner both in quantity and quality in the provision of feed such as oil palm plantation waste.

The results of previous studies indicate that the use of oil palm plantation waste as beef cattle feed has a positive effect on livestock growth and can reduce the provision of forage grass [12]. Therefore, it is necessary to do more research on the use of feed from palm oil plantation waste specific at the farmer level to accelerate the utilization of the potential of this plantation waste. This study aims to determine the effect of using palm oil waste as feed for Bali cattle in the Krida Utama Farmer Group, Margo Mulyo Village, Pondok Kubang District, Bengkulu Tengah Regency.

2. Methodology

The research was conducted from August to November 2018 in Margo Mulyo Village, Pondok Kubang District, Bengkulu Tengah Regency, Bengkulu Province. The stages of implementation in the study were as follows: (1) identification of 26 Bali cattle aged 1.5-2 years without sexing, which were divided into 2 groups to get treatment; (2) preparing palm oil waste feed in the form of solid waste, fermented palm fronds and leaves; (3) feeding according to the given treatment, namely the existing treatment applied by breeders and the introduced treatment (Table 1); (4) data collecting on cattle body weight gain; (5) the data obtained were analyzed using the t test.

Table 1. Oil palm waste feed treatments

Description	Treatments	
	Existing	Introduction
Forage	Forage (field grass) 10% of body weight	Forage (field grass 70% + leaves and palm fronds 30%) 10% of body weight
Additional feed	Rice bran 1% of body weight	Fermented solid waste 1% of body weight
Drugs	Worm medicine	Worm medicine
Vitamins	A, D, E, K	A, D, E, K
Mineral	0,01 g/head	0,01 g/head
Drinking water	Ad libitum	Ad libitum

The variables observed were the weight gain of beef cattle using the SNI 651.4: 2015 [13] method and Body Condition Score (BCS) [14]. The BCS score was estimated based on direct observation using an interval scale of 1-5 from very thin to fat. Bali cattle body weight is calculated by the following formula:

$$\text{Body weight (kg)} = \frac{\text{Chest size}^2 \text{ x Body length}}{11,050}$$

3. Result and Discussion

3.1. Feed nutrient content

The nutrient content of the feed treatment in this study is presented in Table 2.

Table 2. Treatment feed nutrient content

Treatments	Feed ingredients	Water (%)	Crude protein (%)	Energy (Kcal/kg)	Crude fiber (%)	Ash (%)	Ca (%)	P (%)
Introduction	Fermented solid*	9.73	18.46	4405	27.69	17.36	1.79	0.20
	Fermented palm fronds and leaves*	4.25	6.64	4256	41.18	7.81	0.10	0.08
Existing	Rice bran **	1.45	10.93	2017.35	15.07	-	-	-
	Field grass ***	16.85	7.27	-	14.32	-	0.08	0.04

Information * = Results of Proximate Analysis of the Animal Research Institute, Bogor (2018).
** = According to [15]
***= According to [16]

The very important content of feed nutrients in beef cattle feed formulation is crude protein, crude fiber, and metabolism energy. The source of these nutrients comes from forages and concentrates. The fermented solid crude protein content in the introduced treatment was higher than the rice bran in the existing treatment, while the forage protein content in the two treatments was relatively the same. Giving concentrate and forage feed, respectively 1% and 10% of body weight per day, apparently did not meet the existing feed needs. The need for the crude protein in beef cattle rations is 12-13% [13]. This means that the use of solid waste fermentation is in accordance with SNI, while the protein content of rice bran is not in accordance with the SNI. Therefore, fermented solids are more suitable to be used as concentrates.

Making solid fermentation is intended to increase nutritional value, shelf life, and nutrient digestibility in the fattening of Bali cattle. According to [17] the fermentation process is very good to do to improve the nutritional quality of solid feed ingredients of palm oil waste while increasing nutrient digestibility. The solid fermentation process carried out at the research location was able to produce a protein content of 18.46%, which was higher than the non-fermented solid nutrients from the research results of [17] which is only 12.93%. In addition, a fermented feed can improve livestock performance [18–20], improve blood profile [21], and meat quality [19,22].

The crude fiber content needed by beef cattle is 10-40%, the lower the crude fiber content, the better the feed consumed by livestock [23]. Solid waste fermentation contains 27.69% crude fiber, which means that the fermentation process can increase the nutritional value of a material, as a result of the breakdown of complex compounds into simpler compounds to make it easier to digest by beef cattle [17]. From the description above, it can be stated that solid waste fermentation is suitable for use as concentrate feed for beef cattle.

Table 2 also shows that the metabolic energy contained by solid waste fermentation is two times greater than that of rice bran. Metabolism energy functions for the process of breaking down feed nutrients to function for the formation of body tissues or repair damaged body tissues. According to [24] the greater the metabolism energy content in the feed ration, the better it will be for consumption by beef cattle. Fermentation is an effort to increase the nutritional value of feed ingredients so that the quality and availability of nutrients can be increased, including metabolism energy content.

3.2 Average daily gain

Average daily gain (ADG) of beef cattle shows that there are differences between treatments. The use of introduced feed with fermented solid waste was able to increase approximately two times ADG as compared to existing feed with rice bran concentrate (Table 3).

Table 3. Cattle body weight during treatment

Treatment	Beef cattle performance			
	First body weight (kg)	End body weight (kg)	AG (kg)	ADG (kg)
Introduction	174,16	184,02	9,86[a]	0,33[a]
Existing	175,77	180,75	4,99[b]	0,17[b]

The same superscript in the same column shows an insignificant difference (P>0,05)
Different superscripts in the same column indicate significant differences (P<0,05)

The results of the weight statistical analysis showed that there was a significant difference in ADG variable between the two treatments (P <0.05) during the four-month maintenance period. The ADG of introduced technology was able to increase about two times compared to the existing technology with an average of 0.33 kg / head/day and 0.17 kg/head/day, respectively. The use of fermented solid concentrate which has a high crude protein content of 18.46% and a high metabolic energy content of 4405 kcal/kg is able to meet the nutritional needs of beef cattle.

The results of research by [25]) which applied 70% solid waste fermentation to male Bali cattle concentrate can increase ration consumption and digestibility. According to [26] the provision of fermented solid waste concentrate can increase the ADG of cattle to 0.516 kg/head/day. Meanwhile, several research results indicate that feeding beef cattle from palm oil waste can increase the ADG higher than field grass between 0.3-0.52 kg/head/day [27] . ADG research results from the use of introduced feed is still low compared to the results of previous studies. It is assumed that the cattle used are still 1.5 years old, and their growth is relatively slow.

The trend of livestock growth during the research is shown in Figure 1. This trend indicates that livestock growth is better in the introduced feed treatment. Good quality feed in the introduced treatment will accelerate livestock growth [15].

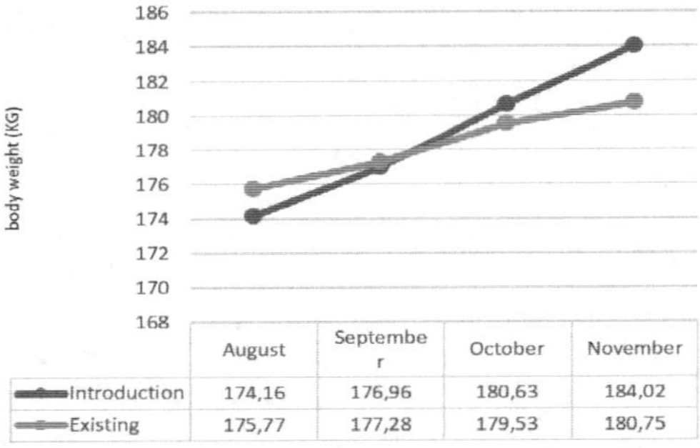

Figure 1. Trends in the growth of research result cattle

3.3. Body condition score (BCS)

The body's energy reserves can be assessed by a visual assessment method known as a body condition score (BCS). The relative score describes the level of muscle and body fat reserves of each cattle. The BCS value between the two treatments at the beginning and at the end was relatively the same with an increase in the score of about 0.3 during 4 months of maintenance (Table 4).

Table 4. Increase in cattle BCS during treatment

Treatment	BCS cattle		
	Early BCS	End BCS	Increased BCS
Introduction	2.77	3.12	0.35
Existing	2.73	3.04	0.31

BCS is a parameter that can be used to evaluate the nutritional status of beef cattle [28]. The initial BCS value for both treatments were 2-3 and increased to 3-4 at the end. The results of this study are also supported by research by [29] at young male Bali cattle. According to [30] states that the ideal livestock BCS score is adjusted to the maintenance objectives. The BCS value between 3-4 is moderate, not thin or fat.

4. Conclusion

The results of this study indicate that introductory feeding in the form of 70% field grass + leaves and 30% palm fronds + fermented solid concentrate can significantly increase ADG compared to the existing treatment. The increase was approximately twofold, respectively 0.33 kg/day/head in the introduced treatment and 0.17 kg/day/head in the existing treatment. However, feeding did not affect the difference in the increase in BCS between the two treatments, namely from 2-3 BCS at the beginning of maintenance to 3-4 at the end of maintenance.

References

[1] Sheil D, Casson A, Meijaard E, Van Noordwjik M, Gaskell J, Sunderland-Groves J, Wertz K and Kanninen M 2009 The impacts and opportunities of oil palm in Southeast Asia: What do we know and what do we need to know? (Bogor: CIFOR)

[2] Pahan. I. 2008 Panduan Lengkap Sawit: Manajemen Agribisnis dari Hulu hingga Hilir Panduan Lengkap Sawit: Manajemen Agribisnis dari Hulu hingga Hilir (Jakarta: Penebar Swadaya)
[3] McCarthy J F, Gillespie P and Zen Z 2012 Swimming Upstream: Local Indonesian Production Networks in "Globalized" Palm Oil Production World Dev. 40 555–69
[4] BPS Provinsi Bengkulu 2020 Provinsi Bengkulu Dalam Angka ed B P Bengkulu (Bengkulu (Id): Badan Pusat Statistik Provinsi Bengkulu)
[5] Khatun R, Reza M I H, Moniruzzaman M and Yaakob Z 2017 Sustainable oil palm industry: The possibilities Renew. Sustain. Energy Rev. 76 608–19
[6] Mayulu H and Sutrisno C I 2016 Beef cattle development policy in Indonesia J. Litbang Pertan. 29
[7] Efendi. Z., Ramon. E. dan Y D 2017 Karakteristik Peternak Sapi Brahman Cross di Perkebunan Kelapa Sawit di Kabupaten Paser Kalimantan Timur Peluang Pengembangan Sapi PotongDengan Perkebunan Kelapa Sawit Di Bengkulu Tengah Provinsi Bengkulu. pp 146–65
[8] Rizali A, Fachrianto F, Ansari M H and Wahdi A 2018 Pemanfaatan limbah pelepah dan daun kelapa sawit melalui fermentasi Trichoderma sp. sebagai pakan sapi potong EnviroScienteae 14 1
[9] Utomo B N and Widjaja E 2004 Limbah padat pengolahan minyak kelapa sawit sebagai sumber nutrisi ternak ruminansia J. Litbang Pertan. 23 22–8
[10] Imsya.A 2008 Konsentrasi N-amonia, Kecernaan bahan kering dankecernaan bahan organik pelepahsawit hasil amoniasi secara in vitro Prosiding Seminar Nasioal Teknologi Peternakan dan Veteriner (Bogor: Pusat Penelitian dan Pengembangan Peternakan) pp 111–5
[11] Nurhayu A, Ishak A B L and Ella A 2013 (Midrib and leaf palm as substituting forages for feed cattle at 108–15
[12] Daliani S D 2010 Pengkajian penggunaan IB pada perbibitan sapi Peranakan Ongole (PO) dan pemanfaatan limbah sawit (solid dan pelepah) untuk efisiensi penggunaan pakan hijauan 50% pada penggemukan sapi PO dalam rangka meningkatkan bobot badan harian > 10% (Bengkulu: BPTP Bengkulu)
[13] Nasional S, Ics I and Nasional B S 2009 Pakan konsentrat - Bagian 2 : Sapi potong
[14] Sukandar A, Purwanto B and Anggraeni A 2008 Keragaan Body Condition Score dan Produksi Susu Sapi Perah Friesian-Holstein di Peternakan Rakyat KPSBU Lembang , Bandung (Performances of Body Condition Score and Milk Production of Holstein- Friesian Dairy Cattles Maintained under Small Dairy Farms in Prosiding Seminar Nasional Teknologi Peternakan dan Veteriner vol 3 pp 86–99
[15] Wibawa A A P, Wirawan I W and Partama I B G 2015 Prospek Penggemukan Sapi di Sekitar Pabrik Kelapa Sawit di Kalimantan Tengah Majalah Ilmiah Peternakan Volume 18 Nomor 1 Tahun 2015 vol 18 (Denpasar) pp 11–6
[16] Utomo B N, Widjaja E and Dara E 2009 Pengaruh pemberian probiotik lokal terhadap PBBH sapi Bali jantan di Kalteng Pengkaj. dan Pengemb. Teknol. Pertan. 12(1) 11–20
[17] Zega A D, Badarina I and Hidayat H 2017 Kualitas Gizi Fermentasi Ransum Konsentrat Sapi Pedaging Berbasis Lumpur Sawit dan Beberapa Bahan Pakan Lokal dengan Bionak dan EM4 J. Sain Peternak. Indones. 12 38–46
[18] Missotten J A, Michiels J, Dierick N, Ovyn A, Akbarian A and De Smet S 2013 Effect of fermented moist feed on performance, gut bacteria and gut histo-morphology in broilers Br. Poult. Sci. 54 627–34
[19] Ahmed S T, Mun H-S, Islam M M and Yang C-J 2014 Effects of fermented corni fructus and fermented kelp on growth performance, meat quality, and emission of ammonia and hydrogen sulphide from broiler chicken droppings Br. Poult. Sci. 55 745–51
[20] Ridla M, Nikmah F K and Ramli N 2014 Performance of mojosari alabio males ducks fed complete ration silage Anim. Prod. 16 176–82
[21] Fazhi X, Lvmu L, Jiaping X, Kun Q, Zhide Z and Zhangyi L 2011 Effects of fermented rapeseed meal on growth performance and serum parameters in ducks Asian-Australasian J. Anim. Sci.

24 678–84

[22] Sutama I, Mudita I M, Suberata I W and Siti N W 2016 Effect Of Dietary Biosupplement Fermented Earthworm Microbes Consortium Inoculant On The Carcass Production And Cholesterol Blood Profile Of Baliness Duck Agric Anim. Sci. 3 1–6

[23] Fitriyani and Asyari H 2017 Kandungan Protein Kasar dan Serat Kasar Pakan Komplit Berbasis Tongkol Jagung dengan Penambahan Azolla sebagai Pakan Ruminansia (Protein Content of Rude and Crude Fiber Based Complete Feed Corn Cob With the Addition of Azolla As Ruminant Feed) J. Galung Trop. 6 12–8

[24] Mariani N P, Mahardika I G, Putra S and Partama I B G 2015 Penentuan Keseimbangan Protein dan Energi Ransum Sapi Bali Jantan Determination of Protein - Energy Ratio at Bali Bull Ration N. P. Mariani, I. G. Mahardika, S. Putra dan I. B. G. Partama J. Peternak. Indones. 17 46–53

[25] Badarina I, Jarmuji J and Gultom D P 2017 Kecernaan Ransum Sapi Bali Dengan Konsentrat Fermentasi Berbasis Lumpur Sawit Dan Bahan Pakan Lokal Agrointek 11 63

[26] Widjaja E and Utomo B N 2006 Prospek Penggemukan Sapi di Sekitar Pabrik Kelapa Sawit di Kalimantan Tengah Semin. Nas. Teknol. Peternak. dan Vet. 110–5

[27] Hendri. Y dan Dewi. R.A 2014 Produksi dan Analisis ekonomi Sapi Bali yang Diberi Pakan Pelepah Sawit di Musim Kemarau J. Pengkaj. dan Pengemb. Teknol. Pertan. 17 81–7

[28] Parish J and Rhinehart J 2008 Body Condition Scoring : Beef cattle Mississippi State Univ. 58–9

[29] Baharun A, L. L Belli H and Hine T M 2017 Charakteristics of Bali Cattle Young Bull Under Traditionally Livestock System in Merbaun Village Regency of Kupang J. Peternak. Nusant. 3 11

[30] Kellogg W 2010 Body Condition Scoring with dairy Cattle Agric. Nat. Resour. 1–6

Etnobotany of medicine plants in the community of Rawang Kao Village, Siak, Indonesia

Sri Wahyuni*, Mar'atul Afidah

Universitas Lancang Kuning, Pekanbaru, Indonesia

*sriwahyunifkip@unilak.ac.id

Abstract. This study aims to determine the plant species used as medicine by the community of Rawang Kao Village, Siak, Indonesia. This research was conducted in Rawang Kao Village, Lubuk Dalam District, Siak Regency, Riau Province in April-May 2021. This study used qualitative descriptive techniques. Samples were taken by snowball sampling technique. Samples in this study were 12 people of Rawang Kao Village who knows about the use of plant as medicine. Data collection techniques used were interview and direct observation to indentify the plant species. Overall, the research results revealed that there are 32 species of plants used as medicine by the people of Rawang Kao Village belong to 21 families. The most widely used plant species as medicine are from Zingiberaceae, as many as 6 species (19 %) and the most widely used plant parts as medicinal ingredients are leaves (56 %). 66 % of plants used as medicine by the people of Rawang Kao Village are planted in the yard of the house. In general, these plants are used as medicine for diabetes, cough, hypertension, and fever.

1. Introduction

Indonesia is a mega-biodiversity country that is rich in a variety of plants and animals. Utilization of plants and animals is carried out in various fields, such as food, industry, and medicine. About 940 species of 30.000 species of plants in Indonesia are known to have medicinal efficacy [1]. Medicinal plants are plants that have medicinal efficacy that can relieve pain, increase endurance, or kill germs [2]. Plant parts commonly used as medicinal ingredients are roots, bark, wood, leaves, flowers, or seeds [3]. Diseases that are treated vary, both external and internal diseases.

The use of plants as medicine is currently growing rapidly. This is due to various supporting factors, such as the availability of raw materials in the form of abundant and diverse biological resources in Indonesia. In addition, the use of plants as medicinal ingredients is considered to have no side effects that have the potential to endanger health. At this time, the tendency of people to use natural products is increasing. According to Kintoko (2006), the history of traditional medicine which has developed into the nation's cultural heritage as well as the global issue of "back to nature" are factors in increasing the market for herbal products [4].

Information related to the use of plants as medicine is passed down from one generation to the next. From ancient times, Indonesian ancestors were familiar with treatment techniques by utilizing plants in the forest and plants around the yard to treat various diseases. Knowledge of medicinal plants is obtained from experience and skills [5].

One of the community groups that use plants as an alternative treatment is the people of Rawang Kao Village, Lubuk Dalam District, Siak Regency. The people of Rawang Kao Village are people of a mixture consisting of several tribes, including the Javanese, Batak, and Malay tribes. However, in general, the majority of the residents of Rawang Kao village are Javanese who still have local wisdom such as the use of plants as traditional medicine which is still maintained until now.

Most of the Rawang Kao Village area is lowland and slightly swampy which is now dominated by oil palm plantations and settlements. This is because the people of Rawang Kao Village are mostly oil palm farmers. However, recently, medicinal plants have begun to be re-developed through several programs such as the KWT (Kelompok Wanita Tani) program. The people of Rawang Kao Village are starting to realize the importance of preserving medicinal plants. Based on this, research on the ethnobotany of medicinal plants in the Rawang Kao Village Community needs to be carried out.

2. Methodology

This research was conducted in Rawang Kao Village, Lubuk Dalam District, Siak Regency in April 2021. The instruments used in this study were interview sheet. The population of this study was the entire community of Rawang Kao Village, Lubuk Dalam District, Siak Regency. The population is a generalization area consisting of objects/ subjects that have certain qualities and characteristics determined by researchers to be studied and then drawn conclusions [6]. The sample is part or representative of the population under study.

This research used snowball sampling technique. Snowball sampling technique is an approach to find key informants who have a lot of information. Using this approach, several potential respondents were contacted and asked if they knew other people with the characteristics referred to for research purposes. Initial contact starts from the base informant (Village Head), then the informant provides recommendations on the names of other informants based on their competence.

In general, good informants are people who are easy to talk to, understand the desired information, and are very happy to provide information to interviewers or researchers. The informants referred to are village heads, community leaders, and local residents who know and use medicinal plants. There are two types of data parameters in the ethnobotany research of medicinal plants in the people of Rawang Ko Village, Lubuk Dalam District, Siak Regency.

a. Main data parameters

The main data parameters are data obtained directly from the original source. The data is primary data collected from research. After the data is collected, it is continued with the preparation in tabular form so that the data is easy to read and understand. The main data obtained are as follows:

- Plant species used as medicine by the people of Rawang Kao Village, Lubuk Dalam District, Siak Regency.
- Plant parts used as medicine by the people of Rawang Kao Village, Lubuk Dalam District, Siak Regency.
- Plant cultivation used as medicine by the people of Rawang Kao Village, Lubuk Dalam District, Siak Regency.

b. Supporting data parameters

Supporting data parameters are data obtained from respondents' personal data in the form of original sources or respondents. In this study, supporting data are directed to respondents who will be interviewed by researchers such as: name, gender, age, education level, and occupation.

3. Result and Discussion

This research was conducted by interviewing the people of Rawang Kao Village who have knowledge of medicinal plants. There were 12 respondents in this research. The respondents consist of males and females with the age of over 20 years. The education level of the respondents varies from elementary school to undergraduate. Most of the respondents' occupations are farmers and housewives. From the results of interviews with respondents, it is known that there are 32 plant species of medicinal plants used by the people of Rawang Kao Village. The following is presented data on medicinal plant species used by the people of Rawang Kao Village:

Table 1. Medicinal Plant Species Used by The People of Rawang Kao Village

No.	Plant Species				Plant Parts
	Common Name	**Latin Name**	**Vern Name**	**Family**	
1	Sirih	*Piper betle*	Sirih	Piperaceae	Leaf
2	Suruhan	*Peperomia pellucida*	Sirih cina	Piperaceae	Leaf
3	Bonglai	*Zingiber montanum*	Bangle	Zingiberaceae	Rhizome
4	Jahe merah	*Zingiber officinale var. rubrum*	Jahe merah	Zingiberaceae	Rhizome
5	Kunyit	*Curcuma longa*	Kunyit	Zingiberaceae	Rhizome
6	Kencur	*Kaempferia galanga*	Kencur	Zingiberaceae	Rhizome
7	Temu hitam	*Curcuma aeruginosa*	Temu ireng	Zingiberaceae	Rhizome
8	Temu kunci	*Boesenbergia rotunda*	Temu kunci	Zingiberaceae	Rhizome
9	Mengkudu	*Morinda citrifolia*	Pace	Rubiaceae	Fruit
10	Serai	*Cymbopogon citratus*	Sereh	Poaceae	Leaf
11	Bawang sabrang	*Eleutherine bulbosa*	Bawang dayak	Iridaceae	Tuber
12	Rumput meranti	*Physalis minima*	Ciplukan	Solanaceae	Fruit
13	Lili putih	*Crinum asiaticum*	Bakung	Liliaceae	Leaf
14	Kelor	*Moringa oleifera*	Kelor	Moringaceae	Leaf
15	Jarak	*Ricinus communis*	Jarak	Euphorbiaceae	Leaf
16	Jarak tintir	*Jatropha multifida*	Tentir	Euphorbiaceae	Sap
17	Jarak hitam	*Ricinus communis var. gibsonii*	Jarak kepyar	Euphorbiaceae	Leaf
18	Sambung tulang	*Euphorbia tirucalli*	Sangkal putung	Euphorbiaceae	Leaf
19	Kembang telang	*Clitoria ternatea*	Kembang telang	Fabaceae	Leaf
20	Turi	*Sesbania grandiflora*	Kembang turi	Fabaceae	Leaf
21	Kejibeling	*Strobilanthes crispus*	Kecibeling	Acanthaceae	Leaf
22	Sambung nyawa	*Gynura procumbens*	Sambung nyowo	Asteraceae	Leaf
23	Pegagan	*Centella asiatica*	Daun tapak kuda	Apiaceae	Leaf
24	Kumis kucing	*Orthosiphon aristatus*	Kumis kucing	Lamiaceae	Leaf
25	Pandan	*Pandanus amaryllifolius*	Pandan	Pandanaceae	Leaf
26	Jeruk nipis	*Citrus aurantifolia*	Jeruk nipis	Rutaceae	Fruit
27	Bunga bintang	*Laurentia longiflora*	Bunga katarak	Campanulaceae	Flower
28	Pasak bumi	*Eurycoma longifolia*	Pasak bumi	Simaroubaceae	Root
29	Brotowali	*Tinospora cordifolia*	Brotowali	Menispermaceae	Leaf
30	Kembang sepatu	*Hibiscus rosa-sinensis*	Bunga raya	Malvaceae	Leaf
31	Rosela	*Hibiscus sabdariffa*	Bunga rosella	Malvaceae	Flower
32	Binahong	*Anredera cordifolia*	Daun binahong	Basellaceae	Leaf

Based on the table and graph above, it is known that 32 species of plants used as medicine by the people of Rawang Kao Village belong to 21 families. The most widely used plant species as medicine are from Zingiberaceae, as many as 6 species (19 %), which are Zingiber montanum, Zingiber officinale var. rubrum, Curcuma longa, Kaempferia galanga, Curcuma aeruginosa, and Boesenbergia rotunda. The use of plants from Zingiberaceae as medicine is mostly done by people from various regions, such as in the people of Rambah Samo District, Rokan Hulu Regency. People from this region use plants from Zingiberaceae for fever, contraception, coughs, bloating, cancer medicine, ulcers, constipation [7].

Meanwhile, the most widely used plant parts as medicinal ingredients are leaves (56 %). Leaves are part of the plant which is almost always abundant in nature because its existence is not affected by the season. Different with fruit and flowers that are available in nature only exists at certain times or seasons. Therefore, the use of leaves for concocting medicine is related to its availability in nature. The advantage of using leaves as medicinal ingredients because the leaves have soft fibers so it's easy extract these substances.

The people of Rawang Kao Village use plants as medicine for various internal and external diseases. The cultivation of plants that are used as medicine is also diverse, including those planted in the yard of the house or growing wild. The following presents data on diseases that the people of Rawang Kao Village believe can be treated by utilizing plants and how to cultivate these plants:

Table 2. Uses of Medicinal Plants and Cultivation Methods

No.	Species	Utility	Cultivation Methods
1	*Piper betle*	Medicine for allergy	Planted in the yard
2	*Peperomia pellucida*	Medicine for ulcer	Growing wild
3	*Zingiber montanum*	Medicine for headache	Planted in the yard
4	*Zingiber officinale var. rubrum*	Medicine for sore throat, cough, and cold	Planted in the yard
5	*Curcuma longa*	Medicine for menstrual pain and fever	Planted in the yard
6	*Kaempferia galanga*	Medicine for hypertension	Planted in the yard
7	*Curcuma aeruginosa*	Medicine for cough, asthma, and malaria	Planted in the yard
8	*Boesenbergia rotunda*	Medicine for cough	Planted in the yard
9	*Morinda citrifolia*	Medicine for diabetes and arthritis	Planted in the yard
10	*Cymbopogon citratus*	Medicine for diabetes and hypertension	Planted in the yard
11	*Eleutherine bulbosa*	Medicine for ulcer, gout, diabetes, and dysentery	Planted in the yard
12	*Physalis minima*	Medicine for flu, cough, diabetes, and lung disease	Planted in the yard and growing wild
13	*Crinum asiaticum*	Medicine for fracture	Planted in the yard
14	*Moringa oleifera*	Medicine for cholesterol, heart disease, and diabetes	Planted in the yard
15	*Ricinus communis*	Medicine for constipation	Planted in the yard and growing wild
16	*Jatropha multifida*	Medicine for wound	Planted in the yard and growing wild
17	*Ricinus communis var. gibsonii*	Medicine for toothache, sprue, and wound	Planted in the yard and growing wild
18	*Euphorbia tirucalli*	Medicine for fracture	Planted in the yard
19	*Clitoria ternatea*	Medicine for fever	Planted in the yard
20	*Sesbania grandiflora*	Medicine for sprue, dysentery, cough, and sore throat	Planted in the yard
21	*Strobilanthes crispus*	Medicine for hemorrhoid and constipation	Planted in the yard and growing wild
22	*Gynura procumbens*	Medicine for hemorrhoid, cholesterol, and gastritis	Planted in the yard and growing wild
23	*Centella asiatica*	Medicine for insomnia, arthritis, and shortness of breath	Growing wild
24	*Orthosiphon aristatus*	Medicine for hypertension and kidney stone	Planted in the yard
25	*Pandanus amaryllifolius*	Medicine for rheumatism	Planted in the yard and growing wild
26	*Citrus aurantifolia*	Medicine for cough	Planted in the yard
27	*Laurentia longiflora*	Medicine for sore eyes	Planted in the yard and growing wild
28	*Eurycoma longifolia*	Medicine for vitality	Growing wild
29	*Tinospora cordifolia*	Medicine for diarrhea, fever, and diabetes	Planted in the yard
30	*Hibiscus rosa-sinensis*	Medicine for fever	Planted in the yard
31	*Hibiscus sabdariffa*	Medicine for hypertension	Planted in the yard

32	*Anredera cordifolia*	Medicine for diabetes, hypertension, and gout	Planted in the yard

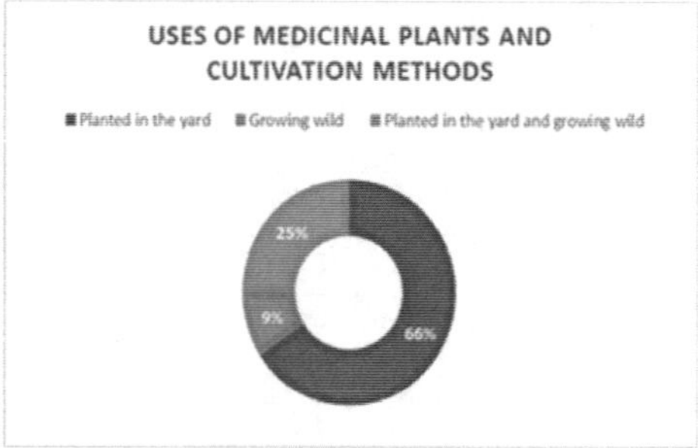

Figure 1. Uses of Medical Plants and Cultivation Methods

From the table above, it is known that medicinal plants can be used to cure various internal and external diseases, such as wounds, ulcer, sprue, to diabetes, fracture, cold, and others. Based on the results of research by Efremila (2015) it is known that ginger (Zingiber officinale) is also used by the Dayak people in Kalimantan as a cold medicine [8]. Ginger is used as an ingredient in herbal medicine because it contains essential oils with chemical compounds active, such as: zingiberin, camphor, lemonin, borneol, shogaol, cineol, fellandren, zingiberol, gingerol, dan zingeron which is efficacious in preventing and treat various diseases [9]. In his book entitled Ragam dan Khasiat Tanaman Obat, Santoso (2008) stated that ginger is efficacious for treating impotence, cough, aches, headache, rheumatism, pain waist, and colds [10].

From the graph above, it is also known that 66 % of plants used as medicine by the people of Rawang Kao Village are planted in the yard of the house and 9,375 % of plants grow wild. Knowledge about plants that can be used as medicine has been passed down from generation to generation in the Rawang Kao Village community. Thus, people have started cultivating medicinal plants in their yards. This proves that the people of Rawang Kao Village still keep the teachings of their predecessors who first used plants as medicine. Of course, this is an example for people in other places to continue to preserve the culture of using plants as medicine so that existing knowledge is not lost in the future.

4. Conclusion

Based on the result of the research, it can be concluded that there are 32 species of plants used as medicine by the people of Rawang Kao Village belong to 21 families. The most widely used plant species as medicine are from Zingiberaceae, as many as 6 species (19 %) and the most widely used plant parts as medicinal ingredients are leaves (56 %). 66 % of plants used as medicine by the people of Rawang Kao Village are planted in the yard of the house. In general, these plants are used as medicine for diabetes, cough, hypertension, and fever.

References

[1] Dorly. 2005. Potensi Tumbuhan Obat Indonesia dalam Pengembangan Industri Agromedisin. Makalah Pengantar Falsafah Sains Sekolah Pascasarjana/ S3. Institut Pertanian Bogor. Bogor.

[2] Darsini, N. N. 2013. Analisis Keanekaragaman Jenis Tumbuhan Obat Tradisional Berkhasiat untuk Pengobatan Penyakit Saluran Kencing di Kecamatan Kintamani Kabupaten Bangli Provinsi Bali. Jurnal Bumi Lestari 13(1): 159- 165.

[3] Adfa, M. 2005. Study Senyawa Flavonoid dan Uji Brine Shrimp Beberapa Tumbuhan Obat Tradisional Suku Serawai di Provinsi Bengkulu. Jurnal Gradien 1(1): 43-50.

[4] Kintoko (2006). Sejarah Pengobatan Tradisional. [Online] Diakses di: books.google.com/Repasetory.Usu.ac.id.chapter%201.pdf [02 Juni 2020].

[5] Sari, L. O. R. K. 2006. Pemanfaatan Obat Tradisional dengan Pertimbangan dan Keamanan.

Majalah Ilmu Kefarmasian III (1): 1-7.
[6] Sugiyono. (2014). Metode Penelitian Pendidikan Pendekatan Kualitatif, Kuantitatif, dan R&D. Alfabeta: Bandung.
[7] Safitri, S., R. Yolanda, dan E. M. Brahmana. Studi Etnobotani Tumbuhan Obat di Kecamatan Rambah Samo Kabupaten Rokan Hulu. Jurnal Mahasiswa Prodi Biologi UPP 1 (1): 110060.
[8] Efremila, E. Wardenaar, dan L. Sisillia. 2015. Studi Etnobotani Tumbuhan Obat oleh Etnis Suku Dayak di Desa Kayu Tanam Kecamatan Mandor Kabupaten Landak. Jurnal Hutan Lestari 3 (2): 234-246.
[9] Goulart, F.S. 1995. Super Healing Foods. Reward Books, a member of Penguin Putnam Inc. New York.
[10] Santoso, H. B. 2008. Ragam dan Khasiat Tanaman Obat. PT Agromedia Pustaka. Yogyakarta.

Can mangrove intensive policies in Indonesia benefit more coastal communities? environmental policy analysis

Anto Ariyanto, Afni Zulkifli*, Jeni Wardi

Universitas Lancang Kuning, Pekanbaru, Indonesia

*afni@unilak.ac.id

Abstract. All This article discusses the government's efforts to restore the environment in coastal areas from the threat of abrasion. The deliberative policy analysis approach was used to analyze the implementation of the Indonesian government's national mangrove labor intensive program policy on Bengkalis Island, Riau Province. The article writing uses a qualitative approach by utilizing descriptive analysis methods. The data collection techniques are interviews, observations, journal articles, online news, official documents, and websites of authoritative institutions. The results of the study show that the implementation of the mangrove labor-intensive national program involves community groups at the site level, this program provides economic, social benefits, and regenerates a culture of mutual cooperation during the Covid-19 pandemic, increases awareness of climate change, and finds several obstacles that need to be addressed. evaluated so that the implementation of this mangrove rehabilitation can run even better.

1. Introduction

Coastal communities are one of the most vulnerable groups affected by climate change. As experienced by the people of Bengkalis Island, Riau Province, Indonesia. In this area which is the foremost island of Indonesia and is only separated by the Melaka strait from Malaysia, there has been a significant reduction in land area, namely an average of 42.5 ha/year due to abrasion [1]. Abrasion that damages ecosystems typical of coastal areas, has resulted in reduced fish catches [2], eliminating agricultural land, plantations, and housing for coastal residents [3]. Collaboration and coordination of work from all levels of government, together with the community, is needed in planning and implementing climate change adaptation and mitigation [4], especially in coastal areas with climate impacts on the coast of Bengkalis Island.

During the Covid-19 pandemic, the Indonesian government made policy breakthroughs with massive community involvement, one of which was through the National Economic Recovery Program (PEN) with an allocation of 4.2% of GDP [5]. This policy is claimed not only for the purpose of environmental recovery, but also for the economic recovery of the community during the pandemic, through community empowerment strategies in management activities and also the utilization of mangrove forests [2]. In 2020 this program covers an area of 15,000 hectares, involving more than 30 thousand people in 50 working days, or reaching more than 1.5 million HOK [6].

Community empowerment through institutional strengthening, policy and community participation has become an important part of restoration activities [7]. With institutional strengthening and community involvement, it is believed that it will increasingly provide increased welfare [8]. Policy support and land use planning need to be integrated in climate change adaptation actions into strategies for increasing the income of coastal communities [9], with the hope of building an economy that is able to face the crisis [10].

The question is whether the implementation of the national mangrove labor-intensive policy which takes place in 2020 and continues in 2021 can provide more benefits for coastal communities, especially during the Covid-19 pandemic? This article will answer it by focusing on the deliberative policy analysis method to analyze policies and their implementation.

2. Methodology

This article was prepared using a qualitative approach with a descriptive method. Data collection techniques were carried out through interviews, observations, journal articles, online news, official documents, and websites of authoritative institutions. The analytical method used is a deliberative approach which emphasizes the participation and involvement of citizens in policy [11] research focus on system discourse to see the extent to which policies contribute to society, and the concept of policy implementation Edward III with a focus on communication or elaboration of coordination between actors, and resources for policy implementation [12].

Referring to these various approaches, the policy analysis in this article leads to the implementation of mangrove labor-intensive programs, the contribution of mangrove labor-intensive policies to coastal communities, as well as supports and obstacles to the implementation of mangrove ecosystem restoration.

3. Result and Discussion

3.1. Implementation of the Mangrove-Intensive National Program (PEN Mangrove)

The implementation of the mangrove labor-intensive program carried out nationally is a breakthrough made by the government to provide economic stimulus to the community, especially during the Covid-19 pandemic. This mangrove restoration activity involves community groups in the village. The total implementation of national mangrove planting in the working area of Indragiri Rokan, covers an area of 692 hectares.

An area of 18 ha is in the Other Use Area (APL) which is a non-forest area, and 674 ha in the Protected Forest (HL) area. Rehabilitation of mangroves in forest and non-forest areas is needed so that they can contribute to ecosystem services and environmental sustainability, in the midst of uneasy pressure due to activities in coastal areas. Table 1 shows the area of mangrove planting in Riau in 2020, carried out in four districts, with the largest area being in Bengkalis Regency.

Table 1. Mangrove Planting Area in Riau Province in 2020

District Location	Mangrove Planting Area (Ha)	Number of Districts	Number of Villages	Number of Groups	Participant
Rokan Hilir	25	1	1	1	51
Bengkalis	319	4	11	12	899
Meranti	55	2	2	2	65
Siak	8	2	3	3	60
Indragiri Hilir	285	4	16	18	461

Source: BPDASHL Indragiri Rokan

The implementation of the national mangrove program begins with the preparation of a technical design, by identifying the location of activities through spatial analysis of forest area maps, national mangrove maps, land cover maps, and abrasion and accretion maps. Furthermore, field checks, measurements, and data collection of community socio-economic information are carried out. The technical draft text will contain information on the name of the implementing group, the location and area of the location, the number and type of seeds, planting patterns, facilities and infrastructure, implementation timetable, budget plan, and location map.

Meanwhile, in 2021, this program will be implemented jointly between the Ministry of Environment and Forestry (KLHK) and BRGM (Peat and Mangrove Restoration Agency). Figure 1 shows the organizational structure of the 2021 national mangrove rehabilitation program.

Source: SK Menteri LHK No 405/2021

For the implementation of the mangrove rehabilitation program, various stages of activities are carried out, starting from field preparation, consisting of institutional preparation or preconditions for the people who will be involved, with outreach activities, group formation, and mentoring. Furthermore, re-measurement of the boundaries of the prospective planting locations, setting of boundary stakes, preparation of seedling points, and installation of stakes. The installation of stakes is carried out as a buffer for propagules seeds to be stronger against currents and tidal waves [13].

The next stage is the provision of seeds or seeds, which can be obtained from seed fertilization, the use of community seeds that are bought back, and the collection of seeds in the form of propagules or other mangrove seeds. For the implementation of planting, must pay attention to factors such as tides, wave season, environmental suitability, as well as the involvement of local communities. These various factors can affect cropping patterns and cropping schedules.

Mangrove planting patterns can be carried out with pure planting patterns, sylvofishery planting patterns, spacing grass planting patterns, plant enrichment patterns, and associated mangrove aquaculture (AMA) patterns. The final stage is payment. Because the implementation of the mangrove rehabilitation program is self-managed, payments are made on an account-to-account basis to the accounts of members of the group implementing the planting, paid according to the number of working days (HOK). Figure 2 shows the stages of implementing the national mangrove.

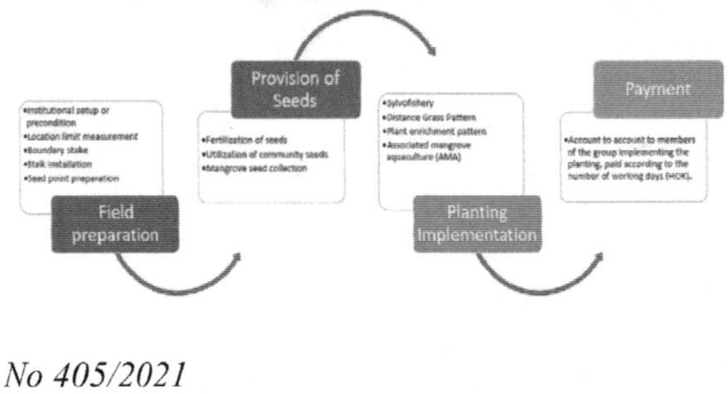

Source: SK Menteri LHK No 405/2021

3.2 Contribution of Mangrove Intensive Policy for Coastal Communities

Nationally throughout Indonesia this activity absorbs 5.9 million workers with an area of 15,000 ha. Components of benefits that can be paid to the community from this program include the provision of salaries (HOK), which is obtained from the manufacture and erection of stakes, the manufacture of signboards and work huts, transportation of seeds and planting, mentoring and supervision. In addition,

IOP Conf. Series: Earth and Environmental Science 1041 (2022) 012059 doi:10.1088/1755-1315/1041/1/012059

the costs borne by the government are also for work operations such as boat rentals, and procurement of seeds.

The community will carry out various restoration activities, including providing seeds or seeds, planting, maintaining, supporting mangrove rehabilitation activities, as well as physical and financial responsibility. All payments are directly given in cash to the account of each group member. This is also to minimize the misuse of the budget. Tables 2 and 3 present data on the national mangrove labor intensive program in Riau Province and community participation in Bengkalis Regency.

Table 2. Describes The National Mangrove Labor-Intensive Program In Riau Province

Year	Large (Ha)	Seeds	HOK	Participant
2020	319	2.498.800	28.797	899
2021	973	3.024.300	44.684	732

Source: BPDASHL Indragiri Rokan

Table 2. Bengkalis Community Involved In National Mangrove Rehabilitation

Village	Group name	Participant	Mangrove rehabilitation area (Ha)
Pangkalan Pinang	Maju Bersama	15	3
Makeruh	Kampung Nelayan	48	30
Teluk Lecah	Bangun Bersama	26	10
Teluk Lecah	Paluh Bakau	40	20
Kembung Luar	Pulau Mandiri	48	12
Pambang Pesisir	Ikatan Pemuda Melayu Peduli Lingkungan III	167	30
Muntai	Ikatan Pemuda Melayu Peduli Lingkungan II	200	80
Muntai Barat	Ikatan Pemuda Melayu Peduli Lingkungan I	254	100
Palkun	Bangkit Bersama	23	12
Sekodi	Berkah Usaha	24	12
Air Putih	Keslimasy	13	4
Pakning Asal	Sepakat	41	6
Total		899	319

Source: BPDASHL Indragiri Rokan

From this national mangrove program, the community receives a budget of Rp. 110 thousand/HOK. The community gets economic benefits, with additional income during the pandemic, between IDR 100,000-150,000 or in a month they can earn around IDR 2,500,000-3,000,000. Several people in one family can be involved in this activity, for a period of 3-4 months. In one month, the income of the people involved in this program can reach Rp. 3 million per person. The poverty line in Riau is IDR 565,937 per capita per month. This value is also above the average monthly wage or net salary of formal workers in Bengkalis Regency in 2020, which is IDR 2,673,959 [14]. This program is recognized by the community as very helpful for the income of the people of Bengkalis Island during the difficult time of the pandemic.

Another benefit that the community feels is mutual cooperation to jointly restore the mangroves in the location where they live. This culture has long been abandoned by the community, due to various factors. However, with the mangrove program, they now gather, discuss and work together to restore the surrounding environment. Mangroves that were previously unnoticed have now become a common asset to be protected. The national mangrove planting also provides a new economic alternative for the people of West Muntai Village, with the opening of the Raja Kecik tourist beach location. This location not only provides economic benefits, but also becomes a place for learning, transferring knowledge about the importance of maintaining the mangrove ecosystem for generations in their village. Mangrove activities also increase public awareness to be directly involved in climate change mitigation and adaptation actions.

3.3. Support and Barriers to Mangrove Intensive Implementation

Support for the implementation of mangroves can be carried out with various policies, starting from the national level that are implemented to the site level. Community groups are assisted and receive direction so that they can independently manage the budget and adjust to the planting period and planting patterns of mangroves. The implementation of mangrove rehabilitation is a multi-stakeholder work. Its implementation requires a strong network between the government, the community and third parties in this case the companion group. The results of previous studies also confirm that the implementation of government programs by involving community participation, the benefits will be felt by all people [15].

However, several changes to the national mangrove work team between 2020 and 2021 caused several obstacles in the field. In 2021, involving BRGM, a self-management contract between the supply of materials and seeds with separate planting, where the supply of materials and seeds uses type 4 swakelola and planting uses type 1 swakelola. This causes the administrative process and integration of work stages to be more complicated and time consuming.

The procurement of materials, seeds, and planting activities should be combined in one self-management contract (type 4), where the payment clause can describe which work components must be transferred to the group account and which must be transferred to the group member's account. This has been practiced in the Mangrove planting labor-intensive activity in 2020 which is managed by the KLHK work unit (BPDASHL).

Another obstacle is the administration which changes frequently, so documents that have been printed and signed must be changed and remade. This makes the administrative process seem convoluted and time-consuming and has the potential to cause conflict between officers and groups in the field. So it is highly recommended that administrative completeness formats be simplified and standardized.

Another obstacle during the implementation of mangrove planting is that the field coordinator and assistants do not fully understand the administration and technical implementation of activities, so that groups often experience confusion during consultations. Administrative and technical strengthening to Field Coordinators and assistants needs to be carried out periodically so that they can understand well the administrative and technical aspects of work and can provide good and clear guidance and instructions to implementing groups.

The climate problem also presents its own challenges to this program, where planting in several locations is carried out by community groups outside the technical design map. Community groups implementing planting need to be given knowledge and skills of participatory mapping using a simple Android-based mapping application, and the rantek map must be revised from real data in the field.

Another obstacle to the implementation of the mangrove planting program, especially in Muntai Barat Village, was that some of the plants had been washed away by the waves of the Melaka Straits, and some of the plants were also attacked by crab pests. For this reason, this mangrove planting policy needs to examine the factors of tidal conditions and sea water currents when determining the location and timing of planting in preparing technical design documents. In order for the planted mangrove seedlings to grow, it is necessary to install a wave barrier and install a plant protector.

Some of the planting sites are mangrove stands that are still close together. In order for the implementation of this program to run according to its objective of restoring the environment, the locations where the stands are still dense and have not been planted can be relocated to more critical areas. Regarding the payment transfer process which takes quite a long time, starting from submitting payments by the group to the process of entering funds into group/member accounts, the payment administration process needs to be simplified, and utilize technology for time efficiency.

Evaluation instruments must also be prepared in a simple, effective, and efficient manner. Executors of evaluation (examination and assessment) of work results need to be clearly appointed and defined. It is recommended to activate and empower the Group Monitoring Team and Facilitators who interact daily on site as evaluators. Technical officers from BRGM, BPDASHL/BKSDA and KPH functioned as supervisors who would guide the evaluators in the evaluation.

4. Conclusion

The national mangrove planting in 2020 and 2021 is carried out not only as a community empowerment program for environmental recovery, but also to help the community's economic recovery during the Covid-19 pandemic. The implementation of the national mangrove on Bengkalis Island involves community groups in all stages of the activity. The community gets more benefits not only in terms of financial income, but also has an impact on increasing social participation, environmental awareness, and forming a climate-resilient community in coastal areas. The implementation of this program is still experiencing various obstacles, especially on administrative issues. For this reason, a more effective and efficient administrative evaluation is still needed so that this community empowerment program can run better.

References

[1] S. Sutikno, "Analisis laju Abrasi Pantai Pulau bengkalis dengan Menggunakan data satelit," Pertem. Ilm. Tah. HATHI XXXI, no. September, 2014.
[2] Ambari, "Abrasi Pesisir Terjadi, Apakah Mengancam Kedaulatan Negara?," Mongabay.co.id, 2019. https://www.mongabay.co.id/2019/06/20/abrasi-pesisir-terjadi-apakah-mengancam-kedaulatan-negara/ (accessed Sep. 12, 2021).
[3] F. Mubarok, "Abrasi Ancam Lahan Perkebunan di Pesisir Pulau Bengkalis," Mongabay.co.id, 2021. https://www.mongabay.co.id/2021/01/22/abrasi-ancam-lahan-perkebunan-di-pesisir-pulau-bengkalis/ (accessed Jul. 14, 2021).
[4] P. Perdinan, "Perubahan Iklim Dan Demokrasi: Ketersediaan Dan Akses Informasi Iklim, Peranan Pemerintah, Dan Partisipasi Masyarakat Dalam Mendukung Implementasi Adaptasi Perubahan Iklim Di Indonesia," J. Huk. Lingkung. Indones., vol. 1, no. 1, 2020, doi: 10.38011/jhli.v1i1.87.
[5] Kamrussamad, Pahlawan-pahlawan Le Parle Covidnomic. Jakarta: PT. Kompas Media Nusantara, 2020.
[6] Admin, "Padat Karya Penanaman Mangrove untuk Pulihkan Ekonomi Nasional," sitinurbaya.com, 2020. http://www.sitinurbaya.com/padat-karya-penanaman-mangrove-untuk-pulihkan-ekonomi-nasional.
[7] M. Mislan and Y. Anwar, "Problems and restoration of cascade Mahakam Lakes in the climate change perspective," in IOP Conference Series: Earth and Environmental Science, 2020, vol. 535, no. 1, doi: 10.1088/1755-1315/535/1/012003.
[8] L. Siwiyanti, A. M, Khairul, L. N, and Arrina, "Peningkatan Kapasitas Lembaga Kemasyarakatan Desa Dalam Meningkatkan Kesejahteraan Ekonomi Masyarakat," J. Masy. Mandiri, vol. 5, no. 3, 2021, [Online]. Available: https://doi.org/10.31764/jmm.v5i3.4413.
[9] A. P. Fischer, "Pathways of adaptation to external stressors in coastal natural-resource-dependent communities: Implications for climate change," World Dev., vol. 108, 2018, doi: 10.1016/j.worlddev.2017.12.007.
[10] S. Syaharani and M. A. Tavares, "Nasib Target Emisi Indonesia: Pelemahan Instrumen Lingkungan Hidup di Era Pemulihan Ekonomi Nasional," J. Huk. Lingkung. Indones., vol. 7, no. 1, 2020, doi: 10.38011/jhli.v7i1.212.
[11] A. Mardiyanta, "Kebijakan Publik Deliberatif : Relevansi dan Tantangan Implementasinya," vol. 24, pp. 261–271, 2011.
[12] L. Agustino, "Analisis Kebijakan Penanganan Wabah Covid-19: Pengalaman Indonesia.," J. Borneo Adm., vol. 16, no. 2, pp. 253–270, 2020, [Online]. Available: https://doi.org/10.24258/jba.v16i2.685.
[13] R. Nopinilianti, B. Burhanuddin, and M. S. Anwari, "Pola Tanam Propagul Dan Bibit Rhizophora Stylosa Griff. Pada Tapak Berlumpur Di Areal Rehabilitasi Mangrove Desa Mendalok," J. Tengkawang, vol. 10, no. 1, 2020, doi: 10.26418/jt.v10i1.39237.
[14] BPS, Provinsi Riau Dalam Angka. BPS Provinsi Riau, 2021.
[15] D.- Kurniasih, "Model Penguatan Kelembagaan Pada Program Sanitasi Lingkungan Berbasis

Masyarakat (Slbm) Di Kabupaten Banyumas (Suatu Pendekatan Collaborative Governance)," Sosiohumaniora, vol. 19, no. 1, 2017, doi: 10.24198/sosiohumaniora.v19i1.7888.

Validation of UV-VIS spectrophotometric method for the determination of Calcium in Maman (cleome gynandra l)

Fathul Jannah[1], S Meri Agritubella[2], Ira Oktaviania[3], Lily Restusari

Poltekkes Kemenkes Riau, Riau, Indonesia

Email: fathul@pkr.ac.id[1], meri@pkr.ac.id[2], ira@pkr.ac.id[3]

Abstract. One of health problems in Indonesia is stunting. Energy and nutrients intake are important factor in the occurrence of stunting conditions. The intake of micronutrients to prevent stunting are vitamin A, zinc, iron, iodine, and several other micronutrients such as calcium and phosphorus are also important in children's linear growth. One of the plants which is a natural resource owned by Provinsi Riau and can be used as an alternative food to obtain calcium is Maman (Cleome gynandra L.). The analytical method used ultraviolet-visible spectrophotometry based on the complex formation between murexide and calcium. The optimum wavelength obtained is 530 nm with a r value = 0.9989. The LOD value obtained was 0.013 ppm and the calcium content in maman leaves was 172.64 ± 4.91 mg/g.

1. Introduction

One of the health problems in Indonesia is the incidence of stunting. In 2018 the prevalence of stunting based on the cross-sectional Indonesian basic health studies (Riskesdas) was 30.8% [1]. The stunting rate in Provinsi Riau according to the results of the Riskesdas 2018 was 27.4%, this value is still higher than the WHO requirement ie 20%. Stunting is the consequence of chronic energy–protein malnutrition. Moreover, other enhancing factors, including infections and diseases, are important in the etiology of stunting [2].

In a situation of insufficient supply of quality food, growth will be seriously hampered by repeated recovery periods The consequences of child stunting can make increased morbidity and mortality, poor child development and learning capacity, increased risk of infections and noncommunicable diseases in adulthood, and reduced productivity and economic capability [3]. Energy and nutrient intake are important factor in the occurrence of stunting conditions. Micronutrient intake that is very important to prevent stunting is vitamin A, zinc, iron and iodine [4]. Several other micronutrients such as calcium and phosphorus also important role in children's linear growth [5,6].

For growth period, a necessity of calcium is important. Lack of calcium in the bones in infants will cause rickets, while in children, lack of deposits can cause stunted growth [7]. The maman plant (Cleome gynandra L) is one of the plants that has a high calcium level. This plant is a plant from South Africa and can grow in tropical and subtropical areas [8]. This plant can be found in Asian countries such as India, Malaysia, Singapore, Thailand and Indonesia.

In Indonesia this plant grows wild like weeds [8]. The calcium content of maman plants in Africa varies depending on where they grow [9]. UV-VIS spectrophotometry is an analytical method that is easy to use. The method provides a simple way to determine the quantity of a little substance. Determination of calcium using a spectrophotometer is based on the formation of a complex between calcium and murexide. Murexid is a compound that has a chromophore group, that is a C=C and C=O double bond which will bind to calcium (Ca).

2. Methodology

2.1 Instrument and material

LaMotte SMART Spectro Spectrophotometer, 350 to 1000nm equipped with 1 cm cells was used for the determination of absorbance. Reference Standard Calcium ($CaCl_2$, Merck), Murexide ($C_8H_8N_6O_6$, Merck), NaOH (merck), nitric acid 65%(p.a) (Sigma Aldrich), aquabidest (waterone). Maman (*Cleome gynandra* L) from Kabupaten Rokan Hilir, Provinsi Riau.

2.2 Stock Standard Solution

Standard stock solution calcium (1000 µg/ml) was prepared by taking accurately weighed 100 mg of $CaCl_2$ into 100 ml volumetric flask and finally volume was made up to the mark with aquabidest. For murexide solution was prepared by taking accurately weighed 50 mg of murexide into 10 ml volumetric flask and finally volume was made up to the mark with aquabidest. 0.5% murexide solution add 25 ml ethanol. This solution stable in 24 hours.

2.3 Destruction Process

Samples (maman leaves) that is mashed with a blender is weighed each 15 grams in erlenmeyer. Sample was destruction by addition of HNO3 65% 25 ml. This solution left for 24 hours. After 24 hours, the solution kept on a hot plate for 1 hour. filtered using filter paper whattman No.42 by eliminating about 5 ml filtrate to saturate the filter paper and then the filtrate was then collected in a bottle is repeated respectively for samples.

2.4 Determination of Maximum Wavelength

The stock concentration of 100 ppm was measured with a spectrophotometer at a wavelength of 400-800 nm.

2.5 Validation of Calcium

Calibration Curve (Linearity)

Series of mixed standards were prepared from the standard stock solution (1000 µg/ml) to five concentrations in the range 0.05 – 0.25 µg/ml.

Accuracy.

The recovery test is carried out using three concentrations (0.05; 0.1, and 0.25 µg/ml). The concentrations obtained next compared to the actual concentrations of the measured standard solution.

Precision.

Precision is measured as relative standard deviation or coefficient of variation. The relative standard deviation value that meets the requirements indicates the precision in a method being carried out. Relative standard deviation measured from six time repeated from one concentration.

Limit of detection (LOD) and Limit of Quantitation (LOQ).

These limit of detection and limit of quantitation can be calculated by the following formula:

$$\text{LOD} : \frac{3\,x\,(\frac{Sy}{x})}{Slope}$$

$$\text{LOQ} : \frac{10\,x\,(\frac{Sy}{x})}{Slope}$$

2.6 Determination of Calcium

The sample solution of the maman (*Cleome gynandra* L) is taken 0.5 ml was put into a 10 ml volumetric flask and added 4 ml NaOH 2N, and added murexide solution 2 ml, finally volume was made up to the mark with aquabidest (Dilution factor =100ml/0.5ml =200 times).

3. Result and Discussion

Maman (Cleome gynandra L) was destruction using wet destruction method. Method uses concentrated HNO3. The use of pure concentrated HNO3 in a close system under elevated temperatur and pressure is well known since 1860. Wet destruction has the advantage of being effective on both inorganic and organic materials. It often destroys or removes the sample matrix, thus helping to reduce or eliminate some types of interferences [11].

Determination of calcium used spectrophometry must using additive reagent for make a complex form of calcium. In this research used murexide solution in alkali condition. Therefor was used NaOH solution, for made alkali condition. The determination of maximum wavelength was using a standard solution of a calcium concentration of 100 ppm, measured in the wavelength range of 400-800 nm using UV-Visible spectrophotometry. The measurement was carried out at specific range from 500 - 560 nm.

Based on results, the maximum wavelength found at 530 nm. Validation of the assay method is conducted to ensure that the analytical methods are accurate and precise to the analyte. A calibration curve for linearity was using a standard solution of calcium with series concentrations of 0.05 μg/ml, 0.1 μg/ml, 0.15 μg/ml, 0.2 μg/ml, and 0.25 μg/ml. From absorbance results made linear regression line equations, which can be used to calculate the calcium in maman. The results of the standard solution of calcium curve are shown in Figure 1 and table 1.

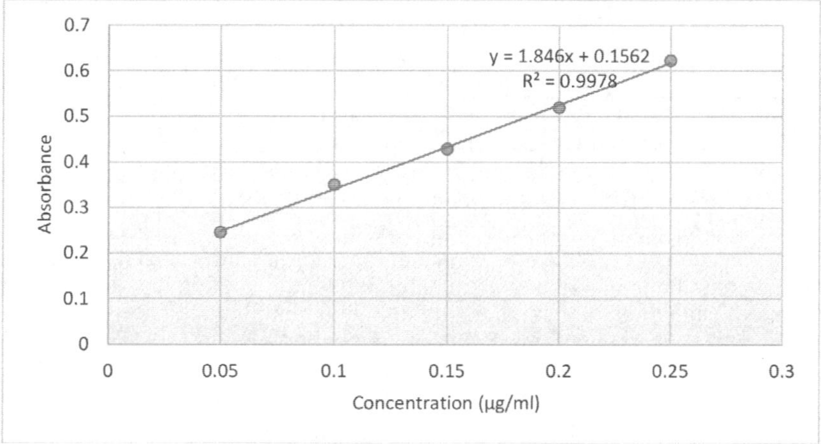

Figure 1. Calcium Standard Curve

The parameter used in linearity is the correlation coefficient (r) in linear regression analysis y = a + bx. The linear regression line equation obtained is y = 1.846x + 0,1562 with a r (correlation coefficient) value 0.9989. The linearity results are shown in Table 1.

Table 1. Linearity of Calcium

Concentration (μg/ml)	Absorbance
0.05	0.2456
0.1	0.3502
0.15	0.4282
0.2	0.519
0.25	0.6227
slope	1.8460
intercept	0.1562
r	0.9989

Accuracy is described from percentage of the recovery. Accuracy testing was used three concentrations in range linearity (low, medium, and high). The range of % recovery found is 91.03 – 98.24% The accuracy results shown in table 2. Requirement for percentage of the recovery for range concentration 1 – 10 μg/ml is 80 – 110% [12]. It indicates that the result of this recovery test meets the predefined accuracy requirements.

Table 2. Percent recovery of Calcium

Absorbance	CM	CA	%recovery
0.2372	0.04386		87.71
0.2513	0.05150	0.05	102.99
0.2323	0.04120		82.41
		average	91.04
0.3593	0.11000		110.00
0.3171	0.08714	0.1	87.14
0.3364	0.09759		97.59
		average	98.24
0.5742	0.22641		90.57
0.6193	0.25085	0.25	100.34
0.6065	0.24391		97.56
		average	96.16

CM : Concentration obtained from measurements (μg/ml)
CA : The actual concentration (μg/ml)

A precision method was carried out using one concentration repeatedly by the same analyst. Thus. the precision can also be described as repeatability or reproducibility. Precision was measured as a relative standard deviation or coefficient of variation. Precision was measured using standard calcium in 0.1 μg/ml. The precision results are shown in Table 3.

Table 3. Precision of Calcium

Replication	Absorbance
1	0.3500
2	0.3546
3	0.3504
4	0.3383
5	0.3621
6	0.3605
Average	0.35265
%CV	2.45

The result shows that the variation of coefficient values of each concentration has 2.45%. Requirement maximum of percent coefficient of variation for concentration 1 μg/ml is 11 [12]. It indicates that the result of precision showed this method reproducible. Limit of Detection (LOD) is the smallest number of analytes in the sample. While the limit of Quantification can be interpreted is the smallest quantity of analytes in a sample. The LOD and LOQ can be calculated statistically from the the standard curve of calcium [13]. LOD results in this study is 0.013 μg/ml and LOQ results obtained was 0.042 μg/ml. Determination of calcium levels was carried out using a sample solution from wet destruction. Sample solution was made in six replications. The results of level measurements are shown in Table 4.

Table 4. Calcium level in Maman Leaves (Cleome gynandra L)

Absorbance	Concentration (μg/ml)	Level (mg/g)
0.3892	0.1262	168.2236586
0.4051	0.1348	179.7052699
0.3985	0.1312	174.8391004
0.3982	0.1311	174.6225903
0.3864	0.1247	166.1563342
0.3949	0.1293	172.2926256
	Average	172.64
	SD	4.91

Quercetin levels can be calculated using the linear regression line y = 1.846x + 0,1562. The measurement results showed that calcium levels were 172 ± 4.91 mg/g in maman leaves. The result showed calcium levels in maman leaves were higher than study from higher than calcium level from maman in Zimbabwe. Calcium levels from that study were ± 18000 g/g [14]. Level calcium in Maman plant from Burkina Faso in Afrika Barat is 440 mg/100 g [15]. The result from that study is lower than level calcium in this study.

4. Conclusion

Based on the results can be concluded that the UV-Vis spectrophotometry method has fulfilled several parameters for the determination of the concentration of calcium in maman leaves (Cleome gynandra L). The result of calcium level was found 172mg/g ± 4.91. r value from linearity test was 0.9989. percentage of recovery have a range 91.03 – 98.24% and the percent coefficient of variation was 2.45. The LOD was 0.013 μg/ml, and LOQ was 0.042 μg/ml.

Acknowledgment

The authors would like to thank the Poltekkes Kemenkes Riau for funding researchers in terms of funds and research facilities so this research can run well.

References

[1] Riset Kesehatan Dasar. 2018. Available online: https://www.litbang.kemkes.go.id/ (accessed on September 2020)

[2] Local Burden of Disease Child Growth Failure Collaborators. Mapping child growth failure across low- and middle-income countries. Nature 2020, 577, 231–234.

[3] Stewart, C P, Iannotti, L, Dewey K G, Michaelsen, K F, and Onyango, A W. 2013. Contextualising complementary feeding in a broader framework for stunting prevention. Maternal & Child Nutrition. 9. 27–45

[4] Souganidis E. 2012. The relevance of micronutrients to the prevention of stunting. Sight and life Magazine. 26(2). 10-18

[5] Stuijvenberg M E, Nel J, Schoeman S E, Lombard C J, Plessis L M, and Dhansay MA. 2015. Low intake of calcium and vitamin D, but not zinc, iron or vitamin A, is associated with stunting in 2-5 years old children. Nutrition; 31:84

[6] Mikhail W Z A, Sobhy, H M, El-sayed, H H, Khairy S A, Abu Salem H Y H, and Samy M A. 2013. Effect of nutritional status on growth pattern of stunted preschool children in Egypt. Academic Journal of Nutrition. 2 (1). 01-09.

[7] Millennium Challenga Account Indonesia. 2014. Stunting dan Masa Depan Indonesia. Available online: www.mca-indonesia.go.id. (accessed on September 2020)

[8] Mishra S S, Moharana S K, and Dash M R. 2011. Review on Cleome Gynandra. International Journal of Research on Pharmacy and Chemistry. Vol1(3). 681-689

[9] Jinazali H, Mtimuni B, and Chilembwe E. 2017. Nutrient Compoisition os Cat's Whiskers

(Cleome gynandra L.) From Different Agro Ecological Zones In Malawi. African Journal of Food Science. Vol 11(1). 24-29.

[10] Hanifah A M. 2019. Analisis Kadar Kalsium (Ca) Pada Susu Sapi Segar Yang Beredar Di Area Madiun Dengan Metode Spektrofotometri UV-Vis. Other Thesis, Stikes Bhakti Husada Mulia Madiun. Available online: http://Repository.Stikes-Bhm.Ac.Id/Id/Eprint/499

[11] Multi Agency Radiological Laboratory Analytical Protocols Manual. 2004. Planning, Assessment, Implementation. Chapter 13: Sample Dissolution. FDA, USGS, NIST.

[12] Gonzales A G, Herrador M A, and Asuero A G. 2010. Intra-Laboratory Assesment of Method Accuracy (Trueness and Precision) by Using Validation Standards. Talanta. 82(5). 1995-1998

[13] Harmita H. 2012. Petunjuk pelaksanaan validasi metode dan Cara Perhitungannya. Pharmaceutical Sciences and Research (PSR). 1(3). pp. 117-135.

[14] Moyo M, Amoo S O, Aremu A O, Gruz J, Subrtova M, Jarosova M, Tarkowski P, and Dolezal K. 2018. Determination Of Mineral Constitunets, Phyrochemicals, And antioxidant Cualities of Cleome Gynandra. Compared to Brassica oleracea and Beta vulgaris. Food Chemistry. The Journal Frontiers In Chemistry. Vol 5. 1-9.

[15] Ouedraogo I W K, Tranchant C C, and Coulibaly Y L B. 2013. Evaluation of Mineral Contents In Cleome gynandra Leaves and Stals From Burkina Faso. Journal of The Cameroon Academy Of Sciences. Vol 11(1). 49-53.

Utilization of bamboo charcoal to decrease tannin content of lindur fruit flour (Bruguiera gymnorrhiza)

Niken Pujirahayu*, Zakiah Uslinawaty, Nuhayati Hadjar, Muhtadir

Department of Forestry, Universitas Halu Oleo, Kendari, Indonesia

*rahayuken08@gmail.com

Abstract. This study aims to determine the correct concentration of bamboo charcoal in the processing of Bruguiera gymnorrhiza fruit to produce mangrove flour with the lowest tannin content. The process of making bamboo charcoal is carried out at the Forestry Laboratory of the Forestry Department, Faculty of Forestry and Environmental Sciences. Meanwhile, the chemical analysis of flour is carried out in the Chemistry Laboratory, Department of Chemistry, Faculty of Mathematics and Natural Sciences, Halu Oleo University Kendari, Southeast Sulawesi. The chemical analysis test was carried out using a completely randomized design method with a one-factor treatment test. This factor is the percentage of bamboo charcoal during boiling and soaking (R0 = 0%; R1 = 15%; R2 = 20%) for the type of mangrove fruit used (P1 = Brugueira gymnorrhiza). The results showed that the provision of bamboo charcoal at the time of boiling and soaking had a significant effect on reducing tannin levels and had a very significant effect on reducing the moisture content and HCN of lindur fruit flour. The best treatment is giving 20% bamboo charcoal with soaking time for 48 hours and changing the soaking water every 24 hours. The best treatment is giving bamboo charcoal 20% with immersion time for 48 hours and changing the immersion water every 24 hours, with the water content of 8.7%, tannin content of 0.063%, and HCN content of 2.093 ppm, where these results meet SNI standards. 01-2997-1996 [1] and ADI (Acceptable Daily Intake) as food ingredients.

1. Introduction

Mangroves have essential ecological and economic functions. One of the economic functions of mangroves is to provide a source of carbohydrates as food for humans. So far, people still do not know about the use of mangrove fruit as a substitute for alternative carbohydrates [2]. One of the mangrove fruits that has potential as a source of food is Bruguiera gymnorrhiza, locally known as Lindur Fruit in Indonesia. These fruits are also cultivated in tropical areas such as South Africa, East Africa, Madagaskar, South Asia, Southeast Asia and Australia, Fruit production begins in two-year-old specimens with 15 cm to 25 cm in length and 2 cm in diametre [3].

The main obstacle that limits the use of mangrove fruit is tannins, which cause a bitter taste in foodstuffs. Lindur is processed into B. gymnorrhiza flour, where pretreatment is required to reduce the content of antinutritional substances, such as HCN and tannin. HCN is a result of an enzymatic reaction after maceration. Tannins are polyphenolic compounds found in most plants and are generally thought to function as chemical defenses against pathogens and herbivory [3].

Excessive consumption of HCN can be toxic because HCN inhibits cytochrome-oxidase in human aerobic respiration, and inhibition of this enzyme can cause asthma, other conditions caused by excessive HCN are headache, dizziness, nausea, vomiting, stomach pains, diarrhoea and sometimes death [4, 5]. Tannins are actually needed to ward off free radicals in the human body, but they can be carcinogens and increase fruit bitterness in excessive concentrations. HCN and tannins are volatile and water-soluble substances [6]. Previous studies recommended using rinse and blanch water to reduce the content of HCN and tannins [7]. However, the water flushing method seems inefficient as it requires repetition [3].

In addition to immersion in water, the process of reducing tannin and HCN levels can also use a mixture of charcoal. The prior study reveales that the use of ash suspension is able to reduce the amount of HCN and tannin by immersing the sample containing HCN and tannin in the ash suspension, because ash contains SiO2, Al2O3, Fe2O3, CaO, K2O, carbon and unburnt residue which can absorb HCN and tannin [3]. Based on this, the use of charcoal can help lower tannin and HCn levels in lindur fruit. In this study tried to use charcoal from bamboo to find out how the steamer against the decrease in tannin and HCN levels in lindur flour.

2. Methodology

The materials used in this study were Bruguiera gymnorrhiza mangrove fruit and bamboo charcoal. The bamboo charcoal used in this study was obtained from homemade products of the Bampu Apus type (While the materials used for analysis were distilled water, filter paper, gelatin solution, saturated NaCl, concentrated H2SO4, acid salt solution, indigocarmine solution, kaolin powder, KMnO4 solution. 0.1 N and a solution of Na2C2O4 (Na-oxalate).

This research was conducted using two stages of research, namely making bamboo charcoal and then soaking and boiling mangrove fruit with bamboo charcoal. Then making flour and testing chemical analysis of lindur flour. Chemical analysis was carried out on three parameters, namely water content, tannin content and HCN content of lindur flour. To determine the effect of charcoal concentration on the parameters of the observations, the method was Completely Randomized Design with one factor treatment test. The factor is the percentage of bamboo charcoal given when boiling and soaking (R0 = 0%; R1 = 15%; R2 = 20%) on the type of mangrove fruit used (P1 = Brugueira gymnorrhiza). Determination of the content of HCN (cyanide acid) was determined by Arianto et al., [8] with modifications

2.1 Bamboo Charcoal preparation

The bamboo used to make charcoal comes from Apus bamboo (*Giganthocloa apus*) and the manufacture of bamboo charcoal is carried out using the following procedure [9].

2.2 Production of Lindur Fruit Flour (Bruguiera gymnorrhiza)

The process of making mangrove flour is as follows:

- Mangrove fruits were sorted as physiologically ripe, namely: those that were 3 months old after flowering with characteristics such as dark green skin, hard flesh, 15 cm to 35 cm in length.
- Peel the outer skin, cut into small pieces (1 cm), then weighed.
- Washed and then boiled in water with a ratio of lindur fruit: water = 1:1.5 (weight/volume) added bamboo charcoal (0%, 15% and 20%) until boiling.
- Removed, washed with clean water, then soaked in water for 48 hours with the replacement of the soaking water every 24 hours. After that, it was dried using direct sunlight for (±) 4 days to dry. The dried fruit was then ground using a dry blender and sieved through an 80-mesh sieve. The flour results were then analysed for water content, HCN and tannins and their yields.

2.3 Determination of Water Content

The method for determining the water content was carried out by the Malangngi et al. [10] method. The water content was determined by weighing 2 g of the extracted test sample. The sample was put into the oven at a temperature of 105 0C for 3 hours then removed from the oven and cooled in a desiccator for 30 minutes, then the weight of the sample was weighed. This treatment was carried out several times until the sample weight was constant. Then the water content is calculated with the following formula:

$$\% \textbf{Kadar Air} = \frac{\textit{Berat Awal} - \textit{Berat Akhir}}{\textit{Berat Awal}} \times \mathbf{100\%} \quad (1)$$

2.4 Determination of cyanide acid (HCN) levels

Determination of the content of HCN (cyanide acid) was determined according to the method of Malangngi et al. [10]. The sample was weighed as much as 20 g which had been mashed, then added 100 ml of distilled water in Erlenmeyer and allowed to stand for 2 hours. Added another 100 ml of distilled water and distilled with steam. The distillate was accommodated in an Erlenmeyer which had been filled with 20 ml of 25% NaOH. After distillation reaches a volume of 150 ml, the distillation process is stopped. After distillation, 5 ml 5% Cl and 8 ml NH4OH were added. The distillate mixture was titrated with 0.02 N AgNO3 solution until turbidity occurred. HCN levels are calculated by the following formula:

$$\mathbf{HCN} = \frac{\mathbf{ml\ AgNO3 \times 0,54}}{\mathbf{Berat\ Bahan}} \times \mathbf{1000\ mg/kg} \qquad (2)$$

2.5 Determination of cyanide acid (HCN) levels

While the total tannin content was determined by Malangngi et al. [10], a total of 0.5 g of the test sample was extracted with 10 ml of diethyl ether for 20 hours, then filtered, and the residue obtained was boiled with 100 mL of distilled water for 2 hours then cooled and filtered. The extract obtained was added with distilled water until the extract volume was 100 mL. A total of 0.1 mL of extract was added with 0.1 mL of Folin Ciocalteu reagent and vortexed, added with 2 mL of Na2CO3, and vortexed again. The absorbance was read at 760 nm after incubation for 30 min at room temperature. The results obtained were plotted against a standard tannic acid curve prepared in the same way. The total tannin content is expressed in mg of tannin/kg extract. Tannin content is calculated using the formula as follows:

$$\%\ \text{Tanin} = \frac{10\ (A-B) \times N \times 0{,}00416}{\text{Sampel (g)}} \times 100\% \qquad (3)$$

The research data were tested using Analysis of Variance (ANOVA) with a significant level of 95% using the SPSS 26 application, and Tuckey's further test was carried out.

3. Result and Discussion

The results of the ANOVA test on *B. gymnorrhiza* fruit flour showed that the concentration of bamboo charcoal given had a very significant effect on the water content and HCN content and had a significant effect on the tannin content of Lindur fruit flour. Then from Tuckey's further test results showed that there were significant differences in water content and HCN levels between the treatment of 0% (control), 15% and 20% bamboo charcoal (Table 1).

Table 1. Water, Tannin and HCN content of *B. gymnorrhiza fruit*/Lindur flour

Treatmant	Water content (%)	Tannin (%)	HCN (ppm)
control	8.7 a	0.063 a	2.093 a
Bambu charcoal (15 %)	6.5 b	0.057 b	1.713 b
Bambu charcoal (20 %)	6.0 c	0.055 b	1.570 c

The value of water content, HCN content and tannins from lindur flour obtained was then compared with the quality standards of SNI 01-2997-1996 flour and Acceptable Daily Intake (ADI) as food ingredients, which are presented in Table 2.

Table 2. Comparison of mean water, tannin and HCN content of lindur flour with SNI and ADI

Variable	results	Maximum standard in food	remarks
Air (%)	8.7	SNI 01-2997-1996 = 12 %	Meet the standard
HCN (ppm)	2.093	SNI 01-2997-1996 = 40 ppm	Meet the standard
Tanin (%)	0.063	ADI = 560 mg/kg	Meet the standard

Boiling and soaking time with bamboo charcoal has benefits in reducing tannin levels of mangrove B. gymnorrhiza fruit. This research proves that the boiling treatment and soaking time allow osmosis and diffusion to occur. This diffusion process occurs at the time of immersion with charcoal content. The higher the charcoal content, the lower the tannin content in the mangrove fruit due to the influence of the diffusion process by charcoal. According to Danarto [11], diffusion can occur if the concentration of the solution in the cell is high, the solution will come out of the cell with a low concentration of solute.

This research shows that in the soaking process, the water conditions change color or foam. This indicates that the diffusion process at the time of soaking occurs with the dissolution of the remaining substances in the fruit. In the process of soaking the water with a low concentration of the solution, it will move into the fruit with a high concentration of solution so that the water content of the fruit is high and slightly expands. One of these soluble substances is tannins because of the nature of the tannins themselves, which are easily soluble in water.

In contrast to the diffusion process, on the contrary, osmosis is moving or moving substances from a solution of low concentration to a solution of high concentration. The process of osmosis occurs during the boiling process, where the substances in the fruit will come out carried by water. On the other hand, unlike the diffusion process, osmosis is moving substances from a low concentration solution to a high concentration solution. The process of osmosis occurs during the boiling process, where the air will carry away the substances in the fruit.

Bamboo charcoal is a solid material with excellent adsorption, especially for materials with a large surface area. Based on the study results, the treatment of giving bamboo charcoal when boiling on mangrove fruit type *B. gimnorrhiza* was proven to reduce water content, HCN, and tannins in flour. The addition of bamboo charcoal concentration in each sample was directly proportional to the decrease in substance content in *B. gymnorrhiza* fruit flour which included water content and tannin content (Table 1).

The water content of lindur fruit that has been known in this study is 71.06%. Based on Table 2, the average moisture content of mangrove fruit meal with concentrations of 0%, 15% and 20% bamboo charcoal administration were 8.7%, 6.5% and 6.0%, respectively. This is in line with previous research conducted by Chrissanty [2]. The decrease in water content occurred due to the provision of bamboo charcoal which contains carbon and has the ability to absorb water from the main ingredient. The decrease in water content causes the sample weight to decrease, so the resulting yield will also decrease. The water content obtained in this study is also lower when compared to Chrissanty's [2] study which obtained the lowest water content of 10.97%.

These data indicate that the moisture content of the lindur fruit flour has met the flour quality requirements because it is lower than the previous study conducted by Kumalaningsih et al. [12], it was stated that the binding and release of HCN and tannins by carbon from lindur fruit would increase. increased as the immersion time also increased. The carbon content in bamboo charcoal is able to bind HCN and tannins from lindur fruit and then release them through the adsorption process, thus the longer the immersion time, the greater the release of HCN and tannin content.

Meanwhile, in this study, it was found that the HCN level was 3,240 ppm. Based on Table 4, the lowest levels of HCN in lindur fruit flour were found in samples with a concentration of 20% bamboo charcoal with a duration of immersion of 48 hours, with an average HCN level of 1.570 ppm and the highest HCN levels in the control sample. with 48 hours of immersion. the average HCN content is 2,093 ppm.

More complete data on the HCN content of Lindur fruit is presented in the appendix. As an anti-nutritional substance, the presence of HCN in food has been determined by the World Food and Agriculture Organization (FAO) where the maximum tannin content that can be present in food is 50 ppm. Meanwhile, based on SNI 01-2997-1996 (Quality Requirements for Cassava Flour) the maximum limit for HCN content is 40.00 ppm [12]. The residual HCN levels obtained in this study indicate that the lindur fruit flour produced meets the requirements as food ingredients because it is present in very small quantities.

The tannin content of lindur fruit flour in this study was 3.7812%. The results of the analysis of tannin levels in Table 1 show that lindur fruit flour has the lowest tannin content when given 20% bamboo charcoal with an average tannin content of 0.05% and a soaking time of 48 hours, while the highest tannin content was in the control sample (0% charcoal). bamboo) with an average tannin content of 0.06%. This result is lower than the results of Chrissanty's study [2] which reported the tannin content of lindur fruit flour was 0.94% with a concentration of 30% rice husk ash in the boiling process. It is suspected that bamboo charcoal is stronger in absorbing tannins than rice husk ash so that different results are obtained. The residual tannin content contained in this lindur fruit flour has met the requirements as food ingredients because the maximum tannin content in foodstuffs determined by the ADI (Acceptable Daily Intake) is 560 mg/kg body weight/day.

Based on the results of the ANOVA test, it showed a significant effect on the decrease in water content, HCN content and tannin content after treatment with bamboo charcoal during boiling and soaking time. It is suspected that the boiling process allows an osmosis process which results in the lifting of the chemical content of lindur fruit by bamboo charcoal which is then carried away by water. While the soaking process with bamboo charcoal resulted in a diffusion process which resulted in lindur fruit with a high concentration of solution being pulled out through the cell cavity by bamboo charcoal which had a low solution content. This is evidenced by the presence of foam in the sample during immersion and changes in the color of the water in the sample for 48 hours of immersion. Photos of the immersion results are presented in the appendix.

The output of ANOVA showed the effect of giving bamboo charcoal on the decrease in water content, HCN and tannins in lindur fruit. Based on subset 1, the average data on the effect of giving bamboo charcoal on moisture, HCN and tannin content with sample codes P1R0, P1R1 and P1R2 were 149.1111, 157.3333 and 184.5556, respectively. While in subset 2 there is sample data P1 which is 3737,0000. From the ANOVA results obtained a significance value of 0.05, which means that the treatment with bamboo charcoal significantly affected the decrease in water content, HCN and tannins of lindur fruit flour.

4. Conclusion

Based on the results of the research that has been obtained, it can be concluded that the provision of bamboo charcoal at the time of boiling and soaking has a significant effect on reducing tannin levels and has a very significant effect on reducing water content and HCN of lindur fruit flour. The best treatment is giving bamboo charcoal by 20% with soaking time for 48 hours and changing the soaking water every 24 hours with the results of water content of 8.7%, tannin content of 0.063 % and HCN content of 2.093 ppm which are these results has met the standards of SNI 01-2997-1996 and ADI (Acceptable Daily Intake) as food ingredients.

References

[1] SNI 01-2997-1996. Tepung Singkong. Standar Nasional Indonesia. Badan Standarisasi Nasional.
SNI 01-3549-2009. Tepung Beras. Standar Nasional Indonesia. Badan Standarisasi Nasional.
SNI 01-3727-1995. Tepung Jagung. Standar Nasional Indonesia. Badan Standarisasi Nasional.
SNI 01-3751-2006. Tepung Terigu. Standar Nasional Indonesia. Badan Standarisasi Nasional.

[2] Chrissanty P A 2011 Penurunan kadar tanin pada buah mangrove jenis Brugueira gymnorrhiza, Rhyzophora stylosa dan Avicennia marina untuk diolah menjadi tepung mangrove Jurnal Industrial 1(1): 31-39.

[3] Amin M N G, Pralebda S A, Hasan M N, Zakariya, Subekti S, Saputra E, Andriyono S, Pramono H and Alamsjah M A 2018 Physicochemical properties of Bruguiera gymnorrhiza flour (BGF) International Food Research Journal 25(5): 1852-1857.

[4] Nhassico D, Muquingue H, Cliff J, Cumbana A and Bradbury J H 2008 Rising African Cassava Production, Diseases Due to High Cyanide Intake, and Control Measures Journal of the Science of Food and Agriculture 88(12):2043 - 2049

[5] Umuhozariho M G, Shayo N B, Msuya J M and Sallah P Y K 2014. Cyanide and selected nutrients content of different preparations of leaves from three cassava species. Afr. J. Food Sci. 8(3): 122-129.

[6] Cumbana A, Mirione E, Cliff J and Bradbury J H 2007 Reduction of cyanide content of cassava

flour in Mozambique by the wetting method Food Chem 101:894–897.
[7] Subandriyo and Setianingsih N I 2016. Extraction Process for Reducing Tannin of Mangrove Fruit
(Bruguiera gumnorrhiza) as a Raw Material for Food Flour. Aquatic Procedia 7(2016) 231 – 235
[8] Arianto A., Nohong B and Nurhaedah 2014 Analisis Kandungan Asam Sianida (HCN) pada Kacang Koro Pedang (Canavalia ensiformis) dengan menggunakan lama Perendaman NaCL yang berbeda Jurnal Galung Tropika 3(3): 186-191.
[9] Suheryanto D 2012 Penelitian pembuatan arang bambu (bamboo charcoal) pada suhu rendah untuk produk kerajinan Jurnal Dinamika Kerajinan dan Batik 32(2): 33-48.
[10] Malangngi L P, Maiske S S dan Jessy J E P 2012 Penentuan Kandungan Tanin dan uji aktivitas Antioksidan ekstrak biji buah Alpukat (Persea Americana Mill.) Jurnal MIPA.
[11] Danarto Y C 2008 Pirolisis sekam padi dengan Katalisator Zeolit Prosiding Seminar Nasional Kimia dan Pendidikan Kimia FMIPA dan UNS.
[12] Kumalaningsih S, Sulistywati dan Wignyanto 2012 Produksi tepung buah Lindur (Bruguiera gymnorrhiza Lamk.) rendah tanin dan HCN sebagai bahan pangan alternatif Jurnal Teknologi Pertanian 13(3): 187-198.

Supplementing safflower oil (*Carthamus Tinctori L*) and inocytol in feed and its effect

Ning Iriyanti*, Ismoyowati, Efka Aris Rimbawanto, Modawy Abdelgader Albasheer Altayb

Universitas Jenderal Soedirman, Purwokerto, Indonesia

*ning.iriyanti@unsoed.ac.id

Abstract. Investigating the effect of safflower oil (Carthamus tinctorius L) and inositol on the quality of Sentul chicken eggs. Eight one 16-week female Sentul chickens were distributed to 81 units of battery cage (3 chickens each). Basal diet was composed of corn, rice bran, soybean kernel, fishmeal, palm oil, calcium car-bonate (CaCO3), topmix, lysine and methionine, as well as safflower oil (Car-thamus tincthorius L) and inositol. The research was conducted in a Completely Randomized Design (CRD) with 9 treatments and 3 replicates (3 chickens/unit). The treatments were R0 = Control/ basal diet + 0% Safflower and 0% Inositol; R1 = Basal diet + 0.5% Safflower oil; R2 = Basal diet + 1.0% Safflower oil; R3 = Basal diet + Inositol 0.5%; R4 = Basal diet + Inositol 1.0%; R5 = Basal diet + 0.5 % Safflower oil + 0.5% Inositol; R6 = Basal diet + 0.5 % Safflower oil + 1.0% Inositol; R7 = Basal diet + 1.0 % Safflower oil + 0.5% Inositol; R8 = Basal diet + 1.0 % Safflower oil + 1.0% Inositol. The data were subjected to analysis of vari-ance (ANOVA), continued with an Honestly Significant Difference test (HSD) using SPSS ver. 17.0 with 5% error. The result showed that supplementing saf-flower oil (Carthamus tinctorius L) and inositol in feed did not significantly af-fect ($P>0.05$) egg weight, egg index, fat and cholesterol content of egg yolk, but significantly affected ($P<0.05$) egg yolk protein, and most significantly affected ($P<0.01$) egg production. Conclusively, supplementing 1% safflower oil in feed could increase egg production, but 1% safflower oil 1% + 0.5% 0.5% inositol could increase protein as well as decreasing fat and cholesterol of egg yolk.

1. Introduction

Sentul chickens are the native chickens of Ciamis Regency, reared for their meat and eggs and recognized as the germplasm of Ciamis Regency as stated in the Letter of Decision of the Minis-ter of Agriculture of Indonesian Republic SK No. 689/Kpts.PD410/2/2013 on establishing Sentul Chickens breed as the native Indonesian chicken breed from Ciamis. Sentul chickens could lay 26 eggs per laying period or approximately 150/year, and they have a faster growth rate as well as better resistance to disease [1].

Sentul chickens is a multifunction chicken, i.e. producing meat and eggs. Sentul chickens could lay 85 to 210 eggs per year. The main components of eggs are water, protein, fat, carbohydrate, vitamin and mineral. According to [2], eggs are composed of shells (11%), white effs (57%) and egg yolk (32%). The optimum egg quality and production is the result of proper maintenance, both the management and quality feed supply. The quality feed contains sufficient protein and energy for basic needs with additional feed supplement or feed additives.

Egg quality is assessed from external and internal factors. The former includes color, weight, index, shell and cleanliness. The latter includes the egg yolk color, Haugh unit, white egg weight, egg yolk weight and pH. Efforts to increase egg quality aim to improve the benefits of eggs through feed manipulation, i.e. supplementing feed additives and/or feed supplement. Additives or supplements could im-prove feed quality and production quality (meat and eggs). The common feed additives for chickens are probiotic, prebiotic, enzyme, herbal substances, vitamin, omega-3 source and other feed with nutritional/functional value.

Safflower oil and inositol are feed supplements incorporated into the feed to increase nutrient metabolism and to obtain optimum performance. Safflower oil is plant oil extracted from the seeds of safflower plants (Carthamus tinctoris L). Safflower oil contains high unsaturated fatty acid (90.33%) and low saturated fatty acid (9.62%). Unsaturated fatty acid in safflower oil mainly consists of 76.22% linoleic acid of the polyunsaturated fatty acid (PUFA) and 13.75 % oleic acid of the monounsaturated fatty acid (MUFA) [3].

Inositol is the group of vitamins B8 and a biomolecule compound with a crucial role in growth, maintenance and reproduction. The symptoms of inositol deficiency include lack of ap-petite and slow growth. Also, inositol contributes to increasing body weight gain; performing biological function; sustaining life, cell growth lipid metabolism and insulin sensitivity; and im-proving the absorption of mineral phosphor.

Also, an increased myo-inositol could improve phytase enzyme derived from microbes in the crop, proventriculus, gizzard, duodenum, jeju-num, ileum, and blood plasma, as well as decreasing the concentration of Ca and P in chicken feed. Inositol is part of the polyhydroxylated sikloalkana, which is known as cyclitol. Inositol is commonly found as phytic acid in plants. Phytic te acid is a subcellular phosphor deposit in the seeds which are bound with the protein [4].

2. Methodology

Eighty-one Sentul chickens aged 16 weeks were reared until 23 weeks old. The basal diet was composed of corn, rice bran, soybean kernel, fishmeal, palm oil, calcium carbonate (CaCO3), top mix, lysine and methionine, and safflower oil (*Carthamus tincthorius L*) and inositol. Cage and equipment for feeding and drinking. A wire cage 35x35x29x56 cm was used to house 2 chickens. Feed was measured using a 2-kg scale with 1g sensitivity and 1200-g capacity. The equipment to analyse cholesterol level was beakers, 10-ml test tubes, and a spectrophotometer. The equipment to analyse fat and protein levels was a desiccator, an oven, a biuret, filter paper and Soxhlet kit, and tools for destruction, distillation and titration.

The experiment was subjected to a Completely Randomized Design (CRD) with 9 treatments and 3 replicates (3 chickens each). The treatments were:

R0 = control/ basal diet + 0% Safflower and 0 % Inositol
R1 = Basal diet + 0.5 % Safflower oil
R2 = Basal diet + 1.0 % Safflower oil
R3 = Basal diet + 0.5 % Inositol
R4 = Basal diet + 1.0 % Inositol
R5 = Basal diet + 0.5 % Safflower oil and 0.5 % Inositol
R6 = Basal diet + 0.5 % Safflower oil and 1.0 % Inositol
R7 = Basal diet + 1.0 % Safflower oil and 0.5 %Inositol
R8 = Basal diet + 1.0 % Safflower oil and 1.0 % Inositol

The diet was composed following iso calorie and iso protein guide. The composition of feed for female Sentul chickens is presented in Table 1, and feed nutrient content in Table 2.

Table 1. Feed composition for female sentul chickens

Feed composition	R_0	R_1	R_2	R_3	R_4	R_5	R_6	R_7	R_8
		%							
Corn	52	52	52	52	52	52	52	52	52
Rice bran	22	22	22	22	22	22	22	22	22
Soybean kernel	11.5	11.5	11.5	11.5	11.5	11.5	11.5	11.5	11.5
Fishmeal	10	10	10	10	10	10	10	10	10
Palm oil	0.5	0.5	0.5	0.5	0.5	0.5	0.5	0.5	0.5
CaCO3	2	2	2	2	2	2	2	2	2
Topmix	1	1	1	1	1	1	1	1	1
Lysine	0.6	0.6	0.6	0.6	0.6	0.6	0.6	0.6	0.6
Methionine	0.4	0.4	0.4	0.4	0.4	0.4	0.4	0.4	0.4
Safflower oil	0	0.5	1	0	0	0.5	0.5	1	1
Inositol	0	0	0	0.5	1	0.5	1	0.5	1
Total	100	100.5	101	100.5	101	101	101.5	101.5	102

Table 2. Feed nutrient for female sentul chickens

Nutrient content	R_0	R_1	R_2	R_3	R_4	R_5	R_6	R_7	R_8
	%								
CP (%)	19.34	19.34	19.34	19.34	19.34	19.34	19.34	19.34	19.34
CFat(%)	11.99	11.99	11.99	11.99	11.99	11.99	11.99	11.99	11.99
CF (%)	10.18	10.18	10.18	10.18	10.18	10.18	10.18	10.18	10.18
ME (Kcal/kg)*	2.853.60	2.860.40	3.867.21	2.853.60	2.853.60	2.860.40	2.860.40	2.867.21	2.867.21
Ca (%)*	2.44	2.44	2.44	2.44	2.44	2.44	2.44	2.44	2.44
P (%)*	0.34	0.44	0.44	0.44	0.44	0.44	0.44	0.44	0.44
Lysine (%)*	1.12	1.12	1.12	1.12	1.12	1.12	1.12	1.12	1.12
Methionine (%)*	0.68	0.68	0.68	0.68	0.68	0.68	0.68	0.68	0.68

Note: (*) Nutrient content was calculated based on the percentage of feed composition applied. The nutrient content was made according to the NRC table [5]. (**) Nutrient content from the analysis of Laboratory of Nutrition and Feed (2020)

This study used Sentul chickens aged 16 weeks and reared for 8 weeks. The samples were collected after 8 weeks of treatment when 50% of the chickens had laid eggs. Three eggs were collected from each of three chickens in the experiment unit. Treatments were offered for 60 days. Data were collected in the beginning, middle and end of the study, which was started at peak production. From each replicate, 9 egg samples were taken, hence 243 eggs in total. The obtained data were subjected to an analysis of variance (ANOVA) following [6]. Further analysis was done in an Honestly Significant Difference test (HSD) using SPSS ver. 17.0 with 5% error. The measured variables were eggs' physical quality (egg production, egg weight, egg index) [7] and chemical quality (protein, fat and cholesterol of egg yolk) ([8]; [9]).

3. Result and Discussion

The result of supplementing safflower oil (*Carthamus tinctorius l*) and inositol in feed and its effect on improving the physical and chemical quality of Sentul chicken eggs are presented in Table 3.

Table 3. The physical and chemical quality of sentul chicken eggs

Treatment	Physical quality of Sentul chicken eggs			Chemical quality of Sentul chicken eggs		
	HDA (%)**	Egg weight (g)^{ns}	Egg index ^{ns}	Egg yolk fat(%) ^{ns}	Protein egg yolk (%)*	Cholesterol egg yolk (mg/g)^{ns}
R_0	56.93±0.06	36.00±2.75	79.05±0.54	29.49±0.49	16.41±0.27	183.20±25.53
R_1	65.44±2.84	37.20±0.23	79.30±3.16	28.44±1.03	16.11±0.51	169.33±5.10

R_2	59.67±1.22	39.18±3.36	78.94±1.68	27.78±0.46	16.53±0.71	151.14±35.63
R_3	48.56±2.15	38.52±4.20	81.92±5.40	28.04±1.04	16.18±0.60	171.41±22.87
R_4	47.59±5.04	37.68±0.09	76.92±4.86	28.13±0.99	15.90±0.57	161.81±28.12
R_5	39.37±5.54	36.67±1.04	81.85±5.74	27.94±0.26	16.11±0.25	161.47±11.46
R_6	47.07±14.20	35.54±2.14	75.86±1.96	29.98±1.50	16.53±0.61	155.04±50.06
R_7	58.41±3.76	37.66±1.01	80.39±8.45	28.69±0.51	17.56±0.16	150.41±46.01
R_8	56.44±2.50	37.33±1.85	77.18±0.87	27.79±0.93	16.85±0.52	158.12±23.78

R_0 = basal diet; R_1= basal diet + safflower oil 0.5%; R_2 = basal diet + safflower oil 1%; R_3 = basal diet + 0.5% inositol; R_4 = basal diet + inositol 1%; R_5 = basal diet + safflower oil 0.5% and 0.5% inositol; R_6 = basal diet + safflower oil 0.5% and inositol 1%; R_7 = basal diet + safflower oil 1% and 0.5% inositol; R_8 = basal diet + safflower oil 1% and inositol 1%
Cholesterol egg yolk

3.1. The physical quality of eggs

The analysis of variance showed that the treatments most significantly affected (P<0.01) egg production (HDA = Hen Day Production) of Sentul chickens, and supplementing 1% safflower oil produced the most egg, i.e. 59.67±1.22%. Egg production is significantly affected by chicken's health, environmental condition, farming management, the development of digestive tracts and reproduction, as well as a balanced supply of feed nutrient.

Nutrients that play a critical role in egg production are energy-protein balance, mineral, calcium, and amino acids especially methionine and lysine, as well as essential fatty acids (oleic, linoleic and arachidonate). Safflower oil contains unsaturated fatty acids that helps the formation of eggs. The main unsaturated fatty acids in safflower oil are oleic and linoleic which contributes around 90% of the total fatty acids. The rest 10% is the saturated fatty acids (palmitic and stearic).

Safflower oil contains 6-8% palmitic acid, 2-3% stearic acid, 16-20% oleic acid and 71-75% linoleic acid [10]. Safflower oil is tasteless and colorless, and the content is similar to that of sunflower oil. Safflower seeds is the source of mineral enriched with mineral (Zn, Cu, Mn, and Fe), vitamin (thiamin and β-carotene), tocopherol, α, β and γ [11]. According to [12], the main carotenoid in safflower oil is β-carotene (0.3-0.35 mg kg^{-1}).

The analysis of variance showed that the treatments did not significantly affect (P>0.05) egg weight and egg index. The range of egg weights in this study was 35.54±2.14 - 39.18±3.36, while the egg index was 75.86±1.96-81.92±5.40. This study saw a numeric increase which demonstrated that safflower oil and inositol can be used to support physiological process in the body. The contributing factors to egg weight and index include genetic factor, sex maturity, age, medicine and nutrient content, especially protein, energy, fatty acids, mineral and vitamin.

Safflower oil contains not only unsaturated fatty acids but also protein, mineral and vitamin that play a critical role in body physiology, especially egg production, egg formation and egg weight. Nutrient content in safflower seeds contained 38-48% oil, 15-22% protein and 11-22% fiber [3], as well as mineral (Zn, Cu, Mn, and Fe) and vitamin (thiamin and β-carotene), tocopherol, α, β and γ [11]. The main carotenoid in safflower oil was β-carotene, i.e. 0.3-0.35 mg kg^{-1} [12]. Fat content in safflower oil are unsaturated fatty acids and free fatty acids. Fat contains water soluble vitamins, namely retinol, calciferol, tocopherol and phylloquinone, as well as unsaturated linoleic fatty acid [3] .

The main unsaturated fatty acids in safflower oil are oleic and linoleic that constitute 90% of the total fatty acid, while the rest 10% consists of saturated fatty acid, namely 6-8% palmitic acid, 2-3% stearic acid, 16-20% oleic acid and 71-75% linoleic acid [10]. Supplementing 0.5%-1% safflower oil and inositol into feed indicated an insufficient protein in lipoprotein process which stimulates estrogen hormone to control egg weight. Estrogen and progesterone hormones stimulated the synthesis of white egg and egg yolk protein. Estrogen stimulates the synthesis of ovalbumin, ovotransferrin and lysozyme protein produced by the tubular glands from magnum. The tubular magnum would secrete most of the protein in white eggs and store it in the form of granule. The secretion of tubular gland is stimulated by progesterone hormone to induce hypothalamus for activating the hormone releasing factor so that LH and pituitary anterior are secreted. Progesterone and androgen would organize the development of oviduct to secrete albumin from magnum [13].

Inositol is the most significant vitamin B-complex as coenzyme to help metabolism process in absorbing nutrient, especially protein. Inositol could improve body immune and handle stress in chicken so that the physical condition remains stable. According to [14], inositol is an important biomolecule for normal body growth, maintenance and reproduction.

Furthermore, inositol is an important vitamin for growth, maintenance and reproduction as well as improving body immune, acting as a natural antioxidant and controlling mineral concentration in the cells [15].

3.2. Chemical qualities of egg yolk

Egg protein was significantly increased ($P<0.05$) 16.11±0.51-17.56±0.16 (R7 = basal diet + safflower oil 1% and 0.5% inositol). Meanwhile, fat (27.78±0.46 - 29.98±1.50%) and cholesterol (150.41±46.01 - 171.41±22.87mg/dl) were not significantly affected ($P>0.05$). However, this study saw a numeric decrease, especially R7 which produced the lowest fat and cholesterol. Omega 3 in safflower oil that consisted of 16-20% oleic acid and 71-75% linoleic acid could help absorb the nutrients, especially the unsaturated fatty acids as the material for lipoprotein.

The unsaturated fatty acid of safflower oil consists of 76.22 % linoleic acid of the polyunsaturated fatty acid (PUFA) and 13.75 % oleic acid of the monounsaturated fatty acid (MUFA) [3]. According to [16], unsaturated fatty acid incorporated in feed would reduce fat deposition in the body due to the decreased fat absorption. Study of [17] showed that feed composition that contained long-chained unsaturated fatty acid could decrease the synthesized fatty acid fat formation, as well as increasing the process of fat breakdown.

Safflower oil would inhibit fat metabolism by limiting the activity of lipoprotein lipase (LPL) in the plasma and adipose tissue and minimize the lipogenesis process. According to [18], fat is stored as triglyceride in adipose tissue. Triglyceride, which is synthesized in the liver or derived from the feed, would be absorbed in the small intestines and converted to the new triglyceride in the cells and bound with phospholipid and apoprotein (chylomicron). Chylomicron would enter lacteal in the lymphatic system then to the blood stream. Triglyceride in the chylomicron would be hydrolyzed by lipoprotein lipase (LPL) and converted into free fatty acids and glycerol which could enter the adipose tissue.

Study of [19] have shown that only free fatty acid could enter the adipose tissue and is esterified into triglyceride in the adipose with the help of the LPL enzyme. The decreased fat content is related to the ability of feed content to inhibit the enzymes for fat metabolism (e.g. acetyl-CoA carboxylase) and affect the fatty acid synthesis in forming the fat [20]. Furthermore, [21] stated that triglyceride converted into chylomicron with VLDL would be absorbed in the lymph as much as 80-90% that was esterified with the long-chained fatty acid. Then, VLDL produced in the liver would transport triglyceride into blood plasma.

According to [22], the triglyceride level would determine fat deposit in the tissue. Fat deposit was also potentially due to the continual synthesis of triglyceride by fatty acids and alpha-glycerol phosphate from glucose because the level of inositol supplement could not optimally increase the insulin sensitivity, which affected the glucose level. According to [23], fatty acid could also synthesize fat from glucose and amino acids. Linoleic acid and oleic acid in feed could affect fatty acid content. In addition, the fat level could escalate due to an increased unsaturated fatty acid profile in the feed. The study of [17] showed that incorporating feedstuff containing unsaturated fatty acid could lower the triglyceride and LDL cholesterol, as well as improving HDL cholesterol in the blood.

Furthermore, according to [24] that linoleic acid would inhibit the absorption of cholesterol by binding fat molecules from feed to minimize its absorbability in the small intestine. The inhibited cholesterol absorption would reduce chylomicron anabolism and reduce triglyceride serum; therefore, the level of cholesterol and triglyceride entering the blood was decreased. It would make VLDL (Very Low-Density Lipoprotein) and LDL anabolism in the serum would decrease.

Linoleic acid would be absorbed and transported through lipoprotein could decrease the biosynthesis of cholesterol and acetyl CoA in the liver. It's been reported in [25] that the decreased cholesterol biosynthesis is due to the inhibited performance of HMG CoA reductase (Hydroxyl methylglutaryl CoA reductase) by cholesterol that entered the remains of chylomicron or LDL which prevented the formation of cholesterol; consequently, the level of LDL and triglyceride were decreased and HDL was increased. The increased HDL would discard the excessive cholesterol in the body, thus reducing cholesterol levels in the meat [26].

Supplementing inositol together with safflower oil in feed would inhibit the activity of acetyl CoA carboxylase which catalyzed acetyl-CoA reaction into fatty acids, thus increasing the fatty acid level. Fatty acid in the form of triglyceride and that bound in albumin are derived either from feed intake or fat synthesis in the liver. Triglyceride form chylomicron or lipoprotein would be hydrolyzed into glycerol and free fatty acid by the lipoprotein lipase (LPL) which was formed by adiposity and secreted into the endothelial cells.

The combination of safflower oil and inositol could improve the performance of Lecithin-cholesterol acyltransferase (LCAT) enzyme and the elements constructing HDL, namely apolipoprotein A-1 and insulin. It's reported by [27] that apolipoprotein A-1 is the main component of HDL that manages the level of cholesterol and allows HDL movement in and out of the blood stream to collect cholesterol excess and decrease VLDL synthesis. Low VLDL would reduce triglyceride hydrolysis, so VLDL is not changed into IDL, thus preventing the second triglyceride hydrolysis which leads to the decreased LDL formation.

According to [28], HDL cholesterol would transport the unused cholesterol together with faces in form of bile salt and neutral steroid hormone; as a result, blood cholesterol is low, leading to decreased egg cholesterol because no excess cholesterol is accumulated in the tissue. Stated that decreased egg cholesterol level is related to decreased blood cholesterol. Fat content in feed contributes to the level of liver fat in chickens; therefore, the higher fat in feed, the higher total fat in the liver, meat and other tissues, like eggs [29].

Most fat is stored in an insoluble form in the adipose tissue where fat metabolism (adiposity) occurs. The precursor for protein in fat formation is triacylglycerol from lipoprotein (VLDL and chylomicron) built in the liver and intestines sent by the blood. Lipoprotein lipase in the surface of blood capillary segmented triacylglycerol into glycerol and fatty acid. Glycerol and fatty acid are transported by adiposity and converted into fat [30]. It's been explained by [31] that inositol plays an important role in inhibiting the activity of fatty acid metabolism, so fatty acid deposition in the tissue is depleting. Inositol could increase the absorption of glucose so that the glucose is not accumulated as fat but used to generate energy.

Supplementing inositol in feed would help decrease meat cholesterol [32] because inositol could increase insulin sensitivity which affects the decreased level of LDL cholesterol. In addition, it was reported by [33] that inositol supplementation in feed could help maintain stable insulin and prevent excessive production by improving insulin sensitivity. An increased insulin level would decrease LDL cholesterol and triglyceride serum as well as increasing HDL cholesterol in the blood.

Inositol together with fatty acids and phosphorus, would stimulate the formation of lecithin [23]. Also, the fatty acid in lecithin would release detach cholesterol from the peripheral tissue which is esterified in the plasma and transported to the liver as HDL; therefore, cholesterol is not accumulated in body tissues due to the oxidized LDL in macrophage [34]. Furthermore, it was explained that insulin is a hormone that plays a role in cell permeability on glucose that is converted to energy [35]. Meanwhile, blood cholesterol would increase when LDL and VLDL increase but HDL decrease due to insulin resistance [36].

Supplementing inositol that belongs to vitamin B6 would affect some biological functions, especially during lipid metabolism which eventually lower the fat level as a transduction signal for insulin [14]. As a sugar alcohol isomer, the inositol also helps the glycolysis process to produce insulin. The liver would store sugar in form of glycogen and bile fluid would emulsify lipoprotein. The liver would store sugar in form of glycogen and emulsify the lipoprotein by the bile fluid.

4. Conclusion

The study concludes that supplementing 1% safflower oil in feed could increase egg production, but 1% safflower oil 1% + 0.5% 0.5% inositol could increase protein as well as decreasing fat and cholesterol of egg yolk.

References

[1] Isyanto A Y, Sudradjat S and Iskandar M 2017 Strategi pengembangan ayam Sentul di Kabupaten Ciamis Mimb. Agribisnis J. Pemikir. Masy. Ilm. Berwawasan Agribisnis 3 1–12
[2] Budiman A, Hintono A and Kusrahayu K 2012 Pengaruh Lama Penyangraian Telur Asin Setelah Perebusan terhadap Kadar Nacl, Tingkat Keasinan dan Tingkat Kekenyalan Anim. Agric. J. 1 219–27
[3] Katkade M, Syed H, Andhale R and Sontakke M 2018 Fatty acid profile and quality assessment of safflower (Carthamus tinctorius) oil J. Pharmacogn. Phytochem 7 3581–5
[4] Widiastoeti D, Santi A and Solvia N 2013 Pengaruh myoinositol dan arang aktif terhadap pertumbuhan planlet anggrek dendrobium dalam kultur in vitro J. Hortik. 22 205–9
[5] NRC 1994 Nutrient requirement of domestic animals Nutr. Requir. Poultry,
[6] Steel R G . and Torrie J H 1995 Principles and Procedures of Statistics, An Approach Biometrics.Principles and Procedures. Translation Bambang Sumantri. (Jakarta: Scholastic PressMain)
[7] Wahju J 2004 Ilmu Nutrisi Unggas. Edisi ke-4 Univ. Gadjah Mada Press. Yogyakarta
[8] AOAC 2005 Official methods of analysis of the Association of Analytical Chemists (Association of Official Analytical Chemists)
[9] Tranggono S B 1989 Kimia lipid PAU Ilmu Pangan dan Gizi, Univ. Gadjah Mada, Yogyakarta
[10] Liu L, Guan L L, Wu W and Wang L 2016 A review of fatty acids and genetic characterization of safflower (Carthamus tinctorius L.) seed oil Org. Chem Curr Res 5
[11] Omidi A H, Khazaei H, Monneveux P and Stoddard F 2012 Effect of cultivar and water regime on yield and yield components in safflower (Carthamus tinctorius L.) Turkish J. F. Crop. 17 10–5
[12] Ben Moumen A, Mansouri F, Richard G, Abid M, Fauconnier M, Sindic M, El Amrani A and Serghini Caid H 2015 Biochemical characterisation of the seed oils of four safflower (Carthamus tinctorius) varieties grown in north-eastern of Morocco Int. J. Food Sci. Technol. 50 804–10
[13] BACON W L, BROWN K I and MUSSER M A 1980 Changes in plasma calcium, phosphorus, lipids, and estrogens in turkey hens with reproductive state Poult. Sci. 59 444–52
[14] Widiastuti E L 2016 Pengaruh Penambahan Sargassum sp. dan Inositol dalam Pakan terhadap Pertumbuhan dan Daya Tahan Juvenil Ikan Gurami (Osphronemus gouramyLac.) J. Natur Indones. 16 72–8
[15] Rahmat D and Wiradimadja R 2011 Pendugaan Kadar Kolesterol Daging dan Telur Berdasarkan Kadar Kolesterol Darah pada Puyuh Jepang (Estimated Cholesterol Levels Meat and Egg Based on Blood Cholesterol on the Japanese Quail) J. Ilmu Ternak Univ. Padjadjaran 11
[16] Wijayanti R and Rahayuni A 2014 Pengaruh pemberian serbuk biji labu kuning (Cucurbita moschata) terhadap penurunan kadar trigliserida darah pada tikus wistar jantan yang diberi diet tinggi lemak
[17] Setiawati T, Atmomarsono U and Dwiloka B 2016 Kadar Lemak Dan Profil Asam Lemak Jenuh, Asam Lemak Tak Jenuh Daging Ayam Broiler Dengan Pemberian Pakan Mengandung Tepung Daun Kayambang (Salvinia molesta) J. Teknol. Has. Pertan. 9
[18] Putri S R and Anggraini D I 2015 Obesitas sebagai faktor resiko peningkatan kadar trigliserida J. Major. 4 78–82
[19] Susanti R and Hidayat E 2016 Profil Protein Susu Dan Produk Olahannya J. MIPA 39 98–106
[20] Nurhayati N, Berliana B and Nelwida N 2020 Massa Protein dan Lemak Daging Dada pada Ayam Broiler yang Mengkonsumsi Ransum Mengandung Bawang Hitam Sains Peternak. J. Penelit. Ilmu Peternak. 18 15–22

[21] Iriyanti Z, Yuwanto T and Sunarjo K 2005 Pengaruh penggunaan minyak lemuru dan minyak kelapa sawit dalam pakan terhadap profil metabolisme lemak pada darah ayam kampung jantan Anim. Prod. 7 67–73
[22] Zahra A A and Edjeng Supriyatna B 2014 Pengaruh pemberian pakan sorgum dan kulit pisang terhidrolisis dengan NaOH terhadap lemak dan kolesterol ayam broiler Agromedia 32
[23] Soeparno S 2011 Ilmu Nutrisi dan Gizi Daging (Yogyakarta, Indonesia: Gadjah Mada University Press.)
[24] Nurillah S 2011 Pengaruh pemberian tepung kaki ayam broiler sebagai subtitusi tepung ikan di dalam ransum terhadar kadar kolesterol dan asam lemak pada kuning telur ayam arab (Gallus turcicus)
[25] Sigarlaki E D and Tjiptaningrum A 2016 Pengaruh Pemberian Buah Naga Merah (Hylocereus polyrhizus) terhadap Kadar Kolesterol Total J. Major. 5 14–7
[26] Sartika R A D 2008 Pengaruh asam lemak jenuh, tidak jenuh dan asam lemak trans terhadap kesehatan Kesmas Natl. Public Heal. J. 2 154–60
[27] Sobari R N 2014 Hubungan Asupan Asam Lemak Jenuh dan Tak jenuh Dengan Kadar Kolesterol HDL Pada Pasien Penyakit Jantung Koroner Di RSUD Dr. Moewardi Naskah Publ.
[28] Mayes P A 2003 Sintesis, pengangkutan dan ekskresi kolesterol Dalam Murray RK, Ganner DK, Mayes PA, Rodwell VW, Ed. Biokimia harper 25th ed. Jakarta EGC
[29] Widyamanda L P, Ismadi V D Y B and Estiningdriati I 2013 Pengaruh penambahan bangle (Zingiber cas-sumunar) dalam ransum terhadap total lipid dan kolesterol hati pada ayam broiler Anim. Agric. J. 2 183–90
[30] Koolman J and Röhm K-H 2005 Color atlas of biochemistry (Thieme Stuttgart)
[31] Lee S A and Bedford M R 2016 Inositol-An effective growth promotor? Worlds. Poult. Sci. J. 72 743–60
[32] Croze M L and Soulage C O 2013 Potential role and therapeutic interests of myo-inositol in metabolic diseases Biochimie 95 1811–27
[33] Bevilacqua A and Bizzarri M 2018 Inositols in insulin signaling and glucose metabolism Int. J. Endocrinol. 2018
[34] Ujani S 2016 Hubungan antara usia dan jenis kelamin dengan kadar kolesterol penderita obesitas rsud abdul moeloek provinsi lampung J. Kesehat. 6
[35] Yuniastuti A, Susanti R and Iswari R S 2018 Efek Infusa Umbi Garut (Marantha arundinaceae L) Terhadap Kadar Glukosa dan Insulin Plasma Tikus yang Diinduksi Streptozotocyn J. Mipa 41 34–9
[36] Afandi M R and Marpaung F R 2019 Correlation Between Apoprotein B/apoprotein Ai Ratio with Homa Ir Value (Homeostatic Model Assesment Insulin Resistance) in Type 2 Diabetes Mellitus J. Vocat. Heal. Stud. 3 78–82

IOP Conf. Series: Earth and Environmental Science 1041 (2022) 012063 doi:10.1088/1755-1315/1041/1/012063

The effects of nano encapsulated *Syzygium Polyanthum* leaves extract inclusion in drinking water

A N 'Afifah, Zuprizal, N D Dono*

Universitas Gadjah Mada, Yogyakarta, Indonesia

*nanungdd@ugm.ac.id

Abstract. A study with six treatments in a Completely Randomized Design was conducted to evaluate the effects of nano encapsulated Syzygium polyanthum leaves extract (SpLE) inclusion in drinking water on small intestinal digesta pH and macromorphology in broiler chickens. The treatments were: drinking water without any addition (negative control), drinking water with 50 ppm antibiotic Tetracycline addition (positive control), or drinking water with1% SpLE, 3% SpLE, 1% nano-capsulated SpLE (N-SpLE), and 3% N-SpLE additions. Each experimental treatment was repeated four times with eight birds per replicate pen. The basal diet and drinking water were provided from days 8 to days 35. One-way ANOVA was used to calculate the effect of the treatments. The difference data in present study were separated with Duncan's new Multiple Range Test. Result showed no treatment effect was detected on macromorphology of all segments on the small intestine. However, drinking water inclusion with 3% SpLE reduced digesta pH ($P<0.05$) in the duodenum and jejunum. Inclusion of 3% N-SpLE reduced digesta pH only in the ileum ($P<0.05$). Reductions of digesta pH in some segments of the small intestinal might shows beneficial effect of drinking water inclusion with Syzygium polyanthum leaves extract on the intestinal health improvement in broiler chickens.

1. Introduction

Phytobiotics are compounds derived from plants added to animal feed to increase livestock productivity through increased digestibility, immunity, nutrient absorption, and elimination of pathogens in the intestines of animals [1]. In the last 20 years, phytobiotics have taken a significant role in animal husbandry as good promoters of growth and can be used to reduce losses caused by antibiotics use reduction or ban [2]. Phytobiotic could be used to serve as feed additives due to lower cost of production, minimum health hazards, reduced risk of toxicity, and environment-friendliness [3].

Recent research about phytobiotic as feed additive have shown encouraging results with regards to weight gain, feed efficiency, intestinal morphology, intestinal microbe, lowered mortality, and increased liveability in poultry [4]. Besides of antimicrobial activity, phytobiotics have a strong antioxidant and anti-inflammatory activities [5]. Phytobiotics can be derived from various ingredients such as extracts of leaves, flowers, fruit, twigs, roots, bark, wood, and plants. The active ingredients contain a wide variety of secondary metabolites with various physiological effects [6]. One of the plants that contain various active compounds so that it has the potential as a phytobiotic is bay leaf (Syzygium polyanthum). Bay leaf (Syzygium polyanthum), which is included in the myrtaceae family, is a spice or medicinal plant that has long been known and developed widely in Indonesia [7]. Bay leaves contain flavonoids, tannins, 0.05% essential oils consisting of eugenol and citral, triterpenoids, alkaloids, and steroids [8].

Bay leaves contain secondary metabolites which have many pharmacological activities in overcoming various diseases [9].

The synergistic effect between secondary metabolites causes pharmacological effects. In addition, secondary metabolites have polyvalent activity, making it possible to overcome various diseases [10]. The active compounds in bay leaves have various pharmacological activities such as; antifungal [11], immunomodulator [12], antimicrobial [13], and antioxidant [14]. However, some of the bioactive compounds in bay leaves are less soluble in water and are easily damaged by influences in the digestive tract environment so that they have low absorption [15].

Various ways to increase the potency of herbal bioactive compounds have been carried out, and one of the formulation strategies that have been developed is to make them in the form of nanoparticles [16]. While the encapsulation process is a technique for coating or coating an active material with a polymer wall layer to produce small micro or nano sized particles [17]. Currently, nanoparticles based on the natural polymer such as chitosan have been widely applied as a drug delivery system due to their specific properties such as biodegradable, mucoadhesive, nontoxic, and enhanced intestinal permeation [18].

Ionic gelation is one of the methods in making nanoparticle preparations. The principle of this method is the occurrence of ionic interactions between the positively charged amino groups in chitosan and the negatively charged polyanions. The most widely used polyanion crosslinker is sodium tripolyphosphate (STPP), because it is non-toxic and multivalent [19]. Therefore, the purpose of this study was to investigate the beneficial effects of nano encapsulated Syzygium polyanthum leaves extract (SpLE) inclusion in drinking water on small intestinal digesta pH and macromorphology in broiler chickens.

2. Methodology

2.1 Animal, diets, and experimental design

The research was conducted in a closed house at the Faculty of Animal Husbandry, Gadjah Mada University, Yogyakarta, Indonesia. One hundred and ninety-two New Lohmann MB 202 broilers aged eight days were plotted into six treatment groups: drinking water without any addition (negative control), drinking water with 50 ppm antibiotic Tetracycline addition (positive control), or drinking water with 1% SpLE , 3% SpLE, 1% nano-capsulated SpLE (N-SpLE), and 3% N-SpLE additions. The treatment was repeated four times with eight birds in each replication cage. The feed formulation and chemical composition for the experimental feed are shown in Table 1. The feed and drinking water were provided ad libitum for 35 days of the experiment.

2.2 Preparation and extraction of bay leaves

Fresh bay leaves are collected from Sleman Regency, Yogyakarta, Indonesia. Leaf samples were dried in an oven at 55 ° C for 2 days. Dry samples were ground using a Thomas-Willey tool (Arthur H. Thomas Co., Philadelphia, USA) with a size of 1 mm. Two hundred grams of mindi leaf powder are put into a beaker and 1000 ml of ethanol (70%) are added. The mixture is macerated for 3 days and stirred every 24 hours. After 3 days, all the macerates were filtered, and the filtrate was evaporated with a water bath. The thick extract is then diluted with distilled water to 2%.

2.3 Preparation of nano-encapsulation bay leaves extract: chitosan: STPP

Nanoencapsulation was processed using the ionic gelation method by mixing bay leaf extract: chitosan: STPP with the optimum formulation (1: 4: 1/90) w/v. Two grams of chitosan were added to 100 ml of acetic acid (1% w/v) and dissolved 0.04 ml of STPP in distilled water and stirred for one hour or until it appeared clear. The bay leaf extract and chitosan were mixed with the optimum formulation (chitosan was dissolved in a pH 4 acetic acid solution) and stirred using a magnetic stirrer for one hour. STPP was then added and stirred for 30 minutes. After the solution became clear, the sediment was separated and the nanoencapsulation solution was used in the in vivo study.

Table 1. Ingredient's composition and nutrients calculation of the experimental basal diet.

Items	Contents
Ingredient's composition, %	
Yellow corn	56.00
Soybean meal	28.50
Meat bone meal	6.00
Rice bran	2.00
Palm oil	5.00
Premix vitamin*	0.20
L-Lysine HCl	0.30
DL-Methionine	0.30
CaCO3	1.50
NaCl	0.20
Total	100.00
Calculated nutrients and energy, %	
Crude Protein	21.26
Metabolizable energy, kcal/kg	3106.44
Crude fiber	3.52
Available phosphorus	0.63
Calcium	1.25
Lysine	1.30
Methionine	0.57

*Premix vitamin (Masamix-Bro) supply per kg diet: Vit. A: 12.500.000 IU; Vit. D3: 2.500.000 IU; Vit. E: 10000 mg; Vit. K3: 2000 mg; Vit. B1: 2000 mg; Vit. B2: 4000 mg; Vit. B6: 1000 mg; Vit. B12: 12000 mcg; Vit.C: 40000 mg; Niacin: 40000 mg; Biotin: 200 mg.

2.4 Intestinal pH and macromorphology
On day 35, one bird was selected with body weight close to the median weight of each cage. The selected bird would be slaughter and surgery according to SNI 99002: 2016 concerning halal slaughter in poultry. Furthermore, the pH and intestinal macromorphology (weight and length) were measured. The digesta pH of 3 intestinal segments (duodenum, jejunum, and ileum) was determined immediately by inserting the tip of a pH meter needle (PT 370, Boeco, Hamburg, Germany). Weight of intestinal was meisured with analytical scales (PM 460, Mettler, Greifensee, Switzerland) and lenght of intestinal was meisured with (Yamayo Phiband, Yamayo, Fukushima, Japan).

2.5 Data analysis
Data were analyzed statistically using one-way ANOVA with a completely randomized design using the SPSS 22 application (SPSS GmbH, Munich, Germany). Data with significant differences with a probability of less than 5% were then tested using Duncan's New Multiple Range Test method.

3. Result and Discussion
There was a significant effect of 1% and 3% SpLE, 1% and 3% N-SpLE supplementation on the pH value of the small intestine (duodenum, jejunum, and ileum). The value of intestinal pH was shown in Table 2. On the other hand, there was no treatment effect that was detected on the macromorphology of all segmen on the small intestine. The value of weight and lenght of the small intestine were shown in Table 3 and Table 4. Table 1.

Table 2. Effect of SpLE and N-SpLE suplementation on intestinal pH[1]

Treatments[2]	Intestinal pH		
	Duodenum	Jejunum	Ileum
Water	5.54±0.48[b]	5.89±0.46[ab]	6.40±0.18[ab]
Tetracycline	4.99±0.55[ab]	6.15±0.47[b]	6.63±0.31[b]
1% SpLE	5.52±0.51[b]	6.01±0.24[b]	6.18±0.30[ab]
3% SpLE	4.81±0.13[a]	5.50±0.18[a]	6.05±0.39[a]

1% N-SpLE	5.56±0.26[b]	6.16±0.09[b]	6.27±0.24[ab]
3% N-SpLE	5.18±0.37[ab]	6.03±0.22[b]	5.98±0.30[a]
SEM	0.13	0.06	0.03
P-*value*	0.075	0.073	0.055

[a,b] Different superscripts within the same column show significant difference (P<0.05).
[1]Means represent 4 pens of 8 birds each per treatment.
[2]Water=negative control (NC), Tetracycline=positive control, NC+1% SpLE, NC+3% SpLE, NC+ 1% N-SpLE, NC+ 3% N-SpLE.

Table 2 shows that the addition of SpLE and N-SpLE can reduce the pH value of all parts of the small intestine (duodenum, jejunum, and ileum). The addition of 3% SpLE decreased (P <0.05) the duodenal pH value. The decrease in pH value was then found in treatment with the addition of Tetracycline, 3% N-SpLE, 1% SpLE, water, and 1% N-SpLE. The addition of 3% SpLE decreased (P <0.05) the pH value of the jejunum. The decrease in pH value was then found in the treatment with the addition of water, 1% SpLE, 3% N-SpLE, Tetracycline, and 1% N-SpLE. The addition of 3% N-SpLE decreased (P <0.05) the pH value of the ileum.

The decrease in pH value was then found in the treatment with the addition of 3% SpLE, 1% SpLE, 1% N-SpLE, water, and Tetracycline. Decreasing pH value will make the digestive tract especially the small intestine more acidic condition. The acidic condition of the small intestine will reduce the growth of pathogenic bacteria, so that it can improve the condition of the digestive tract and its nutrient digestibility which causes the feed rate in the small intestine to get better in the process of nutrient absorption [20][21].

The decrease in pH value with the addition of 3% SpLE in the duodenum was lower than the water treatment, Tetracycline, 1% SpLE, 1% N-SpLE, and 3% N-SpLE. This is probably due to the content of bioactive compounds from bay leaves in the form of high doses of total phenols, flavonoids, tannins and alkaloids. The duodenum is the first part of the small intestine that performs digestion and absorption of nutrients and is located very close to the body wall and is bound to the short mesoduodenum [22].

The decrease in pH value with the addition of 3% SpLE in the jejunum was lower than the water treatment, Tetracycline, 1% SpLE, 1% N-SpLE, and 3% N-SpLE. This is probably due to the high dosage of bioactive compounds in order to maximize the decrease in pH value. Jejunum is the part of the small intestine which is the main absorption site for nutrients[23]. The decrease in pH value with the addition of 3% N-SpLE in the ileum was lower than that of water, Tetracycline, 1% SpLE, 3% SpLE, and 1% N-SpLE treatment. The ileum is the end of the small intestine which functions to absorb some of the end nutrients [24].

Table 3. Effect of SpLE and N-SpLE suplementation on intestinal weight[1]

Treatments[2]	Weight of Intestinal		
	Duodenum	Jejunum	Ileum
Water	7.58±1.53	15.52±4.67	9.66±2.89
Tetracycline	8.81±0.57	15.06±2.30	13.74±2.72
1% SpLE	8.80±2.52	17.88±5.84	12.86±1.31
3% SpLE	7.94±0.33	17.53±2.34	11.43±2.75
1% N-SpLE	8.11±1.29	15.68±2.38	12.62±3.12
3% N-SpLE	8.22±1.29	17.93±2.27	13.32±1.50
SEM	0.20	0.64	0.61
P-*value*	0.865	0.746	0.252

[a,b] Different superscripts within the same column show significant difference (P<0.05).
[1]Means represent 4 pens of 8 birds each per treatment.
[2]Water=negative control (NC), Tetracycline=positive control, NC+1% SpLE, NC+3% SpLE, NC+ 1% N-SpLE, NC+ 3% N-SpLE.

Table 3 shows that the addition of SpLE and N-SpLE has no significant effect on the weight of the small intestine (duodenum, jejunum, and ileum). The absence of the effect of the addition of SpLE and N-SpLE in drinking water to the weight of the small intestine was probably due to the crude fiber content consumed by broiler chickens the same between treatments. The weight of the small intestine is influenced by the nutrient content in the ration, high coarse fiber feed ingredients in the ration can increase the weight of the small intestine [25].

In addition, the absence of the effect of the addition of SpLE and N-SpLE on the weight of the small intestine is suspected because the treatment was not fresh so that the active substance decreased and experienced oxidation which could have a negative effect [26].

Table 4. Effect of SpLE and N-SpLE suplementation on intestinal lenght[1]

Treatments[2]	Lengt of Intestinal		
	Duodenum	Jejunum	Ileum
Water	26.25±2.99	58.75±8.38	60.00±11.46
Tetracycline	28.75±5.12	64.50±6.76	62.50±11.73
1% SpLE	27.88±1.31	4.25±10.24	59.00±9.56
3% SpLE	28.00±1.63	66.00±2.94	59.25±7.93
1% N-SpLE	28.25±1.26	60.50±5.57	59.88±9.63
3% N-SpLE	28.38±2.21	66.03±6.47	68.38±7.18
SEM	0.36	1.23	1.47
P-*value*	0.845	0.614	0.740

[a,b] Different superscripts within the same column show significant difference ($P<0.05$).
[1]Means represent 4 pens of 8 birds each per treatment.
[2]Water=negative control (NC), Tetracycline=positive control, NC+1% SpLE, NC+3% SpLE, NC+ 1% N-SpLE, NC+ 3% N-SpLE.

Table 4 shows that the addition of SpLE and N-SpLE has no significant effect on the length of the small intestine (duodenum, jejunum, and ileum). The absence of the effect of the addition of SpLE and N-SpLE in drinking water to the length of the small intestine was probably caused by the same crude fiber content consumed by broilers between treatments [27]. In addition, improper dosing may cause no real effect. This is in accordance with research conducted by Lisnanti et al. [28] that the addition of ant nest extract (*Myrmecodia sp*) using drinking water media statistically variance also did not significantly affect the intestinal length of broiler chickens. Giving too many doses is likely to cause intestinal inflammation in chickens.

4. Conclusion
Reductions of digesta pH in some segments of the small intestinal might shows beneficial effect of drinking water inclusion with Syzygium polyanthum leaves extract on the intestinal health improvement in broiler chickens.

Acknowledgments
This project was made possible by financial support from the RTA Fund of Gadjah Mada University in 2020.

References
[1] Singh R and Mandal A B 2018 Efficacy of vitamin C in ameliorating ochratoxicosis in broiler chicken Indian J. Anim. Nutr. 35 436
[2] Petričević V, Lukić M, Škrbić Z, Rakonjac S, Dosković V, Petričević M and Stanojković A 2018 The effect of using rosemary (Rosmarinus officinalis) in broiler nutrition on production parameters, slaughter characteristics, and gut microbiological population Turkish J. Vet. Anim.

Sci. 42 658–64
[3] Natsir M H, Hartutik, Sjofjan O and Widodo E 2013 Effect of either powder or encapsulated form of garlic and Phyllanthus niruri L. mixture on broiler performances, intestinal characteristics and intestinal microflora Int. J. Poult. Sci. 12 676–80
[4] Puvača N, Stanaćev V, Glamočić D, Lević J, Perić L, Stanaćev V and Milić D 2013 Beneficial effects of phytoadditives in broiler nutrition Worlds. Poult. Sci. J. 69 27–34
[5] Hashemzadeh-Cigari F, Khorvash M, Ghorbani G R, Kadivar M, Riasi A and Zebeli Q 2014 Effects of supplementation with a phytobiotics-rich herbal mixture on performance, udder health, and metabolic status of Holstein cows with various levels of milk somatic cell counts J. Dairy Sci. 97 7487–97
[6] Utami M M D and Dadik Pantaya 2016 Penggunaan Ekstrak Bawang Putih dalam Pakan terhadap Performans Ayam Broiler Tropis Fase Starter. In: Proc. National Seminar on Research and Community Service Results (Jember, Indonesia) pp 72–5
[7] Utami T P A and Sumekar D W 2017 The Effectiveness Salam' s Leaf (Sizygium polyantha) as Antihypertensive in Wistar' s Rat Majority 6 77–81
[8] Arhiono H N P, Suada I K and Budiasa K 2018 Pengaruh Infusa Daun Salam (Syzygium Polyanthum) Terhadap Kualitas Daging Kambing Pada Suhu Ruang Indones. Med. Veterinus 7 664–74
[9] Heinrich M, Barnes J, Prieto-Garcia J, Gibbons S and Williamson E M 2017 Fundamentals of Pharmacognosy and Phytotherapy E-Book (Elsevier Health Sciences)
[10] Bone K, Simon Mills M and FNIMH M A 2012 Principles and practice of phytotherapy: modern herbal medicine (Elsevier Health Sciences)
[11] Fitriani A, Hamdiyati Y and Engriyani R E R 2012 Aktivitas antifungi ekstrak etanol daun salam (Syzygium polyanthum (Wight) Walp.) terhadap pertumbuhan jamur Candida albicans secara in vitro Biosfera 29 71–9
[12] Sulistiyani, Falah S, Wahyuni W T, Sugaraha T, Tachibana S and Syaefudin 2014 Cellular Mechanism of the Cytotoxic Effect of Extracts from Syzygium polyanthum Leaves Am. J. Drug Discov. Dev. 4 90–101
[13] Warnida H and Sukawaty Y 2016 Efekivitas ekstrak etanol daun salam (Syzygium polyanthum (Wight) Walp.) sebagai pengawet alami antimikroba J. Ilm. Ibnu Sina 1 227–34
[14] Rahman N, Bahriul P and Diah A W M 2014 Uji aktivitas antioksidan ekstrak daun salam (Syzygium Polyanthum) dengan menggunanakan 1, 1-Difenil-2-Pikrilhidrazil J. Akad. Kim. 3 143–9
[15] Aziz T, Yuanita and Susanti 2010 Ekstraksi Eugenol Dari Daun Salam India J. Tek. Kim. Nomor 3 17 17–28
[16] Suryani, Wahyuni, Ariastika D and Rahmanpiu 2016 Formulasi Nanopartikel Kurkumin dengan Teknik Gelasi Ionik Menggunakan Kitosan, Tripolifosfat dan Natrium Alginat serta Uji Stabilitasnya Secara In Vitro Pharmauho 2 17–21
[17] Yunilawati R, Yemirta Y, Cahyaningtyas A A, Aviandharie S A, Hidayati N and Rahmi D 2018 Optimasi Proses Spray Drying Pada Enkapsulasi Antosianin Ubi Ungu J. Kim. dan Kemasan 40 17
[18] Mardliyati E, Muttaqien S El and Setyawati D R 2012 Sintesis Nanopartikel Kitosan- Trypoly Phosphate Dengan Metode Gelasi Ionik : Pengaruh Konsentrasi Dan Rasio Volume Terhadap Karakteristik Partikel Pros. Pertem. Ilm. Ilmu Pengetah. dan Teknol. Bahan 90–3
[19] Gredi J, Taurina W and Andrie M 2017 Analgesic Eff ectivty Of Nanoparticles Chitosan-Ethanol Leaf Extract Papaya (Carica Papaya L .) In White Male Mice (Mus Mucculus) J. ilmu Kefarmasian Indones. 15 228–34
[20] Puspasari D R M M and Istna Mangisah 2014 Pengaruh Level Protein dan Asam Asetat dlam Ransum Terhadap Tingkat Keasaman (pH) Usus Halus, Laju Digesta dan Bobot Badan Akhir Ayam Broiler Anim. Agric. J. 3(3) 3 409–16
[21] Faradila S, Suthama N and Sukamto B 2016 Combination inulin of dahlia variabilis tuber-

Lactobacillus sp optimizing the development of intestinal microflora and growth Pelung-Leghorn chicken crosses J. Vet. 17 168–75

[22] Badrussalam A and Yudiarti T 2020 Pengaruh Penggunaan Aditif Kunyit terhadap Bobot Relatif Organ Pencernaan Ayam Kampung Super Effect of Utilising Turmeric Additive on Relative Weight of Digestive Organs in Super Kampung Chicken Rancangan penelitian 15 273–9

[23] Yunita M N, Agustono B and Saputro A L 2020 Gambaran Histologi Lebar Vili Jejunum Kelinci Jenis Rex dengan Subtitusi Bahan Penyusun Tepung Ikan dengan Tepung Teritip (Cirripedia sp.) J. Nutr. Ternak Trop. 3 30–4

[24] Satimah S, Yunianto V D and Wahyono F 2019 Bobot Relatif dan Panjang Usus Halus Ayam Broiler yang Diberi Ransum Menggunakan Cangkang Telur Mikropartikel dengan Suplementasi Probiotik Lactobacillus sp. J. Sain Peternak. Indones. 14 396–403

[25] Mistiani S, Kamil K A and Rusmana D 2020 Pengaruh tingkat pemberian ekstrak daun burahol (Stelechocarpus burahol) dalam ransum terhadap bobot organ dalam ayam broiler J. Nutr. Ternak Trop. dan Ilmu Pakan 2 42–50

[26] Ketron A C, Gordon O N, Schneider C and Osheroff N 2013 Oxidative metabolites of curcumin poison human type II topoisomerases Biochemistry 52 221–7

[27] Ananda A P, Mangisah I and Wahyono F 2017 Pengaruh Pemberian Tepung Umbi Dahlia, Probiotik Dan Kombinasinya Dalam Ransum Terhadap pH, Bobot Dan Panjang Relatif Saluran Pencernaan Itik Tegal Bul. Ilmu-Ilmu Pertan. 10

[28] Lisnanti E F, Fitriyah N and Anwar M R M 2018 The effect of addition ant nest exstract (Myrmecodiasp) toward the percentage of carcass and long intestine broiler phase finisher 1 60–8

IOP Conf. Series: Earth and Environmental Science 1041 (2022) 012064 doi:10.1088/1755-1315/1041/1/012064

Effects of plant extracts against gastrointestinal nematodes of livestock: a meta-analysis

A Jayanegara*, D M Sinaga, E B Laconi

IPB University, Bogor, Indonesia

*anuraga.jayanegara@gmail.com

Abstract. Infestation of gastrointestinal nematodes to livestock may cause a loss in production performance and under severe condition may lead to death. This study aimed to evaluate the effects of plant extract additions on larval inhibition and mortality of gastrointestinal nematodes. A total of nine articles that comprised of 16 data points were integrated into a database. For each data point, both control and plant extract treatments were specified that included data on number of samples, mean and standard deviation. Effect size of the meta-analysis was based on Hedges' d with the continuous random effects of DerSimonian-Laird method. Results showed that addition of plant extracts had a strong effect to inhibit larval of gastrointestinal nematodes with a cumulative effect size (95% confidence interval) of 36.6 (30.5-42.7, P<0.001). All papers supported such larval inhibition effect of plant extracts except one paper only. Further, plant extracts increased adult mortality of gastrointestinal nematodes with a cumulative effect size of 73.1 (44.6-101.7, P<0.001). In conclusion, addition of plant extracts that rich in bioactive compounds are effective to control the population of gastrointestinal nematodes. The use of plant extracts for such purpose is favorable since it is environmentally friendly.

1. Introduction

Livestock health has always been an important factor to achieve an optimal animal production, and one of which is the health of the digestive tract of livestock. The health of the digestive tract itself is an absence of disease or complaint regarding digestive tract disorders. Accordingly, prevention or avoidance of disease of the digestive tract is an integral part of livestock health, particularly from various pathogens that may cause clinical or subclinical disease, death, morbidity of livestock, and economic loss. Among the gastrointestinal pathogens, infestation of nematodes to livestock may cause a loss in production performance and under severe condition may lead to death [1]. Furthermore, parasitic infection may cause depression of livestock immunity and may lead to the emergence of various other diseases due to subsequent infection by virus, bacteria and/or protozoa.

For several decades, the control of parasites in the gastrointestinal tract has utilized various synthetic anthelmintic compounds. However, increasing resistance of the parasites to the synthetic compounds has emerged due to the intensive use and inappropriate doses [2]. On the other hand, currently there is an increasing consumer demand for healthy and even organic animal products. This has led to the search of potential alternative solutions to control gastrointestinal pathogens, particularly those of natural bioactive compounds such as tannins, flavonoids, saponins and essential oils [3].

In practice, it is often uneasy to use pure compounds of those natural bioactive compounds and therefore the crude extracts of plants are typically used. This study aimed to evaluate the effects of plant extract additions on larval inhibition and mortality of gastrointestinal nematodes by employing a meta-analysis method, in which data are integrated from previously published studies.

2. Methodology

A literature search was performed in order to find various papers investigating the effects of plant extracts on nematodes responsible as parasites in the digestive tracts of ruminants. The search was conducted by using online databases, i.e., Science Direct and Google Scholar. The keywords used in the literature search were “plant extract”, "tannin", "nematode", and "ruminant”. Initially, more than 200 papers were retrieved based on the keywords screening.

After title, abstract and full text evaluation, a total of nine articles [4-12] that comprised of 16 data points were integrated into a database. For each data point, both control and plant extract treatments were specified that included data on number of samples, mean and standard deviation. Effect size of the meta-analysis was based on Hedges’ d with the continuous random effects of DerSimonian-Laird method. Data were presented as forest plots for the parameters of larval population and adult mortality, performed by using the open-source software for meta-analysis in ecology and evolutionary biology (OpenMEE) [13].

3. Result and Discussion

Forest plots on the larval inhibition and adult mortality of gastrointestinal nematodes by plant extracts are presented in Figure 1 and Figure 2, respectively. Addition of plant extracts had a strong effect to inhibit larval of gastrointestinal nematodes with a cumulative effect size (95% confidence interval) of 36.6 (30.5-42.7, P<0.001, Figure 1). All papers supported such larval inhibition effect of plant extracts except one paper only. Further, plant extracts increased adult mortality of gastrointestinal nematodes with a cumulative effect size of 73.1 (44.6-101.7, P<0.001, Figure 2).

The results suggest that plant extracts containing tannins are effective measures to control the population of gastrointestinal nematodes. Tannins have been known to possess toxicity effects on various organisms such as bacteria, protozoa and nematode [14]. Their toxicity effects are typically related to the ability to precipitate proteins, both structural and functional proteins [15]. If tannins bind to digestive enzymes of nematodes, such binding may reduce the activity of the enzymes to degrade macromolecules and, in turn, may decrease the activity and/or population of nematodes. Supporting the present findings, tannins had been reported to reduce nematode population, worm fecundity, fecal egg count and egg output in ruminants [16].

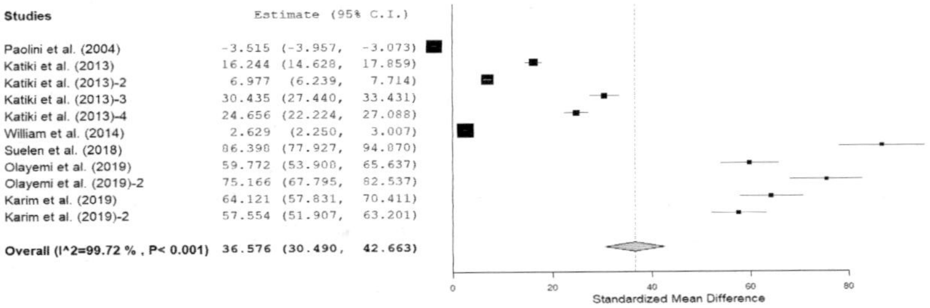

Figure 1. Forest plot on the larval inhibition of gastrointestinal nematodes by plant extracts.

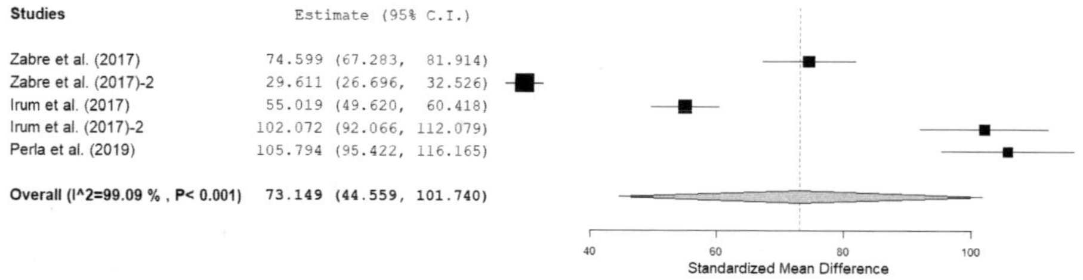

Figure 2. Forest plot on the adult mortality of gastrointestinal nematodes by plant extracts.

4. Conclusion

Addition of plant extracts that rich in bioactive compounds are effective to control the population of gastrointestinal nematodes. The use of plant extracts for such purpose is favorable since it is environmentally friendly.

Acknowledgement

All authors are grateful to the Ministry of Research and Technology, Republic of Indonesia, for providing a research grant through the "Hibah Penelitian Dasar", year 2020, grant number 1/AMD/E1.KP.PTNBH/2020.

References

[1] Haile A, H Hassen, H Gatew, T Getachew, RNB Lobo and B Rischkowsky. 2018. Investigations into nematode parasites of goats in pastoral and crop livestock systems of Ethiopia. Tropical Animal Health and Production. 50:643-650.
[2] Garcia-Bustos JF, BE Sleebs and RB Gasser. 2019. An appraisal of natural products active against parasitic nematodes of animals. Parasites Vectors. 12:306.
[3] Santos FO, APM Cerqueira, A Branco, MJM Batatinha and MB Botura. 2019. Anthelmintic activity of plants against gastrointestinal nematodes of goats: A review. Parasitology. 146:1233-1246.
[4] Paolini V, I. Fouraste and H. Hoste. 2004. In vitro effects of three woody plant and sainfoin extracts on 3rd-stage larvae and adult worms of three gastrointestinal nematodes. Parasitology. 129:69-77.
[5] Katiki LM, JFS Ferreira, JM Gonzalez, AM Zajac, DS Lindsay, ACS Chagas and AFT Amarante. 2013. Anthelmintic effect of plant extracts containing condensed and hydrolyzable tannins on Caenorhabditis elegans, and their antioxidant capacity. Veterinary Parasitology. 192:218-227.
[6] Williams AR, C Fryganas, A Ramsay, I Mueller-Harvey and SM Thamsborg. 2014. Direct anthelmintic effects of condensed tannins from diverse plant sources against Ascaris suum. PLoS ONE. 9:e97053.
[7] Soares SCS, GC de Lima, AC Laurentiz, A Feboli, LA dos Anjos, MSP Carlis, RS Filardi and RS de Laurentiz. 2018. In vitro anthelmintic activity of grape pomace extract against gastrointestinal nematodes of naturally infected sheep. International Journal of Veterinary Science and Medicine. 6:243-247.
[8] Olayemi DO, MM Onakpa and OC Jegede. 2019. Anthelmintic activity of Hymenodictyon pachyanta stem bark extracts against Haemonchus contortus. Folia Veterinaria. 63:30-36.
[9] Karim MA, MR Islam, MA Lovelu, SF Nahar, PK Dutta and MH Talukder. 2019. In vitro evaluation of anthelmintic activity of tannin-containing plant Artemisia extracts against Haemonchus contortus from goat. Journal of Bangladesh Agricultural University. 17:363-368.

IOP Conf. Series: Earth and Environmental Science 1041 (2022) 012064 doi:10.1088/1755-1315/1041/1/012064

[10] Zabre G, A Kabore, B Bayala, LM Katiki, LM Costa-Junior, HH Tamboura, AMG Belem, AL Abdalla, V Niderkorn, H Hoste and H Louvandini. 2017. Comparison of the in vitro anthelmintic effects of Acacia nilotica and Acacia raddiana. Parasite. 24:44.

[11] Irum S, H Ahmed, B Mirza, K Donskow-Lysoniewska, A Muhammad, M Qayyum and S Simsek. 2017. In vitro and in vivo anthelmintic activity of extracts from Artemisia parviflora and A. sieversiana. Helminthologia. 54:218-224.

[12] Acevedo-Ramirez PMC, C Hallal-Calleros, I Flores-Perez, F Alba-Hurtado, MB Mendoza-Garfias, NC del Campo and R Barajas. 2019. Anthelmintic effect and tissue alteration induced in vitro by hydrolysable tannins on the adult stage of the gastrointestinal nematode Haemonchus contortus. Veterinary Parasitology. 266:1-6.

[13] Wallace BC, MJ Lajeunesse, G Dietz, IJ Dahabreh, TA Trikalinos, CH Schmid and J Gurevitch. 2016. OpenMEE: Intuitive, open-source software for meta-analysis in ecology and evolutionary biology. Methods in Ecology and Evolution. 8:941-947.

[14] Jayanegara A, TUP Sujarnoko, M Ridla, M Kondo and M Kreuzer. 2019. Silage quality as influenced by concentration and type of tannins present in the material ensiled: A meta-analysis. Journal of Animal Physiology and Animal Nutrition. 103:456-465.

[15] Makkar HPS. 2003. Effects and fate of tannins in ruminant animals, adaptation to tannins, and strategies to overcome detrimental effects of feeding tannin-rich feeds. Small Ruminant Research. 49:241-256.

[16] Hoste H, F Jackson, S Athanasiadou, SM Thamsborg and SO Hoskin. 2006. The effects of tannin-rich plants on parasitic nematodes in ruminants. Trends in Parasitology. 22:253-261.

Coconut oil effects on in vitro ruminal fermentation and methanogenesis

A Cusiayuni*, RK Nurfatahillah, D Evvyernie, A Jayanegara and KG Wiryawan

IPB University, Bogor, Indonesia

*ayu_cusi@apps.ipb.ac.id

Abstract. Methane emission from ruminants can be mitigated with supplementation of medium chain fatty acids (MCFA) such as lauric acid that present in coconut oil. However, the effects of coconut oil supplementation on in vitro rumen fermentation and methane emission have been varied. This study was conducted to examine the effect of dietary coconut oil supplementation on rumen fermentation and methanogenesis by integrating data from various in vitro studies. A linear regression statistical model approach was applied to a total of 14 studies and 38 data points. The response variables integrated were methane emission, fermentation products, nutrient digestibility, and microbial population. The data obtained were analyzed by using IBM SPSS software version 20. The results showed that methane production was decreased significantly ($P<0.01$) by supplementation of coconut oil. Coconut oil tended to decrease methanogenic archaea, protozoa, total volatile fatty acids and isovalerate proportion ($P<0.1$). However, the oil supplementation did not significantly affect dry matter digestibility, total gas production, N-ammonia, acetate, isobutyrate, valerate, acetate to propionate ratio, population of rumen bacteria, organic matter digestibility, pH, propionate and butyrate. It can be concluded that coconut oil supplementation is able to mitigate methane emission by modifying microbial population structure and fermentation in the rumen.

1. Introduction

Dietary organic matter in the fermentative digestion process by rumen microbes will produce volatile fatty acids (VFA) and a number of gases including carbon dioxide (CO2) and hydrogen (H2). Furthermore, methanogenic archaea will reduce CO2 and H2 to form CH4 or methane gas [1]. The formation of methane gas in the rumen is the final result of fermentation of feed which is released through eructation and feces. Methane gas production from ruminants has been widely studied because it is considered to be one of the largest producers of methane gas and contributes 18% to global warming [2]. Efforts to reduce methane gas production can be solved by supplementing fat in the feed.

Coconut oil is a source of medium-chain fatty acids (MCFA) which contains lauric acid and has been widely studied before to reduce methane gas. According to Kim et al. [4], coconut oil can reduce methane gas in the rumen in vitro by 13% because the value of the acetate: propionate ratio decreases. The effect of coconut oil supplementation on in vitro fermentation parameters varies from several research data that have been conducted. Coconut oil supplementation can be influenced by several factors, one of which is the level of use of coconut oil added to feed. To predict and measure the value of the influence of one independent variable (independent) on the dependent variable (dependent) can use regression analysis.

The independent variable in this study is coconut oil which affects the dependent variable, those are the parameters of fermentation, digestibility, and microbial population. This study was therefore conducted to investigate the effect of coconut oil on rumen fermentation and methanogenesis in vitro by using a simple linear regression mathematical model approach from the research data that have been carried out.

2. Materials and Methods

The data were collected by searching several journals from various sources such as science direct and google scholar about coconut oil effects on in vitro ruminal fermentation and methanogenesis. The database was developed by integrating data from 38 studies that were derived from 14 articles. Various parameter data were recorded in Microsoft Excel 2013 such as oil level (%), protozoa (104/ml), bacterial population (109/ml), methanogenic bacteria (104/ml), total VFA (mmol/l), C2 (%), C3 (%), C4 (%), isoC4 (%), C5 (%), isoC5 (%), C2/C3, pH, NH3 (mmol/l), gas production (ml/g of substrate), methane gas (mmol/d), dry matter digestibility (%), and organic matter digestibility (%).

Table 1. Studies included in the data used to study the effects of coconut oil on fermentation and rumen methanogenesis in vitro

No.	Reference	Method	Diet	Oil level (%)
1	Cieslak et al. [5]	RUSITEC	Hay, wheat meal	0-5
2	Dohme et al. [6]	RUSITEC	Maize silage, concentrate	0-5.4
3	Dohme et al. [7]	RUSITEC	Maize silage, concentrate	0-5.3
4	Dong et al. [8]	RUSITEC	Hay, wheat	0-10
5	Kang et al. [3]	GBI	Cassava chips, rice brand, coconut, palm meals	0-2
6	Kim et al. [4]	GVI	Alfalfa, consentrate	0-3
7	Kongmun et al. [9]	GPT	Roughage, concentrate	0-8
8	Machmuller et al. [10]	RUSITEC	Hay, corn silage, concentrate	0-5
9	Machmuller et al. [11]	RUSITEC	Hay, corn silage, concentrate	0-6
10	Panyakaew et al. [12]	GVI	Corn silage, soybean meal	0-1.5
11	Patra et al. [13]	GPT	Corn silage, alfalfa, hay, concentrate	0-0.62
12	Son [14]	GPT	King grass, lamtoro leaves, and gamal leaves	0-2
13	Sitoresmi et al. [15]	HGT	Grass, concentrate	0-7.5
14	Yabuuchi et al [16]	GBI	Corn grains	0-5

Abbreviations: RUSITEC: rumen simulation technique; GBI: glass bottle incubation; GVI: glass vessel incubation; GPT: gas production technique; HGT: Hohenheim gas test.

Data were analyzed by using analysis of regression in which significance was declared when P-value less than 5%. Statistical analysis was carried out by using IBM SPSS version 20.

3. Result and Discussion

The data used to study the effects of coconut oil on in vitro rumen fermentation and methanogenesis were 38 oil level data with an average use of oil level (%) 2.53 ± 3.11. The results of descriptive statistics for parameters of digestibility, fermentation, and rumen microorganisms can be seen in Table 2.

Table 2. Descriptive statistics of data on the effects of coconut oil on rumen fermentation and methanogenesis in vitro

Parameter	n	Mean	SD	Min	Max
Oil level (%)	38	2.53	3.11	0.00	10.00
Microorganisms					
Protozoa (10^4/ml)	25	3.37	5.27	0.04	25.30
Bacteria (10^9/ml)	22	5.79	6.32	0.35	22.50
Methanogen (10^4/ml)	6	103.28	188.40	0.10	474.80
Rumen fermentation					
Total VFA (mmol/l)	30	91.91	36.86	30.10	182.00

C2 (%)	30	54.22	9.09	32.27	73.82
C3 (%)	29	23.96	5.74	12.37	37.70
C4 (%)	30	14.72	3.58	6.21	21.42
IsoC4 (%)	8	0.85	0.39	0.31	1.33
C5 (%)	12	2.44	0.94	0.75	4.00
IsoC5 (%)	9	0.72	0.57	0.53	2.30
C2/C3	29	2.26	0.53	1.00	3.40
NH_3 (mmol/l)	25	7.78	6.31	0.20	24.10
pH	27	6.63	0.42	5.69	7.20
Gas production (ml/g)	11	96.23	62.91	49.20	228.30
CH_4 gas (mmol/d)	13	5.18	2.87	0.49	10.00
Digestibility (%)					
DMD	9	50.09	17.89	22.60	72.60
OMD	10	41.66	14.23	20.48	58.80

Abbreviations: n: number of observations; SD: standard deviation; Min: minimum; Max: maximum; VFA: Volatile Fatty Acid; C2: acetate; C3: propionate; C4: butyrate; IsoC4: isobutyrate; C5: valerate; IsoC5: isovalerate; C2/C3: acetate/propionate ratio; NH_3: ammonia; CH_4: methane gas; DMD: Dry matter digestibility; OMD: Organic matter digestibility.

Coconut oil supplementation in feed tended to reduce (P<0.10; R2=0.11) the protozoa population in the rumen. Coconut oil which is rich in lauric acid is effective in reducing protozoa [17]. This is also supported by Newbold et al. [18] stated that, MCFA like lauric acid have strong anti-protozoa properties. Protozoa have an important role in rumen methanogenesis because some methanogenic archaea attached to the surface of the protozoa by 20-37%, consequently protozoa defaunation can indirectly reduce the number of methanogenic archaea [19]. Methanogens were also shown tended to decrease (P<0.10; R2=0.53) while the bacterial population were not influenced by the addition of coconut oil.

The effect of coconut oil supplementation in feed can interfere microbial activity on fiber digestibility. The oil will cover the feed fiber causing the rumen microbes cannot degrade the fiber and total VFA production will decrease. However, results of the present study did not support that theory. Coconut oil supplementation in feed tended to reduce (P<0.07; R2=0.11) total VFA production. The mean total VFA concentration was still within the normal range (Table 2). According to McDonald et al. [20], the normal value range of total VFA concentration is between 70-150 mM.

In general, acetate is the largest component of VFA at around 65%, 21% propionate, 14% butyrate [20], 1% isobutyrate, valerate and isovalerate below 3% [21]. Based on the data obtained, the effect of coconut oil is able to reduce the proportion of acetate, isobutyrate, valerate, and the ratio of C2/C3. and increase the proportion of propionate and butyrate (Table 3) although not significant. Isovalerate tended to be significant (P<0.09; R2=0.36) with coconut oil supplementation. This is because isovalerates use hydrogen for their formation [22].

Table 3. Regression equations of the effects of coconut oil on rumen fermentation and methanogenesis in vitro

Parameter	Intercept	Slope	P value	R^2
Microorganisms				
Protozoa (10^4/ml)	4.99	-0.61	<0.10	0.11
Bacteria (10^9/ml)	7.42	-0.59	NS	0.10
Methanogen (10^4/ml)	259.93	-30.61	<0.10	0.53
Rumen fermentation				
Total VFA (mmol/l)	100.78	-4.03	<0.07	0.11
C2 (%)	55.77	-0.60	NS	0.05
C3 (%)	23.23	0.28	NS	0.03
C4 (%)	14.63	0.04	NS	0.00
IsoC4 (%)	0.95	-0.06	NS	0.19
C5 (%)	2.56	-0.07	NS	0.03

IsoC5 (%)	1.96	-0.12	<0.09	0.36
C2/C3	2.37	-0.04	NS	0.07
NH_3 (mmol/l)	9.56	-0.66	NS	0.09
pH	6.56	0.03	NS	0.05
Gas production (ml/g)	102.99	-3.93	NS	0.03
CH_4 gas (mmol/d)	7.43	-0.61	***	0.68
Digestibility (%)				
DMD	51.63	-0.61	NS	0.02
OMD	37.97	1.52	NS	0.09

Abbreviations: This model is very significant (***) at $P \leq 0.01$, significant at $P \leq 0.05$, tends to be significant at $P \leq 0.10$, not significant (NS) at $P \geq 0.10$. VFA: Volatile Fatty Acid; C2: acetate; C3: propionate; C4: butyrate; IsoC4: isobutyrate; C5: valerate; IsoC5: isovalerate; C2/C3: acetate/propionate ratio; NH_3: ammonia; CH_4: methane gas; DMD: Dry matter digestibility; OMD: Organic matter digestibility.

Ammonia concentration (NH3) was not influenced by coconut oil supplementation. Based on the average, the ammonia concentration is still within the normal range as shown in Table 2. The optimum concentration of ammonia in the rumen fluid is within the range of 6-12 mM [20]. The decrease in ammonia concentration in the rumen indicates a decrease in protein degradation by rumen microbes due to oil supplementation in feed [23]. The pH value of the rumen was not influenced by coconut oil supplementation, but the average pH value obtained was still within the normal range as can be seen in Table 2. According to Mottram et al. [24], the range of normal rumen pH values is in the range of 5.5-7.0. The pH of the rumen fluid is a major factor affecting rumen conditions for microbial growth and activity in the rumen [20].

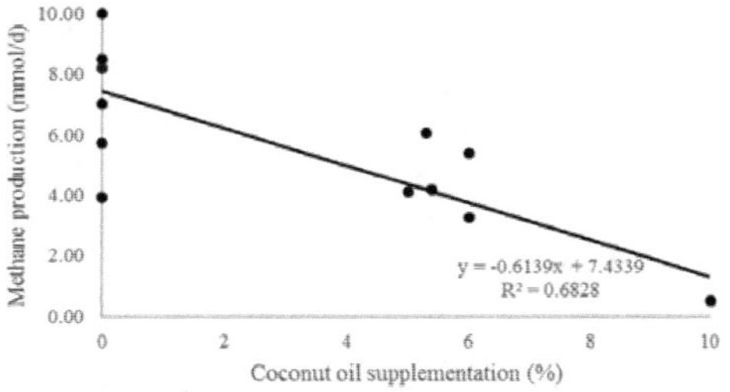

Figure 1. Graph of simple linear regression for methane gas

Coconut oil supplementation had no significant effect on total gas production. Coconut oil supplementation can reduce digestibility in the rumen because coconut oil will cover the fraction of the feed fiber causing rumen bacteria cannot degrade feed fiber and cause total gas production also to decrease [25]. Meanwhile, the methane gas production in the rumen is reduced ($P<0.01$; R2=0.68) by coconut oil supplementation (Figure 1).

According to Tavendale et al. [26], the mechanism of methane gas formation inhibition occurs directly by inhibiting the growth and activity of methanogenic archaea and indirectly by inhibiting fiber digestion thereby reducing hydrogen production. The hydrogen flow to the methanogenesis pathway is diverted to the formation of propionate [16].

Coconut oil supplementation had no effect on the digestibility of dry matter and organic matter. Coconut oil supplementation can disrupt the population and microbial activity in the rumen, consequently the fiber-breaking microbes that are most sensitive to fatty feed will be reduced and feed degradation will decrease [13]. Providing excess fat will cause the fat to cover the fiber fraction in the feed, accordingly the fiber-breaking bacteria cannot degrade feed [27].

4. Conclusion

The effect of coconut oil on rumen fermentation and methanogenesis parameters in vitro from various previous research data has greatly reduced the production of methane gas in the rumen. This decrease was due to the decrease in the population of protozoa and methanogenic archaea. Total VFA and isovalerate also tended to decrease by coconut oil supplementation.

References

[1] Vlaming JB. 2008. Quatinfying variation in estimated methane emission from ruminants using SF5 tracer fechnique. Massey University Thesis, Palmerston North, New Zealand.

[2] Steinfeld H, Gerber P, Wassenaar T, Castel V, Rosales M and Haan CD. 2006. Livestock's Long Shadow, Food and Agriculture Organization of the United Nations, Rome. 390 pages.

[3] Kang S, Wanapat M and Viennasay B. 2016. Supplementation of banana flower powder pellets and plant oil sources on in vitro ruminal fermentation, degestibility, and methane production. Journal of Tropical Animal Health and Production. 48(8):1673-1678.

[4] Kim ET, Park CG, Lim DH, Kwon EG, Ki Ks, Kim B, Moon YH, Shin NH and Lee SS. 2014. Effects of coconut materials on in vitro ruminal methanogenesis and fermentation characteristics. AAJAS. 27(12):1721-1725.

[5] Cieslak A, Szumacher-Strabel M, Szymankiewicz E, Piekniewski M, Oleszak P, Siwinski L and Potkanski A. 2016. Coconut oil reduces protozoa count and methane release during fermentation in a RUSITEC system. Journal of Animal and Feed Science. 15(1):19-22.

[6] Dohme F, Machmuller A, Estermann BL, Pflster P, Wasserfallen A and Kreuzer M. 1999. The role of the rumen ciliate protozoa for methane suppression caused by coconut oil. Letters in Applied Microb. 29(3):187-192.

[7] Dohme F, Machmuller A, Wasserfallen A, Kreuzer M. 2000. Comparative efficiency of various fats rich in medium-chain fatty acids to suppress ruminal methanogenesis as measured with RUSITEC. Canadian Journal Animal Science. 80(3):473-482.

[8] Dong Y, Bae HD, McAllister TA, Mathison GW and Cheng KJ. 1997. Lipid-induced depression of methane production and digestibility in the artificial rumen system (RUSITEC). Canadian Journal of Animal Science. 77(2):269-278.

[9] Kongmun P, Wanapat M, Pakdee P andNavanukraw C. 2010. Effect of coconul oil and garlic powder on in vitro fermentation using gas production technique. Livestock Science. 127(1):38-44.

[10] Machmuller A, Andrea D, Ossowski A, Wanner M and Kreuzer M. 1998. Potential of various fatty feeds to reduce methane release from rumen fermentation in vitro (Rusitec). Animal Feed Science and Technology. 71(1-2):117-130.

[11] Machmuller A, Dohme F, Soliva CR, Wanner Ma and Kreuzer M. 2001.Diet composition affects the level of ruminal methane suppression by medium-chain fatty acids. Australian Journal of Agricultural Research. 52(7):713-722.

[12] Panyakaew P, Goel G, Lourenco M, Yuangklang C and Fievez V. 2013. Medium-chain fatty acids from coconut or krabok oil inhibit in vitro rumen methanogenesis and conversion of non-conjugated dienoic biohydrogenation intermediates. Animal Feed Science and Technology. 180(1-4):18-25.

[13] Patra AK and Yu Z. 2012. Effects of coconut and fish oils on ruminal methanogenesis, fermentation, and abundance and diversity of microbial populations in vitro. Journal of Dairy Science. 96(3):1782-1792.

[14] Putra S. 2006. Pengaruh suplementasi agensia defaunasi dan waktu inkubasi terhadap bahan kering, bahan organik terdegradasi dan produk fermentasi secara in vitro. Animal Production. 8(2):121-130.

[15] Sitoresmi PD, Yusiati LM and Hartadi H. 2009. Pengaruh penambahan minyak kelapa, minyak biji bunga matahari, dan minyak kelapa sawit terhadap penurunan produksi metan di dalam rumen secara in vitro. Bulletin Peternakan. 33(2): 96-105.

[16] Yabuuchi Y, Matsushita Y, Otsuka H, Fukamachi K, Kobayashi Y. 2006. Effect of supplemental lauric acid-rich oils in high grain diet on in vitro rumen fermentation. Animal Science Journal.

77(3):300-307

[17] Machmuller A and Kreuzer M. 1999. Methane suppression by coocnut oil and associates effects on nutrient and energy balance in sheep. Canadian Journal of Animal Science. 79(1):65-72.

[18] Newbold CJ, Fuente G, Belanche A, Ramos-Morales E and McEwan NR. 2015.The role of ciliate protozoa in the rumen. Frontiers in Microbiology. 6: 1313.

[19] Machmuler A, Soliva CR and Kruezer M. 2003. Methane-suppressing effect of myristic acid in sheep as affected by dietary calcium and forage proportion. British Journal of Nutrition. 90(3):529-540.

[20] McDonald P, Edwards RA, Greenhalgh JFD, Morgan CA, Sinclair LA and Wilkinson RG. 2002. Animal Nutrition. 7th Ed, John Wiley Inc, New York. 714 pages.

[21] Hungate RE. 1966. The Rumen and Its Microbes, Academic Press, New York.

[22] Moss AR., Jouany JP and Newbold J. 2000. Methane production by ruminants: its contribution to global warming. Annales de Zootechnie. 49(3):231–253.

[23] Hidayah N. 2014. Ketahanan biohidrogenasi minyak nabati yang diproteksi dengan metode sabun kalsium dan mikroenkapsulasi secara in vitro, Thesis, Bogor Agricultural University, Bogor, Indonesia.

[24] Mottram T, Lowe J, McGowan M and Philips N. 2008. Technical note: a wireless telemetric method of monitoring clinical acidosis in dairy cows. Computers and Electronics in Agriculture. 64(1):45-48.

[25] Hristov AN, Vander Pol M, Agle M, Zaman S, Schneider C, Ndegwa P, VadellaVK, Johnson K, Shingfield KJ and Karnati SKR. 2009. Effect of lauric acid and coconut oil on ruminal fermentation, digestion, ammonia losses from manure, and milk fatty acid composition in lactating cows. Journal of Dairy Science. 92(11):5561-5582.

[26] Tavendale MH, Meagher LP, Pacheco D, Walker N, Attwood GT and Sivakumaran S. 2005. Methane production from in vitro rumen incubation with Lotus pedunculatus and Medicago sativa, and effects of extractable condensed tannin fractions on methanogenesis. Journal of Animal Feed Science Technology. 123:403-419.

[27] Suharti S, Aliyah DN and Suryahadi. 2018. Karakteristik fermentasi rumen in vitro dengan penambahan sabun kalsium minyak nabati pada buffer yang berbeda. Jurnal Ilmu Nutrisi dan Teknologi Pakan. 16(3):56-64.

Effect of palm oil supplementation level on in vitro ruminal fermentability

RK Nurfatahillah*, A Cusiayuni, A Jayanegara, KG Wiryawan, D Evvyernie

IPB University, Bogor, Indonesia

*risankhaerani@gmail.com

Abstract. Palm oil is an abundant source of oil in Indonesia and some other tropical countries. This research aimed to investigate the effect of palm oil supplementation levels on in vitro rumen fermentability by integrating data from various related studies. The data used were 40 data points from 13 studies which reported different levels of palm oil supplementation on in vitro rumen fermentability and nutrient digestibility. Data were tabulated, uniformed for the unit in each parameter, then analyzed using a linear regression method by employing SPSS software version 20. The results showed that palm oil supplementation significantly ($P<0.05$) increased bacterial population and isobutyrate proportion, while reduced protozoa population, and tended to reduce total volatile fatty acids ($P<0.1$). Supplementation of palm oil did not affect dry mater digestibility, organic matter digestibility, methane emission, pH, total gas, ammonia, acetate proportion, propionate, butyrate, valerate and isovalerate. It can be concluded that palm oil supplementation modifies some parameters of in vitro rumen fermentation.

1. Introduction

The process of digestion of feed in ruminants occurs mechanically in the mouth, fermentation by rumen microbes, and hydrolysis by digestive enzymes. In general, the use of oil on ruminants can interfere the rumen microbial activity and have an impact on the fermentation process in the rumen and the final product produced. Oil has a cytotoxic effect on rumen microbes and can inhibit the adhesion of microbial enzymes to feed particles due to oil-covered feed particles [1].

One type of oil that is widely available in Indonesia and several tropical countries and can be used as a feed ingredient is palm oil. Palm oil contains saturated fatty acids (SFA), namely palmitate 44.3%, stearate 4.6%, myristate 1.0%; monounsaturated fatty acids (MUFA), namely oleic acid 38.7%; polyunsaturated fatty acids (PUFA), namely 10.5% linoleic acid and 0.9% others [2].

Palm oil supplementation in ruminants can increase VFA levels and reduce methane gas production [3; 4]. Other research has also resulted in an increase in total VFA accompanied by an increase in acetate, propionate, butyrate, methane gas, ammonia, and a decrease in protozoa population and digestibility [5]. According to [6], the use of palm oil can decrease acetate and increase propionate. Palm oil supplementation does not affect VFA and ammonia levels but can reduce protozoa and methane gas populations [7].

Palm oil supplementation from several studies had different effects on fermentation products in the rumen in vitro. This can be influenced by various factors, one of which is the amount of oil used. The difference in the amount of oil can affect the results produced, so it is necessary to study the relationship between the different amounts of palm oil used on the different effects of these data. This research aimed to investigate the effect of palm oil supplementation levels on in vitro rumen fermentability by integrating data from various related studies using simple linear regression analysis.

2. Materials and Methods

Data were collected from journals that reported the use of various palm oil levels supplementation in the form of Crude Palm Oil (CPO) and Palm Kernel Oil (PKO) on in vitro rumen fermentability. Journal information used in this study is tabulated in Table 1. Data from various publications were tabulated in a database with uniform data units. The data were analyzed using descriptive statistics and simple linear regression using software program of Statistical Product and Service Solutions (SPSS) version 20.

Table 1. Information data were used to assess the effect of palm oil supplementation levels on in vitro rumen fermentability

No	Reference	Oil Source	Diet	Oil Level (%)
1	Dohme *et al.* [8]	CPO	Straw, corn silage, concentrate	0-5.3
2	Dohme *et al.* [8]	PKO	Straw, corn silage, concentrate	0-5.3
3	Hristov *et al.* [9]	CPO	Hay wheat, wheat silage, soybean meal	0-2
4	Yabuuchi *et al.* [4]	CPO	Commercial concentrate, straw	0-0.2
5	Yabuuchi *et al.* [4]	PKO	Commercial concentrate, straw	0-0.2
6	Sitoresmi *et al.* [7]	CPO	Napier, bran	0-7.5
7	Tiven *et al.* [10]	CPO	Napier, bran	0-6
8	Kim *et al.* [11]	CPO	Roughage	0-10
9	Kim *et al.* [11]	CPO	Concentrated	0-10
10	Kim *et al.* [11]	CPO	-	0- 10
11	Hartanto *et al.* [3]	CPO	Commercial pellets	0-3.3
12	Matsuba *et al.* [6]	CPO	Roughage, concentrate	0-15
13	Muktiani *et al.* [5]	CPO	Roughage, concentrate	0-5

CPO (*Crude Palm Oil*); PKO (*Palm Kernel Oil*)

3. Result and Discussion

Data used were 40 data points from 13 studies which reported the use of palm oil at various levels (0-15%) on the characteristics of rumen fermentability in vitro. The results of the descriptive statistics from the data are presented in Table 2.

Table 2. Descriptive statistics of data on the effect of the level of palm oil supplementation on rumen fermentability in vitro

Variabel	n	Min	Max	Average	SD
Protozoa (10^4 ml^{-1})	20	1.27	170.00	34.50	49.36
Bacteria (10^9 ml^{-1})	4	2.80	4.40	3.48	0.81
DMD (%)	4	47.60	61.06	53.44	6.59
OMD (%)	4	50.93	62.16	55.44	4.93
pH	35	5.79	7.16	6.62	0.39
NH_3 (mmol l^{-1})	35	2.12	27.10	14.81	8.83
Total VFA (mmol l^{-1})	38	27.40	171.40	79.82	36.75
C2 (%)	37	11.48	65.40	48.05	13.73
C3 (%)	40	2.87	41.60	19.55	10.94
C4 (%)	40	2.09	20.20	11.12	5.36
IsoC4 (%)	15	0.25	2.53	1.78	0.74
C5 (%)	19	1.05	4.03	2.93	1.02
IsoC5 (%)	17	0.65	5.70	3.32	1.64
C2/C3	18	1.45	8.34	4.21	2.15
Gas (ml g^{-1} substrat)	20	52.20	383.00	131.21	106.82
CH_4 (ml l^{-1} gas)	21	31.32	379.28	199.66	149.32

n: number of observations, SD: standard deviation, Min: minimum, Max: maximum, DMD: Dry matter digestibility, OMD: Organic matter digestibility, CH_4: methane gas, NH_3: Ammonia, VFA: *Volatile Fatty Acid*, C2: acetate, C3: propionate. C4: butyrate; IsoC4: isobutyrate, C5: valerate, IsoC5: isovalerate, C2 / C3: acetate / propionate ratio.

Microbial is the key to the successful use of feed in the rumen [12]. The microbial biomass contained in the rumen is a combination of bacteria, protozoa, fungi, and archae, each of which has a specific function in degrading carbohydrates, proteins, and fats derived from feed [13]. Protozoa are the second-largest microorganism after bacteria and can represent half of the total microbial biomass in the rumen. Increasing the level of palm oil supplementation significantly ($P<0.10$) affected decreasing the rumen protozoa population.

Protozoa do not have the same lipolytic activity as bacteria, as a result, the addition of oil can disrupt the metabolic activity of protozoa and then cause protozoa defaunation [3]. The oil in the rumen will cover the protozoa and are difficult to break down because the protozoa do not produce lipolytic enzymes, this condition can cause the outer surface tension of the protozoa cells to be higher than the inner surface of the cells which results in the protozoa cells experiencing lysis [13].

Increasing the level of palm oil supplementation significantly increased ($P<0.05$) bacterial population. The increase in bacterial population is thought to be due to the defaunation process of protozoa in the rumen. Oil supplementation in the ration can increase the population of rumen bacteria due to reduced bacterial predators, namely protozoa [14]. Protozoa in the rumen are predators of rumen bacteria to meet their protein needs [15] so that a decrease in the population of protozoa in the rumen can give some bacteria the opportunity to develop because it reduces nutrient competition between bacteria and protozoa [16]. Apart from the effects of protozoa defaunation, the total population of bacteria in the rumen is also influenced by rumen pH, type of livestock, frequency of feeding, availability of VFA and ammonia, and the type of feed consumed [17].

Microbial activity in feed fermentation and the chemical composition of feed can affect the digestibility value [18]. The addition of the palm oil supplementation level did not significantly affect the dry matter digestibility and the organic matter digestibility with the resulting slope being negative (Table 3). The difference in the type of ration used (Table 1) and the amount of oil used is thought to influence this value. Research [16] resulted in a reduced digestibility value with the addition of 10% of free oil, while the addition of oil below 10% did not affect the digestibility value.

Dry matter digestibility and organic matter digestibility values used in this study were less than 60% (Table 2), a relatively high digestibility value when the value was more than 60% [16]. Oils that contain unsaturated fatty acids can reduce the digestibility value of fiber due to the inhibition of fiber-degradation by rumen microbes [19]. Digestibility can also be influenced by the physical form of feed ingredients and the feed rate of passage through the digestive tract [20].

Table 3. The regression equation of the level of palm oil supplementation on rumen fermentation and nutrient digestibility in vitro

Variable	*Intercept*	*Slope*	*P value*	R^2
Protozoa (10^4 ml^{-1})	44.07	-4.54	0.05	0.05
Bacteria (10^9 ml^{-1})	2.80	0.25	0.03	0.94
DMD (%)	53.70	-0.13	0.95	0.00
OMD (%)	56.83	-0.67	0.66	0.12
pH	6.66	-0.01	0.54	0.01
NH_3 (mmol l^{-1})	14.48	0.10	0.79	0.00
Total VFA (mmol l^{-1})	90.67	-3.10	0.05	0.11
C2 (%)	50.70	-0.75	0.20	0.05
C3 (%)	19.28	0.08	0.86	0.00
C4 (%)	10.99	0.04	0.86	0.00
IsoC4 (%)	1.25	0.12	0.03	0.27
C5 (%)	2.84	0.12	0.19	0.09
IsoC5 (%)	2.86	0.16	0.13	0.14

C2 / C3	4.30	-0.02	0.88	0.00
Gas (ml g^{-1} substrate)	108.53	5.11	0.35	0.05
CH_4 (ml l^{-1} gas)	201.47	-0.43	0.95	0.00

P≤0.001: very significant, P≤0.05: significant, 0.05≤P ≤0.10: tends to be significant, P> 0.10: insignificant, DMD: Dry matter digestibility, OMD: Organic matter digestibility, CH_4: methane gas, NH_3: Ammonia, VFA: *Volatile Fatty Acid*, C2: acetate, C3: propionate. C4: butyrate; IsoC4: isobutyrate, C5: valerate, IsoC5: isovalerate, C2 / C3: acetate / propionate ratio.

The fermentability of feed in the rumen can be measured from the degree of acidity or pH of the rumen. The addition of palm oil supplementation level does not affect the pH value in the rumen (Table 3). The function of protozoa in the rumen is to maintain pH conditions in neutral conditions by slowing down the fermentation process. Defaunation of protozoa can cause rumen pH to become unstable and tend to decrease [13]. The average pH value of this study is 6.62 which is still within the normal range (Table 2). Normal rumen pH ranges from 5.5 to 7 [13]. The condition of this research was carried out in vitro using artificial saliva, although a decrease in the pH value can still be controlled through the supply of saliva which functions as a buffer during the fermentation process, or through the use of lactate by rumen microbes [17].

Main product of rumen microbial fermentation activity and the main source of energy for ruminants is volatile fatty acids (VFA). The VFA production in rumen fluid can also be used as a measure of feed fermentability [16]. Increasing the level of palm oil supplementation tended linearly to reduce the total VFA (P<0.10). Oil in the rumen can cover the feed and interfere with rumen microbial activity, especially fiber degrading rumen microbes , besides that oil has a cytotoxic effect on rumen microbes [1, 21]. The inhibition of microbial activity in degrading the feed particles is thought to affect reducing the fermentability of feed in the rumen. The VFA components produced from the feed fermentation process in the rumen are acetate (C2) around 65%, propionate (C3) around 21%, and butyrate (C4) around 14% [22].

The proportion of VFA components in the rumen can vary depending on the type of ration given. Addition of palm oil supplementation level had no significant effect on the proportion of acetate, propionate, and butyrate. Fermentability of feed in the rumen also produces VFA components such as isobutyrate (isoC4) around 1%, valerate (C5), and isovalerate (isoC4) below 3% [22; 23]. The proportion of isobutyrate in this study increased significantly (P<0.05) while the proportions of valerate and isovalerate were not significantly affected with the addition of palm oil supplementation levels. Production of isobutyrate and isovalerate uses hydrogen in its formation pathway and increases with the increasing NH3 production [24].

Supplementation level of palm oil did not affect the concentration of rumen ammonia . The concentration of NH3 is strongly influenced by the ability of rumen microbes in degrading feed protein [25]. Palm oil from this study can significantly increase the population of bacteria in the rumen., The increase in bacterial population is thought to affect bacterial activity in degrading feed protein which can increase levels of ammonia in the rumen.

However, in this study there were differences in the feed used (Table 1), these differences could affect the levels of ammonia produced. Increased levels of ammonia can be influenced by types of feed that are high in protein and easy to degrade [22]. The average ammonia levels in this study is 14.81 mM (Table 2) which is in the range of the optimum concentration of ammonia in the rumen fluid, between 6-21 mM [22]. The ammonia formed will be used as the main precursor for protein microbial synthesis in the rumen as long as the carbon skeleton from easily digestible carbohydrates is available [26].

The fermentation process of feed in the rumen also produces gas [27]. The existence of fermentation activity by rumen microbes will produce gas that is formed directly from the fermentation of the substrate (CO2 and CH4) and indirectly from the CO2 that is released from the buffer every time it produces VFA [17]. Palm oil supplementation did not affect the total gas production, although there was a tendency to increase. The increase in gas production in this study is thought to be due to oil which defaunates protozoa and then increases the activity of rumen bacteria in fermenting the feed substrate.

Research by [28] also resulted in an increase in total gas production from cultures that had defaunated protozoa., This increase was accompanied by an increase in the production of CO2 rather than CH4. Apart from the effect of protozoa defaunation, gas production in the rumen can also be influenced by the size of the feed particles, temperature, rumen pH [27], and incubation time for the in vitro scale [29]. The palm oil supplementation level did not affect the production of CH4 in the rumen. Oil supplementation can affect CH4 gas production due to the disruption of protozoa activity in the rumen. Protozoa do not have lipolytic activity, as a result, the addition of oil can disrupt the metabolic activity of protozoa and cause protozoa and bacteria that stick to protozoa to die [3].

Protozoa and methane forming microbes has symbiosis on the transfer of H2 and may affect about 9-25% of the methane formation process or methanogenesis [30], thus reducing the population of protozoa can also reduce the formation of methane gas in the rumen. Apart from the effects of protozoa defaunation, the production of CH4 in the rumen can also be affected by the amount of feed consumption, the rate of fermentation of organic matter, the type of VFA produced, the efficiency of microbial biosynthesis [31], the content of organic matter in feed, and the content of fiber components in the feed [3].

4. Conclusion

Increasing the level of palm oil supplementation in the feed from previous research data can linearly modify several parameters of rumen fermentation in vitro such as bacterial population, protozoa, and concentrations of volatile fatty acids (VFA). Palm oil supplementation can reduce dry matter digestibility, organic matter digestibility, CH4 gas production, propionate, protozoa population and can increase the total production of gas, ammonia, butyrate, valerate, and ratio acetate: butyrate but not linearly.

References

[1] Jenkins TC. 1993. Lipid Metabolism in the Rumen. Journal of Diary Science. 76(12):3851–3863.
[2] Mukherjee S, Mitra A. 2009. Health Effects of Palm Oil. Journal of Human Ecology. 26(3):197–203.
[3] Hartanto R, Cai L, Yu J, Zhang N, Sun L, Qi D. 2017. Effects of supplementation with monensin and vegetable oils on in vitro enteric methane production and rumen fermentability of goats. Pakistan Journal of Agricultural Science. 54(3):693–698.
[4] Yabuuchi Y, Matsushita Y, Otsuka H, Fukamachi K, Kobayashi Y. 2006. Effects of supplemental lauric acid-rich oils in high-grain diet on in vitro rumen fermentation. Animal Science Journal. 77(3):300–307.
[5] Muktiani A, Arifah N, Widiyanto W. 2020. Influence of different vegetable oils on in vitro ruminal fermentability and nutrient digestibility in ettawah crossbred goat. Animal Production. 21(1):22-29.
[6] Matsuba K, Padlom A, Khongpradit A, Boonsaen P, Thirawong P, Sawanon S, Suzuki Y, Koike S, Kobayashi Y. 2019f. Selection of plant oil as a supplemental energy source by monitoring rumen profiles and its dietary application in Thai crossbred beef cattle. Asian-Australasian Journal of Animal Science. 32(10):1511–1520.
[7] Sitoresmi PD, Yusiati LM, Hartadi H, Peternakan F, Mada UG, No JF. 2009. Pengaruh penambahan minyak kelapa, minyak biji bunga matahari, dan minyak kelapa sawit terhadap penurunan produksi metan di dalam rumen secara. Buletin Peternakan. 33(2):96–105.
[8] Dohme F, Machmüller A, Wasserfallen A, Kreuzer M. 2000. Comparative efficiency of various fats rich in medium-chain fatty acids to suppress ruminal methanogenesis as measured with RUSITEC. Canadian Journal of Animal Science. 80(3):473–482.
[9] Hristov AN, Ivan M, McAllister TA. 2004. In vitro effects of individual fatty acids on protozoa numbers and on fermentation products in ruminal fluid from cattle fed a high-concentrate, barley-based diet. Journal of Animal Science. 82:2693–2704.
[10] Tiven NC, Yusiati LM, Rusman, Santoso U. 2011. Ketahanan asam lemak tidak jenuh dalam

Crude Palm Oil terproteksi terhadap aktivitas mikrob rumen domba in Vitro. Media Peternakan. 34(1):42–49.

[11] Kim DH, Mizinga KM, Kube JC, Friesen KG, McLeod KR, Harmon DL. 2014. Influence of monensin and lauric acid distillate or palm oil on in vitro fermentation kinetics and metabolites produced using forage and high concentrate substrates. Animal Feed Science and Technology. 189:19–29.

[12] Suryani N, Suarna I, Budiasa IK. 2018. Imbangan hijauan dan konsentrat dalam ransum terhadap respon fermentasi rumen dan sintesis protein mikroba pedet sapi bali calon induk. Majalah Ilmiah Peternakan. 21(2):60–65.

[13] Yanuartono Y, Nururrozi A, Indarjulianto S, Purnamaningsih H. 2019. Peran protozoa pada pencernaan ruminansia dan dampak terhadap lingkungan. Journal of Tropical Animal Production. 20(1):16–28.

[14] Hanim C, Yusiati LM, Alim S. 2009. Effect of saponin as defaunating agent on in vitro ruminal fermentation of forage and concentrate. Journal of the Indonesian Tropical Animal Agricultural. 34(4):231–235.

[15] Dayyani N, Karkudi K, Zakerian A. 2013. Special rumen microbiology. International Journal of Advanced Biological and Biomedical Research. 1(11):1397–1402.

[16] Suharti S, Aliyah DN, Suryahadi S. 2019. Karakteristik fermentasi rumen in vitro dengan penambahan sabun kalsium minyak nabati pada buffer yang berbeda. Jurnal Ilmu Nutrisi dan Teknologi Pakan. 16(3):56-64.

[17] Suharyono S, NW Hardani S, Wahyono T. 2016. Dinamika hasil fermentasi rumen pada konsentrat yang mengandung suplemen pakan baru (SPB). Jurnal Ilmiah Aplikasi Isotop dan Radiasi. 11(2):99-112.

[18] Lehan M. 2015. Evaluasi nilai kecernaan secara in vitro ransum ternak sapi bali yang disuplementasi dengan probiotik bioplus. Jurnal Peternakan Sriwijaya. 4(1):35–46.

[19] Messana JD, Berchielli TT, Arcuri PB, Reis RA, Canesin RC, Ribeiro AF, Fiorentini G, Fernandes JJ. 2013. Rumen fermentation and rumen microbes in nellore steers receiving diets with different lipid contents. Revista Brasileira de Zootecnia. 42(3):204–212.

[20] Anggorodi R. 1994. Ilmu Makanan Ternak Umum. 5th ed., Gramedia, Jakarta, 216 pages.

[21] Patra AK, Yu Z. 2013. Effects of coconut and fish oils on ruminal methanogenesis, Fermentation, And abundance and diversity of microbial populations in vitro. Journal of Dairy Science. 96(3):1782–1792.

[22] McDonald P, Edwards R, Greenhalgh J, Morgan C, Sinclair L, Wilkinson R. 2002. Animal Nutrition. John Wiley Inc, New York, 693 pages.

[23] Hungate R. 1966. The Rumen and Its Microbes. Academic Press, New York, 533 pages.

[24] Moss AR, Jouany JP, Newbold J. 2000. Methane production by ruminants: Its contribution to global warming. Animal Zootech. 49(3):231–253.

[25] Prayitno RS, Wahyono F, Pangestu E. 2018. Pengaruh suplementasi sumber protein hijauan leguminosa terhadap produksi amonia dan protein total ruminal secara in vitro. Jurnal Peternakan Indonesia. 20(2):116-123.

[26] Purbowati E, Rianto E, Dilaga WS, Sri Lestari CM, Adiwinarti R. 2014. Karakteristik cairan rumen, jenis, dan jumlah mikrobia dalam rumen sapi jawa dan peranakan ongole. Buletin Peternakan. 38(1):21-26.

[27] Tagliapietra F, Cattani M, Bailoni L, Schiavon S. 2010. In vitro rumen fermentation: Effect of headspace pressure on the gas production kinetics of corn meal and meadow hay. Animal Feed Science and Technology. 158(3–4):197–201.

[28] Qin WZ, Li CY, Kim JK, Ju JG, Song MK. 2012. Effects of defaunation on fermentation characteristics and methane production by rumen microbes in vitro when incubated with starchy feed sources. Asian-Australasian Journal of Animal Science. 25(10):1381–1388.

[29] Sugoro I, Wiryawan KG, Astuti DA, Wahyono T. 2015. Gas production and rumen fermentation characteristics of buffalo diets containing by-product from some sorghum varieties. Indonesian

Journal of Animal and Veterinary Science. 20(4):242-249.
[30] Machmüller A, Soliva CR, Kreuzer M. 2003. Methane-suppressing effect of myristic acid in sheep as affected by dietary calcium and forage proportion. British Journal of Nutrition. 90(3):529–540.
[31] Akhadiarto S. 2017. Estimasi emisi gas metana dari fermentasi enterik ternak ruminansia menggunakan metode tier-1 di indonesia. Jurnal Teknologi Lingkungan. 18(1):1-8.

The Composition of Goat Milk Kefir Added with Moringa oleifera Leaf Powder

S R N Endah[1], P D Wulasari[1*], A Nofriyaldi[1], Nurliyani[2] and E Harmayani[2]

[1]Universitas Perjuangan Tasikmalaya, Tasikmalaya, Indonesia
[2]Universitas Gadjah Mada, Yogyakarta, Indonesia

*putridian@gmail.com

Abstract. Goat milk kefir has gained popularity as a functional food. Incorporating natural ingredients with functional properties into kefir making could increase the functional properties in the kefir product. Moringa oleifera leaf has been extensively used for traditional medicine due to its multifunctional properties. This study aimed to investigate the chemical properties of goat milk kefir added with moringa leaf powder. The study was conducted in a Completely Randomized Design, using five treatments and five replicates (three replicates for each treatment). The additional levels of moringa leaf powder were 0, 0.5, 1.0, 1.5, and 2.0 % w/w. The result showed that moringa leaf powder added to kefir making has significantly ($P<0.05$) increased acidity and decreased pH; however, the proximate analysis did not show a significant effect ($P>0.05$). Incorporating 2% moringa leaf powder could increase acidity by 58.61% and decrease pH by 8.53% compared to control. This study found that 2% moringa leaf added to the kefir making produced that the highest acidity (1.115%) and the lowest pH (5.07). The average compositions of total solids, fat, protein, ash, and carbohydrate in this study were 18,17%, 4.939%, 7.22%, 0.988%, and 5.019%, respectively.

1. Introduction

Goat milk products and by-products have gained popularity among consumers because of their small fat lumps that are easier to digest (1). Kefir is one of the by-products that has been extensively consumed for its medicinal value. Kefir is originated from Caucasus Mountain in Rusia, containing some types of lactic acid bacteria (LAB), such as Lactobacillus kefiranofaciens and Lactobacillus parakefiri, as well as the low-alcohol yeast (2). Previous findings reported that consuming kefir is beneficial for type 2 diabetes mellitus patients (Type 2 DM) because it contains anti-diabetes and antioxidants (3-5).

One effort to increase the functional properties of goat milk kefir for Type 2 DM is fortifying/supplementing natural ingredients that contain antioxidant and anti-diabetes properties. Indonesian community has long believed that Moringa oleifera leaf (Moringa oleifera) exhibits anti-diabetes quality because it has high antioxidant properties. The antioxidant properties in moringa oleifera leaf can protect the β-cells from damage due to free radicals; thus, minimizing the risk of hyperglycemia (6, 7). A previous study incorporated Moringa oleifera leaf into fermented milk (yogurt) and reported that the leaf extract did not impose a negative impact on the growth of kefir bacteria (8).

It showed that both goat milk kefir and Moringa oleifera leaf have medicinal values. However, there have been limited references on the functional properties of Moringa oleifera leaf incorporated in the goat milk kefir making and their effects on the levels of acidity, pH, total solids, fat, protein, ash, and carbohydrate. Therefore, this study aimed to investigate the impact of incorporating moringa leaf powder in the goat milk kefir making on the chemical characteristics.

2. Methodology

The main materials in this study were goat milk, kefir grain, and moringa leaf powder. The goat milk is derived from the dairy goat farm "Pakem" Yogyakarta. Kefir grain was collected from Adhigana Multi Makmur, Ltd, Yogyakarta and the Moringa leaf powder was from Nasuha Herb Laboratory, Jakarta.

The research was conducted in a Completely Randomized Design (CRD), using five treatments and three replicates. The goat milk (SK) was added with different levels of moringa leaf powder (BDK), namely T0 (0% BDK); T1 (0.5% BDK); T2 (1.0% BDK); T3 (1.5% BDK) and T4 (2% BDK). The process of manufacturing kefir followed the method by Nurliyani, Sadewa (9) with a slight modification by incorporating milk and moringa leaf powder for pasteurization at 85oC for 15 minutes. A total of 3% of the milk was inoculated to the kefir grain at room temperature and incubated for 18 hours. The samples were collected in a sterile place, and the collected kefir was stored at 4oC for further analysis.

The parameters investigated in the study were acidity, pH value, total solids, fat, protein, ash, and carbohydrate. The pH value was measured using a pH meter to calibrate at buffer pH4. The acidity was expressed in the percentage of lactic acid, determined from NaOH titration, and used phenolphthalein as the indicator (10). The proximate analysis used the AOAC method (11). The result of the proximate analysis was subjected to analysis of variance (ANOVA), and significant effects were examined further using DMRT (Duncan multiple range tests) with a 5% tolerance.

3. Result and Discussion

Table 1 shows that incorporating moringa leaf powder in goat milk kefir making has affected the acidity level ($P<0.05$). The more moringa leaf powder added to the milk, the higher the acidity. Furthermore, 2% of moringa leaf powder produced the most acidity (1.115%) by increasing the initial acidity (control) up to 58.61%. The Standard Codex No. 234 requires kefir product to have at least 0,6% acidity (12).

Therefore, the acidity level in this study 0,703% – 1.115% have met the Codex Standard, but lower than 1.31-1.66% reported by Setyawardani, Sumarmono (13) A previous finding stated that incorporating moringa leaf powder allowed the growth of lactic acid bacteria that also increases the total lactic acid in the kefir (14), and eventually produce a higher acidity than that of control.

On the other hand, Table 1 shows that where moringa leaf powder incorporated in goat milk kefir making has affected the final pH value ($P<0.05$). Hence, the higher the level of moringa leaf powder, the lower the kefir's pH value. Incorporating 2% moringa leaf powder could lower the pH by 8.53% from control (5.07).

This finding was in contrast to Ismael, Farahat (15) that supplementing moringa leaf powder into whey did not affect the pH. The standard pH value for kefir is not determined by the Standard Codex (12); however, the pH value in this study was similar to that referred by Yıldız-Akgül, Yetişemiyen (16) namely 5-5.5. A contributing factor to pH stability is lactic acid bacteria. The more lactic acid bacteria, the lower the pH (17). The correlation between Moringa leaf powder and the increased acidity, as well as decreased pH in the current study, is due to 12% w/w fiber in the moringa leaf powder (18). Fibers are the source of energy for lactic acid bacteria; therefore, the higher level of moringa leaf powder produces higher acidity and lower pH.

Table 1 shows that moringa leaf powder does not affect the total solid ($P>0.05$). The average total solid in this study is 18.17%, which is higher than the previous findings on goat milk, namely 12.09% (19) and 12.37% (20), as well as 9.49% in kefir plus ginger extract (21). The kefir product in this study has met the standard quality of fermented milk products assigned by BSN, i.e., a minimum of 3% (22). The high total solid in this study may be due to the high dry matter in moringa leaf powder incorporated into kefir making. It was in line with Alabi (23), that moringa leaf powder contained 95% dry matter.

Table 1. Acidity, pH, and Total Solids of Goat milk kefir fortified with Moringa leaf powder

Treatments	Acidity (%)	pH value	Total Solid (%)
Milk kefir	0.703 ± 0.597^{a}	5.543 ± 0.23^{a}	18.55 ± 2.509
Milk kefir with 0.5% MOP	0.851 ± 0.629^{b}	5.377 ± 0.2554^{ab}	17.65 ± 0.891
Milk kefir with 1.0% MOP	1.009 ± 0.014^{cd}	5.197 ± 0.643^{ab}	18.47 ± 2.328
Milk kefir with 1.5% MOP	0.893 ± 0.135^{bc}	5.303 ± 0.1778^{ab}	17.42 ± 1.775
Milk kefir with 2.0 % MOP	1.115 ± 0.029^{d}	5.07 ± 0.171^{b}	18.77 ± 0.924

Note: MOP Moringa Oleifera Powder

The same superscripts within row show no significant difference (P<0.05). The characteristics of goat milk compositions fortified with moringa leaf powder. This study found that Moringa leaf powder did not significantly affect (P>0.05) the fat level of kefir; the average fat of kefir fortified with Moringa leaf powder was 4.939% (Table 2). It confirms the required fat levels for kefir products according to Standard Codex, i.e., less than 10% (12). The average fat in this study was higher than that 0.38% in kefir whey (15), 2.09% in kefir made of colostrum goat milk (13), and kefir made of additional ginger extract (3.175) (21).

This finding contradicts the previous findings that Moringa oleifera leaf affected the fat level of whey drink (15). Furthermore, increasing the additional level of moringa leaf powder up to 2% did not affect the fat level. Lactic acid bacteria exhibit the secondary lipolytic activity that breaks milk fat into a simpler compound with the help of the lipase enzyme (24). Moringa leaf powder added to kefir did not affect the activity of lactic acid bacteria of the kefir grain that synthesizes fatty acid from the carbohydrate with the help of the lipase enzyme.

Table 2 illustrates that moringa leaf powder does not significantly affect the final protein (P>0.05). The average protein in this study was 7.22% that met the required kefir quality in the Standard Codex, i.e., a minimum of 2.7% (12). Also, the average protein kefir in this study was higher than 3,20% in kefir plus moringa leaf powder (15) or 3.24% in kefir plus ginger extract (21). The protein content of kefir in the present study was equal to 7.32% in goat milk colostrum kefir (13). Ismael, Farahat (15) stated that incorporating moringa leaf powder would affect the protein level of the kefir product.

However, this study found that moringa leaf powder up to 2% did not affect the level of kefir. The contributing factors to protein levels are protein starter (25) and microorganism (26). The increased protein in the previous study may be due to the high protein concentration in moringa leaf powder (27). Table 2 shows that moringa leaf powder does not affect the ash produced (P>0.05). Ash is the percentage of mineral content in a product (28).

The average ash in this study is 9.88%, which is higher than goat milk kefir in two previous findings, namely 0.77% (19) and 5.24% (20), and 5.43% in kefir plus ginger extract (21). The Indonesian National Standard (SNI) has not set the threshold of total ash in a kefir product. However, the maximum total ash, according to SNI for yogurt product is 1.0%. Therefore, supplementing Moringa oleifera leaf up to 1.5% w/w in this study produced the recommended maximum total ash.

At last, this study found that Moringa leaf powder does not affect (P>0.05) the level of carbohydrate (Table 2), which is 5.019% on average. This level of carbohydrate was not significantly different from previous findings, i.e., 5.53% (29), 5,08% (30) and 3.95% (28). Similarly, Hardiansyah (29) stated that carbohydrate analysis investigates the content of carbohydrates (including fiber) through calculation instead of analysis.

Table 2. Proximate Analysis of goat milk kefir fortified with Moringa leaf powder

Treatments	Fat (%)	Protein (%)	Ash (%)	Carbohydrate
Milk kefir	7.6 ± 0.173	5.303 ± 0.439	0.939 ± 0.143	4.704 ± 2.063
Milk kefir with 0.5% MOP	7.533 ± 0.351	4.679 ± 0.324	0.934 ± 0.048	4.503 ± 0.909
Milk kefir with 1.0% MOP	7.2 ± 0.692	4.921 ± 0.324	0.992 ± 0.131	5.348 ± 2.803
Milk kefir with 1.5% MOP	6.933 ± 0.751	4.791 ± 0.351	0.966 ± 0.081	4.709 ± 2.234
Milk kefir with 2.0 % MOP	6.833 ± 0.577	4.999 ± 0.237	1.107 ± 0.109	5.834 ± 1.551

Note : MOP : Moringa Oleifera Powder

4. Conclusion

Moringa leaf powder incorporated in kefir milk-making could decrease pH value and increase acidity without affecting the compositions. Moringa leaf powder in kefir fermentation also increased the chemical properties and met the standard kefir quality in the Codex.

Acknowledgement

The study was received financial support from the Directorate General of Higher Education of Indonesia (Grant No. 008/SP2H/LT-MULTI/LL4/2020).

References

[1] Attaie R, Richter R. Size distribution of fat globules in goat milk. Journal of Dairy Science. 2000;83(5):940-4.
[2] Leite A, Mayo B, Rachid C, Peixoto R, Silva J, Paschoalin V, et al. Assessment of the microbial diversity of Brazilian kefir grains by PCR-DGGE and pyrosequencing analysis. Food microbiology. 2012;31(2):215-21.
[3] Ostadrahimi A, Taghizadeh A, Mobasseri M, Farrin N, Payahoo L, Gheshlaghi ZB, et al. Effect of probiotic fermented milk (kefir) on glycemic control and lipid profile in type 2 diabetic patients: a randomized double-blind placebo-controlled clinical trial. Iranian journal of public health. 2015;44(2):228.
[4] Alyaqoubi S, Abdullah A, Samudi M, Abdullah N, Addai ZR, Al-Ghazali M. Effect of different factors on goat milk antioxidant activity. Int J ChemTech Res. 2014;6:3091-196
[5] Nurliyani, Eni Hermayani, Sunarti. Nitric Oxide and Inflammatory Cytokine Productions in Diabetic Rats Supplemented with Goat Milk and Soy Milk Kefir. Journal of Food and Nutrition Research. 2015;3(6):384-91
[6] Gupta R, Mathur M, Bajaj VK, Katariya P, Yadav S, Kamal R, et al. Evaluation of antidiabetic and antioxidant activity of Moringa oleifera in experimental diabetes. Journal of Diabetes. 2012;4(2):164-71
[7] Arun Giridhari V, Malathi D, Geetha K. Anti diabetic property of drumstick (Moringa oleifera) leaf tablets. International Journal of Health and Nutrition. 2011;2(1):1-5
[8] Hekmat S, Morgan K, Soltani M, Gough R. Sensory evaluation of locally-grown fruit purees and inulin fibre on probiotic yogurt in mwanza, Tanzania and the microbial analysis of probiotic yogurt fortified with Moringa oleifera. Journal of health, population, and nutrition. 2015;33(1):60-7
[9] Nurliyani, Sadewa AH, Sunarti. Kefir properties prepared with goat milk and black rice (Oryza sativa L.) extract and its influence on the improvement of pancreatic [beta]-cells in diabetic rats. Emirates Journal of Food and Agriculture. 2015;27(10):727
[10] Hashim I, Khalil A, Afifi H. Quality characteristics and consumer acceptance of yogurt fortified with date fiber. Journal of dairy science. 2009;92(11):5403-7
[11] AOAC AoOAC. Official Method of Analysis. . Virginia USA.: Association of Official Analytical Chemists Inc; 2006
[12] Commission CA. Codex standard for fermented milks. Alimentarius Commission No CODEX Stan. 2003:243-2003
[13] Setyawardani T, Sumarmono J, Arief II, Rahardjo AHD, Widayaka K, Santosa SS. Improving composition and microbiological characteristics of milk kefir using colostrum. Food Science and Technology. 2020(AHEAD)
[14] Lengkey H, Balia R. The effect of starter dosage and fermentation time on pH and Lactic acid production. Biotechnology in Animal Husbandry. 2014;30(2):339-47.

[15] Ismael S, Farahat A, Ebrahim Y, Gohari S, Ibrahim G. Fortification Effect of Moringa Oleifera Leaves Powder on Nutritional and Volatile Compounds of Sweet Whey Beverage. Journal of Advanced Studies in Agricultural, Biological and Environmental Sciences. 2016;3(2):1-11.
[16] Yıldız-Akgül F, Yetişemiyen A, Şenel E, Yıldırım Z. Microbiological, physicochemical, and sensory characteristics of kefir produced by secondary fermentation. Mljekarstvo/Dairy. 2018;68(3).
[17] Simova E, Beshkova D, Angelov A, Hristozova T, Frengova G, Spasov Z. Lactic acid bacteria and yeasts in kefir grains and kefir made from them. Journal of Industrial Microbiology and Biotechnology. 2002;28(1):1-6.
[18] Joshi P, Mehta D. Effect of dehydration on the nutritive value of drumstick leaves. Journal of metabolomics and systems biology. 2010;1(1):5-9.
[19] Satir G, Guzel-Seydim ZB. How kefir fermentation can affect product composition? Small Ruminant Research. 2016;134:1-7.
[20] Cais-Sokolińska D, Wójtowski J, Pikul J, Danków R, Majcher M, Teichert J, et al. Formation of volatile compounds in kefir made of goat and sheep milk with high polyunsaturated fatty acid content. Journal of Dairy Science. 2015;98(10):6692-705.
[21] Wulansari PD, Rahayu N. Composition of Cow Milk Kefir Enriched with Ginger Extract (Zingiber officinale). Journal of Livestock Science and Production. 2019;3(2):193-8.
[22] BSN NSAoI. Minuman Susu Fermentasi Berperisa. SNI 7552:2009. Jakarta: Departemen Perindustrian; 2009.
[23] Alabi R. Studies On Traditional VVater Pufification Using Moringa oleifera Seeds. African study monographs. 1994;15(3):135-42.
[24] Garofalo C, Osimani A, Milanović V, Aquilanti L, De Filippis F, Stellato G, et al. Bacteria and yeast microbiota in milk kefir grains from different Italian regions. Food microbiology. 2015;49:123-33.
[25] Zanirati DF, Abatemarco Jr M, de Cicco Sandes SH, Nicoli JR, Nunes ÁC, Neumann E. Selection of lactic acid bacteria from Brazilian kefir grains for potential use as starter or probiotic cultures. Anaerobe. 2015;32:70-6.
[26] Golowczyc MA, Silva J, Teixeira P, De Antoni GL, Abraham AG. Cellular injuries of spray-dried Lactobacillus spp. isolated from kefir and their impact on probiotic properties. International journal of food microbiology. 2011;144(3):556-60.
[27] Madukwe E, Ugwuoke A, Ezeugwu J. Effectiveness of dry Moringa oleifera leaf powder in treatment of anaemia. International Journal of Medicine and Medical Sciences. 2013;5(5):226-8.
[28] Sulistyaningtyas A, Lunggani A, Kusdiyantini E, editors. Kefir Produced from Red Rice Milk by Lactobacillus bulgaricus and Candida kefir Starter. IOP Conference Series: Earth and Environmental Science; 2019: IOP Publishing.
[29] Hardiansyah A. Identifikasi Nilai Gizi Dan Potensi Manfaat Kefir Susu Kambing Kaligesing. Journal of Nutrition College. 2020;9(3):208-14.
[30] Listiawati M, Kurniati T, Windayani N, editors. An effective alternative of pre-dawn meal using milk kefir. Journal of Physics: Conference Series; 2019: IOP Publishing.

The effect of safflower oil (Carthamus Tinctorius L.) and inositol supplementation on egg production

Ismoyowati[1*], N Iriyanti[1] and M Abdelgader[12]

[1]Universitas Jenderal Soedirman, Purwokerto, Indonesia.
[2]Omdurman Islamic University, Khartoum, Sudan.

*ismoyowati@unsoed.ac.id

Abstract. The study aimed to examine the effect of safflower oil and inositol supplementation in a feed on egg production performance and feed conversion ratio of Sentul chickens. The research was conducted using experimental methods based on a completely randomized design (CRD). The materials used were 81 female Sentul chickens aged 16 weeks, which were divided into 9 groups of feed treatment, namely R0: control feed; R1: feed containing 0.5% safflower oil; R2: feed containing 1% safflower oil; R3: feed containing 0.5% inositol; R4: feed containing 1% inositol; R5: feed containing 0.5% safflower oil and 0.5% inositol; R6: feed containing 0.5% safflower oil and 1% inositol; R7: feed containing 1% safflower oil and 0.5% inositol; R8: feed containing 1% safflower oil and 1% inositol. The variables measured consisted of egg production, egg weight, egg mass and feed conversion ratio (FCR). Data were analyzed by analysis of variance. The results showed that the treatment had no significant effect ($P < 0.05$) on all measured variables. The average number of eggs, egg weight, egg mass and FCR for 8 weeks of observation were: 66 ± 9 eggs; 37.31 ± 2.19 g; 16.44 ± 3.31 and 2.26 ± 0.37 The conclusion of this study was the supplementation of safflower oil and inositol with a concentration of 0.5% -1% in feed has not been able to improve the performance of egg production, and the feed efficiency of Sentul chickens.

1. Introduction

Sentul chicken is a native chicken from Ciamis district, including the dual-purpose type, so that it can be used as a meat and egg production [1]. Sentul chicken has the characteristics of fast body weight gain, disease resistance and relatively higher egg production compared to other local chickens. Sentul chickens in one laying period produce eggs as many as 10-18 eggs [2] or about 150 eggs / year [3]. Efforts to improve egg production performance and feed efficiency can be made by improving feed quality. Feed quality improvement can be done by providing feed additives. One function of the feed additive as an immunomodulator to enhance the performance of health in order to increase production. Supplementation of safflower oil and inositol into feed is one of the feed additives that can be added to feed. Feeding efficiency can be measured based on feed conversion.

Safflower oil contains fat-soluble, vitamin retinol, calciferol, tocopherol and phylloquinone as well as unsaturated linoleic fatty acids [4]. The main unsaturated fatty acids in Safflower oil are oleic and linoleic acid. This content accounts for about 90% of total fatty acids and the remaining 10% is saturated fatty acids, namely palmitic and stearic acid [5]. Linoleic fatty acids can control the proteins and lipids needed for follicular development which directly control the production of eggs.

Inositol is a natural nutrient composed of sugar alcohol isomers with a C6 chain and belongs to the B-complex group of vitamins. Inositol is a type of vitamin that the body needs for growth, maintenance and reproduction. Previous studies reported, arginine silicate inositol complex supplementation increased egg production, egg weight, shell weight and shell thickness during the laying period [6]. Based on this background, this study aims to determine the effect of safflower oil (Carthamus inctorius L) and inositol supplementation on egg production performance and feed conversion ratio of Sentul Chicken.

2. Materials and Methods

The research material used 81 female Sentul chickens; 16 weeks old. Chicken rearing was carried out for 12 weeks, and observation of egg production performance and feed conversion was carried out for 8 weeks. The feed additives used as treatment were safflower oil and inositol. Feed ingredients and feed nutrient content are presented in table 1. The research equipment consists of battery cages, feed and drinking water containers and digital scales.

Table 1. Composition of feed ingredients and nutrient content of the feed treatment

Feed ingredients	R_0	R_1	R_2	R_3	R_4	R_5	R_6	R_7	R_8
					%				
Milled corn	52	52	52	52	52	52	52	52	52
Rice bran	22	22	22	22	22	22	22	22	22
Soybean meal	11.5	11.5	11.5	11.5	11.5	11.5	11.5	11.5	11.5
Fish meal	10	10	10	10	10	10	10	10	10
Palm oil	0.5	0.5	0.5	0.5	0.5	0.5	0.5	0.5	0.5
CaCO3	2	2	2	2	2	2	2	2	2
Topmix	1	1	1	1	1	1	1	1	1
Lysine	0.6	0.6	0.6	0.6	0.6	0.6	0.6	0.6	0.6
Methionine	0.4	0.4	0.4	0.4	0.4	0.4	0.4	0.4	0.4
Safflower oil	0	0.5	1	0	0	0.5	0.5	1	1
Inositol	0	0	0	0.5	1	0.5	1	0.5	1
Total	100	100.5	101	100.5	101	101	101.5	101.5	102
				The content of nutrients					
CP (%)	19.34	19.34	19.34	19.34	19.34	19.34	19.34	19.34	19.34
CL (%)	11.99	11.99	11.99	11.99	11.99	11.99	11.99	11.99	11.99
CF (%)	10.18	10.18	10.18	10.18	10.18	10.18	10.18	10.18	10.18
ME (kcal/kg)	2853.60	2860.40	2867.21	2853.60	2853.60	2860.40	2860.40	2867.21	2867.21
Ca (%)	2.44	2.44	2.44	2.44	2.44	2.44	2.44	2.44	2.44
P (%)	0.34	0.44	0.44	0.44	0.44	0.44	0.44	0.44	0.44
Lysine (%)	1.12	1.12	1.12	1.12	1.12	1.12	1.12	1.12	1.12
Methionine (%)	0.68	0.68	0.68	0.68	0.68	0.68	0.68	0.68	0.68

The study was conducted using an in vivo experimental method, with an experimental design, namely a completely randomized design (CRD) consisting of 9 treatments and 3 replications, with each replication consisting of 3 chickens. The treatments given in this study were, R0 = basal feed (0% safflower oil and 0% inositol); R1 = basal feed + 0.5% safflower oil; R2 = basal feed + 1.0% safflower oil; R3 = basal feed + 0.5% inositol; R4 = basal feed + 1.0% inositol; R5 = basal feed + 0.5% safflower oil and 0.5% inositol; R6 = basal feed + 0.5% safflower oil and 1.0% inositol; R7 = basal feed + 1.0% safflower oil and 0.5% inositol and R8 = basal feed + 1.0% safflower oil and 1.0% inositol. The research variables observed were the number of eggs (eggs / head / 8 weeks), egg weight (g), egg mass and FCR.

Measurement of feed consumption is carried out every day. The feed conversion ratio is calculated using the formula: feed consumption (g) / total egg weight (g). The measurement of egg mass is obtained by dividing the total egg weight (g) produced by the chicken population (birds). The data obtained were then entered into data tabulations and analyzed using analysis of variance.

IOP Conf. Series: Earth and Environmental Science 1041 (2022) 012068 doi:10.1088/1755-1315/1041/1/012068

3. Result and Discussion

The results of the research on the effect of safflower oil and inositol supplementation with a level of 0.5-1% are presented in Table 1. Analysis of variance showed that the treatment had no significant effect (P <0.5) on all measured variables.

Table 2. Mean and standard deviation of egg number, egg weight, consumption and feed conversion ratio of Sentul chickens with safflower oil and inositol supplementation in the feed

Treatments	Egg production^{ns}	Egg weights^{ns}	Egg mass^{ns}	Feed consumption^{ns}	FCR^{ns}
R0	58.33±10.02	36.00±2.75	16.28±4.93	4234.37±279.11	2.42±0.14
R1	73.00±9.00	37.43±0.23	19.68±4.23	4315.93±331.69	2.33±0.29
R2	71.00±12.17	39.18±3.36	19.99±3.48	4086.50±213.23	2.19±0.31
R3	65.67±7.77	38.52±4.20	15.85±4.09	3989.58±337.32	2.30±0.24
R4	63.33±10.07	37.68±0.09	16.55±2.73	3971.90±324.19	2.37±0.19
R5	60.67±7.23	36.67±1.04	16.35±3.09	4246.95±363.97	2.27±0.18
R6	68.33±5.51	35.54±2.14	13.03±2.57	3898.29±245.54	2.02±0.21
R7	73.00±6.93	37.66±1.01	16.92±2.31	4221.79±193.73	2.18±0.17
R8	72.33±3.06	37.33±1.84	17.53±1.34	3837.20±207.74	2.29±0.09

Note: R0: feed control; R1: feed containing 0.5% safflower oil; R2: feed containing 1% safflower oil; R3: feed containing 0.5% inositol; R4: feed containing 1% inositol; R5: feed containing 0.5% safflower oil and 0.5% inositol; R6: feed containing 0.5% safflower oil and 1% inositol; R7: feed containing 1% safflower oil and 0.5% inositol; R8: feed containing 1% safflower oil and 1% inositol; ns: non-significant (P <0.05).

Supplementation of safflower oil and inositol in feed by 0.5-1%, has not been effective in increasing egg production and feed efficiency. Safflower oil is an unsaturated fatty acid that controls protein for follicle development and egg weight. Safflower oil supplementation with a percentage of 0.5% and 1% is assumed to not be able to control protein and lipids in the vitellogenin process. This result is consistent with research [7] which reported that chicken feed containing 0-10% safflower seed had no significant effect on egg production, egg mass and feed intake. Layer chicken feed supplemented with two levels (1.5 and 3%) of different oils, namely sunflower, rapeseed, linseed for 12 weeks resulted in relatively equal egg production, egg weight, feed intake, FCR and live weight of the chickens.

However, chickens whose feed had sunflower oil had a lighter yolk color [8]. Inositol supplementation also has not played an effective role in increasing the digestibility of feed protein derived from grains, because it is bound by phytate. Most of the phosphorus in feedstuffs from plants (about 60 to 70%) is in the form of phytate (IP6), which makes it less available to monogastric animals. Phytates are also able to bind nutrients (such as Ca, Zn, Fe, Mg, Cu, starch, lipids, and certain amino acids) so that these nutrients are less available to poultry [9].

The results of the study in Table 2 indicate that there is a tendency to increase the performance of egg production (number of eggs, egg weight, egg mass) and feed efficiency (decrease in the value of the feed conversion ratio) by supplementing with safflower oil and inositol. The use of safflower oil as a source of omega-3 and omega-6 in the ration of laying hens has a good effect on egg production as well as inositol. During the production period chickens need 1-2.5% linoleic acid. Linoleic acid consumed can affect the performance of egg production. If the intake of linoleic acid decreases, it can reduce the performance of egg production [10]. The availability of adequate fat and protein in the ration results in an increase in the maturation rate of follicles, so that the maturation phase can increase the number of follicles [11].

Inositol is one of the vitamins needed for normal body growth, maintenance and reproduction. Inositol can function to break the bonds between phytic acid and proteins, lipids and minerals. Although inositol supplementation has not significantly increased egg production, the results showed that there was an increase in egg production and feed efficiency. Inositol plays a role in the physiology of reproductive hormones and also increases the digestibility of certain feed ingredients from whole grains.

Inositol contained in a feed can reduce levels of anti-nutritional phytic acid found in grains and can break the bonds between phytic acid and protein, lipid and mineral complexes [12]. Inositol with sufficient doses has a positive impact on digestive enzymes, so that the digestibility of feed nutrients increases. High digestibility and absorption of feed nutrients can be used to increase egg production, so that the feed conversion ratio decreases.

4. Conclusion

Safflower oil and inositol supplementation with a level of 0.5-1%, either singly or in combination, has not been able to improve egg production performance and feed efficiency of Sentul chickens.

Acknowledgement

Thank you to the Rector of Universitas Jenderal Soedirman, for funding research skims Institution Research, with a contract agreement letter no. No: T / 318 / UN23.18 / PT.01.03 / 2020.

References

[1] Ermansyah G, W Tanwiriah dan YA Indrawati. 2015. Pengaruh Pemberain Tepung Ampas Tahu didalam Ransum Terhadap Bobot Potong, Bobot Karkas dan Income Over Feed Cost Ayam Sentul. *Students e-Journal* Fakultas Peternakan Universitas Padjadjaran **4**(4):1-6.

[2] Widjastuti T. 2009. Pemanfaatan Tepung Daun Pepaya (Carica Papaya.L L Ess) Dalam Upaya Peningkatan Produksi dan Kualitas Telur Ayam Sentul. *J. Agroland* **16**(3):268-273.

[3] Baktiningsih S, S Mugiyono. dan DM Saleh. 2013. Produksi Telur Berbagai Jenis Ayam Sentul di Gabungan Kelompok Tani Ternak Ciung Wanara Kecamatan Ciamis Kabupaten Ciamis. *Jurnal Ilmiah Peternakan* **1**(3):993-1000.

[4] Katkade, Syed Andhalen,.Dan Sontakke. 2018. Fatty Acid Profile And Quality Assessment Of Safflower (Carthamus tinctorius) Oil. *Journal Of Pharmacognosy And Phytochemisty*. **7**(2):3581-3585.

[5] Khalid N, RS Khan., MI Husaain, M Faroq, A Ahmad, I Ahmed. 2017. A Comperehensive Characterisation Of Safflower Oil For Its Potential Applications As A Bioactive Food Ingredient. *Trend in food science and technology*. **66**(17):176-186.

[6] Sahin K, C Orhan, M Tuzu, JR Komorowski and N Sahin. 2018. Effects of dietary supplementation of arginine-silicate-inositol complex on absorption and metabolism of calcium of laying hens. *PLoS ONE* **13**(1):e0189329. https://doi.org/10.1371/journal.pone.0189329.

[7] Vashan,SJH. N Afzali , M.Mallekaneh , MA Nasseri and A. Allahresani. 2008. The Effect of Different Concentrations of Safflower Seed on Laying Hen's Performance, Yolk and Blood Cholesterol and Immune System. *International Journal of Poultry Science*. **7**(5): 470-473.

[8] Alagawany M, SS.Elnesr, MR. Farag, ME Abd El-Hack, AF.Khafaga, AE Taha, R.Tiwari, MI,Yatoo, P Bhatt, SK. Khurana, and K. Dhama. 2019. Omega-3 and Omega-6 Fatty Acids in Poultry Nutrition: Effect on Production Performance and Health. *Animals*, **9**(573):1-19. doi:10.3390/ani9080573.

[9] Farhadi D, A Karimi, Gh Sadeghi, J Rostamzadeh and MR Bedford. 2017. Effects of a high dose of microbial phytase and myo-inositol supplementation on growth performance, tibia mineralization, nutrient digestibility, litter moisture content, and foot problems in broiler chickens fed phosphorus-deficient diets. *Poultry Science*. **96**:3664–3675 http://dx.doi.org/10.3382/ps/pex186.

[10] Mampioper A, S D Rumetor dan F Pattiselanno. 2008. Kualitas Telur Ayam Petelur yang Mendapat Ransum Perlakuan Substitusi Jagung dengan Tepung Singkong. *Jurnal Ternak Tropika*. **2**(9):42-51.

[11] Suprijatna EU. 2002. Manifestasi Taraf Protein Ransum Periode Pertumbuhan Terhadap Pertumbuhan Organ Reproduksi dan Dampaknya Pada Performan Produksi Ayam Petelur Tipe Medium. *Disertasi* .Program Pascasarjana Universitas Padjadjaran Bandung.

[12] Żyła K, M Mika, R.Duliński, S.Świątkiewicz , J Koreleski, H Pustkowiak, and J Piironen. 2012. Effects of inositol, inositol-generating phytase B applied alone, and in combination with 6-phytase A to phosphorus-deficient diets on laying performance, eggshell quality, yolk cholesterol, and fatty acid deposition in laying hens. *Poultry Science*. **91**:1915–1927. http://dx.doi.org/ 10.3382/ps.2012-02198.

The Anti-Inflammatory Activity of Cherry Leaf Extract (*Muntingia Calabura* L.) Balm Stick

Fitria Nugrahaeni*, Kriana Efendi, Abdul Kholik Aziz

Universitas Muhammadiyah Prof. DR. HAMKA, Jakarta, Indonesia

*fitria.nugrahaeni@uhamka.ac.id

Abstract. In previous studies, cherry leaf extract has been shown to be anti-inflammatory. This means a preparation is needed to deliver the extract. Balm stick is an innovation in a stem-shaped balm that makes it easier to be used so that cherry leaf extract is made into a balm stick preparation. This study aims to determine the anti-inflammatory activity of the balsam stick of cherry leaf extract topically. The study was conducted by varying the concentration of cherry leaf extract at 2.5%, 5%, and 10% and tested its anti-inflammatory activity in male white rats induced by carrageenan. The experimental animals were divided into 5 test groups, whereby the positive control group was given 2.5% hydrocortisone balm stick, the negative control group was given the balm stick preparation and the 3 test groups were given 2.5%, 5%, and 10%. The observations were made using a plethysmometer by looking at the volume of edema in the carrageenan-induced rat paws. Balm sticks with a concentration of 5% and 10% had an inhibitory power of more than 50% with a value of 70.27% and 95.83% while a concentration of 2.5% had an inhibitory power of 44.44%. Balm stick ethanol extract of cherry leaves has anti-inflammatory activity with concentrations of 5% and 10% and demonstrates an increase in anti-inflammatory activity, whereby the greater the concentration, the greater the anti-inflammatory activity produced.

1. Introduction

Advances in science and technology that are increasingly rapid and sophisticated do not shift the role of traditional medicine. In Indonesia, there are about 30,000 types of plants and more than 1,000 species have been known for their benefits as medical plants. One of the medical plants that can be used is cherry [1].

Kersen (*Muntingia calabura* L.) is a fruiting tree that can grow in less fertile soil and is able to tolerate acid, alkaline, and drought conditions. Cherry leaves have various pharmacological effects such as cardioprotective, antipyretic, antioxidant, anti-inflammatory, antidiabetic, antibacterial, and antiulcer [2]. Flavonoids, saponins, and tannins are the ingredients in cherry leaves. The flavonoids contained in cherries are flavones, flavanones, flavans, and biflavans. The content of flavonoids has received much attention as this group of compounds has many activities, one of which is anti-inflammatory [3].

Inflammation is a normal protective response to tissue injury caused by physical trauma, damaging chemical substances or microbiological substances. Broadly speaking, inflammation is divided into two basic patterns; namely acute inflammation and chronic inflammation [4]. Tissue damage due to trauma, microbial invasion, or harmful compounds can cause acute inflammation lasting several days characterized by sub-acute inflammation. Sub-acute inflammation is the period between acute and chronic inflammations [5]. One of the most commonly used methods of Paw edema is based on its ability to inhibit the edema produced in the paws of mice after injection of carrageenan. The effect that can be measured is the volume of edema in the feet of the proinflammatory rats injected before and after using a plethysmometer[6].

The stick balm is an innovation from a stem-shaped balm to make it easier to be used and to avoid hand soiling, so cherry leaf extract is made in a balm stick preparation. The selection of the balm stick formula refers to the formulation carried out by Yati et al. with the concentration that can produce the balm stick with the best physical properties. The formulation of the balm stick consists of wax, fat, and oil[7]. The oil acts as an emollient to give a smooth and soft texture to the skin when applied.

The oil used as a base is olive oil. It is a fatty oil derived from ripe olives (*Olea europaea* L.). Olive oil has many benefits for body health, facial beauty, hair and skin protection, especially due to the high content of oleic acid in olive oil. [8]. The use of olive oil as a base has many advantages with an emollient effect and does not clog pores which makes it unable to cause irritation and is safe to use [9]. This study was conducted to determine the effect of the concentration of balsam sticks of cherry leaf ethanol extract on the inhibition of edema in the feet of male white rats induced by carrageenan.

2. Material and Methods

Materials

Cherry leaf extract (*Muntingia calabura* L.), which was obtained and identified by the Conservation Research Center and Botanical Garden, Indonesian Institute of Sciences in Bogor, Indonesia contains Cera alba (Merck), adeps lanae (Mercks), cetyl alcohol (Mercks), butyl hydroxytoluene (Mercks), olive oil (Bertolli), and carrageenan (Mercks). The comparison material used in this study was hydrocortisone (PT. Kimia Farma).

Mixture Test

This mixture test aims to determine the use of an appropriate surfactant in the formulation of balsam sticks of cherry leaf ethanol extract. The test is carried out with a ratio of 1:1 ethanol extract of cherry leaf and surfactant, which was followed by an observation of its mixability and stability [9].

Balm Stick Making

The balm stick is made with different concentrations of cherry leaf extract as shown in the table below.

Table 1. Formula for Making Balm Stick Preparations

Materials	Functions	Amount			
		F I(%)	F II(%)	F III(%)	F IV(%)
Cherry extract	Active substance	2,5	5	10	-
Cera alba	Hardener	30	30	30	30
Adeps lanae	Fastener	10	10	10	10
Cherry extract	Plasticizer	10	10	10	10
Butyl Hydroxytoluene	Antioxidant	0.1	0.1	0.1	0.1
Span 80	Emulsifier	5	5	5	-
Olive oil	Emollient	Ad 100	Ad 100	Ad 100	Ad 100

The Manufacturing of Anti-Inflammatory Test Materials

As a comparison material (positive control), hydrocortisone powder from Kimia Farma, which was made of 2.5% Stick Balm, was used. 1g of carrageenan powder was then dissolved with 0.9% NaCl to 100 ml in a Beaker glass. It was then stirred until it was dissolved completely so that the concentration obtained is 1% w/v [6].

Grouping and Treatment of Test Animals

The study was conducted experimentally with a completely randomized design, using 25 male white rats which were divided into 5 groups consisting of 5 rats. The division of the groups of rats was as follows: Group I (positive control) was given 2.5% of hydrocortisone stick balm; Group II (negative control) was given a balm stick base (F4); Group III (F1) was given 2.5% of cherry leaf extract balm stick; Group IV(F2) was given 5% of cherry leaf extract balm stick; and Group V (F3) was given 10% of cherry leaf extract balm stick.

After acclimatization for 7 days, the right hind leg of the rats was marked and measured using a plethysmometer as the initial volume. One hour before the induction of carrageenan, the right leg of the rat was smeared with the test preparation for each group 50 times to help penetrate the test preparation through the skin [7]. An hour after the administration of the test preparation, each group was induced by 0.1 ml of 1% w/v carrageenan subplantarly on the paws of the rats that had been marked. Measurements were made every 1 hour for 5 hours. Changes in the level of inflammation that occurred were recorded as the volume of the rats' paws. From the data obtained, the percentage of inflammation inhibition can be determined [6].

The data obtained were tested statistically by using one-way analysis method (One Way ANOVA), to see the effect of the treatment. If it meets the requirements for the ANOVA test and if there is a treatment effect, then it is continued with the Tukey HSD test. If it does not meet the requirements of the ANOVA test, then it is continued with the Kruskall-Wallis test. Meanwhile, if there is an effect of treatment then it is followed by the Mann-Whitney test.

3. Results and Discussion

In this study, the cherry leaves used were obtained from the Research Institute for Spices and Medicinal Plants (BALITRO). The plant determination carried out by the BALITRO laboratory showed that the plant to be used in this study was cherry leaf (*Muntingia Calabura* L.*)*. The yield of cherry leaf extract was 17.92%, with water content of 20.69%, and ash content 8.92%. In addition, the results of the phytochemical screening of extracts were positive for alkaloids, flavonoids, tannins, and saponins [10].

The results of the admixture test showed that both surfactants were mixed with the extract and span 80 was more mixed than tween 80. Therefore, span 80 was used for the formulation. This took place because cherry leaf extract had an HLB that was close to span 80.

The results of the organoleptic test in this study of the three formulations of the balm stick obtained the results of the texture and smell that were not much different from one formula to another. The color had a difference that F3 was more concentrated than F2 and F2 was more concentrated than F1. This shows that the concentration of the extract has no effect on the texture and smell. However, the difference in the concentration of the extract shows that the more concentration of extract used, the more concentrated the color produced on the balm stick preparation.

The results of the homogeneity test in this study of the three balm stick preparations in the three formulas showed a homogeneous arrangement. This demonstrates that all the components used in the balm stick preparation are mixed homogeneously during melting and grinding which means that after printing there are no coarse grains. It can be concluded that there is no effect of the concentration of the extract on the homogeneity of the balm stick preparation.

The pH test aimed to see whether the preparation had a pH suitable for the skin or not. The pH test was carried out using a pH meter. The test results obtained were the pH of the preparation according to a normal skin pH of 4-6 [11]. Therefore, the difference in concentration of cherry leaf extract did not affect the pH of the balsam stick preparations made.

In this melting point test, the results obtained met the requirements of the SNI 16-4769-1998 standard, namely 50-70°C and a good balm stick preparation had a melting point of more than 50% [3]. Each of these formulas had a very high melting point because the formulation of the balm stick had a high melting point and the olive oil base contained oleic acid which was an unsaturated fatty acid composed of 18 C atoms with one double bond [12]. There was an effect of different concentrations of cherry leaf extract on the melting point shown by the one-way ANOVA test ($p<0.05$). This is because the more concentration of cherry leaf extract used, the greater the molecular weight which can affect the higher melting point (Muchson, 2013). Tukey test results showed that formula 1 was comparable to formula 2 and formula 4 ($P>0.05$). Meanwhile, formula 3 was comparable to formula 2 ($P>0.05$) and formula 3 was significantly different from formulas 1 and 4 ($P<0.05$).

According to Hariningsih (2017), the higher the adhesive power of the lipstick, the better the ability of the lipstick to adhere which makes it difficult to be removed. Moreover, the longer the stickiness of a food preparation, the more drugs that can be absorbed into the body. The results of the adhesion test in this study met the requirements of more than 4 seconds (Puspitasari et al., 2018). The results of the one-way ANOVA test showed that there was a significant difference between groups ($P<0.05$), which means that the concentration of cherry leaf extract affected the stickiness. In the Tukey test, F1 was significantly different from F2, F3 and F4 ($P<0.05$).

In the hardness test according to Balsam [3], the preparation is said to be soft if the depth of penetration of the needle is 9-10.5 with a load of 50g. The deeper the needle penetrates, the softer the preparation is. The results of the hardness test obtained from the formulation of the balm stick made from the four formulas met the requirements and were not too soft. The one-way ANOVA test showed that there was a significant difference between groups ($P>0.05$) and the results of the Tukey F1 test were comparable to F4 ($P>0.05$). Nonetheless, F1 was significantly different from F2 and F3 ($P<0.05$). In the meantime, F2 was significantly different from F3 and F4, and F3 was different from F4 ($P<0.05$).

The results of the hardness test are used to find the yield value, namely the pressure required for the preparation to spread. The yield value is inversely proportional to the needle penetration depth. According to Kadu, the greater the needle penetration depth, the smaller the required yield value. This means that the softer the preparation was made, the less effort to apply the preparation which makes it able to spread over the skin [13]. The results obtained by the yield value met the requirements of 100-1000 dyne/cm^2. Hence, itcan be concluded that the preparations made had good dispersion [7].

Table 2. Percentage of Udem Inhibition

Groups		Observation Time Hours				
		T60	T120	T180	T240	T300
				(%)		
Group 1 Positive Control		24,04	69,17	76,66	100	100
	±	±	±	±	±	±
	SD	32,03	41,21	23,13	0	0
Group 3 (2,5%)		18,33	46,10	-13,33	27,78	44,44
	±	±	±	±	±	±
	SD	23,86	24,66	77,36	32,15	44,27
Group 4 (5%)		16,67	25,00	10,84	56,10	70,27
	±	±	±	±	±	±
	SD	20,41	58,46	58,74	11,36	30,27
		23,33	35,00	27,50	84,16	95,83
	±	±	±	±	±	±

Group 5 (10%)	SD	34,05	57,38	40,14	25,42	9,32

Notes :

$\bar{x}$: average inhibition of edema each time

SD : standard deviation of edema inhibition over time

According to Amdeekar, anti-inflammatory treatment is successful if there is a 50% inhibition of inflammation [14]. From the average percent of edema inhibition obtained, it showed anti-inflammatory activity in each group. The anti-inflammatory activity obtained from the test group that had more than 50% anti-inflammatory activity occurred in group 4 (F2 5%) and group 5 (F3 10%). The anti-inflammatory activity in these groups was seen at the 240th and 300th minute observations.

In this study, the concentration of ethanol extract of cherry leaves could provide an anti-inflammatory effect of more than 50%, namely 5% (group 4) and 10% (group 5), by reducing the volume of edema in the feet of rats induced by carrageenan. According to Kuo [12], cherry leaves have anti-inflammatory properties because they contain flavonoids. According to Triswaningsih [10], flavonoids have a benzopyran ring structure that can bind to the cyclooxygenase and lipoxygenase enzymes which makes them able to inhibit the formation of arachidonic acid turning into prostaglandins and leukotrienes which play a role in the inflammatory process. When the formation of arachidonic acid is inhibited by binding to the cyclooxygenase enzyme, inflammation can be thwarted.

The anti-inflammatory activity produced by this cherry leaf extract balm stick is possible by way of the drug being absorbed from the outer layer of the skin by diffusion through the skin barrier, namely 3 skin compartments consisting of the outer surface of the skin, the stratum corneum, and the living tissue beneath. After being applied to the outer surface of the skin, the drug will undergo changes in structure and composition which would determine the bioavailability of the drug. The three main routes of penetration of topical drugs in the intact stratum corneum are the transcellular pathway (through the stratum corneum), the intracellular pathway (diffusion through the lipid matrix between cells), as well as the hair follicle and sweat gland pathway [15].

4. Conclusion

In this study, balsam sticks with ethanol extract of cherry leaves had anti-inflammatory activity at concentrations of 5% and 10%. They showed an increase in anti-inflammatory activity, whereby the greater the concentration, the greater the anti-inflammatory activity produced. This means that the cherry leaf extract balm stick can be used as one of the newest anti-inflammatory drug delivery systems.

References

[1] N. D. Mahmood *et al.*, "Muntingia calabura: A review of its traditional uses, chemical properties, and pharmacological observations," *Pharm. Biol.*, vol. 52, no. 12, pp. 1598–1623, 2014, doi: 10.3109/13880209.2014.908397.

[2] A. D. Puspitasari, D. A. K. Mulangsri, and H. Herlina, "Formulasi Krim Tabir Surya Ekstrak Etanol Daun Kersen (Muntingia calabura L.) untuk Kesehatan Kulit," *Media Penelit. dan Pengemb. Kesehat.*, vol. 28, no. 4, pp. 263–270, 2018, doi: 10.22435/mpk.v28i4.524.

[3] A. Athaillah and S. O. Lianda, "FORMULASI DAN EVALUASI SEDIAAN BALSEM STIK DARI OLEORESIN JAHE MERAH (Zingiber officinale rosc) SEBAGAI PEREDA NYERI OTOT DAN SENDI," *J. Pharm. Sci.*, vol. 4, no. 1, pp. 34–40, 2021, doi: 10.36490/journal-jps.com.v4i1.62.

[4] J. M. Alvarez-Suarez *et al.*, "Anti-inflammatory effect of Capuli cherry against LPS-induced cytotoxic damage in RAW 264.7 macrophages," *Food Chem. Toxicol.*, vol. 102, no. January, pp. 46–52, 2017, doi: 10.1016/j.fct.2017.01.024.

[5] H. N. Pise and S. L. Padwal, "Evaluation of anti-inflammatory activity of nigella sativa: An experimental study," *Natl. J. Physiol. Pharm. Pharmacol.*, vol. 7, no. 7, pp. 707–711, 2017, doi: 10.5455/njppp.2017.7.0204705032017.

[6] D. B. Duarte, M. R. Vasko, and J. C. Fehrenbacher, "Models of inflammation: Carrageenan air pouch," *Curr. Protoc. Pharmacol.*, vol. 2016, no. March, pp. 5.6.1-5.6.9, 2016, doi: 10.1002/0471141755.ph0506s72.

[7] L. P. Dwita, K. Yati, and S. N. Gantini, "The anti-inflammatory activity of nigella sativa balm sticks," *Sci. Pharm.*, vol. 87, no. 1, 2019, doi: 10.3390/scipharm87010003.

[8] A. Widyasanti and J. M. Rohani, "Pembuatan sabun padat transparan berbasis minyak zaitun dengan penambahan ekstrak teh putih," *J. Penelit. Teh dan Kina*, vol. 20, no. 1, pp. 13–29, 2017.

[9] S. E. Priani, W. K. Dewi, and A. Gadri, "Formulasi Sediaan Mikroemulsi Gel Anti Jerawat Mengandung Kombinasi Minyak Jinten Hitam (Nigella sativa L.) dan Minyak Zaitun (Olea europaea L.)," *Kartika J. Ilm. Farm.*, vol. 6, no. 2, p. 57, 2019, doi: 10.26874/kjif.v6i2.143.

[10] D. Triswaningsih, S. Kumalaningsih, Wignyanto, and Pratikto, "Identification of chemical compounds cherry leaves (Muntingia calabura) powder as a natural antioxidant," *Int. J. Agron. Agric. Res.*, vol. 10, no. 5, pp. 84–91, 2017, [Online]. Available: http://www.innspub.net.

[11] F. Nugrahaeni, D. M. Hariyadi, and N. Rosita, "Partition coefficient and glutathione penetration of topical Antiaging: Preformulation study," *Int. J. Drug Deliv. Technol.*, vol. 8, no. 2, pp. 39–43, 2018.

[12] W. L. Kuo, H. R. Liao, and J. J. Chen, "Biflavans, flavonoids, and a dihydrochalcone from the stem wood of muntingia calabura and their inhibitory activities on neutrophil pro-inflammatory responses," *Molecules*, vol. 19, no. 12, pp. 20521–20535, 2014, doi: 10.3390/molecules191220521.

[13] Kadu dkk, "Review on Natural Lip Balm. International Journal of Research in Cosmetic Science.," *Int. J. Res. Cosmet. Sci.*, no. August 2014, p. Halaman 1-2, 2014.

[14] S. Amdekar, P. Roy, V. Singh, A. Kumar, R. Singh, and P. Sharma, "Anti-inflammatory activity of lactobacillus on carrageenan-induced paw edema in male wistar rats," *Int. J. Inflam.*, vol. 2012, 2012, doi: 10.1155/2012/752015.

[15] T. K. Lin, L. Zhong, and J. L. Santiago, "Anti-inflammatory and skin barrier repair effects of topical application of some plant oils," *Int. J. Mol. Sci.*, vol. 19, no. 1, 2018, doi: 10.3390/ijms19010070.

The Effect of HPMC Concentration as a Gelling Agent on Color Stability of Copigmented Blush Gel Extract of Purple Sweet (*Ipomoea Batatas* (L.) Lam.)

Fitria Nugrahaeni*, Nining, Redina Okvianida

Universitas Muhammadiyah Prof. DR. HAMKA, Jakarta, Indonesia

*fitria.nugrahaeni@uhamka.ac.id

Purple sweet potato contains anthocyanin pigments that can be used as natural dyes. However, the use of anthocyanins as dyes has less stable properties. Anthocyanin stability can be increased by the addition of copigments. Copigment that can be used is apple extract at pH 4.6. In this study, a blush was made. One of the blush dosage forms that can be made is gel. Blush gel can provide a cool feeling, easy to wash, soft, and not sticky on the skin. In the creation of gels, the gelling agent is one of the components that affect the characteristics of gel preparations. HPMC is a gelling agent used in the manufacture of blush gel preparations. This study aims to determine the effect of increasing the concentration of HPMC as a gelling agent on the color stability of the copigmented purple sweet potato extract blush gel. The preparations were made in 3 formulas with varying concentrations of gelling agent (F1) 1%, (F2) 1.5%, and (F3) 2%. The evaluation carried out included testing the physical characteristics of the gel, namely organoleptic test, pH test, dispersion test, phase separation test, color stability test, and preference test. The results showed that the evaluation of blush gel preparations met the requirements of pharmaceutical standards with the pH test value of blush gel preparations being 4.72 to 5.84 and the spreadability test value being 5.26 cm to 6.48 cm. The results of the color stability test on the three formulas showed that there were significant differences. The increase in the concentration of HPMC as a gelling agent on the color stability of the copigmented purple sweet potato extract blush gel was unstable during 28 days of storage.

1. Introduction

Blush is a cosmetic preparation that is used to color the cheeks with an artistic touch so that it can improve aesthetics in facial makeup [1]. Blush is available in various color choices, but blush products on the market use synthetic dyes as their dyes [2]. Concerns about the safety of using synthetic dyes encourage the development of natural dyes. Natural dyes were chosen because they are safe if used in the long term. One of the natural dyes that have the potential to replace synthetic dyes is anthocyanins found in purple sweet potato [3].

Purple sweet potato is used as a natural dye because it contains 11.051 mg/100 gram of anthocyanin which is higher than other varieties [4]. In addition, the anthocyanins in purple sweet potatoes can be used as antioxidant compounds that are very useful for the body. The use of anthocyanins as dyes is less stable to oxygen, light, pH, sugar, and temperature [5]. Hence, it is necessary to make efforts to improve its stability. To increase the stability of anthocyanins, it is necessary to add copigmentation [6].

IOP Conf. Series: Earth and Environmental Science 1041 (2022) 012070 doi:10.1088/1755-1315/1041/1/012070

Copigmentation can naturally improve the color of anthocyanins, where the stability and color strength of anthocyanins can be increased by the addition of extracts from different plants that are rich in copigments [7]. One of the compounds that can be used as a copigment is apple extract at pH 4.6 [5]. Apple extract was chosen as a copigment because apples contain the highest amount and activity of antioxidants, as well as the largest amount of flavonoids (quercetin) in apples [8]. With the addition of apple extract at pH 4.6, it is expected to get the desired color, namely a purplish red color [5]. Copigmentation can increase the stability of anthocyanins by way of copigment compounds forming complex bonds with anthocyanins, which then reduces the interaction of anthocyanins with water molecules. The interaction of anthocyanins with water molecules causes anthocyanins to be degraded [9].

One of the blush dosage forms is gel. The gel was chosen because it has the advantage of providing a cool feeling on the skin, easy to wash, soft, and not sticky to the skin [10]. In the gel, there is a gelling agent which is a very important component in the manufacture of gel preparations. HPMC gel base is a gelling agent that is often used in the production of cosmetics and drugs as it can produce a clear gel, easily soluble in water, and has low toxicity [11]. HPMC is a semi-synthetic gelling agent derived from cellulose which is resistant to phenol, can form a clear gel, is neutral, and has a stable viscosity on long-term storage [12]. In addition, HPMC is stable at acidic pH and alkaline pH, which makes it very suitable for anthocyanin compounds which are stable at acidic pH. In this context, it aims to examine the effect of increasing the concentration of HPMC as a gelling agent on the color stability of the pigmented purple sweet potato extract blush gel.

2. Material and Methods

2.1 Preparation Extract

Purple sweet potato (*Ipomoea batatas* (L.) Lam.) powder was obtained from Institut Pertanian Bogor, West Java. 500 grams of purple sweet potato powder was extracted with 3.5 liters of 96% of ethanol. Then, HCl solution was added little by little which makes it to have a pH in the range of 1 to 3. Extraction was carried out by maceration for 3 days, which was followed by another maceration for 2 days while stirring occasionally. The liquid extract obtained was thickened with a rotary evaporator at a temperature of 40 °C to obtain a thick extract [13].

2.2 Phytochemical Screening of Ethanolic Extract

The phytochemical compounds of kencur (aromatic ginger, *Kaempferia galanga*) ethanolic extract, such as anthocyanins were qualitatively identified following standard procedures described in the Harborne and Indonesian Herb Pharmacopoeia [3].

2.3 The Creation of Blush Gel of Purple Sweet Potato (Ipomoea batatas (L.) Lam.)

The ingredients used were weighed. The pigment was prepared by means of purple sweet potato extract, apple extract was added to a pH of 4.6 and then mixed until homogeneous (mixture 1). Gel base was made by dispersing HPMC in Aqua distillata which was then developed for 24 hours. Methyl paraben and propyl paraben were dissolved into propylene glycol, which was then mixed into the gel preparation that had been developed, then grounded until homogeneous [12]. Mixture 1 which was grounded, was then mixed until it was homogeneous.

Table 1. Purple Sweet Potato (*Ipomoea batatas* (L.) Lam.) Blush Gel Formula

Material Name	**Formula (% w/v)**			**Function**
	F1	**F2**	**F3**	
Purple sweet potato extract	12	12	12	Dye
Apple Extract	0,63	0,63	0,63	Copigment
HPMC	1	1,5	2	Gelling agent
Propylene glycol	15	15	15	Humectants
Methyl paraben	0,18	0,18	0,18	Preservative
Propylparaben	0,015	0,015	0,015	Preservative
Aquadest ad	100	100	100	Solvent

2.4 The Evaluation of Purple Sweet Potato (Ipomea batatas L.) Blush Gel Preparation

Organoleptic test was carried out by observing the physical appearance of the purple sweet potato blush gel visually such as smell, shape, and color [14]. The pH test of the gel preparation was measured using a pH measuring device by dipping it into the gel sample. The pH value of a good gel preparation was pH 4.5 to 6.5 and did not irritate the skin [10]. The homogeneity test was carried out by weighing the gel as much as 0.1 grams and was then smeared on a piece of transparent glass and was then observed. Homogeneity was indicated by the absence of coarse grains [14]. Spreadability test of a total of 1 gram of gel preparation was placed on a transparent glass, and was given a weight of 125 grams on top. Afterwards, the diameter formed after 1 minute was measured [15].

2.5 Separation Test Phase

Freeze-Thaw by means of preparations was stored at two different temperatures, namely at a temperature of 4 °C for 48 hours and followed by a temperature of 45 °C for 48 hours (1 cycle). The test was carried out in 6 cycles, whereby each cycle was observed whether there was a phase separation or not in the gel [16].

Centrifugation by weighing as much as 10 grams of each preparation, was put into a centrifugation tube, and then centrifuged at 3750 rpm for 5 hours. Thereafter, an observation was made whether there was separation [17].

2.6 Color Stability Test

A 5 ppm pure anthocyanin solution with 2 solvents was made, namely 90% of ethanol solvent and water. We then measured the maximum wavelength between 400 and 800 nm at t05, t10, and t15, to see if the value was fixed at that time (storing the measuring flask in a cupboard to avoid direct light). It was important to make sure the measured absorbance is between 0.2 and 0.8. After obtaining the data, the measurement conditions (type of solvent and measurement time lag) were determined which were suitable for the color stability test [4].

The reading of the color stability in the blush gel preparation was carried out by weighing the preparation as much as 1 gram and then putting it into a 10 ml volumetric flask dissolved in 90% ethanol to the limit mark. The absorbance was then identified using UV-Vis spectrophotometry at a wavelength of 544 nm with absorbance between 0.2 to 0.8. Observation of color stability was carried out during 28 days of storage [5].

2.7 Data Analysis

The observed data from the dispersion test and color stability test were then analyzed using one-way ANOVA. If there was a significant difference between the formulas, then it was continued with the Tukey HSD test with a 95% confidence level to see if the formula has a significant difference or not.

3. Results and Discussion

Purple sweet potato was macerated using ethanol as solvent. Ethanol solvent was chosen because ethanol is a universal solvent. This solvent can dissolve almost all organic compounds in the sample, both polar compounds and non-polar compounds [2]. Therefore, it is very suitable to find anthocyanins contained in purple sweet potatoes which are polar. In the extraction process, HCl was added so that it became an acidic atmosphere. With an acidic environment, it can denature plant cell membranes and it can dissolve anthocyanin pigments, making them able to leave the cells and prevent flavonoid oxidation [4]. The calculation of the extract yield was then carried out. Yield is the ratio of the weight of the extract obtained with the weight of the simplicia powder. The yield of purple sweet potato extract obtained was 10.5375%.

Anthocyanin qualitative test aims to determine the truth that purple sweet potato extract contains anthocyanins. In the anthocyanin test, purple sweet potato extract was added with HCl. The results of the observations obtained a red color. The red color was formed by flavylium cations, where the number of methoxy groups in the anthocyanin structure was more dominant than the hydroxyl groups. Then, the NaOH solution was added little by little. Observations showed that the green color was slowly fading. At alkaline pH, it turned bluish green due to the carbinol structure of pseudobase [3]. When the pH was higher, the chalcone structure was formed which caused anthocyanins to lose their red color due to the formation of quinonoidal anions [5]. From the observations obtained when compared with the requirements that have been set, it can be concluded that the purple sweet potato extract contained anthocyanin compounds.

The organoleptic test was carried out by observing the physical appearance of the purple sweet potato extract blush gel preparation with the addition of a copigment, namely apple extract at pH 4.6 by visually observing the shape, color, and smell of the blush gel preparation. The results showed a semisolid form, a purplish red color was obtained due to the addition of a copigment, namely apple extract at pH 4.6 and the distinctive odor of the preparation.

In observing the pH of blush gel preparations with the addition of copigment, namely apple extract at pH 4.6, observations were made for 28 days. From these observations, the preparation experienced instability due to the presence of CO2 that was attracted to the preparation, which then reacted with water to release H+. Another aspect that caused changes in the pH value can occur because it was affected by the medium that was decomposed by high temperatures during manufacture or storage time which produces acids or bases. In addition, changes in pH are also caused by environmental factors such as temperature, poor storage, and others [14]. The preparation of purple sweet potato blush gel with the addition of copigment, namely apple extract at pH 4.6, experienced instability during 28 days of storage. However, these preparations still meet the requirements of a good skin physiological pH of 4.5 to 6.5 [10].

In the observation of the homogeneity test of all these formulas, there were no coarse grains placed on a pair of glass plates which were visually observed. When compared with the requirements for good homogeneity, a preparation must show a homogeneous preparation and there are no coarse grains so that it can be concluded that all preparations are homogeneous [18].

In the observation of blush preparations, purple sweet potato extract with the addition of copigment, namely apple extract at pH 4.6 has good dispersion. The requirements for good dispersion for gel preparations are 5 to 7 [14]. It can be concluded that the blush gel preparation of purple sweet potato extract with the addition of copigment, namely apple extract at pH 4.6, met the pharmaceutical requirements, although the dispersion values were different due to differences in the concentration of HPMC. The higher the HPMC concentration, the smaller the dispersion obtained.

Table 2. Spreadability Test Results

Formula	Yield Spread (cm)
F1	6, 48 $\pm$ 0,18
F2	6,18 $\pm$ 0,08
F3	5,26 + 0,08

Based on the results of statistical analysis data, the normality test using the Kolmogrov Smirnov test obtained a significant value > 0.05 which indicated that the data was normally distributed. Thereafter, the homogeneity test was carried out with a significant value > 0.05 which indicated that the data was homogeneously distributed. It was followed by the one-way ANOVA test with a significant value < 0.05 which demonstrated that the data of the dispersion test value of the three formulas had significant differences among the formulas. It was then continued with the Tukey HSD test to see if the formula had a significant difference or not. The Tukey HSD test obtained significant values from the three formulas, namely in F1 there was a significant difference in F3. In F2, there was a significant difference in F3. Furthermore, in F3 there was a significant difference both in F1 and F2.

The freeze-thaw test was carried out to see if there was a phase separation in the blush gel preparation of purple sweet potato extract with the addition of a copigment, namely apple extract at pH 4.6 with two different temperatures. From the observations for 6 cycles, all formulas did not experience phase separation. At 4°C it crystallized so that the consistency of the preparation became thicker. This was because the preparation tended to shrink at low temperatures, making the particles combine to form tighter bonds between themselves. As a result, the viscosity increased and the water rate decreased [12]. However, after being placed at a temperature of 45°C, the preparation returned to its original shape and there was no phase separation. It can thus be concluded that the blush gel preparation of purple sweet potato extract was stable.

Centrifugation is a phase separation test with very fast shaking under the influence of gravity. The preparation was centrifuged at 3750 rpm for 5 hours and the result was equivalent to the effect of gravity for one year [15]. The results of the observations showed that in all the blush gel formulas of purple sweet potato extract with the addition of copigment, namely apple extract at pH 4.6, there was no phase separation. This indicates that the preparation was stable and was expected to be stable equivalent to the effects of gravity for one year. This showcased that the blush gel preparation of purple sweet potato extract with the addition of apple extract copigment at pH 4.6 was stable.

The color stability test aims to measure or see the color stability of each formula. Testing of blush gel preparations of purple sweet potato extract with the addition of copigment, namely apple extract at pH 4.6, was carried out using a UV-Vis spectrophotometer with a wavelength of 544 nm. Anthocyanins could specifically absorb light in the ultraviolet (UV) to violet absorption region, but were stronger in visible light in the spectrum. The presence of conjugated double bonds in the chromophore group contained in the anthocyanin structure makes anthocyanins able to absorb light in visible light. Thus, this is the reason why the color stability test readings were used in a UV-Vis spectrophotometer with a wavelength of 544 nm located in visible light [19].

Anthocyanins as dyes are less stable to oxygen, light, pH, sugar, and temperature [5]. With the addition of apple extract copigment at pH 4.6 in the blush gel preparation, it was expected to increase the color stability of anthocyanins, with the desired color being purplish red. However, based on observations for 28 days, the three formulas experienced changes in absorbance when reading the UV-Vis spectrophotometry. This could be caused by factors that affect color stability, namely pH, light, low temperature, copigment, metal ions, oxygen, enzymes, concentration, and storage time so that anthocyanins were graded. The instability of anthocyanins caused the compound to easily undergo hydrolysis at the glycosidic bond and the aglycone ring became open, thus forming a labile aglycone, as well as a colorless carbinol and chalcone groups [19].

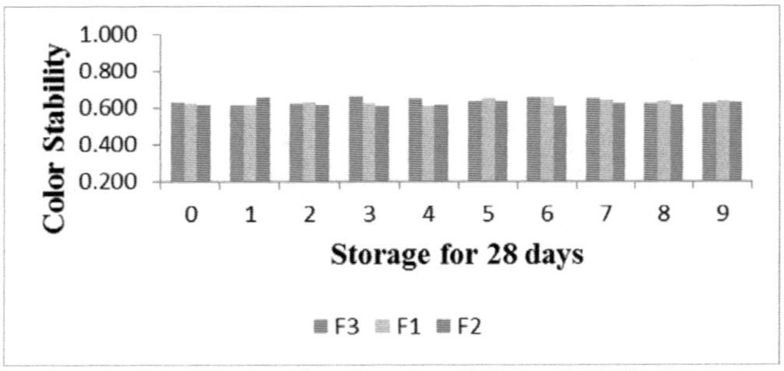

Figure 1. Observation Results of Color Stability Test

Based on the results of statistical analysis data, each formula was tested for normality using the Kolmogrov Smirnov test. A significant value was obtained > 0.05 which indicated that the data was normally distributed. It was then continued with the homogeneity test. The significant value obtained > 0.05 indicated that the data was homogeneously distributed. We then proceeded with testing with the one-way ANOVA parameter which obtained a significant value < 0.05, indicating the color stability value data of the three formulas. There were significant differences among the formulas. We continued with the Tukey HSD test, which obtained a significant value from the three formulas, namely > 0.05, demonstrating that there was no significant difference in the formula during 28 days of observation. It can be concluded that the increase in the concentration of HPMC as a gelling agent on the color stability of the copigmented purple sweet potato extract blush gel did not experience stability during 28 days of storage.

4. Conclusion

Based on the results of the study, the results obtained from the statistical analysis of color stability found that there were significant differences in each formula. It can be concluded that the increase in the concentration of HPMC as a gelling agent on the color stability of the copigmented purple sweet potato (*Ipomoea batatas* (L.) Lam.) blush gel extract did not experience color stability during 28 days of storage.

References

[1] Nurhabibah, A. Najihudin, and D. S. Indriawati, "FORMULATION AND EVALUATION OF BLUSH ON PREPARATIONS FROM THE ETHANOL EXTRACT OF CINNAMON (Cinnamomum burmanni Nees ex Bl) ARTICLE HISTORY," *J. Ilm. Farm. Bahari*, p. 36,40, 2018.

[2] Basito, "Efektivitas Penambahan Etanol 95% dengan Variasi Asam dalam Proses Ekstraksi Pigmen Antosianin Kulit Manggis (Garcinia mangostana L.)," *J. Teknol. Has. Pertan.*, vol. 4, no. 2, pp. 84–93, 2011.

[3] S. Winarti, U. Sarofa, and D. Anggrahini, "EKSTRAKSI DAN STABILITAS WARNA UBI JALAR UNGU (Ipomoea batatas L .,) SEBAGAI PEWARNA ALAMI," *J. Tek. Kim.*, vol. 3, no. 1, pp. 207–214, 2008.

[4] S. Wahyuningsih, L. Wulandari, M. W. Wartono, H. Munawaroh, and A. H. Ramelan, "The Effect of pH and Color Stability of Anthocyanin on Food Colorant," *IOP Conf. Ser. Mater. Sci. Eng.*, vol. 193, no. 1, 2017, doi: 10.1088/1757-899X/193/1/012047.

[5] C. C. Gras, N. Nemetz, R. Carle, and R. M. Schweiggert, "Anthocyanins from purple sweet potato (Ipomoea batatas (L.) Lam.) and their color modulation by the addition of phenolic acids and food-grade phenolic plant extracts," *Food Chem.*, vol. 235, pp. 265–274, 2017, doi: 10.1016/j.foodchem.2017.04.169.

[6] P. Sari, C. H. Wijaya, D. Sajuthi, and U. Supratman, "Colour properties, stability, and free radical scavenging activity of jambolan (Syzygium cumini) fruit anthocyanins in a beverage model system: Natural and copigmented anthocyanins," *Food Chem.*, vol. 132, no. 4, pp. 1908–1914, 2012, doi: 10.1016/j.foodchem.2011.12.025.
[7] M. Hambali, F. Mayasari, and F. Noermansyah, "Ekstraksi Antosianin Dari Ubi Jalar Dengan Variasi Konsentrasi Solven, Dan Lama Waktu Ekstraksi," *J. Tek. Kim.*, vol. 20, no. 2, pp. 25–35, 2015.
[8] L. N. Lestario and S. Andini, "Kopigmentasi Kuersetin Apel (Pyrus malus) terhadap Stabilitas Warna Ekstrak Buah Duwet (Syzygium cumini)," *Pros. Konser Karya Ilm.*, vol. 2, no. 5, pp. 37–42, 2016, [Online]. Available: https://repository.uksw.edu/bitstream.
[9] N. I. Neliyanti, "Ekstraksi dan Uji Stabilitas Zat Warna Alami dari Buah Lakum (Cayratia trifolia (L.) Domin)," *J. Kim. Khatulistiwa*, vol. 3, no. 2, pp. 30–37, 2014.
[10] F. Nugrahaeni, D. M. Hariyadi, and N. Rosita, "Partition coefficient and glutathione penetration of topical Antiaging: Preformulation study," *Int. J. Drug Deliv. Technol.*, vol. 8, no. 2, pp. 39–43, 2018.
[11] N. Noval, R. Rosyifa, and A. Annisa, "Effect of HPMC Concentration Variation as Gelling Agent on Physical Stability of Formulation Gel Ethanol Extract Bundung Plants (Actinuscirpus Grossus)," 2020, doi: 10.4108/eai.23-11-2019.2298326.
[12] S. C. Joshi, "Sol-gel behavior of hydroxypropyl methylcellulose (HPMC) in ionic media including drug release," *Materials (Basel).*, vol. 4, no. 10, pp. 1861–1905, 2011, doi: 10.3390/ma4101861.
[13] R. S. Armanzah and T. Y. Hendrawati, "Pengaruh Waktu Maserasi Zat Antosianin Sebagai," *Semin. Nas. Sains dan Teknol.*, no. November, pp. 1–10, 2016, [Online]. Available: https://jurnal.umj.ac.id/index.php/semnastek/article/view/724/669.
[14] D. M. Hariyadi, N. Rosita, and F. Nugrahaeni, "Formulation, characteristic evaluation, stress test and effectiveness study of matrix metalloproteinase-1 (MMP-1) expression of glutathione loaded alginate microspheres and gel," *Pharm. Sci.*, vol. 24, no. 4, 2018, doi: 10.15171/PS.2018.44.
[15] I. D. K. Irianto, P. Purwanto, and M. T. Mardan, "Aktivitas Antibakteri dan Uji Sifat Fisik Sediaan Gel Dekokta Sirih Hijau (Piper betle L.) Sebagai Alternatif Pengobatan Mastitis Sapi," *Maj. Farm.*, vol. 16, no. 2, p. 202, 2020, doi: 10.22146/farmaseutik.v16i2.53793.
[16] A. Zeinali, T. Edeskär, and J. Laue, "Mechanism of thawing," *Cogent Eng.*, vol. 7, no. 1, 2020, doi: 10.1080/23311916.2020.1716438.
[17] C. M. E. Salenda, P. V. Y. Yamlean, and W. A. Lolo, "PENGARUH VARIASI BABIS KARBOPOL DAN HPMC PADA FORMULASI GEL EKSTRAK ETANOL DAUN TAPAK KUDA (Ipomoea pes-caprae (L.) R. Br. DAN UJI AKTIVITAS ANTIBAKTERI TERHADAP Staphylococcus aureus," *Pharmacon*, vol. 7, no. 3, pp. 249–256, 2018, doi: 10.35799/pha.7.2018.20452.
[18] D. P. Astuti, P. Husni, and K. Hartono, "Formulasi Dan Uji Stabilitas Fisik Sediaan Gel Antiseptik Tangan Minyak Atsiri Bunga Lavender (Lavandula angustifolia Miller)," *Farmaka*, vol. 15, no. 1, pp. 176–184, 2017.
[19] H. E. Khoo, A. Azlan, S. T. Tang, and S. M. Lim, "Anthocyanidins and anthocyanins: Colored pigments as food, pharmaceutical ingredients, and the potential health benefits," *Food Nutr. Res.*, vol. 61, no. September, pp. 0–21, 2017, doi: 10.1080/16546628.2017.1361779.

Sunscreen Factor Formulation and Test of Gel Preparations of 70% Ethanol Extract on Arabica Coffee Leaf (*Coffea arabica* L.)

Sofia Fatmawati*, Fitria Nugrahaeni, Fith Khaira Nursal, Ana Fitriana

Universitas Muhammadiyah Prof. DR. HAMKA, Jakarta, Indonesia

*sofia.fatmawati@uhamka.ac.id

Abstract. Arabica coffee leaf extract (*Coffea arabica* L.) contains phenolic and flavonoid compounds that are potentially active as sunscreen because it might absorb UV (Ultra Violet) A and UV B rays. It is well-known that UV rays exposure causes a negative effect to skin, yet is possible to be avoided by applying sunscreen. This research aims to test the effect of arabica coffee leaf extract on formulation of gel preparations and to discover how its concentration affects the amount of gel preparations' sunscreen factor. There were 4 formulas with different extract concentrations whose characteristics were tested chemically, physically, and they were tested to measure its SPF number. From this test, it was discovered that the gel was qualified and it was plastic thixotropic flow. The measurement of SPF number proposed that F1 could not be categorized as protector because it was only 0.6 while F2, F3, and F4 showed that these formulas could be categorized as minimum protector. The samples were also tested statistically by using One Way ANOVA and Tukey HSD test. The result of the ANOVA test showed the number of sig 0.000 < 0.05 which meant all four formulas were significantly different. The research concludes that the extract of arabica coffee leaf might affect the formulation of gel preparations and its extract concentration might affect SPF number on gel preparations.

1. Introduction

Skin does an important role to protect the body from outside environments such as physical impact and exposure to free radicals. In order to maximize the function of the skin as a protector, it is necessary to give special treatment on the skin. Amongst information about how dangerous UV rays affect our skin, it is also well-known that sunscreen has been trusted to protect our skin from the damage of UV rays[1].

UV rays are rays that are emitted by the sun that can reach the earth surface beside visible light and infrared rays. The wavelength of UV rays is approximately 200-400 nm. The UV spectrum, based on the wavelength, is categorized into three; UV C (200-290), UV B (290-320) and UV A (320-400). Not all of these three UV rays' radiation can reach earth's surface. UV C rays, the one with the strongest energy, cannot reach the surface because of being absorbed in ozone [2].

It is known that UV radiation from the sun causes some damages to human skin. UV A radiation is discovered to cause damages to skin cells and DNA. It also causes photoaging and photocarcinogenesis. These damages can be caused by long term exposure of the sun's rays, even in a small dose. UV A radiation contributes to the decreasing of skin elasticity, the increasing of wrinkles, and the increasing of free radicals that cause acute to chronic changes to skin. Besides, UV A radiation also makes other diseases worsen, such as lupus erythematosus. It also contributes as immunosuppressants that increase cancer cell growth [3]. From another resource, it is stated that the exposure of ultraviolet radiation might cause epidermis damages, which is also called sunburn, pigmentation, wrinkling, premature skin aging, and long term exposure of the sun's rays can also cause mutation of lugature tissue in the stratum corneum[4]

On the other hand, the sun's rays stimulate melanin and pigment as natural sunscreen for human skin. The sun's rays also stimulate hormone protection which allows vitamin D synthesizing to increase skin cell regeneration. It is discovered that the shorter the wavelength, the stronger the energy level, which also causes more damage to the skin (Lu et al., 1996). If the sun's rays expose the skin excessively, it might not resist the negative effects. Therefore, it is necessary to give some protection by using sunscreen. Sunscreen is preparations applied on skin to absorb, to scatter, or to reflect UV rays. It is also applied to help natural protection mechanisms to protect skin from dangerous UV radiation of the sun's rays [6]. It is a product that provides protection from the damage of the sun's rays, but this product is commonly made of synthetic chemicals which allow, in long term exposure, some bad effects to human skin. This research is conducted, to test the formula, so that there would be a chance for natural ingredients to be natural protection which causes less bad effect [7]. The grade of sunscreen's photo-protective is universally measured by Sun Protective Factor (SPF) which determines the increase of the sun's rays' exposure dose with photo-protective products applied without eritema [8].

Coffee is categorized into genus Coffea in the family of Rubiaceae. Genus Coffea L. consists of more than 100 species, but only two species, *Coffea arabica* (arabica coffee) and *Coffea canephora* (robusta coffee), are commercially cultivated on large scales [9]. *Coffea arabica* is a tetraploid species from highland rain forest in the south of Ethiopia. Six of eight chosen species are closely related, including two commercial species *Coffea arabica* dan *Coffea canephora*, as illustrated in the recent phylogeny from genus [10].

Coffee leaf has not been optimally utilized. In Indonesia, coffee leaf is used as a brewed drink named "Aia Kawa" in Sumatera [11]. It is discovered that *Coffea arabica* leaf contains phenolic compounds with a concentration of 17.4% in green leaf and 13.9% in old/pure leaf (Rodrigues et al., 2008). Arabica coffee leaf contains some compounds, such as alkaloid, flavonoid, and phenol which are included into the derivative of hydroxycinnamic acid, caffeine, chlorogenic acid, coumarin, ferulic acid and cinnamic acid. Chlorogenic acid is easily oxidized in aqueous solutions [13] [14].

Phenolic and flavonoid compounds have potential as sunscreen [15]. Flavonoid is commonly able to absorb UV rays, but not able to absorb any radiation on wavelength around 280-315 nm maximally, as other phenylpropanoid compounds. However, compared with other phenylpropanoid compounds, flavonoid compounds will increase significantly, as it is exposed by UV B rays' radiation, including the derivative of hydroxycinnamic acid (for instance: p-Coumaric acid, ferulic acid and caffeine acid). Some research showed that when UV B rays' exposes plants, flavonoid concentration is increased compared with hydroxycinnamic compounds [16].

Sunscreen is applied to help natural protection mechanisms to protect skin from dangerous UV radiation of the sun's rays [6]. It is a product that provides protection from the damage of the sun's rays, but this product is commonly made of synthetic chemicals. In the market, sunscreen is provided in some forms of preparations; salve, cream, gel, lotion, spray, and wax stick [17]. Gel is a semi-solid preparation which consists of suspensions made of inorganic particles in small parts or organic molecules in big parts penetrated by some kinds of fluid [18].

In Indonesia, research related to arabica coffee leaves is limited. Therefore, it is necessary to conduct further research because the resources are overflowing and it should be utilized and developed, for instance, for cosmetic purposes. This research aims to fill the void in the literatures related to arabica coffee leaf by testing how the extract of arabica coffee leaf affects the formulation of gel preparations and discovering how far the extract of arabica coffee leaf concentration affects the grade of Sun Protective Factor of gel preparations.

2. Material and Methods

2.1 Material

The materials used in this study were arabica coffee leaves (Coffea arabica L.) as the active ingredient which was obtained by maceration using 70% ethanol. The chemicals used for the gel formulation were HPMC K100M type 2208 (High Viscosity Grade), Propylene Glycol (Pharmaceutical Grade), Methyl Paraben (Pharmaceutical Grade), Propyl Paraben (Pharmaceutical Grade), Aqua Dest, 70% Ethanol and 70% Ethanol for analysis.

2.2 Plant Determination

The determination of the arabica coffee plant used in the research was carried out at the Herbarium Bogoriense, Botany Division, Biological Research Center, LIPI Cibinong.

2.3 Arabica Coffee Leaf Extraction

Coarse powder of arabica coffee leaves that have been obtained from IPB Bogor was re-pollinated using a grinder or blender. It was then sieved using a mesh sieve no. 40. The extraction method used maceration. The maceration extraction method (cold method) is a better method for extracting cherry leaves compared to the soxhlation method (hot method). This is due to the thermolable nature of flavonoids which will be damaged when extracted using a soxhletation [19]. 1:10 fine powder and macerated solvent. A total of 1300 grams of powder was macerated with 9700 ml of 70% ethanol filter in a dark glass jar at room temperature. It was then left for 2 days and stirred every 8 hours. The maceration results were filtered with a flannel cloth and filter paper. Then, the remaining dregs were re-macerated with 3300 ml of sprinkler for 1 day. Afterwards, it was stirred and filtered again. The filtrate obtained was then concentrated using a rotary vacuum evaporator at a temperature of 50 ° C until a thick extract was obtained. Thick extract of arabica coffee leaves, and the yield was obtained. The yield was calculated based on the ratio of the final weight (weight of the extract produced) to the initial weight (weight of cell biomass used) multiplied by 100% [20].

2.4 Organoleptic Determination of Arabica Coffee Leaf Extract and Determination of Ash Content

Organoleptic determination of arabica coffee leaf extract was carried out by observing the shape, color, and odor of the arabica coffee leaf extract. 2 g of extract was weighed carefully and put in a porcelain crucible that had been annealed,tared, and leveled. Krus was slowly annealed until the charcoal ran out. Afterglow was carried out at a temperature of 600°C for 3 hours, which was then cooled and weighed until a fixed weight was obtained. The ash content was calculated against the material which was dried in the air. If in this way the charcoal cannot be removed, hot water can be added and filtered with an ash-free filter paper. The rest of the paper and filter paper was glazed in the same crucible. The filtrate was put into the crucible, and left to evaporate. Afterglow until the weight was fixed, weighed and calculated.

2.5 Phytochemical Screening Test

Extract was identified of alkaloid, flavonoid, phenol, saponin, tannin, steroid, and triterpenoid contents using Harbonne or Material Medica Method.

2.6 Gel Formulation

Arabica coffee leaf extract gel formula (*Coffea arabica* L.) consisted of arabica coffee leaf extract (*Coffea arabica* L.), HPMC K100M type 2208, Propylene Glycol, Methyl Paraben, Propyl Paraben and Aqua Dest. The gel formulation can be seen in table (1).

Table 1. Formulation Gel Arabica Coffee Leaf Extract

No.	Materials	% (b/v)				Functions
		F1	F2	F3	F4	
1	Ekstract	-	1.5	2	2.5	Active ingredients
2	HPMC	1	1	1	1	*Gelling agent*
3	Propylene Glycol	15	15	15	15	Humectant
4	MethylParaben	0.075	0.075	0.075	0.075	Preservatives
5	Propylparaben	0.025	0.025	0.025	0.025	Preservatives
6	Aqua Dest ad	100	100	100	100	solvent

2.7 Procedure for Preparation Arabica Coffee Leaf Extract Gel

HPMC 4.5 g was developed in a mortar with aqua dest and was then expanded for 24 hours and crushed to form a gel mass (M1). 0.3775 g of methyl paraben was dissolved with a portion of propylene glycol (M2). 0.1125 g of propyl paraben was dissolved with a portion of propylene glycol (M3). The arabica coffee leaf extract was also dissolved with the remaining propylene glycol (M4). M1, M2 and M3, M4 were mixed, crushed, and homogeneous. Furthermore, the evaluation of the determination of the FPS value of gel preparations were carried out.

2.8 Organoleptic Test for Gel

Organoleptic testing was carried out by direct observation of the shape, color and smell of the gel made. Gel was usually clear with a semi-solid consistency[18].

2.9 Homogeneity Test

The homogeneity test was carried out by applying 0.1 g of gel preparation to a piece of transparent glass and observing its homogeneity. The test preparation must show a homogeneous arrangement, indicated by the absence of coarse grains on the object glass [21].

2.10 pH test

The pH test was carried out by turning on the pH meter then dipping the pH meter electrode into the gel formula. It was left for a few moments until the pH meter screen showed a stable number [22].

2.11 Spreadability Test

As much as 1 gram of gel preparation was placed carefully on a glass measuring 20x20 cm, and was given a 125 gram weight on it.It was then measured the diameter formed after 1 minute [23].

2.12 Adhesion Test

A total of 1 gram of gel preparation was flattened on one glass object and covered with another glass object until the two plates were joined. The object glass pair was pressed with a load of 1000 g for 5 minutes. It was then installed on the adhesion test equipment, while simultaneously recording the time it took for the two plates to come off each other [24].

2.13 Viscosity and Flow Properties Test

The viscosity test was performed using the Brookfield Viscometer RV type. The gel preparation was put into a 500 mL beaker glass container, then 'the spindle no. 04 was installed. Afterwards, the spindle was lowered into the preparation to the specified limit. The tool was turned on and set at a speed of 10 rpm, 12 rpm, 20 rpm, 30 rpm and 50 rpm to show a stable number. Then, the results were recorded. To determine the flow properties of the preparation, a curve was made between the shear rate and shear stress [21].

2.14 *Centrifugation Test*

10 grams of each preparation were put into a centrifugation tube and centrifuged at 3750 rpm for 5 hours. It was then observed whether there was separation or not in the gel [21].

2.15 *Freeze Thaw Test*

The phase separation cycle using the freeze thaw method on the gel preparation was carried out in 6 cycles for each formula. Each cycle was observed after 48 hours of storage at 4°C and 48 hours after at 45°C for 24 days. Each cycle was observed whether there was phase separation or not on the gel preparation [21].

2.16 *Determination of the FPS Value of Arabica Coffee Leaf Extract*

The FPS value of the extract in vitro was determined using the Uv-Vis spectrophotometric method developed by Mansur (1986) using equation 3. The extract was diluted with a concentration of 60 ppm, 80 ppm and 100 ppm with 70% pa ethanol and then measured using the Uv-Vis spectrophotometer at a wavelength of 290-320 nm. The determination of the FPS value was carried out in three replications. Then, the data obtained was processed with the Mansur equation (1986). To determine the FPS value of this wavelength, the EE x I value has been described in table 2 [7].

2.17 *Determination of the FPS Value of Arabica Coffee Leaf Extract Gel Preparation*

The FPS value of the extract in vitro was determined using the Uv-Vis spectrophotometric method developed by Mansur (1986) using equation 3. The preparation was weighed as much as 0.02 grams in 5 mL ethanol 70% pa and then measured using a Uv-Vis spectrophotometer at a wavelength 290-320 nm. The determination of the FPS value was carried out in three replications. Then, the data obtained was processed with the Mansur equation (1986). To determine the FPS value of this wavelength, the EE x I value has been described in table 2 [7].

The equation for determining the FPS value can be seen in below:

$$\text{SPF} = \text{CF}\ x \sum_{290}^{320}\ x\ \text{EE}(\lambda)\ x\ \text{I}(\lambda)\ x\ \text{Abs}(\lambda)$$

Based on this equation, there was a variable CF (Correction Factor) = 10, EE (Erythemogenic Effect), I is the simulated intensity of sunlight and Abs is the absorbance of the sample.

The value of EE x I can be seen in table 2.

Table 2. The Value of EE x I

Wave Length (nm)	EE x I
290	0.0150
295	0.0817
300	0.2874
305	0.3278
310	0.1864
315	0.0839
320	0.0180

According to the FDA (Food Drug Administration), the distribution of sunscreen capabilities is Minimum (SPF is between 2-4), Medium (SPF is between 4-6), Extra (SPF is between 6-8), Maximum (SPF is between 8-15), and Ultra (SPF is more than 15) [25].

2.18 Statistical Analysis

Qualitative testing of physical characteristics and determination of the FPS value of gel preparations were analyzed using One Way ANOVA, with a confidence level of 95% ($p > 0.05$). If there WEre significant differences between formulas, we then proceeded with the Tukey HSD test to see which formulas had significant differences.

3. Results and Discussion

The object of this research was fresh arabica coffee plants (*Coffea arabica* L.). Then, the fresh arabica coffee plants were determined in Herbarium Bogoriense, Botany Department of Biology Research Center, LIPI Cibinong. The determination of the plants in this research meant to find out the clear identity of the research object to avoid mistakes in collecting main material for the research. After determined, it was confirmed that the plants were arabica coffee leaf (*Coffea arabica* L.).

Table 3. The Results of Characterization pf Arabica Coffee Leaf Extract

Examination	Results
yield	9.84%
Organoleptic:	
a. form	Thick extract
b. smell	coffee
c. taste	bitter
d. color	Chocolate black
Drying shrinkage	6.61% ± 0.43
Ash content	9.26% ± 0.46

Based on the results of phytochemical screening tests, 70% thick ethanol extract of arabica coffee leaves had alkaloid, phenolic, flavonoid, saponin, tannin and steroid compounds. This is in accordance with previous research [26], [27]. By knowing the truth of its benefits, arabica coffee leaf extract can be formulated into gel preparations that contain antioxidants.

Table 4. Results of Phytochemical Screening Test

Compound	Result
Phenolic	+
Alkaloids	+
Flavonoids	+
Saponins	+
Tannins	+
Steroids	+
Triterpenoid	-

Organoleptic tests included observations on the consistency, color and odor of the gel preparation. The results of organoleptic testing can be seen in table 5. In accordance with the results from the organoleptic test, it was found that all four formulas had semi-solid consistency. For F1, the gel was clear/transparent and odorless because it was not mixed with the extract, while for F2, F3, and F4, they were clear brown and had a typical smell from the extract. The extract concentration affected the color of gel preparations, the higher the concentration, the deeper the brown color.

Homogeneity test aimed to determine the homogeneity of preparations that have been made, homogeneous preparations will produce good quality because it showed the active ingredients of the drug are evenly dispersed in the basic ingredients. Based on the research results, the four formulas produced a homogeneous gel.

Table 5. Results of Organoleptic Tests and Homogeneity Tests

Formula	Organoleptic			Homogeneity
	Consistency	Color	Smell	
1	Semisolid	Transparent	Odorless	Homogeneous
2	Semisolid	Clear brown	Unique	Homogeneous
3	Semisolid	Clear brown	Unique	Homogeneous
4	Semisolid	Clear brown	Unique	Homogeneous

The purpose of this test was to determine that the resulting gel was acceptable for skin pH, because if it is not in accordance with the pH of the skin it can cause irritation. Based on the results of the study, the four formulas met the pH criteria on sunscreen preparations. pH test discovered that all formulas were qualified for pH criterias on sunscreen preparations, which was around 4.5-8.0 (SNI, 1996). The extract concentration affected the acid of the preparations because the extract contained a flavonoid compound which was quite acidic. Therefore, the more the arabica coffee leaf extract concentration, the more the acid it has (Markham, 1988). Arabica coffee leaf extract was 6.16 pH. A product that had too high or too low pH could cause irritation on skin [6]. The graph of the pH test results can be seen in Figure 1.

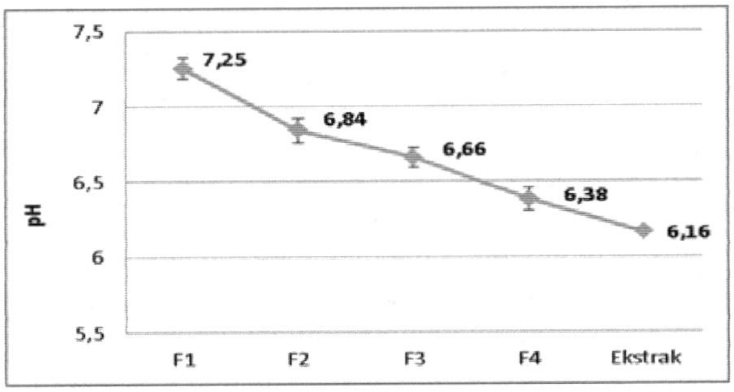

Figure 1. The Results of pH test

The dispersion test was carried out to ensure that the semisolid preparations are able to spread easily without significant pressure, so that they are easy to apply without causing pain and provide comfort to the user. The graph of the results of the scattering power test can be seen in Figure 2. The greater the spreadability of the preparation, the greater the ability of the active substance to spread and contact with the skin. The spreadability is inversely proportional to viscosity, the greater the viscosity of a preparation, the thicker the consistency, so that the resulting spreadability is smaller [30].

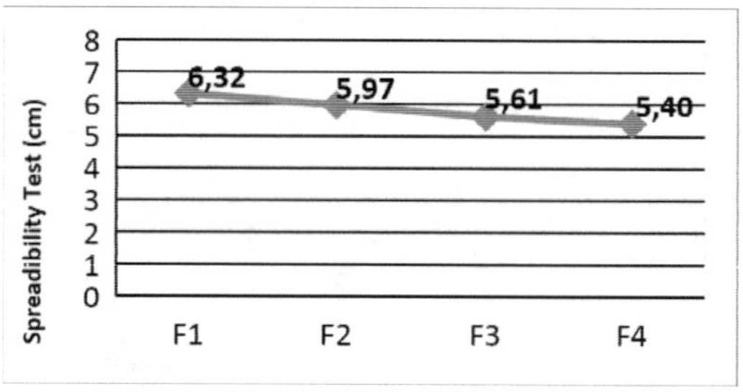

Figure 2. The Results of Spreadability Test

The adhesion test was carried out to determine the ability of the gel to adhere to the skin surface. The graph of the adhesion test results can be seen in Figure 3. Based on the research results, the four formulas met the adhesion criteria. Topical preparations must have sufficient adhesiveness but should not be sticky to the skin as they can reduce comfort during use.

The viscosity test was aimed to determine the thickness of the gel preparation. The higher the viscosity value, the higher the viscosity level of the gel preparation. Viscosity was tested by using *Brookfield* Viscometer RV type. The result revealed that all formulas were qualified for criterias of gel viscosity, which was 2000-4000 *cPs.* [31]. Gel consistency, which was not too liquid or too thick, was a characteristic of the good gel. HPMC concentration in gel preparations was used only 1%, as HPMC K100M is categorized into *High Viscosity Grade* which can be read in *Certificate of Analysis*. Gel viscosity affects the gel itself to disperse and adhere to skin. The higher the viscosity, the more difficult the gel to disperse although it is easier to adhere on skin. The higher the viscosity, the smaller the separation rate of dispersed phase [32]. Of the four formulas, the gel viscosity criteria were met. The graph of the results of the viscosity test can be seen in Figure 4.

All four formulas produced plastic thixotropic flow. It could be seen that the curve was started without getting through point zero (0) but cutting the axis of motion stress (or the outside of the curve was extrapolated cutting the axis) at the certain points named yield grade [33]. It was also shown that the decreased curve was to the left of the increased curve because there was structural change that did not immediately return to the first condition, if the pressure was diminished. Yield grade resulted from F1 was f = 3.496 dyne/cm², F2 was f = 3.097 dyne/cm², F3 was f = 2.979 dyne/cm² and F4 was f = 3.557 dyne/cm². Thixotropic is the characteristic of the flow that is expected for semi-solid preparations because it is expected to be in high consistency in a container. Nevertheless, it is easier to pour when it is given force, making it easier to be applied on skin [34]

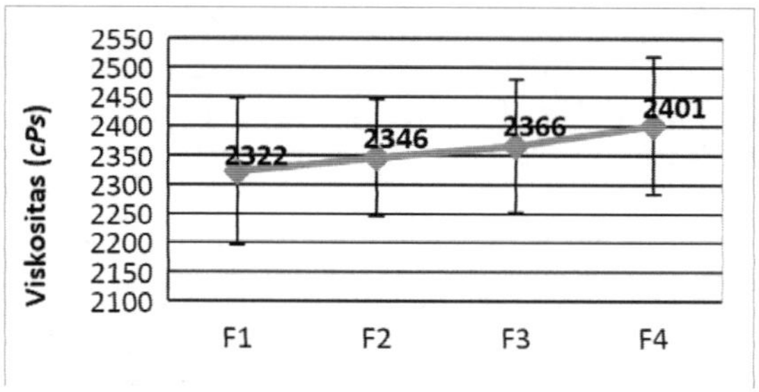

Figure 3. The Results of Viscosity Test

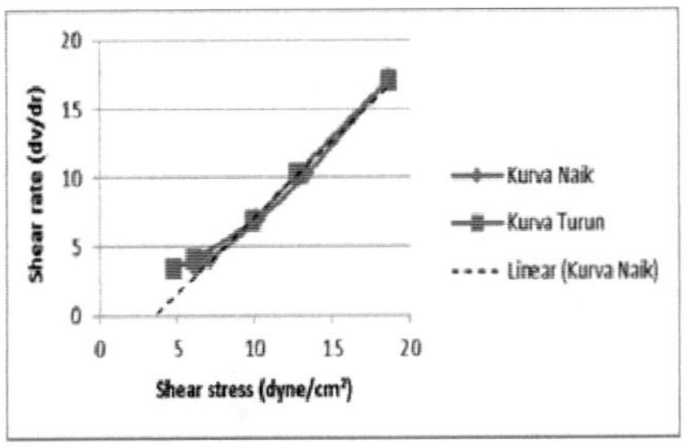

Figure 4. Flow Properties Formula 1

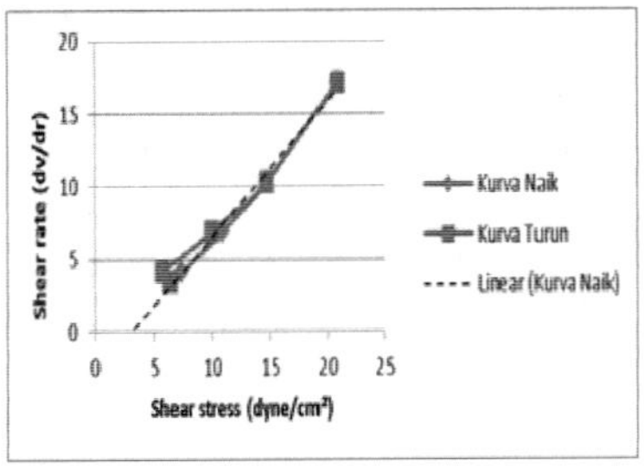

Figure 5. Flow Properties Formula 2

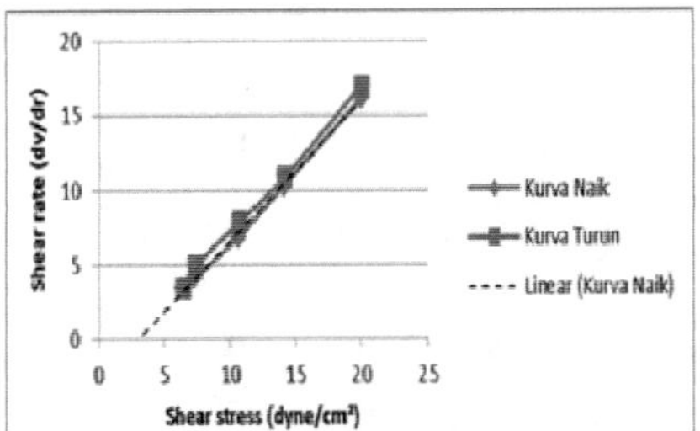

Figure 6. Flow Properties Formula 3

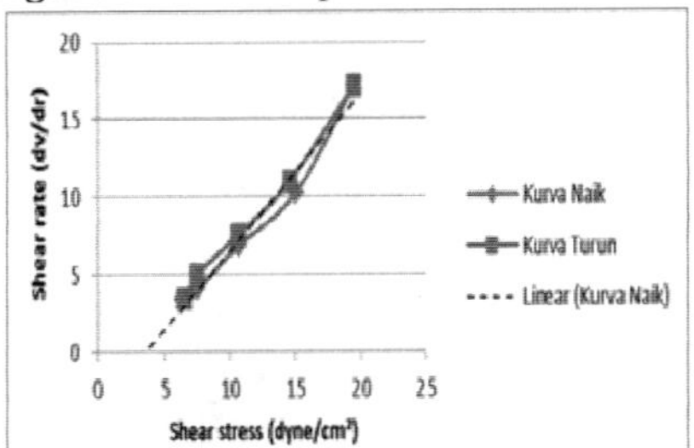

Figure 7. Flow Properties Formula 4

The centrifugation test aimed to observe whether there is a phase separation of the gel preparation and to see the stability of the gel preparation after very strong shaking. Based on the research results, the four formulas did not undergo phase separation or syneresis. The table of centrifugation test results can be seen in table 6.

Table 6. Centrifugation Test

Formula		Results
F1	1	-
	2	-
	3	-
F2	1	-
	2	-
	3	-
F3	1	-
	2	-
	3	-
F4	1	-
	2	-
	3	-

Note : (-) no separation phase

The freeze thaw test is a test that has the purpose of simulating the product during the distribution process in vehicles, which are rarely equipped with temperature control devices. Therefore, this test was carried out at a certain temperature or humidity at a certain time so that the product in its packaging will undergo various changes. If during the six cycles, the preparation is stable or there is no phase that changes, it means that the product is stable during the distribution process [35]. Based on the results of the freeze thaw test, the four formulas were stable as there was no phase separation. The table of freeze thaw test results can be seen in table 7.

Table 7. Results of *Freeze Thaw*

cycle		F1			F2			F3			F4		
		1	2	3	1	2	3	1	2	3	1	2	3
1	4°C	-	-	-	-	-	-	-	-	-	-	-	-
	45°C	-	-	-	-	-	-	-	-	-	-	-	-
2	4°C	-	-	-	-	-	-	-	-	-	-	-	-
	45°C	-	-	-	-	-	-	-	-	-	-	-	-
3	4°C	-	-	-	-	-	-	-	-	-	-	-	-
	45°C	-	-	-	-	-	-	-	-	-	-	-	-
4	4°C	-	-	-	-	-	-	-	-	-	-	-	-
	45°C	-	-	-	-	-	-	-	-	-	-	-	-
5	4°C	-	-	-	-	-	-	-	-	-	-	-	-
	45°C	-	-	-	-	-	-	-	-	-	-	-	-
6	4°C	-	-	-	-	-	-	-	-	-	-	-	-
	45°C	-	-	-	-	-	-	-	-	-	-	-	-

Note: (-) no separation phase

The Sun Protection Factor (SPF) value or FPS value is measured as the ability or effectiveness of a material as a sunscreen. The higher the FPS value, the better the sunscreen will protect against UV rays. The purpose of using sunscreen is based on its ability to absorb, reflect or scatter sunlight [36]. The value of FPS was determined in vitro using the UV-vis spectrophotometry method developed by Mansur (1986). It was done by measuring the absorbance of each sample, namely the extract of arabica coffee leaves and the four formulas for the gel preparation for three replications with a wavelength between 290 - 320 nm where the measurements were described for each 5 nm interval. It is known that the arabica coffee leaf extract has these groups found in flavonoids and phenolic compounds, making it able to absorb radiation energy and produce absorbance values on the uv-vis spectrophotometer.

Based on the results of the study, the arabica coffee leaf extract has a concentration of 60 ppm. Of the three extract concentrations, it has a minimal protection category. The F1 gel preparation was a negative control, which did not contain extracts but produced a low FPS value for sunscreen so that it did not fall into the protection category. F2, F3 and F4 have a minimal protection category. The results of determining the FPS value of extracts and gel preparations can be seen in table 8 and table 9. From the second to fourth formulas, all formulas had minimal protection categories. From the results, it was obtained that gel preparations produce low protection on the skin. This can be overcome by adding other active sunscreen ingredients to support the effectiveness of sunscreens to produce more maximally.

On the SPF grade of the gel preparation that was higher than the extract, from the results of the research of [38] Donglikar and Deore (2017), it was revealed that the phytochemical SPF of the formulation was higher, which indicated the synergy and compatibility of the excipients as well. This result indicated that the prepared formulation has good SPF and good sun protection activity. Moreo et al., (2013) explained that this is possible to happen because it is influenced by the combination and concentration of the gel carrier component, the type of gel, the effect and the interaction of the carrier component such as the humectant used in the formulation. This factor can increase or decrease the UV absorption of sunscreens [39].

Therefore, to develop sunscreens with better safety and high SPF, the formulator must understand the physicochemical principles of not only active UV absorbance, but also vehicle components, such as esters, emulsifiers and emulsifiers used in formulationsThis is because sunscreens can interact with other components of the vehicle, and this interaction can affect the efficacy of the sunscreen.

Table 8. Results of SPF Value Extract

Sample	FPS Va;ue	Protection Category
Ekstract 60 ppm	2.08	Minimum protection
Ekstract 80 ppm	2.49	Minimum protection
Ekstract 100 ppm	3.68	Minimum protection

Table 9. Results of SPF Value Gel Extract

Sample	Value SPF	Protection Category
Formula 1	0.61	-
Formula 2	2.59	Minimum protection
Formula 3	2.95	Minimum protection
Formula 4	3.64	Minimum protection

4. Conclusions

Based on the research results, arabica coffee leaf extract could affect the gel dosage formulation and meet the specified requirements. The results of the ANOVA test showed a grade significance of 0.000 < 0.05, where there were significant differences in the four formulas for gel preparations. This showed that the concentration of arabica coffee leaf extract could affect the SPF value of the gel preparation. In this study, a gel preparation formulation that met the pharmaceutical requirements was obtained by providing an SPF grade with minimal protection.

References

[1] H. Winarsi, *Isoflavon*. Gadjah Mada University, 2005.

[2] COLIPA, *Sun Protection Test Methods*. COLIPA guidelines international, 2006.

[3] K. M. Yuliawati, E. R. Sadiyah, R. Solehati, and A. Elgiawan, "Sunscreen Activity Testing of Robusta Coffee (Coffea cenephora ex Froehner) Leave Extract and Fractions Pengujian Aktivitas Tabir Surya Ekstrak dan Fraksi Daun Kopi Robusta (Coffea canephora)," *Indones. J. Pharm. Sci. Technol.*, vol. 1, no. 1, 2019.

[4] D. RI, *Formularium Kosmetika Indonesia*. Jakarta: Departemen Kesehatan Republik Indonesia, 1985.

[5] M. R. Lu H, Edwards C, Pearse A, "Melanin content and distribution in the surface corneocyte with phototypes," *Br J Dermatol*, vol. 135, pp. 263–267, 1996.

[6] A. D. Puspitasari, D. A. K. Mulangsri, and H. Herlina, "Formulasi Krim Tabir Surya Ekstrak Etanol Daun Kersen (Muntingia calabura L.) untuk Kesehatan Kulit," *Media Penelit. dan Pengemb. Kesehat.*, vol. 28, no. 4, pp. 263–270, 2018, doi: 10.22435/mpk.v28i4.524.

[7] W. A. d. S. Almeida *et al.*, "Photoprotective activity and increase of SPF in sunscreen formulation using lyophilized red propolis extracts from Alagoas," *Rev. Bras. Farmacogn.*, vol. 29, no. 3, pp. 373–380, 2019, doi: 10.1016/j.bjp.2019.02.003.

[8] M. C. Gonçalves *et al.*, "Preliminary studies of the cytotoxicity and photoprotective properties of benzophenone and lactone derivatives," *Rev. Virtual Quim.*, vol. 10, no. 3, pp. 600–608, 2018, doi: 10.21577/1984-6835.20180044.

[9] A. P. Davis, R. Govaerts, D. M. Bridson, and P. Stoffelen, "An annotated taxonomic conspectus of the genus Coffea (Rubiaceae)," *Bot. J. Linn. Soc.*, vol. 152, no. 4, pp. 465–512, 2006, doi: 10.1111/j.1095-8339.2006.00584.x.

[10] H. L. . Hamon, P.; Grover, C.E.; Davis, A.P.; Rakotomalala, J.-J.; Raharimalala, N.E.; Albert, V.A.; Sreenath and E. Stoffelen, P.; Mitchell, S.E.; Couturon, "Genotyping-by-sequencing provides the first well-resolved phylogeny for coffee (Coffea) and insights into the evolution of

caffeine content in its species: GBS coffee phylogeny and the evolution of caffeine content.," *Mol. Phylogenet. Evol*, vol. 109., pp. 351–361, 2017.

[11] D. Ristiana, "Aktivitas Antioksidan Dan Kadar Fenol Berbagai Ekstrak Daun Kopi (Coffea Sp.): Potensi Aplikasi Bahan Alami Untuk Fortifikasi Pangan," *J. Apl. Teknol. Pangan*, vol. 6, no. 2, pp. 89–92, 2017, doi: 10.17728/jatp.205.

[12] O. Rodrigues Salgado, P.; Favarin, J.L.; Leandro, R.A.; Fontão de Lima Filho, "Total phenol concentrations in coffee tree leaves during fruit development," *Sci. Agric*, vol. 65, pp. 354–359, 2008.

[13] C. Manach, A. Scalbert, C. Morand, C. Rémésy, and L. Jiménez, "Polyphenols: Food sources and bioavailability," *Am. J. Clin. Nutr.*, vol. 79, no. 5, pp. 727–747, 2004, doi: 10.1093/ajcn/79.5.727.

[14] I. Tomac and M. Šeruga, "Electrochemical properties of chlorogenic acids and determination of their content in coffee using differential pulse voltammetry," *Int. J. Electrochem. Sci.*, vol. 11, no. 4, pp. 2854–2876, 2016, doi: 10.20964/110402854.

[15] Shovyana HH; A Karim Zulkarnain, "Physical Stability and Activity of Cream W/O Etanolic Fruit Extract of Mahkota Dewa (Phaleria macrocarpha (scheff.) Boerl,) as A a Sunscreen," *YogyakartaFakultas Farm. UGM. Tradit. Med. J.*, vol. 18, no. 2, 2013.

[16] G. Agati and M. Tattini, "Multiple functional roles of flavonoids in photoprotection," *New Phytol.*, vol. 186, no. 4, pp. 786–793, 2010, doi: 10.1111/j.1469-8137.2010.03269.x.

[17] Fields, "Sunscreens:Mechanisms of Action, Use, and Excipients," *Int. J. Pharm. Compd.*, 2008.

[18] A. P. Baskar M, Kiranmathyi B, Sivaraj C, Saraswathi K, "Journal of drug delivery and therapeutics (jddt)," *J. Drug Deliv. Ther.*, vol. 9, no. 3, pp. 661–668, 2019, [Online]. Available: http://dx.doi.org/10.22270/jddt.v9i3.2678.

[19] A. D. Puspitasari and R. L. Wulandari, "Aktivitas Antioksidan dan Penetapan Kadar Flavonoid Total Ekstrak Etil Asetat Daun Kersen (Muntingia calabura)," *J. Pharmascience*, vol. 4, no. 2, pp. 167–175, 2017, doi: 10.20527/jps.v4i2.5770.

[20] R. N. Sani, F. C. Nisa, R. D. Andriani, and J. M. Maligan, "ANALISIS RENDEMEN DAN SKRINING FITOKIMIA EKSTRAK ETANOL MIKROALGA LAUT Tetraselmis chuii Yield Analysis and Phytochemical Screening Ethanol Extract of Marine Microalgae Tetraselmis chuii," *J. Pangan dan Agroindustri*, vol. 2, no. 2, pp. 121–126, 2014.

[21] H. C. A. Loyd V. Allen, Nicholas G. Popovich, "Ansel ' s Pharmaceutical Dosage Forms and," *Int. J. Pharm. Compd.*, p. 537, 2014, [Online]. Available: https://downloads.lww.com/wolterskluwer_vitalstream_com/sample-content/9780781779340_allen/samples/frontmatter.pdf.

[22] S. Septiani, N. Wathoni, and S. R. mita Mita, "Formulasi Sediaan Masker gel Antioksidan Dari Ekstrak Etanol Biji Belinjo," *Fak. Farm. Univ. Padjajaran*, pp. 2–4, 2011.

[23] P. Niyogi, N. Raju, P. Readdy, and B. Rao, "Formulation And Evaluation Of Antiinflammatory Activity Of Solanum Pubescens Wild Extracts Gel On Albino Wistar Rats," vol. 2, no. 3, pp. 484–490, 2012.

[24] L. Allen, *The art, Science, and Technology of Compounding*, Second Edi. Washington DC: American Pharmaceutical, 1998.

[25] V. Damogalad, H. Jaya Edy, and H. Sri Supriati, "Formulasi Krim Tabir Surya Ekstrak Kulit Nanas (Ananas Comosus L Merr) Dan Uji in Vitro Nilai Sun Protecting Factor (Spf)," *PHARMACON J. Ilm. Farm. – UNSRAT*, vol. 2, no. 02, pp. 2302–2493, 2013.

[26] M. Rahmah Nasution and M. Br Manullang, "AKTIVITAS ANTIOKSIDAN SEDUHAN DAUN KOPI KAWA KERING (Coffea arabica L) DENGAN METODE DPPH," *J. Insa. Farm. Indones.*, vol. 3, no. 1, pp. 114–123, 2020, doi: 10.36387/jifi.v3i1.467.

[27] Y. Retnaningtyas, N. Kristiningrum, H. D. Renggani, and N. P. Narindra, "Karakteristik Simplisia dan Teh Herbal Daun Kopi Arabika (Coffea arabica)," *Farm. Jember*, vol. 1, no. 1, pp. 46–54, 2016.

[28] Badan Standardisasi Nasional, "Sediaan Tabir Surya," *Dewan Stand. Nas.*, vol. 16, no. 4399, pp. 1–3, 1996.
[29] Markham KR, *Cara Mengidentifikasi Flavonoid.* Bandung: Penerbit ITB, 1988.
[30] Nutrisia Aquariushinta Sayuti, "Formulasi dan Uji Stabilitas Fisik Sediaan Gel Ekstrak Daun Ketepeng Cina (Cassia alata L.)," *J. Kefarmasian Indones.*, vol. 5, no. 2, pp. 74–82, 2015.
[31] A. Garg, "Spreading of semisolid formulation," USA, pp. 84–102, Sep. 2002.
[32] N. S. Daud, N. S. Daud, L. Ode, and Z. Al, "FORMULASI LOTION TABIR SURYA EKSTRAK ETANOL BERAS MERAH (Oryza nivara)," vol. 1, no. September 2016, pp. 143–150.
[33] Agoes, *Sediaan Farmasi Likuida-Semisolida.* Bandung: Intitut Teknologi Bandung, 2012.
[34] U. Shinde, S. Pokharkar, and S. Modani, "Design and evaluation of microemulsion gel system of nadifloxacin," *Indian J. Pharm. Sci.*, vol. 74, no. 3, pp. 237–247, 2012, doi: 10.4103/0250-474X.106066.
[35] S. Bajaj, D. Singla, and N. Sakhuja, "Stability testing of pharmaceutical products," *J. Appl. Pharm. Sci.*, vol. 2, no. 3, pp. 129–138, 2012, doi: 10.7324/JAPS.2012.2322.
[36] N. Sharon, S. Anam, and Yuliet, "Formulasi Krim Antioksidan Ekstrak Etanol Bawang Hutan (Eleutherine palmifolia L. Merr.)," *Online J. Nat. Sci.*, vol. 2, no. 3, pp. 111–122, 2013.
[37] B. M. Mansur JDS, "Determinação do fator de proteção solar por espectrofotometria," *Mansur MCDA*, 1986.
[38] M. M. Donglikar and S. L. Deore, "Development and evaluation of herbal sunscreen," *Pharmacogn. J.*, vol. 9, no. 1, pp. 83–97, 2017, doi: 10.5530/pj.2017.1.15.
[39] More BH; Sakharwade SN; Tembhurne S; Sakarkar DM, "Evaluation of Sunscreen activity of Cream containing Leaves Extract of Butea monosperma for Topical application," *Int. J. Res. Cosmet. Sci.*, vol. 3, no. 1, pp. 1–6, 2013.

IOP Conf. Series: Earth and Environmental Science 1041 (2022) 012072 doi:10.1088/1755-1315/1041/1/012072

Antioxidant Activity and Sun Protection Factor (SPF) Graded Extract of Katuk Leaves (*Sauropus androgynus* (L.) Merr.)

Sofia Fatmawati*, Ni Putu Ermi Hikmawanti, Akbar Fadillah, Awlia Mustia Putri

Universitas Muhammadiyah Prof. DR. HAMKA, Jakarta, Indonesia

*sofia.fatmawati@uhamka.ac.id

Abstract. Katuk leaf (*Sauropus androgynus* (L.) Merr.) is believed to have medicinal properties, one of which is as an antioxidant. Its efficacy as an antioxidant cannot be separated from the phenolic and flavonoid compounds contained in katuk leaves. This study aims to determine the total phenolic and flavonoid levels as well as the antioxidant activity of graded extracts, namely *n*-hexane, ethyl acetate, and 70% ethanol extract of katuk leaves. The results showed that the total phenolic content and total flavonoid content of each extract were significantly different. Antioxidant activity by the DPPH method was calculated at IC_{50} using quercetin as a comparison. The highest antioxidant potential was shown in the ethanol extract. Each extract has the potential as a sunscreen and ethanol extract provides the highest SPF value.

1. Introduction

Phenol compounds are characterized by the presence of an aromatic ring and one or two hydroxyl groups. Phenol compounds that have more than two hydroxyl groups are called polyphenols, for example the tannin, flavonoid, melanin, and lignin groups [1]. Flavonoid compounds have a characteristic structure of two aromatic rings connected by three C atoms, usually with O atomic bonds in the form of heterocyclic oxygen bonds [1]. Flavonoids are a group of secondary metabolites produced by plants which are included in the large group of polyphenols. Flavonoids have the ability to scavenge free radicals and inhibit lipid oxidation [2].

Sunscreen contains compounds that can protect the skin by absorbing ultraviolet (UV) rays emitted by the sun [3]. Compounds contained in sunscreen can be used to prevent various skin diseases and to protect human skin health from the negative effects of UV rays. Sunscreens are divided into 2 groups, namely physical sunscreens and chemical sunscreens. Chemical sunscreens are sunscreens that absorb ultraviolet light, such as PABA, PABA esters, benzophenone, avobenzone, salicylate, cinnamate and camphor derivatives [4].

Katuk leaves (*Sauropus androgynus* (L.) Merr.) contain secondary metabolites such as tannins, saponins, alkaloids, flavonoids, glycosides, and phenols [5]. The 95% ethanol extract of katuk leaves has a high content of phenolic compounds with 1.49 mgGAE/g fresh weight and 8.71 mgGAE/g dry weight [6]. Antioxidant activity in previous studies stated that the IC_{50} value of katuk leaf methanol extract had a value of 80.81 ppm [7]. Variations in ethanol concentration and extraction methods resulted in different phenolic and flavonoid content in katuk leaves [8], [9]. Multilevel extraction gives better results than individual extraction with each solvent [10].

The aim of the study was to determine the levels of phenolic, flavonoid, as well as antioxidant activity and SPF testing of the *n*-hexane, ethyl acetate, and 70% ethanol extract of katuk leaves from the extraction process with graded maceration, using UV-Vis spectrophotometry method.

2. Materials and Methods

2.1 Preparation Extract

Katuk (*Sauropus androgynus* (L) Merr) leaves were collected and determined at the Institut Pertanian, Bogor, West Java. Simplicia processing begins with fresh katuk leaves taken and cleaned of impurities, then washed with water until clean, drained and finely chopped, and then dried again by aerating. After dried then powdered.

The graded extract was obtained by a stepwise method using solvents with different polarity levels, namely *n*-hexane, ethyl acetate, and 70% ethanol (1:10). The material is soaked for 24 hours, stirring occasionally for the first 6 hours. The residue is then separated from the filtrate. The process is repeated until the solvent is clear (3 repetitions). The residue was then extracted with ethyl acetate and 70% ethanol using the same procedure. Each maserate obtained was then concentrated with a vacuum rotary evaporator at a temperature of 40°C - 50°C to obtain a thick extract.

2.2 Characterization of Extract

The organoleptic test was carried out by observing the shape, color, smell, and taste using the five senses. Each extract was weighed carefully as much as 2 grams, then put into a silicate crucible that had been sized and compared, incandescent in a kiln and the temperature was gradually increased to 600°C (temperature difference of approximately 25°C) until carbon-free. Next, it was cooled in a desiccator and then weighed. The total ash content is calculated against the weight of the test material and is expressed in % w/w. Each 1.0 gram extract was placed in a moisture content balance container which had been previously sized and the initial weight of the extract was seen. Then the extract was dried at a temperature of 105°C at the moisture content balance until a constant extract weight value was obtained, afterward the final weight of the extract was seen. Results are viewed as % MC values [11].

2.3 Phytochemical Screening of Extract

The phytochemical compounds of kencur (aromatic ginger-*Kaempferia galanga*) ethanolic extract, such as phenolic, flavonoid, tannin, alkaloid, and terpene were qualitatively identified following standard procedures describing in the Harborne and Indonesian Herb Pharmacopoeia [12].

2.4 Determination of Total Phenolic Level

From the base solution of 100 ppm gallic acid, several concentrations were made, namely 18 ppm, 30 ppm, 42 ppm, 54 ppm, and 66 ppm. From each concentration of the standard solution of gallic acid, 300 μl was pipetted and then 1.5 ml of *Folin-Ciocalteu* reagent (1:10) was added. After being allowed to stand for 3 minutes, each solution was added with 1.2 ml of 7.5% Na_2CO_3 solution, shaken homogeneously, and allowed to stand in the operating time range at room temperature. All solutions were measured for absorbance at the maximum absorbance wavelength obtained, which was 756.5 nm, then a calibration curve was made for the relationship between gallic acid concentrations. The extract solution obtained was pipette 300 l and added 1.5 ml of *Folin-Ciocalteu* reagent and shaken. It was allowed to stand for 3 minutes, added 1.2 ml of 7.5% Na_2CO_3 solution and allowed to stand again in the operating time range at room temperature. The absorbance of the extract solution was measured by UV-Vis spectrophotometer at the maximum absorbance wavelength. The absorbance of the extract solution was measured by UV-Vis spectrophotometer at a wavelength of 756.5 nm. Performed 3 repetitions [13].

2.5 Determination of Total Flavonoid Level

From 1000 ppm quercetin mother liquor, several concentrations were made, namely 33, 57, 81 and 105 ppm. A total of 0.5 ml of quercetin solution was made with several concentrations of pipette then added with 1.5 ml of methanol and added 0.1 ml of 10% $AlCl_3$ reagent 0.1 ml of sodium acetate (1M) and 2.8 ml of distilled water. Then the solution was shaken and left for an operating time of 60 minutes at room temperature. Measure the absorbance at a wavelength of 434 nm against the standard. Each concentration of 0.5 ml of the test solution was added with 1.5 ml of methanol and added 0.1 ml of 10% $AlCl_3$ reagent 0.1 ml of sodium acetate (1M) and 2.8 ml of distilled water, then allowed to react during the operating time at room temperature. Measure the absorption at the maximum wavelength against the standard [14].

2.6 Antioxidant Activity Test

Weigh 100.0 mg of ethanol extract of katuk leaves from each extraction time dissolved with methanol in a volumetric flask, then diluted to obtain concentrations of 20, 40, 60, 80 and 100 ppm. A total of 0.2 mL of sample solution of each concentration was added with 1 mL of 0.5 mM DPPH and 5 mL of methanol, allowed to stand according to the operating time for 30 minutes and then measured at a maximum wavelength of 515.5 nm. The percentage of radical scavenging activity was calculated by the formula [15].

2.7 Sunscreen Potency Test

The ethanol extract of Arabica coffee leaves was diluted with ethanol to a concentration of 100 ppm. The SPF value was determined by measuring the absorbance of the solution from the extract using a UV-Vis spectrophotometer at a wavelength of 290-320 nm. The sample was dissolved in 5 mL of ethanol p.a. The determination of the SPF value was carried out three times for each sample. The absorbance data obtained were processed by the equation of Mansur (1986) [16].

$$SPF = CF \times EE \times I(\lambda) \times abs(\lambda)$$

The SPF value can be calculated by multiplying the correction factor (CF), the erythema effect spectrum (EE), the intensity spectrum from the sun (I), and also the absorbance (Abs) of the cream sample of coffee leaf ethanol extract. The SPF assessment refers to the provisions of the FDA (Food and Drug Administration), namely minimal protection if SPF 2-4, moderate protection 4-6, extra protection 6-8, maximum protection 8-15, and ultra protection >15 [17].

2.8 Data Analysis

The observed data from the dispersion test and color stability test were then analyzed using one-way ANOVA. If there is a significant difference between the formulas, then it is continued with the Tukey HSD test with a 95% confidence level to see if the formula has a significant difference or not.

3. Results and Discussion

The purpose of using three solvents with different polarities is to obtain active compounds from katuk leaves based on their polarity [10]. In this study, repetition was carried out with the aim of providing sufficient extract weight and being able to attract residual compounds that were still left in the previous iteration. Extraction using solvents with different polarities will produce different polyphenol components so that the antioxidant properties or characteristics of each compound obtained from the extraction are also different [18].

Table 1. Katuk Leaf Extraction Results

Extract type	Average extract weight (g) ± SD	Average extract yield (%) ± SD
n-Hexane	18,05 ± 0,73	7,22 ± 0,29
Ethyl acetate	14,87 ± 1,15	5,95 ± 0,46
Ethanol 70 %	50,27 ± 1,90	20,11 ± 0,76

Table 2. Characteristics of Katuk Leaf Extract

Parameters	Extract type		
	n-Hexane	Ethyl acetate	Ethanol 70%
Organoleptic			
Form	Thick extract	Thick extract	Thick extract
Smell	Typical	Typical	Typical
Favor	Bitter	Bitter	Rather Bitter
Color	Green	Green	Chocolate
Ash Level	2,59 % ± 0,22	0,55 % ± 0,05	6,88 % ± 0,93
Drying Shrinkage	1,08 % ± 0,48	3,04 ± 0,58	3,48 ± 2,41

The yield (Table 1) of the extract obtained from the multilevel extraction resulted in the lowest yield of the ethyl acetate extract, followed by the *n*-hexane extract. The yield obtained in the multistage extraction with 70% ethanol showed the highest yield. Based on the principle of mass transfer of solute into a solvent that is in accordance with characteristics *like dissolves like*, where there is a transfer at the interface layer (solvent and solute) then the solute diffuses into the solvent. The results are different because the ability to attract compounds in each solvent is different too. 70% ethanol extract has the highest yield because 70% ethanol solvent is universal. It can attract non-polar and polar compounds so that 70% ethanol extract is attracted to more secondary metabolites [8].

Organoleptic test on leaf extract was conducted to determine the characteristics of shape, smell, taste, and color. Of the three katuk leaf extracts the difference was only in the extract, the *n*-hexane, and ethyl acetate extracts had a green color while the 70% ethanol had a brown color.

The ash content obtained from this research is *n*-hexane 2.59%, ethyl acetate 0.55%, and ethanol 70% 6.88% as seen in (Table 2). The three extracts had simplicia criteria that were in accordance with the guidelines for a good extract ash content, namely the ash content of not more than 12%. The purpose of ash content is to show the content of organic and inorganic compounds from plants such as alkali metals, alkaline earth metals and heavy metals, as well as to provide an overview of internal and external mineral content until they become extracts.

Drying shrinkage has a goal to give the maximum amount of compounds lost in the drying process. In this study, a moisture balance was used at a temperature of 105°C from the three extracts of *n*-hexane, ethyl acetate, and 70% ethanol. The three results (Table 2) fall into the drying shrinkage limit of no more than 10%.

Table 3. Phytochemical Screening

Detected compounds	Extract type		
	***n*-Hexane**	**Ethyl acetate**	**Ethanol 70%**
Alkaloids			
- Dragendor	-	-	+
- Bouchardat	-	-	+
- Mayer	+	+	+
Phenolic	+	+	+
Flavonoids	+	+	+
Saponins	+	+	+
Tannins	+	+	+
Terpenoids	-	-	-
Steroids	+	+	+

Note: (+) There is a compound (-) There is no compound

The results of phytochemical screening can be seen in (Table 3). Phytochemical screening was carried out to identify the compounds contained in *n*-hexane extract, ethyl acetate extract, and 70% ethanol extract of katuk leaves. Screening is an important step in an effort to reveal the potential of plant resources [1]. The 70% ethanol extract showed positive results for alkaloids, while the ethyl acetate and *n*-hexane extracts were negative.

Determination of total phenolic levels uses the *Folin-Ciocalteu* method because it is easy, cheap, fast to do, and can be done routinely in the laboratory. Polyphenols in plant extracts react with the reagent used, namely *Folin-Ciocalteu* (Blainski et al., 2013). Phenolic compounds react with *Folin-Ciocalteu* reagent only in an alkaline environment so that proton dissociation occurs in phenolic compounds into phenolic ions, hence 7.5% Na_2CO_3 is used to make an alkaline environment. The hydroxyl group in the phenolic compound reacts with the *Folin-Ciocalteu* reagent to form a blue *molybdenum-tungsten* complex which can be detected by a spectrophotometer [19]. Gallic acid (3,4,5-*Trihydroxybenzoic acid*) was used as standard. Gallic acid was chosen as the standard because it represents the general properties of phenolic compounds. In this case gallic acid is a natural phenolic compound that has a strong antioxidant effect [20].

Determination of the maximum wavelength of gallic acid aims to determine at what wavelength the compound reaches the highest absorbance value. The result of measuring the maximum wavelength is 765.50 nm with an absorbance of 0.3445 as the maximum absorbance value. The operating time of the results obtained is stable absorbance at 60 minutes so that it can be concluded that the purpose of determining the operating time is to get the measurement time when the reaction has been running optimally which is characterized by a stable absorbance, so as to maximize the measurement [21].

Various concentrations of gallic acid standards were made because the assay used a calibration curve equation (Table 4), so to make a calibration curve, several concentrations were made in order to obtain a linear equation. The results obtained were at concentrations of 18, 30, 42, 54, and 66 ppm, the data met the requirements of *Lambert Beer*'s law. The calibration curve obtained by the linear regression equation $y = 0.0093.x + 0.1824$ with a correlation coefficient $(r) = 0.9944$.

In the determination of the total phenolic content of katuk leaf extract, as seen in (Table 5) that the 70% ethanol extract had the highest concentration of 16.71 mgGAE/g. The extract was followed by ethyl acetate and *n*-hexane with a concentration of 2.95 mgGAE/g and 0.57 mgGAE/g extract, respectively. Phenolic compounds are extracted well in 70% ethanol because the phenolic content will increase in the extract as the polarity of the solvent increases. Phenolic also tends to be polar so it can dissolve in polar solvents well. These results are similar to other studies which stated polar extracts have the highest phenolic content followed by semi-polar and non-polar extracts. Methanol, a polar solvent, has a higher phenolic content than acetone which is a semi polar solvent [22].

Table 4. Results of Total Phenolic Levels

Extract type	Average Phenolic content (mgGAE/g Extract ± SD)
n-Hexane	0,57 ± 0,02
Ethyl Acetate	2,95 ± 0,07
Ethanol 70%	16,71 ± 0,43

Table 5. Results of Total Flavonoid Levels

Extract type	Average levels of Flavonoids (mgQE/g Extract ± SD)
n-Hexane	55,57 ± 0,11
Ethyl Acetate	88,79 ± 0,73
Ethanol 70%	6,23 ± 0,05

Determination of total flavonoid levels uses quercetin standards. Quercetin is a flavonoid class of the flavonol group [23]. Meanwhile, Quantitative Analysis is conducted by applying the UV-Vis spectrophotometric method (Hanani, 2015). The maximum wavelength resulted from this study was 434 nm. This was due to the reaction of the AlCl3 complex which caused the shift of the wave to become visible, which was indicated by the solution becoming more yellow, and the addition of sodium acetate to maintain the visible wavelength [24]. $AlCl_3$ reagents with flavonoids form complexes between neighboring hydroxyl groups and ketones or with neighboring hydroxyl groups. $AlCl_3$ will react with the ketone group at C4 and the OH group at C3 or C5 in flavone or flavonoid compounds to form a yellow stable complex compound.

The results of the absorbance measurement of the quercetin standard solution to obtain a linear calibration curve obtained results such as (Table 6). These results were entered into the regression equation which then produced linear $y = 0.0024x+0.1897, r = 0.9931$. The flavonoid content was expressed in mgQE/g Extract.

After reading the levels of flavonoids (Table 7) of the graded extract of katuk leaves, the results obtained were *n*-hexane 55.57 mgQE/g, ethyl acetate 88.79 mgQE/g, and 70% ethanol 6.23 mgQE/g. In this study, ethyl acetate had the largest concentration, this is because the content of flavonoid compounds contained in katuk leaves has a low polarity level, namely in the form of aglycones or commonly called aglycone flavonoids. Aglycone flavonoids have less polar properties and tend to be more soluble in chloroform and ether. Flavonoids generally bind to sugars to form glycosides which cause these compounds to be easily soluble in polar solvents, such as methanol, butanol, and ethyl acetate.

In this study, the free radical compound DPPH (2,2-*diphenyl*-1-*picrylhydrazyl*) was used. The results obtained on the measurement of the maximum wavelength is 515.5 nm, the wavelength can be used because the wavelength of the maximum absorbance for measurements with the DPPH method is 515-520 nm [25].

Quercetin was chosen as the standard for comparison because quercetin is a natural secondary antioxidant that has been shown to have free radical scavenging activity. Quercetin, which is a flavonoid group, has several biological activities. These activities can be attributed to the antioxidant properties of quercetin, including its ability to scavenge free radicals [23].

The results of antioxidant activity (Table 8) show that 70% ethanol extract has a lower IC_{50} value than ethyl acetate and *n*-hexane, the smaller the % inhibition obtained, the stronger the antioxidant activity produced [7]. The compounds contained in the ethanol extract are an accumulation of polar, semi-polar, and non-polar compounds. When the extract is macerated in stages, the synergistic function between the compounds will reduce because the components contained in the extract have been separated. The non-polar chemical components were extracted in the *n*-hexane solvent, the semi-polar chemical components were extracted in ethyl acetate, and chemical components that are polar were extracted in 70% ethanol solvent. This causes the antioxidant activity of 70% ethanol extract to be the strongest and at the same time has the least % inhibition, compared to others.

Table 6. Antioxidant Activity Test Results and SPF

Sample Type	Average antioxidant activity ($IC_{50} \pm SD$)	SPF
Quercetin	8,83	-
n-Hexane	88,43 ± 1,20	2,13
Ethyl Acetate	77,65 ± 1,78	5,23
Ethanol 70%	70,33 ± 1,64	2,13

Ethyl acetate extract gave the highest SPF results compared to ethanol and *n*-hexane extracts. The three extracts belonged to the minimum SPF capability (range 2 to 4). Phytochemical compounds in the extract that may play a role in the potential for absorption of ultraviolet light are flavonoids and phenolic compounds [26]. Several groups of active compounds derived from natural ingredients such as flavonoids, tannins, anthraquinones, cinnamates, and glycosides are reported to have the ability to protect against UV rays [27].

4. Conclusions

Based on the results obtained in this study, the determination of total phenolic content had the highest concentration in 70% ethanol extract, which was then followed by ethyl acetate extract and *n*-hexane extract. In the determination of total flavonoid content, the highest yield was in the ethyl acetate extract, then *n*-hexane content, while the 70% ethanol had the lowest content. The results of the antioxidant activity test used the DPPH method. IC_{50} of *n*-hexane extract had the highest concentration in this activity test, followed by ethyl acetate and 70% ethanol extract. The smaller the IC_{50} value, the higher the antioxidant activity produced. Multilevel extraction greatly affects the levels of each extract used. Besides, it will also look for different compounds in the solvent used.

References

[1] P. Tiwari, M. Kaur, and H. Kaur, "Phytochemical screening and Extraction: A Review," 2011.

[2] Z. Zuraida, S. Sulistiyani, D. Sajuthi, and I. H. Suparto, "FENOL, FLAVONOID, DAN AKTIVITAS ANTIOKSIDAN PADA EKSTRAK KULIT BATANG PULAI (Alstonia scholaris R.Br)," *J. Penelit. Has. Hutan*, vol. 35, no. 3, pp. 211–219, 2017, doi: 10.20886/jphh.2017.35.3.211-219.

[3] S. Handani and A. Indyah Sulistyo, "Synthesis of Hydroxyl Radical Scavengers from Benzalacetone and its Derivatives," *J. Phys. Sci.*, vol. 19, no. 2, pp. 61–68, 2008, [Online]. Available: http://web.usm.my/jps/19-2-08/article 19-2-7.pdf.

[4] Y. D. Putri, H. Kartamihardja, and I. Lisna, "Yola et al 2019," *Formulasi dan Eval. Losion Tabir Surya Ekstrak Daun Stevia (Stevia rebaudiana Bertoni M)*, vol. 6, no. 1, pp. 32–36, 2019.

[5] D. Andini, "Potential Of Katuk Leaf (Sauropus androgynus L. Merr) As Aphrodisiac," *J. Major.*, vol. 3, no. 7, pp. 16–21, 2014.

[6] N. Andarwulan, R. Batari, D. A. Sandrasari, B. Bolling, and H. Wijaya, "Flavonoid content and antioxidant activity of vegetables from Indonesia," *Food Chem.*, vol. 121, no. 4, pp. 1231–1235, 2010, doi: 10.1016/j.foodchem.2010.01.033.

[7] C. F. Zuhra, J. B. Tarigan, and H. Sihotang, "Aktivitas Antioksidan Senyawa Flavonoid Dari Daun Katuk (Sauropus Androgunus (L) Merr.)," 2010.

[8] N. P. E. Hikmawanti, S. Fatmawati, and A. W. Asri, "The effect of ethanol concentrations as the

extraction solvent on antioxidant activity of Katuk (Sauropus androgynus (L.) Merr.) leaves extracts," *IOP Conf. Ser. Earth Environ. Sci.*, vol. 755, no. 1, 2021, doi: 10.1088/1755-1315/755/1/012060.

[9] N. P. E. Hikmawanti, S. Fatmawati, Z. Arifin, and . V., "Pengaruh Variasi Metode Ekstraksi Terhadap Perolehan Senyawa Antioksidan Pada Daun Katuk (Sauropus androgynus (L.) Merr)," *J. Farm. Udayana*, vol. 10, no. 1, p. 1, 2021, doi: 10.24843/jfu.2021.v10.i01.p01.

[10] E. C. Jeyaseelan, S. Jenothiny, M. K. Pathmanathan, and J. P. Jeyadevan, "Antibacterial activity of sequentially extracted organic solvent extracts of fruits, flowers and leaves of Lawsonia inermis L. from Jaffna," *Asian Pac. J. Trop. Biomed.*, vol. 2, no. 10, pp. 798–802, 2012, doi: 10.1016/S2221-1691(12)60232-9.

[11] M. Arista, "Aktivitas antioksidan ekstrak etanol 80% dan 96% daun katuk (," *J. Ilm. Mhs. Univ. Surabaya*, vol. 2, no. 2, pp. 1–16, 2013.

[12] S. Winarti, U. Sarofa, and D. Anggrahini, "EKSTRAKSI DAN STABILITAS WARNA UBI JALAR UNGU (Ipomoea batatas L .,) SEBAGAI PEWARNA ALAMI," *J. Tek. Kim.*, vol. 3, no. 1, pp. 207–214, 2008.

[13] Y. Y. Lim and J. Murtijaya, "Antioxidant properties of Phyllanthus amarus extracts as affected by different drying methods," *LWT*, vol. 40, no. 9, pp. 1664–1669, 2007, doi: 10.1016/j.lwt.2006.12.013.

[14] D.-O. Kim, S. W. Jeong, and C. Y. Lee, "Antioxidant capacity of phenolic phytochemicals from various cultivars of plums," *Food Chem.*, vol. 81, no. 3, pp. 321–326, 2003, doi: https://doi.org/10.1016/S0308-8146(02)00423-5.

[15] M. M. Rahman, M. B. Islam, M. Biswas, and A. H. M. Khurshid Alam, "In vitro antioxidant and free radical scavenging activity of different parts of Tabebuia pallida growing in Bangladesh," *BMC Res. Notes*, vol. 8, no. 1, p. 621, 2015, doi: 10.1186/s13104-015-1618-6.

[16] W. A. d. S. Almeida *et al.*, "Photoprotective activity and increase of SPF in sunscreen formulation using lyophilized red propolis extracts from Alagoas," *Rev. Bras. Farmacogn.*, vol. 29, no. 3, pp. 373–380, 2019, doi: 10.1016/j.bjp.2019.02.003.

[17] V. Damogalad, H. Jaya Edy, and H. Sri Supriati, "Formulasi Krim Tabir Surya Ekstrak Kulit Nanas (Ananas Comosus L Merr) Dan Uji in Vitro Nilai Sun Protecting Factor (Spf)," *PHARMACON J. Ilm. Farm. – UNSRAT*, vol. 2, no. 02, pp. 2302–2493, 2013.

[18] R. Y. Senja, E. Issusilaningtyas, A. K. Nugroho, and E. P. Setyowati, "Perbandingan Metode Ekstraksi dan Variasi Pelarut terhadap Rendemen dan Aktivitas Antioksidan Ekstrak Kubis Ungu (Brassica oleracea L. var. capitata f. rubra)," *Maj. Obat Tradis.*, vol. 19, no. 1, pp. 43–48, 2014.

[19] R. Alfian and H. Susanti, "PENETAPAN KADAR FENOLIK TOTAL EKSTRAK METANOL KELOPAK BUNGA ROSELLA MERAH (Hibiscus sabdariffa Linn) DENGAN VARIASI TEMPAT TUMBUH SECARA SPEKTROFOTOMETRI," *Pharmaciana*, vol. 2, no. 1, 2012, doi: 10.12928/pharmaciana.v2i1.655.

[20] Y. M. Huliselan, M. R. J. Runtuwene, and D. S. Wewengkang, "Antioxidant Activity of Ethanol, Ethyl Acetate and n-Hexane Extract from Seswanua Leaves (Clerodendron squamatum Vahl.)," *Pharmacon*, vol. 4, no. 3, pp. 155–163, 2015, [Online]. Available: https://ejournal.unsrat.ac.id/index.php/pharmacon/article/view/8855.

[21] T. G. Arikalang, S. Sudewi, and J. A. Rorong, "Penentuan Kandungan Total Fenolik pada Ekstrak Daun Gedi Hijau (Abelmoschus manihot L .) yang Diukur dengan Spektrofotometer UV-Vis," *Pharmacon J. Ilm. Farm.*, vol. 7, no. 3, pp. 14–21, 2018.

[22] F. Ulfa, A. Anggo, and R. Romadhon, "Uji Potensi Aktivitas Antioksidan Dengan Metode Ekstraksi Bertingkat Pada Lamun Dugong (Thalassia Hemprichii) Dari Perairan Jepara," *J. Pengolah. dan Bioteknol. Has. Perikan.*, vol. 3, no. 3, pp. 32–39, 2014.

[23] M. Materska, "QUERCETIN AND ITS DERIVATIVES: CHEMICAL STRUCTURE AND BIOACTIVITY– A REVIEW," *Polish J. Food Nutr. Sci.*, vol. 58, no. 4, pp. 407–413, 2008, [Online]. Available: http://journal.pan.olsztyn.pl/QUERCETIN-AND-ITS-DERIVATIVES-

CHEMICAL-STRUCTURE-AND-BIOACTIVITY-8211-A-REVIEW,98157,0,2.html.

[24] C. C. Chang, M. H. Yang, H. M. Wen, and J. C. Chern, "Estimation of total flavonoid content in propolis by two complementary colometric methods," *J. Food Drug Anal.*, vol. 10, no. 3, pp. 178–182, 2002, doi: 10.38212/2224-6614.2748.

[25] P. Molyneux, "The Use of the Stable Free Radical Diphenylpicryl-hydrazyl (DPPH) for Estimating Antioxidant Activity," *Songklanakarin J. Sci. Technol.*, vol. 26, no. December 2003, pp. 211–219, 2004, doi: 10.1287/isre.6.2.144.

[26] C. Malsawmtluangi, D. K. Nath, I. Jamatia, E. Z. Lianhimgthangi, and L. Pachuau, "Determination of Sun Protection Factor (SPF) number of some aqueous herbal extracts," *J. Appl. Pharm. Sci.*, vol. 3, no. 9, pp. 150–151, 2013, doi: 10.7324/JAPS.2013.3925.

[27] M. M. Donglikar and S. L. Deore, "Development and evaluation of herbal sunscreen," *Pharmacogn. J.*, vol. 9, no. 1, pp. 83–97, 2017, doi: 10.5530/pj.2017.1.15.

IOP Conf. Series: Earth and Environmental Science 1041 (2022) 012073 doi:10.1088/1755-1315/1041/1/012073

Anti-Inflamatory Activity of Water Extract of *Talinum paniculatum*. (Jact). Gaertn Leaves on Wistar Rat

A Emelda[1]*, Sukmawati[1], A L Dongke[1] and I Marzuki[2]

[1]Universitas Muslim Indonesia, Jl. Urip Sumohardjo KM. 5, Makassar 9023, Indonesia.
[2]Fajar University, Panakkukang-Makassar 90231, Indonesia.

*Correspondence author: andi.emelda@umi.ac.id

Abstract. Inflammation is the body's protective response to tissue injury caused by harmful agents characterized by edema. *Talinum paniculatum* (jact.) Gaertn leaves contain secondary metabolites such as flavonoids, steroids, and tannins known to have anti-inflammatory effects. The research aimed to determine the effective dose of *Talinum paniculatum* (jact.) Gaertn leaves as anti-inflammatory against rats. This research was conducted using rats which were divided into 3 groups: control (prednisone), extract of *Talinum paniculatum* (jact.) Gaertn leaves at doses of 1 g/kgbw and 2 g/kgbw. The width and thickness of the edema were measured for 5 hours using a caliper to calculate the total AUC value. Then it was analyzed statistically using Kruskal-Wallis test followed by Mann-Whitney test. The results of statistical analysis showed that the group at the dose of 2 g/kgbw had an anti-inflammatory effect not significantly different from the control group. It was concluded that the water extract of *Talinum paniculatum* (jact.) Gaertn leaves had anti-inflammatory activity at the effective dose of 2 g/kgbw.

Keywords: Anti-inflammatory, (*Talinum paniculatum* (jact.) Gaertn), Calipers.

1. Introduction

Inflammation is a disease that affects many people, usually characterized by swelling, pain, redness, and heat. Inflammation occurs because there is a normal protective response to tissue injury caused by physical trauma, hazardous chemicals, or microbiological agents. Inflammation is the body's attempt to activate or destroy invading organisms, remove irritants and prepare for tissue repair [1].

Inflammation can be local and systemic, can also occur acutely or chronically, causing pathological abnormalities. Inflammation treatment includes two aspects. The first is to relieve pain as a symptom and the second is to stop the tissue damage process. Reduction of inflammation or inflammatory response using steroids and non-steroidal anti-inflammatory drugs (NSAIDs) can relieve inflammatory reactions well, but long-term use can have side effects. Systemic use of steroids as anti-inflammatory drugs for a long time has side effects in the form of decreased endogenous glucocorticoid synthesis, decreasing the body's immune response to infection, osteoporosis, moon face, and hypertension. Systemic use of non-steroidal anti-inflammatory drugs (NSAIDs) for a long time can also have side effects in the form of digestive tract disorders such as peptic ulcers, analgesic nephropathy, disrupting platelet function, and inhibiting pregnancy induction [2].

The side effects of using these drugs make people want to use drugs that are more effective in healing and have fewer side effects than synthetic drugs, namely traditional medicines [3]. *Talinum paniculatum*

(Jact.) Gaertn contains saponin derivatives, alkaloids, steroids, tannins, flavonoids, provitamin A, fiber, and various important minerals [4].

The leaves of *Talinum paniculatum* (Jact.) Gaertn is empirically effective for treating various diseases such as ulcers, increasing the amount of breast milk production, coughing up phlegm and blood, pneumonia, cold sweat, diarrhea, polyuria, irregular menstruation, vaginal discharge, increase appetite, antibacterial, antiviral, anti-inflammatory, antihistamine and physiologically can improve circulation or blood circulation in the central nervous system or blood circulation to the peripheral nerves [4;5;6;7].

The antioxidant activity of *Talinum paniculatum* leaf infusion, where the results obtained were higher antioxidant activity in fresh leaves extract with 55.03% with an LC50 value of 181.15 μg/ml (DPPH free radical scavenging method), 85.59% (ferric thiocyanate method). Whereas for extraction with hot water, a higher phenolic content was found in the dry leaves extract with 116.11 mg/g extract. The conclusion was that 1 g of extract had antioxidant activity and varied the dosage to experimental mice with doses of 1 g and 2 g [8].

People generally use herbal plants extracted with boiling water for health reasons. When compared to other organic solvents such as ethanol and ethyl acetate, water is more economical, non-toxic, resistant to heating, working more practical, and more able to extract the compounds contained in *Talinum paniculatum* leaves [8].

Based on the description above, this study was conducted to test the anti-inflammatory effect of water extract of *Talinum paniculatum* (Jacq.) Gaertn leaves on wistar rats.

2. Materials and Methods

2.1. Materials

The materials used in this study were aquadest, aluminum foil, water extract of *Talinum paniculatum* (Jacq.) Gaertn leaves, carrageenan lambda (λ), prednisone®, sodium carboxymethylcellulose (Na. CMC) 1% w/v, and sodium chloride (NaCl) 0.9%. Wistar Rat (Ethical clearance: No.:103/A.1/KEPK UMI/IX/2020).

2.2. Methods

This research is experimental research, where the test was carried out using a caliper device. The test animal used was wistar rats which had been adapted and given standard feed for 14 days. Rats were grouped into 3 groups, a control was given prednisone drug and Groups II and III as test group was given water extract of *Talinum paniculatum* (Jacq.) Gaertn 1 g/kgbw and 2 g/kgbw. 10% decocta made by weighing 30 grams in 300 mL aquadest at $90^{\circ}C$ for 30 minutes. The test was carried out by inducing 0.1 mL of 1% carrageenan intraplantarally. The test preparation was administered 1 hour after induction of 1% carrageenan. Measurement of the width and thickness of the edema leg of the test animals was carried out before induction, after induction, and after giving the test preparation at the 1st, 2nd, 3rd, 4th, and 5th hours [9].

3. Result and Discussion

In this study, a research was conducted on the anti-inflammatory test of water extract of *Talinum paniculatum* (jact.) Gaertn leaves on wistar rats. This study aims to determine the anti-inflammatory effect and effective dosage of the extract of *Talinum paniculatum* (jact.) Gaertn leaves as an anti-inflammatory.

This study used calipers by measuring the width and thickness of the rats' feet induced by 0.1 mL of carrageenan 1% w/v on the soles of the feet of experimental rats intraplantarally. The inducer used is the type carrageenan (λ) because it was chosen as an irritant compound that induces cell injury through the release of mediators that initiate the inflammatory process. At the time of the release of inflammatory mediators, maximal edema will happened and lasts for several hours. Besides, this type of carrageenan is easily dissolved in cold or hot water, and the time for dissolving is shorter than other types of carrageenan. Also, the use of carrageen as an inflammation inducer has several advantages including it does not damage tissue, leaves no marks, and provides a more sensitive response to giving anti-inflammatory drugs compared to other compounds [10].

In the process of edema formation, carrageenan will induce cell injury by releasing mediators that initiate the inflammatory process. Edema caused by carrageenan can last for 4 hours and gradually diminishes within 24 hours. Edema occurs due to the release of inflammatory mediators such as histamine, bradykinin, a prostaglandin. Edema caused by carrageenan injection is strengthen by inflammatory mediators especially PGE1 and PGE2 by decreasing vascular permeability. If the capillary permeability decreases, it will cause plasma proteins to go to the injured tissue, causing edema [11].

AUC is obtained from the calculation of the reduction in width and thickness of the edema of each animal. The AUC value informs about the ability of the water extract of ginseng bugis leaves as an anti-inflammatory. The greater the AUC value, the smaller the effect of reducing edema or anti-inflammation, and the smaller the AUC value, the greater the effect of reducing edema or anti-inflammatory [12].

Based on the edema width parameter, the value for edema reduction can be seen in Table 1. From the data, the average width of the edema of the rats' feet can be seen which can be seen in the following figure:

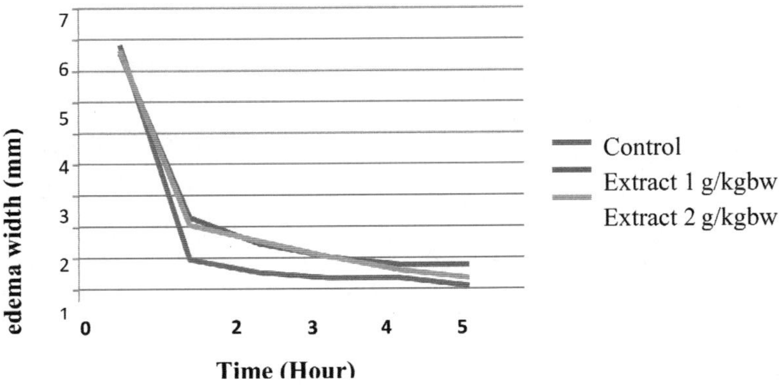

Figure 1. Graphic of the mean value of edema width reduction for 5 hours.

The average result of decreasing the edema width of the rats in Figure 1 shows that all groups experienced a decrease in edema for 5 hours. The prednisone group showed the best reduction in edema width. Prednisone is a corticosteroid class steroid anti-inflammatory drug that inhibits the enzyme phospholipase A2, thereby inhibiting the formation of prostaglandins and leukotrienes. Edema width reduction data is used to determine the total AUC (Area Under Curve) value (Tabel 1).

Table 1. Average AUC (area under curve) avlue of leg width

Treatment Group	Average AUC (mm.hour)
Control	3,5
Extract 1 g/kgbw	7,75
Extract 2 g/kgbw	7,3

Table 1 shows that the positive control group had the lowest average AUC value (3.5 mm.hour), which means that prednisone provided the best anti-inflammatory effect compared to other groups. This was followed by the extract 2 g/kgbw and extract 1 g/kgbw groups with an average AUC value of 7.3 and 7.75 mm.hours respectively.

The results of the calculation of the AUC value for the total width of the rats' feet obtained, were analyzed using the Kruskal-Wallis statistical test and continued with the Mann-Whitney advanced test. Statistical analysis of the AUC value of the total foot width of the rats showed that the positive control group (prednisone) showed no significant difference between the extract group 1 g/kgbw and the extract group 2 g/kgbw with a significant value of $p > 0.05$. This proves that extract 1 g/kgbw and extract 2 g/kgbw have an anti-inflammatory effect which is almost the same as the positive control (Prednisone) in reducing the edema width of rats and also the extract group 1 g/kgbw showed no significant differences in the extract group 2 g/kgbw with a significant value of $p > 0.05$.
This proves that extract 1 g/kgbw and extract 2 g /kgbw have the same effect an anti-inflammatory. Based on the edema thickness parameter, the reduction data can be seen in Table 2. From these data, it is obtained the average thickness of rat foot edema can be seen in the following figure:

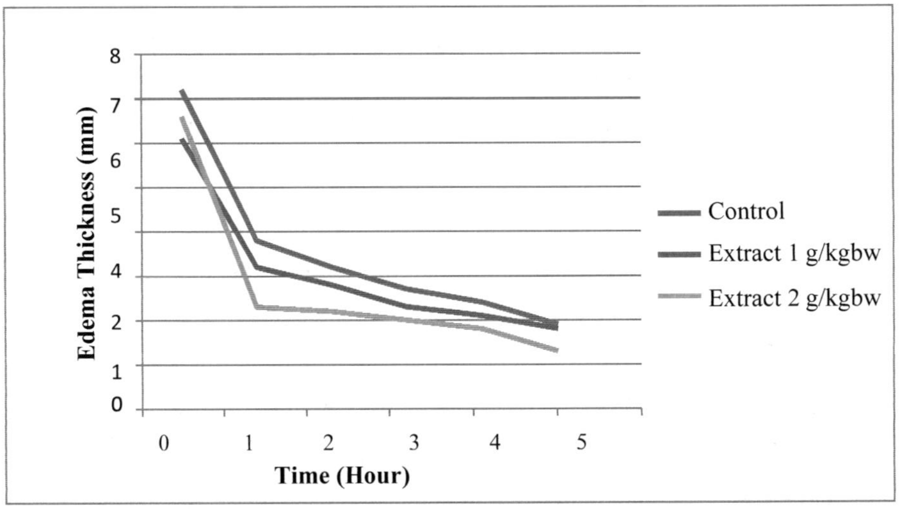

Figure 2. Graphic of the mean value of edema thickness reduction for 5 hours.

Figure 2 describes all groups experienced a reduction in edema thickness up to 5 hours. Group 2 g/ kgbw showed better reduction in edema thickness than other groups. Extract 2 g/kgbw gave a better effect than 1 g/kgbw and control group. Edema thickness reduction data is used to determine the total AUC value then the average value of the total AUC is determined in the followingTable 2.

Table 2. Average AUC value

Treatment Group	Average AUC (mm.Hour)
Control (Prednisone)	13,0
Extract 1 g/kgbw	11,35
Extract 2 g/kgbw	9,0

Table 2 shows that the control group had the lowest average AUC value 9,0 mm.hour, which means that the extract of *Talinum paniculatum* leaves at a dose of 2 g/kgbw gave the best anti-inflammatory effect compared to other groups. This was followed by the 1 g/kgbw group and the control with the average AUC values of 11.35 mm.hours and 13.0 mm.hours, respectively.The results of the calculation of the total AUC value for the thickness of the rats' feet obtained, were analyzed using the Kruskal-

Wallis statistical test and continued with the Mann-Whitney advanced test. Statistical analysis of the AUC value that the control group (prednisone) showed no significant difference between the group 1 g/kgbw and the group 2 g/kgbw with a significant value of $p > 0.05$. This proves that 1 g/kgbw and 2 g/kgbw have an anti-inflammatory effect that is almost the same as the control (Prednisone) in reducing the thickness of rat leg edema and also the extract 1 g/kgbw showed no significant differences in the group. Extract 2 g/kgbw with a significant value of $p > 0.05$. This proves that 1 g/kgbw and 2 g/kgbw have the same effect as an anti-inflammatory.

The effectiveness of the water extract of *Talinum paniculatum* leaves as an anti-inflammatory is thought to be due to the presence of flavonoids in the body which act to inhibit the lipooxygenase enzyme which plays a role in leukotriene biosynthesis. It also inhibits the metabolism of arachidonic acid so that prostaglandin production can be reduced. Flavonoids also inhibit the secretion of lysosomal enzymes which are inflammatory mediators. Inhibition of these inflammatory mediators can inhibit the proliferation of the inflammatory process [13].

4. Conclusion

Based on the research that has been done, it can be concluded that the water extract of *Talinum paniculatum* leaves has an anti-inflammatory effect. The water extract of *Talinum paniculatum* leaves a dose of 2 g/kgbw is an effective dose used as an anti-inflammatory in animal test induced by carrageenan.

5. Reference

[1]. Harvey R A and Pamela C C 2013 Pharmacology Illustrated Review (Jakarta: EGC medical book publisher)

[2]. Goodman G 2003 Basic Therapeutic Pharmacology 10th ed. Volume 2 (Jakarta: EGC Medical Book Publisher)

[3]. Dawud F, Bodhi W and Lolo W A 2014 Test of Anti-Inflammatory Effects of Ethanol Extract of Mahkota Dewa Fruit Skin (Phaleria Macrocarpa Boerl.) Against Foot Edema of Male White Rats. Pharmacon. 3 (1) 8-14.

[4]. Hidayat S 2005 Ginseng, Efficacious Natural Multivitamin (Jakarta: Penebar Swadaya).

[5]. Deliah S 2010 Som Jawa (Talinum paniculatum) Indonesian Ginseng Healing Various Diseases. Industrial Crops Research and Development News. 16 (2) 21.

[6]. Subroto A and Saputro H 2008 Fight Disease with Ant Nests (Jakarta: Penebar Swadaya).

[7]. Emelda A, Fitriana, Adnan RS 2021 Antibacterial Activity of Ethanol Extract and Ethyl Acetate of Ginseng Bugis (Talinum Panicullatum Gaertn.) Leaves Against Staphylococcus Aureus and Escherichia coli Bacteria. Journal of Physics conference series. 1899.

[8]. Lestario L N, Anggelia E C, and Yohanes M 2009 Javanese Ginseng Leaf Antioxidant Activity. Agritech. 29 (2)

[9]. Rahman S, Wati A, and Sukmawati E 2018 Anti-Inflammatory Effects of Ethanol Extract of Cambodian Leaves (Plumeria rubra L.) on White Rats (Rattus Novergicus). Jurnal As-Syifaa. 10 (01) 56.

[10]. Candra A and Santi T D 2017 The Effectiveness of Papaya Leaf Extract (Carica papaya L) as an Anti-inflammatory. Jurnal Aceh Medika. 1 (2) 63-66.

[11]. Corsini E, Paola R D, Viviani B, Genovese T, Mazzon E, Lucchi L, Marinovich M, Galli C L, and Cuzzocrea S 2005 Increased Carrageenan-Induced Acute Lung Inflammation in Old Rats. Immunology. 115 (2) 253-61

[12]. Sutrisna E M, Widyasari D F, and Suprapto 2010 Test of anti-inflammatory effect of ethyl acetate extract of cashew fruit (Anacardium occidentale L.) against edema on the soles of male white rats (Rattus norvegicus) induced by carrageenan. 2 (1) 33-37

[13]. Robbins 2004 Pathology Textbook Edition 7 Volume 1 (Jakarta: EGC medical book publisher)

IOP Conf. Series: Earth and Environmental Science 1041 (2022) 012074 doi:10.1088/1755-1315/1041/1/012074

Assessing food safety compliance for food SMEs in Indonesia

Tri Fajarwaty*, David Jukes

Department of Food and Nutritional Sciences, The University of Reading, United Kingdom

*tri.fajarwaty@gmail.com

Abstract. In Indonesia, SMEs are still being the backbone of the national economy due to their strategic roles in creating economic growth, providing a large share of the job opportunity, encouraging more equal distribution of economic powers, and contributing to the economic development of the community and to the supply of food products for public consumption. SMEs are also have been growing rapidly and mostly is mushrooming in big cities. However, this vital role and rapid growth have not been followed by the development and updating of food safety arrangements, thus many SMEs are unable to comply with the food safety standards and causing various food safety problems. This study assesses the challenges faced by the SMEs in food safety compliance by conducting a literature review and extracting the information from several major/primary papers (academic literature review and gray literature search) to recognize the major gaps and to propose strategic recommendation in the end. It is concluded that in Indonesia, the compliance failures in food SMEs are most likely due to lack of regulatory knowledge and awareness of the importance of food safety principles, lack of proper attention from the government, and the illegal motive to achieve maximum profit with minimum production cost.

1. Introduction

In addition to safeguarding public health, an efficient food control system is critical for ensuring the people's right to safe and high-quality food and preventing outbreaks of food-borne illness that can have far-reaching ramifications on political, social, and economic dimensions. Because of this, governments throughout the world should prioritize the design and execution of a robust food control system that includes all players in the food supply chain. Food laws and regulations, a food control management system, inspection and laboratory services, and information, education, communication, and training are all components of a national food control system. A nation's ability to safeguard public health, avoid foodborne disease, and boost export and import/international trade is also a part of this [1,2].

National food control systems vary widely per country, as do the organizational structures that underpin them, with each system bringing with it advantages and drawbacks. Multiple agencies or a single entity might be responsible for food safety oversight. Food safety is overseen by one agency in a single agency system, however in multiple agency systems, food safety regulation is shared across numerous agencies and may be separated into national, state, and local authorities. It is important to note that the integrated food control system contains certain phases of operations from policy development to education and training that are carried out in an integrated fashion with effective coordination and collaboration [1,3].

Regardless of whether a country is developed or developing, creating a national food control system is difficult. Small and medium-sized businesses, or SMBs, are a key source of difficulty for the food industry (SMEs). As the number of small and medium-sized firms (SMEs) has risen rapidly around the globe, food safety regulations in many countries have not kept pace, and as a result, many SMEs are unable to meet national and international food safety requirements [4,5]. As a growing country, Indonesia has been plagued by numerous and complex food safety issues created by SMEs for many years because of its vast population and other variables, such as economic and political. It is possible that Indonesia's national food control system does not adequately address the food safety demands of small and medium-sized enterprises.

As a result, while many studies have been done on the function of the national food control system in ensuring food safety and quality, only a small number have examined how this national food control operates on small and medium-sized enterprises (SMEs). Consequently, this study's general purpose is to describe Indonesia's Food Safety Controls in particular the position and problems of the system to manage small and medium-sized businesses (SMEs).

2. Methodology

For this article, we conducted a literature review and extracted material from various major/primary documents, including FAO technical guidelines, electronic sources, as well as journals, to better comprehend and explain Indonesia's food safety control system. Food safety issues in both developed and developing nations are discussed here.

3. Findings

3.1 Classification of Small and Medium Sized Enterprises (SMEs)

Taylor [6] said that there is no universally accepted definition of a small business. But they might be categorized based on the number of workers they employ as well as their revenue and profit margins. This study uses the term "SMEs" to refer to micro, small, and medium-sized firms. Small-scale producers in Indonesia are referred to using a variety of terminologies and classifications. In accordance with the Micro, Small, and Medium-Sized Enterprises Act of 2008 [7]:

- Micro enterprise is characterized to have a net worth of at most IDR 50,000,000, excluding land and buildings; or having annual sales turnover of up to IDR 300,000,000 ;
- Small enterprise is characterized to have a net worth of more than IDR. 50,000,000 up to IDR 500,000,000 excluding land and buildings; or has an annual sales turnover of more than Rp. 300,000,000 up to a maximum of Rp 2,500,000,000.
- Medium enterprise is characterized to have a net worth of more than IDR. 500,000,000 up to IDR 10,000,000,000 excluding land and buildings; or has an annual sales turnover of more than Rp. 2,500,000,000 up to a maximum of Rp 50,000,000,000".

Statistics Based on the number of employees, Indonesia [8] categorized small and medium enterprises. Small and medium-sized firms have between 5 and 99 employees, whereas micro and home enterprises have between 1 and 4 employees. A term used by FAO/WHO [9] to describe small food producers was "small and/or less developed food business (SLDB)." Food safety management system status, not output volume, is what the phrase 'less developed firms' refers to, according to the report. From street food sellers to the home food industry (HFI) (Industri Rumah Tangga Pangan/IRTP), Indonesia's SLDB encompasses a wide range of businesses. SMEs in Indonesia may have a broad range of definitions, which might lead to confusion and misinterpretation of their authority.

3.2 Profiles and food safety aspects of SMEs

Despite the fact that more than 98% of Indonesia's SMEs are still micro-scale, SMEs account for more than 99 percent of the country's enterprises in all sectors, including food, and about 120 million people are employed by them [10]. Agriculture and food businesses make up more over half of all national businesses, according to country data from the Ministry for Cooperatives and Small Business [11]. As of 2019, there are roughly 65 million small and medium-sized enterprises (SMEs) in Indonesia, according to the Ministry of Cooperatives and SMEs [10]. As a result, it may be estimated that there are over 30 million food businesses in Indonesia.

SMEs in Indonesia's food industry have been described by Puspa and Kühl [12] as having a number of common traits (home-made food producers). According to the authors, the following features are critical to recognizing food safety issues in this industry and establishing successful food safety programs in the future. The following is a list of the features:

a. Structure

There are many misconceptions about the informal sector, which includes food SMEs like street food sellers and "warungs," which are generally seen as difficult to manage and falling behind in economic growth.

b. Identification

As many street food sellers and "warungs" as there are, their formal documentation and identification is lacking. As of this writing, there is no official statistics on the actual quantity of these creatures. Since there is no licensing or registration required to conduct business, this is the primary reason. In addition, many street food sellers are seasonal.

c. Assets

Most street food sellers have very little money and rely on their own or their employees' fundamental talents to make a living. There is a good chance that the owner and/or employees of street food vendors are either ignorant or have just a minimal formal education. SME food safety programs are typically hampered by a financial and human resources burden.

d. Infrastructure

'Warungs' and'street food sellers' often operate in a small space. In most cases, they do not have access to drinking water or power. For mobile street food sellers, the lack of potable or clean water to wash their hands and clean the dishes increases the danger of foodborne illness. It is one of the biggest obstacles to increasing food safety in small businesses.

The government encourages the development and existence of local small businesses while simultaneously preventing the loss of life and money due to contaminated food. Developing a food safety policy and strategy that encourages SMEs to grow while protecting public health is essential for achieving this goal [9]. Although small and medium-sized enterprises (SMEs) are often regarded as beneficial to the society, they are also widely regarded as the primary cause of foodborne disease. In the majority of foodborne outbreaks caused by street meals, raw food was of low quality, vendors did not conduct proper food safety and hygiene, and basic hygiene amenities such as running water, washing facilities and toilets were lacking.

Poor hygiene and sanitation conditions, as well as low levels of knowledge, contribute to Indonesia's foodborne disease issue, which is worsened by the fact that the majority of Indonesians still live-in poverty. According to Vollaard et al. [16], a study on Jakarta's street food and paratyphoid and typhoid illness was shown to be connected. There are several variables that contribute to bacterial contamination of street food, including poor food handling techniques by street food sellers such as infrequent hand washing, direct food contact, and the usage of unsanitary ice cubes.

As a result, food sellers were well aware of transmission factors, but they were unable to put this information into reality in their businesses.

There are many incidents of foodborne illness in Indonesia, including those caused by the consumption of SMEs goods, such as street-food sold near schools, according to an annual survey by the National Agency of Drug and Food Control Republic of Indonesia. A study by Dewanti-Hariyadi et al. [17] found that instances of food poisoning in Indonesia are more likely to go unreported. The lack of a robust foodborne surveillance and reporting system is likely to blame. Indonesia's foodborne disease rate may be substantially greater than the data gathered by the government indicates. Small and medium-sized businesses (SMEs) are also notorious for having a poor degree of food safety compliance. Boric acid, formaldehyde and illegal/non-food-grade colorants such as Methanyl yellow and Rhodamine B are still widely used in Indonesia.

In addition, the recommended amount of food additives, such as sweeteners, is frequently exceeded It was discovered that more than 70% of Indonesian households that produce food have not properly applied Good Manufacturing Practices (GMP) in their production process during a National Agency for

Drugs and Food Control Republic of Indonesia inspection in 2019 [18]. A guideline issued by NADFC for food SMEs, GMP or Cara Produksi Pangan yang Baik (CPPB), must be followed to guarantee that the food they produce is safe for consumption. The PIRT (Pangan Industri Rumah Tangga) certificate and label need GMP compliance as the major criteria. More than 20% of the street food consumed by schoolchildren (PJAS) exhibited a high degree of microbiological contamination, excessive use of food additives (sweetener and preservative), abuse of non-food grade chemical additives (Rhodamine B, Formaldehyde, borax, and Methanyl yellow) [18]. The lack of GMP awareness, lack of food safety expertise and competence, and a lack of financial assistance might all be contributing reasons to the problems faced by Indonesian food SMEs in using GMP.

3.3 Challenges in the food safety compliance for SMEs

"Assuring Food Safety and Quality: Guidelines for Strengthening National Food Control Systems" was released by the FAO/WHO in 2003 as an updated guideline that intends to encourage national authorities, particularly in developing countries, in strengthening their food control systems. Basic to the National Food Control System are food law and regulation, management of the control of the food supply chain, inspection and testing, and IEC (informal and formal) as a foundation for a system that safeguards the public's health by preventing the consumption of unsafe foods, promotes fair trade practices in the food industry, and contributes to economic development [1].

Othman [19] found that national food safety programs in Southeast Asia often lacked a number of critical components. There is a knowledge of the impact of hazardous food on the health and economic well-being of the country, as well as the necessity for more food safety inquiry and research. Lack of cooperation among food safety agencies as well as weak methods for identifying food safety issues as well as insufficient food policies are all factors that need to be addressed.

Indonesia's National Food Control System has met all of the FAO/WHO guidelines for a strong national food control system, such as the existence of a food regulatory framework that defines roles and responsibilities, provides guidance for best practices, establishes consequences for non-compliance and is in harmonization with international and regional standards. As evidenced by the significant number of fraudulent acts and tainted and inferior food items identified in small businesses (SMEs), this regulatory framework has proved troublesome.

There has never been a comprehensive and reliable study undertaken in Indonesia to uncover the causes of the various breaches and non-compliance committed by SMEs. A lack of regulatory knowledge and awareness about the importance of food safety principles and an illegal desire to maximize profit with the lowest possible production costs is the most likely reason for food SMEs' failure to comply with regulatory requirements [20], according to Sugiyono et al. [20]. According to Rahayu et al. [21], SMEs' personnel lack basic understanding of food sanitation and hygiene, which are the most fundamental elements of food safety.

It's possible that SMEs workers have a lower degree of education than the majority of Indonesians, based on statistics from Statistics Indonesia [22]. Only nine persons in the 100-person population who are at least 15 years old have graduated from college or university. Indonesia's small and medium-sized enterprises (SMEs) face several obstacles to food safety compliance, including a lack of funding from the government for food safety regulatory bodies [18]. The inadequate implementation of regulations was due to a lack of government resources for food safety supervision.

More than half of small and medium-sized businesses (SMEs) have difficulty complying with regulations because of their small size and lack of help from the government, according to a study published in the Journal of Business Ethics. When it comes to food safety rules and regulations in Asia-Pacific area, there are several issues to overcome, including the form of legislation, the difficulty of implementing and socializing it, as well as the food chain's complexity. [23] There are several major barriers to regulatory compliance in SMEs in developed countries such as the United Kingdom: lack of knowledge (regulation aspects and food safety requirements and principles); distrust in the relevance and enforcement of certain regulations; the lack of proper labor; and a lack of money to implement food safety practices.

Yapp and Fairman [24] and the OECD [25] agree that non-compliance or compliance failures in the target group (SMEs) are influenced by: (1) lack of regulatory knowledge and understanding (SMEs unable to comply with regulations because they do not comprehend with what is required); (2) compliance failures related to the willingness of SMEs to comply with the regulations, whether because it is too costly or because they do not want to be seen as a burden.

Furthermore, Indonesia's food regulation system has failed to take into account various weaknesses in the enforcement, monitoring, and follow-up of SMEs, notably street food, as well. Despite the fact that street food contributes much to local economic growth, it is often overlooked and undervalued by the government. As an example, in Indonesia, food safety license or registration for small and medium-sized enterprises (SMEs) is required before to distribution of certain types of food, but only for processed goods.

However, according to Jayasuriya [26], registering or licensing street food is an useful tool in managing street food because it allows SMEs to have access to food safety IECT programs and monitoring, which are critical for ensuring the safety of street food. Street food sellers in Indonesia are usually on the move, making it impossible for the government to institute a registration system. Street sellers in various locations of Indonesia, such as Jakarta, have recently been given additional regulations by the local government, such as Governor Regulation Number 10 of 2015 on the organization and empowerment of street vendors. According to the law, street sellers must be registered and identified. Additional details are included in the rule. Where street vendors might operate in more sanitary circumstances, including food safety.

4. Conclusion

Indonesia's food safety issues, like those in any other developing country, are complicated and multilayered. The small and medium-sized businesses (SMEs) are a key source of compliance issues. Compliance failures in food SMEs are attributed to a lack of regulatory knowledge and awareness, as well as a desire to maximize profits while minimizing costs, a lack of government resources for food safety control, and the absence of a registration system for street vendors. A more constructive attitude toward SMEs is needed by the government in order to obtain a better approach to controlling and strengthening food safety in small businesses.

References

[1] Food and Agriculture Organization of The United Nations and World Health Organization 2003 Assuring Food Safety and Quality : Guidelines for Strengthening National Food Control Systems. (Great Britain: FAO/WHO)

[2] Neeliah SA, Goburdhun D and Neeliah H 2009 National food control systems: lessons from Mauritius. Ecology of Food and Nut. 48(2) 137–56

[3] Neeliah SA, Goburdhun D 2007 National food control systems: a review. Food Reviews Int. 23(1) 35-51

[4] Mwamakamba L, Mensah P, Takyiwa K, Darkwah-Odame J, Jallow A and Maiga F 2012 Developing and maintaining national food safety control systems: Experiences from the WHO African region. African Journal of Food, Agriculture, Nutritition and Development 12(4) 6291-304

[5] Fernando Y, Ng H. H and Walters T 2015 Regulatory incentives as a moderator of determinants for the adoption of Malaysian food safety system. British Food Journal 117(4) 1336-53

[6] Taylor E 2001 HACCP in small companies: benefit or burden? Food Control 12(4) 217-22

[7] Law No 20/2008 on Micro, Small, and Medium enterprises

[8] Statistics Indonesia n.d. Ekonomi dan Perdagangan (Economics and Trade) https://www.bps.go.id/Subjek/view/id/9#subjekViewTab1|accordion-daftar-subjek2.Viewed 10.08.21

[9] FAO/WHO 2006. FAO/WHO guidance to governments on the application of HACCP in small

and/or less-developed food businesses. FAO Food and Nutrition Paper No. 86
[10] Ministry of Cooperatives and SMEs 2021 The development of SMEs and Large Enterprises 2012-2013. Data of Micro, Small, and Medium Enterprises (MSMEs) https://kemenkopukm .go.id/uploads/laporan/1617162002_SANDINGAN_DATA_UMKM_2018-2019.pdf.
[11] Lestari S H 2007. Kajian efektivitas model penumbuhan klaster bisnis UKM berbasis agribisnis. Pengkajian Produk Unggulan Dalam meningkatkan Ekspor UKM dan Pengembangan Ekonomi Lokal 1-25 Viewed 10.08.21
[12] Puspa J and Kuhl P 2007 Food quality and safety measures and controls: The future challenge for Indonesia. International Agricultural Research for Development University of Kassel-Witzenhausen and University of Gottingen
[13] Akinbode SO, Dipeolu AO, and Okuneye PA 2011 Willingness to pay for street food safety in Ogun State, Nigeria. Journal of Agricultural & Food Information 12(2) 154 - 66
[14] Mensah P, Yeboah-Manu D, Owusu-Darko K and Ablordey A 2002. Street foods in Accra, Ghana: how safe are they? Bulletin of the World Health Organization 80 546 – 54
[15] Khairuzzaman M, Chowdhury FM, Zaman S, Al Mamun A, Bari ML. 2014 Food safety challenges towards safe, healthy, and nutritious street foods in Bangladesh. International Journal of Food Science 1-9
[16] Vollaard AM 2004 Risk factors for typhoid and paratyphoid fever in Jakarta, Indonesia. JAMA 291(21) 2607
[17] Dewanti-Hariyadi R and Purnomo E. Global food safety hand book case study : Indonesia Available online: https://seafast.ipb.ac.id/publication/journal/food-safety-in-indonesia.pdf.
[18] NADFC 2019 Yearly Report
[19] Othman NM 2007 Food safety in Southeast Asia: challenges facing the region. Asian J. of Agric. and Dev. 4 83-92
[20] Sugiyono et.al and Simorangkir T (ed.) 2011 The compilation of laws regarding food (food safety) Badan Pembinaan Hukum Nasional Kemenkum HAM RI Jakarta
[21] Rahayu WP, Nababan H. and Hariyadi P 2012. Food safety for Food Security in increasing the competitiveness of micro, small and medium enterprises to strengthen the national economy Widyakarya Nasional Pangan dan Gizi X
[22] Statistics Indonesia 2020 Statistics of Education 2020 https://www.bps.go.id/publication/2020 /11/27/347c85541c34e7dae54395a3/statistik-pendidikan-2020.html
[23] Prabhakar SVRK, Sano D and Srivastava N 2010 Food safety in the Asia-Pacific region: Current status, policy perspectives, and a way forward. Sustainable consumption and production in the Asia-Pacific region: Effective responses in a resource constrained world, 215-238.
[24] Yapp C and Fairman R 2006 Factors affecting food safety compliance within small and medium-sized enterprises: implications for regulatory and enforcement strategies. Food Control 17(1) 42-51
[25] OECD 2000 Reducing the risk of policy failure: challenges for regulatory compliance
[26] Jayasuriya DC. 1994 Street food vending in Asia: some policy and legal aspects. Food Control 5(4) 222-6

IOP Conf. Series: Earth and Environmental Science 1041 (2022) 012075 doi:10.1088/1755-1315/1041/1/012075

In-silico screening of inhibitor on protein epidermal growth factor receptor (EGFR)

Jeremi Ongko, Jesica Viona Setiawan, Alfina Gracia Feronytha, Agnes Juliana, Andy Effraim, Mariana Wahjudi, Yulanda Antonius*

Universitas Surabaya, Surabaya, Indonesia

*yulandaantonius@staff.ubaya.ac.id

Abstract. The screening process to detect early-stage lung cancer is injurious to a patient's survival. Fortunately, there are natural compounds that have been acknowledged to possess anticancer properties, work as the protein binding inhibitors of lung cancer promotors: EGF and EGFR. The study aims to identify inhibitors of EGFR protein binding. Assessments were accomplished based on several parameters related to EGFR proteins, such as pathways, protein activity, conformational changes, and numerous information using the STRING database and KEGG pathway database. Ten inhibitor compounds that expressed highest activity were selected for further analysis were: (20R,22R)-5beta,6beta-Epoxy-4beta,12beta,20-trihy-droxy-1-oxowith-2-en-24-enolide, irinotecan, flavopyridol, teniposide, exatecan, daphnoretin, indirubin, topitecan, wentilactone, and evidiamine. The native ligand Lapatinib was used as positive control in this analysis. The analysis was accomplished by molecular docking using Vina 4 in the PyRx software. Interactions between the ligands and residues were investigated using LIGPLOT+ 2.2. The In-silico analysis of the ten candidate compounds revealed that (20R, 22R)-5beta, 6beta-Epoxy-4beta, 12beta, 20-trihydroxy-1-oxowith-2-en-24-enolide expressed the lowest binding energy value, which is -10.4 kcal/mol, indicated the closest binding energy value to Lapatinib as the control. Based on the interaction of amino acids, (20R,22R)-5beta, 6beta-Epoxy-4beta, 12beta, 20-trihydroxy-1-oxowith-2-en-24-enolide has excellent potential to be utilized as next inhibitor com- pound candidates for EGFR protein, because it binds to the Lys745 residue. It mirrors the positive control and has a binding energy on the range of the specified acceptable parameters.

1. Introduction

Epidermal Growth Factor Receptor (EGFR) protein belongs to the ErbB receptor family and a cell surface receptor for members of the epidermal growth factor family of extracellular ligands. EGFR has a key role in signal transduction processes by regulating major cellular functions, such as cell proliferation and apoptosis. The EGFR protein is a tyrosine kinase receptor with a size of 170 kDa located at the surface of the cell and activated by binding the specific ligands, including Epidermal Growth Factor (EGF). EGFR has four functional domains, namely: extracellular ligand-binding domain, transmembrane domain, intracellular tyrosine kinase domain, and C-terminal regulatory do- main.

The C-terminal regulatory domain has several phosphorylated tyrosine kinase domains that specializes in ligand binding. After binding to the ligand, dynamic conformational changes will occur in the extracellular and intracellular domains of the kinase receptor eventuating to transphosphorylation of tyrosine residues in the C-terminal regular tory domain.

This will provide a docking location for downstream molecules that leads to the activation of multiple pathways, such as mitogen-activated protein kinase (MAPK), phosphatidylinositol 3-kinase/AKT, signal transduction and the activation of the STAT3 & STAT5 transcription pathways. Activation of these pathways will eventuate in the blocking process of apoptosis, proliferation, invasion, and metastasis is all important for the cancer phenotype [1]. EGFR is a transmembrane receptor protein tyrosine kinase that is expressed on the epithelial, mesenchymal, and neurogenic tissues. Overexpression of EGFR is implicated in the pathogenesis of many tissue malignancies, including non-small cell lung cancer.

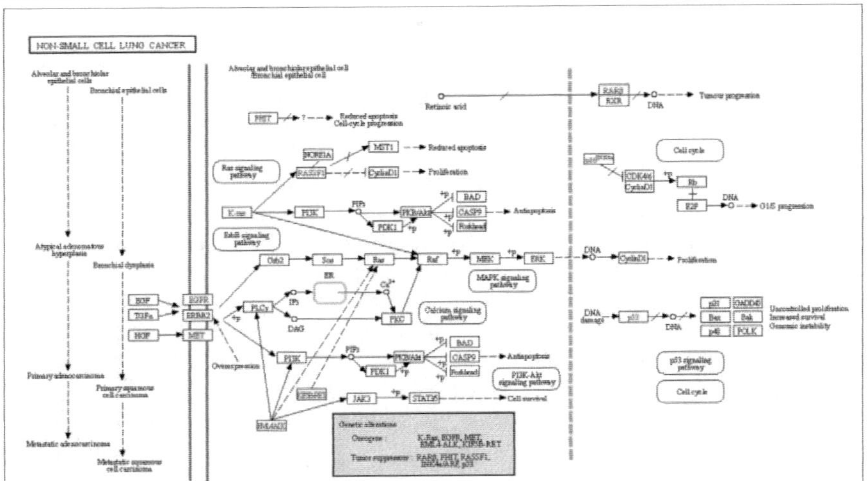

Figure 1. Non-small Lung Cancer Pathway from KEGG

When a specific ligand such as EGF and TGFα binds to EGFR, the EGFR which should functions normally will undergo a conformational change and phosphorylation occurs in the intracellular domain which eventuate the downstream signal transduction from numerous pathways such as Raf1-extracellular signal-regulated kinase, PI3K/AKT, signal transduction, and the activation of the STAT transcription factor. Depending on the pathway (Figure 1), result of the signal could be an uncontrollable cell proliferation as the result of disrupted of apoptotic mechanism [2]. Lung cancer is a complex type of disease and is one of the main causes of cancer mortality in the world [3].

The cancer survival rate can be elevated, depends on the stage that detected at initial medical diagnosis. Therefore, the screening process aims to detect early-stage lung cancer [4]. There are several acknowledged compounds that naturally available and have potential activities as the binding inhibitor between EGF and EGFR protein. Thanks to the discovery of the anticancer properties in these compounds, the therapeutic drugs for cancer derived from these compounds had been successfully manufactured.

For example, there are ethyl acetate and ethanol extracts which contains friedelan-3-one, stigmast-5-en-3-ol, palmitic acid, kaur-16-en-18-oic acid, and benzoic acid that discovered from Thai herbal plants such as Bridelia ovata, Croton oblongifolius, and Erythrophleum succirubrum. This signifies the urgent necessity to perform the screening process of natural compounds that importantly involve in the treatment or prevention of lung cancer by inhibiting EGF binding with EGFR protein to instigate apoptosis [5]. Therefore, this research aimed to identify the compounds that have potential as candidates of EGFR protein inhibitor.

2. Methodology

2.1 Preparation of EGFR Protein
An assessment was accomplished based on several parameters regarding EGFR protein activities such as pathways, protein activity, conformational changes, and other supplementary parameters. The assessment supported by the STRING database (https://stringdb.org/), the KEGG pathway database (https://www.genome.jp/ kegg/path- way.html), and other scientific research reports. The study revealed that the EGFR protein with the code PDB: 1XKK binds to the ligand of the Lapatinib compound as a compound that has been used for the lung cancer treatment [6]. The protein was further modified using ChimeraX to remove water molecules, free compounds, Lapatinib ligands, and provide polar hydrogen bonds to these proteins.

2.2 Literature Study of EGFR Protein Inhibitor Candidates and Compounds Preparation
Studies have also been accomplished. The potential compounds that capable to inhibit the activity of EGFR protein by replacing the attachment of specific ligands that trigger lung cancer has been investigated through various journals and traced into PubChem (https://pubchem.ncbi.nlm.nih.gov/) to obtain their structure. Forty-five inhibitor candidate compounds were selected for further analysis and one positive control were used to redock. The positive control in this analysis was Lapatinib, it was the native ligand and conformed inhibitor compound in the activity of EGFR protein [6]. The 3D structure of the compounds was stored in .SDF form and then was submitted into the PyRx program to be measured by application's Openable to determine its minimum energy and minimize the structure of the ligand until it was ready for dock ing.

2.3 Candidate Compound Screening for EGFR Protein Inhibitor (Molecular Docking)
Screening was accomplished by molecular docking using Vina 4 within the PyRx program version 0.8 [7]. Docking location for center position X: 18.3392, Y: 35.8643, Z: 38.1427 and grid box dimensions X: 16.4623 Å, Y: 13.3738 Å, Z: 20.3377 Å. The analysis is considered success if the binding energy results' value id exceed or draw close to the binding energy results of the Lapatinib as control compound.

2.4. Visualization and Amino Acids Interaction Analysis of EGFR Proteins with Candi- date Compounds
The EGFR protein was visualized with ligands of secondary metabolite com- pounds that passed the predetermined parameters using the Chimera X [8], to inspect the position of the hydrogen bond between the protein and the candidate compound. The visualized compounds were stored in .PDB form. Then each compound was investigated using LIGPLOT+ software version 2.2 to examine the interaction between protein amino acids and ligands [9].

3. Result and Discussion

The reason of the using Lapatinib as the control as candidate comparison is the ability to elevate cytotoxicity against lung cancer cells. Lapatinib exerts its activity intra- cellularly by competing with ATP for the ATP-binding domain in the cytoplasmic tail of the tyrosine kinase receptor. Many natural secondary metabolite compounds derived from various plants have anticancer properties [10]. The analysis was accomplished based on the binding energy between the EGFR protein and the 10 compounds to determine excellent candidates for attachment inhibitor of specific ligands to EGFR pro teins as lung cancer origin.

The molecular docking analysis was run using PyRx program by comparing the binding energy score of 10 compounds against the binding energy score of the Lapatinib control compound, which is -11.0 and setting a minimum score parameter of -10.0 to ensure an effect is produced. The results of the candidate compounds obtained are as follows:

Table 1. Screening Results of Secondary Metabolite Candidates to Control (Lapatinib) based on the Binding Energy Score

No.	Name of Compound	PubChemID	Binding Energy (kcal/mol)	Model 3D Ligands
1	Lapatinib (Positive control)	208908	-11.0	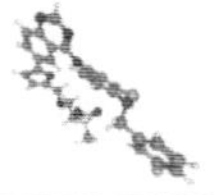
2	(20R,22R)-5beta,6beta-Epoxy- 4beta,12beta,20-tri-hydroxy-1-oxowith-2-en-24-enolide	46872824	-10.4	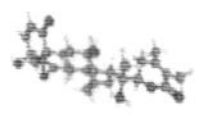
3	Exatecan	151115	-10.2	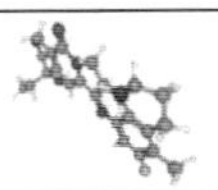
4	Irinotecan	60838	-10.1	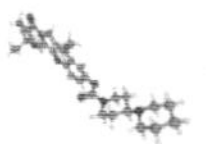

Based on the results of the binding energy analysis between candidate compounds and EGFR protein using PyRx, six candidate compounds were considered to be potential inhibitors. They obtained the highest binding energy value among the 10 compounds; the values were close to the binding energy value of Lapatinib. Those six candidate compounds are: a) (20R,22R)-5beta,6beta-Epoxy-4beta,12beta,20-trihydroxy-1- oxowith-2-en-24-enolide with $\Delta G = -10.4$ kcal/mol, b) Exatecan with $\Delta G = -10.2$ kcal/mol, c) Irinotecan with $\Delta G = -10.1$ kcal/mol, d) Flavopyridol with $\Delta G = -10.1$ kcal/mol, and e) Teniposide with $\Delta G = -10.1$ kcal/mol.

Meanwhile, the binding energy value between Lapatinib and EGFR protein is -11.0 kcal/mol. The binding energy demonstrates the affinity/bond between the candidate compound (ligand) and the EGFR protein. When the smaller/negative binding energy is obtained, the stronger/stable the bonds are formed [11]. Furthermore, the results of the docking were visualized between the six best candidate compounds and Lapatinib with EGFR protein using Chimera X.

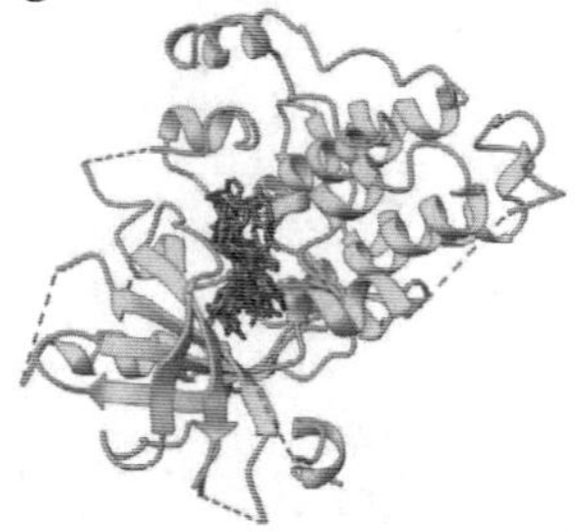

Figure 2. Protein - Ligand Visualization Results based on Docking Results between EGER (PDB: 1XKK)

All the ligands above bind at the same location and a little is seen in some compounds that have hydrogen bonds with proteins while some such as teniposide and irinotecan do not have bonds with proteins. The interaction between amino acid of EGFR protein with ligands of the five candidate compounds and Lapatinib was analyzed using the LIGPLOT+ program.

Table 2. Results of the Interaction of EGFR Protein Amino Acids with Ligands

Name of complex	*Hydrogen Bond*	*Hydrophobic Bond*
EGFR – Lapitanib	Thr790: 3.04 Å Lys745: 3.02 Å	Met1002, Leu792, Gly796, Met793, Ala743, Leu844, Gln791, Met766, Thr854, Cys775, Phe856, Asp855, Leu858, Asn842, Gly721, Val726, Ser720, Arg841, Leu718, Gly719
EGFR – (20R,22R)-5beta,6beta- Epoxy-4beta,12beta,20- trihydroxy-1- oxowith-2-en-24- enolide	Asp855: 2.88 Å Lys745: 3.18 Å	Leu718, Ile744, Leu844, Thr790, Ala743, Leu777, Ile789, Arg841, Thr854, Leu788, Val726, Gly719
EGFR – Irinotecan	-	Leu718, Gly796, Leu844, Thr790, Ala743, Leu777, Cys775, Met766, Arg776, Phe856, Thr854, Asp855, Lys745, Val726, Gly719
EGFR - Flavopiridol	Met793: 2.93 Å	Cys797, Leu718, Met1002, Gly796, Leu792, Ala743, Leu844, Lys745, Thr790, Asp855, Gly719, Thr854, Gly721, Val726, Ser720
EGFR - Teniposide	-	Arg841, Cys797, Leu718, Met1002, Gly796, Gly719, Ala743, Thr854, Leu844, Leu777, Thr790, Arg776, Cys775, Asp855, Met766, Lys745, Val726, Ser720, Gly721, Leu799
EGFR – Exatecan	Asp855: 3.30 Å	Leu799, Asp800, Arg841, Met793, Leu844, Leu792, Ala743, Thr854, Val726, Gly721,

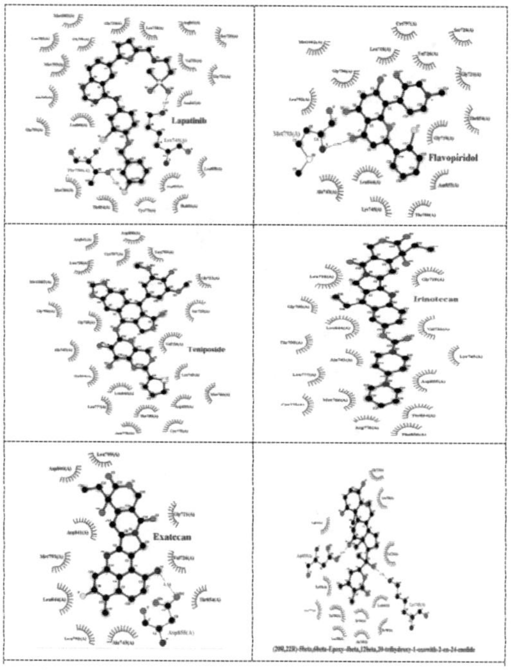

Figure 3. Interaction of amino acid residues within EGFR protein with potential ligands

The observation of hydrogen bonds and amino acids that interact between EGFR proteins and ligands of candidate compounds can be done using the LIGPLOT+ program. The interaction analysis between EGFR proteins with these ligands revealed that the Lapatinib formed two hydrogen bonds with Thr790 and Lys745 from the EGFR ATP binding pocket. The amino acids in EGFR that interact with Lapatinib are Met1002, Leu792, Gly796, Met793, Ala743, Leu844, Gln791, Met766, Thr854, Cys775, Phe856, Asp855, Leu858, Asn842, Gly721, Val726, Ser720, Arg841, Leu718, and Gly719.

Irinotecan and Teniposide compounds were unable to form hydrogen bonds with EGFR. It is probably due to a different active site that influenced the different inhibition path from the binding pocket. In case of the flavopyridol candidate compound, it has one hydrogen bond with Met793 from EGFR at a distance of 2.93 Å. Exatecan compounds has one hydrogen bond with Asp855 from the EGFR at a distance of 3.30 Å.

Then the (20R,22R)-5beta,6beta-Epoxy- 4beta,12beta,20-trihydroxy-1-oxowith-2-en-24-enolide compound formed two hydrogen bonds with Asp855 and Lys745. Lys745 were the same amino acid that Lapatinib, as the positive control, have interaction with. Therefore, with the interaction with Lys745, the (20R,22R)- 5beta,6beta-Epoxy-4beta,12beta,20-trihydroxy-1-oxowith-2-en-24-enolide compound have a potential as an EGFR protein inhibitor that could be further developed. These results indicate that (20R,22R)-5beta,6beta-Epoxy-4beta,12beta,20-trihydroxy-1-oxowith-2-en-24-enolide compounds are the most excellent candidate as the EGFR protein inhibitor, which has similar mechanism pathway with Lapatinib.

4. Conclusion

Based on literature review and in silico analysis, this research obtained the top ten candidate compounds based on the lowest binding affinity and sorted it into the five most excellent candidate compounds. The binding energy analysis discovered that (20R,22R)-5beta,6beta- Epoxy-4beta,12beta,20-trihydroxy-1-oxowith-2-en-24-enolide compounds have the lowest binding energy value, which is -10.4 kcal/mol, which has the closest value to the binding energy of Lapatinib (positive control). The interaction of amino acids confirmed that, (20R,22R)- 5beta,6beta-Epoxy-4beta,12beta,20-trihydroxy-1-oxowith-2-en-24-enolide is the most promising compound to be further developed as EGFR protein inhibitor opposing lung cancer development, as it binds to the Lys745 residue, copying the Lapatinib mechanism resulting binding energy within the specified acceptable parameters.

Acknowledgment

Authors thank to Faculty of Biotechnology, University of Surabaya for supporting this research.

References

[1] Nwizu, Tobenna and Salgia R 2012 Molecular Biology of Lung Cancer and Therapeutic Implications Emerging Cancer Therapeutics 3(1):1–10.

[2] Bethune G, Bethune D, Ridgway N and Xu Z 2010 Epidermal growth factor receptor (EGFR) in lung cancer: an overview and update Journal of Thoracic Disease 2(1):48–51.

[3] Parris B, O'Farrell H, Fong K and Yang I 2019 Chronic obstructive pulmonary disease (COPD) and lung cancer: common pathways for pathogenesis. Journal Of Thoracic Disease, 1(1), S2155-S2172.

[4] Wu G X and Raz D 2016 Lung Cancer Screening Cancer Treatment and Research 170:1-23.

[5] Poofery J, Khaw-on P, Subhawa S, Sripanidkulchai B, Tantraworasin A, Saeteng S, Siwachat S, Lertprasertsuke N and Banjerdpongchai R 2020 Potential Thai Herbal Extracts on Lung Cancer Treatment by Inducing Apoptosis and Synergizing Chemotheraphy Molecules 25(231):1–30.

[6] Elrayess R, Aziz Y M A, Elgawish M S, Elewa M, Yassen A S A, Elhady S S, Elshihawy H A and Said M M 2020 Discovery of Potent Dual EGFR/HER2 Inhibitors Based on Thiophene Scaffold Targeting H1299 Lung Cancer Cell Line Pharmaceuticals,14(1), 9.

[7] Dallakyan S and Olson A J 2015 Small-Molecule Library Screening by Docking with PyRx.

Methods in molecular bology 1263: 243-250.
[8] Pettersen E F, Goddard T D, Huang C C, Meng E C, Couch G S, Croll T I, Morris J H and Ferrin T E 2021 UCSF ChimeraX: Structure visualization for research ers, educators, and developers Protein Science, 30(1): 70–82.
[9] Laskowski R A and Swindells M B 2011 LigPlot+: multiple ligand-protein interaction diagrams for drug discovery J. Chem. Inf. Model., 51:2778-2786.
[10] Nirmala M J, Samundeeswari A and Sankar P D 2011 Natural plant resources in anti-cancer therapy-A review. Research in Plant Biology 1(3):1-14.
[11] Laksmiani N L, Paramita N L P V and Wirasuta I M A G 2016 In vitro and in silico anti-oxidant activity of purified fractions from purple sweet potato ethanolic extract. International Journal of Pharmacy and Pharmaceutical Sciences 8(8): 177-181.

Effect of Various Feed Additives on Carcass and Meat Quality of Two Different Strains of Chickens

E Tugiyanti* and E Susanti

Universitas Jenderal Soedirman, Purwokerto, Indonesia

*elly.tugiyanti@unsoed.ac.id

Abstract. Assorted feed additives for broiler chickens, including prohibited antibiotics, are widely sold in the market. The objective of incorporating supplement in feed is to improve broilers' performance and meat quality. This study conducted an experiment to 60 DOC broilers from each Cobb and Lohmann strains. The experiment was conducted in a 2x4 factorial CRD. The treatments were feed additives, i.e. antibiotics, probiotics, acidifiers and phytobiotics. Each treatment was repeated 3 times. The measured variables were carcass and non-carcass percentage as well as physical quality of broiler meat. Analysis of variance showed that feed additives did not significantly affect ($P>0.05$) carcass and non-carcass percentage and meat physical quality of the two broiler strains. Conclusively, probiotics, acidifiers and phytobiotics are the potential alternatives as a substitute for antibiotics for Cobb and Lohmann broiler chickens as reflected from the carcass percentage and meat quality.

1. Introduction

Broiler chicken is one of the fastest sources of meat growth in the world, and in Indonesia broiler chickens are included in poultry which meat is mostly consumed by humans. Indonesia's climatic conditions, which have high temperature and humidity, cause many disease problems in broiler chickens. Therefore, an antibiotic is used to prevent disease. Its development growth promoter antibiotics (AGPs) are frequently used to increase the bodyweight of poultry. However, there has been a growing concern about the negative effects of AGP and its residues in meat product as well as antibiotic resistance to the consumers.

The impact AGP on the emergence of antibiotic resistance in zoonotic pathogenic bacteria in the gut microbial community of poultry [1]. The prohibition of AGP is regulated in Law Number 18/2009 junto Number 41/2014 concerning Livestock and Animal Health. Article 22 paragraph 4c stated, "Everyone is prohibited from using feed mixed with particular hormones and/or antibiotics as feed additives". AGP incorporated in feed would impact positively on chickens' performance, but it brings negative effects to human health. The antibiotic residue in chicken eggs and muscle tissue [2] would make the human body resist some types of antibiotics.

The commonly used AGP for the poultry industry includes Avilamycin, Flavophospholipol, Enramycin, Monensin, Penicillin, Virginiamycin, Tetracycline, Erythromycin, Salinomycin, and Bacitracin methylene disalicylate (BMD). Accordingly, farmers start to find alternatives for antibiotics, such as probiotic, acidifier and phytobiotic. These substances do not leave residual traces in meat; hence, healthier as a feed additive [3][4][5].

However, the effects of probiotic, acidifier and phytobiotics on chickens are varied. One contributing factor is the strains of broiler chicken. The large variety of broiler strains allows farmers to select ones with high-performance ones and compatible with environmental conditions. Different genetic qualities in each strain result in different abilities to respond to the environment, hence different growth rates [6]. Further effects are evident on the final weight, carcass weight and meat quality [7]. Accordingly, it is important to investigate the effect of broiler stains and feed additive on broiler carcass and meat quality

2. Methodology

This study used sixty DOC of each Cobb and Lohmann strains. The basal feed used commercial feed contained 20-22% crude protein, 5-6% crude fat, 5% crude fiber (maximum), 8% ash (maximum), 12% water (maximum), 0.8-11% Ca and 0.5% feed additives (minimum) namely antibiotic, probiotic powder, acidifier powder, and mixed garlic, turmeric, and ginger powder. Feed additives were given through the basal feed. A total of 24 slated broiler cages were made of bamboo and wood, measuring 80 cm x 80 cm per unit.

This experimental study was conducted in Completely Randomized Design (CRD) with the factorial pattern. The first factor is the broiler strains (Cobb and Lohmann) and the second factor is feed additives (powder) including 0.125% antibiotic, 0.2% probiotic, 0.8% acidifier and phytobiotic (as well as mixed 0.25% garlic, 2.1% turmeric, and 2.1% ginger powder). The levels of each feed additive are adjusted to the levels recommended on the label of each feed additive, except for the levels of ginger, turmeric and garlic based on the best results of previous studies [8][9]. The eight combined treatments were Cobb broiler given antibiotic, Cobb broiler given probiotic, Cobb broiler given acidifier, Cobb broiler given phytobiotic, Lohmann broiler given probiotic, Lohmann broiler given acidifier, Lohmann broiler given phytobiotic.

These eight treatments were repeated three times, and each unit was filled with 5 chickens, hence 120 broilers in total. The chickens were harvested at the age of 35 days and then they were slaughtered by cutting three channels, namely the respiratory tract, and left and right blood vessels in the neck. Then the chicken undergoes processing, namely scalding, plucking, removal internal organ except the giblet (liver, heart, and gizzard), and eviscerating to obtain the carcass and non-carcass. The meat that was analyzed to obtain physical quality of the meat was the breast meat. The physical quality of meat observed consisted of water holding capacity, cooking loss, and meat tenderness. The collected data were subjected to Analysis of Variance (ANAVA), and any significant effect would be subjected to an Honestly Significant Difference test (HSD).

3. Result and Discussion

3.1 Carcass and non-carcass Percentage

Table 1. The average carcass percentage of Cobb and Lohmann Strains

Treatments	Carcass percentage (%)		Non-carcass percentage (%)	
	Cobb	Lohmann	Cobb	Lohmann
Basal feed+antibiotic	74.03±3.01	76.91±5.74	28.98±4.37	20.43±3.51
Basal feed+probiotic	74.53±1.63	74.72±1.92	25.79±1.07	25.28±1.92
Basal feed + acidifier	75.97±0.75	76.06±2.07	24.03±0.75	23.94±2.07
Basal feed + Phytobiotic (garlic+ turmeric + ginger)	75.96±3.99	77.48±5.14	24.04±3.99	25.85±0.87

The carcass and non-carcass percentage of Cobb and Lohmann strain in this study were 74.03–75.97% and 74.72–77.48%, respectively. This result confirmed the previous studies, i.e. 69.16% of Cobb strain [7] and 63.00–66.16% of Lohmann strain [10].

Furthermore, the non-carcass percentage of Cobb (24.03–28.98%) and Lohmannn (20.43–25.85%) in this study was within the normal range as per the previous study namely 22.26-24.13%[10]. Analysis of variance showed that the interaction between different feed additives and broiler strains did not significantly affect ($P>0.05$) the carcass percentage and non-carcass percentage. It demonstrated that any broiler strains respond similarly to antibiotics, probiotic, acidifier and phytobiotic treatments.

Therefore, probiotic, acidifier and phytobiotic can substitute antibiotics that have been prohibited for animal feed additives. Additionally, the relatively similar genetic quality between Cobb and Lohmann allows farmers to use Lohmann and Cobb interchangeably for their livestock farming. A previous study[11] reported that strains with different characteristics would produce relatively similar carcass and non-carcass percentage when fed on quality feed. Furthermore, [12] and [13] reported that acidifier and phytobiotic play a similar role to that of antibiotics.

Acidifier improves digestibility by increasing the performance of digestive enzymes, decreasing intestinal pH and maintaining the balance microbes in the digestive system. Meanwhile, phytobiotic which is produced from the plant's secondary metabolite (either contain nutrition, no nutrition even anti-nutrition) and incorporated to the feed can increase livestock productivity by improving feed characteristics, increasing the health of the digestive tract by controlling bacteria as well as improving production performance and the quality of livestock products.

Similarly, probiotic bacteria can improve livestock performance by exhibiting a competitive exclusion in the intestines. Besides, probiotic bacteria will modulate body immune of broiler chickens by improving the goblet cells, the inflammatory mitigation and body immune status. Garlic, ginger and turmeric all contain essential oils that act as natural antibiotics. In addition, garlic contains the bioactive substance allicin, while ginger contains oleorosin and gingerol, phenolic compounds, antioxidants.

The bioactive substances in turmeric contain curcuminoid compounds consisting of curcumin, desmetoxicumin as much as 1-5% and bisdesmetoksikurcumin as much as 1-5% and other useful substances such as essential oils consisting of ketones sesquiterpenes, turmerone, tumeone 60%, zingiberen 25%, felandren, sabinen, borneol and sineil. Accordingly, different feed additives would result in a relatively similar carcass and non-carcass percentage.

3.2 Meat quality of different strains of broiler chickens

Table 2. Water holding capacity, cooking loss, and meat tenderness of broiler chickens consuming additives-fortified feed

Treatments	pH		Cooking loss		Meat tenderness	
	Cobb	Lohmann	Cobb	Lohmann	Cobb	Lohmann
Basal feed + antibiotic	6.21±0.19	5.88±0.03	31.56±0.89	28.88±0.44	5.49±1.40	6.08±1.35
Basal feed + probiotic	6.17±0.14	5.94±0.11	28.17±1.96	28.55±1.63	4.13±1.02	4.74±0.67
Basal feed + acidifier	6.02±0.09	5.87±0.13	31.81±1.21	28.75±3.15	5.99±2.00	4.87±1.44
Basal feed + Phytobiotic (garlic+ turmeric + ginger)	6.00±0.14	6.02±0.02	29.39±1.08	28.01±0.39	5.24±0.76	5.61±1.47

Analysis of Variance showed that no interaction was observed between feed additives and broiler strains, demonstrated by the non-significant effect on pH, cooking loss and meat tenderness. The contributing factor to this is the similar function of antibiotic, probiotic, acidifier and phytobiotic to inhibit pathogenic bacteria and improve body immune of the livestock[14]; therefore, the chickens are always healthy, on top-performance and with normal glycolysis process which affect the quality of the broiler meat [3][15].

Meat pH in this study was 6.00 – 6.21, which is relatively comparable to 6.00 reported by Irma[16]. The effect of phytobiotics is not significantly different on the quality of meat, because the bioactive substances, essential oils and antioxidants contained in phytobiotics can prevent the pH from dropping too quickly. It is due to the donation of H + ions contain in antioxidants.

An antioxidants are effective in influencing the process of depleting glycogen reserves into lactic acid and preventing the oxidation process by free radicals as well[17]. The rate at which muscle pH decreases which will rapidly result in a low water-binding capacity, because of the increased contraction of the actomyosin that was formed, thus squeezing the liquid out of the meat. The meat tenderness is one of the factors which determine meat quality. Turmeric has curcuminoid as antibacterial agent, it also stimulated bile duct to produce more bile liquid by cholekinetic and choleretic way.

In liver, lipid metabolism undergoes well, result high ATP to produce more amino acid to develop muscle. Therefore, broilers have more muscle and a little fat. It also optimized gastrointestinal works, increase appetite and growth rate. The garlic and ginger have similar potential as antibacterial agent like turmeric, but the effect in broiler are lower than turmeric [3]. Furthermore, cooking loss and meat tenderness is affected by several factors, including fat content and pH[18].

4. Conclusion

a. This study concludes that different feed additives and broiler strains produced similar carcass percentage and meat quality.
b. Probiotic, acidifier and phytobiotic are the potential substitute for the prohibited antibiotics.

Acknowledgement

The researchers express their gratitude to the Jenderal Soedirman University for the research grant up to completion.

References

[1] Salaheen S, Kim S W, Haley B J, Van Kessel J A S and Biswas D 2017 Alternative growth promoters Modulate broiler gut microbiome and enhance body weight gain *Front. Microbiol.* **8** 1–11 10.3389/fmicb.2017.02088

[2] Ngangguk C A, Detha A I and Wuri D A 2013 Assessment of tetracycline residues in the meat ofbroilers, free-range broilers and rejected layer chickens sold in Kupang city *J. Kaji. Vet.* **2** 175–81 https://doi.org/10.35508/jkv.v2i2.1002

[3] Pertiwi H and Dadi T B 2020 Phytoadditive Suplementation to Improve Production Performance of Broiler Replacing Antibiotic Growth Promoters (AGPs) in Indonesia -a Review *Sylwan* **164** 398–406

[4] Lillehoj H, Liu Y, Calsamiglia S, Fernandez-Miyakawa M E, Chi F, Cravens R L, Oh S and Gay C G 2018 Phytochemicals as antibiotic alternatives to promote growth and enhance host health *Vet. Res.* **49** 1 18 10.1186/s13567-018-0562-6

[5] Gunal M, Yayli G, Kaya O, Karahan N and Sulak O 2006 The effects of antibiotic growth promoter, probiotic or organic acid supplementation on performance, intestinal microflora and tissue of broilers *Int. J. Poult. Sci.* **5** 149–55 10.3923/ijps.2006.149.155

[6] Kalia S, Bharti V K, Gogoi D, Giri A and Kumar B 2017 Studies on the growth performance of different broiler strains at high altitude and evaluation of probiotic effect on their survivability *Sci. Rep.* **7** 1–8 10.1038/srep46074

[7] Risnajati D 2017 Comparison of Final Weight, Carcass Weight and Carcass Percentage of Various Broiler Strains *Sains Peternak.* **10** 11–4 10.20961/sainspet.v10i1.4808

[8] Natsir M H and Muharlien E W dan 2016 The use of a combination of turmeric flour (Curcuma domestica) and ginger (Zingiber officinale) in the encapsulated and non-encapsulated forms on the characteristics of the intestines and intestinal microflora of broilers *Bul. Peternak.* **40** 1–10 *ISSN-0126 4400* https://doi.org/10.21059/buletinpeternak.v40i1.8890

[9] Nuningtyas Y F 2014 Pengaruh penambahan tepung bawang putih (Allium sativum) sebagai aditif terhadap penampilan produksi ayam pedaging *Ternak Trop.* J.Tropical Animal Production **15** 65–73

[10] J Sibarani V D Y and L D M 2014 Percentage of carcass and non-carcass and abdominal fat of broiler chickens treated with citric acid acidifier in double step down feed *Anim. Agric. J.* **3** 273–80 http://ejournal-s1.undip.ac.id/index.php/aaj

[11] Melly Pratiwi, Muh. Amrullah Pagala A S A 2016 Carcass production and abdominal fat of Cobb and Lohmann strains broiler chickens were fed different diets *JITRO* **1** 1–17 DOI: http://dx.doi.org/10.33772/jitro.v3i1.986

[12] Maya Septiana O sofjan dan M H N 2014 Effects of Addition of Natural Acidifier and Phytobiotic Mixtures in Non and Encapsulated Forms in Commercial Feed on Quality of Laying Chicken Eggs Animal science Faculty University of Brawijaya 1–11

IOP Conf. Series: Earth and Environmental Science 1041 (2022) 012076 doi:10.1088/1755-1315/1041/1/012076

[13] Cecep Hidayat R 2019 Review: Opportunities for Developing Phytogenic Feed Additives as Substitute for Antibiotics in Broiler Rations in Indonesia *J. Ilmu dan Teknol. Peternak. Trop.* **6** 188–213 10.33772/jitro.v6i2.7139

[14] Akhadiarto S 2014 Effect of addition of local probiotics in the diet of broiler chicken performance **16** 16 22 https://doi.org/10.29122/jsti.v16i1.3397

[15] Husein Abdurrahman Z and Yanti Y 2018 Overview of the Effect of Probiotics and Prebiotics on Chicken Meat Quality *J. Trop. Anim. Prod.* **19** 95–104

[16] Irmayani 2019 Analysis of Microbial Contamination and pH Value of Broiler Chicken Meat on Lakessi Traditional Markets in Parepare City *J. Galung Trop.* **8** 1–8 DOI: http://dx.doi.org/10.31850/jgt.v8i1.431

[17] Tugiyanti E, Ibnu H.S, Novie A.S E S dan S M 2016 Pengaruh pemberian tepung daun sukun ke dalam pakan terhadap kualitas daging itik tegal jantan Umur 9 minggu *Prosiding Seminar Nasional Peternakan 2016 Optimalisasi Sumberdaya Lokal Peternakan Rakyat Dalam Mendukung Pengembangan Badan Usaha Milik Rakyat (BUMR)* pp 174–82

[18] Tugiyanti E, Yuwanta T, Zuprizal and Rusman 2013 Improving performance, meat quality and muscle fiber microstructure of native Indonesian muscovy duck through feed protein and metabolizable energy *Int. J. Poult. Sci.* **12** 653–9 10.3923/ijps.2013.653.659

IOP Conf. Series: Earth and Environmental Science 1041 (2022) 012077 doi:10.1088/1755-1315/1041/1/012077

Microbiological profile of concentrated yoghurt manufactured from low and full fat milk during storage

T Setyawardani*, J Sumarmono and D R Kusuma

Universitas Jenderal Soedirman, Purwokerto, Indonesia

*triana.setyawardani@unsoed.ac.id

Abstract. The microbiological profile illustrates the existence and growth of microflora in fermentation products, such as concentrate yogurt. The study aimed to identify and evaluate microflora in yogurt paste observed from the total bacteria, total lactic acid bacteria (LAB), total yeast, LAB viability, and level of acidity. Concentrate yogurt was made of full-fat milk and low-fat milk that was fermented and hung for 24h and stored at a low temperature for observation on day 0,15, 30, 45, and 60. The data were subjected to a Generalized Linear Model (GLM) followed by the Duncan test. The result showed significant ($P<0.05$) changes in the total bacteria, total LAB, total yeast, LAB viability dan acidity (pH) of the fermentation products; however, no significant difference ($P>0.05$) was identified regarding the use of raw material. Storage made the fermented products suffer from the decreased total bacteria, LAB dan its viability, but there was an increase in the total yeast and pH. Conclusively, storing concentrate yogurt in a cool place for 60 would decrease the microflora profile (total bacteria, LAB dan LAB viability) but increase the total yeast and pH value.

1. Introduction

Concentrated yogurt, a fermented product, is yogurt that undergoes further process (hanging) to decrease water content and increase total solid. In general, fermentation product could support growth and improve health. The traditional method to make yogurt paste is by straining the yogurt at a particular time and temperature.

On an industrial scale, concentrate yogurt manufacture uses advanced technologies, such as centrifugation, recombination, and ultrafiltration. Compared to regular yogurt, concentrate yogurt has a higher total solid (+ 23-25 g/lt), white and creamy color, soft texture and a slight syneresis [1, 2]. The main ingredient for yogurt paste is full-fat milk, but some research modifications used low-fat milk. The low-fat yogurt paste is a suitable low-calory diet for consumers.

An indicator to determine the shelf life of a product is by evaluating the microflora during storage. During storage, concentrated yogurt would undergo some microbiological changes in the total LAB, LAB viability, and acidity. Concentrate yogurt contains less water than yogurt, and it affects total microbes and viability. Product acidity affects the growth of spored microbes and contaminant bacteria. The latter would change the characteristics of the products during storage.

Moreover, non-starter LAB may grow during storage, particularly in most types of cheese (Mannu et al., 2000). Different main ingredients (full fat and low-fat milk) would affect the growth of microflora in the stored products. Low-fat ingredients have different abilities to capture the living and growing microflora before the products are being stored. The processing method also affects the total microflora; a previous study reported that coliform is present in Labneh product made using the cloth-bag method (Al-Kadamany et al., 2003). This study aimed to evaluate the microbiological profile of concentrate yogurt made of full-fat and low-fat ingredients saved in cold storage [2].

2. Materials and Methods

The study used an experimental method in a 2x5 factorial-Completely Randomized Design (CRD) and three replicates. The first factor is the type of milk (F= full fat; L= Low-fat), and the second factor is storage time (P1= no storage; P2= 15 days; P3=30 days; P4= 45 days; and P5= 60 days). The measured variables were total bacteria, total LAB, total yeast, LAB viability, and pH value. The data were subjected to the GLM procedure (Generalized Linear Model), with 5 % significance.

2.1 Concentrate yogurt production

Full-fat and low-fat milk (2.5L each) was pasteurized separately and incorporated with 5% (v/v) yogurt starters *L. bulgaricus, S. thermophilus,* and *L. acidophilus* in each treatment unit. Following this process was incubation at 37°C for 4h. The indicators of yogurt products are sour taste and the thick, soft texture. The yogurt was processed into concentrated yogurt using the straining and hanging method at 5°C for 24h. At last, the finish concentrated yogurt was stored at 5°C for 60 days and evaluated every 15 days for its microbiological profile.

2.2 The procedure for calculating total microflora

The yogurt paste sample was homogenized, taken 1ml, and mixed with a 9-ml solvent for the first dilution. The solution was diluted until 10^9, and the last three series of dilutions were grown in a media on a petri dish using the pour plate method. The microbes developed in different media: bacteria in NA media (Nutrient Agar), LAB in MRSA, and yeast in PDA (Potato Dextrose Agar). All microbes were incubated at 37°C.

3. Result and Discussion

3.1 Total bacteria and yeast

The present study used different main ingredients (full-fat and low-fat milk) and found that the concentrated yogurt contained an equal number of total bacteria regardless of the type of milk. However, different storage time resulted in varied total bacteria and yeast across treatments (P<0.05).

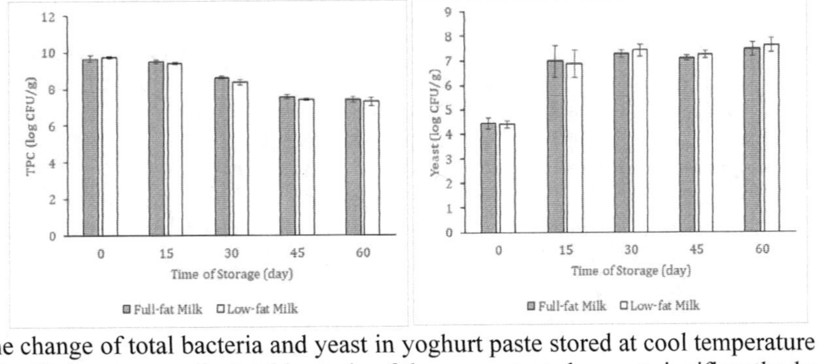

Figure 1. the change of total bacteria and yeast in yoghurt paste stored at cool temperature for 60 days.

During the 60-day storage, the total bacteria of the concentrated yogurt significantly decreased, and the total yeast increased. The microbial profile of the fermentation product (in the storage) would experience some changes and growth according to the regular pattern of microbial growth. Before being stored, the highest average of total bacteria was 9.7033±012 CFU/g and gradually decreased. LAB, the producer of metabolite. i.e., lactic acids as the dominant total bacteria.

In the storage, the total bacteria started to diminish according to the growth phase; therefore, at the end of the stage, the total bacteria reached the ultimate decrease. The acid-producing metabolites allowed the yeast to survive the highly acidic environment and break down peptides, thus producing an alkali condition that increases the pH of the product. Total bacteria of the yogurt decreased during 14-day storage at 7oC, indicating bacteria's durability against acidity and cold/condensation.

Some yeasts that contaminated the products were Trichosporon, brassicae, Cryptococcus curvatus, Kluyveromyces, marxianus, Trichosporon cutaneum, Debaryomyces, hansenii, Pichia farinose, Candida blankii, and Geotrichum candidum [1]. At the end of storage day, the pH value increased due to an extensive metabolism activity of yeast and mold using lactic acid as the source of carbon and the proteolysis process that released most of the alkali components[3].

3.2 Total and LAB viability

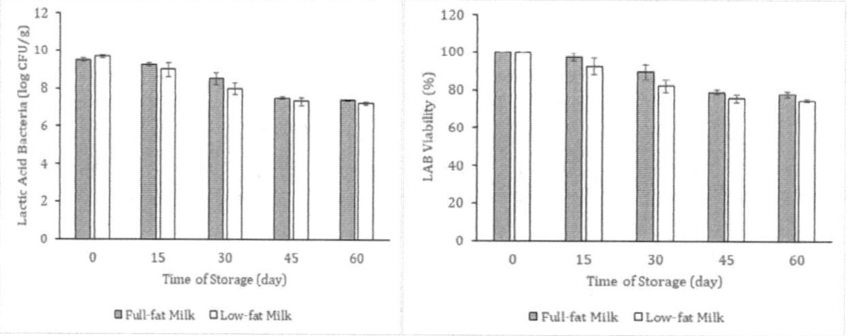

Figure 2. The change of total LAB and viability while concentrate yogurt was stored for 60 days in a cool temperature

The contributing factors to the viability of probiotic LAB in food products include the strains, pH, hydrogen peroxide and soluble oxygen, the produced metabolite concentration (lactic acid and acetic acid), the media as a buffer, storage temperature, and additional ingredients [4, 5].

The result showed that the total lactic acid bacteria (LAB) of the concentrated yogurt were not affected by the type of milk (full-fat and low-fat) but affected by the storage time ($P<0.05$). Meanwhile, LAB viability was not affected by the types of the main ingredients and the storage time. The total LAB decreased while the concentrated yogurt was in store. At the end of 60-day storage, the average of the total LAB was 7.32 log CFU/g. This decrease was followed by the declining viability and the lowest (76.28%) was observed in the concentrated yogurt stored for 60 days.

Accordingly, the longer the storage, the more inhibited the LAB growth because the psychotropic and mesophilic yeast also developed. Salinity and acidity are the inhibiting factors to LAB growth when the product is stored. Non-starter LAB that could better survive in low pH and high saline environment also develop in concentrated yogurt during storage [6].

Previous studies on the fermentation product of goat milk yogurt added with fruits reported a decreased viability during the four weeks of storage. The viability of *B. animalis Bb-12* and *Propionibacterium jensenii* 702 remained at 10 7 and 10 8 CFU, respectively, compared to *L. acidophilus* La-5 that declined to 10 5 CFU at the end of storage time [7].

3.3 pH value of concentrate yogurt

The produced lactic acid determines the shelf life of a product and its product sensory properties [8].

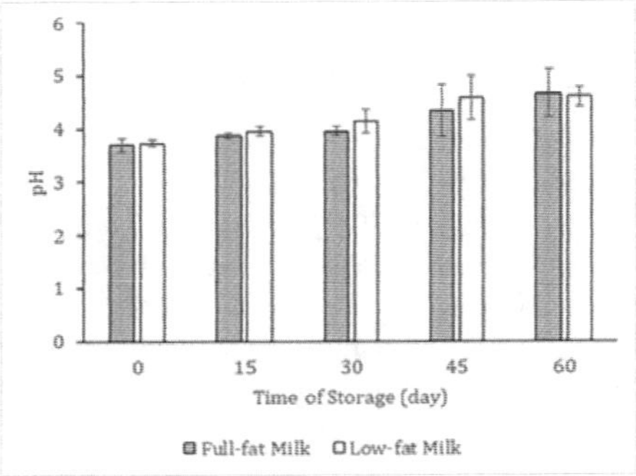

Figure 3. The change of pH value when the yoghurt paste is stored for 60 days at cool temperature

Concentrate acidity is an indicator of the sensory and microbiological properties of a fermentation product. The range of pH value is 3.7 to 4.6, where the longer the storage time, the higher the pH. Therefore, the storage time significantly affected ($P<0.05$) the pH value of a product. A significant increase in pH was observed in 45-60 days of storage. Furthermore, different pH values across treatments showed varied levels of acidity; thus, the lower the pH, the more acidic the product.

Besides metabolic acids, the other contributing factors to the decline of pH are the release of whey and the increased solubility of calcium phosphate mycelium that made the soluble calcium in the curd/solid missing. The produced lactic acid determines the shelf life of a product and its product sensory properties [9]. In general, the pH of yogurt that is stored before 45 days is under the casein isoelectric point. The isoelectric point is a condition where negative casein is depleting; therefore, it decreases the repulsion electrostatics among the casein molecules.

Additionally, the pH value reflects the accumulated metabolite produced by LAB during the fermentation. Labneh is a fermentation product with a low pH however, it allows yeast and molds to grow although the product is in cool storage. It explains that the shelf life of concentrated yogurt should not exceed two weeks. The result of studies on fermentation product confirms those of cheese product that storing cheese for 60 days would significantly affect the metabolic products, thereby decreasing the pH value [10].

4. Conclusion

The microflora profile of yogurt paste stored in cold temperature for 60 days undergo some changes, namely the decreased total bacteria, LAB and LAB visibility, as well as the increased total yeast and pH value.

References

[1] Nsabimana, C., B. Jiang, and R. Kossah, Manufacturing, properties and shelf life of labneh: a review. International Journal of Dairy Technology, 2005. 58(3): p. 129-137.

[2] Al-Kadamany, E., et al., Estimation of shelf-life of concentrated yoghurt by monitoring selected microbiological and physicochemical changes during storage. LWT-Food Science and Technology, 2003. 36(4): p. 407-414.

[3] Setyawardani, T., W.P. Rahayu, and N.S. Palupi, Physicochemical and Stability of Goat Cheese with Mono and Mixed Culture of Lactobacillus plantarum and Lactobacillus rhamnosus. Anim. Prod., 2016. 1(2): p. 36-42.

[4] Fonteles, T.V., et al., Optimization of the fermentation of cantaloupe juice by Lactobacillus casei NRRL B-442. Food and Bioprocess Technology, 2012. 5(7): p. 2819-2826.

[5] Costa, M.G.M., et al., Sonicated pineapple juice as substrate for L. casei cultivation for probiotic beverage development: process optimisation and product stability. Food Chemistry, 2013. 139(1-4): p. 261-266.

[6] Marino, M., M. Maifreni, and G. Rondinini, Microbiological characterization of artisanal Montasio

cheese: analysis of its indigenous lactic acid bacteria. FEMS Microbiology Letters, 2003. 229(1): p. 133-140.
[7] Ranadheera, C.S., et al., Probiotic viability and physico-chemical and sensory properties of plain and stirred fruit yogurts made from goat's milk. Food Chemistry, 2012. 135(3): p. 1411-1418.
[8] Guggisberg, D., et al., Rheological, microstructural and sensory characterization of low-fat and whole milk set yoghurt as influenced by inulin addition. International Dairy Journal, 2009. 19(2): p. 107-115.
[9] Park, Y., J. Lee, and S. Lee, Effects of frozen and refrigerated storage on organic acid profiles of goat milk plain soft and Monterey jack cheeses. Journal of dairy science, 2006. 89(3): p. 862-871.
[10] Setyawardani, T., et al., The Physicochemical and Sensory Qualities of Goat Cheese with Indigenous Probiotic Starter at Different Temperatures and Storage Durations. Animal Production, 2018. 19(3): p. 197-205.

Effect of dietary supplementation with flying fish (exocoetidae) silage on the egg quality and duck performance

L Ambarwati* and S P Syah

Universitas Sulawesi Barat, Majene, Indonesia

*lilisambarwati@unsulbar.ac.id

Abstract. This study aimed to evaluate the effect of dietary supplementation with flying fish silage on egg quality and duck performance. This study was design using a completely randomized design (CRD), 4 treatments of dietary supplementation (P0= 7%, P1= 8%, P2= 9%, P3= 10%) with 5 replications of laying duck. The observed variable of the egg quality was egg weight, egg yolk index (EYI), and Haugh unit (HU), and the duck performance were daily weight gain (DWG) and feed conversion ratio (FCR). The data were analyzed using ANOVA and followed by Duncan's test. The results showed that the flying fish waste silage have no significant effect on egg quality and duck performance (P> 0.05). The EYI ranged from 0.790 to 0.816, HU ranged from 95.04 to 99.09, egg weight ranged from 63 to 66.92 g, DWG ranged from 3.34 to 5.42, and FCR ranged from 1.85 to 2.26. In conclusion, dietary supplementation with flying fish waste silage up to a level of 10% was not able to decrease egg quality and duck performance.

1. Introduction

The duck population in Indonesia from year to year has a relatively slow population increase compared to other poultry. The duck population in 2019, with a total of 61,221,313 ducks, only increased by 2.80% (1,669,600 ducks) from the previous year (59,551,713 ducks). In West Sulawesi, the population of duck was 396,484, up 1.65% (6,459 duck) from the previous year (390,025 duck). Furthermore, the National duck egg production in 2019 was 352,939.13 Tons. It only increased 4.26% (14,431.55 Tons) from the previous year (338,507.58 Tons). Meanwhile, the duck egg production of West Sulawesi in 2019 was 2,510.97 tons, an increase of 1.65% (40.87 tons) from the previous year (2,470.10 tons) [1].

One of the most important that determines the productivity of ducks is the availability of nutrients in the feed. The high price of feed is one of the reasons why duck farmers prefer to shepherd their ducks rather than being kept in cages. Meanwhile, high FCR is also the reason duck farmers do not get maximum profit. The high cost of feed ingredients for protein sources, especially animal origin, has to be found a solution. Fish meal that is still imported can be an opportunity for the West Sulawesi region to produce fish feed ingredients with almost the same quality as imported fish meal.

Flying fish (Exocoetidae), commonly called as Tuing-tuing fish by the local community, are the catch of fishermen with abundant production every year and are characteristic of West Sulawesi. This fish production is very abundant around April to November. Since it is an abundance, many of them will rot, because they are not sold, so that they become a source of environmental pollution. The abundance of flying fish production in West Sulawesi could be a solution to the problem of availability of protein feed substitutes for fish meal as duck feed through the use of flying fish silage.

Flying fish contains 78.18% of water, 18.88% of crude protein, 0.13% of Fat, 0.28% of crude fiber, 0.32 of BETN (Non-Nitrogen Free Extract), and 1.21% of ash. Fish silage is a liquid product made from

whole fish or fish processing industry waste that is liquefied to resemble slurry by enzymes contained in the fish themselves through the help of acids or microbes that are intentionally added [2].
The current study on the use of alternative source of protein in duck feed to improve the performace and the egg quality of laying ducks has been done [[3][4] [[5][[6][7]. The study on the use of flying fish as a source of protein for fliying fish meal has never been carried out.

This research was conducted to determine the effect of dietary supplementation with flying fish silage on egg quality, which includes of the egg weight, EYI, HU, and duck performance could also be observed by DWG and FCR.

2. Materials and Methods

2.1 Cages and Animal Preparation

A total of 40 laying ducks, at 24 weeks with the live weight of 1500 to 1600 g, were used. The experimental ducks were obtained from Breeding Farm, Lembang, Majene, West Sulawesi. Metabolic cages made from wood and bamboo slats, with each plot of cage dimension was 30 x 30 x 40 cm with a battery system and were equipped with a feeding containing and an infused drinking water supply system. The experimental ducks were placed into 40 plots of cage for 14 days acclimatization. Each cage was filled with one laying duck.

2.2 Dietary Treatments

The duck feed and drinking water were provided ad libitum. The duck feed used during the study was mesh formulated with flying fish silage, adapted to the nutritional requirements of laying ducks (> 24 weeks). The feed ingredients of laying ducks were showed in Table 1.

Table 1. Feed ingredients and nutrient content of the treatments for laying period of duck (24 – 35 weeks)

Feed ingredients (%)	Feed supplemented with flying fish silage (%)			
	P0	P1	P2	P3
Corn	40	40	40	40
Bran	25,7	23,7	23,2	23
Soybean Meal	10	10	9,25	8,7
Tofu Dregs	12	10.5	10	9
Fish Meal	7	7	7	7
Fish Silage	0	3.5	5.25	7
Oil	3	3	3	3
Limestone	0.7	0.7	0.7	0.7
Top Mix	0.5	0.5	0.5	0.5
Lisin	0.6	0.6	0.6	0.6
Methionin	0,3	0,3	0,3	0,3
Salt	0.2	0.2	0.2	0.2
Total	100	100	100	100
	Nutrient content (%)			
Crude protein	17.07	17.68	18.45	18.70
Metabolic energy (kcal/kg)	2765	2743	2762	2742
Crude Fiber	7.56	7.02	8.64	8.16
Fat	6.70	6.80	7.54	8.18
Ash	9.35	6,60	8.09	8.16
BETN*	60.68	61.08	58.05	53.34
Water	16.79	17.90	16.92	17.07

Note: Calculations from the NRC Table (1994) and laboratory analysis of animal feed nutrition, Hasanuddin University (2020). P0 = basal feed (control), P1 = basal feed + 5% flying fish silage, P2 =

basal feed + 7.5% flying fish silage, P3 = basal feed + 10% flying fish silage. *Non-Nitrogen Free Extract.

2.3. Research Design

This study using a randomized completed design with 4 treatments and 5 replications of laying duck. The treatments were: P0 (basal feed/control), P1 (basal feed + 5% flying fish silage), P2 (basal feed + 7.5% flying fish silage), P3 (basal feed + 10% flying fish silage).

2.4. Variables and Sampling

The variable which observed in this study were egg weight, egg yolk index (EYI), Haugh unit (HU), daily weight gain (DWG), and feed convention ratio (FCR). Egg weight was recorded every day and carried out using a digital weight scale (Camry model: EK3650/EK3651, Hongkong) with an accuracy of 0.1 g. EYI was calculated as the ratio of egg yolk height and yolk diameter. HU was calculated by the Haugh formula: $HU = 100 \log (H + 7.57 - 1.7W^{0.37})$, which H is albumen height and W is egg weight. The egg yolk height, yolk diameter, and the albumen height were measured by using a digital micrometer. DWG was calculated by reducing between the final body weight and the initial body weight then divided by the maintenance period of the duck. FCR was calculated as the ratio of the number of feeds consumed every week with the weight gain of ducks each week.

2.5. Statistical Analysis

The data obtained were analyzed using analysis using one-way ANOVA by the general linear model procedures in IBM® SPSS® Statistics Subscription Ver 1.0.0.1461. If there is a significant effect of treatments, was further analyzed by the Duncan's multiple range test among mean values. Significant difference was evaluated at the level of $P < 0.05$.

3. Result and Discussion

The data results of egg quality and duck performance with the addition of flying fish silage in each treatment were shown in Table 2.

Table 2. The egg quality (egg weight, EYI, and HU) and duck performance (DWG and FCR) by dietary supplementation with flying fish silage

Variables	Treatments			
	P0	P1	P2	P3
EYI	0.392 ± 0.01	0.388 ± 0.01	0.407 ± 0.02	0.414 ± 0.03
HU	98.87 ± 0.11	95. 04 ± 8.77	99.03 ± 0.07	99. 09 ± 0.07
Egg weight	63 ± 8.93	65 ± 7.04	65 ± 7.91	66.92 ± 5.40
DWG	3.38 ± 1.08	4.25 ± 1.25	4.69 ± 1.39	5.42 ± 2.83
FCR	1.97 ± 0.43	2.26 ± 0.39	2.00 ± 0.39	1.85 ±0.26

Note: P0 = basal feed (control), P1 = basal feed + 5% flying fish silage, P2 = basal feed + 7.5% flying fish silage, P3 = basal feed + 10% flying fish silage. EYI = egg yolk index, HU = haugh unit, DWG = daily weight gain, FCR = feed conversion ratio.

3.1 Egg Quality

The addition of flying fish silage to duck basal feed did not have a significant effect on EYI of duck egg in each treatment ($P> 0.05$). Table 2 shows that the addition of flying fish silage to a level of 10% was not able to increase the EYI value of duck eggs. The presence of formic acid and probiotics in flying fish silage has not yet been able to stimulate nutrient absorption and metabolism, so that the blood flow and material deposition cannot be accelerated.

This resulted in the formation of immature yolk, so that the increase in the IKT of duck eggs was not significant. The addition of formic acid and the growth of natural probiotic microbes in the process of flying fish silage production, principally, it is done to increase the digestibility of feed ingredients, increase nutritional value, and prevent less optimal use of feed in the intestines. The natural ingredients are able to stimulate the uptake of glucose from the intestines and increase the uptake of glucose by the liver cells from the circulatory system. Glucose is then processed into several forms, namely oxidized as an energy source, stored in the liver as glycogen or converted into fatty acids as raw material for egg yolk precursor biosynthesis [8].

The EYI, at the time the egg was laid, was 0.45 then decreased to 0.30 if the eggs were stored for 25 days at 25°C [9]. According to BSNI [10], EYI of first quality was 0.458 – 0.521, quality second was 0.394 – 0.457, and third quality was 0.330-0.393. In this study, the EYI of P3, P2 and P0 treatment were included in the second quality. Purwati *et al.* [11] reported that EYI with 1-day storage period ranged from 0.33 to 0.51 with a mean of 0.42. In this study, the treatment with high protein content (P3) has not been able to increase the yolk index. The higher the protein and fat content in the feed, the higher the egg yolk index [4].

The result from this study, showed that the HU of egg duck were not significant (P>0.05) different among the groups (Table 2). It was assumed that because the content of ovomucin in albumin was very viscous and the albumin level was getting higher. Nuraini *et al.*[4] reported that ovomucin plays a very important role in water binding to form the albumin gel structure. If the ovomucin nets are many and strong, the albumin will more viscous, causing the albumin viscosity to be high as shown in the HU indicator. The HU value of newly laid egg is 100. HU egg with the best quality value is above 72, while the HU value of rotten eggs is under 50 [12].

That was indicated that the HU value eggs of this study in all treatments was classified as good quality with an average HU value of 95.04 - 99.09. According to Moghadam *et al* [13], the ovomucin content in egg white highly affects the value of HU. Albumin that contains less ovomucin will cause it too watery faster. Otherwise, if ovomucin in the albumen is high, it will cause more viscous and higher of egg albumen, so that the HU value also will be high.

The average of egg weight in this study ranges from 63 to 67 g where there was no significant (P>0.05) different among the groups (Table 2). This was assumed because the protein and energy content in the feed were almost same. The use of flying fish silage in basal feed was unable to increase protein levels and metabolic energy in the duck feed. Nutritional deficiencies in animal diet can cause a decrease in egg production and will reduce weight gain of poultry [14]. Protein and amino acids (especially methionine) in feed are the most important food substances in controlling egg size, other than genetics and body size of poultry. Methionine content is the main limiting factor for amino acids in the formation of egg weight. Giving methionine in the same treatment may influence the height ovomucin in albumin, increasing the egg white height, causing the average weight of the egg yolk is lower, resulting in lower egg weight [13].

The results of this study were in line with study conducted by Sumiyati [3], it was reported that there was no significant effect of adding cassava leaf meal and golden snail to laying duck feed on egg weight. The treatment only increased feed consumption an average of 184.6 g/bird. Furthermore, Fouad [15] reported that the addition of methionine with a level of 2 to 5 g / kg in the feed of ducks aged 28 weeks did not affect eggs weight, FCR and HU. The difference result on weight, quality, and performance may be related to age, stress, adaptation period, source of feed ingredients and the amount of feed consumption [16].

3.2 Duck Performance

The study result showed that, there was no significant different (P>0.05) of FCR among the group (Table 2). The average FCR ranged from 1.85 to 2.26. The results of this study were lower than previous studies by Suswoyo and Rosidi [17] reported that FCR of Tegal ducks in the production period ranged from 3.16 to 4.83.

In addition, it was lower than Sumiati *et al.* [3] with an FCR value ranged from 4.14 to 4.18. Furthermore, [6] reported that the addition of feed which containing 10% of lebium fish waste flour in the diet has significant effect on the FCR in 16 week old male local ducks. In the other hand, this study result in line with Shabani *et al.* [18] reported that the addition of fish waste silage at every 120 g/kg of feed has not been able to increase FCR of broiler chickens. Moreover, Fouad *et al.* [15] reported that The addition of methionine 2 - 5 g/kg at laying duck had no significant effect on FCR.

The smaller the FCR, the better the feed efficiency, while the greater the FCR, the worse the feed efficiency. The amount of FCR could be used as an overview of production efficiency of laying duck [19]. The poor FCR of ducks is because ducks have a habit of always drinking so that a lot of feed is wasted, and cause feed quickly decompose and inviting flies. Suswoyo dan Rosidi [17] stated that in general, when the ducks eat, the feed will be scattered or wasted when the ducks are moved from the drinking area and a lot of the feed is dissolved in the feed place when drinking. The poor FCR of ducks is also possibly caused by the inability of ducks to control the amount of feed consumption which is regulated by the amount of energy consumption which the chickens can do.

The addition of flying fish silage on the basal feed has no significant effect (P>0.05) on DWG. There was no different of DWG among the group (Tabel 2). This study result in line with Fouad *et al.* [15] and Zang *et al.* [16], however, it was different with [15] reported that the addition of methionine was able to increase egg production and body weight gain in 60 week old Shaoxin laying ducks during the 6 week trial period. Shabani *et al.* [18] reported that the addition of fish waste silage in every 120 g/kg of feed could increase DWG in broilers.

The value of DWG was smaller because the absorption of nutrient was not optimal du to crude protein linear with crude fiber content. This causes a longer process of feed digestion, and absorption of nutrients to produce weight gain becomes less. The formic acid which presents in silage cannot activate digestive enzymes to work optimally. Moreover, the probiotics which present in silage also have not worked optimally to expand the intestinal villi and facilitate absorption of nutrients.

The formic acid serves to increase the digestibility of feed ingredients, increase the nutritional value of feed ingredients, and prevent clumping. The *Lactobacillus casei* which present in fish silage will help the digestion of fish [18]. The present of lactic acid bacteria in digestive tract is able to maintain the balance of microflora ecosystem in the intestinal. Furthermore, the lactic acid bacteria could produce a lactic acid which beneficial to inhibit the growth of pathogenic microbial.

4. Conclusion

The addition of flying fish silage to duck basal feed up to 10%, has not been able to improve egg quality and duck performance. It is necessary to do further study by adding the level of flying fish silage above the level of 10% so that the performance and quality of duck eggs could be better.

Acknowledgement

The authors would like to thank The Ministry of Research, Technology and Higher Education of the Republic of Indonesia for supporting the research funding through Beginner Lecturer Research Schemes with contract number: T/019/UN55.C/KP.00.05/2020.

References

[1] Direktorat Jenderal Peternakan dan Kesehatan Hewan, 2017 *Statistik Peternakan dan Kesehatan Hewan 2017/ Livestock and Animal Health Statistics 2017* .

[2] Hany Handajani, Sri Dwi Hastuti S, 2013 Penggunaan berbagai limbah asam organik dan bakteri asam laktat terhadap nilai limbah ikan *Depik* **2**, 3 p. 126–132.

[3] Sumiati S Darmawan A and Hermana W, 2020 Performances and Egg Quality of Laying Ducks Fed Diets Containing Cassava (Manihot esculenta Crantz) Leaf Meal and Golden Snail (Pomacea canaliculata) *Trop. Anim. Sci. J.* **43**, 3 p. 227–232.

[4] Susan Nuraeni. M. Anwar Djaelani S dan K, 2019 Nilai Haugh Unit (HU), Indeks Kuning Telur (IKT) dan Ph Telur Itik Pengging Setelah Pemberian Tepung Daun Kelor (Moringa oleifera Lam .) *Bul. Anat. dan Fisiol.* **4**, 2 p. 107–115.

[5] Palupi R Lubis F N Rismawati R Sudibyo I and Siddiq R A R, 2018 Effect of Indigofera zollingeriana Top Leaf Meal Supplementation as Natural Antioxidant Source on Production and Quality of Pegagan Duck Eggs *Bul. Peternak.* **42**, 4 p. 301–307.

[6] Daud M Fuadi Z and Mulyadi M, 2020 Performan dan Produksi Karkas Itik Lokal dengan Pemberian Ransum yang Mengandung Limbah Ikan Leubiem (Canthidermis maculata) *J. Agripet* **20**, 1 p. 9–16.

[7] Darmawan, A. Sumiyati H W, 2016 Kualitas fisik telur itik Magelang yang diberi ransum mengandung tepung daun Indigofera sp. dan minyak ikan lemuru *Bul. Makanan Ternak, 2016* **103**, 1 p. 11–19.

[8] Kasiyati Djaelani M A and . S, 2019 Effect of Supplementation of Moringa oleifera Leaf Powder on Reproductive Performance and Ovarian Morphometry of Pengging Ducks *Int. J. Poult. Sci.* **18**, 7 p. 340–348.
[9] Yuanta T, 2004 *Dasar Ternak Unggas* Yogyakarta: Kanisus.
[10] Badan Standarisasi Nasional, 2008 *SNI 3926:2008 Telur Ayam Konsumsi* .
[11] Purwati D Djaelani M A and Yuniwarti E Y W, 2015 Indeks Kuning Telur (IKT), Haught Unit (HU) dan Bobot Telur pada Berbagai Itik Lokal di Jawa Tengah *J. Biol.* **4**, 2 p. 1–9.
[12] Ismoyowati I Indrasanti D and Sulistyawan I H, 2018 The Differences of Feed Quality and Egg Production Performance of Tegal and Magelang Ducks on Farming in Central Java *Bul. Peternak.* **42**, 3 p. 197–202.
[13] Nassiri Moghaddam H Kazemi Fard M Agah M J Hosseini S J and Mirakzehi M T, 2012 Effect of different levels of methionine, protein and tallow on the productive performance and egg quality of laying hens in the late-phase production *Brazilian J. Poult. Sci.* **14**, 2 p. 149–158.
[14] Olarotimi O J and Adu O A, 2017 Potentials of non-conventional protein sources in poultry nutrition *Arch. Zootec.* **66**, 255 p. 451–457.
[15] Fouad A M *et al.*, 2016 Effects of dietary methionine on performance, egg quality and glutathione redox system in egg-laying ducks *Br. Poult. Sci.* **57**, 6 p. 818–823.
[16] Zhang L B and Guo Y M, 2008 Effects of liquid DL-2-hydroxy-4-methylthio butanoic acid on growth performance and immune responses in broiler chickens *Poult. Sci.* **87**, 7 p. 1370–1376.
[17] Suswoyo I and Rosidi, 2017 Produksi Telur dan Konversi Pakan Itik Tegal Terkurung Kering Yang Diberi Kombinasi Probiotik dan Vitamin C November p. 704–709.
[18] Shabani A Boldaji F Dastar B Ghoorchi T and Zerehdaran S, 2018 Preparation of fish waste silage and its effect on the growth performance and meat quality of broiler chickens *J. Sci. Food Agric.* **98**, 11 p. 4097–4103.
[19] Imawan M R Sutrisna R and Kurtini T, 2016 Pengaruh Ransum Dengan Kadar Protein Kasar Berbeda Terhadap Pertumbuhan Itik Betina Mojosari *J. Ilm. Peternak. Terpadu* **4**, 4 p. 300–306.

Production profile and microbiological properties of locally produced pork sausage in the Philippines

MM Dealino[1,2] and CM Bueno[2*]

[1]College of Agriculture, Southern Luzon State University, Lucban, Quezon, Philippines
[2]College of Agriculture and Food Science, University of the Philippines, Los Baños, Laguna, Philippines

*cmbueno@up.edu.ph

Abstract. This study evaluated the micro-enterprise production practices and microbiological properties of Philippine pork sausage. It was noted that most sausage manufacturers that fall under the said business category require workers to take a bath prior to operation and wear aprons while few require hand washing and use of hairnet and facemask. No one used hand gloves. Meat materials are sourced locally and undergo no washing prior to grinding which are mostly facilitated at room temperature. Only few of the respondents have designated processing area while others process in-store or at the back of their house. All the respondents use stainless steel mixing basin and manual stuffing with 46.67% using stainless steel table. Processors use hog casing and/or artificial edible casing. Results of the microbial analysis showed all sausage samples were beyond the limit for coliforms while only 26.77% were below the limit set for Escherichia coli. Salmonella had 93.33% incidence while all samples were tested positive for Staphylococcus aureus. Results showed that micro-enterprise production process of local pork sausage may compromise the safety of the consuming public and comprehensive evaluation by the local government unit is necessary to comply with the Philippine Food Safety Act of 2013.

1. Introduction

Fresh sausage is a highly perishable type of sausage which should be handled appropriately. This type of sausage is not heat treated and is sold in raw state, thus, must be cooked prior to serving [1]. This product may last for 2-4 days if stored in the refrigerator and for 1-2 months in the freezer [2]. A famous Philippine pork sausage or locally known as longanisa or chorizo, is an example of a fresh sausage with high volume manufactured by micro-enterprise. It was first made in 1739 by Juan Suarez in which he tried to copy the sausage "sorizo de bilbao" from Acapulco, Mexico by using local ingredients (salt, oregano extract, atsuete, garlic and others). The product is displayed and hanged in front of the stores along the streets in the town area [3].

As a native delicacy in the different areas in the Philippines, it is then important to assess the production aspect, and microbiological safety since most of this product is manufactured by the micro-enterprises. There are limited studies done on the evaluation of this local sausage pertaining to the above-mentioned parameters and the safety of the consuming public must be assured, thus this study.

2. Materials and Methods

2.1 production profiling of the local pork sausage
An on-the-spot visit to 15 of the total 17 local pork sausage producers (88%) in the island of Luzon was conducted in coordination with the local government unit representative. An unannounced assessment of sausage manufacturing by these micro-enterprises was facilitated through an assisted survey and actual observation from 6:00 to 10:00 in the morning. Actual observation on the production of the local pork sausage was facilitated to document the processing practices of the product.

2.2 Microbiological properties of the local pork sausage
2.2.1 Procurement of sample materials
Sausage samples (about 2.0 kg from each respondent) were procured on the same day from 7:00–9:00 am from 15 local pork sausage processors representing 88% of the total manufacturers in the identified study area.
2.2.2 Microbial analysis
The samples were subjected to microbiological analyses on the day of collection. The E.coli and coliform count were noted using the Pour Plate Method (Chromocult Agar, 35°C, 24 hrs) [4]. Presence of Salmonella was analyzed through conventional method (Hektoen Agar- Bismuth Sulfite Agar – XLD Agar, 35°C, 24 hrs) [5] while the presence of Staphylococcus aureus was determined using the conventional method (Baird Parker Agar, 37°C, 24 hrs, Bactident Coagulase 37°C, 24 hrs) [6].
2.2.3 Statistical analysis.
Frequency distribution of the data gathered were computed and assessed. Sample values, percentages and averages of the data were determined and interpreted.

3. Result and Discussion

3.1 Production Profiling of the Local Pork Sausage
3.1.1 Personnel Hygiene and Personal Protective Equipment
Personnel hygiene practices.
Every production day prior to sausage processing, the respondents required their personnel to take a bath (93.33%), hand wash (26.67%) and cut fingernails (6.67%) as shown in Table 1. Bathing regularly is an important part of proper personal hygiene to remove dirt and odor from head to toe. Meanwhile, hand washing and use of hand sanitizers are preventive measure for food contamination [7], thus, it is important to know when to wash hands while processing food.

Table 1. Personnel hygiene practices*

Hygiene Practice	Distribution Of Respondents
	Percentage (%)
Bathing	93.33
Hand Washing	26.67
Cutting Of Fingernails	06.67

*Multiple answers

Food handlers should religiously practice hand washing after touching bare human body parts other than clean hands and clean exposed parts of the arm; using the toilet; coughing, sneezing, using a handkerchief or disposable tissue, eating or drinking; and handling soiled equipment or utensils. In addition, hand washing should be done during food preparation as frequent as necessary to prevent contamination when changing task such as switching from raw to ready-to-eat product. More so, hand washing should be done before wearing working gloves and after engaging in other activities that contaminate the hands.

Fingernails of food handlers should be trimmed and without nail polish nor with artificial fingernails. Since human is both a reservoir and vector of microorganisms that can contaminate food directly or indirectly, it is very important that personnel in food processing has good personal hygiene.

Personal protective equipment.
Wearing personal protective equipment (PPE) is a must when it comes to handling food. This is done for food safety reasons and to avoid contamination in any way. Table 2 showed that producers required their workers to wear aprons (66.67%), hairnets (33.33%), facemasks (26.67%), white rubber boots (6.67%) and none of them require wearing of gloves. Hairnet would keep the hair from falling, thus avoiding contamination and would keep hair strands from ending up on the product. Facemask would keep saliva from spilling when processor talk and would prevent contamination when sneezing or coughing.

Wearing a clean apron on top of clean working clothes would minimize contamination. Wearing of safety shoes or white boots is often practiced in commercial food processing plant to serve as protection against slippery floor and to prevent cross contamination, provided that they are worn inside the production area only. Gloves are worn to avoid food contamination through the hands of the food handlers. No bare hands contact with ready-to-eat foods such as meat-deli is mandated by the U.S. FDA Food Code [8]. Single use of gloves is recommended.

Table 2. Producers using Personal Protective Equipment (PPE)*

Personal Protective Equipment	Distribution Of Respondents
	Percentage (%)
Apron	66.67
Hairnet	33.33
Hand Gloves	00.00
Rubber Boots	06.67
Face Mask	26.67

*Multiple answers

It was observed that there are manufacturers allowing their personnel to wear sleeveless shirt during hot weather and baseball cap. Wearing sleeveless shirt is prohibited since this type of clothes exposes the armpit where hair grows and is prone to sweating. Food processing establishments should require their personnel to change clothes and wear uniforms inside the processing area to avoid contaminants from the clothes worn outside. Moreso, the use of baseball cap is discouraged since it will not contain the processors' hair properly to keep it from ending in the food, not like that of hairnets, and sweat accumulates in the cap from long use without washing.

It was also observed that some workers were wearing earrings, ring and wristwatch. These accessories should not be worn when working with food for these may incidentally fall on the product and become contaminant, and these also harbor bacteria and body fluids which will contaminate the foods [7].

3.1.2 preparation of the meat prior to sausage processing
Handling the meat material for sausage making is very important since it is highly perishable and will contribute most in the microbial load of the finished products. It was noted that microorganisms commonly enter meat preparation from meat, spices, and other ingredients [9].

Table 3 showed the meat material came directly from slaughterhouse (20%), wet market (26.67%) or delivered by suppliers (53.33%). All hogs are slaughtered in the town's slaughterhouse. Quality of meat depends on the subsequent handling from hauling live animals until slaughtered up to its form for meat preparation. From the abattoir, animals must be rested and fasted before they are slaughtered. The higher the stress level of the live animal prior to slaughtering, the poorer the meat quality. [10]. Water source of the abattoir is also a factor contributing to microbial quality since it comes into contact with meat during washing.

The meat from the wet market is more exposed to environmental contaminants. The carcasses brought to the wet market did not undergo the chilling process and are hanged in wholesale parts, fabricated to retail cuts resulting to very high microbial loads. It was revealed that pork meat samples from the public market stalls had high microbial plate count and found positive for E.coli [11]. Rapid chilling of carcass after washing is important in inhibiting the rapid growth of microorganisms [11], and that most of the pathogenic microorganisms grow rapidly at the temperature 40-140°F which is the so called "temperature danger zone" [13].

Table 3. Meat material source, form, and preparation of meat

Meat Source	Distribution Of Respondents	
	Frequency	Percentage (%)
Slaughterhouse	3	20.00
Market	4	26.67
Supplier	8	53.33
Total	15	100.00
Forms Of Meat When Bought		
Ground Meat	9	60.00
Whole Meat	6	40.00
Total	15	100.00
Preparation Of Meat Before Grinding		
Washing	0	00.00
No Washing	15	100.00
Total	15	100.00

Sixty percent (60%) of the processors bought their raw meat material in ground form, while 40% buy it whole or intact and do their own grinding. Those ground meat from the suppliers have greater microbial contamination than the intact meat bought from the wet market since additional handling activities were done prior to delivery providing more opportunity for contamination. The grinder used may have been operating the whole day at room temperature in an unsanitized environment, using contaminated containers and other food contact areas.

In addition, ground meat, which is more prone to microbial multiplication due to greater surface area, is held much longer at higher temperature giving greater chance for microbial growth and multiplication. It was stated by all the respondents that the meat for sausage processing is not washed prior to grinding. Manufacturers believed that if the meat is washed prior to grinding, the sausage will spoil easily, and some respondents stated that the sausage will harden. In reality, washing prior to grinding will greatly reduce the microbial load of meat. Their belief that meat will harden due to washing has no basis at all, in fact, hardness is no longer an issue in ground meat due to disruption of meat structure during grinding.

Meat holding temperature.

Table 4 shows the conditions where meat is held prior to processing. Delivered ground meat was placed in stainless basin at room temperature (93.33%), in which one of them placed it directly in the mixing equipment. Only 6.67% placed the ground meat in an air-conditioned room.

Table 4. Conditions where meat is held prior to processing

Condition	Distribution of Respondent	
	Frequency	Percentage (%)
Refrigerated Temperature	0	00.00
Room Temperature	14	93.33
Air-Conditioned Temperature	1	06.67
Freezing Temperature	0	00.00
TOTAL	15	100.00

The producers stated that they immediately process the meat after delivery. Placing ground meat at room temperature will favor greater microbial growth and multiplication compared to those placed in an air-conditioned room. Refrigeration will retard the growth of microorganisms in ground meat.

3.1.3 Processing of Sausage

Processing is crucial in keeping the microbial load of the sausage low. Through the processing environment, equipment and handlers, microorganisms can make their way into the sausages which can have a significant impact on its microbiological status. The series of activities done in processing would reveal how safe a processed food is [9].

Production Area.

Presented below (Table 5) are the locations where sausages were processed. The respondents processed their sausages in an area inside their outlet stores (40%), in a designated processing area (26.67%), inside their houses (20%), and outside the house (13.33%).

Since, the sausage is under the cottage industry level or micro-enterprise, most of the producers do not have appropriate processing facility. Processing the sausage in an open area or in an inappropriate location does not assure food safety due to numerous opportunities for microbial contaminants. There is also high incidence of contamination from the environmental pollutants, pets, playing kids, unnecessary person in the production area and others.

Table 5. Location of sausage production area

Location	Distribution Of Respondents	
	Frequency	Percentage (%)
Designated Processing Area	4	26.67
Outlet Store	6	40.00
Inside The House	3	20.00
Outside The House	2	13.33
Total	15	100.00

Processing

Sausage ingredients were mostly mixed by bare hands, 93.33% (Table 6). Manufacturers claimed that bare hands mix the sausage thoroughly and for ease of stuffing. Bare hands without proper hand washing and disinfection will cause contamination of the products. Hands are important in transmitting foodborne pathogens which concentrate more around and under the fingernails [14].

Processors hands should be cleaned properly by rinsing under clean, running warm water with sufficient amount of recommended cleaning compound, rubbed together vigorously for about 20 seconds, rinse again under clean, warm running water; sanitized and thorough drying to avoid recontamination.

Table 6. Mixing method, storage, and table used in sausage manufacture

Methods	Distribution Of Respondents	
	Frequency	Percentage (%)
Bare Hands	14	93.33
Hands With Gloves	0	00.00
Use Of Meat Mixer	1	06.67
Total	15	100.00
Place Where Sausage Mixtures Are Held Prior To Stuffing		
Table	14	93.33
Refrigerator	0	00.00
Stuffing Equipment	1	06.67
Total	15	100.00
Table Used In Sausage Making		
Stainless Steel Table	7	46.47
Kitchen Table	7	46.67
Plastic Tables	1	06.67
Total	15	100.00

Proper hand washing reduces microbial load but friction and water were found to have the most important roles [14]. Bare hands can be used in processing food provided that proper hand washing, and sanitation will be done. Wearing hand gloves must be done to protect the food product from being contaminated by avoiding touching anything except the food being processed. However, it was noted that food workers only wear hand gloves to protect their hands from getting dirty. The use of properly cleaned and sanitized meat mixer assures the production of good quality sausage. It will minimize time and effort in processing and would keep the mixture from processors' hands, thus may minimize contamination.

The producers place the sausage mixtures on top of the table (93.33%) and few (6.67%) directly placed it in the sausage stuffing equipment. The practice of not placing the sausage mixture under low temperature creates an environment for rapid microbial growth. They do not place it in a refrigerator since the large mixing basin does not fit in the refrigerator.

All the manufacturers use stainless steel basin and stuffing funnel. However, only 46.67% of them use stainless steel tables, others use kitchen table with plastic sheet covering (46.67%) and plastic tables (6.67%). Stainless steel material is recommended in food processing due to its durability, strength, rust resisting property and ease of cleaning. Though stainless-steel materials are resistant to stain and corrosion and does not easily breed bacteria, it must be cleaned properly and regularly to avoid build-up of contaminants. Using kitchen table and plastic table are inappropriate since the materials are not of food grade which may contaminate the products.

Sausage Casing

Table 7 shows the different casing materials used in stuffing sausage. The manufacturers use either natural casing from hog small intestines (40%), edible collagen casing (40%) or uses both natural and edible casings (20%). Natural casing which is typically from hog and sheep intestines has been traditionally used for centuries in stuffing sausages and is most preferred by consumers since it is from natural material.

However, natural casing produce less uniform size of the finished product and it is inherently contaminated with microorganisms, subject to rancidity and less durable while artificial edible casing made of collagen has uniform size and shape; and has less microbial issues. Some of the respondents prefer to use both types of casings.

IOP Conf. Series: Earth and Environmental Science 1041 (2022) 012079 doi:10.1088/1755-1315/1041/1/012079

Table 7. Sausage casing used

Casing	Distribution Of Respondents	
	Frequency	Percentage (%)
Natural	6	40.00
Artificial	6	40.00
Natural And Artificial	3	20.00
Total	15	100.00

3.2 Microbiological Properties of the Local Pork Sausage

Testing the microbiological properties of the finished products provides a way of assessing how well the slaughter and the production processes were executed to control contamination [25]. The numbers of bacteria (the total viable count) on the surface of carefully handled carcasses may be up to 103 to 104 organisms per cm2 but with poor hygiene this could rise to more than 106 organisms per cm2 (Warriss, 2000).

The microorganisms of concern for meat and meat products are coliform, E.coli, Salmonella and Staphylococcus aureus [26] causing meat borne illnesses/spoilage and found their way into the product due to cross contamination from ingredients, equipment, materials and food handlers. Exogenous contamination control is difficult, thus, consumers encounter meat spoilage and meat poisoning. Table 8 presents the microbial properties of sausage samples from the participating respondents.

Coliforms

All samples tested were beyond the acceptable limit for coliforms of <10cfu/g. Coliform bacteria indicate the potential presence of disease-causing bacteria in water [15]. This implies that the water used in sausage production from the preparation of raw materials up to processing is contaminated. Coliform testing is especially significant for the dairy and meat industries, where food products are exceedingly susceptible to contamination, and it became a priority due to the numerous possibilities of contamination.

These bacteria are common in the environment such as in soil and in the intestines of animals which are generally not harmful. Hence, regular coliform testing for food producers and processors is important [16]. In sausages, coliforms should be absent in encased prepared meat containing minced meat [17].

Table 8. Microbial properties of the sausages from different producers in Southern Luzon, Philippines

Samples	Coliform Count**, Cfu/G (Limit: <10 Cfu/G)	*E.Coli* Count, Cfu/G (Limit: 5,000 Cfu/G)	*Salmonella* Detection (Presumptive Test)	*S. Aureus* Detection
A	1.1×10^{5}	$1.3 \times 10^{4**}$	+	+
B	$>6.5 \times 10^{6}$*	$1.9 \times 10^{5**}$	+	+
C	$>6.5 \times 10^{6}$*	$1.4 \times 10^{5**}$	+	+
D	1.3×10^{5}	$2.0 \times 10^{5**}$	+	+
E	7.2×10^{5}	$1.2 \times 10^{5**}$	+	+
F	1.3×10^{4}	2.8×10^{3}	+	+
G	6.2×10^{4}	2.5×10^{3}	+	+
H	2.5×10^{5}	$8.5 \times 10^{4**}$	+	+
I	2.3×10^{5}	$6.6 \times 10^{4**}$	+	+
J	5.7×10^{5}*	$1.2 \times 10^{4**}$	+	+
K	1.4×10^{5}	$7.9 \times 10^{3**}$	-	+
L	6.1×10^{5}*	$1.5 \times 10^{4**}$	+	+
M	3.0×10^{5}*	$5.8 \times 10^{3**}$	+	+
N	1.1×10^{4}	<250*	+	+
O	3.1×10^{5}*	3.1×10^{2}	+	+

Note * - Estimated Plate Count ** - Beyond the Limit/Acceptable level

Escherichia coli (E.coli)
The *E.coli* count ranges from <250 to 2.0 x 10^5. The study revealed that only 4 of the 15 samples (Samples F, G, N and O) were within the allowable level of 5,000cfu/g. This bacterium is commonly used as fecal indicator organism. It is a bacterial contaminant of sausage and other meat products that cause serious diarrheal illness which sometimes resulting in complications that can lead to death. The results imply that there were improper hygiene and handling during the processing of the local pork sausage [17] since it was reported from the first part of the study that most of the respondents do not require hand washing and checking of the fingernails before starting the processing of the sausage, and all of the respondents do not require use of hand gloves in processing.

Also, workers do not wear appropriate PPE, such as face masks and hairnets, while processing the sausage. Another is the practice of not washing the meat before grinding after long time of dripping from the slaughterhouse. The principal sources of microbial contamination were the raw materials and food handlers [18]. This pathogen is killed by mild heat and inactivated at temperatures higher than 70°C, hence, pasteurization can totally inhibit *E.coli.* [19].

Salmonella
In this study, 14 of the total 15 samples from different producers were tested positive for *Salmonella*. This pathogen is the most frequently reported bacterial cause of foodborne illnesses and is particularly problematic as a foodborne pathogen because it has the ability to persist in many different conditions [20]. Human is a vector of transmission of this microorganism since they grow best in body temperature.

The result may again be attributed to lack of proper hygiene, poor implementation of using appropriate PPE, improper handling of materials and ingredients, and other causes [17]. This bacterium causes the disease *salmonellosis*, a gastrointestinal illness which typically occurs 12–72 hours upon ingestion of the contaminated food [20]. This illness is usually characterized by acute onset of fever, abdominal pain, diarrhea, nausea and sometimes vomiting, that lasts for 2–7 days [20, 21]. Symptoms may be mild but, in some cases, dehydration can become severe and life-threatening especially to children and elderly. *Salmonella* is heat labile making it easy to eliminate if proper cooking will be done. The cooking temperatures of 62.8°C can destroy *Salmonella* [20].

Staphylococcus aureus
This study reported that 100% of the sample sausages are tested positive for *S. aureus* which may be accounted to improper hygiene and sanitation of workers. Incomplete and improper wearing of PPE and jewelries/accessories which were observed from the first part of the current study accounted for the presence of *S. aureus* in all samples tested.

It was reported that human is a major source of contamination in food poisoning caused by *S. aureus* [22] that is why personnel hygiene is very important in food processing. Negligence to wash hands after visiting the bathroom is one of the poor personal hygiene practices which may result in up to 107 pathogens under the fingernails of the food handler [19]. Contamination of food may occur through direct contact with human, through respiratory tract droplets or by skin fragments [23], [24]. This bacterium produces enterotoxins which is responsible for foodborne illnesses. *S. aureus* can also tolerate salt up to 20% concentration [23].

Although *S. aureus* is readily killed at cooking and pasteurization temperatures, increased amount of toxin from this bacterium can cause serious illness. In all sausage products, proper hygiene, handling, and storage procedures are essential to control contamination [17].

4. Conclusion
In production profiling it was documented that micro-enterprise pork sausage producers do not implement strict rules and regulations in processing such as wearing of appropriate personal protective equipment, use of jewelry and workers personal hygiene. Most of these manufacturers do not have appropriate processing area and equipment, thus, hygiene and sanitation of the processing areas and materials used are compromised.

A high load of microbial contaminants in the sausage production was noted. The values from the results of Coliforms and E. coli. and the presence of Salmonella spp and S. aureus could be alarming for the consumers. Those pathogens may pose serious illnesses when contamination is significantly high especially when the product is not sufficiently cooked to eliminate the microorganisms of concern.

From the profiling, the attention of the producers of the local pork sausage should be called upon regarding the current processing procedure being implemented resulting to product contamination. Training-seminar can be conducted for these micro-enterprises for them to realize the importance of proper meat handling and processing on their product quality as well as to comply with the food safety standards that the country is currently strengthening.

Acknowledgement
The authors would like to acknowledge the Commission on Higher Education Kto12 Program for the financial support.

References

[1] Heinz G and Hautzinger P. Meat processing technology for small- to medium scale producers. Food and Agriculture Organization of the United Nations Regional Office for Asia and The Pacific, Bangkok. 2007. p456.

[2] Boyer R and Mckinney J. Food storage guidelines for consumers. 2013. [cited 2020 October 1]. In: Virginia Cooperative Extension. Available from: https://www.pubs.ext.vt.edu/content/dam/pubs_ext_vt_edu/348/348-960/FST-286.pdf.

[3] Dela Cruz TV. Lucban's Original Langgonisa. Siyasig Southern Luzon State University Official Research Journal. 2006; 2(1) 68-75.

[4] BacteriologicaL Analytical Manual (BAM). Enumeration of the Escherichia coli and the Coliform Bacteria. 2002; cited 15 May 2018. Available from https://www.fda.gov/Food/FoodScienceResearch/LaboratoryMethods/ucm064948.htm.

[5] Bacteriological Analytical Manual (BAM). Salmonella detection. 2011; cited 15 May 2018. Available from https://www.fda.gov/Food/FoodScienceResearch/LaboratoryMethods/ ucm070149.htm.

[6] Bacteriological Analytical Manual (BAM). Staphylococcus aureus-detection. 2001; cited 15 May 2018. Available from https://www.fda.gov/Food/FoodScienceResearch/Laboratory Methods/ucm071429.htm.

[7] Fraser AM. Food Safety: Safe Food Handler. Department of Family and Consumer Sciences. NC State University, Raleigh, NC. 2003; cited 10 May 2018. Available from http://www.foodsafetysite.com/resources/pdfs/EnglishServSafe/ENGSection4.pdf.

[8] Food and Drug Administration Food Code. 2017; cited 10 May 2018. Available from https://www.fda.gov/food/fda-food-code/food-code-2017.

[9] Melngaile A, Ciekure E, and Valcina O. 2014 Microbiological quality of meat preparations and meat products. In: Proceeding of the 9th Baltic Conference on Food Science and Technology. Jelgava, Latvia, 8-9 May 2014. pp61-65.

[10] Warris PD, Brown SN, Adams SJM, and Corlett IK. Relationships between subjective and objective assessments of stress at slaughter and meat quality in pigs. Meat Science. 1994; 38(2) 329-340.

[11] Yandug RS III, Ventura DC JR, Ybañes RHD and Ybañes AP. Microbial plate count and detection of Escherichia coli in Pork Meat Samples from Stalls in a Public Wet Market in Cebu, Philippines. International Research Journal of Interdisciplinary & Multidisciplinary Studies. 2016 March; 2(2): 65-72.

[12] Doyle EM. Survival and Growth of Bacterial Pathogens on Raw Meat During Chilling. Food Research Institute, University of Wisconsin-Madison. February 2002; cited 15 April 2017. Available from http://smas.chemeng.ntua.gr/miram/files/publ_292_11_2_2005.pdf.

[13] Food Safety and Inspection Service. Food Safety Education. USDA FSIS; Updated 28 Jun 2017. Cited 10 May 2018. Available from https://www.fsis.usda.gov/wps/portal/fsis/topic s/food-safety-education/get-answers/food-safety-fact-sheets/safe-food-handling/danger-zone-40-f-140-f/ct_index#:~:text=%22Danger%20Zone%22%20(40%20%C2%B0F%20%2D%20140%20%C2%B0F)&text=Bacteria%20grow%20most%20rapidly%20in,called%20the%20%22Danger%20Zone.%22&text=Never%20leave%20food%20out%20of%20refrigeration%20over%202%20hours.

[14] Stagnitta PV, Micalizzi B, and Stefani De Guzmán AM. Prevalence of some bacteria yeasts and molds in meat foods in San Luis, Argentina. J Publ Health. 2006; 14(3): 141–144.

[15] Swistock, Bryan R., Clemen S, Stephanie and William Sharpe. Coliform Bacteria. PennState Extension.

Accessed April 15, 2018. Available from https://extension.psu.edu/coliform- bacteria.
[16] Feng P, Weagant SD, Grant MA, Burkhardt W. Bacteriological Analytical Manual (BAM) Chapter 4: Enumeration of Escherichia coli and the coliform bacteria. US FDA. 2002; Updated October 2020. Cited 15 May 2018. Available from https://www.fda.gov/food/ laboratory-methods-food/bam-chapter-4-enumeration-escherichia-coli-and-coliform- bacteria.
[17] Association of Food and Drug Officials. Safe Practices for Sausage Production - Distance Learning Course Manual. Version 1. April 2014; Cited 15 April 2018. Available from http://www.afdo.org/resources/Documents/Committee%20Reports%202016-2017/Safe-Practices-Sausage-Production_04-14.pdf.
[18] Cordoba MG, Cordoba JJ, Jordano R. Evaluation of microbial hazards during processing of Spanish prepared Flamenquin. J Food Prot. Jun 1998;61(6):693-699.
[19] Gungori E and Gokoglu N. Determination of Microbial Contamination Sources at a Frankfurter sausage processing line. February 2010; Turkish Journal of Veterinary and Animal Sciences 34(1):53-59.
[20] Bedale W and Milkowski AL. Salmonella Fact Sheet, American Meat Science Association. 2015. Cited 5 April 2018. Available from https://meatscience.org/docs/default- source/publications-resources/fact-sheets/salmonella-fact-sheet-2015.pdf?sfvrsn=0.
[21] World Health Organization. Salmonella (non-typhoidal). 20 February 2018. Cited 10 April 2018. Available from http://www.who.int/mediacentre/factsheets/fs139/en/.
[22] Nel S, Lues JFR, Buys EM, Vente RP. The personal and general hygiene practices in the deboning room of a high throughput red meat abattoir. Food Control. 2004; 15: 571-578.
[23] Dodd C, Aldsworth TG and Stein RA, editors. Foodborne Diseases. 3rd ed. 2017. London: Academic Press; 2017. p576.
[24] Morris JG Jr and Potter ME. Foodborne Infections and Intoxications. 4th ed. 2013. London: Academic Press; p568.
[25] Food Standards Agency. Meat Industry Guide: Chapter 13 Microbiological Criteria; September 2017. Cited 10 April 2018. Available from https://www.food.gov.uk/sites/ default/files/media/document/chapter13-microbiological-criteria-final_version-2_2.pdf.
[26] Warriss PD. Meat Science: An Introductory Text. 2nd ed.. 2009. United Kingdom: CABI Publishing. p248.

Farming Sustainability: A Sensitivity Analysis

Y N Wakhidati*, M Sugiarto, H Aunurrohman, S Mastuti and J A Suryani

Universitas Jenderal Soedirman, Purwokerto, Indonesia

*yusmi.wakhidati@unsoed.ac.id

Abstract. The sustainability of broiler farming cannot be separated from the sensitivity of business feasibility. The very high price fluctuation since before the pandemic until now has become a threat to business sustainability. This study reveals the characteristics of broiler farming and the impact of price fluctuations on the feasibility of broiler farming in Banyumas Regency, Indonesia. This study used a survey method. 60 broiler farmers who are members of the contract growing were randomly selected as respondents. Data were analyzed using descriptive statistic. The impact of price fluctuations was calculated from a business feasibility simulation of changes in product selling prices. The results showed that the average broiler farm size was 5,444 head/period and cultivated 5 times in one year. The average investment was IDR 154,735,835.48 and earned profit IDR 70,045,228.33 per year with a positive NPV and IRR was 33.36 percent. Sensitivity analysis with a scenario of a decrease in selling prices by 3 percent shows that farmer profits fall to IDR 36,459,511.28 per year, with a negative NPV and an IRR of 7.57 percent, below the reference interest of 9.94 percent.

1. Introduction

Banyumas Regency is one of the regencies in Central Java which is the center for broiler chicken cultivation. This district occupies a strategic position in western Central Java. The climatic and topographic conditions of Banyumas Regency are very supportive of the cultivation of broiler chickens. This is indicated by the positive trend of increasing broiler population in the last three years. The broiler population in 2017, 2018 and 2019 were 7,537,900 heads, 8,208,000 heads and 11,741324 heads respectively[1].

This population development is indicated by the increasing interest of breeders in following a partnership pattern. [2] Through a partnership scheme, breeders are guaranteed the availability of production facilities in various business scales, both large and small [3]. This cooperation model also provides certainty for the selling price of chickens. This is expected to reduce risks and support the sustainability of broiler livestock business [4]. Even so, the selling price of livestock has very high fluctuation. Changes in selling prices certainly greatly affect profits and business sustainability.

Business sustainability cannot be separated from business feasibility analysis. Business feasibility analysis reflected in NPV and IRR illustrates the ability of businesses to generate profits and the rate of return on investment invested [5]. Even though farmers follow a partnership pattern, the contract price received by the farmer is determined based on only one production period. This condition make farmers getting different price levels in one year of production. This study will reveal the characteristics of broiler chicken farming in Banyumas Regency and the impact of changes in contract prices on business feasibility.

2. Materials and Methods

The research method used was a survey of broiler breeders in partnership in Banyumas Regency. A total of 60 breeders were randomly selected to be respondents. The data collected are investment data and production data and analyzed using descriptive analysis. The analysis used to determine the feasibility of a business is the NPV and IRR analysis with the interest of rate (9.4%). To determine the sensitivity of business feasibility, a contract price reduction simulation is performed to produce a negative NPV and an IRR below the reference interest rate.

$$NPV = \sum_{t=0/1}^{n} \frac{Bt - Ct}{(i+i)^t}$$

NPV was calculated using the following formula [6]
Explanation:
Bt = revenue at year t
Ct = costs at year t
i = discounted factor
t = year
n = project duration
A project or investment is feasible if NPV is positive.

IRR was calculated using the formula:

$$IRR = i_1 + \frac{NPV_1}{NPV_1 - NPV_2} x(i_2 - i_1)$$

Explanation:
i_1 = the discounted factor that yields a positive NPV
i_2 = the discounted factor that yields a negative NPV
NPV_1 = positive NPV
NPV_2 = negative NPV
A project or investment is feasible if the IRR is higher than the interest rate in effect.

3. Result and Discussion

3.1 Respondents' characteristic

The characteristics of livestock business will present data, business scale, length of raising livestock for one period and the number of raising livestock periods in one year. The summary of the characteristics of livestock farming is presented in table 1 below.

Table 1. Characteristics of livestock businesses

	Small (<4999)	Medium (5000-7499)	Large (> 8000)	Average
Livestock as the main business (%)	64.86%	83.33%	100%	75%
Scale enterprises (head/period)	3,027	6,042	12,909	5,442
Lenght time of raising livestock for one period (days)	38	37	37	37
Raising livestock periods in one year (times)	6	5	5	5

Broiler chicken farming in Banyumas Regency is not a secondary business for farmers. This is evident from the results of the analysis which shows that 75% of farmers state that raising is their main job. All large-scale farmers make this business as their main job. On a smaller business scale, some farmers consider this business as a secondary business to earn additional income. On the larger business scale, raising livestock would be needed more time and attention from farmers. All farmers who have large scale of business category devote their time to raising broilers.

The average number of birds in one period of raising in the category of small, medium and high scale are 3.027 heads: 6,042 heads and 12,909 heads respectively. Lenght time of raising livestock for one periode are 37 days and there was 5 times of raising livestock periods in a year. Farmers need moret time to prepare cages and equipment so that an average of 5 times period a year is considered the most ideal [7].

3.2 Feasibility analysis

The business feasibility analysis in this study includes the results of the analysis regarding the amount of investment, the average contract price, the average revenue, production costs, profits and the results of the NPV and IRR calculations. The results of the business feasibility analysis are listed in the table below.

Table 2. Average investment costs, production costs, revenues, profits, NPV and IRR

	Low (<4999)	Middle (5000-7499)	High (> 8000)	Average
Investment (IDR / bird)	5,760.06	5,422.82	4,855.73	5,684.64
Average contract price (IDR / kg)	17,935.27	17,438.00	17,937,55	17,845.30
Revenue (IDR / bird)	34,501.24	39.929.35	41,777.01	41,220.47
Cost of Production (IDR / bird)	31,760.05	37,098.61	39,897.49	38.647.17
Profit (IDR / bird)	2,741.19	2,830.74	1,879.52	2,573.30
NPV (with interest of rate 9.4%)	84,401,849.17	160,849,464.17	147,166,303.87	111,198,188.87
IRR	30.87	32.98	38.38	33.36

The investment cost is the cost incurred by the farmer for making the cage and equipment. The larger the business scale, the greater the investment required. In calculating production costs, investment costs will be calculated as fixed costs [7]. When viewed from the need per livestock, the larger the business scale have the lower the fixed cost per head [8]. The investment cost per livestock for small, medium and large enterprises are IDR 5,760.06; IDR 5,422.82 and IDR 4,855.73 respectively.

The contract prices received by farmers ranged from IDR 17,438.00 to IDR 17,937.55 for each period. With this average price level, the income received varies depending on the live weight of the livestock. The results of the analysis showed that small, medium and large-scale breeders received revenues of IDR 34,501.24 ; IDR 39,929.35; and IDR 41,777.01 respectively. Although large-scale breeders receive the highest revenue, they also incur the highest production costs so that the profits they receive are the smallest. Production costs per head for small, medium and large-scale breeders are IDR 31,760.05; IDR 37,098.61 and IDR 41,777.01 respectively. Profits obtained from tips of livestock raised in small, medium and large businesses are IDR 2,741.19; IDR 2,830.74 and IDR 39,897.49.

The feasibility analysis shows that an interest rate of 9.4% gives a positive NPV and an IRR above the reference interest rate with an average of 33.36%. This shows that the broiler chicken farm is feasible to run with an average contract price of IDR 17,845.30.

Table 3. Sensitivity Analysis: 3% price fall

	Low (<4999)	Midle (5000-7499)	High (> 8000)	Average
Contract price (decreased by 3%)	17,397.21	16,959.00	17,399.42	17,309.94
Revenue (IDR / year)	33,468.35	38,733.64	40,526.89	39,986.618
Cost of production (IDR / year)	31,760.05	37,098.61	39,897.49	38,647.18
Profit (IDR / year)	1,708.30	1,635.027143	629.40	1,339.44
NPV (9.4%)	13,179,820.48	23,706,791.42	-159,176,768.69	-16,313,493.35
IRR	24.15	25.25	-	7,57

Sensitivity analysis is conducted to see changes in business feasibility when the contract price decreases by 3%. The results of the analysis show that when the contract price drops by 3% to IDR 17,309.94, the average NPV becomes negative, and the IRR becomes 7.57%. This price fluctuation greatly affects business feasibility [9], especially for large scale breeders. The results of the NPV calculation on this scale are negative, even though the breeders still get profits, this profit cannot cover the investment costs.

In contrast to the small and medium scale businesses, a 3% reduction in contract prices still results in a positive NPV and an IRR above the reference interest rate. This occurs because large-scale breeders experience a decrease in revenue while production costs remain unchanged. This result is different from previous research which states that a 5% change in price still gives a positive NPV and an IRR above the reference interest rate and is still worth working on [10].

4. Conclusion

Broiler chicken farming is very sensitive to changes in selling prices. Any change in contract prices will greatly affect business feasibility and business sustainability. This very sensitive nature of the business threatens the business continuity in the future. A mechanism is needed to regulate contractual agreements between breeders and partner companies so that there is certainty in prices that can ensure the continuity of the breeders' business.

References

[1] Mastuti S, E Yuwono, R Widiyanti, NN Hidayat dan L Setiana. 2020. Analisis Trend Populasi dan Zooning Pengembangan Ternak Ayam Niaga Pedaging di Kabupaten Banyumas in Proc. Seminar Teknologi dan Agribisnis Peternakan VII-Webinar: Purwokerto 27 Juni 2020. ASBN:978-602-52203-2-6 (in Indonesian with abstract in English).

[2] Haryuni N and Z Fanani. 2017. Study of Feasibility on Broiler Business Development. Journal of Development Research. 1(2):63-67.

[3] Kumar M and SR Torane. 2019. Financial Feasibility of Broiler Production in Case of Non Contract and Contract Farms. Economic Affairs. 64(2): 395-400.

[4] Mendes AS, DC Gudoski, AF Cargnelutti, EJ Silva, EH Carvalho and GM Morello. 2014. Factors that Impact the Financial Performance of Broiler Production in Southern States of Parana Brazil. Brazillian Journal of Poultry Science. 16 (1): 113-120.

[5] Al-Sharafat A and TM Al-Fawwaz. 2013. Economic Analysis of Different Broiler Farm Capacities: A CAse Study of Jordan. International Journal of Business and Mangement. 8(5):41-47.

[6] Firdaus M and L Komalasari. 2010. Feasibility Analysis of Integrated Broiler Production. Media Peternakan. 33 (3):182-188.

[7] Balamurugan V and M Manoharan. 2013. Cost and benefit of investment in integrated broiler farming-A case study. International Journal of Current Research and Academic Review. 2(4):114-123.

[8] Singh VP, VK Sharma, MS Sidhu and HS Kingra. 2010. Broiler Production in Punjab. Agricultural Economics Research Review. 23 (July-December): 315-324.
[9] Amalia FR, A Daryanto and H. Rujito. 2015. Comparative Feasibility Analysis of Modern and Traditional Systemof Broiler Chicken Farm Business. Indonesian Journal of Business and Entrepreneurship. 1(2): 90-95.
[10] Gandhy A and D Sutanto. 2017. Analisi Finansial dan Sensitivitas Peternakan Ayam Broiler PT Bogor Eco Farming Kabupaten Bogor. Jurnal Optima. 1(10):1-11.

Improving genetic quality of local livestock for achieving food security

E Kurnianto

Universitas Diponegoro, Semarang, Indonesia

*kurniantoedy17@gmail.com

Abstract. Indonesia has local livestock breeds, both ruminant and non-ruminant. The existence and preservation of those animals is very important. The government, the concerned parties and farmers of community member are required to be responsible for maintaining the productivity of livestock. The quality genetic could be improved by applying breeding and selection. There are three efforts can be used to improve the livestock population and productivity, those are artificial insemination, embryo transfer, and separation technology of spermatozoa X and Y. The objective to be achieved is to help meet the need of food, especially animal protein, of the domestic community, in addition to the welfare of the community.

1. Introduction

Mutual attention between government, concerning parties and the communities is very important in preserving livestock. The attention of the three parties becomes strong in animal conservation effort. The conservation is carried out for the livestock that are feared to be extinct due to uncontrolled use. The community's need for nutritious food demands the fulfillment of both qualitative and quantitative need. Improper government import taps can decline the number of local livestock to cover the needs of these communities. Within the different species used for food production, only a few breeds are developed towards high quality breed fitting in high input systems. Conservation of farm animal genetic resources become important not only for their biological role, but also for their economic, socials and cultural values [1].

It is known that the performance of livestock is influenced by genetic and environmental factors. To obtain the superior livestock, selection program should be done. Selection of livestock must be supported by a good environment. No matter how good the genetic quality of the livestock is, if it is not supported by a good environment, then the performance is not optimal [2] [3]. The high quality of livestock will lead to good performance and productivity. Thus, it can help meet people's nutritional needs.

2. Discussion

2.1 Livestock genetic resources

Animal Genetic Resources, hereinafter referred to as AGR are animals or their genetic material, but not including fish of their genetic material, which contain units that function as carries heredity trait, both with actual and potential value, which can be used to create creates breeds. We should know the differences about origin, local and introduced or exotic genetic resources. It has been stated in the

Government Regulation of Republic of Indonesia No 48 of 2011 that original livestock genetic resource is livestock genetic resource originated purely from Indonesia.

Then, the local livestock genetic resource is livestock genetic resource resulting from a cross or introductions that have been adapted and reproduced in their environment. Meanwhile, introduction (exotic) livestock genetic resource is livestock from abroad, either already or that has not been proven to adapt to the environment in Indonesia. Livestock Breed is a group of livestock which has distinctive phenotypic characteristics, and these characteristics can be inherited to their offspring.

Furthermore, a livestock line is a group of individual livestock in one family which has certain characteristics which are used for breeding or breeding purpose. Genetic diversity can be understood more clearly as presented in Figure 1 [4].

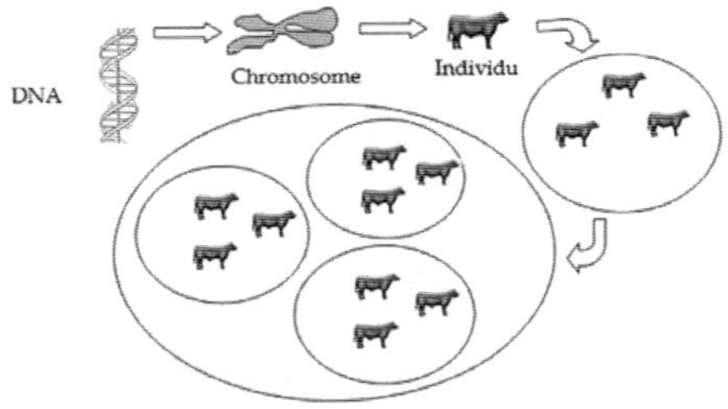

Figure 1. Genetic diversity between individuals in a population or between populations within species that start from differences in gene structure.

Determination of line animal genetic resource is government recognition of animal genetic resources that already exist in a germplasma source of livestock breed area that has been raised by breeders from generation to generation and belongs to the community.

2.2 Conservation

An integration of the local animal and introduction animal would be beneficial to breeding between those animals for create a new breed. When kept is live animal, more than one location is needed. Indonesia has many species of local livestock, both ruminant and non-ruminant. The effort of researchers from both research institutes and universities are continuously made to maintain the existence of existing livestock and seek recognition of new breeds of livestock. The performance of local livestock in several breeds is not inferior to introduced livestock. For example, some reproductive traits of local livestock are better than those of (Bos taurus and Bos indicus). Thus, there is a compelling reason to breed and preserve it.

Livestock conservation must be conducted within the livestock own environment [2]. A sufficient population size for each genotype is necessary. It has been known that the preserved genotype will always be the result of combined effects of genetical and environmental variables, the latter being peculiar to the breeding environment. The breeding environment where the livestock has evolved or environments that are thought to be similar with the original one will enable them to live and reproduce.

One way to conserve the genetic resources of livestock is by conservation of individual either in situ or ex situ. In situ conservation or on-farm conservation might be the appropriate way for local animal to be conserved whereby populations are raised by the farmers in their own environment. The most effective on-farm conservation is that which is nationally promoted and combines elements of existing national programs.

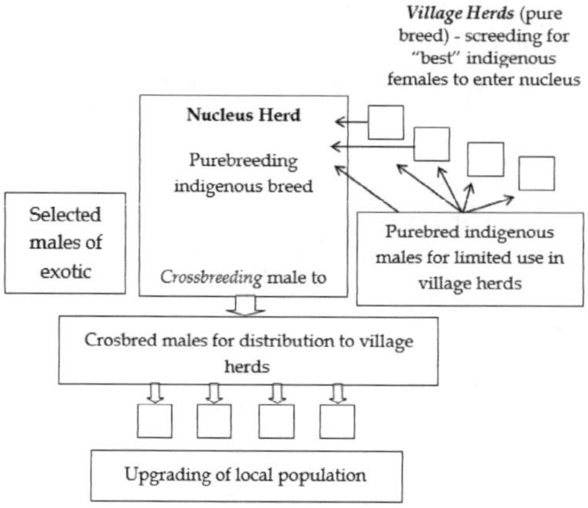

Figure 2. Open nucleus herd breeding scheme – Base for conserving a local breed and upgrading local population [5]

A local livestock breeding program is basically no different from a typical livestock breeding program. To achieve success in the livestock improvement program, at least 4 components must be fulfilled [6], those are 1) A data recording system, 2) Methods and tools to estimate the value of breeding, 3) A system for evaluating livestock to be used as parent to produce the next generation through selection and mating system, and 4) Structure to transmit genetic advantage from the breeding population to the production population. The specialty in improving local livestock is that the purity of the breed must be maintained through in situ conservation efforts. Crosses may be done ex situ and only for economic purposes. It was suggested by [7] [8] that the steps in preserving and developing local livestock through purification of the breed are: 1) Identification of livestock population, 2) Determination of genetic traits, 3) Determination of production parameters, 4) livestock breeding in situ, and preservation of ex situ genetic materials. Figure 2 present open nucleus herd breeding scheme – base for conserving a local breed and upgrading local population.

As presented in statement above that the parties are responsible in preserving the local animal/ livestock. Figure 3 present the mutual collaboration among them. Determination of line animal genetic resource is government recognition of animal genetic resources that already exist in a germplasma source of livestock breed area that has been raised by breeders from generation to generation and belongs to the community. In situ conservation is very useful for maintaining the presence of local livestock in the local area that is already suitable for the life of this livestock. Development of ex situ breeding program for phenotypic characteristics can create o long-term and sustainable conservation program.

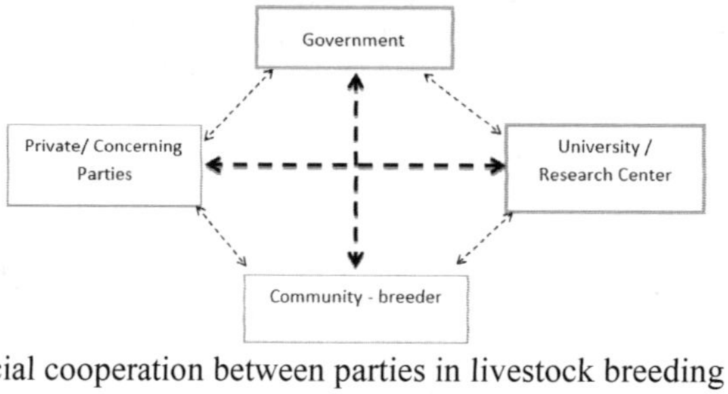

Figure 3. Beneficial cooperation between parties in livestock breeding and preservation

Livestock preservation efforts are carried out by the government in accordance with its authority who cares about the existence and development of livestock. Components of government, society and concerned parties who have succeeded in developing certain livestock bred are proud of the wealth of the area. Livestock commodities and the number of breeds that have been determined by the Minister of Agriculture is presented in Table 1.

Preservation of animal genetic resources is supervised by the government. Of course, the community is a component of actors in the field who supports this government program. Geographical distribution and population of both males and females should be detected in an inventory of livestock genetic resources that are indicated too unsafe. Management of genetic resources is carried out in stages starting from the national to provincial and district level.

Table 1. The Distribution of local livestock decided by Minister of Agriculture

Animal	The Number of Breed
Sheep	≥ 7
Goat	≥ 10
Sapi	≥ 11
Buffalo	≥ 8
Horse	≥ 4
Itik	≥ 13
Ayam	≥ 8
Total	≥ 61

2.3 Scope of use

Livestock genetic resources are utilized in scope of rearing and breeding. The process of raising local and original livestock must be based on the welfare of the animal. The existence of livestock raised by the farmer, community or institution remains under government supervision. The government provides certain incentive to people who successfully breed local livestock. The community is prohibited from breeding livestock businesses which can reduce the existence of native and local livestock.

Animal genetic resources and the breeding arrangement is aimed to:

a. Guarantee the preservation and use in a sustainable manner
b. Ensuring the maximum and sustainable availability of qualified germplasma
c. Providing data on genetic resources of livestock.

2.4 Testimony

Improving the genetic quality of local livestock is carried out programmatically through cooperation of various component. This quality improvement is for local livestock whose less secure status, which has high social value for genetic diversity, and has high economic value.

In the context of improving the quality and breeding of the black Kejobong goat, in the involvement of the related parties is a must. Figure 4 presents a scheme of effort to achieve the target of qualified black Kejobong goat. Assistance efforts by the sub-service for animal husbandry at Purbalingga Regency Agriculture Service are always carried out through field workers. Meanwhile, community service activities in the form of counseling are also often done, including by university. The most important thing and key word is coordination and communication between parties, so that the contribution of each party is clear and does not overlap.

2.5 Improving genetic quality

Genetic diversity is potentially threatened by many factors, those were changes in producer preferences of socio-economic factors, a range of disaster (drought, diseases epidemics, etc.). Statement of [1] said that the development of technology could change in decline of diversity. It is especially in the incorrect use of technology, for example the use of artificial insemination using frozen semen of introduction breed to native livestock reding the performance of these native livestock. We must be careful in using technology. The government should give appreciation or incentive to farmer groups for maintaining the existence of local livestock.

The awareness of the significance of conservation and sustainable use of AGR is often limited at the policy level. There is an increased emphasis on biotechnology, and less attention is paid to more holistic

IOP Conf. Series: Earth and Environmental Science 1041 (2022) 012081 doi:10.1088/1755-1315/1041/1/012081

breed improvement activities involving design of breed program, establishment, support and testing of alternative, and the involvement of local farmers [9].

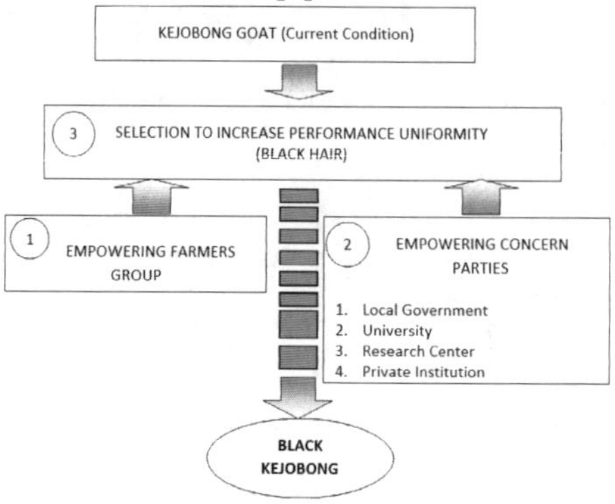

Figure 4. An example of achieving qualitative of kejobong goat thought empowering

Breeding can be implemented if the limited population is:

a. Breeding to increase the population.
b. Conducting selection to select good-looking and high-productivity livestock. If the diversity decreases, then mating with other animals that are not related to increase the diversity, so that selection can be carried out again.
c. In situ conversation.
d. Development of conservation and livestock conservation are (ex situ).

The requirement for improving genetic quality is:

a. Data recording for qualitative and quantitative traits.
b. Molecular data (blood plasma gene, DNA).
c. Data analysis regarding molecular gene expression to phenotype.
d. Controlled mating system within breeds.
e. Selection program for obtaining good livestock having high productivity.

At the community level, the application of technology to accelerate population increase is artificial insemination (AI) growth is technology. Meanwhile, at the government level, embryo transfer is more appropriate. By using artificial insemination, one cow produces one calf. However, in embryo transfer, one cow can produce 17 embryos than can be deposited in recipient parent [10]. The other method for increasing livestock population is separation technology of spermatozoa X and Y. The presence of spermatozoa in formation process of sex has an important meaning to obtain a certain sex [11].

2.6 Animal product – nutritional need

The National Forum for Food and Nutrition (Widya Nasional Pangan dan Gizi) XI 2018 formulated a number of recommendations, one of which is nutritional adequacy rate as a national reference. It was stated for this forum that improving public nutrition and increasing food safety and quality insurance are two very important things.

It was stated in Regulation of Minister of Health of the Republic of Indonesia Number 28 of 2019 concerning the recommended nutritional adequate rate that the protein need that should be met for the people of Indonesia per capita per day aged 10-70 years old for men are 50-75 g, and women are 55-65 g. In 2019, the need for animal protein from livestock has been met [12]. Indonesian people have many choices, such purebred and local chicken, duck meat, mutton-lamb, and beef/buffalo. Meat originating from local livestock only fulfills part of the community's needs.

During the last 5 years, the national consumption of beef per capita has reach 2.6 kg/year or 7.12 g/day, in which it was 64% supplied locally, equivalent to 2,500,000 live cows, while 36% are imported, equivalent to 1,450,000 live cows. It means that the domestic animal population and productivity should be increased because of still many importations. The government is fully responsible for this problem. Nationally, the special effort for cows to be required for pregnancy (UPSUS SIWAB) program from 2017 to the present has shown a very encouraging realization of more than 10 million of acceptors or 105.5%, pregnancies of 76.3% and calve birth of 71.9%. If the livestock by farmers are of good quality, then the livestock productivity will be higher.

References

[1] Oldenbroek, K. 2007. Utilization and Conservation of Farm Animal Genetic Resources. The Netherlands, Wageningen Academic Publishers. Tisdell, C. 2003. Socioeconomic causes of loss of animal genetic diversity: analysis and assessment. Ecological Economics. 45(3):365-376.
[2] Falconer, D.S. and T.F.C. Mackay. 1997. Introduction to Quantitative Genetics. Fourth Ed. Longman.
[3] Lynch, M. and B. Walsh. 1998. Genetics and Analysis of Quantitative Traits. Sinauer Association, Inc.
[4] Bagley, M.J., S.E. Franson, S.A. Christ, E. R. Waits and G.P. Toth. 2002. Genetic Diversity as An Indicator of Ecosystem Condition and Sustainability: Utility for Regional Assesments of Stream Condition in the Eastern United States. EPA/600/R-03/056, September 2002. U.S. Environmental Protection Agency, Cincinnati, OH.
[5] Philipsson, J., J.E.O. Rege and A.M. Okeyo. 2006. Sustainable breeding programmes for tropical farming system. In: Animal Genetics Training Resource, Version 2. (eds. J.M Ojanggo, B. Malmfors and A.M. Okayo). International Livestock Research Institute, Nairobi, Kenya, and Swedish University of Agricultural Sciences, Uppsalla, Sweden.
[6] Dekkers, J.C.M, J. P. Gibson, P. Bijma and A.M. van Arendonk. 2004. Design and Optimisation of Animal Breeding Progammes. Lecture Note, Iowa State University.
[7] FAO . 2007. The state of the World's Animal Genetic Resources for Food and Agriculture. Food Agriculture Organization of the United Nations. Rome, Italy.
[8] Da Silva Mariante, A. and A.R. de Bem. 1992. Animal Genetic Resource Conservation Programme in Brazil. Animal Genetic Resource Information No. 10
[9] FAO. 2004. Conservation strategies for animal genetic resources, by D.R. Notter. Background Study Paper No. 22. Commission on Genetic.
[10] Subchan, F.A., R. Handarini dab S.W. Siswanti. 2016. Perbedaan waktu penyuntikan hormone FSH terhadap respon superovulasi sapi Angus. Jurnal Peternakan Nusantara. 2(2):159-166.
[11] Sophian, E. and F. Sofiati. 2016. Peranan bioteknologi reproduksi dalam peningkatan kualitas ternak. Biotrends. 7(1):42-47.
[12] TROBOS, Media Agribisnis Peternakan. 2019. Terpenuhi, Kebutuhan Protein Hewani dalam Negeri. Trobos Online.

Current goat milk production, characteristics, and utilization in Indonesia

J Sumarmono

Universitas Jenderal Soedirman, Purwokerto, Indonesia.

*juni.sumarmono@unsoed.ac.id

Abstract. In Indonesia and some southeast Asian countries, the number of goats raised for milk production is growing. In these countries, goat milk is the second most produced and consumed milk after cow milk. In Indonesia, dairy goats' development is promising, but the farming of dairy goats and marketing milk products is still scattered and unstructured. The milk is produced from varieties of goat breeds, including Peranakan Etawa, Saanen, Nubian, and their crosses. Goat milk has unique and distinct characteristics and is an excellent source of nutrients for humans. Compared to cow milk, it has smaller fat globules, more homogenous, shorter rennet coagulation time, and softer curd. Goat milk utilization includes fluid milk for direct consumption; frozen fresh milk; dried or powdered milk; fermented milk products such as yogurt, kefir, including its derivatives such as cosmetics, concentrated yogurt, yogurt cheese, ice cream, shakes; cheeses such as fresh cottage-type cheese, acid-coagulated cheese and mozzarella; and also, other traditional products such as sweets & candies and caramels. The development of dairy goats and goat milk consumption is driven by several factors, including the low cost for start-up and production, consumers' demand, alleged health properties of goat milk, and varieties of products.

1. Introduction

Indonesia's rising population needs a sufficient supply of food, both plant-based and animal-based food. By June 2020, the total population has reached 268.583.016 people. Concerning milk-based foods, in 2019, data from National Statistical Bureau (BPS) showed that milk consumption in Indonesia was 16.23 kg/capita/year, with a national demand of 4.3 million tons per year for milk and milk products. The overall production of fresh milk was 996,442 tons, which accounted for just 22 percent of the total demand. Cow milk predominates in the domestic production of fresh milk.

Global production of non-bovine milk reached 133 million tons per year, representing more than 17 percent of all milk production. Goat milk production is projected to be 12.2 metric tons per year or around 2 percent of total milk production [1]. Although the marketing and distribution system is less structured than in European countries, most goat milk production is approximately 52 percent, and consumption is in Asia. Asian countries with largest producers of goat milk are India, Bangladesh, and Pakistan; African countries are Sudan, South Sudan and Somalia; and European countries are France, Spain, and Greece [2].

Following the continuous rise in dairy goats' global population, milk demand will continue to increase [3]. In many Southeast and East Asian countries, the need for high prices of goat milk is encouraged by the belief that goat milk has health and medicinal properties [4]. The lower cost also contributes to the increased interest in dairy goat farming for start-up and production compared to dairy cattle farming.

Goat milk is considered to have lower allergy potential but more excellent digestibility compared to cow milk. Furthermore, greater interest is given to the existence of health-promoting compounds in goat milk. Goat milk can be processed as a carrier for functional ingredients, such as prebiotic ingredients or probiotic bacteria, into various products [2].

2. The fourth industrial revolution (4IR)

Sustainable dairy goat farming has been developed in some Asian countries, including Indonesia, Malaysia, and Thailand, in partnership with plant-based agro-industries such as palm oil, cocoa, coffee, and rubber plantations [3]. Over the past year, Indonesia's total goat population has increased by more than 600 thousand heads, reaching 18,975,955 heads by the end of 2019. Peranakan Etawa (PE), Saanen, Anglo Nubian, dan their crosses (Saanen x PE or Anglo Nubian x PE) are well-known goat breeds for milk production by farmers [5]. Farmers mainly raised Peranakan Etawa and Jawarandu goats for meat production, but many farmers milked the does for around 0.5 to 1.5 liters per day (Picture 1).

Dairy goats (dual purposes) and goat milk are raised and produced in several different areas in Central Java province, including Banyumas, Cilacap, Purworejo, Sragen, and Tegal. In Baturraden, Banyumas District, Central Java Province, there is a National Breeding Centre for Peranakan Etawah and Saanen Goats. Goat milk is also produced by farmers in the provinces of West Java (Bandung, Sumedang, Sukabumi, Garut), Yogyakarta (Sleman, Kulonprogo), East Java (Malang, Blitar, Banyuwangi, Lumajang), Padang (Nagari Bukik Batubuah), Lampung (Kota Metro), North Sumatra (Asahan), and Aceh (Banda Aceh). However, there is no official data available on the total domestic goat milk production.

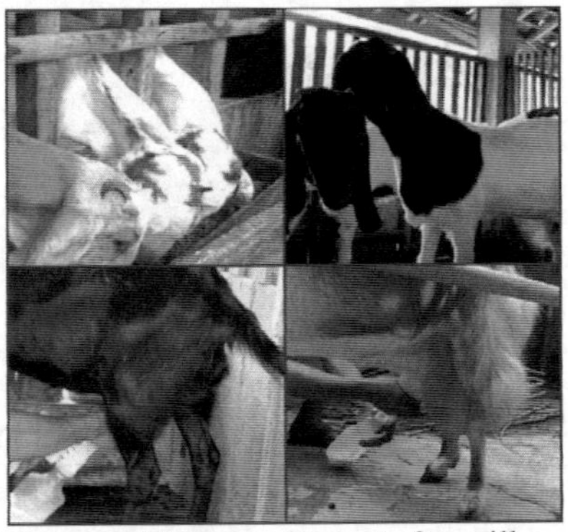

Figure 1. Peranakan Etawa and Jawarandu goats for milk production in Indonesia.

Milk production varies greatly, from as low as 0.5 L/d to more than 2 L/d. On average, goat milk production in Indonesia is 1.06 L/d or 191.8 L per lactation (Table 1), which is relatively low compared to the milk production of dairy goats in other countries. The average milk production of goats in Mexico is reported at 1.5 L/d [6], and Saanen goats in Brazil is documented at 3.21 kg/d [7]. As in other dairy animals, many factors influence goat milk production such as breed, feed and feeding, the extent of lactation, and the shape of udder [5, 8-10].

Table 1. Milk production of some goat breeds in Indonesia

Breeds	Average Milk Production (L/d)	Peak Production (L/d)	Total Milk Production (L)*
Peranakan Etawah	0.86	1.22	154.3
Peranakan Etawah	0.70	n/a	126.1
Sapera	1.47	2.19	264.6
Anglo Nubian	1.19	1.98	214.2
Senduro	1.09	n/a	196.6
Average	*1.06*	*1.80*	*191.8*

3. Milk Composition

Undoubtedly, goat milk is a rich source of nutrients such as carbohydrates, fats, proteins, minerals, and vitamins, vital for normal physiological processes and muscle building in humans. This is why goat milk has been a significant part of human's diets for a long time. Goat milk is used as a complementary food to reduce malnourished children in Indonesia [13] and in other parts of the world [14]. Goat milk provides protein, fat, and other nutrients similar to cow milk (Table 2).

The average composition of fresh goat milk in Indonesia is presented in Table 3. On average, fresh goat milk contains 13.51 percent total solids, 3.2 percent proteins, 4.77 percent fats, 4.96 percent lactose, and 0.80 percent minerals or ash. As reviewed by [2], casein is a primary protein in goat milk (74%), followed by whey protein (17 percent) and non-protein nitrogen component (9 percent). Variation in the relative proportion of milk protein can be attributed to genetic polymorphisms within goat populations. Goat milk is also reported to have a higher proportion of small-sized casein micelles than cow milk, leading to the better digestibility of fresh goat milk and its products.

Table 4 shows that the predominant fatty acids in goat milk are saturated fatty acids (SFA). The primary fatty acids in goat milk were stearic and palmitic, and oleic acid was the primary unsaturated fatty acid. In contrast, linoleic acid was the primary polyunsaturated fatty acid, and capric acid was the principal medium-chain triglyceride [15]. Various factors affect goat milk composition, such as genetics, stage of lactation, season, type of diet, and health [16].

Table 2. Comparative composition of goat and cow milk

Composition	Goat Milk		Cow Milk	
	[17]	[18]	[17]	[18]
Protein (g/100 g)	3.48	2.85	2.82	3.27
Fat (g/100 g)	5.23	3.58	3.42	3.49
Lactose (g/100 g)	n/a	4.13	n/a	4.52
Casein (g/100 g)	n/a	2.14	n/a	2.55
Saturated fatty acids (mg/100g)	3683.10	n/a	2436.41	n/a
Polyunsaturated fatty acids (mg/100g)	213.25	n/a	109.32	n/a
Total essential amino acids (mg/100g)	1639.84	n/a	1298.36	n/a
Total non-essential amino acids (mg/100g)	1840.15	n/a	1522.58	n/a

Table 3. Goat milk composition of several breeds of Indonesian goats

Components	Peranakan Etawah [19]	Peranakan Etawah x Anglo Nubian [20]	Peranakan Etawah x Saanen [20]	Average
Total solids, %	13.67	13.17	13.68	13.51
Protein, %	3.60	2.98	3.02	3.20
Fat, %	4.42	4.88	5.01	4.77
Ash, %	0.80	n/a	n/a	0.80
Lactose %	n/a	4.97	4.94	4.96

Table 4. Fatty acids composition of goat milk

Fatty Acids	Organic Farming [21]	Conventional Farming [21]	Conventional Farming [15]
Caprylic acid (C8:0)	1.53	1.04	1.76
Capric acid (C10:0)	3.69	2.66	5.07
Lauric Acid (C12:0)	5.51	4.07	1.89
Myristic acid (C14:0)	7.49	6.01	4.64
Palmitic acid (C16:0)	24.68	24.89	15.71
Stearic acid (C18:0)	19.76	24.87	0.69
Palmitoleic acid (C16:1)	0.39	0.49	12.14
Oleic acid (C18:1)	29.9	31.44	19.46
Linoleic acid (C18:2)	2.62	3.7	3.81
Linolenic acid (C18:3)	0.48	0.33	0.62

Saturated fatty acids (SFA)	62.67	38.66	45.06
Monounsaturated fatty acids (MUFA)	30.29	31.93	20.24
Polyunsaturated fatty acids (PUFA)	3.09	4.03	4.73
Unsaturated fatty acids (UFA)	33.38	35.96	24.97
PUFA/SFA	0.05	0.11	0.10
UFA/SFA	0.55	1.04	0.55

4. Milk Characteristics

Fresh milk from Peranakan Etawah goats has a pH between 6.4 to 6.6, with a specific density between 1.0278 to 1.0289 [22]. Goat milk is also characterized as having goaty flavor or "prengus" which due to particular flavor components. Too strong goaty flavor limits the acceptability of goat milk to consumers [23].

A recent market survey shows that most consumers (68.4 percent) in Medan city, North Sumatra have a positive understanding of goat milk in terms of nutritional content and aroma and are thus happy to drink goat milk. On the contrary, 31.6 percent of consumers have a negative opinion and have a low preference for goat milk [24].

To some extent, goat milk differs from cow milk and is known to have unique characteristics [25-28], which can be summarized as follow:

a. Goat milk shows higher digestibility, alkalinity, buffering capacity than cow milk
b. For individuals with cow milk allergies and lactose sensitivity, goat milk is a better option.
c. Goat milks differ in the proportions of the four significant caseins (αs-1, αs-2, β, and κ), and the micellar size, which is related to genetic polymorphisms,
d. Goat milk yogurt gel is weaker than cow's milk gel, which can be attributed to low casein content, α-casein ratios, and micellar size.
e. Heat stability of goat milk is considerably lower than cow milk, which may relate to the high content of ionic calcium and low micellar solvation in goat milk, hence affecting thermal instability
f. The occurrence of genetic polymorphisms in goat milk proteins causes differences in the proportions of caseins α-s1 and α-s2 among individuals and breeds of goats

In addition, a report showed that goat milk products have different textural properties compared to cow and sheep milk products [25]. Goat milk products have lower hardness, consistency, adhesiveness, extrusion force, stability, and higher syneresis than cow milk. A proteomics analysis demonstrates that goat milk is sensitive to heating. Heating increases the digestibility of goat milk proteins. The high-temperature short-time method has different effects on milk proteins than the use of the low-temperature long time method. [29].

5. Milk Utilization

In addition to liquid milk ready for consumption, there are various goat milk products, including powdered/dried milk, yogurt, kefir, cheese, and other products such as sweets, caramels, soap, etc. Most products are manufactured by households, small and medium enterprises, and also members of cooperative enterprises.

5.1 Liquid milk

The most prevalent use of goat milk is liquid milk prepared for consumption. Pasteurized goat milk is available in 250 mL bottles in different flavors, such as original, chocolate, and strawberry flavors [30]. It is also a common practice to sell fresh unpasteurized goat milk in frozen forms. Many goat farmers members of a cooperative combine production, processing, and marketing their own pasteurized goat milk products. Pasteurized goat milk is further processed into milkshakes of iced fruity milk. Marketing is done either by direct selling or online.

5.2 Powdered/dried milk

Fresh goat milk is made into powdered or dried milk in some regions [31, 32]. Small enterprises such as CV. Bumiku Hijau and CV. Sahabat Ternak, as well as members of cooperative enterprises in Sleman, the Special Area of Jogjakarta, manufacture powdered goat milk. The powdered goat milk produced requires 4 to 6 hours of pan drying of fresh milk with added sugar, followed by powder-milling of dried milk [32].

5.3 Fermented milk

One right way to expand dairy goat farming is to process fresh goat milk into functional food products, especially fermented products. Fermented milk products, such as yogurt and kefir, are widespread and popular in Indonesia. Goat milk yogurt is made by adding commercial yogurt culture and 4-6 hours fermentation. Drink- or stirred-type yogurt is more commonly accepted by consumers than set-type yogurt. Processing fresh goat milk into fermented milk products is usually done by micro and small enterprises. Yogurt can be further processed into a more nutrient-dense product known as concentrated yogurt [33].

Goat milk kefir is a result of 12 to 24 hours fermentation of goat milk by kefir microflora, either raw or pasteurized milk. Although kefir production is relatively new in Indonesia, it has gained enormous popularity in recent years. Kefir made from goat milk contains lactic acid bacteria and yeast of 7.17 ± 0.92 log CFU/ml and 6.76 ± 0.39 log CFU/ml, respectively [34]. Due to its perceived beneficial properties for wellbeing, kefir from goat milk has gained more recognition and market demand.

5.4 Cheese

To date, the commercial production of goat cheese in Indonesia is unknown. The majority of local cheeses is originated from cow milk. This might be a result of the scattered availability of goat milk and low market demand. Cheese production requires a higher amount of goat milk compared to other products such as yogurt and kefir. Nevertheless, the characteristics of soft cheese produced from Peranakan Etawah goat milk has been reported [35]. The cheese was characterized as having a soft texture with a pH of 4.6, total titratable acidity of 0.5%, and a mild goaty flavor. The composition of the cheese was moisture 57%, crude protein 15%, total fat 22%, and mineral 2.5%.

In addition, the cheese contained at least 108 probiotic lactic acid bacteria, which was added as the culture starter. In addition, Setyawardani [36] reported having successfully stored artisanal goat cheese containing probiotic lactic acid bacteria for 60 days. Interest to producing different varieties of cheeses is increasing, such as fresh cottage-type cheese, mozzarella, cream cheese, and acid-coagulated cheese. Increased goat milk production will eventually lead to an increased interest in the production of goat cheese, particularly the mozzarella cheese.

5.5 Other products

In addition to liquid milk, powdered milk, and fermented milk, there are several other goat milk products manufactured by households and members of cooperative enterprises, such as crackers [24], milk soap [37], sweets/candies [38], caramels [39], kefir soap [40], and kefir cream/ mask [41]. These products are mostly artisan or traditional products, but they significantly contribute to the increasing demand for and utilization of goat milk.

6. Conclusion

Goat milk is gaining more attention and premium price in Indonesia. Most goat milk is produced by dual purposes breed of goats, which in general have lower milk production than dairy goats. Goat milk is known to have distinctive characteristics, composition, and improved health properties than cow milk. In Indonesia and neighboring countries, the large markets for nutritious and functional foods, the health benefits of goat milk and its products, and the connection of goat milk with natural and sustainable agriculture can be used as a factor for the further growth of goat milk and goat milk production.

References

[1] C. S. Ranadheera, C. A. Evans, S. K. Baines, C. F. Balthazar, A. G. Cruz, E. A. Esmerino, M. Q. Freitas, T. C. Pimentel, A. E. Wittwer, and N. Naumovski. 2019. Probiotics in goat milk products: Delivery capacity and ability to improve sensory attributes. Comprehensive reviews in food science and food safety, 18(4): 867-882. https://doi.org/10.1111/1541-4337.12447.

[2] S. Verruck, A. Dantas, and E. S. Prudencio. 2019. Functionality of the components from goat's milk, recent advances for functional dairy products development and its implications on human health. Journal of functional foods, 52: 243-257. https://doi.org/10.1016/j.jff.2018.11.017.

[3] B. A. Miller and C. D. Lu. 2019. Current status of global dairy goat production: an overview. Asian-Australasian Journal of Animal Sciences, 32(8): 1219. https://dx.doi.org/10.5713%2Fajas.19.0253.

[4] J. B. Liang and P. Paengkoum. 2019. Current status, challenges and the way forward for dairy goat production in Asia–conference summary of dairy goats in Asia. Asian-Australasian Journal of Animal Sciences, 32(8): 1233. https://dx.doi.org/10.5713%2Fajas.19.0272.

[5] S. Rusdiana, L. Praharani, and S. Sumanto. 2015. Kualitas dan produktivitas susu kambing perah persilangan di Indonesia. Jurnal Penelitian dan Pengembangan Pertanian, 34(2). http://dx.doi.org/10.21082/jp3.v34n2.2015.p79-86.

[6] A. M. Galina, J. Pineda, R. I. H. Piedrahita, P. Vázquez, G. Haenlein, and J. Olmos. 2019. Effect of Grazing on the Fatty Acid Composition of Goats Milk or Cheese. Advance Dairy Research, 7(3): 1-8. DOI: 10.35248/2329-888X.19.7.227.

[7] L. F. P. Soares, A. CorrÊA, M. Nunes, A. F. D. Souza, F. F. R. D. Carvalho, M. Maciel, and D. O. Vale. 2020. Milk production and the feeding costs of lactating Saanen goats fed diets containing spineless cactus. Revista Caatinga, 33(2): 550-554. http://dx.doi.org/10.1590/1983-21252020v33n227rc.

[8] Y. Y. Suranindyah, B. P. Widyobroto, and S. D. Astuti. 2020. Impact of Extended Lactation on Fatty Acid Profile and Milk Composition of Dual Purpose Tropical Goat. Pakistan Journal of Biological Sciences: Pjbs, 23(2): 113-118. https://doi.org/10.3923/pjbs.2020.113.118.

[9] D. N. Febriana, D. W. Harjanti, and P. Sambodho. 2018. Korelasi ukuran badan, volume ambing dan produksi susu kambing Peranakan Etawah (PE) di Kecamatan Turi Kabupaten Sleman Yogyakarta. J. Ilmu-ilmu Peternakan, 28(2): 134-140. http://dx.doi.org/10.21776/ub.jiip.2018.028.02.06.

[10] N. Ali, N. Munawarah, and N. Sofyan. 2017. Pengaruh pemberian ampas tahu terhadap produksi air susu dan pertambahan berat badan kambing Peranakan Etawa (PE). Jurnal SAINTEK Peternakan dan Perikanan, 1(1): 23-26. https://doi.org/10.31605/saintek%20perikan.v1i1.59.

[11] J. Palayukan, A. Murthadho, N. D. Putra, A. Furqon, A. Ridhowi, T. E. Susilorini, and S. Suyadi. 2020. Interaksi Antara Bentuk Ambing dan Paritas Terhadap Produksi Susu Kambing Senduro. TERNAK TROPIKA Journal of Tropical Animal Production, 21(1): 57-62. https://doi.org/10.21776/ub.jtapro.2020.021.01.7.

[12] S. D. Widyawati, R. F. Hadi, and A. Hanifa. 2019. The Effects of Linseed Supplementation in Ration on Milk Production and Quality of Lactating Ettawa Crossbreed Dairy Goats. IOP Conf. Series: Earth and Environmental Science, 292: 012004. https://doi.org/10.1088/1755-1315/292/1/012004.

[13] E. Sebtiarini, E. R. Dewi, and L. Dewanti. 2016. Goat Milk Utilization as Complementary Feeding of Children after Six Month to Reduce Undernourished Rate in Indonesia. Nursing and Health, 4: 24-28. DOI: 10.13189/nh.2016.040202.

[14] G. F. W. Haenlein. 2017. Why Does Goat Milk Matter?-A Review. Nutrition & Food Science International Journal, 2(4): 87-90. DOI: 10.19081/NFSIJ.2017.02.555594.

[15] J. Sumarmono, M. Sulistyowati, and Soenarto. 2015. Fatty Acids Profiles of Fresh Milk, Yogurt and Concentrated Yogurt from Peranakan Etawah Goat Milk. Procedia of Food Science, 3(Elsevier): 216-222. https://doi.org/10.1016/j.profoo.2015.01.024.

[16] G. Getaneh, A. Mebrat, A. Wubie, and H. Kendie. 2016. Review on goat milk composition and its nutritive value. Journal of Nutritions and Health Science, 3: 401-409. https://doi.org/10.15744/2393-9060.3.401.

[17] P. Paengkoum, X. T. , S. Paengkoum, and C. Yuangklang. 2018. Feeds and feeding of dairy goat in Asia. in The 4th International Asian-Australasian Dairy Goat Conference 17-19 October 2018. Tra Vinh, Vietnam.

[18] S. Stergiadis, N. P. Nørskov, S. Purup, I. Givens, and M. R. F. Lee. 2019. Comparative nutrient profiling of retail goat and cow milk. Nutrients, 11(10): 2282. https://doi.org/10.3390/nu11102282.

[19] T. W. Murti, G. R. T. Wijakangka, G. M. Wisudanto, and W. T. Nugraha. 2018. Physico-chemical and mineral content of goat milk products in Sleman district of special region of Yogyakarta, Indonesia. in The 4th International Asian-Australasian Dairy Goat Conference, 17-19 October 2020. Tra Vinh, Vietnam.

[20] L. Praharani, S. Supriyati, and R. Krisnan. 2015. Milk Quality of Anglo Nubian X Etawah Grade Goats and Saanen X Etawah Grade Goats at First Kidding Period. in The 6th International Seminar on Tropical Animal Production, 20-22 October. Universitas Gadjah Mada, Yogyakarta.

[21] V. Wanniatie, M. B. Sudarwanto, T. Purnawarman, and A. Jayanegara. 2019. Chemical compositions, contaminants, and residues of organic and conventional goat milk in Bogor District, Indonesia. Veterinary World, 12(8): 1218. 10.14202/vetworld.2019.1218-1224.

[22] N. Ratya, E. Taufik, and Arief, II. 2017. Karakteristik kimia, fisik dan mikrobiologis susu kambing Peranakan Etawa di Bogor. Jurnal Ilmu Produksi dan Teknologi Hasil Peternakan, 5(1): 1-4.

[23] S. S. Sakr. 2017. Reducing Goat milk Flavor by Beta–Cyclodextrin (β-CD) and its Effect on Texture and Microstructure in Soft Goat Cheese. Journal of Food and Dairy Sciences, 8(4): 179-183. https://dx.doi.org/10.21608/jfds.2017.38197.

[24] N. A. Siregar, A. T. Susanti, A. A. Nasution, S. H. Sitindaon, D. J. S. Siregar, N. Lubis, S. P. Ginting, and Y. L. Henuk. 2020. Consumer preference on milk crackers made from goat milk based on organoleptic test in Medan, Indonesia. IOP Conference Series: Earth and Environmental Science, 454: 012016. https://doi.org/10.1088/1755-1315/454/1/012016.

[25] V. B. da Silva and M. P. da Costa. 2019. Influence of Processing on Rheological and Textural Characteristics of Goat and Sheep Milk Beverages and Methods of Analysis, in Processing and Sustainability of Beverages, A M Grumezescu and A M Holban, Editors. Woodhead Publishing: United Kingdom. p. 373-412.

[26] Y. W. Park. 2017. Goat milk–chemistry and nutrition, in Handbook of milk of non-bovine mammals, Y W Park, G F W Haenlein, and W L Wendorff, Editors. Wiley Blackwell: Oxford, UK. p. 42-83.

[27] E. Moschopoulou, L. Sakkas, E. Zoidou, G. Theodorou, E. Sgouridou, C. Kalathaki, A. Liarakou, A. Chatzigeorgiou, I. Politis, and G. Moatsou. 2018. Effect of milk kind and storage on the biochemical, textural and biofunctional characteristics of set-type yoghurt. International Dairy Journal, 77: 47-55. https://doi.org/10.1016/j.idairyj.2017.09.008.

[28] Y. W. Park. 2007. Rheological characteristics of goat and sheep milk. Small Ruminant Research, 68: 73-87. https://doi.org/10.1016/j.smallrumres.2006.09.015.

[29] D. Chen, X. Li, X. Zhao, Y. Qin, J. Wang, and C. Wang. 2019. Comparative proteomics of goat milk during heated processing. Food Chemistry, 275: 504-514. https://doi.org/10.1016/j.foodchem.2018.09.129.

[30] N. Lailia, M. Rondhi, and D. Soejono. 2020. Analisis Rantai Pasok dan Strategi Pengembangan Susu Kambing Pasteurisasi di Goatzilla Farm & Cafe. Forum Agribisnis (Agribussines Forum), 10(March): 11-26. https://doi.org/10.29244/fagb.10.1.11-26.

[31] Y. Agustina. 2020. Analisis Pengendalian mutu bahan Baku Utama Susu Kambing Bubuk di Bumi Haijau Yogyakarta. Jurnal Teknosains, 9(2): 127-138. https://doi.org/10.22146/teknosains.42139.

[32] W. T. Nugraha. 2017. Gambaran Kondisi Wilayah dan Jalur Tata Niaga Susu Kambing di Kabupaten Sleman. Journal of Livestock Science and Production, 1(1): 24-30. http://dx.doi.org/10.31002/jalspro.v1i1.450.

[33] J. Sumarmono, T. Setyawardani, and A. H. D. Rahardjo. 2019. Yield and Processing Properties of Concentrated Yogurt Manufactured from Cow's Milk: Effects of Enzyme and Thickening Agents. IOP Conference Series: Earth and Environmental Science, 372: 012064. https://doi.org/10.1088/1755-1315/372/1/012064.

[34] T. Setyawardani and J. Sumarmono. 2015. Chemical and microbiological characteristics of goat milk kefir during storage under different temperatures. Journal of the Indonesian Tropical Animal Agriculture, 40(3): 183-188. https://doi.org/10.14710/jitaa.40.3.183-188.

[35] J. Sumarmono, T. Setyawardani, and S. A. Santosa. 2019. Effect of Storage Conditions on The Characteristics and Composition of Fresh Goat Cheese Containing Probiotics. Animal Production, 21(1): 56-63. http://dx.doi.org/10.20884/1.jap.2019.21.1.776.

[36] T. Setyawardani, J. Sumarmono, and K. Widayaka. 2019. Effect of cold and frozen temperatures on artisanal goat cheese containing probiotic lactic acid bacteria isolates (Lactobacillus plantarum TW14 and Lactobacillus rhamnosus TW2). Veterinary World, 12(3): 409. https://dx.doi.org/10.14202%2Fvetworld.2019.409-417.

[37] S. Akhiriani and N. M. Indrayani. 2018. Peningkatan Produktivitas Agroindustri Sabun Susu Kambing Etawa "AFINDA". Warta Pengabdian, 11(4): 121-138.

[38] Y. A. Widanti and S. Sutardi. 2020. Pendampingan pengembangan produk permen susu di desa Balerante Jawa Tengah. Jurnal Masyarakat Mandiri, 4(5): 1030-1039. https://doi.org/10.31764/jmm.v4i5.3242.

[39] R. W. Arief, N. Santri, and R. Asnawi. 2018. Pengenalan pengolahan susu kambing di kecamatan Sukadana

kabupaten Lampung Timur. Jurnal Teknologi & Industri Hasil Pertanian, 23(1). http://dx.doi.org/10.23960/jtihp.v23i1.45-56.

[40] P. D. Wulansari and D. Wijayanti. 2020. Karakteristik fisik, kimia dan mikrobiologi sabun kefir susu kambing. Jurnal Pangan dan Agroindustri, 8(3): 145-153. http://dx.doi.org/10.21776/ub.jpa.2020.008.03.4.

[41] F. Ferawati, E. Erpomen, Y. F. Kurnia, R. Reswati, and K. Khalil. 2018. Implementasi teknologi fermentasi susu kefir pada peternakan sapi perah di kota Payakumbuh. Jurnal Hilirisasi IPTEKS, 1(4): 297-305.

IOP Conf. Series: Earth and Environmental Science 1041 (2022) 012083 doi:10.1088/1755-1315/1041/1/012083

The impact of agricultural waste on river water quality of kreo watershed in Semarang city

W Setyaningsih[1,2], R S Sanjaya[1*]

[1]Universitas Negeri Semarang, Semarang, Indonesia
[2]Universitas Diponegoro, Semarang, Indonesia

*sanjayarois@gmail.com

Abstract. Agricultural waste, especially from the use of fertilizers and pesticides, has become a serious problem in several countries because it has caused a decrease in river water quality. The land use and livelihoods of the people in the Kreo watershed are oriented towards agricultural activities. This study aims to (1) analyze the pattern of agricultural land use in the Kreo watershed and (2) analyze the impact of agricultural waste on river water quality in the Kreo watershed. The method used is descriptive quantitative. Analysis of the impact of agricultural waste on river water quality was carried out by laboratory testing of river water samples that referred to the criteria for water quality standards based on class according to Government Regulation of the Republic of Indonesia Number 82 of 2001 for class I river designation. Laboratory tests of water samples used parameters TSS, TDS, Nitrate, BOD, and COD. While the analysis of the status of water quality using the STORET method. The results showed that the pattern of use of rice fields was clustered upstream and in the middle, farmland was clustered in the downstream and mixed gardens were spread out in the upstream, middle, and downstream areas. The status of river water quality in the upstream and middle parts of the Kreo watershed has been lightly polluted, while the downstream still meets the quality standards. The river water will be harmful to human health if it is intended as a raw source of drinking water.

1. Introduction

Indonesia is a country known as an agricultural country, but nowadays most of its agricultural land has turned into critical land due to the excessive use of chemical fertilizers and pesticides. Agricultural waste originating from the use of fertilizers and pesticides has polluted land, water and environmental resources, even around 85% of the water that irrigates rice fields in Java Island contains 54 mg/L nitrate or about 20% higher than the specified quality standard (Sutriadi and Sukristiyonubowo, 2013). Improper management of agricultural activities will have an impact on decreasing the quality of river waters (Berka et al., 2001).

Residues from agricultural activities will cause degradation of land and river waters (Lawniczak et al., 2016). This is because residues from agricultural activities contain various kinds of pollutants such as sediment, nitrogen, and pesticides which are carried along with surface runoff (Lenat, 1984). The high intensity of agricultural activities in an area can be reflected by the high concentration of nutrients in river waters (Heathwaite & Johnes, 1996). One way to identify the impact of intensive agricultural activities is from the high concentration of nitrate in river waters (Howden et al., 2013). The high concentration of nitrate comes from the excessive use of inorganic fertilizers or nitrogen fertilizers by farmers (Hamilton et al., 1993).

The Kreo watershed has a hilly topography and has experienced a decline in the quality of land and water resources caused by several triggering factors. The most dominant trigger factor is the influence of agricultural activities, namely the way farmers cultivate the land, high rainfall, and the expansion of agricultural land in protected areas (Setyaningsih et al., 2019). Land use and people's livelihoods in the Kreo watershed are still oriented to agricultural activities, especially in the upstream and middle parts (Muyassaroh and Juhadi, 2015).

Waste from agricultural activities from the use of fertilizers and pesticides is a non-point source of pollutants and is one of the causes of pollution in river water (Taylor et al., 2016). In addition, solid waste from agricultural activities such as plastic wrap for fruits, mulch, polybags and so on also affects the quality of river waters. According to Central Java Governor Regulation Number 156 of 2010 the allocation of river water use in the Kreo watershed is designated as a class I river, namely as a raw source of drinking water.

In addition, for the long-term plan the use of Jatibarang Reservoir water in the Kreo Watershed will be used as a source of clean water for Regional Drinking Water Company Semarang City (Hidayah et al., 2016). Therefore, studies related to monitoring the quality of river water in the Kreo watershed are very important to do. The purpose of this study was to (1) analyze the pattern of agricultural land use in the Kreo watershed of Semarang City and (2) analyze the impact of waste from agricultural activities on the condition of river water quality in the Kreo watershed in Semarang City.

2. Methodology

The location of the research was carried out in the Kreo watershed in the city of Semarang which is one part of the sub-watershed of the Garang watershed system. The topography of the area is a hilly area. Astronomically, the research location is located at 7o0'43" - 7o6'23" south latitude and 110o18'28" - 110o22'42" east longitude. While administratively the Kreo sub-watershed covers 3 sub-district administration areas which include Gunungpati District, Mijen District, and Ngaliyan District. The research location can be seen in Figure 1.

This research is descriptive quantitative research. Analysis of agricultural land use patterns using digitization techniques from Google Earth images recorded in December 2020 using GIS software. While the analysis of the impact of agricultural waste on river water quality is carried out by laboratory testing of river water samples which refers to the criteria for water quality standards based on class according to Government Regulation of the Republic of Indonesia Number 82 of 2001 for class I rivers. River water sampling technique refers to Indonesian National Standard 6989.57-2008 concerning Surface Water Sampling Methods and is based on the results of consultations at the Health Laboratory and Medical Device Testing Center of Central Java Province.

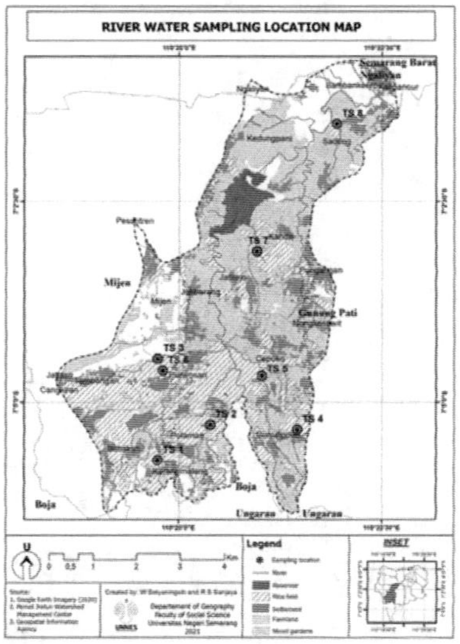

Figure 1. Research location and river water sampling

Data collection techniques through interviews with farmers, field observations, and documentation. A sampling of river water quality tests was carried out at 8 observation stations and the day after fertilization. Analysis of river water quality in this study used physical parameters (TSS and TDS) and chemical parameters (Nitrate, BOD, and COD). The location of river water sampling in this study is presented in Figure 1 and Table 1 above.

Table 1. Location of river water sampling

Sample number	Location		
	Latitude	Longitude	Village
1	7°5'43.35"S	110°19'44.27"E	Karangmalang
2	7°5'16.84"S	110°20'23.01"E	Polaman
3	7°4'27.91"S	110°19'44.42"E	Purwosari
4	7°5'20.51"S	110°21'26.93"E	Gunungpati
5	7°4'40.52"S	110°21'1.10"E	Cepoko
6	7°4'36.74"S	110°19'48.12"E	Purwosari
7	7°3'7.55"S	110°20'57.48"E	Kandri
8	7°1'32.44"S	110°21'56.35"E	Sadeng

Determination of water quality status in this study used the STORET method (Sulthonuddin et al., 2019). Based on the Decree of the Minister of the Environment Number 115 of 2003, the STORET method is one of the most commonly used methods to determine the status of water quality (Rintaka et al., 2019). In principle, the use of the STORET method is to compare water quality standards according to their designation (Afriani et al., 2021).

Determination of river water quality status of the Kreo watershed in Semarang City using the STORET method is carried out by assigning a score to each water quality parameter (physics, chemistry, biology). If the measurement results meet the water quality standard (measurement result quality standard) then it is given a score of 0. If the measurement result does not meet the water quality standard (measurement result > quality standard) then it is given a score as in Table 2. below.

Table 2. Determination of the value system to determine the status of water quality

Number of parameters	Score	Parameters		
		Physics	Chemical	Biology
< 10	Maximum	-1	-2	-3
	Minimum	-1	-2	-3
	Average	-3	-6	-9
≥ 10	Maximum	-2	-4	-6
	Minimum	-2	-4	-6
	Average	-2	-12	-18

Source: Decree of the Minister of Environment Number 115 of 2003

Then the negative number of all parameters is calculated, and the quality status is determined from the number of scores obtained using the value system. Meanwhile, the classification of water quality status based on the STORET method is divided into 4 classes which are presented in table 3. below.

Table 3. Classification of water quality using the STORET method

Class	Score	Information
A	0	Meet quality standards
B	-1 s/d -10	Lightly polluted
C	-11 s/d -30	Medium polluted
D	≥ -31	Heavily polluted

Source: Decree of the Minister of Environment Number 115 of 2003

3. Result and Discussion

3.1 Agricultural land use patterns

The total area of agricultural land in the Kreo Watershed in Semarang City is 2,577.79 hectares or 67.02% of the total area of the Kreo watershed in Semarang City. In general, the use of agricultural land in the Kreo watershed is classified into 2 forms of utilization, namely wetland agriculture in the form of rice fields and dry land agriculture in the form of farmland and mixed gardens. The composition of agricultural land use at the research site can be seen in table 4. below.

Table 4. Patterns of agricultural land use in the Kreo Watershed, Semarang City

No.	Land Use	Area (Ha)	Percentage (%)
1.	Rice field	950,72	24,72
2.	Farmland	537,64	13,98
3.	Mixed gardens	1089,43	28,32

Source: Google Earth Image analysis and data processing (2021)

The use of rice fields in the Kreo watershed of Semarang City has a clustered pattern in the upstream and middle parts, but the area of rice fields is more dominant in the upstream. The proportion of rice fields in the upstream area covers 71.48% of the total area of rice fields in the Kreo Sub-watershed of Semarang City, and as much as 24.65% of rice fields in the middle and 3.8% in the downstream. The land use of the farmland in the research location has a pattern that spreads in the upstream and middle parts and tends to cluster in the downstream.

The proportion of farmland in the downstream area covers 63.38% of the total farmland area in the Kreo Watershed of Semarang City, and as much as 25.15% of farmland is concentrated in the middle and 11.48% in the upstream. Mixed gardens have a pattern that spreads in the upstream, middle, and downstream areas, but the mixed gardens are more concentrated in the middle. The proportion of mixed garden area in the middle part covers 45.80% of the total mixed garden area in the Kreo Sub-watershed of Semarang City, and the middle reaches 32.71% while in the downstream reaches 21.48%.

3.2 Agricultural solid waste

Solid waste in this study is plastic waste that is difficult to decompose from the remnants of the use of mulch, polybags, plastic wrappers, and plastic sacks of fertilizer packaging that can be seen with the naked eye. Based on observations at several locations on agricultural land, it was found that there were plastic remnants from the use of mulch and plastic bags of fertilizer packaging found on agricultural land as shown in Figure 2. and Figure 3. below.

Figure 2. Plastic waste left over from mulch on agricultural land and remaining plastic sacks of fertilizer packaging (Research documentation, 2021)

On agricultural land with an intercropping system between rice plants and fruit trees and in orchards, farmers tend to use plastic wrap to maintain fruit quality when harvesting and avoid pests. When it is harvest time, most of the farmers will save the plastic wrap which is still suitable for use in the next planting period. Meanwhile, some farmers will immediately dispose of the plastic wrapping waste by collecting it and then burning it or dumping it directly on agricultural land. The plastic wrap used at the research location is presented in Figure 4. below.

Figure 3. Use of plastic wrap at the research site (Research documentation, 2021)

Solid waste from agricultural activities such as plastic wrap, residual use of mulch and polybags and plastic sacks of fertilizer packaging tends to accumulate in river waters where the flow is calm and will disrupt the flow of river water. At several locations of river conditions observation stations found solid waste in the form of plastic waste around the river. The solid waste comes from agricultural activities from the remnants of the use of mulch and polybags used by farmers.

This is indicated by the use of land around the river which is dominated by agricultural land in the form of farmland and mixed gardens. In addition, the location of the river waters is quite far from residential areas due to the sloping topography of the area and quite difficult to access. The environmental conditions around the river are presented in Figure 5. below.

Figure 4. Remaining use of mulch and polybags at the research site (Research documentation, 2021)

Based on the real condition of agricultural land and river waters at the research site, it can be said that solid waste (plastic) from agricultural activities has polluted river waters in the Kreo Watershed, Semarang City. This can be seen from several points of observation of river conditions, there is plastic waste from the rest of the use of fertilizers, mulch, and polybags in river bodies and the environment around the river.

3.3 Agricultural liquid waste

Liquid waste from agricultural activities cannot be seen with the naked eye, therefore the identification of contaminants in the form of liquid waste in this study is from the use of fertilizers and pesticides used by farmers. As many as 71% of farmers choose to use chemical fertilizers, 6% of farmers use manure, and 22% of farmers use a mixture of both chemical fertilizers and manure. The composition of fertilizer use in the research location is presented in Table 5. below.

Table 5. Agricultural Liquid Waste

Types of fertilizer		Frequency (%)	Location						
			Karang malang	Polaman	PPurwosari	Gunungpati	Cepoko	Kandri	Sadeng
Dung		6	2	1	0	1	0	2	0
Chemical	Urea	51	7	7	13	14	2	5	3
	NPK	15	2	1	4	5	1	1	1
	Pusri	3	0	1	2	0	0	0	0
	TSP	3	0	0	3	0	0	0	0
Compost		0	0	0	0	0	0	0	0
Mix (dung and chemical fertilizer)		22	4	2	7	1	3	4	1

The types of fertilizers that are mostly used by farmers in the Kreo sub-watershed of Semarang City are urea (51%) and NPK (15%). Gunungpati and Purwosari villages are the areas with the highest use of Urea and NPK fertilizers. In addition, some farmers also use Pusri (3%) and TSP (3%) fertilizers in their farming activities. The amount of fertilization of agricultural land in the Kreo Sub-watershed of Semarang City, especially in wetland agriculture (rice fields) is generally carried out 2 times per planting period. As many as 18% of farmers in the research location only applied 1 time fertilization with dung and 12% of farmers applied 2 times fertilization in one planting period.

Meanwhile, 82% of farmers do 2 times fertilization with chemical fertilizers in one planting period. Then only 4% of farmers fertilize 3 times in one planting period, and 8% of farmers do 1 time fertilization in one growing season. The composition of the amount of fertilization per planting period is presented in Table 6. below.

Table 6. Amount of fertilization per planting period

Location	Dung			Chemical fertilizer		
	1 time	2 times	3 times	1 time	2 times	3 times
Karangmalang	4	1	0	1	13	0
Polaman	1	2	0	0	10	1
Purwosari	8	2	0	1	27	1
Gunungpati	1	1	0	1	18	0
Cepoko	1	2	0	0	5	1
Kandri	1	4	1	0	9	1
Sadeng	2	0	0	5	0	0
Frequency (%)	18	12	1	8	82	4

Source: Interview and data processing of researchers, 2021

Based on the results of interviews, around 92% of farmers in the research location used chemical pesticides in overcoming and eradicating pests that often attack their agricultural land. For example, the pesticides most often used by farmers are Dencis, Starban, Postin, Pastak, Regent, and so on. Around 7% of farmers do not eradicate pests on the agricultural land they cultivate due to financial constraints and 1% of farmers choose to eradicate pests with natural ingredients using water from tobacco stew to reduce expenses. The pest control system in the Kreo watershed is presented in Table 7. below.

Table 7. Pest eradication system at the reseach site

Location	Pest Eradication		
	Natural ingredient	Chemical Pesticides	Not etadicated
Karangmalang	0	14	1
Polaman	0	12	0
Purwosari	0	29	0
Gunungpati	1	20	0
Cepoko	0	6	0
Kandri	0	11	1
Sadeng	0	0	5
Frequency (%)	1	92	7

Source: Interview and data processing of researchers, 2021

3.4 Water Sample Laboratory Test

Sampling of river water was carried out the day after fertilization with the assumption that after fertilization activities were the peak of residue accumulation in agricultural land, especially from residues of nitrogen fertilizer use. The results of laboratory tests of river water samples at the research location are presented in table 8. below

Table 8. Laboratory tests of river water samples

Parameters	Unit	Quality standards	TS 1	TS 2	TS 3	TS 4	TS 5	TS 6	TS 7	TS 8
TDS	mg/L	1000	97	143	122	99	101	120	69	98
TSS	mg/L	50	3	1	12	3	6	5	23	11
BOD	mg/L	2	2,9	0,9	3,8	4	1,6	6,8	4,6	1,6
COD	mg/L	10	9	3	9	12,6	3	18	15	3
Nitrat (NO3) sebagai N	mg/L	10	0,58	2,64	2,84	0,92	1,48	1,06	4,8	1,4

Source: Laboratory Test, 2021

Based on the results of laboratory tests of river water samples in the Kreo Watershed, Semarang City, from 8 observation stations, the results obtained varied on each parameter. The concentrations of BOD and COD parameters at several locations have exceeded the class I river designation quality standards. Meanwhile, the TDS, TSS, and Nitrate parameters still meet the specified quality standards. Comparison of the concentrations of each parameter of river water quality test in the Kreo watershed, respectively, is presented in the following graph.

a. TDS

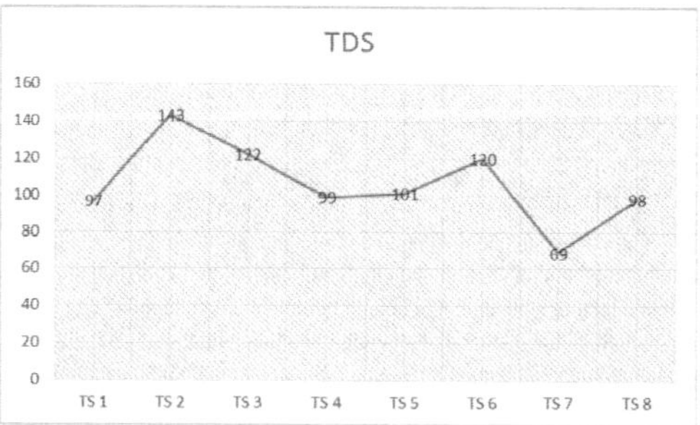

Figure 5. Trends in TDS concentration at the study site (Laboratory tests and research data processing, 2021)

The highest concentrations of TDS were found in TS2, TS3 and TS6. This is influenced by the condition of the river current which is very calm so that it allows the accumulation of dissolved solids in the water to remain stable and not affected by the flow of river currents. The high concentration of TDS in TS3 and TS6 is thought to have come from the use of fertilizers because the water sampling coincided with the farmer's fertilization schedule at that location. The lowest TDS value is found in TS7, this is due to land use along the right and left of the river in the form of mixed gardens which are unproductive agricultural land and are underutilized and managed by farmers.

The pattern of TDS concentration in river waters in the Kreo Watershed of Semarang City has a decreasing trend from upstream to downstream. The highest concentration of TDS is located upstream. This is influenced by differences in the level of slope from upstream to downstream. The upstream has a more sloping terrain and the potential for erosion is higher, besides that the concentration of agriculture is located in the upstream area of the Kreo watershed so that the utilization of agricultural land in the upstream part is more intensive than the middle and downstream parts of the Kreo watershed.

b. TSS

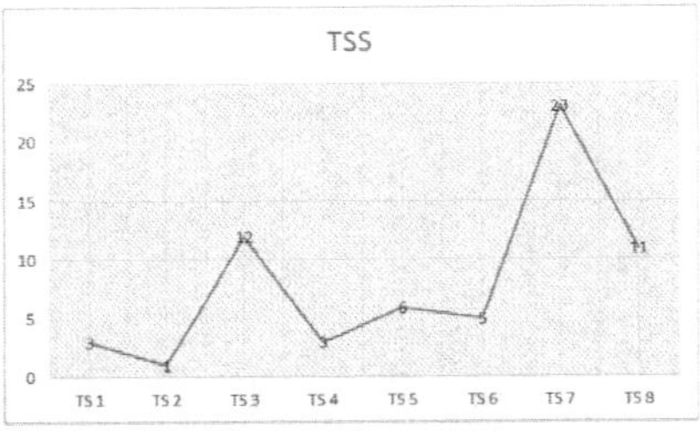

Figure 6. Trends in TSS concentration at the study site (Laboratory tests and research data processing, 2021)

TSS is closely related to topographic conditions, rainfall and soil characteristics (Danz et al., 2013). The highest concentration of TSS is located in TS7. This is influenced by the location around the river water sampling which has a Mediterranean soil type with a moderate level of sensitivity to erosion (erodibility). So that the soil particles from the surrounding agricultural land will be very easily carried to the river body through irrigation channels and surface runoff. Meanwhile, the lowest TSS concentration was in TS2-Polaman. This is influenced by soil characteristics that have low sensitivity to erosion with reddish brown latosol soil types. In addition, around the river the condition of agricultural land tends to be flat. So that the soil particles that are eroded and carried along to the river body tend to be smaller.

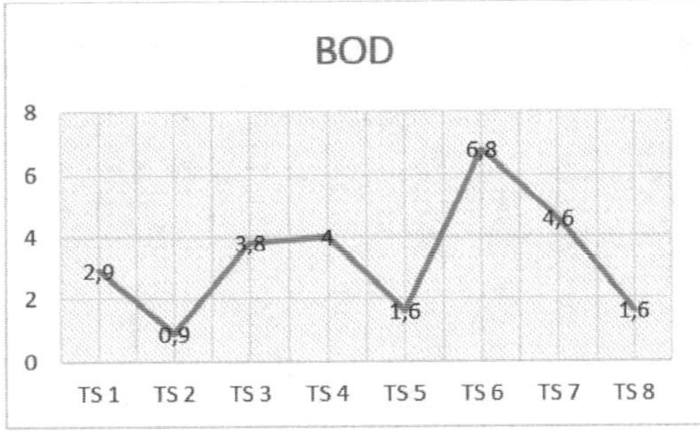

Figure 7. Trends in BOD concentration at the study site (Laboratory tests and research data processing, 2021)

c. BOD

Biochemical Oxygen Demand (BOD) is generally used to measure the contamination of biodegradable organic materials in water (Udeigwe and Wang, 2010). Several previous studies such as the research of Dudley and Karr, (1979); Schreiber and Neumaier (1987); Bishop et al., (1999) show that agricultural activities have contributed to organic waste which is indicated by the high concentration of BOD in river waters.

Based on the results of laboratory tests, 5 of the 8 points of river water sampling locations showed BOD values that exceeded the class I water quality standard because the BOD concentration was more than 2 mg/L. The BOD value from upstream to downstream varies, however, the highest BOD value is located upstream, namely at TS6 which reaches 6.8 mg/L. In addition, TS3 which is located close to TS6 and is only 280 meters away also has a BOD value that exceeds the quality standard. The high BOD value in TS3 and TS6 was caused by the use of manure used by farmers.

In addition, sampling on TS3 and TS6 was carried out on the day of fertilization by farmers. As many as 25% of farmers in this location use livestock manure when fertilizing in addition to continuing to use chemical fertilizers as the main fertilizer.

d. COD

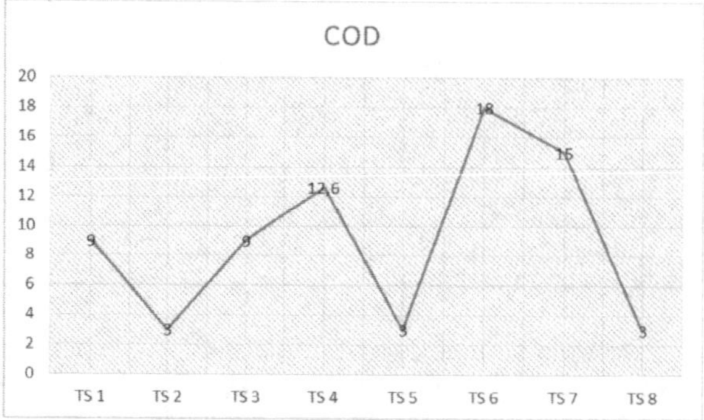

Figure 8. Trends in COD concentration at the study site (Laboratory tests and research data processing, 2021)

Chemical Oxygen Demand is the amount of chemical oxygen needed to oxidize organic materials present in water through chemical reactions because it is difficult to destroy biologically (Singh et al., 2011). The highest COD concentrations were in TS6, TS7, and TS4. The BOD value always follows COD, if the BOD value is high, the COD value will also be high, and vice versa. According to Tchobanoglous et al. in Nurjanah et al. (2017) suggested that BOD is part of COD.

Therefore, location points with high BOD values will also have higher COD values than other points with lower BOD levels. The COD value will increase along with the increase in organic matter entering the river body. The high demand for oxygen to oxidize pollutants will also affect the oxygen requirements of organisms living in river waters. COD concentrations that exceed the quality standard will disrupt the sustainability of river water ecosystems, especially in aquatic biota because the oxygen demand in river waters is not fulfilled.

e. Nitrat

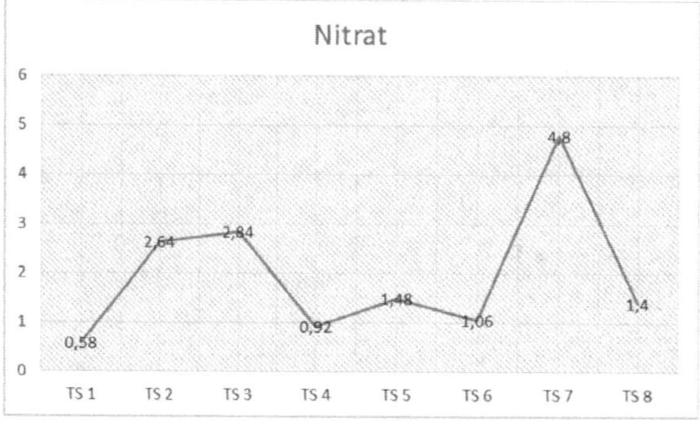

Figure 9. Trends in nitrate concentration at the study site (Laboratory tests and research data processing, 2021)

The concentration of nitrate at the study site varies from upstream to downstream. In the upstream and midstream, the nitrate value tends to be higher. This is influenced by differences in productivity levels in managing and utilizing agricultural land from upstream to downstream. In the upstream and midstream, agricultural activities in the Kreo watershed are more intense than in the upstream. Therefore, the nitrate concentration in the upstream and middle of the Kreo watershed is higher.

Although the nitrate value at all sampling locations still showed numbers below the quality standard, the nitrate value could be said to be high because in one area of agricultural land only 1 person was fertilizing so that the accumulation of residue from the use of nitrogen fertilizers was not as much as if fertilization was carried out simultaneously between one farmer to another.

Kreo River Water Quality Status

Waste from agricultural activities, both solid waste and liquid waste is one of the factors as a contributor to pollutants in rivers in the Kreo Watershed, Semarang City. The results of laboratory tests on river water samples showed that the concentrations of BOD and COD at several sample points had exceeded the class I river water quality standards.

Meanwhile, the TDS, TSS, and Nitrate parameters still met the quality standards. This is influenced by the agricultural system that takes place at the research site. The mechanism of agricultural activities in the Kreo watershed of Semarang City is not agriculture which is carried out simultaneously. The system of agricultural activities from the beginning of the planting period, fertilizing, spraying pesticides, and harvesting in one area of agricultural land/rice fields is not carried out simultaneously (simultaneously) between one farmer and another, but in these activities there are intervals or lags of several days.

Central Java Governor Regulation Number 156 of 2010 concerning Water Designation and Water Quality Management of the Garang River, stipulates that the rivers in the Kreo watershed are designated as class I rivers, namely as drinking water raw water, but in facts the rivers in the Kreo watershed are used to irrigate agricultural crops, fisheries, and tourism facilities for the Jatibarang Reservoir.

Therefore, the analysis of the STORET method to determine the status of water quality in each class of water use designation such as class II, class III, and class IV is very necessary. Information on the status of water quality in each designation class can be used as consideration in the watershed management plan. The results of the analysis of river water quality status based on each designation class are presented in Table 9. below.

Table 9. River water quality status of the Kreo Watershed in Semarang City

Sample Location	Score	Class I	Score	Class II	Score	Class III	Score	Class IV
TS 1	-2	Lightly polluted	0	Meet quality standards	0	Meet quality standards	0	Meet quality standards
TS 2	0	Meet quality standards	0	Meet quality standards	0	Meet quality standards	0	Meet quality standards
TS 3	-2	Lightly polluted	-2	Lightly polluted	0	Meet quality standards	0	Meet quality standards
TS 4	-4	Lightly polluted	-2	Lightly polluted	0	Meet quality standards	0	Meet quality standards
TS 5	0	Meet quality standards	0	Meet quality standards	0	Meet quality standards	0	Meet quality standards
TS 6	-4	Lightly polluted	-2	Lightly polluted	-2	Lightly polluted	0	Meet quality standards
TS 7	-4	Lightly polluted	-2	Lightly polluted	0	Meet quality standards	0	Meet quality standards
TS 8	0	Meet quality standards	0	Meet quality standards	0	Meet quality standards	0	Meet quality standards

Source: Laboratory tests and data processing (2021)

Analysis of the status of class I river water quality shows that from 8 sampling locations, 5 samples were lightly polluted, namely at observation stations TS1, TS3, TS4, TS6, and TS7. In the upstream part of the Kreo watershed, the type of land use is dominated by wetland agriculture (rice fields), so that the upstream part is the center of agricultural activities in the Kreo watershed. Analysis of the status of river water quality from 6 samples taken upstream shown in TS1 to TS6 shows that as many as 4 samples of class I river water have been lightly polluted. Therefore, river water in the upstream part of the Kreo watershed will be dangerous if used as a source of drinking water. Contamination of pollutants in the upstream river is influenced by intensive agricultural activities, especially from the use of fertilizers and pesticides used by farmers.

The use of agricultural land in the middle part of the Kreo watershed is dominated by dry land agriculture in the form of mixed gardens and farmland as well as wetland agriculture (rice fields). In the middle of the river water sampling is done at the observation station TS7 which is one of the input sources (inlet) to the Jatibarang Reservoir. In the future, Jatibarang Reservoir water will function as a source of clean water for Regional Drinking Water Company Semarang City Semarang City (Hidayah et al., 2016). Based on the analysis of the water quality status of class I river designation, it shows that the condition of the river in TS7 has been lightly polluted. Thus the river water will be harmful to human health if used as a source of drinking water.

Based on laboratory tests of water samples at the observation station TS8 which is located downstream of the Kreo watershed which is the output of the Jatibarang Reservoir and various flows from other tributaries downstream, it shows that the condition of the river water quality status still meets the quality standards. This is influenced by the type of agricultural land use in the downstream which is dominated by dry land agriculture in the form of farmland and mixed gardens. The productivity of dryland agriculture in the Kreo watershed is lower than that of wetland agriculture. In addition, the use of fertilizers and pesticides in dry land agriculture is not as intense as in wetland agriculture. Therefore, the contamination of liquid waste from agricultural activities in the downstream is not as much as in the upstream and middle parts.

The quality and status of river water quality in the upstream, middle, and downstream parts of the Kreo watershed have different results. In the upstream and in the middle the water quality status of the river has been lightly polluted, while in the downstream the water quality status still meets the quality standard for class I rivers. This is inseparable from the way farmers manage their agricultural land. Although the existing agricultural systems in the Kreo watershed from upstream, middle, and downstream have homogeneous characteristics, the schedule when fertilizing and spraying pesticides in one agricultural area or rice field is not carried out simultaneously between one farmer and another.

The laboratory test of nitrate parameters which is an indication of pollutants from the use of chemical nitrogen fertilizers (Urea and NPK) in one plot of rice fields with one farmer who fertilizes turns out to produce the highest nitrate of 4.8 mg/L a day after fertilization from the specified quality standard. by 10 mg/L in class I rivers. If all farmers in the area apply fertilization at the same time, there will be an increase in nitrate accumulation in river waters near the agricultural land and the potential for pollution from agricultural waste will be higher.

At this time the water quality status of the Kreo watershed, especially in the upstream and middle parts, has been lightly polluted, if fertilization and pesticide spraying activities are carried out simultaneously, the potential for river pollution in the Kreo watershed can increase to moderately to heavily polluted. In addition to having a negative impact on human health if the water is used as a source of drinking water, the entry of pollutants into river waters will also have a negative impact on the life of biota and river aquatic ecosystems in the Kreo watershed.

4. Conclusion

The results of the study show that the pattern of rice field use tends to cluster in the upstream part and a small portion in the middle. The pattern of land use in the farmland spreads in the upstream and middle parts but tends to cluster in the downstream. While the pattern of mixed garden land use spreads in the upstream and downstream parts, but in the middle it tends to cluster.

Agricultural waste, both solid and liquid waste, has polluted river waters in the Kreo Sub-watershed, Semarang City. The results of laboratory tests of river water samples showed that the concentration of BOD and COD parameters at several sample locations had exceeded the class I river quality standards, while the TDS, TSS, and Nitrate parameters still met the quality standards. Analysis of the STORET method shows that the status of water quality in the upstream and middle parts of the Kreo Sub-watershed of Semarang City has been lightly polluted, while the status of water quality in the downstream still meets the quality standards.

Acknowledgement
The author expresses his gratitude to M. Lutfi Wirawan and Allesandro Aryo Setyaki who have helped the author in sampling river water and the author expresses his deep gratitude to the Central Java Province Health Laboratory and Medical Device Testing Center which has provided direction in taking river water samples in this research.

References
[1] Afriani, S., Agustina, S., Karina, S., Irwan, I., dan Kazrina, C. S. M. 2021. The assessment of water quality by STORET method in the northern waters of Banda Aceh. In IOP Conference Series: Earth and Environmental Science (Vol. 674, No. 1, p. 012062). IOP Publishing.
[2] Berka, C., Schreier, H., dan Hall, K. 2001. Linking water quality with agricultural intensification in a rural watershed. Water, Air, and Soil Pollution, 127(1), 389-401.
[3] Bishop, C. A., Mahony, N. A., Struger, J. N. G. P., Ng, P., dan Pettit, K. E. 1999. Anuran development, density and diversity in relation to agricultural activity in the Holland River watershed, Ontario, Canada (1990–1992). Environmental Monitoring and Assessment, 57(1), 21-43.
[4] Danz, M. E., Corsi, S. R., Brooks, W. R., dan Bannerman, R. T. 2013. Characterizing response of total suspended solids and total phosphorus loading to weather and watershed characteristics for rainfall and snowmelt events in agricultural watersheds. Journal of Hydrology, 507, 249- 261.
[5] Dudley, D. R., & Karr, J. R. 1979. Concentration And Sources Of Fecal And Organic Pollution In An Agricultural Watershed 1. JAWRA Journal of the American Water Resources Association, 15(4), 911-923.
[6] Hamilton, P. A., Denver, J. M., Phillips, P. J., dan Shedlock, R. J. 1993. Water-quality assessment of the Delmarva Peninsula, Delaware, Maryland, and Virginia-Effects of agricultural activities on, and distribution of, nitrate and other inorganic constituents in the surficial aquifer. US Geological Survey open-file report, 93, 95-p.
[7] Heathwaite, A. L., dan Johnes, P. J. 1996. Contribution of nitrogen species and phosphorus fractions to stream water quality in agricultural catchments. Hydrological processes, 10(7), 971-983.
[8] Hidayah, S. N., Widyorini, N., dan Purnomo, P. W. 2016. Analisis Kesuburan Perairan Waduk Jatibarang Berdasarkan Distribusi Dan Kelimpahan Bakteri Heterotrofik. Journal of Management of Aquatic Resources, 5(4), 443-452.
[9] Howden, N. J., Burt, T. P., Worrall, F., Mathias, S. A., dan Whelan, M. J. 2013. Farming for water quality: balancing food security and nitrate pollution in UK river basins. Annals of the Association of American Geographers, 103(2), 397-407.
[10] Keputusan Menteri Negara Lingkungan Hidup No. 115 Tahun 2003 Tentang Pedoman Penentuan Status Mutu Air.
[11] Lawniczak, A. E., Zbierska, J., Nowak, B., Achtenberg, K., Grześkowiak, A., dan Kanas, K. 2016. Impact of agriculture and land use on nitrate contamination in groundwater and running waters in central-west Poland. Environmental monitoring and assessment, 188(3), 172.
[12] Lenat, D. R. 1984. Agriculture and stream water quality: a biological evaluation of erosion control practices. Environmental management, 8(4), 333-343.
[13] Muyassaroh, N. L., dan Juhadi, J. 2015. Pengaruh Kondisi Sosial Budaya Terhadap Kerusakan Lahan Di Das Kreo Kota Semarang Dan Sekitarnya. Geo-Image, 4(1).
[14] Peraturan Pemerintah RI Nomor 82 Tahun 2001 Tentang Pengelolaan Kualitas Air dan Pengendalian Pencemaran Air.

[15] Peraturan Gubernur Jawa Tengah Nomor 156 Tahun 2010 Tentang Peruntukan Air dan Pengelolaan Kualitas Air Sungai Garang di Provinsi Jawa Tengah.
[16] Rintaka, W. E., Hastuti, A. W., Susilo, E., dan Radiarta, N. 2019. The used of storet index to assess water quality in perancak estuary, Bali, Indonesia. In IOP Conference Series: Earth and Environmental Science (Vol. 246, No. 1, p. 012012). IOP Publishing.
[17] Schreiber, J. D., dan Neumaier, E. E. 1987. Biochemical oxygen demand of agricultural runoff (Vol. 16, No. 1, pp. 6-10). American Society of Agronomy, Crop Science Society of America, and Soil Science Society of America.
[18] Setyaningsih, W., Sriyono, S., dan Benardi, A. 2019. Kajian Kerusakan Lahan Di Daerah Aliran Sungai (DAS) Kreo Akibat Pembangunan Pemukiman Di Sekitar Waduk Jatibarang Kota Semarang. Media Komunikasi Geografi, 19(2), 177-186.
[19] Singh, R., Kumar, A., Kirrolia, A., Kumar, R., Yadav, N., Bishnoi, N. R., dan Lohchab, R. K. 2011. Removal of sulphate, COD and Cr (VI) in simulated and real wastewater by sulphate reducing bacteria enrichment in small bioreactor and FTIR study. Bioresource technology, 102(2), 677-682.
[20] Sulthonuddin, I., Hartono, D. M., Utomo, S. W., dan Said, C. A. A. 2019. Water Quality Assessment of Cimanuk River in West Java Using STORET Method. In IOP Conference Series: Earth and Environmental Science (Vol. 239, No. 1, p. 012007). IOP Publishing.
[21] Sutriadi, M. T., dan Sukristiyonubowo, S. 2013. Pencemaran Nitrat pada Air Sungai Sub DAS Klakah, DAS Serayu di Sistem Pertanian Sayuran Dataran Tinggi. Jurnal Tanah dan Iklim, 37(1), 35-44.
[22] Taylor, S. D., He, Y., dan Hiscock, K. M. 2016. Modelling the impacts of agricultural management practices on river water quality in Eastern England. Journal of Environmental Management, 180, 147-163.
[23] Udeigwe, T. K., dan Wang, J. J. 2010. Biochemical oxygen demand relationships in typical agricultural effluents. Water, Air, & Soil Pollution, 213(1), 237-249.

Utilization of plastic waste as an eco-friendly construction material

Zainuri*, G Yanti, S W Megasari

Universitas Lancang Kuning, Pekanbaru, Indonesia

*zainuri@unilak.ac.id

Abstract. Plastic paving blocks use concrete sand and plastic waste as binding materials. This study aims to calculate the potential for reducing plastic waste in nature if paving blocks are mass produced. The study was carried out using an experimental method, using a laboratory as a place of testing. If the reduction of plastic waste is considered, the 60% plastic:40% concrete sand variation with a compressive strength value of 13.92 MPa is included in grade C. The conclusion of this study is the use of plastic waste for the manufacture of plastic paving blocks with a job mix of 60% plastic.:40% sand as much as 1 m3 of 1,225.6 kg/m3 or can reduce plastic waste by 0.098 %/m3/day and if a manufacturer produces 10 m3 in a day, then the potential for reducing plastic waste is 7.3536 tons/day.

1. Introduction

Plastic waste is disposed every day. Low-income people tend to be less concerned about the environment and have low awareness of disposing waste, so they are contributors of large amounts of plastic waste, even they realized it or not [1]. The right solution to overcome the problem of plastic waste is recycling [2], [3]. Plastic waste is increasing due to the increasing trend of bottled water (type of PET plastic) consumption and leaving tap water consumption which gives a bad perception and doubts its safety for consumption [4]. The increasing use of bottles as containers for bottled water has led to the idea of using plastic bottle waste as a substitute for brick walls with some advantages such as saving energy, reducing CO2 emissions, making bricks, and reducing the use of more expensive cement [5].

Plastic waste can still be reused into the same or even different products than before. The nature of plastic that is not easily biodegradable or durable can be reused for household products, such as plates, cups and plastic spoons. The disadvantages of these recycled plastic products are the color are less bright and after some time they emit an unpleasant aroma. Tableware or houses with colors and aromas that do not support are less desirable. Therefore, the use of plastic waste for other products is a breakthrough so that the waste can be utilized and not continuously pollute the environment.

Plastic waste can be used as a substitute or additional material in construction materials. In road construction, for example, the use of plastic waste, asphalt and aggregates can improve performance and strength as well as cost savings by reducing the use of asphalt [6]. In addition, PET plastic, quarry dust and fly ash can be the main ingredients in the manufacture of paving blocks with increased strength [7]. By utilizing plastic waste in the manufacture of one of the building materials, it means that the use of plastic waste is not limited to making handmade as stated in [8].

If previous studies used plastic waste from the PET (Polyethylene Terephthalate) [9], this time the waste used is HDPE (High Density Polyethylene) plastic. HDPE plastic waste usually comes from bottles that are rather thick and colored. Physical appearance is tougher than the type of PET or PP. The difference in physical appearance raises the thought of trying it in making paving blocks. Does the use

IOP Conf. Series: Earth and Environmental Science 1041 (2022) 012084 doi:10.1088/1755-1315/1041/1/012084

of HDPE type of plastic waste have a good binding capacity for the product and can produce products that comply with SNI quality standards? These considerations are the background of this study.

After confirming that the manufacture of paving blocks by utilizing plastic waste from the HDPE type has the strength according to SNI quality standards. This study aims to calculate how much the potential for reducing plastic waste is if the product is mass-produced. The strength of the proven product is a consideration for local manufacturer to produce and market them actively. If the public understands that the plastic waste is still have economic value, then segregation efforts starting from the household can be carried out, so it is expected to further reduce the plastic waste that is wasted in nature.

2. Methodology

Paving blocks are usually used to cover the ground surface to make it look neater and more comfortable to pass. Conventional paving blocks are made of cement, fine aggregate and added a little water to assist in the binding process. Different from the usual paving blocks, plastic paving blocks that utilize plastic waste use concrete sand and HDPE (High Density Polyethylene) plastic waste as binding materials. Bonding uses fire as a heating element to melt the plastic.

The equipment used is still simple such as scales, gas stove, skillet and spatula. Equipment that is also important is the mold. Cube-shaped mold as required by SNI with the size of the mold for the sample maker (5 x 5 x 5) cm. The study uses an experimental method that uses a laboratory as a place of testing. Some of the tests carried out in the laboratory are compressive strength and product absorption tests. A series of tests that can be performed after the sample is made.

There are 9 variations of paving block samples made. The variation is determined based on the planned weight by dividing the variation of plastic waste from 10%, 20%, 30%, 40%, 50%, 60%, 70%, 80% and 90% of the total weight of the material used. The steps for making samples are weighing the material according to the planned variation; melt the plastic waste in the skillet and add the concrete sand; stir until mixed using a spatula; put the mixture into the iron mold; removing the sample from the mold; samples were cooled at room temperature for 28 days. The formulations used in the test are:

- Compressive strength test [10] :
  $fc' = \frac{P}{A}$ (1)
- Product absorption test [11] :
  $Absorbsi(\%) = \frac{A-B}{B} \times 100\%$ (2)

The next step is to determine the best variation from several samples according to SNI quality standards and finally calculate the reduction of plastic waste. Physical quality standards [12] are:

Table 1. Paving block Physical Strength.

Quality	Usage	Compressive Strength (MPa) Average	Compressive Strength (MPa) Min	Porosity Average (%)
A	Road Pavement	40	35,0	3
B	Car Parking Lot	20	17,0	6
C	Pedestrian Area	15	12,5	8
D	City Park	10	8,5	10

3. Result and Discussion

Plastic paving blocks have almost the same strength as conventional paving blocks [13]. The strength of paving blocks can be increased by increasing the quantity of plastic so that the compressive strength value can be greater than conventional paving blocks [14]. Making paving blocks using plastic waste is more efficient and economical [15]. The percentage of plastic waste that is better used as a cement substitute is 5% – 13% in the concrete mix [16].

Samples that have been left for 28 days at room temperature, without special treatment, are ready to be tested. Although there is no provision for when the plastic paving block sample is eligible to be tested, the material used is one of the concrete forming materials, so it is determined to follow the test age provisions according to the concrete regulations, that is after the sample is 28 days old. When the age of the sample is 28 days, it is expected that the bonding that occurs between the materials has been perfect, and the sample has reached its maximum strength.

The weight of each sample is weighed and then the average weight is determined for each variation of the sample made. The sample size (5 x 5 x 5) cm, the weight obtained in the range of 120 grams to 200 grams. The more the amount of concrete sand used, the heavier the product will be. After the sample is weighed, it is tested to determine the value of compressive strength and porosity. The results of the weighing and testing are listed in Table 2 below.

Table 2. Average of Compressive Strength and Porosity of Paving Block

No	Sample Variation Plastic : Sand	Average Weight (gram)	Average Compressive Strength (MPa)	Average Porosity (%)
1	10% : 90%	-	-	-
2	20% : 80%	193,00	5,36	4,8
3	30% : 70%	198,20	10,40	1,1
4	40% : 60%	173,80	14,48	0,3
5	50% : 50%	164,60	13,36	0,3
6	60% : 40%	153,20	13,92	0,2
7	70% : 30%	136,80	11,36	0,4
8	80% : 20%	126,20	10,72	0,7
9	90% : 10%	116,60	8,88	0,9

With the most plastic variations, which is 80% of the weight of all materials, the average weight obtained is 126.20 grams. The smaller the variation of plastic in the paving block mixture, the average weight of the sample increases. This condition is slightly deviated from the 20% plastic variation: 80% concrete sand which has an average weight of 193.00 grams smaller than the 30% plastic variation. This shows that 20% of plastics are not able to bind 80% of concrete sand perfectly. A weak bond makes a lot of concrete sand loose and the sample is hollow, so the sample becomes less dense. As a result, the sample weight is smaller than previously estimated.

The weak bond from the sample variation of the 20% plastic : 80% concrete sand directly affects the strength of the product. The very low average compressive strength value of 5.36 MPa is below the SNI quality standard so that it can be said that this variation is not suitable for use in the manufacture of paving blocks made from plastic waste.

The heavier the product, the compressive strength value tends to be higher. With the lowest average weight in the variation of 80% plastic: 20% sand concrete, the compressive strength value obtained is 10.72 MPa which places the product at quality D. The compressive strength value continues to increase placing the product at quality C which is better than quality D. The highest compressive strength value of 14.48 MPa was achieved in the sample with a variation of 40% plastic: 60% sand concrete. In the variation of 30% plastic: 70% concrete sand, the compressive strength value experienced a considerable decrease, namely 10.40 MPa which puts the product at D quality. The decrease in the compressive strength value occurred due to the reduced binding due to the smaller portion of plastic.

The porosity test showed that all samples made were within the safe threshold for all qualities in SNI. Only samples with a variation of 20% plastic: 80% sand concrete did not reach the strength of A quality with the highest porosity value of 4.8%. Although the porosity value from sample variation of 20% plastic : 80% concrete sand almost reached 5.0%, the product still met the B quality requirements while the others were at A quality. Thus, it can be said that all samples made with 8 variations of composition different materials are declared to have complied with the requirements of the porosity value of the product.

Selection of the optimal job mix depends on the needs. When the potential for reducing plastic waste is considered, those that use a large variety of plastics are preferred. Although the variation of 40% plastic: 60% concrete sand has the highest compressive strength value, the preferred variation is from the 60% plastic : 40% concrete sand with a compressive strength value of 13.92 MPa slightly lower but both are in C quality.

The variation of 70% plastic: 30% concrete sand and 80% plastic: 20% concrete sand uses more plastic than the variation of 60% plastic : 40% concrete sand. However, the quality of the product achieved is based on the lower compressive strength value, namely D quality. However, a better quality will be chosen in deciding which job mix to use.

With consideration of product strength and plastic waste reduction, the 60% plastic variation: 40% concrete sand was chosen as the job mix plan. Calculation of material requirements for the job mix variations are:

The average weight of the job mix for the mold (5 x 5 x 5) cm with a paving block sample of 153.20 grams is:

Volume per mold = 0,05 m x 0,05 m x 0,05 m
= 0,000125 m^3
Plastic waste weight = 153,20 gr x 60%
= 91,92 gr
= 0,09192 kg
Concrete sand weight = 153,20 gr x 40%
= 61,28 gr
= 0,06128 kg

A sample measuring (5 x 5 x 5) cm requires 91.92 grams of plastic waste. To make 1 m^3 of paving block, the job mix is calculated as follows:

The weight of the job mixes for 1 m^3 is:
Amount per 1 m^3 sample = 1 m^3: 0,000125 m^3
= 8.000 buah
Plastic waste weight = 0,09192 kg x 8.000 bh
= 735,36 kg
Concrete sand weight = 0,06128 kg x 8.000 bh
= 490,24 kg

The standard mold for paving blocks is square with a size of (20 x 10 x 8) cm. With 1 m^3 of job mix, ± 625 paving blocks can be made. The weight of plastic waste needed to produce 1 m^3 of paving blocks is 735.36 kg of plastic waste. When a manufacturer who produces paving blocks decides to produce paving blocks made of plastic, then the plastic waste from the type of HDPE needed will have a higher economic value. Scavengers and sorters will immediately separate plastic waste from the intended type and sell it to manufacturer who produce paving blocks made from plastic waste.

The advantage of paving blocks made of plastic is a product that is more resistant and not easily cracked or broken. If supported by adequate equipment such as mixing machines and printers, then one manufacturer can produce 10 m^3 in a day. Given the product can be immediately released from the mold because of the fast-cooling factor. With a variation of 60% plastic: 40% concrete sand, if one manufacturer produces 10 m^3/day, then the need for plastic waste for this activity is calculated as follows:

Paving block weight per 1 m^3 is:
= (1 m^3 : 0,000125 m^3) x (153,20 gr/m^3 : 1.000 gr/kg) = 1.225,6 kg/m^3
Plastic waste needs per day:
= 60% x 1.225,6 kg/m^3 x 10 m^3 = 7.353,60 kg = 7,3536 ton
Plastic waste needs per year:

= 365 day/year x 7,3536 ton = 2.684,064 ton/year

Percentage of plastic waste reduction per m^3 of paving block:

$$= \frac{(60\% \text{ x } 1.225,6 \text{ kg/m3 : } 1.000 \text{ kg/ton})}{7.486,80 \text{ kg/ton}} \times 100\% = 0,00982209\ \%/m^3/\text{day}$$

Urban waste products recorded in several big cities in Indonesia are shown in Table 3 below [17].

Table 3. Volume of waste transported per day by type of waste and city.

City	Inorganic Waste (ton)	
	2018	2019
Banda Aceh	85,20	88,49
Medan	745,00	750,00
Padang	189,00	192,00
Pekanbaru	228,25	353,76
Palembang	336,54	357,12
Bandar Lampung	15,12	19,17
Pangkal Pinang	26,70	51,02
DKI Jakarta	3.407,87	3.519,14
Bandung	513,28	533,59
Yogyakarta	101,6	98,35
Surabaya	761,57	754,92
Banjarmasin	118,30	126,16
Manado	350,00	450,00
Kendari	71,60	56,00
Gorontalo	50,77	48,65
Ambon	46,05	37,60
Ternate	13,70	15,01
Manokwari	33,28	35,82
Total	7.093,83	7.486,80

If producing just 1 m^3 of paving block has the potential to reduce plastic waste by 0.01% (0.00982208 % /m^3/day), then one manufacturer who produces 10 m^3/day contributes 0.098 % (0.00982209 % x 10 m^3/day). day). You can imagine and calculate how much plastic waste will be reduced if this product is mass produced.

Previous studies have calculated the amount of waste that can be reduced by 3.9816 tons/day for each standard production of 5 m^3/day paving blocks with a variation of 70% PET plastic: 30% sand [18]. The contribution in reducing PET type plastic waste is almost the same as the results of this study using HDPE plastic if the calculation is equated in 1 m^3. Another study calculated the reduction of waste in producing 1,000 paving blocks by 600 kg [13]. Proving that the use of plastic waste in the production of paving blocks makes a real contribution to the reduction of plastic waste that can pollute the environment.

4. Conclusion

The conclusion of this study is the use of plastic waste for the manufacture of plastic paving blocks with a job mix of 60% plastic: 40% sand as much as 1 m3 which is 1,225.6 kg/m3 or can reduce plastic waste by 0.098 %/m3/day and if one manufacturer only produces 10 m3 in a day, so the potential for reducing plastic waste is 7.3536 tons/day.

Acknowledgements

The author would like to thank LPPM Universitas Lancang Kuning for the financial support for this research. The authors also express their gratitude to the Laboratory of the Civil Engineering Study Program, Universitas Lancang Kuning, which has provided this research facility.

References

[1] P Rajkumar 2015 A Study on the Plastic Waste and Environmental Degradation ABC Journal of Advanced Research 4 9-15
[2] W d'Ambrieres 2019 Plastics Recycling worldwide: Curent Overview and Desirable Changes Field Actions Science Reports Special Issue 12-19
[3] H Alhazmi, F H Almansour and Z Aldhafeeri 2021 Plastic Waste Management: A Review of Existing Life Cycle Assessment Studies Sustainability 13 1-21
[4] I Gambino, F Bagordo, B Coluccia, T Grassi, G D Filippis, P Piscitelli, B Galante and F D Leo 2020 PET-Bottled Water Consumption in View of a Circular Economy: The Case Study of Salento (South Italy) Sustainability 12 1-13
[5] M Jalaluddin 2017 Use of Plastic Waste in Civil Constructions and Innovative Decorative Material (Eco-Friendly) MOJ Civil Engineering 3 359-68
[6] R Manju, S Sathya and K Sheema 2017 Use of Plastic Waste in Bituminous Pavement International Journal of ChemTech Research 10 804-11
[7] C Nivetha, M Rubiya, S Shobana, R G Vaijayanthi, M E Viswanathan and R Vasanthi M E 2016 Production of Plastic Paver Block From the Solid Waste (Quarry Dust, Flyash & PET) ARPN Journal of Engineering and Applied Sciences 11 1078-79
[8] P Purwaningrum 2016 Upaya Mengurangi Timbunan Sampah Plastik Menjadi Produk dan Jasa Kreatif Jurnal Teknik Lingkungan 2 141-47
[9] Zainuri, G Yanti, S W Megasari dan D Setiawan 2020 Paving Block Ramah Lingkungan Dengan Pemanfaatan Limbah Plastik Laporan Akhir Penelitian Utama Pekanbaru
[10] SNI 03-1974-1990 1990 Tata Cara Pengujian Kuat Tekan Beton Badan Standarisasi Nasional Jakarta
[11] SNI 03-6433-2000 2000 Standar Nasional Indonesia Pengujian Absorbsi Badan Standarisasi Nasional (BSN) Jakarta
[12] SNI 03-0691-1996 1996 Persyaratan Mutu Bata Beton (Paving Block) Badan Standarisasi Nasional Jakarta
[13] J Ghuge, S Surale, B M Patil, and S B Bhutekar 2019 Utilization of Waste Plastic in Manufacturing of Paver Blocks International Research Journal of Engineering and Technology (IRJET) 6 1967-70
[14] A S Moon, M A.Riyaz, N P Chautre, M S I Ansari, R M Piprewar, and R S Balpande 2020 Utilization of Waste Plastic in Paver Blocks International Research Journal of Engineering and Technology (IRJET) 7 2225-29
[15] S Dinesh, A Dinesh and K Kirubakaran 2016 Utilisation of Waste Plastic in Manufacturing of Bricks and Paver Blocks International Journal of Applied Engineering Research 11 364-68
[16] M Prasad and D Jaysawal 2017 Use of Plastic Waste in Concrete Mix International Research Journal of Engineering and Technology (IRJET) 4 1914-17
[17] BPS-Statistics Indonesia 2020 Environment Statistics of Indonesia 2020: Water and Environment Badan Pusat Statistik Jakarta
[18] Zainuri 2021 Handling of Plastic Waste in Paving Block Production Jurnal Teknologi Lingkungan 22 170–77

Biogas as methane mitigation path on dairy cattle waste management

A Herliatika* and Y Widiawati

Institute of Research and Development of Agricultural, Bogor, Indonesia

*tikaagustinherlia@gmail.com

Abstract. Greenhouse gas (GHG) caused global warming. The GHG emissions in ruminants come from enteric fermentation and manure management. Some of manure management of large ruminants (dairy cattle and beef cattle) are Biogas and manure stacking. Biogas in manure management is the most suggested, since beside reduce the GHG emission, it's also produced an alternative energy source. Biogas system had been applied by many sectors including small holder farmers. There is no data available on the reduction value of GHG emission resulted from Biogas system. Therefore, the equation of methane mitigation from biogas was needed to predict the potential mitigation from this activity. This study aimed to create a simple equation to predict methane mitigation from manure management activity of dairy cattle, also to predict added value as an energy source in manure management. A correction factor of emission from dry stacking of manure compared to emission from biogas is also required in the calculation. Equations formed from the study are EB= (940,57 CO2-eq /AU/year) x N, EBB= 130,73 CO2-eq/AU/year x N, and EDS= 0,659281 x EB or EDS= 4,74336 x EBB. These equations used to predict greenhouse gas production in the manure of dairy cattle. The final result showed that manure management of dairy cattle should be done by using Biogas. Gases produced in biogas can be used as alternative energy by burning it and contribute to reducing the GHG production in manure management.

1. Introduction

GHG has been increasing the earth's temperature and harming the earth's condition. There are some paths to minimize the negative effect of GHG, which in turn also leads to an increase in the cost. Cost for minimizing the negative effect of methane emissions (Marginal Abatement Cost) normally is below 20€/t CO2-eq but predicted more in agriculture caused by most of the livestock found on an extensive farm and some of them are unproductive animals, especially in developing countries [76].

Methane mitigation can be used to minimize the Marginal Abatement Cost by minimizing total GHG emissions. Livestock emissions came from enteric methane and manure management. Process of manure decomposition produce some amount of CH4 and N2O. There are common manure managements in farmers, such as Fresh stacking, dry stacking, spread in lands, and biogas. Among them, biogas is one of manure management which is giving added value as a source of energy alternative. This management usually for manure management of large ruminants (dairy cattle and beef cattle).

Dairy cattle cultivation usually carried out in long term compared to fattening of beef cattle. This condition leads to dairy cattle manure potentially use as a sustainable source of biogas. Another path to manage dairy cattle manure is dry stacking [98], Both paths need to be evaluated to looking at which path is more efficient than others. There is no correction factor information if amount of manure management replaced from dry stacking to Biogas. This correction factor uses to predict mitigation efficiency if some manure manages through Biogas.

This study aimed to create a formula used to predict GHG production in manure management of dairy cattle. Also, to give some considerations about the best strategies path used to mitigate methane emission in dairy cattle manure management.

2. Materials and Methods

manure management, Dry Stacking Emissions during a year which let the dry manure stack all days, and biogas which place manure into a digester to conduct the fermentation activity. GHG emissions from biogas will calculate before while the methane production still storage in digester or other tank and after gas burning while the methane production uses to produce energy.

A correction factor of GHG emissions uses dry stacking during a year compared to biogas, both before and after gas burning, will calculating the proportion of difference total GHG emissions from Biogas (Before and after gas burning) and total GHG from dry stacking with total GHG emissions from Biogas (Before and after gas burning). Formula to calculate the correction factor of GHG emissions from dry stacking compares to GHG emissions from Biogas showed in Formula 1.

$$\text{Correction factor} = (1 - (\frac{(\text{Biogas GHG emissions} - \text{Dried Stacking Emissions})}{\text{Biogas GHG emissions}}))......................(\text{Formula 1})$$

If Biogas GHG emissions > Dry Stacking Emissions, then the value of Correction factor nilai< 1
If Biogas GHG emissions < Dry Stacking Emissions, then the value of Correction factor nilai> 1

3. Result and Discussion

3.1 Developing a formula used to GHG emissions predicted in manure management of dairy cattle

Manure management in dairy cattle is 50% through dry stacking and 50% through biogas. Evaluation of both management needs to be done to know the efficiency of both paths in minimized the negative effect of GHG. Direct emission of CH4 was produced in fresh manure, meanwhile, emissions of N2O, CH4, and CO2 were produced during storage. Dry stacking of manure focusing on N2O emissions, even though some CH4 also produced, meanwhile the number is very low so it can be ignored in the calculating process.

In Biogas, GHG emissions during storage, focusing on CH4 and CO2 emissions. Mitigation in Biogas aimed to minimize N2O production during dry stacking storage in a year, also reallocate some CH4 emissions to produce alternative energy. Dry stacking storage shifted the manure management application from dry stacking storage to Biogas will reduce or minimized N2O emitted during the dry storage as well as use CH4 produced for energy alternative.

Formula use on GHG emission calculation in this study restricted only in dairy cattle manure storage . GHG emissions from fresh manure aren't predicted in this formula. Global emissions of manure management CO2 and manure management N2O are higher on Dairy cattle than beef cattle. Emissions of manure management CO2 for dairy cattle and beef cattle are 3,8% and 1,4%, respectively. Emissions of manure management N2O for dairy cattle and beef cattle are 5,4%vs 3,6%, respectively

3.2 Calculation of CH_4 and CO_2 conversion volumes against both gas weights

Moleculer Weight (Mr) of 1kg of CH_4 is 16 with n (1000g/16) = 62,5mol, in room temperature 270C (300K), and pressure 1 atm is 0,0821, so V value through equation PV=nRT, is 1539,375L, so every 1L of methane equal with 0,65g of methane, or 1m^3 of methane is equal with 650g (0,65kg) of methane. Mr of 1kg of CO_2 is 44 with n (1000g/44) = 22,73mol, in room temperature 270C (300K), and pressure 1 atm is 0,0821, so V value through equation PV=nRT, is 559,84L, so every 1L of CO_2 is equal with 1,79g of CO_2, or 1m^3 of CO_2 is equal with 1790g (1,79kg) of CO_2.

3.3 GHG emissions from dry stacking manure during a year

N_2O production in manure management calculated using the equation in modified by Puslitbangnak 2016 in Formula 2:

N_2OD (mm) = [$N_{(T)}$ x $Fex_{(h)}$ x $BW_{(t)}$] x $FE_{t(s)}$ x 44/28 x 365 ....(Formula. 2)

Description:
N_2OD (mm) = N_2O emissions from manure (kg N_2O/year)
$N_{(T)}$ = The population of animals T in 1 year (head)
$Fex_{(h)}$ = excretion of N every 1000kg of body weight (BW)
$BW_{(t)}$ = Average of BW (kg/head)
$FET_{(s)}$ = The emission factor of N_2O in the manure of animal T with S manure management (kg N_2O-N/kg N)
44/28 = Molecule weigh of N_2O/ Molecule weigh of N2
365 = Total days in a year (1/year)

The emissions value of N_2O (kg N_2O/year) can be convert to N_2O (kg $CO_{2\text{-}eq}$/year) after multiplying with 265 [65]. Value Fex(h) on dairy cattle is 0,47 (kg N/1000kg BW); The value of FET(s) on dairy cattle which manure stacking in drying is 0,02. So, 1 head of dairy cattle with BW 432,07kg/head can produce 2,34 kg N_2O/AU/year or equal with 620,1 kg $CO_{2\text{-}eq}$ /AU/year if the manure stack dried during a year.

3.4 GHG emission in biogas before gas burning

Predicted biogas production from 5 heads of dairy cattle with BW 432,07kg is 1m^3/day [9]. Total methane in biogas purified using nylon 5µm for 20 minutes is 67,5971% [7]. So1 head of dairy cattle (1AU) can produce 0,2 m^3/year biogas equal with 73 m^3/AU/year which contains 49,35 m^3/AU/year methane. 1 m^3 of methane equal with 0,65kg of methane, so 1 head of dairy cattle during a year can produce 32,08kg of CH_4. The emission of CH_4 (kg of CH_4/year) can be converted into CH_4 (kg $CO_{2\text{-}eq}$/year) then multiplying with 28 [5]. So, 1 head of dairy cattle can produce 898,24kg $CO_{2\text{-}eq}$/AU/year and added with CO_2 emission in biogas if the manure management doing through biogas.

Almost 100% of biogas contains Methane and CO_2 if methane contained in the 73 m^3/AU/year biogas is 49,35 m^3/AU/year then almost 23,65 m^3/AU/year is CO_2, 1m^3 of CO_2 is equal with 1,79kg of CO_2, so CO_2 produce by 1 head of dairy cattle in a year is 42,33kg CO_2/AU/year. Then the total of GHG emission in manure management through biogas before gas burning is 940,57 $CO_{2\text{-}eq}$/AU/year (Summation from 898,24kg $CO_{2\text{-}eq}$/AU/year and 42,33kg CO_2/AU/year).

3.5 GHG emission in biogas after gas burning

The total of methane contained in biogas, it can be burn is 2005mol (weight of CH_4 is 32,08kg equal with 32080g with Mr 16). Burning of 2005mol methane can produce 2005mol of CO_2. The residue of the burning is 49383,15L of CO_2 (n=2005mol; P=1atm; R=0,0821; and T=270C equal with 300K using formula V=nRT/P). So, emission of CO_2 after gas burning is 88,40kg CO_2/AU/year (49383,15L x 1,79kg/m^3).

Total gas emissions from manure management activity using 1 head of dairy cattle during a year got from summation of CO_2 emission in biogas and CO_2 produce after gas burning. It is 130,73 CO_2/AU/year (42,33kg CO_2/AU/year and 88,40kg CO_2/AU/year).

3.6 The correction factor of GHG emission from dry stacking against biogas

Differences production of $CO_{2\text{-}eq}$/AU/year in dry stacking and biogas before gas burning is 340,23kg $CO_{2\text{-}eq}$/AU/year higher on dry stacking management. So, the correction factor is 1-0,340719 or 0,659281. Differences production of $CO_{2\text{-}eq}$/AU/year in dry stacking and biogas after gas burning is 489,37 $CO_{2\text{-}eq}$/AU/year higher on dry stacking management. So, the correction factor is 1+3,74336 or 4,74336.

3.7 Formulas to predict GHG emissions on dairy cattle manure

Depends on all equation before, so formulas for GHG emissions prediction on dairy cattle are (GHG emissions in $CO_{2\text{-}eq}$/AU/year)

EB = 940,57 (kg $CO_{2\text{-}eq}$/AU/year) x N ....(Formula. 3)

EBB = 130,73 ($CO_{2\text{-}eq}$/AU/year) x N ....(Formula. 4)

EDS = 0,659281 x EB ....(Formula. 5)

or EDS= 4,74336 x EBB ....(Formula. 6)

Descriptions:

EB = GHG emissions from biogas before gas burning (kg $CO_{2\text{-}eq}$/year)

N = Total animals in a year (AU); 1 AU calculating depends on BW of mature dairy cattle (432,07kg); for dairy cattle>2th (1AU); 1-2th (0,5AU); and <1th (0,25AU) [10]

EBB = GHG emissions from biogas after gas burning (kg $CO_{2\text{-}eq}$/th)

EDS = GHG emissions from dry stacking (kg $CO_{2\text{-}eq}$/th)

3.8 Using the formula to predict GHG emissions on the manure of dairy cattle

Data need for predict the GHG emission on manure of dairy cattle is Data of the population (head) between 2015-2019 and the percentage of animals classified based on age as a report survey in 2017 at [3]. GHG emissions calculation predicted in 3 provinces (East of Java, Central of Java, and West of Java). The results show in Table 1.

Table 1. GHG emissions predicted in manure management of dairy cattle

Province	2015	2016	2017	2018	2019
		Population (head) [1]			
West Java	116400	119595	115827	118800	120719
Central Java	134670	137334	138560	154202	139111
East Java	255947	265002	273881	295809	287482
Indonesia	518649	533920	540441	581822	561061
		Population (AU) [2]			
West Java	90050	92522	89607	91907	93391
Central Java	104184	106245	107193	119295	107620
East Java	198007	205012	211881	228845	222403
Indonesia	401240	413054	418099	450112	434051
		EB (ton$CO_{2\text{-}eq}$/AU/year) [3]			
West Java	84698	87023	84281	86445	87841
Central Java	97992	99931	100823	112205	101224
East Java	186239	192828	199289	215245	209186
Indonesia	377394	388506	393251	423362	408255
		EBB (ton$CO_{2\text{-}eq}$/AU/year) [4]			
West Java	11772	12095	11714	12015	12209
Central Java	13620	13889	14013	15595	14069
East Java	25885	26801	27699	29917	29075
Indonesia	52454	53999	54658	58843	56743
		EDS predicted using EB (ton$CO_{2\text{-}eq}$/AU/year) [5]			
West Java	55840	57373	55565	56991	57912
Central Java	64604	65882	66471	73974	66735
East Java	122784	127128	131387	141907	137912
Indonesia	248809	256134	259263	279114	269155
		EDS Predicted using EBB (ton$CO_{2\text{-}eq}$/AU/year) [6]			
West Java	55840	57373	55565	56991	57912
Central Java	64604	65882	66471	73974	66735
East Java	122784	127128	131387	141907	137912
Indonesia	248809	256134	259263	279114	269155

1 Animal Population based on value in [3]; 2 AU calculation using calves, heifer (young), and cows percentage on survey in 2017 [3]; 3 Calculating use Formula. 3; 4 Calculating use Formula. 4; 5 Calculating use Formula. 5; 6 Calculating use Formula. 6

3.9 Biogas is a source of friendly environmentally alternative energy

Manure storage as biogas potentially develops in our country. Compare to dry stacking which contribute to higher GHG emissions, biogas contributes some alternative energy as a result of methane burning. The gas-burning activity also contributes to GHG emissions in CO_2. Burning 1mol of methane can produce 1 mol of CO_2, which has reactivity 28 times lower than methane. Depends on [1], as much as $1m^3$ of methane is equal with 11,17kWh. Mature dairy cattle can produce 49,35 m^3/AU/year of methane (previous calculation).

So, it means that 1 ST of dairy cattle can produce 551,24kWh/AU/year. If it used to light a lamp that needs 30 Watt, so in a year, manage the manure of 1 ST dairy cattle can lighting 18374 lamps for 1 hour. The problem found to implemented biogas as a mitigating activity and producing alternative energy is installation cost for build biogas is too expensive. Also, the small amount of manure produced as a result of a farmer only has a few dairy cattle. Regarding that situation, making portable tank biogas [2] which easy and cheap, can be adopted. This portable biogas also matches small scale farming which only has animals in a small amount.

4. Conclusion

Greenhouse gas emissions in manure management of dairy cattle are higher on biogas before gas-burning compared to dry stacking, but after gas burning the GHG emissions, It's in reverse. GHG emissions from biogas before gas-burning predicted using formula EB= 940,57 (CO2-eq/AU/year) x N and GHG emissions from biogas after gas-burning predicted using formula EBB= 130,73 (CO2-eq/AU/year) x N. GHG emissions from dry stacking during a year predicted using EDS= 0,659281 x EB or EDS= 4,74336 x EBB. Manure management through biogas potentially reducing GHG emissions, also providing friendly environmentally alternative energy.

References

[1] Aji KP, AN Bambang. 2019. Konversi energi Biogas menjadi energi listrik sebagai alternati energi terbarukan dan ramah lingkungan di desa langse kecamatan margorejo kabupaten pati. In: Proc. Seminar Nasional Teknoloogi Industri, Lingkungan dan Infrastruktur (SENTIKUIN) Volume 2 tahun 2019, Fakultas Teknik Uniersitas Tribhuwana Tunggadewi. Malang, 24 Agustus 2019. Pp:B4.1-B4.7.

[2] Atmodjo T, D Rosadi, Hardoyo. 2014. Perancangan tangki biogas portable sebagai sarana produksi energi alternatif di peesaan. Widyariset. vol.17(3):409-416.

[3] Direktorat Jenderal Peternakan dan Kesehatan Hewan. 2019. Statistika Peternakan dan Kesehatan Hewan. Ditjen PKH Kementerian Pertanian RI, Jakarta. 221 pages.

[4] Gerber PJ, H Steinfeld, B Henderson, A mottet, C Opio, J Dijkman, A Falcucci, G Tempio. 2013. Tackling climate change through livestock – A global assessment of emissions and mitigation opportunities. Food and Agriculture Organization of the United Nations (FAO), Rome

[5] IPCC (Intergovernmental Panel on Climate Change). 2006. IPCC Guidelines for National Greenhouse Gas Inventories, Prepared by the National Greenhouse Gas Inventories Programme. Eggleston HS, Buendia L, Miwa K, Ngara T, Tanabe K, editors. Institute for Global Environmental Strategies, Kanalawa. 10.87 pages.

[6] IPCC (Intergovernmental Panel on Climate Change). 2014. IPCC Fifth Assessment Report. www.ipcc.ch

[7] Isaksson, LH, AG Sanabria, Z Klimont, P Rafaj, W Schopp. 2020. Technical potentials and costs for reducing global anthropogenic methane emissions in the 2050 timeframe-results from the GAINS model. Environ Res. Commun. Vol.2(2020):1-21.

[8] Saleh A, MWK Planetto, RD Yulistiah. 2016. Peningkatan presentase metana pada biogas menggunakan variasi ukuran pori membran nilon dan variasi waktu purifikasi. J Teknik Kimia. Vol.22(4):35-44.

[9] Widiawati Y, A Herliatika, Zuratih, RA Saptati. 2019. Emisi dari subsektor Peternakan. Dalam: Metode Penilaian adaptasi dan Inventarisasi Gas Rumah Kaca Sektor Pertanian. Agus F, editor.

Badan Penelitian dan Pengembangan Pertanian, Jakarta. 154 Pages.
[10] Widyastuti FR, Purwanto, Hadiyanto. 2013. Biogas melalui pemanfaatan limbah padat pada peternakan sapi perah bangka botanical garden pangkalpinang. Metana. Vol.9(2):19-26.
[11] Yendraliza Y, M Magfirah, M Rodiallah. 2018. Struktur populasi dan potensi kecamatan benai kabupaten kuantan singingi untuk pengembangan sapi potong. J Ilmiah Ilmu-Ilmu Peternakan. Vol.21(2):70-77

Does rebound effect influence the factors of carbon emission in Indonesia? Kaya index and LMDI decomposition

Eka Sudarmaji[1,2*], Noer Azam Achsani[1], Yandra Arkeman[1], Idqan Fahmi[1]

[1]IPB University, Bogor, Indonesia
[2]Universitas Pancasila, Jakarta, Indonesia

*esudarmaji@univpancasila.ac.id

Abstract. Increased energy efficiency led to increased economic growth, the energy used in the economy, and overall production. Unfortunately, the rebound effect and carbon emissions were adverse effects on the final result of energy efficiency. Indonesia relied on new energy efficiency technologies to reduce its energy intensity while avoiding the rebound effect. Hence, the other strategy to stimulate increased energy efficiency in the economic sector was an essential part of government policy overcoming resource constraints. This paper used the logarithmic-mean Divisia index decomposition and KAYA identity to recognize the determinant factors of carbon emissions and rebound effect changes in Indonesia. According to the research findings, substantial energy rebounds in Indonesia have revealed that energy efficiency gains in Indonesia may be related to the rebound effect. Indonesia sought to go in this direction, with the national energy policy aiming to lower energy efficiency by 1% each year to stimulate energy savings in all sectors. This paper advocated enhancing energy efficiency, notably in the building sector's plan to encourage energy efficiency measures.

1. Introduction

Energy efficiency is one indication that includes both energy consumption and the forces that drive energy use. Energy indicators, in particular, were helpful because they relate energy use to important activity metrics such as GDP and output value. Energy consumption increase has been related to economic expansion, products and services, population building, and transportation. Thus, energy indicators are a valuable tool for policymakers and could also predict future development in energy use. A variety of factors influenced the relationship between energy and economic growth. To provide a fair understanding of the country's aggregated indicators as a whole, energy used and operation was needed. The most popular reasons for total energy usage in the economy were increased activity and economic development.

Using the Logarithmic Mean Divisia Index (LMDI) method to decompose the extended Kaya identity, this paper examined the driving factors of carbon emissions in Indonesia from 1971 to 2017. It was in response to a query about how carbon abatement has progressed in Indonesia over the past 36 years. What would happen if the current pattern continued? The model for evaluating efficiency was also discussed. We identified the root cause of dramatically rising CO2 emissions in the last 36 years based on the findings.

However, many energy efficiency solutions were commonly acknowledged to fail to cut energy usage by the amount expected by simple engineering models. One reason for this was that such advancements reduce the cost of energy services, which boosts demand for those services. It was referred to as the direct rebound effect. Even if demand for energy services remained constant, there were various reasons why the overall energy savings for the economy may be smaller than basic calculations indicate. For example, if electric vehicle vehicles cut fuel consumption, the savings could be spent on other goods and services that also required energy to supply.

The difference between the prospective and actual energy savings from an energy efficiency increase was sometimes referred to as the rebound effect (RE). Potential energy savings were estimated ex-ante on the assumption that there were no behavioral responses to efficiency improvements. In contrast, actual energy savings were measured or estimated ex-post and included one or more previously described behavioral responses. A rebound effect, for example, meant that potential energy savings had been "taken back." There were two types of rebound effects: direct and indirect. However, the evidence on rebound effects was limited and unclear, but direct effects are more robust than indirect effects.

Direct rebound effects came from increasing energy service consumption, such as lighting, whose effective price has decreased due to enhanced energy efficiency. As a result, energy consumption would be higher than in a scenario in which none of these behavioral responses occur. At the same time, the direct rebound impact may be inversely related to the level of energy service usage because the marginal utility from that service decreased with higher consumption.

These effects were determined by the scale of the energy-related costs and their 'visibility to the consumer. Indirect rebound effects resulted from re-spending the cost savings from energy efficiency improvements on other goods and services that also required energy to provide and thus contribute to greenhouse gas (GHG) emissions. Most empirical research estimates of indirect rebound impacts were 'economy-wide,' referring to energy use and emissions across the national economy and frequently along worldwide supply chains. Therefore, these effects were often estimated using input-output (I-O) models calibrated to the national economy's structure.

2. Literature Review

The rebound effect occurs when technical improvements make equipment more energy efficient. The cost per service unit of the equipment would decrease as the equipment became more energy efficient. As a result, prices tend to fall, making people consume more productive services and increased demand. Part of energy conservation was symbolized as a reflection effect. In developed countries, critics of energy efficiency programs said that the rebound effect occurred when consumers used energy-efficient technology more than efforts to reduce those energy cost savings.

According to the literature, there are three sorts of rebound effects from energy efficiency gains: direct, indirect, and economic (Greening, Greene, & Difiglio, 2000). Improvements in energy efficiency may also boost economic development, leading to an increase in energy consumption. There have been few studies on the repercussions of increased energy efficiency in underdeveloped countries. However, minimal empirical data suggested that direct rebound effects may be discovered.

There are a few methods to calculate Energy Rebound Effects or testing Jevon's paradox or Khazzoome Brookes Postulate, i.e., Computable General Equilibrium, LMDI, Cobb-Douglas, and Input-output methods. Chong, Liu, Ma, Li, Li, & Song (2017) and González, Landajo & Presno (2014) used the LMDI to analyze the energy consumption in uni-european and China. LMDI method can be found at Cansino, Collado, Merch (2019) for analyzing the backfire effect or Jevon's paradox in Spain caused by energy efficiency actions; Wang, Gao, Tang, Yuan, and Zuo (2018) investigated the rebound impact of energy consumption in China's three industrial sectors.

The addicting LMDI and extended KAYA identity were utilized in this paper to represent the various consequences of energy consumption fluctuations. First, using the same equation, decomposition analysis was done to break down variations in energy use into various pre-defined components. Upon decomposition index analysis, the author investigated how the rebound effects influenced CO2, GDP, Energy Intensity, and Carbon Intensity.

IOP Conf. Series: Earth and Environmental Science 1041 (2022) 012086 doi:10.1088/1755-1315/1041/1/012086

3. Methodology

Due to the flexibility of acceptance, ease of use, and a relatively low data requirement, IDA was more widely accepted as a decomposition tool. Ang (2015) summarized the IDA methods in terms of their advantages and disadvantages. Ang (2015) then advocated the Logarithmic Mean Divisia Index (LMDI) for general use. Several studies had also applied extended Kaya identity using LMDI (C. Ma & Stern, 2008; Wang, Liu, Zhang, & Song, 2014; Zhang, 2019). M. Ma and Cai (2018) and M. Ma, Cai, and Cai (2018) have conducted studies combining Kaya identity and LMDI for decomposition to total energy-related CO2 in the construction sector.

This decomposition ' effect' led to a change in the carbon reduction expressed in the Kaya. The famous KAYA equation formula was as follows:

$$\text{PEC} = \text{POP x } \frac{GDP}{POP} \times \frac{Primary\ Energy}{GDP} \times \frac{Energy\ Consumption}{Primary\ Energy} \quad (1)$$

Description

- PEC = Primary Energy Consumption
- GDP = Gross Domestic Product – Using IPP
- Energy Intensity = Primary Energy/GDP
- Consumption Intensity = EC/Primary Energy

To forecasting CO2 – this paper adds CO2 emission in the LMDI model, the equation formula as follows:

$$\text{CO2} = \text{Pop x } \frac{GDP}{POP} \text{ x } \frac{Primary\ Energy}{GDP} \text{ x } \frac{CO2}{Primary\ Energy} \quad (2)$$

Use addictive LMDI Analysis.

$$\left(\Sigma L(POP^T, POP^0)Ln\left(\frac{POP_{effect^T}}{POP_{effect^0}}\right)\right) + \left(\Sigma L(GDP^t, GDP^0)Ln\left(\frac{GDP_{effect^T}}{GDP_{effecr^0}}\right)\right) + \left(\Sigma L(PEC^T, PEC^0)Ln\left(\frac{PEC_{effect^T}}{PEC_{effect^0}}\right)\right) + \left(\Sigma L(CO2^T, CO2^0)Ln\left(\frac{CO2_{effect^T}}{CO2_{effect^0}}\right)\right) \quad (2a)$$

Rebound Method

$\Delta E_{intensity}$ was a proxy for technological improvement or technological change (Cansino, Roman-Collado, & Merchan, 2019); hence, energy intensity could change energy consumption and change economic growth. However, on the other hand, improvements in $E\Delta_{intensity}$ could be a backfired effect, commonly called the Rebound effect (RE), on energy consumption.

$$\Delta E_{Energy\text{-}efficiency} = \Delta E_{tot} - \Delta E_{Pop} - \Delta E_{GDP} \quad (3)$$
$$\gamma^1_{total} = \gamma^1_{GDP} + \gamma^1_{POP} + \gamma^1_{intensity} \quad (4)$$

From equation (4), we concluded that if $\gamma^1_{total} \geq 1$ expresses strong decoupling efforts, it implied that the country's energy consumption decreases while the national economy grows. Moreover, if $0 < \gamma^1_{total} < 1$, it expresses weak decoupling efforts so that the inhibiting effect of country energy consumption was more vulnerable than the effect of economic growth. Finally, if $\gamma^1_{total} \leq 0$, it expresses that there were no decoupling efforts $\frac{Energy\ Consumption}{Y=GDP} = \Delta EI = \frac{\Delta PEC}{PEC}$,

When energy intensity decreased by 1%, it would result in the same 1% reduction in energy costs. In other words, the resulting energy saving of 1% would be the same as the decrease in energy intensity or can be written, as follows:

$$\frac{\Delta PEC}{PEC} 1_{\mathrm{GDP}} = \frac{\Delta EI}{EI} \tag{5}$$

However, the expected energy saving of 1% did not necessarily occur because, at the same time, the rebound effect increased energy consumption. Therefore, it could be concluded that Expected Energy Saving (EES) was not the same as Actual Energy Saving (AES) because of the rebound effect. The formula of Rebound Effect (RE) can be written as follow:

$$RE = \frac{\frac{\Delta PEC}{PEC} 1y - \frac{\Delta PEC}{PEC}}{\frac{\Delta PEC}{PEC} 1y} = 1 - \frac{\frac{\Delta PEC}{PEC}}{\frac{\Delta PECI}{PECI}} = \frac{\frac{\Delta GDP}{GDP}}{-\frac{\Delta EI}{EI}} = \frac{\frac{\Delta GDP}{GDP}}{\frac{\Delta GDP}{GDP} - \frac{\Delta PEC}{PEC}} \tag{6}$$

$$\frac{\frac{\Delta PEC}{PEC} 1GDP - \frac{\Delta PEC}{PEC}}{\frac{\Delta PEC}{PEC} 1GDP}$$

Expected Energy Saving Target Ratio (ESTR) difference with Actual Energy Saving (AES) could be said to be the difference. The condition Of Zero RE = 0, would occur when the decline of 1% energy intensity would cause an increase of 1% energy consumed in generating the output economy as it did before. The addicted LMDI decomposition was used to get four aspects to represent the various consequences of energy consumption changes: population effect, GDP growth effect, energy intensity effect, and CO2 intensity effect. The LMDI became popular when (International Energy Association) IEA used LMDI to forecast CO2, followed by most researchers in energy.

First, they used to use the Laspeyres index in early 1990, then AMDI. The Data of CO2 gas emission, GDP, population, Primary energy consumption was taken to decompose the GDP-effect, Growth-effect, CO2-effect & Energy Intensity effect, consisting of 3.128 observation data. The data came from IEA and comprised data for Indonesia from 1971-2017.

4. Result and Discussion

4.1 Indonesia's energy situation

Table 1. Delta CO2, Population, GDP, Energy Intensity and CO2 effects

No	Year	**Delta CO2**	**Population Effects over CO2**		**GDP Effects over CO2**		**Energy Intensity Effects over CO2**		**Carbon Intensity Effects over CO2**	
1	1971-75	27,178	8,443	31,1%	13,506	49,7%	(8,420)	-31,0%	13,649	50,2%
2	1976-80	68,611	13,388	19,5%	30,795	44,9%	(11,152)	-16,3%	35,580	51,9%
3	1981-85	41,300	17,222	41,7%	16,615	40,2%	1,441	3,5%	6,022	14,6%
4	1986-90	102,272	20,621	20,2%	50,911	49,8%	11,899	11,6%	18,841	18,4%
5	1991-95	139,427	27,892	20,0%	97,644	70,0%	(30,454)	-21,8%	44,345	31,8%
6	**1996-97**	43,53	**6,514**	**15,0%**	**14,182**	**32,6%**	**(5,801)**	**-13,3%**	**28,635**	**65,8%**
7	**1997-98**	13,36	**6,782**	**50,8%**	**(74,242)**	**-555,6%**	56,858	**425,5%**	**23,965**	**179,3%**
8	1998-03	131,485	37,522	28,5%	62,169	47,3%	(1,635)	-1,2%	33,430	25,4%
9	2003-08	80,624	45,331	56,2%	139,482	173,0%	(111,451)	-138,2%	7,262	9,0%
10	2008-13	137,719	49,551	36,0%	162,889	118,3%	(107,578)	-78,1%	32,857	23,9%
11	2013-17	156,394	42,295	27,0%	135,835	86,9%	(53,450)	-34,2%	31,713	20,3%

IOP Conf. Series: Earth and Environmental Science 1041 (2022) 012086 doi:10.1088/1755-1315/1041/1/012086

When oil prices surged in the 1970s, Indonesia's energy demand skyrocketed, resulting in a windfall in export profits. Between 1968 and 1981, exports led to high GDP rates of more than 7% on average. However, due to declining oil prices, growth slowed to an average of 4.5 percent per year between 1981 and 1988. Economic reform occurred during the 1980s, including a controlled depreciation of the rupiah to boost export competitiveness and banking sector deregulation.

As a result, from 1989 to 1997, international investment poured into Indonesia, notably in the export-oriented manufacturing sector. In 1998, real GDP fell by 13.1 percent, and the economy hit a low point in mid-1999, with real GDP growth of 0.8 percent. By that time, Indonesia's real GDP growth had reached 6% in 2012, falling to 5.1 percent in 2004 and 5.6 percent in 2005. The administration then made efforts to relax foreign direct investment regulations in order to revive the economy. Consequently, Indonesia managed to boost GDP growth to slightly more than 5% in 2016–17. As a result, Indonesia's primary energy consumption rose significantly, with an annual growth rate of 5,157 percent on average between 1971 and 2017.

Indonesia increased its GDP growth by about 5% in 2016. Indonesia's demand for primary energy has risen at an average annual rate of 5.157% over the last forty years, from 1971 to 2017. all of the direct energy supply was greater than 10,462.6 PJ Indonesia's emissions have increased just 2048 tons per year since 1971 due to moderate economic growth and only moderate energy intensity improvement. Total CO2 emissions in Indonesia spanned the 46 years from 1971 to 2017 and 940 metric tons in that time frame. Indonesia has set a target of reducing its emission intensity by 29.41% by 2030. However, Indonesia's emissions would continue to rise rapidly due to the business climate change it will face in the coming decade.

For the Indonesian leaders, keeping the country's energy security had become extremely difficult. As can be seen in Figure 1, the primary energy source folds 49 times. GDP (or economic) components and population growth were the key factors behind the rise in energy demand. Using decomposition analysis, it could be seen that the GDP components were playing a more critical role in spiking energy demand and energy intensity. The 1997-to-1998 Asian crisis was the only time when GDP and energy intensity components decreased. An example of the Decomposition for Indonesia was depicted in the graph in the figure below.

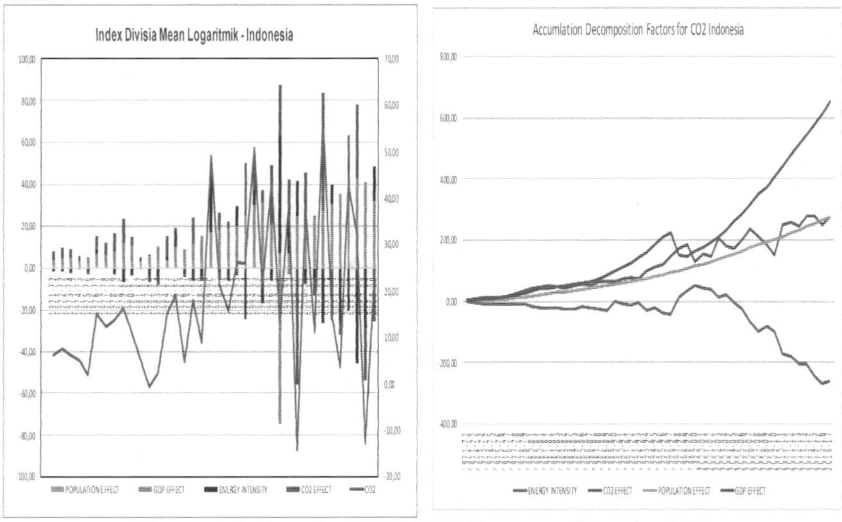

Figure 1. Indonesia CO2 LMDI Decomposition Analysis

4.2 Decomposition analysis

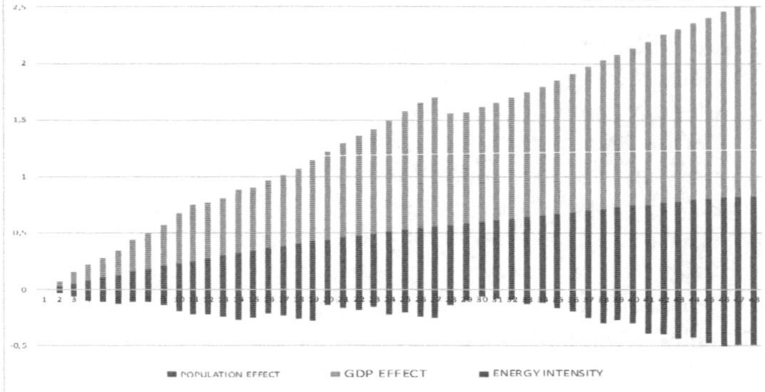

Figure 2. Indonesia Decomposition Growth Factors 1971-2017 With 1971 = 100

Ensuring a reliable and adequate energy supply in Indonesia had become increasingly challenging for Indonesia's leaders. Whereas primary energy supply folds more than 49 times, the determinant factors leading to rising demand for energy were caused by GDP's components and population growth components. Through decomposition analysis, Fig. 1 and table 1 clearly showed that the GDP's components had a more significant role in encouraging energy demand and energy intensity components to play their function in soaking demand for energy throughout 1971-2017.

Decreased part of the GDP and energy intensity components occurred only in the Asian Crisis, which began in 1997-1998. The weakening of the energy intensity components occurred from 1985 to 1990 due to the drop in oil prices. Oil prices, for example, fell from USD 69.97 per barrel in September 1985 to USD 31.11 per barrel in February 1986. As a result, Indonesia's GDP growth rate fell by 2.1 percent, 7.3 percent, and 7.8 percent, respectively, in 1985, 1986, and 1987. The second drop happened during the Asian Crisis, which lasted from 1997 to 1998.

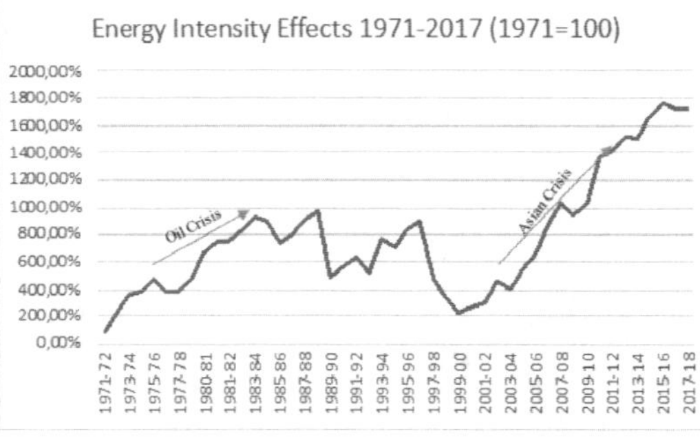

Figure 3. Accumulation Energy Intensity 1971-2017

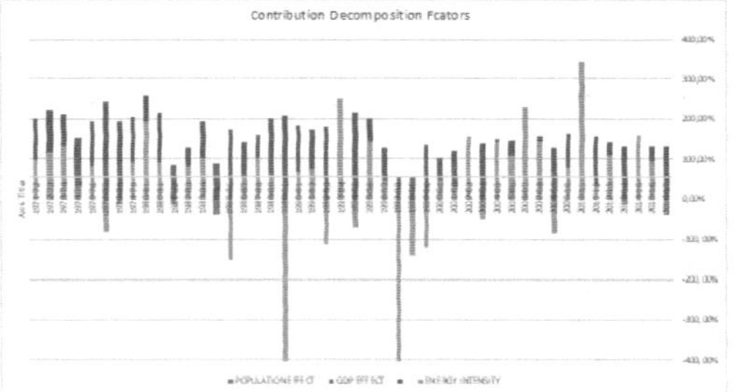

Figure 4. Indonesia Decomposition factors: Pop-effect, GDP-effect, EI-effect & CO2-effect 1971-2017

IOP Conf. Series: Earth and Environmental Science 1041 (2022) 012086 doi:10.1088/1755-1315/1041/1/012086

4.3 Rebound Analysis

Table 2. Five stated of rebound effects

Status of Energy Rebound		
No	State of Energy Rebound (RE)	d(EI) in Energy Use from 1% of Energy Intensity
1	Super Conservatism (RE<0%)	d(EI) < - 1%
2	Zero (RE=0)	d(EI) = - 1%
3	Partial (0% < RE < 100%)	-1% , d(EI) < 0%
4	Full (RE-100%)	d(EI) = 0%
5	Back Fire (RE>100%)	d(EI) > 0%

Source: Brockway 2t al, 2017

Brockway et al. identified five possible states of rebound effects, see table 2 above. Based on those possible stated, we investigated how CO2 changes over time to attribute changes to Population, Gross Domestic Product, Energy Intensity (Energy per unit GDP), and Carbon Intensity (CO2 per unit energy).

Table 3. Rebound calculation 1971-2015

Year	D(PE)	D(EI)	% Rebound	Remarks	Year	D(PE)	D(EI)	% Rebound	Remarks
1971-72	4,1%	-3,0%	-38,5%	Super Conservation	1994-95	10,3%	2,0%	607,0%	Back Fire
1972-72	4,6%	-3,5%	-29,5%	Super Conservation	1995-96	3,6%	-4,1%	10,8%	BackFire
1973-74	3,7%	-3,9%	6,4%	BackFire	1996-97	3,4%	-1,4%	-148,7%	Super Conservatio n
1974-75	4,0%	-1,0%	-302,2%	Super Conservation	1997-98	-2,2%	12,0%	81,8%	Back Fire
1975-76	4,5%	-2,3%	-101,3%	Super Conservation	1998-99	4,8%	4,0%	218,6%	BackFire
1976-77	11,1%	2,3%	575,0%	BackFire	1999-00	8,4%	3,2%	361,4%	BackFire
1977-78	7,0%	0,3%	2836,7%	BackFire	2000-01	2,3%	-1,4%	-63,8%	Super Conservatio n
1978-79	4,7%	-2,7%	-74,2%	Super Conservation	2001-02	3,6%	-0,8%	-332,4%	Super Conservatio n
1980-81	4,1%	-5,8%	28,9%	BackFire	2002-03	0,2%	-4,7%	95,7%	BackFire
1980-81	5,2%	-2,6%	-99,8%	Super Conservation	2003-04	6,5%	1,4%	554,5%	Back Fire
1981-82	2,5%	0,2%	1163,1%	Back Fire	2004-05	1,5%	-4,0%	61,6%	BackFire
1982-83	1,8%	-2,3%	20,1%	BackFire	2005-06	2,3%	-3,2%	29,2%	BackFire
1983-84	3,9%	-2,9%	-34,0%	Super Conservation	2006-07	-0,4%	-6,7%	106,0%	BackFire
1984-85	3,5%	1,1%	422,6%	BackFire	2007-08	1,7%	-4,1%	57,3%	BackFire
1985-86	10,3%	4,3%	338,0%	BackFire	2008-09	7,1%	2,3%	401,7%	BackFire
1986-87	3,3%	-1,6%	-103,2%	Super Conservation	2009-10	4,0%	-2,2%	-83,7%	Super Conservatio n
1987-88	2,7%	-3,1%	11,1%	BackFire	2010-11	-3,9%	- 10,4%	137,2%	BackFire
1988-89	5,7%	-1,7%	-231,0%	Super Conservation	2011-12	5,0%	-1,0%	-376,3%	Super Conservatio n
1989-90	21,2%	13,6%	256,4%	BackFire	2012-13	2,3%	-3,2%	27,0%	BackFire

1990-91	4,9%	-2,0%	-148,0%	Super Conservation	2013-14	5,3%	0,3%	1731,9%	BackFire
1991-92	4,2%	-2,2%	-93,1%	Super Conservation	2014-15	0,2%	-4,6%	95,8%	BackFire
1992-93	9,8%	3,2%	405,4%	BackFire	2015-16	2,2%	-2,7%	20,6%	Back Fire
1993-94	0,1%	-7,4%	98,2%	BackFire	2016-17	6,3%	1,2%	626,2%	BackFire

Table 3 found that Indonesia's rebound effect mirrors its energy consumption, shown in Fig. 4. The most noticeable rebound effect took place in the year 1977-1978. It happened due to the rise in oil prices in the 1970s. Hence, it resulted in an export revenue windfall that sustained high economic growth rates, with an average growth of over 7% between 1968 and 1981. In the middle of 1979, oil prices rose due to a decline in oil output following the Iranian revolution and global oil supplies. The price of crude oil had more than doubled to $39.50 per barrel over the next 12 months. Moreover, it has contributed to a high rate of economic growth for Indonesia.

Based on table 3 and Figure 5, Indonesia had experienced more than 16 times super conservation and 30 times backfire effect, taking the rebound measurement (Saunders, 2015). The results show that, between 1971 and 2017, the energy rebound effect amounted to an average of 198.4%, implying that Indonesia could not merely rely on technical means to reduce energy consumption and emissions. The average rebound effect has been around 249.44%, with an upward trend in the last decade, indicating that the expected energy savings have been missed. The results show that GDP was mainly responsible for increasing CO2. Economic instruments should also be used as complements to ensure energy conservation and emission reduction results.

According to the International Energy Agency, energy and carbon intensity have improved in recent years, reverting to levels not seen since the 1990s. GDP growth has been lower since the global financial crisis, and this trend has continued to the present. Furthermore, the present COVID-19 epidemic was causing the world economy to slip back into a slump. Three interconnected impacts – somewhat reduced economic growth, increased energy intensity, and increased carbon intensity – resulted in a slower increase in world CO2 emissions.

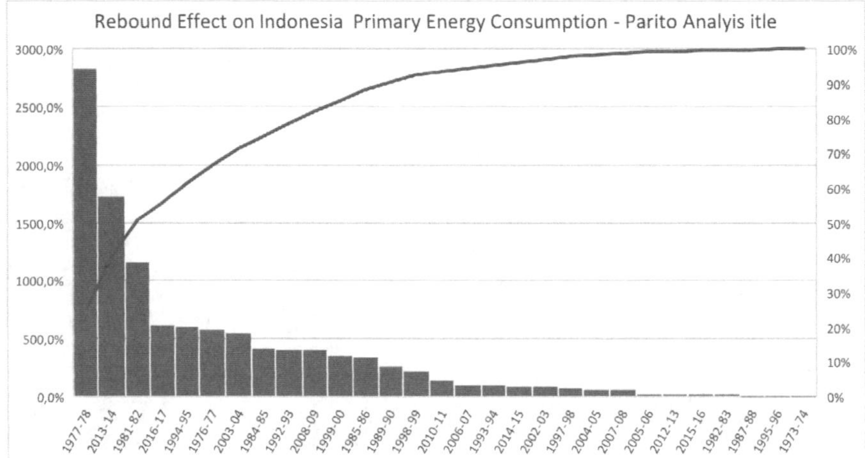

Figure 5. Indonesia Decomposition factors: Pop-effect, GDP-effect, EI-effect & CO2-effect 1971-2017

4.4 Decomposition of changes in CO2 emissions

From 1971 to 2017, Indonesia has seen substantial emissions increase of about 20.48 metric tons per year, driven by strong economic development and modestly improved energy intensity. Total CO2 emissions in Indonesia from 1971 to 2017 were 941.40 metric tons over 46 years. Indonesia promised to lower its emissions intensity by 29.41% by 2030. Emissions in Indonesia may continue to rise sharply as a result of economic activity.

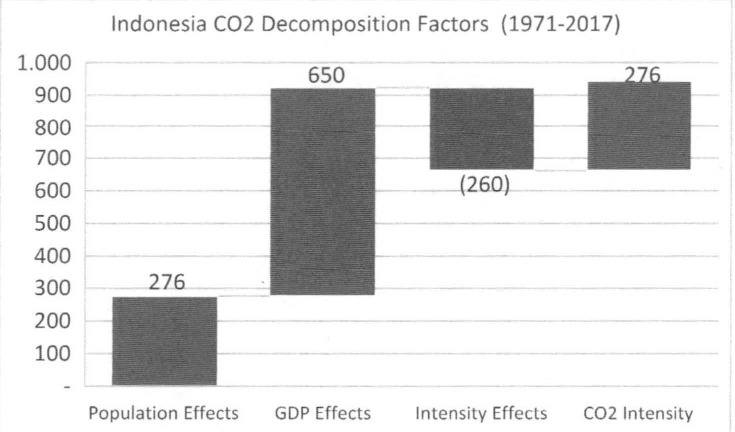

Figure 6. Indonesia Accumulation Decomposition Effect 1971-2017

As shown by urbanization, the population impact was a factor that exacerbated the growth in CO2 emissions. Household consumption was one of the variables that led to CO2 emissions. Fortunately, throughout the previous decade, the outcome demonstrated that Indonesia's GDP effect was taking off due to the government increasing private sector growth to diversify the petroleum economy. According to the findings, the GDP impact was the most compelling element in the yearly rise in CO2 emissions, followed by the population effect in Indonesia (see Fig. 6). This conclusion is consistent with previous research on the influence of GDP, as measured by the proportion of output in GDP.

The reduction in total CO2 emissions is attributable to the energy intensity effect. Energy intensity has reduced CO2 emissions in Indonesia by increasing the technical aspect. The energy intensity effect has been linked to lower CO2 emissions. This research discovered that, based on the percentage of carbon intensity impact on CO2 emissions during the previous decade, Indonesia was a CO2 emission-reduced country. The total of these four variables equaled the total of CO2. Again, it was shown that Kaya's index could identified that the GDP and Population were the primary driving force of CO2 emissions, and rebound effects contributed indirectly.

As suggested by this research, now is a good time for Indonesia to increase renewable energy deployment. The government is responsible for supporting the Energy Efficiency Saving (Retrofitting) projects in Indonesia (Nasip & Sudarmaji, 2018). The retrofitting programs are doable technically; hence the government must unlock the energy efficiency industry through incentives policies or social finance instruments such as project support from banks and pension funds (Sudarmaji & Ardianto, 2020). Other aspects of the government's energy conservation policy can be made more effective with the national nudges.

Similarly, nudges are accepted in several different industries besides manufacturing. Sudarmaji & Thalib, 2020 found that social norms and restraints have influenced electric use in rural areas. Nudging energy conservation solutions were included in the 'nudges.' It usually occurred in many countries nudging can be used in many industries. Carbon dioxide emissions can lower the energy needed in future energy-conservation projects (Krstic & Krstic, 2015).

5. Conclusion

The present COVID-19 epidemic was causing the world economy to slip back into a slump. Three interconnected impacts – somewhat lower economic growth, improved energy intensity, and improved carbon intensity – contributed to a slower increase in global CO2 emissions in Indonesia. According to the research findings, substantial energy rebounds in Indonesia have revealed that energy efficiency gains in Indonesia may be related to the rebound effect. Indonesia sought to go in this direction, with the national energy Policy (KEN) aiming to lower energy efficiency by 1% each year to stimulate energy savings in all sectors. In the low carbon scenario, the industrial sector will dominate other sectors by 2050, accounting for 37% of total output. Electricity consumption will be more dominating, accounting for 33% of total demand. It comprises the implementation of suitable policy actions to reduce CO2 emissions.

According to this paper, the increase in CO2 emissions was mainly caused by economic expansion, followed by population growth (urbanization). This paper advocated enhancing energy efficiency, notably in the financial sector's plan to encourage energy efficiency measures. Further research may look at whether Indonesia's low-carbon economic objective was technically feasible. Our findings also show how significant the rebound impact was in Indonesia, which might be a crucial component and driving factor for economic growth. It raised the question of whether climate-related measures may constrain future economic development. This article now proposes that now is a favorable moment for Indonesia to accelerate the deployment of renewable energy sources.

References

[1] Ang, B. W. (2015). LMDI decomposition approach : A guide for implementation. Energy Policy, 86, 233–238.

[2] Cansino, J. M., Roman-Collado, R., & Merchan, J. (2019). Do Spanish energy efficiency actions trigger JEVON'S paradox? Energy, 181, 760–770.

[3] Krstic, B., & Krstic, M. (2015). Models of Irrational Behaviour of Household and Firm. Ekonomika, 61(4), 1–10. https://doi.org/10.5937/ekonomika1504001K

[4] Ma, C., & Stern, D. I. (2008). Biomass and China's carbon emissions: A missing piece of carbon decomposition. Energy Policy, 36, 2517–2526.

[5] Ma, M., & Cai, W. (2018). What drives carbon mitigation in the Chinese commercial building sector? Evidence from decomposing an extended Kaya identity. Science of the Total Environment, 634, 884–899.

[6] Ma, M., Cai, W., & Cai, W. (2018). Carbon reduction in China's commercial building sector: A bottom-up measurement model based on Kaya-LMDI methods. Energy, 165, 350–368. Retrieved from https://doi.org/10.1016/j.energy.2018.09.070

[7] Nasip, I., & Sudarmaji, E. (2018). Managing Tax Dispute Due to IFRS-16 on the Retrofits Implementation in Indonesia. International Journal of Engineering and Technology (IJET), 7(3.21), 200–208.

[8] Saunders, H. D. (2015). Recent Evidence for Large Rebound : Elucidating the Drivers and their Implications for Climate Change Models. The Energy Journal, 36(1), 23–48.

[9] Sudarmaji, E., & Ardianto, Y. (2020). Alternative Financing for Energy Efficiency Saving in Indonesia Under LCCA Analysis. International Journal of Psychosocial Rehabilitation, 24(6), 3178–3186.

[10] Sudarmaji, E., & Thalib, S. (2020). To Nudge Or Not To Nudge Households: Energy Efficiency Case In Indonesia. International Journal of Production Economics, 24(6), 3225–3241.

[11] The ASEAN Secretariat. (2019). ASEAN Integration Report 2019. Jakarta, Indonesia.

[12] Wang, W., Liu, X., Zhang, M., & Song, X. (2014). Using a new generalized LMDI (logarithmic mean Divisia index) method to analyze China's energy consumption. Energy, 67, 617–622.

[13] Zhang, Y. (2019). Energy Rebound Effect Analysis Based on Technological Progress. In IOP Conference. Series Earth and Environmental Science 300.

IOP Conf. Series: Earth and Environmental Science 1041 (2022) 012087 doi:10.1088/1755-1315/1041/1/012087

Study of phytoplankton biology index and water quality parameters of kali Surabaya River

O H Cahyonugroho[1,2*], S Hariyanto[1*], G Supriyanto[1]

[1]Universitas Airlangga, Surabaya, Indonesia.
[2]Universitas Pembangunan Nasional Veteran Jawa Timur, Surabaya, Indonesia.

*sucipto-h@fst.unair.ac.id

Abstract. Phytoplankton is used as an indicator of water quality because of its short life cycle and fast response to changes in the aquatic environment, especially in river ecosystems. In addition, the existence of industrial, agricultural, and community activities around the river will cause changes in the water quality of the waters. This study aim was to analyze the distribution of the phytoplankton community based on the biological index related to the water quality parameters of the Kali Surabaya river. The study was carried out in January – February 2021. The study sample used a random sample method which was divided into four observation station. The samples obtained were identified by the type of phytoplankton and tested for water quality parameters (pH, temperature, brightness, strong currents, dissolved oxygen, biological oxygen demand, nitrate, phosphate, and chlorophyll-a). The results showed the phytoplankton abundance index in the Kali Surabaya river is low with a range of 485,49–626,21 cell/L, the diversity index (H') is moderate with a range of 1.446–1.703, uniformity index (E) in the range of 0,61–0,626, and dominance index (D) with a range of 0,293–0,368. The relationship between the biological index of phytoplankton with the water quality of the Kali Surabaya River that the water quality parameters that had more influence on the diversity index, uniformity index, and dominance index were pH, dissolved oxygen, and nitrate.

1. Introduction

Rivers as open water ecosystems that are often referred to as public waters are strongly influenced by the surrounding environment. Changes in water quality conditions in river flow impact waste from existing land use [1]. Kali Surabaya River is the main river of the Brantas watershed downstream, which is the main source of water needs as well as a domestic and industrial waste disposal site in the Surabaya area. The main designation upstream is as a source of irrigation water for about 800,000 hectares of agricultural area. Another designation in the upstream is used as raw water for domestic purposes for residents of Surabaya and surrounding areas. In some locations, the water of the Kali Surabaya River is used for fish cultivation of approximately 4,000 hectares of fishery area [2].

The Kali Surabaya River is part of the Brantas river that flows from Lengkong Baru Dam into the jagir floodgates of Surabaya. According to the Government Regulation Number 82 of 2001 and the Governor Regulation of East Java No. 2 of 2008 on water quality management and water pollution control in East Java Province, Kali Surabaya River includes class I river, which is water whose allocation can be used for drinking water raw water so that it is always controlled by water quality.

One way to control water quality can be done by looking at biological parameters using phytoplankton as an indicator. Phytoplankton plays a vital role in water, its ecological function as a primary producer, and the beginning of the chain in the food web causes phytoplankton to be often used as a scale of fertility measure or bioindicator of water. The increasing nutrient content in the waters will directly affect the phytoplankton community and the surrounding environment. The complexity of the food chain and food web and the importance of phytoplankton as a primary producer will considerably influence the dynamics of river ecosystems. The more stable the aquatic environment, directly it will strengthen the stability of plankton communities. The presence of phytoplankton dramatically affects fish and aquatic larvae [3]. The distribution of aquatic environmental communities in rivers is analyzed through biological indices, namely, the diversity index, dominance index, and phytoplankton uniformity index.

A good quality water is a basic human need. The presence of water is not only determined by its amount, but also its quality. In everyday life, good quality water determines both humans and other living things [4]. Water quality can be assessed physically and chemically, such as temperature, pH, brightness, dissolved oxygen (DO), biological oxygen demand (BOD), nitrate, phosphate, and optimal balance between oxygen and carbon dioxide [4].

Until now, monitoring water quality conditions and information regarding phytoplankton biology indexes used as bioindicators of quality and pollution of the Kali Surabaya river have not been carried out. Therefore, this study was conducted to analyze the distribution of phytoplankton communities using biological indices related to the water quality parameters of the Kali Surabaya River.

2. Methodology

The study was conducted in The Kali Surabaya Watershed from January to February 2021. Phytoplankton sampling and water samples were conducted at four stations. Location of station 1 through station 4 as in figure 1.

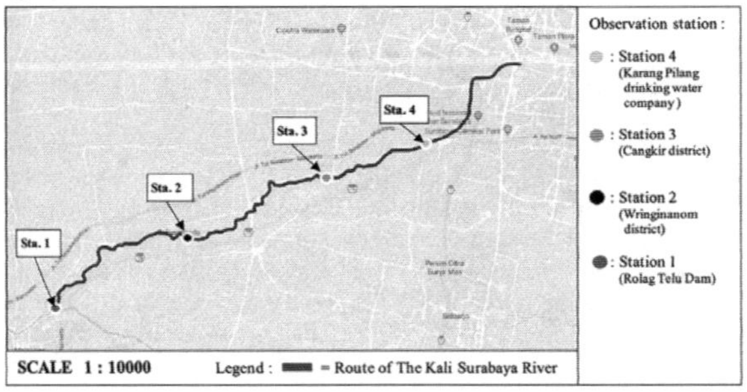

Figure 1. Locations of study area in the Kali Surabaya River

The phytoplankton sampling procedure was carried out by taking river water using a 5 L bucket and filtering it using a 35 mikron net plankton size of 100 L, after that it was put into a 100 mL bottle, then 12 mL of 4% formalin solution was added to the bottle. The filtered water containing phytoplankton was then identified in the Laboratory of the Faculty of Marine and Fisheries Sciences, Malang, based on the manuals Yamaji (1979) and Newel & Newel (1977).

To determine the distribution of phytoplankton communities and quantitative analysis of abundance using the Lackey Drop Macrotransect Counting method. While to calculate the biological index of phytoplankton, namely the diversity index, uniformity index, and dominance index using the Shannon-Wiener method. And for measuring water quality parameters (pH, temperature, brightness, strong currents, DO, BOD, nitrate, phosphate, and chlorophyll-a) using the following analytical methods which is temperature (SNI 06-6989.23-2005), brightness (SNI 7644-2010), strong currents (SNI 7644-2010), pH (SNI 7644-2010), dissolved oxygen (DO) (SNI 06-6989.14-2004), biological oxygen demand (BOD) (SNI 6989.72-2009), nitrate (NO3-N) (SNI 6989.79-2011), phosphate (PO4) (SNI 06-6989.31-2005), and chlorophyll-a (SNI 06-4157-1996).

The data obtained will be processed using statistical methods of principal component analysis (PCA) with Minitab 18 software. The above statistical method is intended to find out the main components of the Kali Surabaya river water quality parameters.

3. Result and Discussion

3.1 Abundance and biological index of phytoplankton.

The results of the calculation of phytoplankton abundance index in all observation Station as shown in table 1. obtained a range between 485,49 – 626,21 cell/L, which is relatively low. Abundance with a value of < 1.000 cell/L is classified as low, abundance between 1.000 – 40.000 cell/L is relatively moderate, and abundance > 40.000 cell/L is high. The average results of the diversity index (H') across all observation stations ranged from 1,446 – 1,703. It shows that the phytoplankton community in Kali Surabaya River has moderate diversity, moderate biota community stability, and moderate polluted water quality.

Table 1. Abundance Index and Phytoplankton Biology Index in Kali Surabaya River

Biology Index	Study's Station			
	1	2	3	4
Abundance Index (cell/L)	557,717	535,523	626,212	485,491
Diversity Index (H')	1,703	1,494	1,446	1,496
Uniformity Index (E)	0,626	0,571	0,561	0,576
Dominance Index (D)	0,293	0,342	0,365	0,368

The criteria for H' values range from 1 to 3 then the stability of the biota community is moderate. If the diversity index is between 1 - < 2,5, it falls into moderate diversity and moderate pollution. The greater the value of H' indicates the more diverse life in these waters. This condition is a better place to live. The calculation of the average value of the uniformity index (E) in all observation station ranges from 0,561 – 0,626. From the range seen, the value of E < 1 indicates that the biota rate between common genera or the uniformity structure between species is low.

The average value of the dominance index (D) in all observation stations ranges from 0,293 – 0,368. If the dominance value is close to the number 1, then in the structure of the biota community observed, some species significantly dominate other species. The species Nitzschia sp. found in all observation stations indicates that the river is in eutrophic – hyper-eutrophic conditions [6]. The more domineering composition of Bacillariophyte indicates that the waters are moderately polluted and have optimal nutritional conditions [7].

The rainy season plays a major role in determining the hydrological properties and dominance of certain types of plankton photos through the entry of allochthonous nutrients into rivers that may have stimulated the growth, reproduction and survival of phytoplankton populations, leading to an increase in their cell density [8].

3.2 Water quality parameters in the Kali Surabaya River

Depiction of temporal trend of the water quality parameters of the Kali Surabaya river, whose value affects the diversity, distribution, and pattern of phytoplankton abundance presented in table 2.

Table 2. Water quality parameters in Kali Surabaya River

Parameter	Unit	Water quality standards	Station 1	Station 2	Station 3	Station 4
pH	-	5,0 – 9,0	9,475	9,331	9,263	9,294
Temperature	^{O}C	± 5	27,781	27,594	28,088	27,638
Brightness	cm	-	9,375	12,000	6,063	5,563
Strong currents	m/dt	-	1,045	1,197	1,047	1,103
DO	mg/L	> 0	5,994	4,538	3,775	3,694
BOD	mg/L	12	16,134	18,549	14,987	13,522
Nitrate	mg/L	20	6,007	4,649	4,473	5,154

Phosphate	mg/L	5	0,366	0,327	0,366	0,291
Chlorophyll-A	mg/m^3	-	0,354	0,428	0,379	0,273

At shown in table 2. the measurements of acidity degrees (pH) at all observation stations range from 9,263 – 9,475. It shows that the pH of the Kali Surabaya river is stable and viable for the river organism. Fluctuations pH is greatly influenced by the process of respiration carried out by phytoplankton. The more carbon dioxide produced from the respiration process, the lower the pH. However, conversely, if photosynthesis activity is higher, it will cause a higher pH [9]. On temperature measurements ranging from 27,59O – 28,09OC. It shows that the temperature in the Kali Surabaya river has an optimal temperature for phytoplankton growth. The optimal temperature for phytoplankton ranges from 25O – 30OC [10].

The temperature of the waters is affected by the intensity of light entering the water. On brightness, measurements range from 5,56 - 12 cm. It shows the sun's sheer power into the river when sampling is low due to the brightness level of < 50 cm. Brightness is a water condition that describes the optical properties of water that are determined based on the amount of light absorbed and emitted by materials contained in the water. On the current strength (currents speed) measurements range from 1,045 – 1,197 m/s. It shows that the current strength of the Kali Surabaya river belong to the strong current velocity. This is based on the criteria for currents < 0,1 m/s including very weak currents speed, currents 0,1 – 1 m/s including medium currents speed and currents > 1 m/s including strong currents speed.

In the measurement of dissolved oxygen (DO) ranging from 3,694 - 5,994 mg/L. It shows that the dissolved oxygen contained in the Kali Surabaya river is low. Although phytoplankton can still live well, it is seen that quite a lot of phytoplankton types are found in the Kali Surabaya river. Plankton can live well at more than 3 mg/L oxygen concentrations [11]. In BOD measurements range from 13,522 - 18,549 mg/L. It shows that the concentration of BOD kali Surabaya river is classified as high more than water quality standards or water quality status Kali Surabaya river is heavily polluted with a pollution index that exceeds 10 mg/L.

The tendency to increase BOD concentration in the rainy season can be caused by the deposition of domestic wastewater from the residential, the possibility of industrial wastewater disposal [12]. Nitrate measurements range from 4,47 – 6,00 mg/L. It shows that nitrate sources from the mainland in the form of waste containing nitrates and the concentration of nitrates of the Kali Surabaya river is still fairly normal. The permissible nitrate concentration in the waters is 10 mg/L. Nitrates are one of the essential nutrients needed by all water biota, including phytoplankton. Nitrate levels of less than 0,44 mg/L are also a limiting factor for phytoplankton growth [13].

In phosphate measurements ranging from 0,29 – 0,366 mg/L. It shows that the phosphate concentration of the Kali Surabaya river is still good in supporting phytoplankton life due to the presence of organic matter that enters the river and undergoes a decomposition process that produces nutrients. The main sources of nutrient input in the waters are fertilizer runoff from agricultural land, the use of detergents containing phosphates, erosion of soil containing nutrients, as well as domestic and industrial waste disposal [14].

Phosphate levels below 0,09 mg/L can be a limiting factor for phytoplankton growth [15]. On chlorophyll-a measurements range from 0,27 – 0,43 mg/m3. It shows that the condition of the Kali Surabaya river is in a good category and belongs to the oligotrophic water type. It is based on the criteria of waters with chlorophyll-a < 15 mg/m3 categorized into good condition, chlorophyll-a levels of 15 – 30 mg/m3 of the moderate category, and levels of chlorophyll-a > 30 mg/m3 of bad categories [16].

3.3 Correlation between biology index, phytoplankton class, and water quality

Analysis of the principal components in figure 2. shows that most of the contributing parameters are in the first component, amounting to 48,2% of the total variance, while the second component contributes 26,7% of the total variance. So that the overall contribution of the main component constituents reached 74,9%.

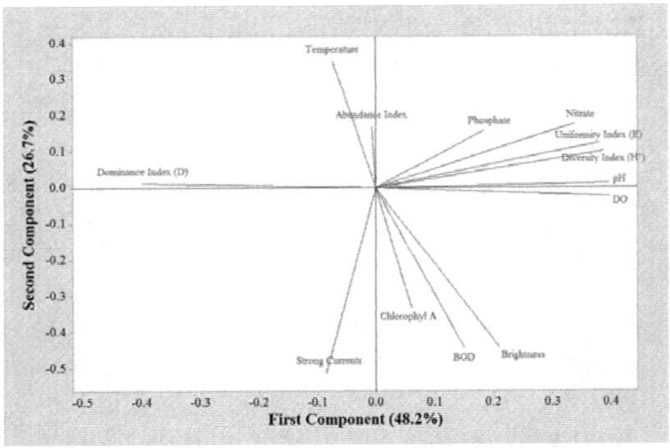

Figure 2. The principal components analysis on the phytoplankton biological index with the water quality of the Kali Surabaya River.

The parameters that contribute to the first component are pH (0,397), DO (0,398), nitrate (0,38), diversity index (0,387), uniformity index (0,380) and dominance index (-0,398). The results showed that the increase in pH, DO, and nitrate values, will be followed by an increase in the diversity index, uniformity index, and dominance index in the Kali Surabaya river. In the second component, the contributing parameters are temperature (0,350), brightness (-0,443), strong currents (-0,512), BOD (-0,445) and chlorophyll-a (-0,334).

4. Conclusion

The phytoplankton abundance index in the Kali Surabaya river is relatively low with a range of 485,49 – 626,21 cell/L, the diversity index (H') is relatively moderate, the stability of the biota community is relatively low, and the quality of moderately polluted water with a range between 1,446 – 1,703, the uniformity index (E) indicates the distribution of biota among other species, namely Nitzschia sp. with a range of dominant indexes between 0,293 – 0,368 which is close to the number 1.

Measurements of water quality parameters, Kali Surabaya shows that the water of the Kali Surabaya river is still feasible for organisms such as phytoplankton to develop/survive due to stable water conditions. However, it is proved by the value of pH, temperature, strong currents, brightness, DO, nitrate, phosphate, and chlorophyll-a, which is still below the water quality standard except for the BOD value above the water quality standard. This causes the Kali Surabaya river into the criteria of water quality status to be heavily polluted with a pollution index of > 10 mg/L.

The analysis of the principal components of the phytoplankton biology index with Kali Surabaya river water quality, water quality parameters that are more influential on diversity index, uniformity index, and dominance index are pH, DO, and nitrate.

Acknowledgments

The authors thank to the Directorate of Research and Community Service, Deputy for Strengthening Research and Development of the Ministry of Research and Technology / National Research and Innovation Agency who has provided support through Postgraduate Research Grants - Doctoral Dissertation Research Scheme, with Contract No. 489 / UN3.15 / PT / 2021.

References

[1] Kahirun S L O, Surya R A, Erif L O M, Yasin A and Ifrianty 2019 Indicators of river water quality by using macroinvertebrates in wanggu river. J Ecogreen, 5 63–67.
[2] Satmoko Y and Nusa I S 2019 Surabaya river water quality condition. case study: improving raw water quality of PDAM Surabaya. J Teknologi Lingkungan 20 19–28.

[3] Sirait M, Rahmatia F, and Pattulloh 2018 Comparison of diversity index and phytoplankton dominance index on the ciliwung river Jakarta. J Kelautan 11 75–79.
[4] Khairuddin, Yamin M and Syukur A 2016 Analysis of water quality kali ancar using macroinvertebrate bioindicators. J Biologis Tropis 16 10-22.
[5] Dema V G A 2019 Analysis of river water quality in kali surabaya river flow. J Swara Bhumi 5 1–6.
[6] Soeprobowati T R and Suedy S W A 2011 Rawakening lake phytoplankton community. J Sains dan Matematika 19 19-30.
[7] Basavaraja D, Narayana J, Puttaiah E T and Prakash K 2012 Phytoplankton species diversity indices in anjanapura reservoir, Western Ghat region, India. J Environ. Biol. 34 805–10.
[8] Kim T W, Najjar, R G and Lee K 2014 Influence of precipitation events on phytoplankton biomass in coastal waters of the eastern united states. Global Biogeochemistry Cycles 28 1–13.
[9] Damar A, Colijn F, Hesse K J and Wardiatno Y 2012 The eutrophication states of jakarta, lampung, and semangka bays: nutrient and phytoplankton dynamics in Indonesian tropical waters. J Tropical Biol. Conservation 9 61-81.
[10] Kadir M A, Damar A, and Krisanti M 2015 Spatial and temporal dynamics of zooplankton community structure in jakarta bay. J Ilmu Pertanian Indonesia 20 247-56.
[11] Wijayanti 2011 Diversity of plankton species in different places environmental conditions in rawa pening Semarang regency. Thesis. IKIP PGRI Semarang: Semarang.
[12] Yudo S, and Said N I 2019 Surabaya river water quality conditions, case study: improvement of raw water quality PDAM Surabaya. J Teknologi Lingkungan 20 19-28.
[13] Sudarsono 2012 Phytoplankton community structure in the waters of sei jang village, bukit bastari district of Tanjungpinang city. Thesis. Water Resources Management Study Program. FKIP. Raja Ali Haji Maritime University, Riau Islands.
[14] Alvarez-Vázquez L J, Fernández F J, and Martínez A 2014 Optimal control of eutrophication processes in a moving domain. J Franklin Institute 351 4142-82.
[15] Yuliana 2012 Implications of changes in nutrient availability to the rapid development (blooming) of phytoplankton in the waters of Jakarta bay. Dissertation. Graduate Faculty of Bogor Agricultural Institute, Bogor.
[16] Fitra F, Zakaria I J and Syamsuardi 2013 Primary productivity of phytoplankton in bungus bay. J Biol. 2 59-66.

IOP Conf. Series: Earth and Environmental Science 1041 (2022) 012088 doi:10.1088/1755-1315/1041/1/012088

Factor and perception analysis of the impact of boat parking on river pollution

B Rahman*, M S Aryanto

Urban Regional and Planning, Sultan Agung Islamic University, Semarang, Indonesia

*bobyrahman@unissula.ac.id

Abstract. This paper examines factors affecting river pollution from boat parking activities and the community perception of these factors. Uses a quantitative methodology, utilizes questionnaires from people active in the Silugonggo River, using technical factor analysis to find factors that influence river pollution, followed by perception analysis techniques to analyze the community's social view of river pollution from boat parking. The results showed that the number of a boat parked, and the length of parking were factors affecting river pollution. The results of community perceptions also confirm these results, by giving a negative view of social environmental conditions on these two factors.

1. Introduction

The North Coast of Java or often called "Pantura" has a long coastal area. It is one of the areas that have a good level of productivity [1]. This factor is formed because the coastal area of the north coast has many river estuaries so that it can raise to potential such as high organic matter content. In addition, this productivity is formed due to the high diversity of coastal ecosystems such as mangroves, ponds, and river estuaries that can be used as an access to and from rivers and the sea. [2]. The diversity of northern coastal ecosystems has led to many activities, especially economic ones that can have an impact on the quality of the coastal environment

One of the rivers in the northern coastal area is the Silugonggo River. This river is in Juwana District, Pati Regency, Central Java Province. The Silugonggo River is the largest in Juwana District, which has a variety of activities such as boat parking, toilets, fish auction, and industry, so it has the potential to cause pollution to the quality of its river water [3]. Several results of testing the water quality of the Silugonggo River showed that the results of pollution exceeded the water quality standard [3 – 5]. The results of the primary survey related to the pH level were also at pH 8 and 9 or above the water quality standard.

From the various activities, there is a very dominant activity physically on the Silugonggo River, the activity is boat parking. Boat parking activities have the potential to pollute rivers, one of which is pollution through spills or seepage of oil from ships [6]. In addition, the length of parking can also cause the fall of used boat luggage into the river which can pollute the river in color and smell. Based on the problems above, a study related to the significant determinants factor of boat parking in affecting the quality of the Silugonggo river is conducted.

Several studies discuss boat parking on rivers along with river pollution. Previous research conducted by Jumardin (2018) [7] analyzed how boat activities can pollute solid waste. More in the field of policy,

IOP Conf. Series: Earth and Environmental Science 1041 (2022) 012088 doi:10.1088/1755-1315/1041/1/012088

research by Faturachman (2015) [8] analyzed that one aspect of ship safety is how to manage the potential for ship pollution to the sea. Research on boat parking on the Silugonggo River is related to the potential for oil seepage of parked ships that can cause pollution [9] and the impact of river parking on river functions [10]. This research completes the previous research by fulfilling certain part that has not been discussed. It is the main factor and community perception as a social point of view in the field.

2. Methodology

The method used in this study was a quantitative method with factor analysis techniques. Factor analysis is an analysis technique that can provide an overview of the correlation between the observations of several independent indicators so that it can provide the results of the analysis of the covariance of several unobserved basic variables [11 – 12].

The perception was another analysis technique used in this study. Perception draws dimensions about social conditions [13]. This provides an overview of the community's social view of the existing field conditions, whether it is seen as negative or positive [14]. Perception can also describe the psychological condition of the existing community related to community and environmental conflicts [15].

Data collection in this study was carried out by distributing questionnaires. Questionnaires were distributed to people who were considered close to activities on the Silugonggo River. These communities are fishermen, boat drivers, or boat owners. The number of respondents in this study reached 466 respondents.

Table 1. Variable of Analysis Factor Parameter

Variable	Parameter
Boat size	1 = Less than 80GT, 2 = 80-150GT, 3 = more than 150GT
Number of Boat Parking	1 = < 50 boats, 2 = 50-100 boats, 3 = > 100 boats
Parking Length	1 = < 5 days, 2 = 5-8 days, 3 = > 8 days
Parking Length by size	1 = less than 80GT, 2 = 80-150GT, 3 = more than150GT
Parking density	1 = does not interfere with the flow, 2 = moderate interfere with the flow, 3 = high interfere with the flow

Table 2. Variable of Community Perception Parameter

Variable	Parameter
Number of Parking Affects River Pollution	1 = does not affect, 2 = affects, 3 = highly affects
Parking length causes water color to change	1 = Does not cause the water color to change, 2 = causes the water color to change, 3 = highly causes the water color to change
Parking length causes water to smell	1 = Does Not Cause Water to Smell, 2 = Causes Water To Smell, 3 = Highly Causes Water To Smell

3. Result and Discussion

The study is located downstream of the Silugonggo River in Juwana District, which starts from the Juwana Bridge to Serapat Island with a distance of 2.2 km. Downstream of the Silugonggo River has various widths and depths, which are between 60-100 meters wide and 8-15 meters deep. Its location which is directly adjacent to the Java Sea makes the Silugonggo river a center of fishery activity for the Juwana sub-district and its surroundings. This causes several negative impacts for the river and the community around the river. The large number of large fishing boats that are parked irregularly causes small fishermen not to be able to go to sea because the access of heading to the sea is hampered by parking of fishing boats.

In addition, the large number of boats parked cause the flow of the river to the sea to be blocked, causing sedimentation in the river. Furthermore, this is exacerbated by the number of burnt shipwrecks which worsen river sedimentation and pollute the river

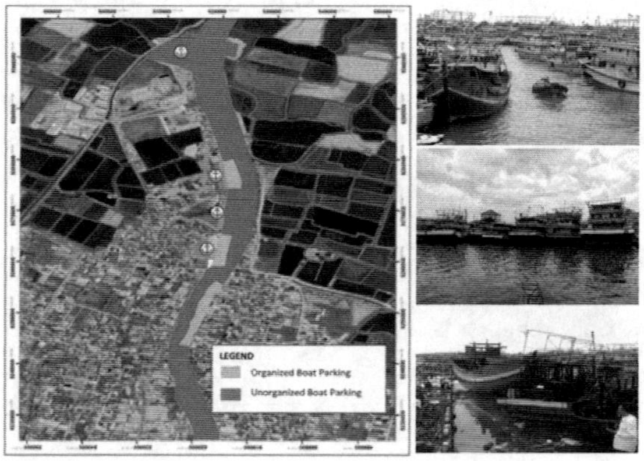

Figure 1. Study Area Silugonggo River and Parking Activity
Source : google earth and author document, 2020

Factor analysis is a reduction effort to focus more on the problem factors, so that the handling can be seen more comprehensively. The connection and the focus of the handler can be carried out more clearly. However, before reading the final results of factor analysis, it is necessary to conduct tests related to factor analysis and the materials used. These tests provided information regarding whether factor analysis is appropriate or not when used.

Table 3. KMO and Bartlett's Test

Kaiser-Meyer-Olkin Measure of Sampling Adequacy.		.589
Bartlett's Test of Sphericity	Approx. Chi-Square	243.846
	Df	10
	Sig.	.000

From table 3, the KMO value or measure is used to determine whether the factor analysis process can be carried out or not. KMO values above 0.5 indicated that the factor analysis process can be carried out, while KMO values below 0.5 indicate that the factor analysis process cannot be carried out [16]. Judging from the results, the number 0.589 or above 0.5 indicated the factor analysis process can be carried out.

Table 4. Anti-image Matrices

		Boat_Size	Number of_Boat_Parking	Parking Length	Parking Density	Parking length according to size
Anti-image Covariance	Boat_size	.817	-.212	.007	-.119	-.206
	Number of _boat_parking	-.212	.852	.123	-.195	.066
	Parking_length	.007	.123	.816	-.157	-.263
	Parking_Density	-.119	-.195	-.157	.836	-.088
	Parking Length_by_size	-.206	.066	-.263	-.088	.780
Anti-image Correlation	Boat_size	.613[a]	-.254	.008	-.144	-.258
	Number of _boat_parking	-.254	.510[a]	.148	-.231	.080
	Parking_length	.008	.148	.564[a]	-.190	-.330
	Parking_Density	-.144	-.231	-.190	.655[a]	-.110
	Parking Length_by_size	-.258	.080	-.330	-.110	.590[a]

a. Measures of Sampling Adequacy(MSA)

The Anti-image Matrices table shows that the KMO MSA value for each variable lies in the diagonal values of the anti-image correlation matrix [17]. The KMO MSA value of a variable below 0.5 should be discarded, then re-do the factor analysis process without involving the discarded variable. After that, re-check the KMO MS for each variable, whether it meets the requirements (above 0.5) or not. Because it is more than 0.5, then all variables can be continued or do not need to be discarded.

Table 5. Communalities

	Initial	Extraction
Boat_size	1.000	.546
Number of _boat_parking	1.000	.723
Parking_Length	1.000	.674
Parking_Density	1.000	.490
Parking_Length_by_Size	1.000	.633

Extraction Method: Principal Component Analysis.

From the table of the communalities value, the value of the number of boats parking is 0.723. This value can mean the factors that have been formed (2 factors based on the principal component analysis method) can explain the variance of the Number of boats Parking by 72.3%. The factors that have been formed can explain the variance of parking length by 67.4% and Parking Density by 63.3% and so on.

Table 6. Variance Explained

Component	Initial Eigenvalues			Extraction Sums of Squared Loadings			Rotation Sums of Squared Loadings		
	Total	% of Variance	Cumulative %	Total	% of Variance	Cumulative %	Total	% of Variance	Cumulative %
1	1.800	36.008	36.008	1.800	36.008	36.008	1.542	30.843	30.843
2	1.266	25.314	61.321	1.266	25.314	61.321	1.524	30.478	61.321
3	.781	15.627	76.948						
4	.601	12.021	88.969						
5	.552	11.031	100.000						

Extraction Method: Principal Component Analysis.

From table 6, Eigenvalues include one approach that can be used to determine the number of factors to be maintained in the analysis. The recommendation is that the eigenvalue of a factor that is greater than 1 must be maintained if the eigenvalues reach 1,800 and 1,266 or more than 1. The first factor can explain 36.008% of the variance factor and the second factor can explain 25.314% of the variance factor.

Table 7. Component of Matrixa

	Component 1	Component 2
Boat Size	.664	.324
The number of Boats Parking	.360	.770
Parking Length	.549	-.610
Parking Density	.677	.177
Parking length by Size	.685	-.405

Extraction Method: Principal Component Analysis.
a. 2 components extracted.

The loading factor is the correlation value between the original value and the factor value. The factor loading value can be used to determine whether a variable will be included in one of the factors. High factor loading is above 0.4 [19]. So that the factors of boat size, parking length, parking density and parking length by size can be grouped into a high correlation. As for the division of factors, we can see the closeness of the component values. When looking at component 2, the proximity of the factor is more readable, namely factor 1, the boat size, the number of parking boats, and the parking density which have a positive value. While the parking length and the parking length by size have a close value which is indicated by a negative value.

	Component	
	1	2
Boat Size	.703	.228
The number of Boats Parking	.794	-.303
Parking Length	-.029	.821
Parking Density	.610	.343
Parking Length by Size	.212	.767

Extraction Method: Principal Component Analysis.
Rotation Method: Varimax with Kaiser Normalization.
a. Rotation converged in 3 iterations.

Figure 2. Rotated Component of Matrixa

To recheck the division, it can be seen from the rotated component matrix, or try to roll back the possible component matrix. The results still show the same distribution of factors. So, when observed based on the class division, it can be concluded that the number of boat parking represents the size of the boat parking. The number of boat parking and parking density becomes the first factor named the number of boat parking. Meanwhile, the parking length and the parking length by size have a relationship in time. So, the second factor is the parking length.

From the results of the analysis, it is found two main factors causing river pollution by boat parking, namely the factor of the number of boat parking and the parking length. Both of these factors will be tested for the impact of aspects of social views through public perception. The results from the picture can show that the community, which consists of fishermen and boat owners, gives a critical perception regarding boat parking and the environment.

On the factor of the number of boats parking, the public has a perception that the number of boats must begin to be regulated because 79.8% of the community views that the number of boat parking greatly affects river pollution. Not much different from the factor of parking length, public perception also gives the same view. The parking length is considered by the community to influence river pollution, both on changes in river color and on the smell of river water.

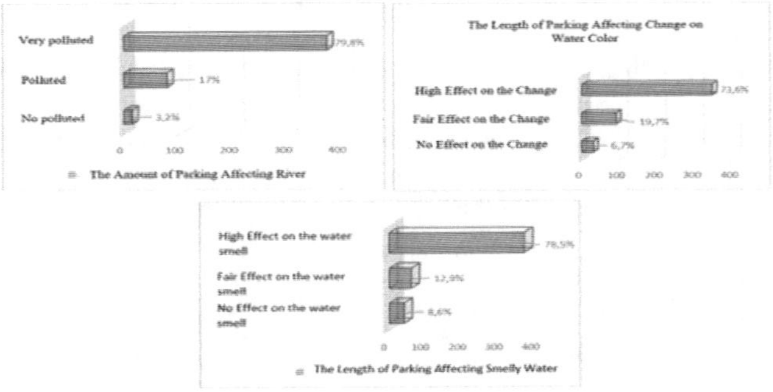

Figure 3. Perception About Parking Activity on Silugonggo River

The results of factor analysis and perceptions as social views in society are formed due to several facts in the field. The number of boats parking makes the river flow delayed, has an effect on sedimentation and becomes a place for gathering garbage, thus causing local scale pollution. While the parking length has the potential to cause oil spillage into the river, causing river pollution which affects changes in the color and smell of water in the river.

Regulation is needed on river parking activities. The length of parking is something that must be considered more from the side of the arrangement. because based on the results of the analysis of boat parking provides the dominant factor for the potential for ship parking, with the simulation the longer the ship parking lasts, the more potential for river pollution by ships.

4. Conclusion

This study looks back at the main factors that can be a general concern in the arrangement of boat parking and looks at the condition of the community's view of the existing factors. The main contribution of this research is that we identify the number of boat parking, and the parking length is general factors that can be the main consideration in reducing river water pollution by boat parking activities. These factors have also received a negative social field view from public perception. So, it has a high value for better handling. These results imply that if the government plans to maintain the sustainability of the Silugonggo River through parking arrangements, it must minimize the factors that affect river pollution, namely the number of boat parking and the parking length. By understanding the factors affecting river pollution, the government can make parking arrangements in a better direction, so that it can improve the social view of the community, especially fishermen, boat owners related to parking activities on the Silugonggo River.

References

[1] A A Saraswati 2004 Konsep Pengelolaan Ekosistem Pesisir (Studi Kasus Kecamatan Ulujami, Kabupaten Pemalang, Jawa Tenggah Tek.Lingk vol 5 no 3 pp 205–211.
[2] M Zainuri 2010 Kontribusi Sumberdaya Fitoplankton Terhadap Produktivitas dan Keseimbangan Ekosistem dalam Pengelolaan Wilayah Pesisir. Pengukuhan Guru Besar Universitas Diponegoro Semarang 78 Semarang, pp. 1.
[3] L Chauro Aina, E Rita, and F Kaswinarni 2017 Biomonitoring Pencemaran Sungai Silugonggo Kecamatan Juwana Berdasarkan Kandungan Logam Berat (Pb) Pada Ikan Lundu BIOMA J. Ilm. Biol vol 5 no 2 pp 1–11.
[4] S E Arvianto, A Satriadi, and G Handoyo 2016 Pengaruh Arus Terhadap Sebaran Sedimen Tersuspensi Di Muara Sungai Silugonggo Kabupaten Pati J. OSEANOGRAFI vol 5 no 1 pp 116–125.
[5] G Hidayah, S Y Wulandari, and M Zainuri 2016 Studi Sebaran Klorofil-a Secara Horizontal di Perairan Muara Sungai Silugonggo Kecamatan Batangan, Pati Bul. Oseanografi Mar vol 5 no 1 pp. 52.
[6] Y Darlan and U Kamiludin 2016 Penelitian Lingkungan Pantai Dan Logam Berat Perairan Pariaman - Padang-Bungus Teluk Kabung Sumatera Barat J. Geol. Kelaut vol 6 no 1 pp 12–22.
[7] I Jumardin, A Djabier, Bustamin, and Y Herlambang 2018 Bagaimana Efektivitas Penerapan Marpol Pada Kapal-Kapal Di Pelabuhan Pantoloan Venus vol 6 no 11 pp 61–70.
[8] D Faturachman, Muslim and S Agung 2015 Analisis keselamatan transportasi penyeberangan laut dan antisipasi terhadap kecelakaan kapal di Merak-Bakauheni Flywheel J. Tek. Mesin Untirta vol. 2 no 1 pp 14–21
[9] C S Mardiantari 2017 Aspek Hukum Perlindungan Lingkungan Laut Dari Kegiatan Kapal Perikanan Di Juwana Kabupaten Pati Provinsi Jawa Tengah Thesis on Universitas Atma Jaya Yogyakarta.
[10] M. S. Aryanto 2021 Pengaruh Parkir Kapal Perikanan Terhadap Fungsi Sungai Thesis on Universitas Islam Sultan Agung.
[11] F. Rachman 2019 Factor analysis Factor analysis Int. J. Res. Sci. Innov vol VI no IX pp 212–216.
[12] A M Tiro, Sukarna, and Aswi 2006 Analisis Faktor. Makasar: Andira Publisher.
[13] W Klein, M R Dove and A J Felson 2021 Engaging the unengaged: Understanding residents' perceptions of social access to urban public space Urban For. Urban Green Vol. 59 no. January, p. 126991.

[14] V D Simonchuk and N V Romasheva 2021 Public perception of carbon capture (utilization) and storage projects: world experience and the situation in Russia in E3S Web of Conferences, vol 266 pp 1–10.
[15] L Zhang, G Huang, Y Li, and S Bao 2021 A Psychological Perception Mechanism and Factor Analysis in Landsenses Ecology: A Case Study of Low-Carbon Harmonious Discourse Int. J. Environ. Res. Public Health vol 18 no 13 pp 6914.
[16] N K Malhotra and D F Birk 2015 Marketing Approach: An Applied Approach Updated Second European Edition London: Prentice-Hall.
[17] A Field 2009 Discovering Statistics Using SPSS ISM (London, England) Introducing Statistical Methods Series London: SAGE.
[18] F Hair Jr, Joseph, W C Black, B J Babin and R E Anderson 2010 Multivariate Data Analysis, vol. 7. New Jersey: Prentice-Hall.

AUTHOR INDEX

Afifah, A N ... 419
Abdelgader, M ... 448
Achsani, Noer Azam ... 575
Adli, D N ... 63
Afidah, Mar'atul ... 386
Agritubella, S Meri ... 399
Altayb, Modawy Abdelgader Albasheer ... 411
Ambarwati, L ... 515
Ambarwati, Sri ... 158
Anditya, C ... 273
Andiyan, Andiyan ... 242
Anggono, T ... 273
Antonius, Yulanda ... 498
Arief, Yanuar Z. ... 139
Ariyanto, Anto ... 131, 392
Arkeman, Yandra ... 575
Arlenny ... 139, 173
Aryanto, M S ... 591
Asteria, D ... 295
Athaillah, Teuku ... 29
Aunurrohman, H ... 531
Azhar, T ... 215
Aziz, Abdul Kholik ... 452
Bachtiar, M. ... 124
Budianto, A D ... 63
Bueno, CM ... 521
Cahyanto, T A ... 145
Cahyonugroho, O H ... 585
Cendrawati, D G ... 273
Cheris, Rika ... 329
Cusiayuni, A ... 430, 436
Darma, R ... 55
Dealino, MM ... 521
Dewi, W N ... 268
Dhiniati, F ... 324
Diansari, L E ... 324
Dikman, D.M. ... 191
Dinata, A ... 324
Djatmiko, O. E. ... 211
Dongke, A L ... 487
Dono, N D ... 419
Efendi, Kriana ... 452
Efendi, Z ... 380
Effraim, Andy ... 498
Einstein, A ... 211
Elsera, M ... 262
Emelda, A ... 487
Endah, S R N ... 443
Eteruddin, Hamzah ... 139
Evvyernie, D ... 430, 436
Fadillah, Akbar ... 478
Fadlan, Arif ... 47
Fahmi, Idqan ... 575
Fajarwaty, Tri ... 492
Fajri, I ... 235
Fardiaz, D ... 197
Fatmawati, Sofia ... 465, 478
Febriana, Dela ... 313
Febriyanti, Rizki ... 1
Feronytha, Alfina Gracia ... 498
Firdaus, F ... 191
Fitriana, Ana ... 465
Fitriani, K ... 350
Gani, S A ... 235
Guntoro ... 131, 370
Hadjar, Nuhayati ... 405
Hambali, Erliza ... 8
Hamzah ... 131
Handoko, Tito ... 334
Hariyanto, S ... 585
Harmayani, E ... 443
Hartono, B ... 96
Hasanti, Fathimah ... 139
Hastuti, D.R.D. ... 55
Hati, D L ... 197
Herawati, H ... 112, 124
Herdiansyah, H ... 289, 295
Herliatika, A ... 569
Hidayat, Fahmi ... 47
Hidayat, N ... 74
Hikmawanti, Ni Putu Ermi ... 478
Idris, Iswadi ... 301
Imbayarwo-Chikosi, V E ... 80
Indradji, M ... 343
Indrasanti, D ... 343
Insusanty, Enny ... 363
Iriani, ES ... 22
Iriyanti, N ... 448
Iriyanti, Ning ... 411
Irsyad, M I A ... 273
Ishak, A ... 380
Islami, Nur ... 252
Ismoyowati ... 411, 448
Jannah, Fathul ... 39, 399
Jayanegara, A ... 426, 430, 436
Jukes, David ... 492

Juliana, Agnes 498
Kabullah, Muhammad Ichsan 280
Kamsiati, E. 112
Kamsiati, E 124
Kodir, Abdul 228
Krishnadas R 118
Kurnianto, E 536
Kushandajani 222
Kusnadi, H 380
Kusnadi 1
Kusuma, D R 510
Kusumawati, Yuly 47
Laconi, E B 426
Latifah, Eva Umi 47
Linawati 96
Lisnawita 370
Listyorini, Haniek 242
Madusari, Sylvia 313, 357
Maimun 235
Majid, M S A 215
Malik, H. 308, 319
Marzuki, I 487
Mastuti, S 531
Matitaputty, PR 203
Megasari, S W 563
Mugabe, P H 80
Muhammad, Juandi 252
Muhtadir 405
Mulyadi, Aras 334
Mulyawanti, I. 22
Munasik 74
Murni, Yetty 158
Mutamima, A 268
Mutia, Tika 334
Nasip, Ismiriati 158
Natsir, M H 63
Naufal, M. 308, 319
Nawari 252
Neyman, S N 145
Ngongo, Y 203
Nining 458
Nofriyaldi, A 443
Nofrizal 252
Nopryanti 363
Novita, Lidya 301
Nufutomo, T K 350
Nugrahaeni, Fitria 452, 458, 465
Nugroho, Yoga 29
Nuningtyas, Y F 63
Nur, S. 211
Nurbaety, N 74
Nurdin, R 215
Nurfatahillah, RK 430, 436
Nurliyani 443
Nursal, Fith Khaira 465
Oktaviania, Ira 399
Okvianida, Redina 458
Ongko, Jeremi 498
Pamungkas, D. 191
Panjaitan, N V 152
Pari, Gustan 8
Pattipeilohy, Mery 102
Pelamonia, Alamanda 102
Perwitasari, F. D. 88
Prayitno, Pranoto Hadi 242
Pujirahayu, Niken 405
Pundoko, I 152
Purnama, I 268
Purnama, Ig. L. S. 308, 319
Purnama, Indra 36
Putra, A. R. S. 88
Putra, Ardi 334
Putra, R 350
Putri, A.S. 191
Putri, Asgami 285
Putri, Awlia Mustia 478
Qadri, Muhibbul 29
Radiati, L E 197
Rahim, A. 55
Rahman, B. 591
Rahmawan, H 145
Ramon, E. 380
Ratnawaty, S. 203
Razali 235
Renganathan R 118
Restusari, Lily 399
Rimbawanto, Efka Aris 411
Riyanta, Aldi Budi 1
Rizqiana, S 68
Rosanti, Aulia Dewi 47
Rosetia, Amanda 329
Rosyida, Fatiya 228
Rosyidah, Masayu 242
Ruindungan, D 152
Rumbayan, M 152
Ruslan, I 273
Rz, I O 184
Rz, Ira Oktaviani 39
Salman, D. 55
Samnuzulsari, T. 262
Samsi, M 343
Sanjatmiko, P 289
Sanjaya, R S 550
Santosa, L. W. 308, 319

Santoso, S. 55
Sanusi 235
Sari, Gresita Fitaloka 14
Sari, L R H 197
Sari, Lala Novita 357
Sari, V I 374
Sari, Vira Irma 313, 357
Sari, Vonny Indah 36
Sartiyah, S. 215
Sasmitaloka, K S 22
Setiawan, David 131, 173, 370
Setiawan, Jesica Viona 498
Setyaningsih, W 550
Setyawardani, T 510
Sinaga, D M 426
Siswanto, Hadi 139
Siswati, Latifa 131, 285, 363
Sjofjan, O 63
Soemantri, A S 124
Solikin, N 96
Sudarmadji 308, 319
sudarmaji, Eka 158
Sudarmaji, Eka 575
Sufiriyanto 343
Sugiarto, M 211, 531
Sugiono 96
Suhartati, F M 68
Sujianto 334
Sukmawati 487
Sukoco, H 145
Sulistiowati, R 262
Sumarmono, J 510, 542
Supratiwi 222
Supriyanto, G 585
Suryani, Ani 8
Suryani, J A 531
Suryapratama, W 68
Susanti, E 74, 505
Susi, Neng 36, 363
Sutisna, N 289
Suwignyo, B. 88
Syafrani, S 268
Syah, S P 515
Syahza, Almasdi 252
Tanjung, Ardyanto 228
Thamrin 252
Tinov, MY Tiyas 334
Tugiyanti, E 505
Uslinawaty, Zakiah 405
Uthia, R. 184
Uthia, Rahimatul 39
Wahjudi, Mariana 498
Wahjuni, S 145
Wahyuni, S 262
Wahyuni, Sri 386
Wahyuningtyas, Neni 228
Wakhidati, Y N 531
Wakhidati, Y. N 211
Wardani, Anggita R.K 47
Wardi, Jeni 392
Widayanti, SM 22
Widiati, R. 88
Widiawati, Y 569
Widiyawati, R. 191
Widyanarko, U. 308
Wiryawan, KG 430, 436
Wulandari, WA 380
Wulasari, P D 443
Yandra, Alexsander 39
Yanti, G 563
Yanti, Rina Novia 8
Yasid, Hamdan 285
Yudiatmaja, W E 262
Yuhanah, Yuhanah 242
Yuliasari, S 380
Yunanto, M K 262
Yusuf, R. 235
Yuswardi, Yuswardi 242
Yuvendius, H 374
Yuvendius, Hazra 36
Yuwanto 222
Yuwono, E. 343
Zainuri 374, 563
Zamzami 370
Zezai, T A 80
Zondra, E. 374
Zulkarnain 334
Zulkifli, Afni 392
Zuprizal 419
Zuvara, R A 295